Annual Review of Astronomy and Astrophysics

EDITORIAL COMMITTEE (1999)

KENNETH BRECHER
GEOFFREY BURBIDGE
DAVID J. HELFAND
JOSEPH S. MILLER
ALLAN SANDAGE
FRANK H. SHU
SCOTT D. TREMAINE
GEORGE WALLERSTEIN

RESPONSIBLE FOR THE ORGANIZATION OF VOLUME 37 (EDITORIAL COMMITTEE, 1997)

KENNETH BRECHER
GEOFFREY BURBIDGE
ANNE P. COWLEY
DAVID J. HELFAND
ALLAN SANDAGE
ANNEILA I. SARGENT (GUEST)
FRANK H. SHU
DAVID STEVENSON (GUEST)

Production Editors: CHRISTINE MCGEEVER, LARISA NORTH
Subject Indexer: CHERI WALSH

ANNUAL REVIEW OF ASTRONOMY AND ASTROPHYSICS

VOLUME 37, 1999

GEOFFREY BURBIDGE, *Editor*
University of California at San Diego

ALLAN SANDAGE, *Associate Editor*
Observatories of the Carnegie Institution of Washington

FRANK H. SHU, *Associate Editor*
University of California at Berkeley

www.AnnualReviews.org   science@annurev.org   650-493-4400

ANNUAL REVIEWS
4139 El Camino Way • P.O. Box 10139 • Palo Alto, California 94303-0139

ANNUAL REVIEWS
Palo Alto, California, USA

COPYRIGHT © 1999 BY ANNUAL REVIEWS, PALO ALTO, CALIFORNIA, USA. ALL RIGHTS RESERVED. The appearance of the code at the bottom of the first page of an article in this serial indicates the copyright owner's consent that copies of the article may be made for personal or internal use, or for the personal or internal use of specific clients. This consent is given on the condition that the copier pay the stated per-copy fee of $8.00 per article through the Copyright Clearance Center, Inc. (222 Rosewood Drive, Danvers, MA 01923) for copying beyond that permitted by Section 107 or 108 of the US Copyright Law. The per-copy fee of $8.00 per article also applies to the copying, under the stated conditions, of articles published in any *Annual Review* serial before January 1, 1978. Individual readers, and nonprofit libraries acting for them, are permitted to make a single copy of an article without charge for use in research or teaching. This consent does not extend to other kinds of copying such as copying for general distribution, for advertising or promotional purposes, for creating new collective works, or for resale. For such uses, written permission in required. Write to Permissions Dept., Annual Reviews, 4139 El Camino Way, P.O. Box 10139, Palo Alto, CA 94303-0139 USA.

International Standard Serial Number: 0066-4146
International Standard Book Number: 0-8243-0937-5
Library of Congress Catalog Card Number: 63-8846

Annual Review and publication titles are registered trademarks of Annual Reviews.
∞ The paper used in this publication meets the minimum requirements of American National Standards for Information Sciences—Permanence of Paper for Printed Library Materials. ANSI Z39.48-1992.

Annual Reviews and the Editors of its publications assume no responsibility for the statements expressed by the contributors to this *Annual Review*.

Typeset by Techbooks, Fairfax, VA
Printed and Bound in the United States of America

PREFACE

This volume was planned at a meeting held on April 12, 1997, in Pasadena, California. Those who attended the meeting included Geoffrey Burbidge (Editor), Allan Sandage and Frank Shu (Associate Editors), Sam Gubins (Editor-in-Chief, Annual Reviews), Ken Brecher, Anne Cowley, and David Helfand (Editorial Committee members), Naomi Lubick (Production Editor), and guests Anneila Sargent and David Stevenson.

In the Preface to Volume 36, I pointed out that 32 articles were scheduled for this volume. In fact, only 14 are contained here. For the next volume, Volume 38 (2000), 24 articles are scheduled.

Once again, I wish to thank the Authors, the Associate Editors, the original Production Editor, Christine McGeever, and the current Production Editor, Larisa North, for all the work that has gone into the production of this volume.

Geoffrey Burbidge
Editor
March 1999

Contents

ADVENTURES IN COSMOGONY, *A. G. W. Cameron*	1
A CRITICAL REVIEW OF GALACTIC DYNAMOS, *Russell M. Kulsrud*	37
FREQUENCY ALLOCATION: THE FIRST FORTY YEARS, *Brian Robinson*	65
REFERENCE FRAMES IN ASTRONOMY, *K. J. Johnston and Chr. de Vegt*	97
PROBING THE UNIVERSE WITH WEAK LENSING, *Yannick Mellier*	127
THE HR DIAGRAM AND THE GALACTIC DISTANCE SCALE AFTER HIPPARCOS, *I. Neill Reid*	191
NUCLEOSYNTHESIS IN ASYMPTOTIC GIANT BRANCH STARS: RELEVANCE FOR GALACTIC ENRICHMENT AND SOLAR SYSTEM FORMATION, *M. Busso, R. Gallino, and G. J. Wasserburg*	239
PHYSICAL CONDITIONS IN REGIONS OF STAR FORMATION, *Neal J. Evans II*	311
HIGH-ENERGY PROCESSES IN YOUNG STELLAR OBJECTS, *Eric D. Feigelson and Thierry Montmerle*	363
SOURCES OF RELATIVISTIC JETS IN THE GALAXY, *I. F. Mirabel and L. F. Rodríguez*	409
THE FIRST 50 YEARS AT PALOMAR: 1949–1999: THE EARLY YEARS OF STELLAR EVOLUTION, COSMOLOGY, AND HIGH-ENERGY ASTROPHYSICS, *Allan Sandage*	445
ELEMENTAL ABUNDANCES IN QUASISTELLAR OBJECTS: STAR FORMATION AND GALACTIC NUCLEAR EVOLUTION AT HIGH REDSHIFTS, *Fred Hamann and Gary Ferland*	487
ORIGIN AND EVOLUTION OF THE NATURAL SATELLITES, *S. J. Peale*	533
FAR-ULTRAVIOLET RADIATION FROM ELLIPTICAL GALAXIES, *Robert W. O'Connell*	603
INDEXES	
Subject Index	649
Cumulative Index of Contributing Authors, Volumes 26–37	675
Cumulative Index of Chapter Titles, Volumes 26–37	679

Related Articles

From the ***Annual Review of Earth and Planetary Sciences***, Volume 27, 1999:

Ups and Downs in Planetary Science, Carolyn S. Shoemaker

Hubble Space Telescope Observations of Planets and Satellites, Philip B. James and Steven W. Lee

Kuiper Belt Objects, David Jewitt

From the ***Annual Review of Nuclear and Particle Science***, Volume 48, 1998:

Current Quests in Nuclear Astrophysics and Experimental Approaches, F. Käppeler, F.-K. Thielemann, and M. Wiescher

ANNUAL REVIEWS is a nonprofit scientific publisher established to promote the advancement of the sciences. Beginning in 1932 with the *Annual Review of Biochemistry*, the Company has pursued as its principal function the publication of high-quality, reasonably priced *Annual Review* volumes. The volumes are organized by Editors and Editorial Committees who invite qualified authors to contribute critical articles reviewing significant developments within each major discipline. The Editor-in-Chief invites those interested in serving as future Editorial Committee members to communicate directly with him. Annual Reviews is administered by a Board of Directors, whose members serve without compensation.

1999 Board of Directors, Annual Reviews

Richard N. Zare, Chairman of Annual Reviews
  Marguerite Blake Wilbur Professor of Chemistry, Stanford University
Peter F. Carpenter, *Founder, Mission and Values Institute*
W. Maxwell Cowan, *Vice President and Chief Scientific Officer, Howard Hughes Medical Institute, Bethesda*
Sandra M. Faber, *Professor of Astronomy and Astronomer at Lick Observatory, University of California at Santa Cruz*
Eugene Garfield, *Publisher*, The Scientist
Samuel Gubins, *President and Editor-in-Chief, Annual Reviews*
Daniel E. Koshland, Jr., *Professor of Biochemistry, University of California at Berkeley*
Joshua Lederberg, *University Professor, The Rockefeller University*
Gardner Lindzey, *Director Emeritus, Center for Advanced Study in the Behavioral Sciences, Stanford University*
Sharon R. Long, *Professor of Biological Sciences, Stanford University*
Michael Peskin, *Professor of Theoretical Physics, Stanford Linear Accelerator Ctr.*
Peter H. Raven, *Director, Missouri Botanical Garden*
Harriet A. Zuckerman, *Vice President, The Andrew W. Mellon Foundation*

Management of Annual Reviews

Samuel Gubins, President and Editor-in-Chief
Richard L. Burke, Director for Innovation
John W. Harpster, Director of Sales and Marketing
Steven J. Castro, Chief Financial Officer
Gwen Larson, Director of Production

Annual Reviews of

Anthropology	Fluid Mechanics	Physiology
Astronomy and Astrophysics	Genetics	Phytopathology
Biochemistry	Immunology	Plant Physiology and Plant
Biomedical Engineering	Materials Science	Molecular Biology
Biophysics and Biomolecular Structure	Medicine	Political Science
	Microbiology	Psychology
Cell and Developmental Biology	Neuroscience	Public Health
	Nuclear and Particle Science	Sociology
Earth and Planetary Sciences	Nutrition	SPECIAL PUBLICATIONS
Ecology and Systematics	Pharmacology and Toxicology	Excitement and Fascination of
Energy and the Environment	Physical Chemistry	Science, Vols. 1, 2, 3, and 4
Entomology		

For the convenience of readers, a detachable order form/envelope is bound into the back of this volume.

ADVENTURES IN COSMOGONY

A. G. W. Cameron
Harvard-Smithsonian Center for Astrophysics, 60 Garden Street, Cambridge, Massachusetts 02138; e-mail: acameron@cfa.harvard.edu

Key Words autobiography, nucleosynthesis, stellar astrophysics, planetary sciences, scientific advising

■ **Abstract** I was born and educated in Canada, obtaining my PhD in experimental nuclear physics. When I learned that technetium had been found in stellar spectra, I taught myself some astrophysics and began to study stellar nucleosynthesis. This is an account of those studies and of the pathway through much of theoretical astrophysics and planetary physics that was a natural outgrowth of the pursuit of nucleosynthesis problems. I also discuss my experiences in government service and in academia, in organization of conferences, in governmental advising, and in academic administration. In particular, I emphasize the logical connections among the various scientific themes that I have pursued.

Introduction

To celebrate my 60th birthday a conference was held and a *festschrift* volume was prepared. The editors, exhibiting great modesty, declared themselves incapable of summarizing my career for this volume and asked me if I would do so myself. I agreed to do this (Cameron 1986a), and the resulting chapter contains an account of my life until then that concentrates on a chronological account of how my scientific interests developed and ranged over a variety of subjects. It also attempts to mention the various students and postdocs who have collaborated with me.

In this article I give a brief autobiographical sketch and follow it with an account of my approach to various scientific topics. For these topics I attempt to remember the events that interested me and how, in many cases, questions in one area prompted investigations in another area. I also give an account of my other professional experiences, including administration, scientific advice giving, conference organization, and the like.

The principal motivation here is to provide my colleagues with a coherent account of these scientific interests and also to give historians of science a glimpse of the connectedness of much of my research. I have lived long enough to see how at least one historian of science has treated some of my work. I refer to Steve Brush, who has recently written histories of the development of topics such as nucleosynthesis, the formation and evolution of the solar nebula, and the origin of the moon (Brush 1996a,b,c).

In seeing my own work covered in these histories, I have become aware of the different points of view held by scientists and historians: The scientist is always seeking to discover the truth about scientific questions, whereas the historian largely regards the truth as unknowable and seeks to record what the scientific beliefs were at different times. But the search for truth usually takes place within a rational context of scientific motivations, and when scientists write papers describing their final conclusions, they frequently suppress discussions of their motivations and of the blind alleys encountered along the way. So I think that personal histories written by scientists make a useful complement to the accounts of historians. But there is the danger that the scientist may underplay or not mention at all the various errors made along the way, particularly if they were embarrassing. In addition to that, memories are unintentionally selective. So these are my caveats about what follows.

Autobiographical Sketch

I was born in Winnipeg, Canada, on June 21, 1925. Both of my parents were of Scottish descent. My mother's family name was Bell, and her ancestors immigrated to Ontario a few generations earlier. Her father led a rather colorful life. He ran away from school as a teenager to join an expeditionary force to put down an uprising by Indians and French-Indian half-breeds in Manitoba. Thereafter he made his life on the prairies, living part of the time with the Indians and organizing parties of them to hunt meat for the construction crews building the Canadian Pacific Railway. Eventually he settled in Winnipeg as treasurer of the Winnipeg Grain Exchange.

My father was born in London, England. His father, in turn, spent most of his life in Malmesbury as a school headmaster, and earlier generations were heavily involved with the Presbyterian Church in Scotland. My father was trained as a chemist. He spent a brief period of time working with Soddy, an early investigator of radioactivity (my father even wrote a brief book on the subject). Then he turned his attention to biochemistry and accepted a position at the Manitoba Medical College, a branch of the University of Manitoba. He spent the remainder of his life there as head of the biochemistry department.

So I grew up in an environment in which scholarly and professional work was valued. I was told that around the age of four I addressed all men as "Doctor," clearly an early attempt at forming a hypothesis based upon limited data.

I attended a private school for boys in Winnipeg, where I was sometimes at the head of the class, sometimes second. I was particularly keen about science and mathematics, more so than the other boys. It was probably good for me that I was required to participate in sports, although I was not enthusiastic about them. Later I did my undergraduate work at the University of Manitoba, majoring (not a term used by Canadians) in physics and mathematics.

My PhD studies were carried out at the University of Saskatchewan. There they had a 28 MeV betatron, which we used to study photonuclear reactions. My thesis advisor was Leon Katz, and we worked together to develop a method to solve photonuclear activation curves. The betatron generated a *bremsstrahlung* spectrum of photons extending to the electron energy that generated the spectrum, and so the number of reactions induced in a target bombarded with two *bremsstrahlung* spectra at adjacent energies differed by the difference in the two spectra; this difference was itself continuous in energy and corresponded to a difference in the total activation in the target per unit time (or dose). Nowadays it would be a simple matter to compute these differences by computer. Then we did it graphically by hand and set up a method to solve for the photonuclear cross sections as functions of energy called the photon difference method (Katz & Cameron 1951).

For a while we had a small industry going, generating cross section curves spanning the giant dipole resonance and determining the characteristics of the resonance as a function of atomic number. Later we found that we had sufficient energy stability in the betatron so that for light elements like oxygen we could measure an activation curve in small energy steps and find kinks in it corresponding to discrete energy levels in the target nucleus (Katz et al 1954).

The S-Process

It was now 1952, and I did not have to think hard about a post-doctoral position, because I had become fairly well known in the photonuclear field, and I was given an unsolicited offer of an assistant professorship at Iowa State College (as it was then) at Ames, Iowa. There they had just obtained a 70 MeV synchrotron, and it was attractive to contemplate extending my photonuclear work to higher energy. However, when I arrived I found that the only activity going on with that machine was an attempt to increase the beam current, which was too small to do very much of interest. That kind of engineering did not interest me at all, so I mainly became involved in helping a graduate student analyze some photographic emulsions in a study of higher energy photoproton reactions (Hoffman & Cameron 1953).

Iowa State housed the Ames Research Center of the United States Atomic Energy Commission. One day, not long after I arrived, I was browsing through its reading room, and I came across the current issue of *Science News Letter*, now *Science News*, in which I saw a story reporting that an astronomer, Paul Merrill, had discovered lines of the unstable element technetium in red giant stars of class S. I found this very exciting. This meant that in the recent past there had been a big flood of neutrons in the interior of that class of star. Where had they come from? That moment marked an instant turning point in my career.

I had long had an interest in astronomy but had never had an opportunity to take a course in it (what was called an astronomy course at the University of Manitoba was simply a navigation class in which the only useful thing I learned was spherical trigonometry), and Iowa State did not have any such courses. So I bought all the graduate-level astrophysics texts I could find, subscribed to the *Astrophysical Journal*, and started some intense reading.

When I had reached the point where I felt comfortable calculating thermonuclear reaction rates, I started looking for Merrill's neutrons. In ordinary hydrogen burning, the only neutron production occurs in rare reactions of deuterium with itself, and the few neutrons thus produced are immediately gobbled up by capture on hydrogen. So that could not be it. After hydrogen burning comes helium burning, and for that one must start looking at reactions in carbon. ^{12}C? No. ^{13}C? Ahah! ^{13}C$(\alpha,n)^{16}$O is exothermic! And so are similar reactions in nuclei consisting of several alpha-particles and a neutron. But the ^{13}C case was particularly interesting because that is the fastest such reaction taking place on the nuclei involved in the carbon cycle. Unfortunately that nucleus also has a small abundance in the cycle, but it was not clear to me how many neutrons were needed because Merrill had not given technetium abundances for his S stars. This led to my first paper in astrophysics (Cameron 1955).

By now I had lost all enthusiasm for pursuing photonuclear reactions and felt the need to find a job where I could pursue these new interests. I thought I needed a nuclear physics environment rather than an astrophysical one because the major tasks would involve nuclear physics, and the relevant astrophysical environments in which these new nuclear reactions would occur would not be known for a long time. I had already spent some time at the Canadian Atomic Energy Project at Chalk River, Ontario, and that was a logical place to apply. The application was accepted and off I went.

In order to work in this new field of nuclear astrophysics, as I had started to call it, it was clear that a lot of nuclear reaction cross sections would be needed. Sometimes the relevant experimental measurements were available, but more often there were no useful measurements and a cross section would have to be estimated. For reactions on the relatively light nuclei, where the thermonuclear rates were frequently dominated by individual resonances, I spent a lot of time with my slide rule calculating rates for nuclei where resonance positions were known but other important data were not. If a spin and parity was known, one still had to guess a reduced width, or strength, of that resonance, usually using some sort of average value that could easily be off by an order of magnitude. This may sound like rather useless information, but thermonuclear reaction rates are so sensitive to energy and temperature that such estimates were quite often considerably better than having no information at all.

Things were better for heavier nuclei, for there nuclear level densities become sufficiently great that it is possible to average them for many spins and parities, and if one has a reasonable knowledge of the nuclear masses, energy densities, and nuclear particle and radiation widths, the cross sections can be calculated with probable errors of a factor two or three. For the purpose of understanding general processes in nuclear astrophysics that is generally satisfactory.

Over the next few years I devoted quite a bit of effort to devising improved semi-empirical methods for calculating nuclear masses, nuclear level densities, and nuclear radiation widths (toward the end with Arnold Gilbert), and these proved to be very useful tools for doing nuclear calculations within astrophysical contexts (Cameron 1957a,b, 1958a).

At this point it may be interesting to recount my early experiences with calculating machinery. At Chalk River I wanted to solve the differential equations for helium burning, forming ^{12}C, destroying it to make ^{16}O, and destroying that to make ^{20}Ne. One of my colleagues wired up a plugboard with germanium diodes for an IBM accounting machine in the Chalk River accounting department, so that the machine became a card-programmed calculator. The machine worked with mechanical relays and was very slow. You could do the equivalent of DO loops but no IF branching. That was good enough for the helium-burning problem. Looping was achieved by preparing small packets of cards to program the loop, and then replicating these packets a large number of times. I would leave many trays of these cards prepared in the accounting room, and ask the accounting people to please keep the hopper filled. The results of the calculations were punched out on these cards as they passed through the machine.

Then one day the head of the accounting office told me that the IBM office in Ottawa would have a new IBM 650 computer on display in their office for a couple of weeks, and perhaps I would like to go down there with some of them to see a demonstration. The IBM 650 was a two-address calculator that operated from a magnetic drum storing 2000 words. I said I would like to go if I could try writing a program to run on it. So along came the instruction book from IBM, and I wrote a simplified version of an s-process network with the nuclei in the network compressed into about ten groups. The accountants punched the cards for me and off we went to Ottawa. I had even tried to be sophisticated by optimally locating the instructions around the drum to minimize access time. In Ottawa we fed my cards into the machine, and the impossible happened: The program ran perfectly!

I had used as starting abundances the solar system abundances that had recently been published by Suess and Urey. This calculation revealed for me the essential features of the s-process: the preferential capture of the neutrons on the iron abundance peak and the increase of the heavier element abundances by three orders of magnitude (Cameron 1957c,d, 1958b). I was absolutely delighted with this and asked if I could come back next week to try to run a complete s-process. It happened that the staff was planning a late-night farewell party for one of their number, and permission was given to come that afternoon. So I expanded the program to the full network, one mass number per nucleus, and once again the cards were punched. Back in Ottawa, the cards went into the machine, and the program failed to run. So I had to learn console debugging in real time. After finding four errors, I was able to run the program. And it ran. And it ran. And it ran. Thank goodness for the party: I was able to go down to the party repeatedly to ask them to find me some extra boxes of blank cards to use to take output. It was nearly dawn before I arrived back at Chalk River.

Shortly after this Chalk River purchased a Datatron 205, which was easier to use than the IBM 650. The Datatron 205 used decimal machine coding and had a drum that stored 4000 words. I was to be the principal user of this machine for the next three years. It was generally in use by others for short calculations during the day. But late in the afternoon, I would think about what I wanted to calculate next and code the problem. Toward the end of the workday I would go to the computation building to punch my cards. Then I'd feed them in, debug to the degree necessary, and head to the cafeteria for supper. If all seemed well with the run, I'd leave instructions to the operators about how to take the problem off the machine the next morning, and I would head home.

The relatively long run times I could get on weekends were particularly useful. However, sometimes when I would come in on Saturday or Sunday to check on the calculation, I would find the computer stopped. I would try three remedies, in the following order: (*1*) give the machine a kick; (*2*) look for a pattern of lights in the displays on the electron tube modules that would be a forbidden combination and, if found, rummage around in the customer engineer's office to find a replacement package; (*3*) call the customer engineer at home to come in and fix things (this did not make me very popular with the customer engineer).

The relatively simple problems I was able to compute in this manner led to a fairly large number of publications during 1959.

Advanced Stages of Thermonuclear Reactions

At that time hydrogen and helium burning were fairly well understood from the nuclear physics point of view (apart from a few uncertain parameters), but the details of their participation in stellar evolution were less well understood. How the more advanced stages fit into stellar evolution was a mystery. It seemed to me appropriate to proceed entirely on a nuclear physics basis, which meant considering a gas processed through hydrogen and helium burning, and containing initially solar abundances of the heavier elements, and heating and squeezing it to see what would happen.

But what would happen if one squeezed but did not heat? Clearly red giant stars were headed in that direction in their helium cores. When the matter became dense enough, the thermal energy would become negligible and the gas would form a crystalline lattice, but the ions would retain significant zero-point energies and motions, and there would be a significant probability that nuclear coulomb barriers could thus be penetrated. The penetrability would increase with increasing density.

I made some simple estimates of the high-density-induced reaction rates for helium and heavier nuclei to find when energy generation rates might blow up such degenerate cores. Thermonuclear reactions are driven by a high temperature. But at Chalk River there was no such thing as an English-Greek dictionary, so how would I find an appropriate name for the equivalent high density process? Fortunately I remembered that surfaces of constant thermodynamic properties of various kinds all began with the prefix "iso-," and indeed the dictionary produced "isopycnal" for surfaces of constant density. So that was how pycnonuclear reactions received their name (Cameron 1959a).

The process of squeezing and heating would lead through a series of stellar burning stages. First would be carbon burning. This was fairly straightforward. The main complications came from the emission of protons and alpha-particles in the carbon reactions with itself. These would generally interact with the lightest nuclei present in the mixture, and thus a nuclear reaction network would be generated in the light nuclear region extending from carbon itself to a little beyond magnesium (Cameron 1959b). Oxygen burning (with itself) behaved somewhat similarly, except that neutrons were also released, and the network of reactions that needed to be considered was greatly extended and started with the previous products of carbon burning. The neutron captures generated a modified form of the s-process (Cameron 1959c).

However, a new kind of reaction started to become important at this time: the photonuclear reaction, the reverse of a particle capture reaction. In a neutron capture reaction, the neutron would enter a nucleus to form a compound state, and photons would be emitted in a complicated cascade through the intermediate levels approaching the ground state.

The photonuclear process was the reverse of that: A statistical equilibrium would be set up among the lower-lying states of a nucleus with equal rates of formation and destruction of each state. As the temperature rose, the states in the vicinity of the neutron separation energy would become excited, and neutrons would be emitted. The rate at which this emission would happen could be calculated by a detailed balance argument. If you knew the rate of neutron capture you could compute the rate of its inverse.

As an aside, one consequence of maintaining statistical equilibrium among the lower excited states of a nucleus is that it will be common for those states to be able to undergo beta decay or electron capture to a different nuclear ground state more massive than the ground state of the nucleus considered. That ground state can in turn decay back to a lower state of the original parent nucleus. I called this the photobeta process, but the name did not stick (Cameron 1959d). The process as described is one form of the URCA process, in which large numbers of neutrinos and antineutrinos will be emitted. It can also be important in the s-process by speeding up the decay of nuclei along the neutron capture path for which a beta decay is energetically possible but highly forbidden.

The first important application of photonuclear processes was in neon burning, which occurred between carbon and oxygen burning. In this case an alpha-particle would be ejected. Most of the time it would be captured again on an oxygen, which would result in no net change, but occasionally the capture would take place on neon to form magnesium, and in this way the neon would be gradually depleted. The second way in which this showed up was in the photodisintegration along the s-process path in oxygen burning; certain nuclei with low neutron binding energies would become photodisintegrated, thus establishing some road blocks along the way (Cameron 1959c).

The net result of carbon and oxygen burning was to form nuclear products in the range magnesium to sulfur. The final stage of energy generation transforms this group of nuclei into another group centered around ^{56}Fe. This process is strictly driven by photodisintegrations. Notice that the average mass number doubles here. In order to do this, half of the original nuclei must be completely disintegrated into neutrons, protons, and alpha-particles, and these in turn must be captured on the other remaining nuclei. To accomplish this the temperature must be raised above at least four billion degrees K.

I had followed all of this during 1958 and 1959, at least crudely. The paper describing the approach to nuclear statistical equilibrium and the formation of the iron peak was returned to me with referee's comments to be answered just at the time that I had left Chalk River to spend a year at Caltech. At that time I had become so immersed in other problems that I never got around to completing a revision of this paper, and it was eventually incorporated into my series of Yale lecture notes, to be described later (Cameron 1963).

Neutron Stars

The formation of the iron equilibrium peak is the end of the line for the thermonuclear evolution of a stellar interior. If the temperature is raised further, the equilibrium peak shifts downward to helium, absorbing a great deal of energy and causing the star to collapse. At the time that I was thinking about this problem, most astrophysicists considered the collapse to be an unsolved problem not worth worrying about because there were no relevant observations. There was no reason for this, because the fundamental physics of the problem were in place. Oppenheimer and Volkoff had shown us how to calculate the structures of neutron stars, and Oppenheimer and Snyder had told us how to calculate the collapse into a black hole. But the details of the relevant physics could be improved in many ways.

Oppenheimer and Volkoff, in their neutron star calculations, had treated the neutrons effectively as heavy electrons, meaning that they had ignored the nuclear forces between the neutrons. This was of considerable practical importance for the collapse of a star, because the Oppenheimer-Volkoff neutron star models had a maximum mass of 0.7 solar masses, whereas Chandrasekhar's white dwarf models had a maximum mass of 1.4 solar masses, only slightly reduced by the effects of electron captures at the center. That meant that if a star had enough mass to surmount the Chandrasekhar peak, it would also surmount the Oppenheimer-Volkoff peak, so that all collapses would produce black holes and there would be no neutron stars. This became a compelling question for me.

There are many components to the forces between nucleons, but in general there is a net attraction between them at longer distances, and a repulsion between them at short distances. This is why nuclear matter tends to have a constant density inside nuclei, except in the surface regions. In constructing models of neutron stars I used a potential that had been used for interactions in nuclear matter, and the resulting models had maximum masses considerably greater than the Chandrasekhar limit (there was an error in these calculations that did not invalidate this result). Thus it appeared that neutron stars could be formed in nature. At the same time, it also appeared inevitable that a sufficiently massive star would be able to collapse to form a black hole (Cameron 1959e).

But there were other aspects of neutron stars that I found fascinating. As the mass of such a star increases, the fermi levels of the fermions in the interior increase, until soon the fermi energies of the nucleons exceed the mass energy differences between them and other hyperons into which they could be transformed. Of course, any such transformation would also have to conserve momentum, but conceptually this is always possible at an energy threshold if two such nucleons meet head-on and each makes a transformation. In this way it became clear that a neutron star interior was likely to become a hyperonic zoo. The details of the zoo will remain somewhat obscure for some time because the nuclear forces between the different hyperonic species are not known in detail, and these forces determine the actual transformation thresholds.

These calculations formed a systematic campaign for understanding the nuclear consequences of squeezing and heating a body of hydrogen and helium gas through the range of conditions in a stellar interior. I did other nuclear astrophysical calculations in the following years, particularly related to the r-process, but I was already being diverted into thinking about other aspects of astrophysics and about planetary sciences.

Meteorites and Elemental Abundances

The key to sorting out the individual processes of nucleosynthesis was the fundamental paper by Suess & Urey (1956), published at an optimal time for me. They utilized elemental abundance data from a variety of sources, some astronomical, but mainly from meteoritical analyses, to construct a table of the abundances of the elements. The data they had to work with were very crude in most cases, so they needed a scheme to smooth these data and to interpolate the most uncertain elements. They recognized that the abundances varied in a remarkably smooth way, particularly among the heavy elements and especially among the rare earths. Therefore, their empirical smoothing and interpolation procedure consisted of insisting that the isotopes of odd mass number should vary smoothly in abundance, and they adjusted the elemental abundances to make this happen. This was the abundance distribution utilized by the Burbidges, Fowler, and Hoyle in their classification of nucleosynthesis processes (Burbidge et al 1957). I used it as well, with similar results.

But like most empirical procedures, the process was not perfect. The abundance table could be improved by using the identified characteristics of the s- and r-processes among the heavy elements, quasi-statistical-equilibrium abundances in the silicon to calcium region, and judiciously applied smoothing of odd mass number abundances elsewhere (Cameron 1959e). During the next two decades I produced a series of versions of the abundance table, applying these criteria, but also incorporating the better abundance determinations in meteorites, particularly carbonaceous chondrites, that were often produced in response to the need for improved abundances. Finally Ed Anders took over the task of improving this table, incorporating a lot of new measurements of his own. During this period the measured data became much better. I had tried to standardize on the Type I carbonaceous chondrites as seeming to represent the least fractionated source of abundance data. The most recent table of Anders & Grevesse (1989) contains very little data that are interpolated rather than measured.

An accidental but nevertheless significant byproduct of these tasks was that I became deeply involved in trying to understand meteorites: their compositions, their discrete components such as inclusions and chondrules, their ages, their thermal histories, and their origins. Without some understanding of these things I could not have confidence in using meteorites as abundance sources.

But it is impossible to discuss such questions in isolation. The formation of the meteorites reflects a number of important processes that occurred during the formation of the solar system and, in turn, the formation of the solar system involves many questions about the interstellar medium, star formation, and galactic history. And we come full circle to the advanced stages of stellar evolution, supernovae, the production of extinct radioactivities, and their introduction into the forming solar system. The interconnections among these topics have dominated my research interests for the last three decades.

Extinct Radioactivities and Solar System Formation

Jesse Greenstein had told me in the late 1950s that I would be welcome to spend a year at the California Institute of Technology on his Air Force contract. In 1959 I felt the time had come to take him up on this offer, my purely nuclear physics approach had run its intended course. I felt the need to immerse myself in an astrophysical environment, at least for a while. I went to Caltech with a fairly open mind regarding what I wanted to do. That did not include observing. My new colleagues at Caltech were quite surprised, because most people going there were anxious to take advantage of the facilities. But I did not see an observational program that would directly answer the physical questions in which I was interested, so I felt my time was best spent studying in the astrophysical library.

I found Fritz Zwicky to be very cordial toward me. For years Fritz had been pushing his ideas about neutron stars to anyone who would listen and had been universally ignored. I believe that part of the problem was his personality, which implied strongly that people were idiots if they did not believe in neutron stars. He was right, of course, but that was not the way to make friends and influence people. So he was delighted that someone had independently come to believe in his point of view.

Fritz gave me a diagram showing how the principal spectral features in a long-observed Type I supernova varied with time and asked me if I could make anything out of it. Apart from identifying the strongest feature as a helium line, I got nowhere with the problem and concluded that I was not really suited for that sort of thing.

I had not been at Caltech for long when John Reynolds (1960) at Berkeley announced the discovery of the extinct radioactivity ^{129}I (via its decay product, ^{129}Xe) in meteorites. This extinct radioactivity had been sought for some time, but its actual discovery triggered another major shift in my research interests. This nuclide was a product of the r-process and thus would be produced in supernovae. The key factor was the abundance of the decay product in meteorites (5×10^{-5} relative to the stable isotope ^{127}I), which implied important things about both the galactic environment and the formation of the solar system. Compared with a continuous production of r-process nuclei, the indication here was that the ^{129}I had been isolated for of order 10^8 years before winding up in primitive solar system material. I was inspired to think about the formation of the solar system and, to some degree, to think about the time distribution of supernova activity within the galaxy.

For a long time the problem of the origin of the solar system was considered a curious intellectual puzzle containing no real scientific content. People had long since noticed a few interesting regularities about the solar system. The angular momentum vectors associated with spin and orbital motion tended to be aligned, and the orbital radii of the planets increased in a roughly geometrical fashion. These were the facts that a successful theory of the origin of the solar system was expected to explain. The tools of the trade were various artificial constructs consisting of massive bodies passing close to the Sun, or of contracting gas clouds shedding rings of gas from time to time. No attempt was made to relate such scenarios to prior states of matter in the galaxy that could evolve into any of these scenarios. In other words, the entire subject was considered rather disreputable. People thought that you had to work backward from the present state of the solar system to determine its earlier states.

My approach to the subject was quite the reverse, no doubt influenced by my nuclear astrophysics background. The presence of ^{129}I in the early solar system, together with the general theory of stellar nucleosynthesis that had developed, indicated that 4.6×10^9 years ago nucleosynthesis was occurring, contaminating the interstellar medium with new elements, and interstellar clouds thus contaminated were contracting to form new stars.

A clear similarity existed between the supernova implosion of a massive star and star formation. In the case of a massive star, the central region contracted and heated until disintegration of the iron peak elements occurred, decreasing the adiabatic index below 4/3 and initiating a dynamic collapse. In the case of an interstellar cloud (this was before we knew about cores in the clouds) contraction and heating would occur until molecular hydrogen started dissociating and the adiabatic index fell below 4/3, similarly initiating a dynamic collapse. In the case of the cloud, the collapse would lead through the ionization of hydrogen atoms and the single and double ionization of helium atoms. The radial collapse would be modified by the angular momentum to form a central star surrounded by a disk. Apart from the details of needing to get the hydrodynamics right, the problem would then be to evolve this system forward in time toward the present.

The community now follows this approach, and it has proven much more productive than attempts to figure out the evolution in reverse.

A Time of Transition

At the end of my year at Caltech I returned to Chalk River, but it was clear that I had outgrown the place. I now needed an environment suited to supporting my research in nuclear physics, in astrophysics, and in the geosciences and planetary sciences. Also, the space age had begun, and all sorts of new initiatives in space research were opening up. The United States was involved in an intense international competition in this area, and resources were becoming available to do a lot of very interesting things.

Where did Canada stand in all this? I was appointed to a committee of the Canadian National Research Council to consider that question and make recommendations. The contrast with the United States was stark. The newly-formed NASA was planning to measure particles and fields throughout interplanetary space, to send missions to the planets, and to develop space observatories for astronomy. The Canadians thought that they would be able to send a few rockets to probe the upper atmosphere. That would hardly be an effort commensurate with the relative populations and economies of the two countries. I was terribly disgusted with this situation and very quickly decided that my future lay south of the border.

Thus it was that in the spring of 1961 I was one of the initial people hired when the Goddard Institute for Space Studies opened in New York City. GISS was attractive to me in many ways. It was part of a NASA initiative to establish good working relationships with academic people working in any of the space sciences, and in the general New York area there were many first-class universities with a desire to do things in the fields that were opening. GISS supported both senior and junior postdoctorals and academics on sabbatical or other short visits. I took a leading role in organizing scientific conferences held at GISS on quite a variety of topics, mostly in areas where I had some expertise or was trying to acquire some. With our resources we were able to support the preparation of proceedings of these meetings, which was helpful in giving GISS a distinctive academic flavor.

I developed relationships with several of the institutions in the area. GISS was physically located in buildings adjacent to or rented from Columbia University, and although I did not have an academic appointment at Columbia I did supervise several Columbia students from the physics department. I did a little teaching at New York University and supervised some of their graduate students in physics. I also had closer relationships with the physics department at Yale and at the Belfer Graduate School of Science of Yeshiva University, which I shall now describe.

In the spring of 1962 Vernon Hughes, chair of the physics department at Yale, approached me with the news that some of the more junior graduate students in his department were interested in possibly doing theses in astrophysics. He wondered if I would be interested in supervising them, in which case a visiting lectureship could be set up, I could come up to Yale for the day from New York about every other week, and this would also give me a chance to give some courses in the subject during these visits. So I went to New Haven to interview the students, of whom there were four. Three who agreed to try out this plan (the fourth switched to plasma physics) were Dave Arnett, Carl Hansen, and Jim Truran.

Since summer was fast approaching and I would not start seeing them on a regular basis until the fall, I thought about giving them a research problem that would cause them to read widely among astrophysical topics during the summer, but from which I would not expect quick solutions. So I asked them to calculate the chemical evolution of the galaxy. This task was critically dependent on the rapidly emerging theory of stellar nucleosynthesis, combined with the current status of stellar evolution calculations and early ideas about mixing processes within the galaxy. The first paper to emerge from these studies did not appear until three years later, but it was a good training exercise. There were so many unknowns and arbitrary assumptions in the exercise that I always referred to it as a game, not to be taken too seriously (initially, anyway). But gradually others treated the game quite seriously, particularly Beatrice Tinsley, and a recognized topic within astrophysics emerged.

The first three years of my Yale appointment were particularly memorable for me. I had become involved in several branches of science that had traditionally stood on their own with few connections to neighboring disciplines. These branches of science included nuclear physics, astrophysics, geophysics and planetary sciences, and meteoritics. For the first year of lectures I concentrated the course on nuclear astrophysics because the students would need this background for their thesis work. In my day at Yale I would give a two-hour lecture that would be tape recorded. I would then take the tape recording with me back to New York, where it would be transcribed at GISS. The tape and its transcript would the be turned over to the students at Yale; Arnett, Hansen, and Truran would then take turns in writing the lectures into sets of notes, incorporating material from the lectures and from reference material. The notes would then come back to New York to be reproduced and freely distributed, not only within the Yale physics department but to anyone who wanted them.

In the subsequent two years I decided to attempt something more ambitious. The objective was to describe the universe as seen through the eyes of a physicist. In this case "the universe" meant everything from the center of the Earth to the cosmological horizon. That meant I had to learn a great deal of new physics, and I found the lecture courses of Landau and Lifshitz to be ideal for that purpose. This was my real training for the interdisciplinary studies that would occupy me for the next three or four decades. Once again Arnett, Hansen, and Truran worked hard to turn the lectures into sets of notes. Again these were widely distributed. I had quite a few suggestions that I turn these notes into books, but that would have been an undertaking of sufficient magnitude to greatly interfere with the research I was doing. However, I still find these notes to be a valuable resource.

I carried on with my Yale appointment for another three years, involving a second set of graduate students. However, I did not again challenge myself with anything quite so ambitious. Meanwhile, I was undergoing further institutional transitions at home.

It is said that all institutions, particularly those involving a civil service, start out in a flexible way, but with the passage of time rigid procedures develop to make life less than fully enjoyable for those subjected to them. I had become an American citizen by this time, and thus was a regular member of the civil service. Since I had previously been a Canadian civil servant at Chalk River, in a sense I should have become adjusted to the evils of bureaucracy. But in all my interactions with universities while in New York, I had come to appreciate and value the greater sense of personal and academic freedom in such institutions.

In 1965 I was invited to become an adjunct professor of space physics at the Belfer Graduate School of Science at Yeshiva University in New York. A Jewish undergraduate institution, Yeshiva was nearly a century old and had ambitious objectives. It had developed a large medical school of excellent reputation, and the Belfer Graduate School of Science was a newer but parallel attempt to build an institution of similar quality in the physical sciences. I had long since learned to admire the qualities of scholarly dedication common among the Jewish people, and so I agreed to the adjunct appointment and to teach a course in space physics in the fall of 1965 at Belfer. A year later I was invited to join the faculty at Belfer full time. By now I had become irritated with the growing rigidities of the civil service, so I readily agreed to this new appointment. However, what I primarily shed were the civil service restrictions, since I agreed to continue contributing intellectually to GISS by spending a couple of days a week there. My Belfer students all had office space at GISS too, so they were not neglected.

In some respects my life now became a little more simplified. At the peak I was advising about a dozen graduate students from four institutions. Over the next two or three years this number diminished to about five students at just one institution (Belfer), and I had more time to think about science and to work on some of my own problems. Computing styles had changed. At GISS there was a central IBM computer, state-of-the-art for those days but very slow by contemporary standards. People programmed in Fortran, and you handed in decks of punched cards to be run by the computing staff and waited for the output to come back so you could discover your next bug and correct it. Interactive computing it was not. I was quite happy to channel problems requiring computing through a student or postdoc for a while, but then I would be asked to look through a lot of Fortran code to try to spot an error, and I would feel handicapped because I had not programmed in Fortran until then. So I set myself a programming problem in order to learn Fortran properly (calculating the structure of the primitive solar nebula), and although this took care of my learning problem, I badly missed the interactive character of the computing. So for a while I continued to rely mainly on students and colleagues.

The peak in government funding of science occurred about the time I moved to Belfer. Thereafter there was much more competition for these funds. The competition did not affect me in a significant way for quite a while, but funding for research in basic physics, chemistry, and mathematics became much tighter than funding for the space sciences (we were still carrying out the lunar landings), and this turn of events severely affected my colleagues at Belfer. Internal institutional funds became scarce and services suffered. It slowly became clear to me that the bright intellectual promise of the middle 1960s for the future of the Belfer School of Science would fade away in the 1970s. My future had better lie elsewhere. Since I was getting tired of the crime, grime, incivility, and congestion in Manhattan, I felt the elsewhere should not be there either. I was to be among the earliest of the faculty to leave Belfer, but a general exodus soon took place, and in a few years Yeshiva closed Belfer. It is a measure of the quality of my colleagues there that in the next few years many of them were elected to the National Academy of Sciences.

In 1972 George Field was designated the new director of Harvard College Observatory and was given an initial year to contemplate what new initiatives to undertake there. There had been a great deal of friction between HCO and the Smithsonian Astrophysical Observatory, which shared the observatory facilities in Cambridge. SAO had consisted originally of only a small group in Washington, but had moved to Cambridge to encourage its growth and diversification, with a Harvard professor (Fred Whipple) as director. Sputnik proved to be the magic propellant for this process: Whipple established a world-wide satellite tracking network, and SAO bulged at the seams. Under the circumstances, friction was inevitable. Early in Field's year of contemplation, the decision was made to merge the directorships of the two observatories under a single director, who would be a Harvard professor (and, initially, Field).

Thus it was that in the fall of 1972 Field asked me if I would be interested in a Harvard appointment. His motivation was that he was thinking of the need for an umbrella organization that would hold the two observatories together intellectually, and he was thinking of appointing a series of associate directors in the different subfields of astronomy as part of this umbrella organization. He had in mind appointing me as associate director for planetary sciences. I agreed to this proposal, and thus was set in motion the complex procedures involved in appointing a Harvard professor. The umbrella organization was to be called the Center for Astrophysics. During the next few months I talked to George on a number of occasions as his plans took shape. I moved my office to Cambridge in the summer of 1973, bringing to an end my years of transition. I have been in Cambridge for 26 years.

Organization of Conferences

Before returning to a discussion of my research programs, I want to describe some of the nonresearch aspects of my career. I have already mentioned my role in organizing a series of scientific meetings at GISS. These meetings were done with the encouragement of the director, Robert Jastrow, and were made possible with his discretionary funds. Some of these meetings were intended to review the status of an established field in which we had an interest, such as stellar evolution. Others were intended to explore new frontiers, such as meetings on the origin of the solar system, or on the origin of atmospheres and oceans. But we also had the ability to react quickly to important new developments, such as in our meeting on pulsars shortly after their discovery.

I was involved in organizing other conferences during these same years. Because of my nuclear astrophysics research, I was a regular participant in the Gordon Conference on Nuclear Chemistry during those years in which nucleosynthesis was one of the conference topics (1956, 1958, and 1961). These meetings were wonderful occasions. They brought together theorists such as myself with people working on meteorites. After the Suess-Urey paper on abundances (Suess & Urey 1956), this topic had become a prime source of interest for me. At the 1961 conference quite a few of the meteoriticists expressed a wish for more frequent meetings that would explore their field more thoroughly. Accordingly, Ed Anders and I agreed to try to establish a new meeting topic within the framework of the Gordon Research Conferences.

I took primary responsibility for negotiations with the Gordon Conference organization, and I found its officers to be interested and helpful. Thus it was that the first Gordon Conference on the Chemistry and Physics of Space was held in the summer of 1962 in Tilton, New Hampshire, with me as chairman and Anders as vice-chairman. The intent was to bring together scientists from the astrophysical and planetary science (including meteoritical) communities in the hope that the mutual interactions would produce interdisciplinary research. These objectives were partly successful: During the next decade only a few astrophysicists attended this conference, but in general the meteoriticists and planetary scientists loved it. Their fields were enjoying an influx of new resources as NASA prepared for the lunar landings, and new instrumentation was yielding important measurements of excellent accuracy. But at the end of the decade lunar samples were returned, and all of a sudden everyone had far too many meetings to go to, and the Gordon Conference on the Chemistry and Physics of Space suffered. I chaired the conference for a second time in 1970, but it died three years after that.

However, this was not to be my last involvement with the Gordon Research organization. At the end of the 1980s NASA began a possible new initiative to study the origin of the solar system as a scientific field, and in fact to consider planetary systems in general as part of this field. I participated in most of the organizing meetings, and when the program itself was approved and initial funds provided, I chaired its steering committee. It was clear to the steering committee that general meetings bringing together people from all the disciplines in the Origins program were desirable, and something along the lines of a Gordon Conference would be a good way to do this. Because I was the one on the committee with experience in dealing with the Gordon Conference organization, it was logical that I should be the one to explore that possibility. I found the organization more bureaucratic (because it now had a considerably bigger program) than it had been, but the arrangements for the new conference went smoothly enough. Thus it was that in the summer of 1991 I was the first chairman of the Gordon Conference on the Origins of Solar Systems.

During the 1960s, I had been an interested participant in the first two conferences on relativistic astrophysics, which were held in Texas in 1963 and 1964. There was no meeting in 1965, and it appeared to me that the organizers had run out of steam. Because this was a worthwhile activity, and because I was going to become a full-time member of the staff at Belfer in 1966, it seemed to me that hosting a third meeting in this series would be a way to put Belfer on the map, so to speak. I proposed this idea to the previous organizers. Their response was enthusiastic, and with their help we did it in New York City in early 1967. The local arrangements were ably handled by my secretary. The meeting itself was held in a hotel in central Manhattan, and the participants were transported to a reception at Yeshiva. This was thus the third "Texas Symposium," and it established the pattern for moving the site of the meeting to a different location every second year, soon to include overseas locations.

I became involved in a number of theoretical astrophysics workshops starting in 1971 when David Pines organized a Summer Study Group on Pulsars as part of the summer program at the Aspen Center for Physics. This occasion was most enjoyable and productive, fostering recreation with interactive discussions and lectures, and some of us started to think about repeating this kind of activity. So Stirling Colgate and I sought the participation of George Field (who was not at the meeting) in making a proposal to the Aspen Center for Physics to hold a three-week workshop in astrophysics each year and in proposing to NASA that it fund participation in the workshop. Permission and funding were obtained, and the astrophysics activity has been a part of the Aspen Center since then. I went regularly to the early workshops and then occasionally until 1985.

Another outcome of the discussions Stirling Colgate and I had during the Aspen workshop in 1971 was the organization of several theoretical astrophysics workshops in different places, the objective being to bring the free-wheeling spirit of the Aspen style to other institutions. The organization mostly fell to me, and my procedure was to find a co-organizer, usually at the destination institution. Six of these workshops were held, one at Green Bank, one in Santa Cruz, and four at the headquarters of the Kitt Peak National Observatory in Tucson. For the latter four we found a sponsor in Leo Goldberg, director of Kitt Peak, who thought it would be intellectually stimulating to bring theorists to interact with the observers in Tucson. Leo provided us some travel support. The sessions at Kitt Peak were held in the central courtyard. These were stimulating occasions, but I did not try to do them after 1975, and no more workshops of this kind were held.

My conference organizing activities continued for a short time after I moved to Harvard, but then declined. During my years in Cambridge I have been involved in a small amount of such organizing, but not on a scale comparable to my earlier efforts. This change seems to be correlated with the availability of personal computers, so that it became more efficient for me to write my own letters and papers using a word processor, and my need for a secretary decreased to the point where I could no longer justify the salary involved in competition with scientific expenditures. At that point the amount of personal effort involved in organizing a conference became a barrier I seldom wanted to surmount.

Planning for Science

Space Science is Big Science, and it costs Big Public Money. As such it requires Big Justification, involving Big Planning and Big Advice. I have been involved in several stages of this process.

I became a NASA employee in 1961, and as such I could be only a member of NASA internal committees for the purpose of giving advice to NASA. But my first taste of the planning process was through my attendance at part of the 1962 Space Sciences Summer Study of the Space Science Board, held in Iowa City under the chairmanship of Jim Van Allen. I was there not as a participant, but as a resource person. I had voice but not vote. That same year I became a member of the Astronomy Subcommittee of the NASA Space Sciences Steering Committee. There I had voice and vote. This was a useful introduction for me to NASA operations, and it was useful to visit several NASA centers for its meetings.

In 1965, while still a NASA employee, I again attended the Space Sciences Summer Study of the Space Science Board, again with voice but not vote. But that year I also became a consultant to the United States Atomic Energy Commission at Livermore, California, in connection with its Project Plowshare, and to the NSF Advisory Panel on Atmospheric Sciences. So my knowledge of government advisory activities was expanding rapidly.

After 1966 I was no longer a NASA employee. Starting in 1969, I began a series of advisory activities that became quite intensive. In that year I served on the Astronomy Survey Committee (the Greenstein Committee) of the National Academy of Sciences, on its Steering Committee and as chairman of its Space Astronomy Panel, and in parallel with that as a member of the Space and Planetary Physics Panel of the Physics Survey Committee of the Academy. These activities extended through 1971. That year marked the beginning of my participation in a series of internal NASA planning meetings for the exploration of the outer planets. Several committees were working simultaneously toward the end of the process in 1974, and in 1974 and 1975 I chaired a small committee composed of the chairs of the other committees to produce an integrated plan for outer planets exploration. I attribute my rapid and extensive involvement with these various activities to the intensive learning process I had undergone in connection with my Yale lectures and with the interdisciplinary diversification of my interests. These experiences gave me a broad overview of the interesting problems in many fields of space science, and so did the variety of committee activities themselves. This overview and breadth of interests meant that I did not have any particular axes to grind and people could view me as an impartial chairman.

In 1973 I became a member of the Committee on Planetary and Lunar Exploration (COMPLEX) of the Space Science Board (SSB) of the Academy. At that time the SSB issued annual reports to which the committees of the Board contributed. The SSB and its committees played a rather passive role in this process, listening to NASA present its plans for space flight activities, and then generally blessing them. This passive role was about to change.

In 1974 I became a member of the SSB itself. More importantly, Jerry Wasserburg was brought in to chair COMPLEX, of which I remained a member. Jerry started by pointing out that NASA could claim the blessing of the SSB for any activities that it wanted to carry out, as long as NASA called those activities by the label the SSB or its committees had blessed. The reason for this was that the SSB was endorsing the code names of the flight programs that NASA wanted to fly, but it was not defining the contents of those programs, which could thus change and often did.

Jerry proposed that COMPLEX should define the scientific objectives of a program but not any program names by which such a program had been advocated. A simple illustration might be the following: According to the old way, COMPLEX might say, "we endorse the NASA Mariner Jupiter Orbiter program." But it would be better to say, "during the next decade NASA should explore the Jupiter system, characterizing the planet's atmosphere by measuring the composition and structure to the following accuracies..., and measuring the properties of the major satellites to the following accuracies..., and mapping the intensities of the system's particles and fields to the following accuracies..." No mission code name would be mentioned.

A first step in this direction was to systematize planetary investigations. We defined three evolutionary stages in such investigations. The first stage is reconnaissance, in which there are typically one or a few planetary flybys and some basic measurements are made. The second stage is exploration, which would typically involve one or more orbiters, and the third stage is detailed investigation, likely to combine orbiters with landers. For each target system a systematic strategy of study can thus be defined, and COMPLEX would formulate its recommendations in these strategic terms. The time period allowed for the strategy to be carried out was usually a decade, which would leave NASA free to formulate its tactics within its fiscal and technological constraints, and its success (or lack thereof) would be judged by how well the strategic objectives had been met.

During the 1974–75 academic year, COMPLEX followed these principles in formulating a strategy for exploration of the outer planets, and the next spring its recommendations were incorporated in its report to the SSB. Usually in the past the SSB had picked samples of its committee reports to incorporate in its own annual report. This time it faced a more contentious proposition. Wasserburg demanded the SSB review the COMPLEX report in detail and incorporate the whole report as part of its own annual report. Needless to say, that meeting was rather wild.

Other committee chairmen complained that endorsing the entire COMPLEX report would exhibit undue favoritism for COMPLEX, and in any case there was insufficient time for such a review. After a lot of screaming and shouting the proposition to do the review passed by a slim margin, provided the Board would be given the next morning a list of the actual recommendations contained within the COMPLEX report (and there were a lot of them). Evidently Jerry did not get much sleep that night, but in the morning the list was there, and the Board accepted as its policy these recommendations after the usual minor wordsmithing.

This marked a major turning point for the Board. There quickly came a consensus that all of the committees should undertake to produce strategies for research in their disciplinary areas. It was clear that these exercises would take longer than a year to prepare, and so the idea of the SSB annual report was abandoned. Henceforth the Board would produce specialized reports, many of which would be strategy documents. COMPLEX itself foresaw the need for two more documents to round out its scientific overview: strategies for the inner solar system and for minor bodies.

The committees were, in general, slow to come to grips with the task of preparing a strategy. The planetary exploration model did not transfer easily to many other disciplines where research was not programmatic—astronomy, for example—where one has to think about how to prioritize instrumentation because observational programs are either individualistic or involve small consortia. The committee chairmen had difficulty understanding how to distinguish strategic concepts from NASA program planning. So the Board and its committees experienced a lot of confusion. But COMPLEX was able to make rapid progress on a strategy for the exploration of the inner solar system, still guided by Jerry Wasserburg.

In the spring of 1976 I was elected to the National Academy of Sciences. Shortly thereafter I was asked to become chairman of the SSB. I was hesitant about that because I knew it would consume a great deal of time and effort. But I recognized that the logical candidate for the job should be someone who was an SSB member, who was an Academy member, and who had gone through the COMPLEX experience. So I felt that I had to do it.

During the first year of this chairmanship, I estimate that I spent more than half the time on SSB affairs. In subsequent years it was about a third of the time. Of course, impartiality required that I resign from COMPLEX. But I made a point of attending one meeting of each committee each year.

The evening before each SSB meeting I held a meeting of the committee chairmen to discuss everyone's progress and to make plans. From time to time I attended meetings of the SSB's sister boards: the Space Applications Board and the Aeronautics and Space Engineering Board. I was ex officio a member of the NASA Advisory Council. I met periodically with the NASA administrator but saw more of the associate administrator for space science. On a few occasions I met with the President's science advisor.

When Congress was considering the annual budget proposals for NASA I would sometimes give testimony at the senate or house hearings to present the SSBs views and try to explain our strategies. I attended summer studies. In short, I was busy.

In many of the committees progress in strategy preparation was slow or stalled, often because the chairmen were not quite sure how to proceed or how to formulate the tasks for the committees to perform. Often this meant waiting for a committee chairmanship to reach its end and then to make a judicious choice of the successor with strategy building in mind. In several cases I was able to choose a COMPLEX veteran who had become familiar with the nuts and bolts of forming strategies. But as my own three-year term came to an end there remained a great deal to do to finish these strategies. I asked a few SSB members to form a panel to recommend the next chairman, and then I kept clear of their deliberations to avoid any hint of trying to influence their choice. That was clearly a mistake. They informed me that they had consulted widely among their colleagues and there was a very clear consensus that I needed to stay on for a second term to get the strategies finished. I was rather dismayed, but felt I had to consent.

When my second term came to an end in 1982, the strategies were either all done or in their last stages of preparation. I felt rather worn out by all the advice-giving exercises in which I had been engaged. I was pleased that after that I was not often called upon to sit on advisory committees, and I felt that I had done my duty to participate in the public side of science. NASA showed its appreciation by giving me its Distinguished Public Service Medal.

Local Administration

I mentioned that when the Center for Astrophysics was set up in 1973 I came in as associate director for planetary sciences. Actually this position did not involve a great deal of administration because the job was mainly one of coordination of activities in the two observatories. Each observatory retained its own budget under control of the director, George Field, and he chose not to delegate much authority to spend the funds to the associate directors. Research activities were largely funded by grants and contracts anyway, usually under control of the principal investigators. So to a considerable extent the associate directors formed an advisory panel to the director. I was happy enough with that procedure because it did not eat into my research time too much. For me the most distasteful aspect of the job was the fact that I was actually administering the work of a number of SAO employees, and so I had to subject them to an annual evaluation process, which I regarded as demeaning to professional peers in sharp contrast to the collegiality of the relationships on the Harvard side.

Field actually felt the need for advice from a smaller, more informal, less constrained group, and thus over time there developed a "kitchen cabinet" consisting of Riccardo Giacconi, Herb Gursky, and me, often attended by John Gregory, the chief administrator on the SAO side. This cabinet met more often than the associate directors met and tended to deal with the more contentious issues facing the center.

In 1976, about the time that I was becoming chairman of the Space Science Board, the dean asked me to become chairman of the astronomy department. Unlike the situation in many astronomical institutions, the department was a separate entity from the observatory and was administered separately, with the professors responsible for educational activities and making recommendations, through the chairman, to the dean of the faculty of arts and sciences (FAS). It had its own budget, which was negotiated each year with the dean's office. I felt that a term as department chairman was unavoidable in the long run, so I might as well serve my turn. I accepted the appointment.

At that time the department had an unfortunate reputation for neglecting its graduate students, who were taking too long to get their degrees. Over a period of time as chairman and later as chair of the departmental committee on academic studies, I managed to improve their lot significantly. We went through a series of graduate and undergraduate curriculum reforms. The committee on academic studies set stricter timetables for the progression through graduate studies.

Probably the most useful innovation was the establishment of individual thesis advisory committees (TACs) for each student after he or she had finished the first stage of graduate work, consisting of a research project. Several faculty members would sit on a TAC, and frequently one or more Smithsonian scientists and someone from outside the Center for Astrophysics. A TAC would normally meet twice a year to hear a report from a student and to give advice, including whether it was reasonably feasible that a given research project could be completed in the estimated time. The TAC could meet more frequently if needed in difficult cases. The students have generally valued the TAC experience.

Before the establishment of the Center for Astrophysics, Harvard College Observatory had its own deliberative body called the Observatory Council, consisting of both professors and research appointees. The council had fallen into disuse under the center. However, the directorships of HCO and SAO are not collegially equivalent: SAO is a line organization and HCO is not. Thus in time the staff members on the Harvard side felt that they were not receiving as much information about Harvard-side expenditures as they were accustomed to, probably because SAO was now a considerably bigger organization and took a much larger share of the director's time. A grass-roots effort led to the revival of the Observatory Council. It was agreed that when desirable the Council would meet, with the department chairman also chairing the Council.

Another way in which I managed to improve the lives of the graduate students was to use gift and discretionary money to buy them their own Unix computer. Things change slowly at Harvard. The reaction from the dean's office was, "but graduate students have always paid to have their theses typed; why should you be using departmental money to let them do it themselves?" Within the next two or three years they came to realize why it was appropriate.

The normal term of a department chairman is three years. My term was extended twice, in one- and two-year increments, to a total of six years. Thus it happened to coincide with my chairmanship of the SSB. I had managed to drop the associate directorship for planetary sciences a little earlier. But as chair of the Committee on Academic Studies I continued to be involved in departmental administration for a few years under my successors as department chairmen. After that I was mostly free of departmental administrative duties. But I did serve another term as associate director within the center, under its newer director, Irwin Shapiro. This time I was in charge of theoretical astrophysics. This experience was similar to my previous one, except that the Center for Astrophysics was now quite a lot larger, primarily because of growth on the SAO side, and there was more business to be transacted at the meetings.

The Evolution of Computing Technology

I have mentioned earlier in this account my experiences in mechanizing my computing at Chalk River and my learning experience in the "hand in your card deck" mode at GISS. As long as I was at Belfer I had ready access to the computer at GISS. But when I went to Harvard I found that the central computer was an SAO machine from the Control Data Corporation (CDC) (the CDC 6400), and you had to pay for your computing time from grant funds. You still handed in your card deck, and the computing rates were high, because the machine was very expensive and had high maintenance costs. So I looked for a way to recover the style of individual interactive computing that I had grown to love at Chalk River.

I found it in a Hewlett-Packard programmable calculator that used interpreted Basic. It also had a 5 MB hard disk attached. The calculator had a speed comparable to the Datatron machine I used at Chalk River and considerably more storage and was sufficient for the relatively simple calculations I was doing at the time because I had the machine to myself and I could run it 24 hours per day.

But simple calculations usually lead to more extensive ones, and after two or three years I began to get frustrated with the slow speed of the machine I had, and I began to look at other options. It was now the mid-1970s, and minicomputers had appeared on the market and had become popular, particularly for controlling research equipment. But they also ran Fortran and you could get hardware floating point units. I considered Digital Equipment Corporation (DEC) minicomputers, but was bothered by the small exponent range of the floating point numbers, which was rather unsuited to many astrophysical calculations.

I found that Data General (DG) minis were better in this respect, with a larger exponent range, and so I wound up getting a DG Nova 3. A little later we added a Nova 4. Each machine had a working memory of 256 KB, and we eventually got a couple of 75 MB hard disks, each the size of a washing machine. Each machine could also support two users (or alternatively, two working programs). This was a significant step forward. My postdocs compared the execution times of some of their programs on the CDC 6400 and our Novas, and found that when fully loaded our machines had a throughput well over half that of the 6400 at a tiny fraction of the cost. In the late 1970s we were running stellar evolution programs with lots of bells and whistles, as the jargon put it, including using equations of state with nonideal gases. The nice thing was that we could run the machines full time without worrying about running out of dollars in a computing account. And it was interactive computing.

Also in the late 1970s the microcomputer revolution began. It was an ideal time for me because by now my postdocs and I were competing for time on the Novas, and it was difficult to find the time for mundane tasks such as preparing letters or scientific manuscripts. I began to invest in relatively inexpensive microcomputers for this purpose, thus unloading some of the work from the Novas. These early machines ran an early and very simple operating system called CP/M, which was a precursor to DOS.

The stage was now set for what I think of as the big computer war within the Center for Astrophysics. The CDC 6400 was nearing the end of its useful lifetime. The aging components were difficult to replace and maintenance costs were becoming unbearable. It was an important matter for the associate directors to consider.

The SAO computer manager wanted to upgrade to a newer CDC machine. John Gregory, the head of the SAO administration, also favored this approach and began negotiations with CDC. Some other people began to wonder, "There are rumors about getting a Boston-area Cray to be shared among the universities here. Couldn't we replace the 6400 with part of the time on such a Cray?" I said, incredulously, "You want to do the SAO accounting on a part-time Cray?" I maintained that the cost/performance ratio was much lower for minicomputers than on CDC mainframes and that we were entering a new era in which computing would be cheap and distributed among many machines, to judge from my own experience. The net result of this debate was a decision by the associate directors to set up a computing committee, and each was asked to appoint a representative to the committee. I appointed myself.

I found that committee's deliberations to be educational. A number of minicomputer companies made presentations, and we visited some of them to see demonstrations. We prepared benchmark programs representing the variety of astrophysical computations typically being done on the 6400 and saw them run on the competing machines. The CDC entry with a performance comparable to the 6400 also ran the benchmarks. The CDC salesmen were shocked to find themselves challenged from below by what they regarded as inferior technology, but nevertheless technology with a lower cost/performance ratio. Late in the game DEC informed us under nondisclosure that they were bringing out a new machine, a kind of superminicomputer, called a VAX 11/780. We benchmarked this machine and found it the best of the lot, with a performance comparable to that of the 6400, but with one glaring problem: the limited floating point exponent range. But DEC apparently wanted to generate forward momentum for their new machine from the prestige of a sale to SAO, so they promised to provide a hardware option with a much larger exponent range. At that point the battle was clearly over, although it took a while for the grumbling to die out. The purchase of the VAX proved to be the first step in the proliferation of many computers throughout the Center for Astrophysics.

Although the Center was now set in a new direction, the decision to purchase the VAX had little effect on my own computing. We owned the Nova minis, and just as we had a throughput comparable to that of the 6400, it was also comparable to that of the VAX, and the computing charges to use the VAX included paying for the salaries of the computing staff, which was thus more expensive than computing our way. The situation continued until the middle of the 1980s.

When Willy Benz came to Harvard and wanted to work with me to run his smoothed particle hydrodynamics (SPH) program, I bought a Sun workstation that would be capable of running the software. This move produced a surprise of its own: the program had been running on Cray XMP machines at Los Alamos (I'll describe it in greater detail in a subsequent section). We found that it lost only a factor 7 in speed on the Sun. The basic reason for this is that the program did not vectorize efficiently and therefore ran inefficiently on the Cray. But this was a big jump up in computing power for us.

My computing experiences since then have led me through a sequence of Sun and DEC Unix workstations, and more recently Intel workstations running Windows NT. The prices of such workstations have tended to remain about the same or to slowly decline, and the performance to be expected from the machines has doubled every year or two. Thus computing is powerful, interactive, and distributed, and the desired goals of my computing history have been achieved. Nevertheless, I look forward to the yearly improvements. Productivity continues to increase quite rapidly.

Nuclear Astrophysics

In the account of my experiences in New York and afterwards, one underlying feature is the broadening of my scientific interests that developed in response to scientific challenges and to the needs of the advice-giving processes. Thus an account of my scientific history becomes complicated because I tended to oscillate between astrophysical and planetary science problems and often these problems were interrelated. Given the question of whether I should follow a chronological account and jump back and forth among different fields, or take broad fields one at a time and jump back and forth over the years, I have chosen to do the latter.

When I started thinking about thermonuclear reactions in stars in 1953, the main sequence and immediate postmain sequence nuclear transformations in stars like the sun had been explored, but beyond that the subsequent course of stellar evolution was a complete blank. While at Chalk River I was able to find the stellar burning pathways through to the three possible endpoints of stellar evolution, at least in principle. But I was very far from determining how the stars follow such pathways, in whole or in part, in the course of their evolution. Such questions required the combinations of expertise involved in writing a stellar evolution code, and, perhaps even more importantly, the use of hydrodynamics to study explosive nucleosynthesis and to understand the highly complex phenomena associated with supernova explosions. The Suess and Urey paper of 1956 allowed nuclear physicists to determine the principal processes that had been involved in the formation of the elements, but beyond exploring how the nuclear processes themselves would go, one could say very little about the stellar environments in which they took place.

Thus, in retrospect, it seems that 1960 was a good year for me to begin the process diversifying my interests into other areas, particularly related to the formation of the solar system. Over the subsequent quarter century I have from time to time worked on many other problems involving nucleosynthesis, usually in collaboration with a student or postdoc, but it has helped that the theoretical understanding of stellar evolution and supernovae has advanced as far as it has. More detailed calculations can be made in application to specific stellar scenarios. But none of this work has really occupied me full time.

One problem I was unable to solve fully plausibly was the environment in which the r-process operated. The nuclide abundance peaks associated with the r-process indicated that the mean capture path associated with this process lay well to the neutron-rich side of the valley of beta stability, so that the beta decay lifetimes would be small fractions of a second, and clearly some kind of supernova environment would have to be involved. Although with various collaborators I have suggested a number of different r-process scenarios, none of them proven to be fully satisfactory. The topic remains in unsatisfactory condition.

During the 1960s and 1970s my work in nuclear astrophysics consisted largely of refinements, extensions, and variations on the previous work. New nuclear mass formulas were generated (Cameron & Elkin 1965), and their parameters were used to improve the semi-empirical fits to nuclear level densities (Gilbert et al 1965, Gilbert & Cameron 1965) and nuclear radiation widths, thus improving the tools being used to estimate neutron capture cross sections and other nuclear reaction rates (Truran et al 1966). These tools also let me explore equilibrium conditions at high density (Tsuruta & Cameron 1965).

A fair amount of work was done on the properties of neutron stars, even before they were positively identified as pulsars in 1968 (Tsuruta & Cameron 1966a, b). It appeared that a number of interesting phenomena might be associated with vibrations of neutron stars, but I lost interest in that possibility when Bill Langer showed in his thesis that the transformations between internal hyperons due to the vibrations would damp the vibrations in a matter of seconds (Langer & Cameron 1969). My thinking about the rotational properties of neutron stars did not lead me to predict the actual properties found for pulsars. I found it amusing that at a scientific meeting in Bombay in late 1968, Henk van de Hulst remarked that theorists had never succeeded in predicting completely new types of astrophysical objects. But by the time he had returned home from that meeting that statement was no longer true, since the period decay of the Crab pulsar had been found and had removed any doubt about the reality of neutron stars.

My interest in higher densities led me to consider some variations on the URCA process in which a nucleus alternates between beta decay and electron capture, thus emitting both neutrinos and antineutrinos and cooling the material. One way to do this is to locate the nucleus close to a boundary in degenerate material where electron capture on a particular nuclear species changes the composition near the stellar center. Purely thermal broadening of the Fermi surface can open phase space so that both processes are possible. But the processes happen more dramatically if the transforming nuclei are physically transported repeatedly across the composition boundary. This can happen in density vibrations in which the Fermi energy is raised and lowered (Tsuruta & Cameron 1970), and also in convection processes in which the interior is stirred (Tsuruta & Cameron 1976).

Solar Nebula Processes

I mentioned earlier that initially I started thinking about the origin of the solar system in terms of the collapse of an interstellar cloud. Given that the cloud had some angular momentum (if nothing else, due to the constraints of magnetic lines making the cloud corotate with its motion around the center of the galaxy), the collapse might be expected to form a central object in hydrostatic equilibrium surrounded by a circumstellar disk. The disk would be in nearly Keplerian rotation about the central protostar.

In order to understand what would happen in this kind of system, some sort of models were required. The simplest approach would be to decouple the calculation of the vertical structure of the disk from the calculation of the radial structure. When I was at GISS learning to program in Fortran I experimented with this approach to calculating disk structures, and later Milton Pine expanded on this in his thesis (Cameron & Pine 1973). The results of these numerical exercises were useful for a large number of back-of-the-envelope calculations on the behavior of matter in the solar nebula, particularly on the role of turbulence in promoting the growth of the original interstellar grains in the nebula into larger structures (Cameron 1973).

No attempt was made to evolve the structures by estimating the transport of radiation or angular momentum (and hence mass), but it was clear that significant cooling would occur in times ranging from years in the inner nebula to perhaps thousands of years in the outer nebula. These first attempts to quantify the conditions in the early solar nebula had the unfortunate effect of making meteoriticists think of the nebula as a "hot nebula," and thus to assume that the solid materials that they dealt with in their meteorites had to condense from the initial gas in the nebula in a distinct condensation sequence. These ideas persisted for several years. Even today many meteoriticists speak of "conditions in the solar nebula" without specifying the evolutionary stage to which their remarks should apply.

Meanwhile, during the preceding decade at GISS and Belfer I had been collaborating with Dilhan Ezer in a variety of stellar evolution calculations emphasizing the premain sequence stages of stellar evolution. The starting point for a study of solar evolution would be the condition that all the energy released by gravitational contraction would go into the dissociation of hydrogen molecules and the ionization of hydrogen and helium. This condition established an upper limit to the initial radius of the sun by neglecting the loss of energy by radiation from the solar material on the way toward this state (Ezer & Cameron 1965). Other studies during this time explored a wide range of stellar masses (Ezer & Cameron 1967a, b), premain sequence evolution with mass loss (Ezer & Cameron 1971a), and prenucleosynthesis stars composed only of hydrogen or of hydrogen and helium (Ezer & Cameron 1971b). The latter studies were extensions of my earlier program to calculate the sequence of nuclear burning stages starting with a pure hydrogen gas.

Later, at Harvard, I tackled the more realistic problem of forming sun-like stars from a solar nebula. First, postdoc Richard Epstein developed an equation of state good for lower temperatures and nonideal gases. His successor, James "Jas" Mercer-Smith, then modified a stellar evolution code to add material to the surface of the model at a rate that would be consistent with the transport of mass through the solar nebula by some kind of dissipation process, so that we grew the model from a small initial mass seed. In this manner we discovered a phenomenon that so far no one else has bothered to incorporate in premain sequence studies. In the central entropy effect, the small seed will have a low entropy, and the entropy of the surrounding layers will rise because of the increasing amounts of gravitational energy that is released by the accretion. Hence deuterium ignition starts off center, and it takes of order 10^8 years for thermal conduction in the center to raise the temperature there to the ignition point of deuterium. During this time the Sun has a high-density semi-degenerate core and a somewhat expanded envelope (Mercer-Smith et al 1984). Only people doing radially symmetric hydrodynamic collapse calculations have found this effect, but they usually do not report it in their papers.

In this early approach to modeling the solar nebula it was necessary to identify processes that would lead to dissipation in the gas and that would be sources of viscosity, so that the Lynden-Bell and Pringle prescription for the evolution of such disks could be applied. I assumed that turbulence would do this job, arising both from thermally driven convection within the disk (due to vertical transport of energy in at least some parts of the disk) and from mechanically driven mixing due to the infall of gas clumps onto the disk as part of the molecular cloud collapse that itself would become irregular due to turbulence in the cloud.

More recently it has become evident that magnetic instabilities (the Balbus-Hawley effect) will inevitably play an important role in parts of disks with significant amounts of ionization. But by far the most important source of dissipation and angular momentum transport now appears to be gravitationally driven spiral density waves within the disk, which can operate with great efficiency. These density waves have been successfully modeled with three-dimensional hydrodynamics, but the dissipation rates are so sensitive to the extent by which the mass surface density of the disk exceeds a critical threshold that we do not have good analytical expressions to simplify the treatment and we have to parameterize the transport rates (Cameron 1995).

During the last three decades a great deal of progress was made in understanding the accumulation of the terrestrial planets, due in particular to Viktor Safronov and George Wetherill. Meanwhile, I thought a bit about the formation of the giant planets.

It was important to have better models of the interiors of the giant planets. The early work in this direction was done by Hubbard (1969, 1970), whose approach was to make numerical models and to fit them to the planetary parameters by varying the hydrogen/helium ratio in the interior. I found this quite disturbing from a cosmogonic point of view, since a large planet such as Jupiter should represent a good sample of the solar composition where the hydrogen/helium ratio was fairly well determined. Thus this ratio should agree with that in the Sun for the giant planets (at least in bulk, allowing for possible gravitational settling of helium). I thought that a better approach would be to fix the ratio in the giant planets and to vary the size of the core, which would be composed of elements heavier than helium, to fit the planetary parameters. This approach was followed by Morris Podolak in his thesis and subsequently by Podolak & Cameron (1974, 1975).

But I also thought that the giant planets, particularly Jupiter, might have formed as a result of gravitational instabilities in the solar nebula. In our modeling exercises we had found that the amount of heavier elements in the giant planets provided planetary cores of 10 to 20 Earth masses (in Jupiter a solar composition would give a core of only about one Earth mass). I investigated these "giant gaseous protoplanets" (often called GGPPs in the literature), with the idea of understanding how they could pick up the additional heavy elements as a result of infall (DeCampli & Cameron 1979, Slattery et al 1980). But then David Stevenson persuaded me that such infalling heavy elements would dissolve in the gas phase before they could penetrate to the core. So I abandoned the GGPP idea, and thought about the cores as forming by basically the same mechanisms as the terrestrial planets and gravitationally capturing gas from the solar nebula, an idea that I had previously explored (Perri & Cameron 1974).

The formation of a massive gaseous envelope as the core grew in mass had been investigated by Fausto Perri for his thesis, who found that there eventually occurred an instability in such a system whereby the gas would rapidly accumulate onto the core (not quite a dynamical collapse). This picture has dominated thinking in the field for a number of years, until things have reversed themselves again. A few years ago Willy Benz found, in the course of numerical hydrodynamic simulations of the solar nebula and the resulting spiral density waves, that there was a tendency to have a few local instabilities in the density waves, leading to gravitational collapse. Benz' finding has been confirmed with more detailed computations by Nelson et al (1998). So GGPPs are now back in favor. My estimates of the rate of radial migration of solids in the nebula (Cameron 1995) should allow the planetary cores to grow to a reasonable size before the gas becomes dense enough to dissolve the incoming material.

Time Constraints and Triggered Star Formation

Meteoritic abundances of the elements played a key role in the early determination of the nucleosynthesis processes. Currently the refined isotopic analyses of meteoritic components are revealing a much more complex and interesting situation concerning astrophysical and solar system environments. Bulk samples of matrix material, and chondrules and inclusions, which are themselves bulk samples, sometimes reveal isotopic abundance patterns for certain elements that deviate significantly, and in different ways, from normal solar system material. The most spectacular anomalies have been found in individual interstellar grains extracted from meteoritic matrices, which clearly sample individual stars (see for example Bernatowicz & Zinner 1997). But sometimes bulk samples also have large anomalies. Particularly interesting are the fractionated and unknown nuclear (FUN) inclusions in which many different elements exhibit these anomalies. Guy Consolmagno and I were able to show that the anomalies in the heaviest elements can be interpreted most simply as a change in the abundance ratios of the s-process and r-process components of the elements involved, and in a self-consistent manner (Consolmagno & Cameron 1980).

It is my own belief that these are inclusions from other young stellar systems that had not yet been incorporated into meteorites when they were caught up in bipolar outflows from those systems and thrown into interstellar space, subsequently to become part of the solar nebula. This young field of research is very promising.

Extinct radioactivities form a different kind of isotopic anomaly, because they can arise in material with a normal solar system composition, but in which the material involved in the extinct radioactivity was transported from its point of nucleosynthetic origin to the solar nebula in a time not longer than several half-lives of the radioactivity.

I described the impact the discovery of extinct ^{129}I had on the direction of my career. Many other extinct radioactivities are now known. From the point of view of timing the formation of the solar system, the most important of these are the shortest-lived ones, which happen to have the lowest mass numbers: ^{26}Al, ^{36}Cl, ^{41}Ca, ^{53}Mn, and ^{60}Fe. All except the fourth of these can be formed by neutron capture processes, and at one time I thought that AGB red giant stars might have injected them into the solar nebula. Later I came to favor production of all five of these extinct radioactivities in collapse-type supernovae. Here the matter of timing becomes crucial.

If a core in a molecular cloud forms at a density of about 10^{-21} gm/cm^3 (about 10^3 atoms/cm^3), then a uniform sphere starting in free fall at this density takes about two million years to complete its collapse. A problem then arises with the presence of ^{41}Ca in the solar nebula, since it has a half-life of 10^5 years. It already requires about four or five half-lives to travel from the site of the supernova to the position of a molecular cloud core, for if it were closer the core would be shredded to pieces and could not form the solar system (Vanhala & Cameron 1999). Therefore the abundance of ^{41}Ca would be depleted by a factor of more than 10^7 before the core could finish free-fall collapse and the radioactivity could be incorporated into a meteoritic inclusion. This abundance is already much smaller than the abundance observed in inclusions.

But matters are worse if the cloud core starts from a condition of rough pressure balance against gravity, for then the collapse time is increased by another significant factor. This consideration provided a compelling motivation for suggesting that the supernova shock carrying the extinct radioactivities triggered the cloud core into collapse. Numerical simulations of such collapses indicate that the total time required following this triggering event is only a few half lives, which then would be consistent with observation.

Jim Truran and I had already suggested the idea of a supernova trigger for formation of the solar system in 1977, and this idea attracted some initial interest (Cameron & Truran 1977). General isotopic anomalies in solar system material had provided the motivation at that time, but additional evidence was not forthcoming, and the idea became dormant. In 1995 I was part of a group that revived the idea of the trigger, based on an analysis of reported gamma rays from the Orion Nebula (Cameron et al 1995), but I now believe that this was the wrong reason for such a revival. The hydrodynamic simulations of the triggered collapse put the scheme on a more solid footing (Vanhala & Cameron 1999). Nevertheless, many quantitative issues remain to be solved in such a scenario, particularly concerning the details of the mechanism that can mix the supernova material into the collapsing core material.

The Formation of the Earth-Moon System

I have left this topic until the end because of a logical discontinuity with the remainder of the research topics on which I've worked. The other research topics were generally initiated as a result of questions arising in my other work. That is not the case here, although indeed the processes leading to the origin of the Moon must be integrated smoothly into the general picture of the origin of the solar system itself. That has not been the case historically, where the question of lunar formation has in the past been discussed independently of all such considerations. Of course, scientific attention was sharply focused on the lunar origin picture during the period of Project Apollo.

It was maintained by Harold Urey and by many other people who argued for manned exploration of the Moon that such an endeavor would allow us to determine the correct theory among the three theories of lunar origin: the Moon formed by rotational deformation and fission of the Earth, the Moon and the Earth formed in mutual orbit, and the Moon formed elsewhere and was captured by the Earth.

Each theory had its partisan followers. As the period of the lunar landings approached, a psychologist named Ian Mitroff obtained a NASA grant to interview the "moon scientists" during that time to determine their prior opinions about many aspects of lunar science including the question of lunar origin. His major objectives were to study the psychological aspects of the ways in which the lunar scientists formed their opinions, and to find out how these opinions would evolve in the light of the arrival of pertinent lunar data.

Somehow it was determined that I was a lunar scientist, and so I got onto Mitroff's list. Mitroff (1974) found that consensus had been reached about such straightforward questions as the ages of the lunar rocks, but on indirect questions such as the mechanism of the origin of the Moon, there was no consensus at all. The fission hypothesis had lost a little plausibility, but otherwise few opinions had been modified. In fact, many of the lunar scientists had relatively little interest in the question of origin. So it was clear that going to the Moon did not help us quickly identify the correct theory of lunar origin. The data did not point unambiguously at any of the theories; each of them had difficulties. Harold Urey's hope was not to be fulfilled.

Shortly after I moved to Harvard, I happened to be thinking about the large amount of angular momentum in the Earth-Moon system—it is much higher than normal for a terrestrial planet. I wondered how massive an object would be required such that, if it struck the Earth a glancing blow and stuck, it would give the Earth a rotation having the Earth-Moon amount of angular momentum. It is a simple high school physics problem, so it did not take long to determine that the object would have to have a mass of about one-tenth that of the Earth, or about the mass of Mars. That determination suggested an interesting scenario.

I had brought Bill Ward to Harvard as a postdoc, and the two of us played around with the idea for a while. In 1974 I went to a meeting on planetary satellites, and there I listened to Bill Hartmann (with Donald Davis) present a paper suggesting that a collision with the Earth might throw up enough material to form the Moon. Following the paper I remarked that I had been working on the problem of such a collision, but that in order to account for the angular momentum of the Earth-Moon system the collider would have to have at least the mass of Mars. This phrase has caught the public imagination and is usually quoted in connection with this problem, but most people have ignored the fact that this number is a lower limit. Bill Ward and I first presented the details of this requirement at a Lunar and Planetary Science meeting in 1976 (Cameron & Ward 1976), and we presented a follow-up analysis of the disk such a giant impact would generate at the same meeting a couple of years later (Ward & Cameron 1978).

There followed an extended quiet period in which nothing seemed to be happening regarding this problem, during which the amount of activity associated with lunar sample analysis gradually declined. However, a meeting was called in Kona, Hawaii, in 1984 to discuss the origin of the Moon. In preparation for that I made some simple calculations that used "particle-in-cell" hydrodynamics to study a giant impact. The principal conclusion from that exercise was that the material placed in orbit that would be available to form the Moon came predominantly from the impactor rather than the protoearth, contrary to what the geochemists tended to expect (Cameron 1986b). The Kona meeting was a lively affair and is celebrated in the field as the launching event where a bandwagon formed for the giant impact theory and most people hurried to jump on board.

A few months afterwards I gave a colloquium concerning the origin of the Moon at Los Alamos, and indicated the need for supercomputer calculations to study the giant impact. Afterward two people came to talk to me about this. One was a Swiss postdoc, Willy Benz, who had developed a smoothed particle hydrodynamics (SPH) code (then a relatively new technique) and who thought that it might be well suited to a giant impact study. The other was one of my former postdocs, Wayne Slattery, who was interested in working on the problem using one of the regular hydrodynamics codes, and who had access to large amounts of time on classified machines. A perfect combination! Our collaboration was established immediately.

Oral reports of early results were given in 1985 and a series of published papers on the giant impact started in 1986 (see references in Cameron 1997). In 1986 Willy Benz came to Harvard as a postdoc and later was appointed an assistant professor. Some of the computing continued at Los Alamos and the rest of it was done on several Unix workstations that we purchased in the next few years.

Giant impact calculations have formed a substantial part of my computing activities since then. In the earlier years an individual simulation on a small machine would take many weeks or months for a simple case involving about 3000 particles in the SPH code. Under such conditions these calculations were mainly sampling exercises to see how much material would be placed in orbit, whether it contained significant amounts of iron (undesirable) and how it was divided between the protoearth or the impactor sources. I tended to run cases with unreasonably large amounts of angular momentum (like twice that in the Earth-Moon system today) in order to get good yields to study.

Much more recently, with the development of inexpensive PCs, it became possible to make more rapid progress by carrying out systematic surveys of giant impact parameters, using 10,000, 20,000, and even 100,000 particles in the runs. All the original runs assumed that the giant impact occurred after the accumulation of the Earth was essentially complete (except for the addition of the impactor mass). However, the current best parameter fits require the use of larger impactors, and also that the impact occurred when the Earth was only about half accumulated. Also, I am now collaborating with Robin Canup, who takes my SPH output and integrates the paths of the orbiting particles much farther forward in time using a symplectic N-body integrator. In this way the combined results now go through the giant impact stage and then the lunar accumulation stage until the orbiting material either accumulates onto the Moon, escapes, or is deflected onto the Earth. Thus for the first time we are able to get complete simulations of a given case (Cameron & Canup 1998). Quite a bit of work remains to complete the parameter survey, however.

My Research Career in Retrospect

As I look back on the account of my research career, I am struck by how fortunate I have been in the timing of my research opportunities. My training was in nuclear physics, and the field of nuclear astrophysics opened up just at the right time for me. I had to teach myself astronomy and astrophysics to take advantage of the opportunity, but I was by then a physicist and this was an interesting application of the basic knowledge I had acquired.

Nowadays both physics and astrophysics are much broader fields, and I do not think that the average astronomer or astrophysicist has enough time to get the breadth of training at the graduate level that would really be desirable to get a good overview of all of the physical phenomena that contribute to the complexities of the various fields of the astrophysical sciences. Things have been rendered still more complex because we are seeing an intellectual merging of the geosciences with astrophysics. Again, at an early stage, I had to teach myself a lot of geophysics and even some mineralogy and geology.

Physicists often exhibit a great deal of arrogance, since they consider that they are practicing the queen of the sciences. But I have learned that there is a great deal of value in what people in the other sciences have to tell me about their observations and conclusions. I reserve the option of reinterpreting what they have told me so that it fits together with whatever else I know about the subject. Sometimes this process will cause me to change my mind about something I thought I already understood.

The core of my intellectual approach to trying to understand the universe is to seek consistency everywhere. Of course ugly facts are always coming along to spoil beautiful theories, but sometimes the revised theories that incorporate the ugly facts are even more beautiful. Sometimes they make you realize that nature is complex and you do not really understand it. But I have a reputation for frequently changing my mind and that is caused by the constant search for consistency. I am counting on that to sustain me and keep me mentally alive as I head toward retirement. But although I will have a pension to replace my salary, I want to keep the computers running as I try to resolve yet another inconsistency, as long as I am physically able to do so.

Visit the Annual Reviews home page at http://www.AnnualReviews.org

LITERATURE CITED

Anders E, Grevesse N. 1989. Abundances of the elements—meteoritic and solar. *Geochim. Cosmochim. Acta* 53:197

Bernatowicz TJ, Zinner E. eds. 1997. *Astrophysical Implications of the Laboratory Study of Presolar Materials*. Woodbury, NY: Am. Inst. Phys.

Brush S. 1996a. *Nebulous Earth*. Cambridge, UK: Cambridge Univ. Press

Brush S. 1996b. *Transmuted Past*. Cambridge, UK: Cambridge Univ. Press

Brush S. 1996c. *Fruitful Encounters*. Cambridge, UK: Cambridge Univ. Press

Burbidge EM, Burbidge GR, Fowler WA, Hoyle F. 1957. Synthesis of the elements in stars. *Rev. Modern Phys.* 29:547

Cameron AGW. 1955. Origin of anomalous abundances of the elements in giant stars. *Astrophys. J.* 121:144

Cameron AGW. 1957a. Nuclear radiation widths. *Can. J. Phys.* 35:666

Cameron AGW. 1957b. A revised semiempirical atomic mass formula. *Can. J. Phys.* 35:1021

Cameron AGW. 1957c. Nuclear reactions in stars and nucleogenesis. *Pub. Astron. Soc. Pacific.* 69:201

Cameron AGW. 1957d. Stellar evolution, nuclear astrophysics, and nucleogenesis. Chalk River Report CRL-41

Cameron AGW. 1957e. A revised table of abundances of the elements. *Astrophys. J.* 129:676

Cameron AGW. 1958a. Nuclear level spacings. *Can. J. Phys.* 36:1040

Cameron AGW. 1958b. Nuclear astrophysics. *Annu. Rev. Nucl. Sci.* 8:299

Cameron AGW. 1959a. Pycnonuclear reactions and nova explosions. *Astrophys. J.* 130:916

Cameron AGW. 1959b. Carbon thermonuclear reactions and the formation of heavy elements. *Astrophys. J.* 130:429

Cameron AGW. 1959c. Neon and oxygen thermonuclear reactions. *Astrophys. J.* 130:895

Cameron AGW. 1959d. Photobeta reactions in stellar interiors. *Astrophys. J.* 130:452

Cameron AGW. 1959e. Neutron star models. *Astrophys. J.* 130:884

Cameron AGW. 1963. *Nuclear Astrophysics*, comp. WD Arnett, CJ Hansen, JW Truran. New Haven: Yale Univ. (Lecture notes)

Cameron AGW. 1973. Accumulation processes in the primitive solar nebula. *Icarus* 18:407

Cameron AGW. 1986a. Some autobiographical notes. In *Cosmogonical Processes*, ed. WD Arnett, CJ Hansen, JW Truran, S Tsuruta. Utrecht: VNU Sci. Press

Cameron AGW. 1986b. The impact theory for origin of the moon. In *Origin of the Moon*, ed. WK Hartmann, RJ Phillips, GJ Taylor. Houston: Lunar Planet. Inst.

Cameron AGW. 1995. The first ten million years in the solar nebula. *Meteoritics* 30:133

Cameron AGW. 1997. The origin of the moon and the single impact hypothesis V. *Icarus* 126:126

Cameron AGW, Canup R. 1998. The giant impact occurred during earth accretion. *Lunar Planet. Sci. XXIX* p. 1062. Houston: Lunar Planet. Inst. (CD-ROM)

Cameron AGW, Elkin RM. 1965. Role of the symmetry energy in atomic mass formulas. *Can. J. Phys.* 43:1288

Cameron AGW, Höflich P, Myers PC, Clayton DD. 1995. Massive supernovae, Orion gamma rays, and the formation of the solar system. *Astrophys. J.* 447:L53

Cameron AGW, Pine MR. 1973. Numerical models of the primitive solar nebula. *Icarus* 18:377

Cameron AGW, Truran JW. 1977. The supernova trigger for formation of the solar system. *Icarus* 30:447

Cameron AGW, Vanhala H, Höflich P. 1997. Some aspects of triggered star formation. In *Astrophysical Implications of the Laboratory Study of Presolar Materials*, Woodbury, NY: Am. Inst. Phys. Conf. Proc. 402, ed. TJ Bernatowicz, E Zinner, p. 665. Woodbury, NY: Am. Inst. Phys.

Cameron AGW, Ward WR. 1976. The origin of the moon. *Lunar Science VII,* Houston: Lunar Sci. Inst.

Consolmagno GJ, Cameron AGW. 1980. The origin of the FUN anomalies and the high temperature inclusions in the Allende meteorite. *Moon Plan.* 23:3

DeCampli WM, Cameron AGW. 1979. Structure and evolution of isolated giant gaseous protoplanets. *Icarus* 38:367

Ezer D, Cameron AGW. 1965. A study of solar evolution. *Can. J. Phys.* 43:1497

Ezer D, Cameron AGW. 1967a. Early and main sequence evolution of stars in the range 0.5 to 100 solar masses. *Can. J. Phys.* 45:3429

Ezer D, Cameron AGW. 1967b. Evolution of stars of low mass. *Can. J. Phys.* 45:3461

Ezer D, Cameron AGW. 1971a. Pre-main sequence stellar evolution with mass loss. *Astrophys. Space Sci.* 10:52

Ezer D, Cameron AGW. 1971b. The evolution of hydrogen helium stars. *Astrophys. Space Sci.* 14:399

Gilbert A, Cameron AGW. 1965. A composite nuclear level density formula with shell corrections. *Can. J. Phys.* 43:1446

Gilbert A, Chen FS, Cameron AGW. 1965. Level densities in lighter nuclei. *Can. J. Phys.* 43:1248

Hoffman MM, Cameron AGW. 1953. Angular and energy distributions of photoprotons from aluminum and tantalum. *Phys. Rev.* 92:1184

Hubbard WB. 1969. Thermal models of Jupiter and Saturn. *Astrophys. J.* 155:333

Hubbard WB. 1970. Structure of Jupiter: chemical composition, contraction, and rotation. *Astrophys. J.* 162:687

Katz L, Cameron AGW. 1951. The solution of x-ray activation curves for photonuclear cross sections. *Can. J. Phys.* 29:518

Katz L, Haslam RNH, Horsley RJ, Cameron AGW, Montalbetti R. 1954. Fine structure in the $C^{12}(\gamma,n)C^{11}$ and $O^{16}(\gamma,n)O^{15}$ activation curves. *Phys. Rev.* 95:464

Langer WD, Cameron AGW. 1969. Effects of hyperons on the vibrations of neutron stars. *Astrophys. Space Sci.* 5:213

Mercer-Smith JA, Cameron AGW, Epstein RI. 1984. On the formation of stars from disk accretion. *Astrophys. J.* 279:363

Mitroff II. 1974. *The Subjective Side of Science: A Philosophical Inquiry into the Psychology of the Apollo Moon Scientists.* Amsterdam: Elsevier

Nelson AF, Benz W, Adams FC, Arnett W. 1998. Dynamics of circumstellar disks. *Astrophys. J.* 502:342

Perri F, Cameron AGW. 1974. Hydrodynamic instability of the solar nebula in the presence of a planetary core. *Icarus* 22:416

Podolak M, Cameron AGW. 1974. Models of the giant planets. *Icarus* 22:123

Podolak M, Cameron AGW. 1975. Further

investigations of Jupiter models. *Icarus* 25: 627

Reynolds JH. 1960. Determination of the age of the elements. *Phys. Rev. Lett.* 4:8

Slattery WL, DeCampli WM, Cameron AGW. 1980. Protoplanetary core formation by rain-out of minerals. *Moon Plan.* 23: 381

Suess HE, Urey HC. 1956. Abundances of the elements. *Rev. Mod. Phys.* 28:53

Truran JW, Hansen CJ, Cameron AGW, Gilbert A. 1966. Thermonuclear reactions in medium and heavy nuclei. *Can. J. Phys.* 44: 151

Tsuruta S, Cameron AGW. 1965. Composition of matter in nuclear statistical equilibrium at high densities. *Can. J. Phys.* 43:2056

Tsuruta S, Cameron AGW. 1966a. Some effects of nuclear forces on neutron star models. *Can. J. Phys.* 44:1895

Tsuruta S, Cameron AGW. 1966b. Cooling and detectability of neutron stars. *Can. J. Phys.* 44:1863

Tsuruta S, Cameron AGW. 1970. URCA shells in dense stellar interiors. *Astrophys. Space Sci.* 7:374

Tsuruta S, Cameron AGW. 1976. The URCA process in convective cores. *Astrophys. Space Sci.* 39:397

Vanhala HAT, Cameron AGW. 1999. Numerical simulations of triggered star formation I. collapse of dense molecular cloud cores. *Astrophys. J.* In press

Ward WR, Cameron AGW. 1978. Disc evolution within the roche limit. *Lunar Planet. Sci. IX* Houston: Lunar Planet. Inst.

A CRITICAL REVIEW OF GALACTIC DYNAMOS

Russell M. Kulsrud
Princeton University Observatory, Peyton Hall, Princeton, New Jersey

Key Words MHD dynamos, galactic magnetic fields, alpha effect, flux expulsion, small-scale fields

■ **Abstract** I present the general picture of how galactic magnetic fields grow in disks according to the alpha-Omega dynamo theory. Emphasis is placed on following the lines of force during the dynamo process. The dynamo equation is presented together with a simple growing solution for the galactic disk. Then I take up the various critical questions that have been raised concerning the galactic dynamo theory. These are (1) the importance of the escape of flux from the disk in order for the magnetic field to grow; (2) the physics of turbulent diffusion and its mixing of field lines together so that the rms field is possibly greater than the mean field; (3) whether magnetic reconnection plays a role in the galactic dynamo; (4) whether small-scale fields can grow large enough to swamp the dynamo. Then I discuss the possible seed fields from which the dynamo starts and their relation to the primordial hypothesis. Finally I take up the question of the final evolution of the galactic field after the alpha effect saturates. My conclusion is that all these problems warrant attention but none of them seem to be serious enough to cast any real doubt on the dynamo as the most likely generator of galactic fields.

1. INTRODUCTION

The origin and evolution of the large-scale galactic magnetic field (Zweibel & Heiles 1997) is a challenging theoretical problem. It has been under discussion for over 40 years, since the magnetic field was first taken seriously by Fermi (1949). The early work supposed that the galactic field had its origin prior to the formation of the galaxy, or at least the galactic disk (Alfven 1950, Biermann 1953, Hoyle 1958, Hoyle & Ireland 1960, 1961). However, it has been pointed out by Parker (1969) that such a field might be expelled by dynamic motions in less than a billion years. This then led to the suggestion that a dynamo mechanism driven by cyclonic turbulence and differential rotation of the interstellar medium could sustain the field and even could generate it from a small seed field to its present state (Parker 1970a, 1971a, Vainshtein & Ruzmaikin 1971, 1972). The theory was modeled after parallel theories of dynamos in the earth and the sun (Parker 1955a,b).

At the present time the dynamo appears to be the most likely explanation for the existence of the galactic field. It has been formulated in a precise way through the mean field dynamo equation whose solutions have been developed over the years. The resulting field appears to correspond to the magnetic field of our galaxy and others (Kronberg 1994, Beck 1996). On the other hand, a number of criticisms have been leveled at the dynamo theory. Attempts to answer them have led to an increased understanding of the physics of dynamo action on the scale of the galaxy. These criticisms primarily concern the intense development of small-scale fields. If these fields are not limited by saturation, then they would grow to a large enough amplitude to control the turbulence and stop the action of the dynamo (Kulsrud & Anderson 1992, 1993, Anderson & Kulsrud 1993, Cattaneo & Vainshtein 1991, Vainshtein & Cattaneo 1992, Cattaneo 1994, Gruzinov & Diamond 1994, 1995, 1996).

It appears, however, that the small-scale fields do saturate at levels that need not interfere with the general dynamo (Kulsrud & Anderson 1992, Pouquet et al 1976, Chandran 1996, 1997b).

There are other serious physical problems, however, that are still unresolved. They concern the expulsion of flux from the galactic disk and the general concept of turbulent diffusion of larger scale fields. Since these are still unresolved, there is still some doubt as to the validity of the dynamo theory as the source of galactic fields.

Questions of a more astrophysical nature are:

1. What are the sizes of the various coefficients in the mean field dynamo equation, i.e. the intensity of the cyclonic motions that transform the toroidal magnetic field into the poloidal field?
2. What is the magnitude of the turbulent diffusion that leads to the expulsion of unwanted flux and the smoothing of the remaining large-scale magnetic field?
3. What are the initial conditions of the galactic field, the weak field that is amplified by the dynamo?
4. Finally, what is the behavior of the magnetic field when it is strong enough to produce a back effect on the turbulence that limits further growth?

In this review I first briefly summarize the creation and evolution of the magnetic field as viewed from the nonprimordial theory as I understand it, and as it now appears to be accepted. Throughout this review I make the assumption of flux freezing of magnetic field lines. Flux freezing can be broken by fast magnetic reconnection if this process is possible on the galactic scale. My prejudice is that it is not. Therefore, by and large in this review I assume flux freezing to be exact. The various difficulties this assumption entails in galactic dynamo theory bring out the important role that fast magnetic reconnection could play if it existed. It turns out that the absence of magnetic reconnection is constraining on the power of the dynamo, but it is still possible that galactic magnetic fields could be generated even in its absence.

After the qualitative discussion I review the elements of this theory in detail: the derivation of the dynamo equation, its solutions and an estimate of its physical parameters, the seed fields from which the dynamo starts, and its final behavior. Next, I take up the main criticism, the behavior of the small-scale fields, and indicate why it may not be so important. I attempt to present a physical picture based on the concept of flux freezing of the lines of force and indicate where the dynamo theory still seems weak from this point of view. Finally, I indicate how these remaining criticisms may be met.

The theory of the galactic magnetic field has been thoroughly discussed in the excellent book by Ruzmaikin, Shukurov & Sokoloff (1988b, referred to as *RSS* in the remainder of this article). This book represents the best systematic account of the dynamo theory as applied to the galactic magnetic fields. An excellent elementary account for the general reader is Ruzmaikin et al (1988a). The physics questions are laid out most clearly in the book by Parker (1979, referred to as *Par* in this review). The simplest and most direct presentation of the dynamo theory is given in Moffatt (1978), while the mathematically rigorous treatment is in the book by Krause & Radler (1980). The most recent general review of the dynamo is that of Beck et al (1996). The whole question of the validity of dynamo as a source of the galactic magnetic field was set forth in an admirable presentation by Field (1995).

2. THE ORIGIN OF GALACTIC MAGNETIC FIELDS ACCORDING TO DYNAMO THEORY

Let us start with the supposition that there is no magnetic field present in the universe at the time of recombination. That is, no process in the universe produced a magnetic field of interest prior to this time. (This seems a reasonable assumption, since any such field would make the task of explaining the galactic field that much easier.)

Then, during the formation of the galaxy, there would be considerable turbulence that would lead by the Biermann battery to extremely weak fields, coherent on the size of the protogalaxy (Biermann 1950, Kemp 1982, Kulsrud 1997, Kulsrud et al 1997a). Further, some dynamo action would amplify this magnetic field to appreciable strength. Collapse into the disk would further increase this field.

Once the disk has been formed and star formation commences, interstellar turbulence is generated by stellar winds and stellar explosions (McIvor 1977, Higdon 1984). The turbulent motions are altered into cyclonic motions by the Coriolis forces associated with galactic rotation (Parker 1955a, Par pp. 559–78, RSS pp. 162–166).

Such cyclonic motions have the capability of rotating any toroidal field into a poloidal field direction, and any poloidal field into the toroidal direction (Parker 1955a). This happens on scales comparable to the size of the largest turbulent eddies, so in general, many small poloidal or toroidal loops are formed. The general turbulence leads to mixing and smoothing of these loops until larger scale poloidal and toroidal fields are formed.

In addition, the differential azimuthal rotation produces a toroidal field from any poloidal field. This latter process is more powerful than the cyclonic motions, but only works to create toroidal not poloidal flux.

Thus, the dynamo has twofold action: it converts poloidal to toroidal flux by differential rotation and reconverts toroidal flux to poloidal flux by cyclonic motions. The latter action is termed the α effect. All the cyclonic motions, those with upward and downward motions and those above and below the galactic plane, produce a poloidal field with the same sense.

Now, it is easy to show with boundary conditions allowing the escape of magnetic flux that the combined effects of α and Ω lead to amplification of the field. However, if the boundary conditions are such that all the flux is confined to the disk, then the total radial flux does not change in time and no amplification occurs. In fact, the toroidal field near the central plane of the disk has one sign, while that near the edges of the disk has the other. Thus, a most important feature of the dynamo is the expulsion of the reversed flux near the boundary of the disk, leaving behind only flux with the direction of that near the central plane.

The action of α and Ω is to double the positive and negative fluxes and separate them with respect to height. The doubling rate is proportional to the geometric mean of the Ω (differential rotation) and α (cyclonic) effects. This must occur faster than the turbulent diffusion rate or else the field decreases. Also, the flux must be expelled at a rate comparable to or faster than diffusion if the field is to grow.

Now the process is linear in the field strength, so there are various, dynamo normal modes. The different modes have different numbers of nodes in Z, the coordinate perpendicular to the galactic disk. However, the mode with no nodes generally grows much faster than the others. Because the disk is thin, the primary variation of the mode is with Z. The mode is local in horizontal coordinates to lowest order, so the field may grow faster at certain radii than others. However, smoothing in radius by horizontal diffusion controls the relative amplitudes, and one finds that the primary normal mode in Z breaks up into distinct eigenmodes having different radial dependences. There is a fastest growing one of these, and this determines the final shape of the generated galactic field.

In summary, there are four important ingredients to the galactic dynamo: (1) differential rotation producing the Ω effect, (2) cyclonic motions producing the α effect, (3) escape of flux with wrong sign, and (4) turbulent diffusion.

All these effects arise naturally in the galactic disk except the third. The escape is helped by cosmic ray pressure and, when the field becomes strong enough, magnetic buoyancy (Parker 1966, 1967, 1971a, 1992, Par pp. 341–356). However, because of the large electrical conductivity of the interstellar medium, a flux line cannot escape entirely. The escaped part of the line must remain connected to the remainder of the line. It can, however, effectively stretch to infinity in the vertical direction, removing itself from consideration. It must be remembered, however, that the part of the line to escape is initially loaded with interstellar matter that cannot escape because it is held down by gravity (Vainshtein & Rosner 1991). This matter must slip down the escaping line, and onto the piece of the lines closer to the central plane pointing in the "correct" direction. This important escape process has not been sufficiently examined in the context of the dynamo. It needs much more careful consideration.

As long as the field energy is small compared with the energy in cyclonic motions and turbulence, the evolution of the field is linear. One can break up the evolution into normal modes of different growth and expand the field in these normal modes. The amplitude of these modes is given by the seed fields that are present initially, after the galactic disk formed. Then the fastest growing mode determines the character of the field only after enough time elapses for the number of e-foldings of the fastest modes to differ by more than unity from the next fastest.

By making use of the observed properties of the interstellar medium the e-folding time of the fastest mode can be estimated to be of order several hundred million years. Thus, one would expect twenty or thirty growth times to elapse for this mode. A respectable seed field cannot be too small, or else the field cannot reach its present value of 3 microgauss. In fact, the seed fields should be at least 10^{-13}–10^{-12} gauss (Field 1995).

It has been suggested that such seed fields might arise from fluxes observed in supernova remnants or thrown out by stars. The Crab nebula has a coherent field of order 10^{-4} gauss over a radius of 1 parsec, which would give a mean field about 3×10^{-14} gauss when averaged over the scale of the dynamo mode (3 kpc radial extent). However, during one growth time, 5×10^8 years, there would be $N = 5 \times 10^6$ supernovae, each contributing a field of this size so that one would have a root mean square seed field of $\sqrt{N} \times 3 \times 10^{-14}$ gauss, which would provide a sufficient seed field. However, by linearity the present field would be the sum of 5×10^6 incoherent dynamo modes or 5×10^6 incoherent amplitudes. The root mean square field would thus be 2×10^3 times larger than the average field unless flux freezing was broken by rapid reconnection. Thus, the stellar fields would seem to provide poor seed fields. It is expected that the fields resulting from protogalactic activity would provide a stronger, more coherent seed field, and the smaller stellar fields would not lead to such a coherent field.

Lastly, what is the long-term fate of the galactic field? When the field becomes strong, it will resist the cyclonic motions, and after being twisted from an azimuthal direction to a poloidal direction, it will twist back. Also, the mixing velocities will develop a springiness that will resist their mixing the field. However, the differential rotation, which is driven by much stronger gravitational forces, characterized by the high kinetic energy densities of rotation, will probably be unaffected. Thus, the final behavior will be a battle between weakening of the field by escape of flux by ambipolar diffusion driven by magnetic buoyancy and cosmic ray pressure, and strengthening of the field by differential rotation. Most of the dynamo action is suppressed at this time.

Such an evolution would be reminiscent of that of a primordial field suffering differential rotation (Howard & Kulsrud 1997).

It is interesting to reflect on how the lines of force behave under the above story. If the conductivity is infinite, the number of lines of force cannot change nor the lines break (Alfven 1950). We start off with a weak coherent field with relatively few lines of force, say, all azimuthal. After we finish we must have the same number of lines but a much stronger field. How does the dynamo accomplish this?

Let us start with a toroidal field in the positive direction about the galaxy. Tracing through our story, a cyclonic motion makes a poloidal loop from the toroidal field. The radial component of the part of the loop near the central plane has one sign and the different part of the same loop near the edge of the disk has the opposite sign. The differential rotation shears the two parts of the line in different toroidal directions. Thus, the same line now goes back and forth, twice in the same direction, the positive direction, and once in the negative direction. With Ω decreasing with radius, one finds that the part of the line near the central plane has the positive direction, and the part near the edge the negative direction. (Parker 1971a). Then the negative part of the line is expelled out of the disk, so that the same line now goes forward a certain distance, leaves the disk in the vertical direction, then goes back the same distance in the extragalactic (or halo) region, comes back to the disk, and goes forward again, thus doubling the flux in the disk.

After an amplification by a factor A, the line makes A excursions forward in the toroidal direction interspersed with leaving the disk after each forward motion and going backward an equal distance, returning to the disk at each starting point. After it has done this A times, the line now counts A times in the positive direction, and it can proceed forward.

This picture could be simplified by magnetic reconnection, if it were fast enough, which could break the line in A parts. Then one need not worry about the intermediate steps. However, this picture brings out the importance of flux expulsion.

After this survey of the whole picture of the dynamo action, let us turn to a more quantitative description of all these processes.

We first consider the central dynamo. Then we consider the small-scale problem. Next we consider the seed fields. Finally, we discuss the saturated galactic field.

3. THE DYNAMO EQUATION

3.1 Derivation of the Dynamo Equation

In order to judge more quantitatively whether the cyclonic motions can amplify a given galactic field (arising initially from a seed field), it is necessary to represent the dynamo action by a reasonable differential equation. Such a differential equation was first derived by Steenbeck et al (1966) and then by Parker (1970b), Vainshtein (1970), Vainshtein & Ruzmaikin (1971, 1972), and Vainshtein & Zeldovich (1972). Its derivation is presented in many places (Moffatt 1978, Par pp. 567–578, RRS pp. 177–181, Vainshtein 1970, Kulsrud & Anderson 1992). To bring out the physics, we sketch a simple one here.

The idea is that the magnetic field **B** can be written as

$$\mathbf{B} = \bar{\mathbf{B}} + \mathbf{b}, \tag{1}$$

and the velocity field **U** as

$$\mathbf{U} = \bar{\mathbf{V}} + \mathbf{v}. \tag{2}$$

It is imagined that $\bar{\mathbf{B}}$ and $\bar{\mathbf{V}}$ represent ensemble averages over an ensemble of many realizations of the velocity field **U**. It is further assumed, at least at first, that **B** is so weak that the velocity fields in the different members of the ensemble are unaffected by it. (Any velocities produced by Lorentz forces are too small to have any effect on **B**.) This is the kinematic assumption. Now, **B** satisfies the standard magnetic differential equation

$$\frac{\partial \mathbf{B}}{\partial t} = \nabla \times (\mathbf{U} \times \mathbf{B}) + \lambda \nabla^2 \mathbf{B}, \tag{3}$$

where $\lambda = \eta c/4\pi$ is the magnetic diffusivity. The units of the resistivity η are chosen so that $\lambda = 10^7/T^{3/2}$ cm^2/sec, where T is the temperature in units of 10^4 degrees kelvin (=1 ev). It is clearly seen that this last term is negligible, for $T = 1$ ev, for any scale greater than 10^{12} cm if the time evolution is shorter than a Hubble time, so that generally in the galactic dynamo the λ term can be dropped.

The velocity field in the galaxy consists of $\bar{\mathbf{V}}$, the differential rotation of the galaxy, and **v**, the turbulent motions generated by stars and supernovae, the turbulent and cyclonic motions. The scale of variation of these turbulent velocities is generally less than 100 parsecs, while the scale of the magnetic field of interest, $\bar{\mathbf{B}}$, is longer than a kiloparsec. Breaking the equation into a mean part and a random part, we get

$$\frac{\partial \bar{\mathbf{B}}}{\partial t} = \nabla \times (\bar{\mathbf{V}} \times \bar{\mathbf{B}}) + \nabla \times (\overline{\mathbf{v} \times \mathbf{b}}), \tag{4}$$

and

$$\frac{\partial \mathbf{b}}{\partial t} = \nabla \times (\mathbf{v} \times \bar{\mathbf{B}}) + \nabla \times (\mathbf{v} \times \mathbf{b}) - \nabla \times (\overline{\mathbf{v} \times \mathbf{b}}), \tag{5}$$

and we drop the $\bar{\mathbf{V}} \times \mathbf{b}$ term in Equation (5) as it can be removed by a gallalean transformation. The second equation is to be solved for each realization of the turbulent motions, and the ensemble average of $\mathbf{v} \times \mathbf{b}$ is to be calculated at each point.

To carry this out, a crucial assumption is made that has been widely criticized in the past few years (Kulsrud & Anderson 1992). This assumption is that one may neglect the last, $\mathbf{v} \times \mathbf{b}$, terms in Equation (5). The supposition is that **b** is small compared with $\bar{\mathbf{B}}$ so that this term is negligible. Then we have

$$\frac{\partial \mathbf{b}}{\partial t} = \nabla \times (\mathbf{v} \times \bar{\mathbf{B}}), \tag{6}$$

and thus **b** only exists on the small scale of **v** (<100 parsecs in the galactic case). This assumption makes possible the explicit representation of the dynamo by the dynamo equation. A further simplification is also made, that the decorrelation time of the velocity field τ is short compared with the rate of evolution of the mean field (see Chandran 1997a). With these assumptions it is possible to derive an evolution equation for $\bar{\mathbf{B}}$ in terms of any statistics of the turbulence.

Now, the statistics of the turbulence must include both random turbulent and cyclonic motions. Because the scale of interstellar turbulence is not really small compared with the thickness of the disk of the galaxy, and because the cyclonic motions are dominated by rotation about the vertical direction, we cannot really assume homogeneity and isotropy of the turbulence with any accuracy. However, because of the analytic simplicity gained from such an assumption, we do make it anyway in our derivation. The more general case is discussed in Krause & Radler (1980).

The most general isotropic homogeneous correlation tensor for the velocity field is (Par pp. 524–529)

$$\langle v_i(r_0, t) v_j(r + \rho, t + \tau) \rangle = \left[\left(\delta_{ij} - \frac{\rho_i \rho_j}{\rho^2} \right) f(\rho) + \epsilon_{ijk} \rho_k g(\rho) \right] h(\tau). \quad (7)$$

The first term on the right is the normal correlation tensor for isotropic mirror symmetric incompressible turbulence, while the second, ϵ_{ijk}, represents the cyclonic motions.

Let us assume $\mathbf{B} = \bar{\mathbf{B}}$ at time t and advance $\mathbf{B}$ forward a short time Δt but with $\tau \ll \Delta t$. We neglect $\bar{\mathbf{V}}$ in Equation (4) and restore it later. To first order,

$$\mathbf{b}(\mathbf{r}, t') = \int_t^{t'} \nabla \times [\mathbf{v}(\mathbf{r}, t'') \times \bar{\mathbf{B}}(\mathbf{r}, t)] \, dt''. \quad (8)$$

Then one can advance $\mathbf{B}$ to next order by iterating again

$$\mathbf{B}(\mathbf{r}, t + \Delta t) = \nabla \times \int_t^{t+\Delta t} [\mathbf{v}(\mathbf{r}, t') \times \mathbf{b}(\mathbf{r}, t')] \, dt'. + \bar{B}(r, t). \quad (9)$$

We have left out the $\nabla \times (\mathbf{v} \times \bar{\mathbf{B}})$ term since its average is zero.

Then substituting Equation (8) for $\mathbf{b}$, taking the ensemble average of this over all members of the ensemble between t and $t + \Delta t$ (for fixed members before t), and then finally over all members before t, we get

$$\left\langle \frac{\partial \mathbf{B}}{\partial t} \right\rangle \Delta t = \left\langle \nabla \times \int_t^{t+\Delta t} dt' \int_t^{t'} dt'' [\mathbf{v}(\mathbf{r}, t') \times (\nabla \times [\mathbf{v}(\mathbf{r}, t'') \times \bar{\mathbf{B}}(\mathbf{r}, t)])] \right\rangle. \quad (10)$$

(The corrections from assuming $\mathbf{B}(t) = \bar{\mathbf{B}}$ are negligible if $\tau \ll \Delta t$.)

Since $\Delta t \gg \tau$, the time integration on the right-hand side can be done, and is $\Delta t \int h(t) = \Delta t \tau$, so that

$$\frac{\partial \bar{\mathbf{B}}}{\partial t} = \tau \langle \nabla \times [\mathbf{v}(\mathbf{r}, t) \times (\nabla' \times [\mathbf{v}(\mathbf{r}', t) \times \bar{\mathbf{B}}])_{r'=r}] \rangle. \quad (11)$$

Then, substitution from Equation (7) for the correlation tensor enables us to evaluate this expression. Restoring the $\bar{\mathbf{V}}$, we can write

$$\frac{\partial \bar{\mathbf{B}}}{\partial t} = \nabla \times (\alpha \bar{\mathbf{B}}) + \beta \nabla^2 \bar{\mathbf{B}} + \nabla \times (\bar{\mathbf{V}} \times \bar{\mathbf{B}}), \tag{12}$$

where α depends on g and β depends on f.

By introducing the mean helicity of the turbulence $\langle \mathbf{v} \cdot \nabla \times \mathbf{v} \rangle$, which is expressible in terms of g, and the kinetic energy $\langle \mathbf{v}^2 \rangle$, which is expressible in terms of f, we can show that

$$\alpha = -\frac{\tau}{3} \langle \mathbf{v} \cdot \nabla \times \mathbf{v} \rangle, \tag{13}$$

and

$$\beta = \frac{\tau}{3} \langle \mathbf{v}^2 \rangle. \tag{14}$$

The β term in Equation (12) is the diffusion tensor and represents the mixing of the field (Parker 1971b, 1973, Par pp. 490–511). It is the same result as one gets from mixing of any scalar field, and can be thought of as resulting from a random walk of various field lines with mean step size $v\tau$, and rate of steps $1/\tau$. During the steps the field strength of the line does not change, and the resulting mean field $\bar{\mathbf{B}}$ at any point is the mean of all the field strengths of all the flux lines brought to that point by the ensemble.

The first term, the α term, represents the effect of the cyclonic motions on the mean field. Again, one can model the effect of the field lines as the result of random steps taken by the field lines. In one step a given line of force of the mean field is rotated by $\tau(\nabla \times \mathbf{v})$ about the step direction $\mathbf{v}\tau$ so that the mean field is turned and displaced by an amount $\tau \mathbf{v} \cdot (\nabla \times \mathbf{v})$ per unit time.

Let the galactic mean field $\bar{\mathbf{B}}$ be in the toroidal direction at some time. Then after a sufficiently long time, the pieces of the field line get displaced both upward and downward. As these pieces are rotated into the poloidal directions, the pieces displaced upward are rotated in one direction, and the pieces displaced downward in the opposite direction. Since α is of constant sign above the central plane, a poloidal field of one sign gets produced near the central plane and of the opposite sign near the edge of the disk. Then the differential rotation of the galaxy acts on this intermediate field, through the $\nabla \times (\bar{\mathbf{V}} \times \bar{\mathbf{B}})$ term, so as to yield a change in the original magnetic field that strengthens it near the central plane and weakens it near the disk boundary.

In this way one sees that the dynamo equation well represents the general physical picture we have of the operation of the galactic dynamo in amplifying any initial field. It further provides the ability to judge the rate of growth of the field. On the other hand, the method of derivation we have sketched here also enables us to appreciate the assumptions made in the galactic dynamo equation.

Before proceeding, it is necessary to point out the symmetry of the α coefficient. Motions away from the galactic plane rotate one way above the plane (they rotate backward relative to the galactic rotation), so that $\alpha > 0$ for $z > 0$. Below the plane they rotate the same way (backward) so that $\alpha < 0$ for $z < 0$. It turns out that this is just the condition for the fastest growing mode to have the fields symmetric in Z.

3.2 Solution of the Dynamo Equations

Let us assume that the galaxy is, in the mean, axisymmetric (ignoring turbulent motions and spiral structure). Introduce cylindrical coordinates R, Θ, and Z, so that B_R and B_Z are components of the poloidal field and B_Θ is the toroidal field. Then components of the dynamo equation are

$$\frac{\partial B_R}{\partial t} = -\frac{\partial}{\partial Z}(\alpha B_\Theta) + \beta(\nabla^2 B)_R, \tag{15}$$

$$\frac{\partial B_\Theta}{\partial t} = R\frac{d\Omega}{dR}B_R - \frac{\partial}{\partial R}(\alpha B_Z) + \beta(\nabla^2 B)_\theta, \tag{16}$$

$$\frac{\partial B_Z}{\partial t} = \frac{\partial}{R\partial R}(R\alpha B_\Theta) + \beta(\nabla^2 B)_Z. \tag{17}$$

The α term in the second equation is considerably smaller than the Ω term, so we drop it. The first two equations then decouple from the last equation, which we ignore.

Now the galactic disk is very thin compared with its radius, so that **B** and α vary much more rapidly in Z than in R. Thus, to lowest order, we drop the radial derivatives, and we get

$$\frac{\partial B_R}{\partial t} = -\frac{\partial}{\partial Z}(\alpha B_\Theta) + \beta\frac{\partial^2}{\partial Z^2}B_R, \tag{18}$$

$$\frac{\partial B_\Theta}{\partial t} = -\Omega B_R + \beta\frac{\partial^2}{\partial Z^2}B_\Theta. \tag{19}$$

We have assumed that $\Omega \sim 1/R$ so that $Rd\Omega/DR = -\Omega$, a relation well satisfied near the sun in our galaxy.

Because α is odd and the product $\partial/\partial Z$ and α is even, the equations are symmetrical with respect to Z. Therefore, there are both symmetric and antisymmetric solutions of B_R and B_Θ.

The symmetric solution,

$$B_\Theta(Z) = B_\Theta(-Z), \quad B_R(Z) = B_R(-Z), \tag{20}$$

is called the quadrupole solution. It is the most unstable mode in the galactic disk.

The antisymmetric dipole solution,

$$B_\Theta(Z) = -B_\Theta(-Z), \quad B_R(Z) = -B_R(-Z), \tag{21}$$

is of less importance for galaxies.

To show that these lead to a growing solution, let us assume the dissipation, β, is small, take $\alpha = \alpha_0 Z/h$, and B_R and B_Θ proportional to $e^{\gamma t}$. Then from Equation (18), $B_R = -(\alpha_0/h\gamma)B_\Theta$, and from Equation (19), $B_\Theta = -(\Omega/\gamma)B_R$, so that $\gamma^2 = \Omega\alpha_0/h$ and $B_R = -\sqrt{\alpha_0/h\Omega}B_\Theta$, where B_R and B_Θ are constant in Z. Since $\alpha_0 \approx 1.0$ km/sec (Parker 1971a), and $h\Omega \approx 10$ km/sec for $h = 300$ pc, $B_R \approx 0.3 B_\Theta$ and the field is primarily a toroidal field. The growth time $1/\gamma \approx 10^8$ years. On the other hand, $(\beta/h^2)^{-1} \approx 2 \times 10^8$ years. Although this is smaller, it is clear that a more careful analysis is needed to realistically establish that the growth is positive. Nevertheless, the double-reacting $\alpha\Omega$ mechanism leads to a growth if dissipation is negligible. This relation between B_Θ and B_R is characteristic of the behavior of the actual solution near $Z = 0$.

It is clear that if no flux can escape from the disk, the magnetic field cannot grow. Integrate Equation (15) between $-h$ and h,

$$\frac{d}{dt}\int_{-h}^{h} B_R dZ = \left[\beta\frac{dB_R}{dZ} - \alpha B_\theta\right]_{-h}^{h}. \tag{22}$$

The right-hand side just represents the rate of passage of radial flux through the disk boundaries. If this vanishes, then $\int B_R dZ$ is a constant in time and γ must be zero. Thus, flux escape is vital to dynamo growth.

It is generally assumed that the field is zero or negligible outside the disk and that the boundary conditions are $B_R(\pm h) = B_\Theta(\pm h) = 0$. These are termed vacuum boundary conditions, and are standardly assumed in most thin disk calculations (see Brandenburg et al 1992). If there is indeed a growing mode, the B_R flux must increase, and the term on the right-hand side of Equation (22) represents the rate of loss of negative B_R flux. It is the case (see RSS Figure VII.1.c) that B_R is negative (positive) near $Z = h, (-h)$, and its slope is positive (negative).

The problem of flux loss, which has been emphasized by Parker, will be discussed below.

Dynamo growth is a competition between the product of α/h and Ω, and β^2/h^4. Indeed, the equations can be made dimensionless by the transformations $Z = Z'h$, $B_R = \sqrt{\alpha_0/h\Omega}B_R'$. With this substitution, only the dimensionless parameter,

$$D = -\frac{\Omega h^3 \alpha_0}{\beta^2}, \tag{23}$$

enters into the equations. This parameter is called the dynamo number. α is generally expressed as $\alpha_0 \alpha'(Z)$ where α' gives the functional dependence of α on Z and is of order one. The critical value of D for instability is between 6 and 10 for a reasonable α' profile such as $\sin Z$ and for vacuum boundary conditions.

One finds that γ can be written

$$\gamma = \gamma'\frac{\beta}{h^2}, \tag{24}$$

where for vacuum boundary conditions γ' is a function of D. The actual numerical results depend on the shape assumed for α'. Choices are for $Z' > 0$, (1) $\alpha' = 1$, (2) $\sin \pi Z'$, and (3) Z'. The critical values for D for these three cases are $-6, -8, -13$, respectively (RSS Figure VII.2). More negative values of D lead to growth. An approximate expression for γ for larger D and case (2) is

$$\gamma = \sqrt{\frac{\Omega\alpha_0}{h} - \frac{\pi^2}{4}\frac{\beta}{h^2}}, \tag{25}$$

(RSS Figure VII.2). This result corresponds closely to our approximate result above.

The best analytical solutions of the dynamo equation in the thin disk limit are given for case (3) by Soward (1978). Good numerical solutions for arbitrary thickness are to be found in Stix (1975, 1978), Walker et al (1994), Brandenburg et al (1992, 1993), Moss et al (1992, 1993), and Beck et al (1994). The solution of the dynamo equation in a thin disk has been carried out locally since Z variations are large compared with radial variations. However, since the growth rate depends on α, β and $R d\Omega/dR \approx -\Omega$, all of which change by finite amounts in radius, one would expect such a local analysis to give a growth rate depending on R, $\gamma(R)$. If there is large amplification, the field amplitudes at different radii would soon be very different and radial diffusion would become important.

A more careful analysis leads to the actual radial envelope of the eigenmodes and the mean growth of such modes. One finds that the fields are strongest where $\gamma(R)$ is largest. This analysis is carried out with numerical examples in RSS pp. 191–200 and also in Soward (1978). It is found that this leads to an eigenmode more restricted in radius than one might have expected at first. According to RSS, the mode is appreciable in an annulus of 2 or 3 kiloparsecs radial thickness about the galactic center, and tends to concentrate inward from the solar radius. (See Figure VII.5 of RSS.) These solutions depend on the actual variation of α, which is only estimated by them.

3.3 The Physical Parameters of the Dynamo Equation

The physical parameters Ω and β are fairly well known. The parameter α is more difficult to evaluate. It has been extensively investigated for certain models of cyclonic motion, but, up until recently, it has been difficult to construct a sufficiently good model for cyclonic motion. The fanning out of the turbulent cell as it rises owing to decreasing density is somewhat problematical because of the inhomogeneity of the interstellar medium. Further, the statistical properties of the turbulence are also uncertain because of the uncertainties in their sources.

The situation has been significantly improved by six recent papers of Ferriere (1992a,b, 1993a,b, 1996, 1998). In these papers she makes a realistic model for supernova explosions. Starting with a purely toroidal field, she calculates the direct effect of each supernova on it, determining the loop of poloidal flux each supernova produces. Then summing over an assumed height and frequency distribution of supernova explosion centers, she arrives at a much more precise value for α. Of particular significance in her work is the entrance of a new parameter, v_{esc}, the mean vertical velocity associated with supernovae remnants. This parameter has special importance for the flux escape problem.

Ferriere also considers the effect of groups of supernovae going off together, the superbubbles (McCray & Snow 1979, Spitzer 1990). Although rarer than supernovae, they actually make a larger contribution to α than that from the sum of individual supernova. The results of her investigation on α are summarized in the final paper (Ferriere 1998). The Z dependence she finds for α extends to much greater heights than previously supposed, making it necessary to reevaluate the dynamo growth rates.

Making use of the values obtained for α from Ferriere's earlier work (1992a,b, 1993) and of the approximate expression for γ, Equation (25), Field (1995) reevaluates the numerical growth rate for the galactic dynamo. He finds a growth time of 400 million years if one neglects the β effect in Equation (25). Field's growth time gives 25 e-folding times in 10^{10} years. This corresponds to 7 powers of ten, so the seed field must be larger than about 3×10^{-17} gauss for the galactic dynamo to amplify it to the present size field. Thus, the seed fields that form the initial conditions for the dynamo must be much stronger than that obtained from a simple battery mechanism.

On the other hand, the reality of substantial growth for the galactic dynamo seems to be well established by these results, provided that the dynamo can survive the criticisms leveled against it.

4. CRITICISMS OF THE DYNAMO THEORY

There have been several serious criticisms of the dynamo theory. These criticisms all have their root in the very strong flux freezing of the galactic plasma. Flux freezing seems an impossible condition to avoid (unless possibly rapid magnetic reconnection occurs; this possibility is discussed below). First, we assume absolute flux freezing. Then, as discussed in the introduction, lines cannot break and the only possibility for flux amplification is a toroidal stretching of each line by an amount A equal to the amount by which B is amplified. Imagine the field originally in the positive toroidal direction. Then this toroidal stretching must involve a back-and-forth motion of the final lines. These motions must occur a number of times equal to the amplification factor. But these motions are just what the galactic dynamo accomplishes, with those portions of the field lines that are of the wrong toroidal sign expelled from the galaxy.

4.1 The Escape of Field Lines

The rate of escape of the negative portions of the field lines is represented by the surface terms $\beta \partial B_R/\partial Z$ and $\beta \partial B_\Theta/\partial Z$, the first of which is in the right-hand side of Equation (22). In the framework of the mean field theory, this is apparently accomplished easily by the imposition of the vacuum boundary conditions. However, from the point of view of flux freezing, there is a serious difficulty: These parts of the line of force are loaded with interstellar plasma, and the removal of this much flux (equal to the retained flux minus the tiny initial flux) from the disk should correspond to removal of one half of all the interstellar matter, an obvious absurdity. The resolution of the problem is: As the flux is lifted out of the galaxy, the lines are not horizontal, but form arcs along which the interstellar plasma slips, transferring itself from the parts of the flux lines being expelled to the other parts of the flux lines, those parts with the correct sign of toroidal field. (The lifting force is primarily due to cosmic ray pressure when the field is weak.)

Parker has emphasized that this should be happening to all of the flux lines and is the chief way for the cosmic rays to escape from the galaxy (Parker 1966, 1967, 1992, Par pp. 341–356). As the flux escapes, it is inflated to infinity by the cosmic ray pressure. As the plasma slips downward along the lines, it unloads the lines leading to faster escape. However, in its downward motion, it interacts with the cosmic rays so that the relative velocity is at most a few times the Alfven speed (Kulsrud & Pearce 1969). The rate of escape of the flux is governed by the rate of mass slippage along the lines.

Parker (1971a, 1992) attempts to relate this escape physics to the boundary conditions of the mean field dynamo. The relation is fairly close, as can be seen from the following estimates.

If the magnetic field grows in 500 million years, an amount of flux equal to the total flux, and thus an amount of matter equal to the entire interstellar medium, must pass through the galaxy in this time. This requires a net downward vertical velocity of $v_z \approx \gamma h \approx 0.5$ km/sec. This matter must return to the galactic by sliding down the field lines with a parallel velocity comparable to the Alfven speed of, say, 30 km/sec, and must go a distance several times the thickness of the disk. This yields a net downward velocity of about 10 km/sec. Thus, the matter returns at a little faster net speed than that at which it leaves. Hence, if the density on the escaping lines is greater than one tenth of that in the disk, then the lines can unload their plasma fast enough to escape.

Once the pieces of the lines of force are unloaded, they should float to infinity under the influence of cosmic ray pressure. Whether they reconnect in such a way as to detach themselves from the disk or not does not seem to be critical. If they do escape to infinity stretching themselves indefinitely, one would end up with the vertical magnetic flux comparable to the horizontal flux since an equal amount of flux has been removed. Thus, the mean vertical field should be $h/\pi R \approx 0.01$ times the horizontal field. In any event, the picture of the behavior of the dynamo in terms of flux lines appears to be essentially equivalent to that in terms of the dynamo equation.

4.2 Diffusion, the β Factor

The β factor in the dynamo equation appears to be important through its relation to flux escape, but not so important numerically as far as Z diffusion of flux in the disk. Since the solutions require B_R to change sign over the disk, there must be some mixing of the positive and negative lines of force (RSS Figure VII.1). Once they are mixed, it is difficult to see how they could unmix.

One can describe this situation very crudely as follows: During one e-folding of the dynamo, a fraction of flux of opposite signs of order $\delta = (\beta/h)[B_R(h)/B_R(0)]$ is mixed. If the mixing is thorough, then after one e-folding of B in any small area, the flux is predominantly in the positive direction $B_\theta > 0$, but it consists of a proportion $1 - \delta$ of positive lines and δ of negative lines. After the next stage of amplification, because the dynamo cannot distinguish between the positive and negative tubes of force, the fraction of positive flux is $(1 - \delta)^2$ and the fraction of negative flux 1 minus this. After n e-foldings occur, the proportions are $(1 - \delta)^n$ and $1 - (1 - \delta)^n$. If $n\delta \gg 1$, then the rms field is about $e^{n\delta} = A^\delta$ times the mean field where A is the amplification factor. Thus, in order that the magnetic field pressure not be too much larger than the field pressure, inferred, say, from the Faraday rotation results, it is necessary to require that A not be too large. Since $\beta/\gamma h^2 \approx 1$, and $B_R(h)/B_R(0) \approx 0.2$, we would have $\delta \approx 0.2$ and only a few e-foldings would be allowed.

The easiest way out of this problem is simply to let the tubes of opposite sign reconnect (Parker 1957, Sweet 1958, Ji et al 1998, Uzdensky et al 1996, 1997, Kulsrud 1998). This would convert the field energy into heat and lower the magnetic pressure so that the fraction of the area occupied by negative flux would be small. Let us see if this is possible.

An estimate of the cross-sectional area of each tube might be as follows. Each turbulent eddy has a radius $d \approx \lambda_{eddy}$. The first mixed tubes would likely be of this size. After each mixing they would be reduced by one half so that the final cross-section of the "negative" tubes would be $\lambda_{eddy}/2^n = \lambda_{eddy}/A$, where A is the amplification factor of the field. If A were 10^4, then the radii of the cross-section (the "size") of each tube would be 1 pc. The space between tubes would be occupied by reduced field, so that the tubes would press against each other, squeezing out the plasma at the rate V_A/L, where L is the length of the field line between places where it escapes from the disk. If V_A has its current value of 3×10^5 cm/sec, and L is 300 pc, then the time for them to come together would be a logarithmic factor times $L/V_A \approx 10^8$ years. As the tubes get closer, the current density rises until the Ohmic heating gets large enough to balance the plasma pressure loss due to the plasma escape between the tubes. The pressure would then settle down to a large enough value to balance the magnetic pressure. At this time the thickness of the layer between the tubes, δ, would be the Sweet (1958), Parker (1957) thickness $\delta \approx L/S^{1/2}$, where S is the Lundquist number $LV_A/\eta \approx 10^{19}$, $\delta \approx 10^{-10.5}$ and the reconnection velocity with which the tubes merge would be $V_A/L \approx 10^{-4}$ cm/s. The time to completely merge one negative tube with a positive tube at this velocity would be $T_{rec} = dS^{1/2}/V_A \approx 3 \times 10^{20}$ sec.

The reconnection would be faster if the current density were so high that the relative drift velocity, V_D, of the electrons relative to the ions were comparable to the sound speed. But $V_D = cB/4\pi n_e e\delta = 4 \times 10^{1.5}$ cm/sec, a value considerably lower than the sound speed. Therefore, the resistivity should be normal, and there should be no reason for the reconnection to proceed any faster.

Thus, because of mixing, either one must make sure that escape is fast enough in the dynamo solution to avoid serious mixing, or one can only allow a limited amplification of the mean field.

4.3 Small-Scale Fields

Perhaps the most vehement objections to the dynamo have their root in the problem of small-scale fields. These objections are based on an old estimate of Zeldovich (1957) for the behavior of two-dimensional fields. He estimates the small-scale energy to be larger than the large-scale energy by the magnetic Reynold's number, $R_M = LV/\lambda$.

From the general concept of flux freezing we have B/ℓ a constant for any piece of a field line, so as lines are stretched by turbulence, the magnetic field should intensify. The rate of stretching of field lines is essentially the turnover rate, γ, of the eddy stretching them (Batchellor 1950).

Let us assume that turbulence has a Kolmogoroff spectrum $I \sim k^{-5/3}$ where I is the kinetic energy per unit k. Then the rms value of $\tilde{v}_k$ at any scale k^{-1} is

$$\tilde{v}_k \approx \tilde{v}_0 (k_0/k)^{1/3}, \qquad (26)$$

where k_0 and $\tilde{v}_0$ are the wave number and velocity of the largest eddy (at the outer scale). The turnover rate γ_k of eddies of scale k^{-1} is

$$\gamma_k \approx k\tilde{v}_k \approx k_0 \tilde{v}_0 \left(\frac{k}{k_0}\right)^{2/3}. \qquad (27)$$

From this we see that the smallest eddies turn over the fastest, and the magnetic field is amplified most rapidly by eddies at the inner scale with wave number $k_{max} \approx R_e^{3/4} k_0$, where R_e is the hydrodynamic Reynold's number, $R_e = \ell_0 \tilde{v}_0/\nu$, with $\ell_0 = k_0^{-1}$, and ν is the kinematic viscosity. This result is derived in Kulsrud & Anderson (1992), and their numerical estimate shows that $1/\gamma_{max} \approx 10{,}000$ years, a time very short compared with the large-scale dynamo growth time of 10^8 years.

Thus, the intense activity of magnetic fields on small scales must be understood. They could potentially swamp the turbulence in a very short time. The small-scale fields have been discussed in the galactic context by Biermann & Schluter (1951), Parker (Par pp. 519–521), Kazantsev (1968), Ruzmaikin & Shukurov (1982), Kulsrud & Anderson (1992, 1993), Anderson & Kulsrud (1993), Kulsrud et al (1997b), Molchanov et al (1985), and Vainshtein (1980), and in a more general context by Kraichnan (1976a,b), Kraichnan & Nagarajan (1967), Nagarajan (1971), Pouquet et al (1976), Cattaneo & Vainshtein (1991), Vainshtein & Cattaneo (1992), Cattaneo (1994), and Gruzinov & Diamond (1994, 1995, 1996).

A summary of the galactic situation is as follows: Divide the small scales of the magnetic turbulence into two parts: the small scales, between the inner and outer turbulent scales, and the ultra-small scales, those smaller than the turbulent inner scale. A simple calculation shows that the mean free path of a plasma is smaller than the inner scale by a factor $R_e^{1/4}$ ($\approx$10 for the interstellar medium). Therefore, it is not appropriate to apply the ordinary resistive MHD equations below this scale, since magnetic perturbations are collisionlessly damped (Barnes 1968, Kulsrud et al 1997b). However, the magnetic differential Equation (3) is still valid. For simplicity, we will ignore this problem and consider the ultra–small-scale magnetic fields to be governed by MHD equations.

Now, as noted above, the small-scale energy grows at the turnover rate of the smallest eddy, γ_k (i.e. with a timescale of 10^4 years). Equating this rate to the resistive dissipation rate, $k_\eta^2 \lambda$, one finds they are equal at wave number $k_\eta = (\gamma/\lambda)^{1/2}$, which corresponds to 10^9 cm. For comparison, the inner scale is about 10^{16} cm. Thus, there is a range of 10^7 for the ultra-small scales.

Now, Kulsrud & Anderson (1992) have shown that when the magnetic energy is so small that one can ignore back reaction, the magnetic energy propagates downward toward the resistive scale at the rate $5\gamma/4$, while the amplitude at any fixed k grows at the rate $3\gamma/4$. Thus, the total energy grows with the rate (band width $5\gamma/4$) $\times$ (growth rate at fixed k, $3\gamma/4$) $= 2\gamma$. This is consistent with estimates of the growth of the total magnetic energy. However, when the magnetic energy is concentrated at very small scales, k, it exerts a large Lorentz force $kb^2/4\pi$, which is larger the smaller the scale. This relatively large force produces appreciable small-scale velocities, which when combined with the large-scale shear velocity of the eddy, just cancels its amplification effect on the ultra–small-scale fields. In this way, the flow of energy to small scales is blocked and the magnetic fluctuations at very small scales are saturated, (see Par pp. 519–521, Kulsrud & Anderson 1992).

The saturation condition for magnetic fluctuations b_k at scale k given by

$$\frac{k^2 b_k^2}{4\pi \rho} = \gamma_{max}^2, \qquad (28)$$

a constant. From this, one sees that the saturation limit on $\tilde{v}_k$ decreases with increasing k. When the field has just saturated at some scale k, it continues to grow at smaller k until this scale also saturates. (In fact, because of this saturation, the resistive scale k_η is never reached unless the initial field is extremely small, $< 10^{-25}$ gauss.)

There is an apparent inverse cascade, although in fact no magnetic energy is exchanged between scales. In fact, the peak of energy moves to larger and larger scales, until the inner scale λ_{min} is reached. At this point the small scale magnetic energy is concentrated at λ_{min} and satisfies

$$k\tilde{v}_k \approx \gamma_{max} \approx k\tilde{v}_k \qquad (29)$$

with $k = 1/\lambda_{min}$, or, in other words, there is equipartition of the small-scale magnetic energy with the kinetic energy of the smallest eddy.

Thus, a careful consideration of the behavior of ultra–small-scale magnetic fields shows that they saturate with a total energy much smaller than the total turbulent energy. Their net effect on the turbulence is negligible.

These results are contrary to the criticisms of the dynamo theory leveled by Cattaneo & Vainshtein (1991) and Gruzinov & Diamond (1994, 1995, 1996). These criticisms are not identical.

Cattaneo & Vainshtein (1991) claim that the small-scale magnetic energy increases until all the energy is concentrated at the resistive scale and the resistive damping is in balance with production of magnetic fluctuations by the interaction of a turbulent eddy with the large-scale field. That is, their basic equation is

$$\frac{\partial \tilde{\mathbf{b}}}{\partial t} = \nabla \times (\tilde{\mathbf{v}} \times \bar{\mathbf{B}}) + \eta \nabla^2 \mathbf{B}, \tag{30}$$

and the steady state solution of this leads to $\tilde{b} \approx R_M^{1/2} \bar{B}$. Cattaneo & Vainshtein (1991) then claim that the dynamo is saturated by small-scale fields, and turns off, if the initial large-scale field satisfies $\bar{B}^2/4\pi > \rho \tilde{v}_0^2 / R_M$. To derive this result it is necessary to assume that the turbulent velocities are unaffected, and no other velocities are developed until $\tilde{b}^2/4\pi = \rho \tilde{v}_0^2$, and full equipartition is reached. But as noted above, in the case of the galactic dynamo, velocities on the ultra-small scale are produced by much smaller fields, and these velocities stop the growth of magnetic fluctuations long before $\tilde{b}^2/4\pi = \rho \tilde{v}_0^2$. It also should be remarked that in Equation (30), the term $\nabla \times (\tilde{v} \times \tilde{b})$ is neglected, although this actually has no bearing on the final result.

The criticism of Gruzinov & Diamond (1994, 1995, 1996) is a little different. They do not get such a large suppression of the dynamo as do Cattaneo & Vainshtein (1991). They do, however, neglect any self-saturation of the small-scale fields. By taking into account the vector character of the magnetic fields they find that the small-scale fields are proportional to the magnetic Reynold's number times $\mathbf{B}_0 \cdot \langle \mathbf{v} \times \mathbf{b} \rangle = \mathbf{B} \cdot \mathbf{E}$ where $\mathbf{E}$ is the electric field driving the dynamo. Thus, as the small-scale magnetic fields suppress the dynamo they reduce themselves, and the final suppression of the α effect of the dynamo is much less. In fact, they find that $\alpha = \beta \bar{\mathbf{B}} \cdot \nabla \times \bar{\mathbf{B}}$ in the limit of large Reynold's numbers, the contribution from $\mathbf{v} \cdot \nabla \times \mathbf{v}$ being suppressed entirely. The physical significance of this is not clear, but since the basic estimate for b leaves out the all-important self-saturation, the result does not seem relevant for the galactic dynamo.

In summary, magnetic fields on all scales smaller than the inner scale saturate at a level corresponding to equipartition with the smallest eddy.

This is not the end of the story by any means. The energy at larger scales, those between the inner and outer scales, continues to grow. This behavior has been examined by Pouquet et al (1976), and also by Chandran (1996, 1997b) making use of turbulence techniques developed by Kraichnan (1965), the direct interaction approximation (DIA) and its various modifications. The DIA is an attempt to understand turbulence by taking into account all the three mode interactions. These three mode interactions arise from nonlinear terms in the MHD equations. These interactions are themselves modified by all the other modes. As an example, the presence of magnetic fields gives a springiness to the plasma that produces a finite frequency in the magnetic modes. Then, in the three-mode interaction, if the frequencies of the three modes do not match properly, very little energy is exchanged.

Pouquet et al (1976) have simplified the DIA equations by a method developed by Orszag (1972) that introduces an artificial decorrelation time into the three mode interactions. (This is termed the eddy damped quasilinear normal Markorian approximation, EDQNM.) This decorrelation represents as closely as possible the modification of this interaction by other modes. Chandran (1997b) has carried out an integration of the full DIA equations for the beginning of the same inverse cascade problem treated by Pouquet et al (1976) in the EDQNM approximation. He shows that the results are nearly the same, thus justifying their approximation. Chandran (1997b) has introduced a different approximation to the DIA equations, the realizable Markovian closure, or the RMC approximation introduced by Bowman et al (1993). He has reinvestigated the inverse cascade process and obtained conclusions very close to those of Pouquet et al (1976).

These calculations are quite complicated and difficult to carry out and are not well understood physically. However, the results are simple to express. Starting with saturation of the magnetic energy at the inner scale, the magnetic energy continues to grow at larger scales. It grows at each scale between the inner and the outer scales until equipartition is reached. When equipartition is reached, one can think of the hydrodynamic eddies as being replaced by hydromagnetic modes. The eddy motion no longer occurs freely, but is affected by the magnetic forces and becomes oscillatory. These are, of course, not hydromagnetic waves as one usually understands them, since there is no uniform field to carry them, but the frequencies are equivalent to hydromagnetic waves with the uniform field replaced by the rms field.

With time, the magnetic energy inverse cascades to larger and larger scales, reaching equipartition at increasing scales until the outer scale is nearly reached. However, fortunately for the dynamo, equipartition is not attained at the very largest outer scale. This is probably because kinetic energy is constantly being injected hydrodynamically at this scale. The hydrodynamic motion will perturb the magnetic field and try to increase it to equipartition. However, before this can happen the injected energy is transferred to the next smaller scale. Thus, the largest eddy is not strongly affected by the small-scale magnetic fields. Since the mean field dynamo is primarily driven by the largest eddy and its cyclonic motion, the α and β coefficients are not changed very much. Hence, the conclusion appears to be that the total effect of all the small-scale fields on the dynamo is not very important.

5. SEED FIELDS

The dynamo equations are linear in $\bar{B}$ (except for any saturation effects). Thus, at the beginning there must be a non-zero initial field for the dynamo to act on. This initial field is termed the seed field from which the galactic field grows. Various ways to produce the seed field have been suggested.

One suggestion is that the seed fields arise from extra terms in the MHD equations leading to a battery mechanism (Biermann 1950, Par p. 235, RSS p. 255, Kemp 1983, Kulsrud 1997, Kulsrud et al 1997a). Other methods to start the field have also been proposed. (Harrison 1970, Mishustin & Ruzmaikin 1971). By these processes one could start from zero field and grow the field to a certain critical value. When this value is reached, the dynamo becomes more important than the battery and the subsequent growth is by the dynamo.

A second suggestion is that the seed field came from stars and supernova remnants (Michel & Silk 1973, Rees 1987, RSS p. 255). Strong magnetic fields are observed in supernova remnants such as the Crab nebula, and these could provide the seed fields. (These fields themselves have an origin, a similar battery, and dynamo generation in the stars.)

A third suggestion is that since the galactic disk formed by collapse of the protogalaxy, the seed field came from the protogalaxy. There are battery and dynamo mechanisms driven by turbulence in the protogalaxy which could generate substantial magnetic fields, and these could be transferred by collapse into the galactic disk (Pudritz & Silk 1989, Poezd et al 1993, Beck et al 1994, Kulsrud et al 1997a).

Finally, one could regard a primordial field from the early universe as a sort of seed of large strength (Alfven 1950, Piddington 1970, 1981, Howard & Kulsrud 1997).

The seed fields cannot be too small since the dynamo has only a finite number of e foldings, even setting aside the question of too much mixing discussed in a previous section. In fact from Field's estimate of 25 e foldings (Field 1995), the initial seed field for the dynamo modes must be larger than 3×10^{-13} gauss. From our concerns about mixing, it might have to be considerably larger.

5.1 The Biermann Battery Mechanism

In Ohm's law for a plasma there is an additional electron pressure term that is responsible for the battery.

The equation of motion of the electrons is

$$n_e m_e \left(\frac{\partial \mathbf{v_e}}{\partial t} + \mathbf{v_e} \cdot \nabla \mathbf{v_e} \right) = -\nabla p_e - n_e e \left(\mathbf{E} + \frac{\mathbf{v_e} \times \mathbf{B}}{c} \right) + \nu_{ei} n_e m_e (\mathbf{v_i} - \mathbf{v_e}), \quad (31)$$

where $n_e, m_e, \mathbf{v_e}, p_e$ are the electron density, mass, velocity, and pressure, ν_{ei} is the electron ion collision frequency, and $\mathbf{v_e} - \mathbf{v_i}$ is the relative electron ion drift velocity. Express the latter in terms of the current $\mathbf{j} = n_e e (\mathbf{v_i} - \mathbf{v_e})/c$ and the resistivity $\eta = \nu_{ei} mc/ne^2$, divide the equation by $n_e e$ and replace v_e by v. This yields Ohm's law in the familiar form

$$\mathbf{E} + \frac{\mathbf{v} \times \mathbf{B}}{c} = -\frac{\nabla p_e}{n_e e} + \eta \mathbf{J}. \quad (32)$$

(The curl of this equation gives $\partial \mathbf{B}/\partial t$.)

The first term on the right-hand side of Equation (32) is the Biermann battery term. If $\mathbf{B} = 0$ initially, it will cause $\mathbf{B}$ to grow linearly until the dynamo term $\nabla \times (\mathbf{v} \times \mathbf{B})$ becomes equal to the battery term. (If $v \approx c_s$, the sound speed, then this happens when the gyroradius becomes smaller than the scale heights of the pressure and density.) With the extra term, $\mathbf{B}$ satisfies an equation similar to the equation for vorticity $\omega = \nabla \times \mathbf{v}$. In fact, if they are both initially zero, one can show that the relation

$$\Omega = \frac{e\mathbf{B}}{Mc} = -\omega \qquad (33)$$

holds, where M is the ion mass and Ω is the vector ion cyclotron frequency (Kulsrud et al 1997a). This relation is valid until either the viscosity or the resistivity becomes important. Thus, a rough value for the initial field can be obtained from a rough value for the vorticity. Taking $\omega \approx v/L \approx (100 \text{ km/sec})/10 \text{ kps} = 3 \times 10^{-16} \text{ sec}^{-1}$, one gets a value of 3×10^{-20} gauss. This value gives too small a seed field by itself.

5.2 Stellar Seed Fields

A popular choice for seed fields has centered on magnetic fields from stars and supernovae. (Michel & Silk 1973, Rees 1987, RRS p. 255). However, seed fields directly from stars are also quite small when one takes into account the tremendous adiabatic decrease associated with flux freezing. It seems more promising to make use of fields from supernova remnants. Take the magnetic field of 10^{-4} gauss observed in the Crab nebula as an example. It is thought to be due to the wind-up of the pulsar magnetic field by rapid rotation of the pulsar. After expansion to 100 parsecs, say, the field would be reduced by 10^4 to 10^{-8} gauss.

It is proposed to take the seed field as the random sum of the fields from all the supernova remnants taking place in the range of the principal dynamo mode $\Delta R \approx 3$ kpc in one dynamo-mode growth time. The field from each individual supernova remnant should be expanded as a series in the complete set of dynamo modes. One obtains the initial amplitude of the principal normal mode by multiplying the spatial extent R_{sn} of each supernova remnant by the spatial variation of the mode and taking the random sum. If N is the number of supernova and B_{sn} is the field of one, then one gets

$$B_{seed} = \sqrt{N} \left(\frac{4\pi R_{sn}^3}{3}\right) \left(\frac{1}{\text{vol. gal}} \frac{R}{\Delta R}\right) \left(\frac{R_{sn}}{\Delta R}\right) B_{sn}. \qquad (34)$$

The second and third factors give the ratio of the remnant volume to that of the dynamo mode. The fourth factor takes into account that the supernova remnant flux is of both signs, so that only the mode gradient leads to a non-zero contribution to the mode. Taking one supernova per every 30 years in the galaxy, $\Delta R/R = 1/3$, and a growth time of 500 million years, we have $N \approx 5 \times 10^6$. The resulting seed field is

$$B = 3 \times 10^{-12} \text{ gauss},\tag{35}$$

a possibly adequate seed field.

However, one objection to this seed field is that the dynamo operates on any field independent of its sign. Thus, it would amplify N flux tubes of one sign and $N - \sqrt{N}$ of the opposite sign, and keep these fields well mixed. We have shown that reconnection should not be able to merge these tubes of force to any extent, and thus the rms field should be $\sqrt{N} \approx 2.5 \times 10^3$ larger than the mean field.

It is difficult to see how this can be avoided unless some way can be found to make reconnection go much faster than the Sweet (1958) Parker (1957) rate.

5.3 Protogalactic Seed Fields

Recent cosmological numerical simulations for the formation of galactic structure have led to insight into the shear turbulence present in protogalaxies (Kulsrud et al 1997a). The magnetic differential equation, with its battery term, was included in these simulations and it has been shown that the Biermann battery term leads to magnetic field of order 10^{-21}–10^{-20} gauss.

However, the turbulence has been shown by the numerical simulations to have a Kolmogoroff spectrum, and to be in equipartition with the gravitational binding energy of the protogalaxy. On the basis of these facts it is possible to show that during the collapse of the protogalaxy by a factor of only 2 in radius, the magnetic field is easily amplified to equipartition with the smallest eddy (a value of $B \approx 10^{-9}$ gauss). This scale of the smallest eddy is less than a thousandth of the radius of the galaxy. However, if the inverse cascade continues to larger scales after this first equipartition is reached, the magnetic field will evolve to a larger scale and a stronger value. This value and scale are larger and more coherent than that obtained from supernova remnants. It thus seems very likely that the resulting field would provide a natural seed field for the galactic dynamo. Dynamo calculations have already been carried out based on this assumption to show that the strong fields in damped Lyman alpha clouds could be explained by the dynamo (Pudritz & Silk 1989, Poezd et al 1993, Beck et al 1994).

The scale and initial field strength produced by the protogalaxy can be determined only after a better understanding of the inverse cascade phenomena is achieved. However, it seems very likely that the protogalaxy is a leading candidate to provide the seed field.

5.4 The Primordial Field

The boldest and most controversial suggestion for a seed field is that the magnetic field was already present in full strength when the galaxy formed, and needed no further amplification by a dynamo (Alfven 1950, Biermann 1953). In fact, it would already be strong enough to saturate any dynamo action. Any tendency toward cyclonic motions by Coriolis forces would be balanced by the tension force of the primordial field. Any mixing velocities would stretch the field lines and they would bounce back before mixing.

The primordial hypothesis for the origin of the galactic field has been supported by Alfven (1950), Fermi (1949), Hoyle (1958), Hoyle & Ireland (1960, 1961), Piddington (1970, 1981), Kulsrud (1990), and many others. Of course, it transfers the solution of the origin problem to an earlier epoch. However, it is certainly a feasible hypothesis, although not as philosophically satisfying as the dynamo solution. There is really no reason why a strong field could not be present in the disk from the very beginning. There are even several observations supporting it.

The first observation is that stellar formation which seems to depend on the magnetic field strength has not changed very much over the age of the galactic disk.

A second observation is that the relative isotopic abundance of Li as evidenced by early halo stars seems to have changed relatively rapidly in early stages (Fields et al 1995, Lemoine et al 1997). This result is symptomatic of the production of Li7 by cosmic ray spallation with cosmic ray densities comparable to those of the present. Without strong fields, cosmic rays would escape faster and would have lower densities.

A third observation consists of fragmentary evidence that some damped Lyman alpha systems produce rotation measures in quasars that are stronger than quasars not behind such systems, indicating these systems have substantial magnetic fields. (Wolfe et al 1993, Kronberg et al 1992, Perry 1994). Some of these systems are so young as to not have rotated once. This would indicate that very little dynamo action had occurred.

Parker and others have presented arguments against the primordial hypothesis as the sole mechanism for galactic fields (Parker 1969, 1970a, Par p. 519, Rosner et al 1988, Wielebinski 1989). The first of their arguments is that any such fields would rapidly escape the disk by turbulent diffusion. This escape is difficult since these fields are anchored in the interstellar medium, and could not totally escape without taking the bulk of the interstellar medium with them (Vainshtein & Rosner 1991).

The second argument against the primordial origin is that any primordial field would end up tightly wrapped by differential galactic rotation reversing every 100 parsecs in radius. A careful analysis of this wrapup has been carried out by Howard & Kulsrud (1997). They have shown that the field indeed ends up reversing rapidly with radius. However, they have also shown that if the primordial field was not completely uniform over the size of the galaxy, the reversals would not cancel out and there would be a mean field, since field of one direction would have a different flux than the other. It is difficult to see how this model could be contradicted by observations. This picture of the tightly wrapped field is reminiscent of a well-mixed dynamo field.

A reasonable approach to a primordial field, if one grants its existence, is to include it in the dynamo theory as an initial condition.

6. SATURATION OF THE DYNAMO

The galactic magnetic field is at present near a saturated value. If it became any stronger, then the disk would become thicker since the field is already nearly strong enough to support the interstellar medium against gravity.

The dynamo is linear and has exponential growing modes. Thus, it is expected that some nonlinear effects would saturate dynamo action. Since a strong field will resist cyclonic motion, it is natural that the dynamo saturates when it is strong enough to limit the α effect. One could expect this to happen when the Coriolis force $\approx \rho \Omega v_0$ during cyclonic motion is comparable to the magnetic tension force $B_0^2 \Delta\theta / 4\pi r$ where $\Delta\theta$ is the angle of rotation of the cyclonic cell, and r is the size of the cell. Now, $\Delta\theta \approx \Omega\tau \approx \Omega r/v_0$, so that cyclonic motion is suppressed if $B_0^2/8\pi \approx \rho v_0^2/2$, i.e. at equipartition. This is about the observed size of the mean field. If B^2 is larger, then the galactic dynamo should be suppressed. Parker (1992) has argued that, making use of magnetic reconnection, one can avoid this suppression, and continue to amplify the field. He feels this is necessary because the magnetic field continues to be lost by buoyancy, mixing, and perhaps ambipolar diffusion (Parker 1967). We have argued against such loss because the field lines are anchored in the disk. However, if flux freezing is broken by ambipolar diffusion (a type of buoyancy), then there could be loss.

Let us ask: What is the present state of the evolution of the galactic field, or more precisely, what happens after α and β are reduced by a saturated field? The Ω effect is not reduced, so that the field lines will continue to be stretched in the toroidal direction as long as there is a radial magnetic field. Curiously, this is precisely the problem considered in Howard & Kulsrud (1997) in the context of the primordial origin. In this paper they neglected any dynamo actions and included only ambipolar diffusion and differential rotation. They evolved the magnetic field from a simple, nearly uniform state (the primordial hypothesis). However, there is little difference between this and starting from a dynamo mode. The magnetic field strengthens until ambipolar diffusion forces the field lines out of the disk. This weakens the radial field and reduces the stretching. The toroidal field strength rises linearly to a maximum value dependent on initial conditions, and then decays as $1/\sqrt{t}$, always remaining equal, at time t, to that field that can diffuse half the flux out of the disk by ambipolar diffusion in the same time t. For the present time the result is about 2 microgauss, depending on the ambipolar diffusion properties of the clouds. If the field weakens further, the dynamo action will, of course, recommence.

7. CONCLUSION

The $\alpha - \Omega$ dynamo theory is a beautiful and satisfactory theory for the origin and sustainment of a strong coherent galactic magnetic field. It is possible to interpret it in terms of twisting and stretching of magnetic field lines. This picture indicates the vital importance of expelling those parts of the field lines pointing in the wrong direction. Each field line of an initially weak field is imagined to double back and forth many times. It moves forward a certain distance, then leaves the disk and goes far away, moving in a reverse direction, and then returns to the disk at a smaller toroidal angle. It repeats this process as many times as are necessary to achieve amplification. It need not and probably will not break.

Thus, a single line is made to go in the toroidal direction multiple times, consistent with flux freezing, by the $\alpha - \Omega$ dynamo.

The expulsion of the field is a crucial problem that has not been sufficiently addressed. It appears that escape is possible at a rate fast enough to satisfy the vacuum mean field dynamo boundary conditions. However, if the escape does not happen rapidly enough, the mixing of plus and minus field lines can be considerable, leading to very large field energies for a given mean field.

The problem of the effect of small-scale fields on the dynamo is probably exaggerated. If one treats the ultra–small-scale fields carefully, it appears that they are saturated at energies too small to have an effect. Energy at the inner scale seems to inversely cascade up to just short of the outer scale, so that the turbulent spectrum consists of one large hydrodynamic eddy with a tail of hydromagnetic waves. The eddy is the currently driven convective cyclonic cell that is continually being generated by supernovae and supernova remnants. It is important and basic to properly understand all the various properties of small-scale fields, but the essential result would seem to be that small-scale fields have a negligible effect on the dynamo.

The question of the seed field is an open one. Magnetic fields generated by the Biermann battery or the Harrison or Mishustin-Ruzmaikin effect seem to be too weak to amplify sufficiently for the dynamo. On the other hand, the protogalaxy with its strong turbulence is able to generate very strong seed fields on quite substantial scales. Since the disk comes from the protogalaxy, these seed fields seem unavoidable.

The question of stellar fields as the sole seed fields seems problematical since they would lead to strongly interwoven plus and minus fields, with an rms field more than a thousand times larger than the mean field. This is unavoidable since the linear dynamo theory cannot distinguish between plus and minus initial fields. One would not have to worry about such fields if the protogalaxy provides stronger seed fields, stronger than the rms fields from supernova remnants and stars.

The hypothesis of a primordial field is physically possible and the objections directed against it can be ruled out. However, if it is not too strong, then it merely provides a reasonable seed field for the galaxy. The strength of such a field can be considered a parameter in the theory of the galaxy field. At the moment, no plausible theory for the generation of such a primordial field of any magnitude has been advanced. The only limits on a universal field are its effect on the black body radiation. However, this is a weak limit which is probably not of interest. It is interesting that because of the effects of intergalactic fields on the super high energy cosmic rays, the solution may emerge from this direction.

Finally, consider the evolution of the saturated galactic field, when it is strong enough that the α and β effects are suppressed, but not the Ω stretching effect. The field will continue to grow linearly until limited by ambipolar diffusion, which will eject the radial component of the field (as well as the toroidal component). The field will then weaken till the α and β effects are reestablished, and the whole thing could repeat.

In conclusion, we seem to be in pretty good shape with the $\alpha - \Omega$ dynamo as the engine driving the galactic field. The chief problem still remaining is the speed of expulsion of flux. The remaining problems surrounding the dynamo certainly need more investigation, but do not seem to be threatening.

ACKNOWLEDGMENTS

I would like to thank Reinar Beck, Steve Cowley, Pat Diamond, George Field, Gene Parker, Bob Rosner, Sam Vainshtein, and Ellen Zweibel for many useful discussions of the galactic dynamo problem. This work was partially supported by NASA Grant No. NAGW 2419 and by DOE under contract No. DE-AC 02-76-CHO-3073.

Visit the Annual Reviews home page at http://www.AnnualReviews.org

LITERATURE CITED

Alfven H. 1950. *Cosmical Electrodynamics*. Oxford: Clarendon

Anderson S, Kulsrud R. 1993. In *Theory of Solar Planetary Dynamos*, ed. MRE Proctor, PC Mathews, AM Rucklidge, pp. 1–7. England: Cambridge Univ. Press

Barnes A. 1967. *Phys. Fluids* 10:2427–36, 437

Batchellor GK. 1950. *Proc. Roy. Soc. London*. A201:405–16

Beck R, Brandenburg A, Moss D, Shukurov A, Sokoloff D. 1996. *Annu. Rev. Astron. Astrophys*. 34:155–206

Beck R, Poezd AD, Shukurov A, Sokoloff DD. 1994. *Astron. Astrophys*. 289:94–100

Biermann L. 1950. *Z. Naturforsch* 5a:65–71

Biermann L. 1953. *Annu. Rev. Nucl. Sci*. 2:335–64

Biermann L, and Schluter A. 1951. *Phys. Rev*. 82:863–68

Bowman J, Krommes J, Ottaviana M. 1993. *Phys. Fluids* 5:3558–89.

Brandenburg A, Donner KJ, Moss D, Shukurov A, Sokoloff DD. 1992. *Astron. Astrophys*. 259:453–46

Brandenburg A, Donner KJ, Moss D, Shukurov A, Sokoloff DD. 1993. *Astron. Astrophys*. 271:36–50

Cattaneo F, Vainshtein SI. 1991. *Ap. J. Lett*. 376:L21–24

Cattaneo F. 1994. *Ap. J*. 434:200–95

Chandran BDG. 1996. Phd Dissertation Princeton University

Chandran BDG. 1997a. *Ap. J*. 482:156–66

Chandran BDG. 1997b. *Ap. J*. 485:148–58

Fermi E. 1949. *Phys. Rev*. 75:1169–74

Ferriere K. 1992a. *Ap. J*. 389:286–96

Ferriere K. 1992b. *Ap. J*. 391:188–98

Ferriere K. 1993. *Ap. J*. 404:162–84

Ferriere K. 1993. *Ap. J*. 409:248–61

Ferriere K. 1996. *Astron. Astrophys*. 310:438–55

Ferriere K. 1998. *Astron. Astrophys*. 335:488–99

Field GB. 1995. In *The Physics of the Interstellar Medium and Intergalactic Medium.* ed. A Ferrara, CF McKee, C Heiles, PR Shapiro, pp. 1–14, *Astron. Soc. Pac. Conf. Ser.* Vol. 80

Fields BD, Olive KA, Schramm DN. 1995. *Ap. J.* 435:185–202

Gruzinov A, Diamond P. 1994. *Phys. Rev. Lett.* 72:1651–53

Gruzinov A, Diamond P. 1995. *Phys. Plasmas* 2:1941–46

Gruzinov A, Diamond P. 1996. *Phys. Plasmas* 3:1853–57

Harrison ER. 1970. *MNRAS* 147.279–86

Higdon JC. 1984. *Ap. J.* 285:109–23

Howard A, Kulsrud R. 1997. *Ap. J.* 483:648–65

Hoyle F. 1958. In *La Structure et l'Evolution de l'Universe*, pp. 53–73, ed. R Stoops. Brussels: Institut de Solvay

Hoyle F, Ireland JG. 1960. *MNRAS* 120:173–86

Hoyle F, Ireland JG. 1961. *MNRAS* 122:35–39

Ji H, Yamada M, Hsu S, Kulsrud R. 1998. *Phys. Rev. Lett.* 80:3256–59

Kazantsev AP. 1968. *Sov. Phys. JETP* 26:1031–34

Kemp JC. 1982. *Publ. Astron. Soc. Pac.* 94:627–33

Kraichnan RH. 1964. *Phys. Fluids* 7:1163–77

Kraichnan RH. 1976a. *J. Fluid Mech.* 75:657–76

Kraichnan RH. 1976b. *J. Fluid Mech.* 77:753–68

Kraichnan RH, Nagarajan S. 1967. *Phys. Fluids* 10:859–70

Krause F, Radler K-H. 1980. *Mean Field Electrodynamics and Dynamo Theory.* Berlin: Akademie-Verlag. Oxford: Pergamon

Kronberg PP. 1994. *Rep. Prog. Phys.* 57:325–82

Kronberg PP, Perry JJ, Zulkowski ELH. 1992. *Ap. J.* 387:528–35

Kulsrud RM. 1990. In *Galactic and Extragalactic Magnetic Fields, IAU Symp. 140*, ed. R Beck, PP Kronberg, R Wielebinski. p. 527

Kulsrud RM. 1997. In *Critical Dialogues in Cosmology*, ed. N Turok, pp. 328–40. Singapore: World Scientific

Kulsrud RM. 1998. *Physics of Plasmas* 5:1599–1606

Kulsrud RM, Anderson SW, 1992. *Ap. J.* 396:606–30

Kulsrud RM, Anderson S. 1993. In *Theory of Solar Planetary Dynamos*, ed. MRE Proctor, PC Mathews, AM Rucklidge, pp. 195–201. England: Cambridge University Press

Kulsrud RM, Cen R, Ostriker JP, Ryu D. 1997a. *Ap. J.* 480:481–91.

Kulsrud RM, Cowley S, Gruzinov A, Sudan R. 1997b. *Phys. Rep.* 283:213–26

Kulsrud RM, Pearce WP. 1969. *Ap. J.* 156:445–69

Lemoine M, Schramm DN, Truran JW, Copi CJ. 1997. *Ap. J.* 478:554–62

McCray R, Snow TP, 1979. *Annu. Rev. Astron. Astrophys.* 17:213–39

McIvor I. 1977. *MNRAS* 178:85–100

Michel FC, Yahil A. 1973. *Ap. J.* 179:771–80

Mishustin IM, Ruzmaikin AA. 1971. *Sov. Phys. JETP* 34:233–35

Moffatt HK. 1978. *Magnetic Field Generation in Electrical Conducting Fluids.* Cambridge: Cambridge University Press

Molchanov SA, Ruzmaikin AA, Sokoloff DD. 1985. *Sov. Phys. Uspekhi* 28:307–27

Moss D, Brandenburg A. 1992. *Astron. Astrophys.* 256:371–74

Moss D, Brandenburg A, Donner KJ, Thomasson M. 1993. *Ap. J.* 409:179–89

Nagarajan S. 1971. *Solar Magnetic Fields, IAU Symp. 43*, ed. R Howard, pp. 487–504

Orszag SA, Patterson GS. 1972. In *Lecture Notes in Physics vol. 70*, ed. JB Keller, JS Pakadakis, pp. 127. New York: Springer-Verlag

Parker E. 1955a. *Ap. J.* 121:491–507

Parker E. 1955b. *Ap. J.* 122:293–314

Parker E. 1957. *Phys. Rev.* 130:830–36

Parker E. 1966. *Ap. J.* 145:811–33

Parker E. 1967. *Ap. J.* 149:517–34

Parker E. 1969. *Ap. J.* 157:1129–35

Parker E. 1970a. *Ap. J.* 160:383–404
Parker E. 1970b. *Ap. J.* 162:665–73
Parker E. 1971. *Ap. J.* 163:255–78
Parker E. 1971b. *Ap. J.* 163:279–85
Parker E. 1973. *Astrophys. Space Sci.* 22:279–91
Parker E. 1979. *Cosmical Magnetic Fields: Their Origin and Their Activity.* Oxford: Clarendon Press (referred to as Par)
Parker E. 1992. *Ap. J.* 401:137–45
Perry JJ. 1994. In *Cosmical Magnetism. Contributed Papers in Honor of Professor L. Mestel*, ed. D Lyndon-Bell, pp. 144–51. Cambridge: Inst. Astron.
Piddington JH. 1970. *Aust. J. Phys.* 23:731–50
Piddington JH. 1981. *Ap. J.* 247:293–99
Poezd A, Shukurov AM, Sokoloff D. 1993. *MNRAS* 264:285–87
Pouquet A, Frisch U, Leorat J. 1976. *J. Fluid Mech.* 77:321–54
Pudritz RE, Silk J. 1989. *Ap. J.* 342:650–59
Rees M. 1987. *QJRAS* 28:197–206
Rosner R, Ducca 1988. In *The Center of the Galactic IAU Symp. 136*, ed. M. Morris, pp. 319–28
Ruzmaikin AA, Shukurov AM. 1982. *Astrophys. Space Sci.* 82:397–407
Ruzmaikin A, Shukurov AM, Sokoloff D. 1988a. *Nature* 336:341–47
Ruzmaikin AA, Shukurov AM, Sokoloff DD. 1988b. *Magnetic Fields of Galaxies.* Dordrecht: Kluwer (referred to as RSS)
Soward AM. 1978. *Astron. Nachr.* 299:25–33
Spitzer L. 1990. *Annu. Rev. Astron. Astrophys.* 28:71–101
Steenbeck M, Krause F, Radler K-H. 1966. *Z. Naturforsch.* 21a:369–76
Stix M. 1975. *Astron. Astrophys.* 42:85–89
Stix M. 1978. *Astron. Astrophys.* 68:459
Sweet PA. 1958. In *Electromagnetic Phenomenoa in Ionized Gases Proc. IAU Symp. 6*, ed. B Lehnert, pp. 123–34
Uzdensky D, Kulsrud R, Yamada M. 1996. *Phys. of Plasmas* 3:1220–33
Uzdensky D, Kulsrud R. 1997. *Phys. of Plasmas* 4:3960–73
Vainshtein SI. 1970. *Sov. Phys. JETP* 31:87–89
Vainshtein SI. 1980. *Sov. Phys. JETP* 52:1099–1107
Vainshtein SI, Cattaneo F. 1992. *Ap. J.* 393:165–71
Vainshtein SI, Rosner R. 1991. *Ap. J.* 376:199–203
Vainshtein SI, Ruzmaikin AA. 1971. *Sov. Astron.* 15:714–19
Vainshtein SI, Ruzmaikin AA. 1972. *Sov. Astron.* 16:365–67
Vainshtein SI, Zeldovich YaB. 1972. *Sov. Phys. Uspekhi* 15:159–72
Walker MR, Barenghi CF. 1994. *Geophys. Astrophys. Fluid Dynamics* 76:265–81
Wielebinski R. 1989. In *The Interstellar Medium in Galaxies*, ed. HA Thronsen, JM Shull, pp. 344–69. Dordrecht: Kluwer
Wolfe AM, Turnshek DA, Lanzetta KM, 1993. *Ap. J.* 404:480–510
Zeldovich YaB. 1957. *Sov. Phys. JETP* 4:460–62
Zweibel EG, Heiles C. 1997. *Nature* 385:131–36

FREQUENCY ALLOCATION: The First Forty Years

Brian Robinson

Research Fellow Emeritus, Australia Telescope National Facility, Epping, N.S.W. 2121 Australia; e-mail: brobinso@ozemail.com.au

Key Words Frequency management, IUCAF, world radio conferences, radio interference, mobile satellites, radio astronomy

■ **Abstract** In 1960 ICSU set up an Inter-Union Commission (IUCAF) on the Allocation of Frequencies for Space Research and Radio Astronomy, to keep key parts of the radio spectrum clear for passive, scientific use. IUCAF represents URSI, IAU and COSPAR at World Radio Conferences (WRCs) convened by the International Telecommunications Union (ITU) in Geneva; the WRCs establish the international law which governs users of the radio spectrum. This review recounts many serious threats posed to passive scientific research by commercial and military operations, particularly those involving radio emissions from aircraft and spacecraft. The continual conflict between commercial greed and scientific curiosity has often put the future of radio astronomy, space research, and earth exploration in jeopardy. The conflict increases as we move into the Information Age.

1. THE BEGINNINGS: 1952–1959

This review recounts the role of the Inter-Union Commission on the Allocation of Frequencies (IUCAF), and its predecessor in the International Scientific Radio Union (URSI), to keep some parts of the radio spectrum clear for passive, scientific use. IUCAF is responsible for protecting the frequencies needed by radio astronomy, space research, and environmental monitoring of the Earth. The 40-year struggle has a parallel in the continuing fight of astronomers to keep some sky dark, despite street lights, advertising, aircraft, spacecraft, and recurrent ideas for space advertising.

1.1 Interference

In the late nineteenth century, attempts to detect radio waves from extraterrestrial sources were made by Ebert, Jansen, Lodge, and others. Lodge found that a sensitive coherer was subject to "too many terrestrial sources of disturbance in a city like Liverpool to make the experiment feasible" (Smith 1960). So there has been awareness of "terrestrial sources of disturbance" since the earliest experiments.

Interference played a major part in Karl Jansky's 1932 discovery of radio waves from the galaxy. It was a result of his investigation, for Bell Telephone Labs, of the sources of interference experienced on 20.5 MHz radio links, using a directional antenna to scan in azimuth. The arrival of one type of interference advanced by four minutes each day, which Jansky identified as a sidereal rate (Jansky 1979).

1.2 World War II: New Technology

Between 1932 and World War II, radio astronomy had been an individual, amateur effort: Grote Reber was building receivers and a steerable dish in the backyard of his house in Wheaton, Illinois.

The development of radar in the United Kingdom from 1935, and the enormous growth of radar and communications technology from 1939 to 1945 trained many people in microwave techniques and gave them advanced hardware. During the war J.S. Hey had identified radiation from a sunspot, and observed emission from Cygnus A and radio reflections from meteor trails.

The scene was set for great expansion of radio astronomy after the war, and radio observatories began at Sydney and Cambridge. The Radiophysics Laboratory in Sydney had been the center for radar development in the South Pacific, and when the war ended the group remained intact, looking for other applications of the new radio technology.

Martin Ryle had been involved in British radar research and development, particularly countermeasures in aircraft. At the end of the war Ryle was put in a major's uniform and sent to Germany to steal Nazi technology. Ryle came back to England with airborne radar equipment and two Wurzburg radar antennas, which Ryle, Graham Smith, and Bruce Elsmore set up as an interferometer at Cambridge. Other important groups formed in Europe: Jodrell Bank, Leiden, Meudon, Onsala, and others. Wurzburg antennas appeared everywhere. In the United States work began at Harvard University and the Naval Research Laboratories, located in Washington, D.C., to follow up the early work at Sydney and Cambridge.

However, the amazing growth of technology during the war equally set the stage for remarkable developments in radio communications, broadcasting, and navigation, both civil and military. The potential for interference to radio astronomy from radar and communications was unavoidable. In the last forty years we have seen a burgeoning use of the radio spectrum for every conceivable purpose. Conflict arose between those who want to transmit and those who want to listen to emissions from space—understandably so because the emissions from space are, to traditional communications engineers, unbelievably weak.

1.3 International Cooperation in Radio Astronomy

From the time of the 1905 Berlin Conference, there had been many international efforts to carve up the radio spectrum between different users. The International Telecommunications Union (ITU) was born in 1919, as a branch of the League of Nations, to wield the carving knife. In 1959 the ITU (now a United Nations body) called a World Conference to re-allocate the radio spectrum and to allocate bandwidths that were inaccessible at the time of the 1937 ITU World Conference in Cairo but became accessible with technology that was brand new at the time of the 1947 Atlantic City World Conference.

Radio astronomers were quick to prepare for and take part in the 1959 World Conference in Geneva. Radio astronomy observations of "radio sources" had already been carried out at frequencies from 18 MHz (Shain 1951) to 24 GHz (Piddington and Minnett 1949). In 1951 the first radio spectral line, atomic hydrogen, had been discovered from the Milky Way at 1420.405 MHz (Ewen & Purcell 1951).

The ITU has a technical arm that was then known as Comite Consultatif International des Radiocommunications (CCIR). Recommendations from CCIR study groups carried great weight at ITU world conferences. Balth van der Pol from Holland played an important part in CCIR as director from 1949 to 1957, as did F.L.H.M. Stumpers, R.L. Smith-Rose and Fred Horner from the United Kingdom, and John P. Hagen from the United States. From 1953 important work was done in London by a CCIR study group responsible for radio astronomy. A history of the CCIR from 1929 to 1989 can be found in Kirby (1978, 1989) and Stumpers (1989). Kirby and Struzak (1994) give a retrospect on the CCIR view on frequency allocations for radio astronomy.

Union Radio Scientifique Internationale   Radio astronomy had found an international home in URSI as Commission V. The new commission was founded in 1948 with the title "Extraterrestrial Radio Noise." By the time of its first meeting at the 1950 URSI General Assembly in Zurich, the title of Commission V had been changed to "Radio Astronomy." At the first session, on September 13, 1950, the death of Karl Jansky in February was announced. Commission V next met in Sydney in 1952, in The Hague in 1954, and in Boulder in 1957.

CCIR met in Warsaw in 1956 and drew up Recommendation No. 173 on Protection of Frequencies Used for Radio Astronomical Measurements (Smith-Rose 1956). "Rec. 173" was referred to Commission V of URSI, and the 1957 General Assembly of URSI in Boulder set up Sub-Commission Ve on Frequency Allocation, charged with protection of the radio spectrum for radio astronomy and preparing recommendations to be submitted to the ITU. The initial membership (Haddock 1957) was F.T. Haddock (Chair), E.J. Blum, A. Hewish, V.V. Vitkevitch, W.N. Christiansen, A.H. de Voogt, T. Hatanaka, L. Erikson, F. Becker, A.E. Covington, G. Righini, J. Tuominen, O.E.H. Rydbeck, R. Coutrez and A.P. Mitra.

Sub-Commission Ve under F. T. Haddock, then J.W. Findlay (from 1958), took part in the submissions made by CCIR (at its Plenary in Los Angeles in April 1959) to the 1959 ITU World Conference. Activity by national governments was also required—the ITU is a United Nations body, and only U.N. member countries can vote on proposals. The involvement of governments is crucial because delegates from URSI and IUCAF can speak at conference sessions, but never vote.

The World Conference lasted for three months, and the case for radio astronomy was supported in rotation by J.H. Oort (IAU President), B. van der Pol, H.C. van de Hulst, J.F. Denisse, J.W. Findlay, C.L. Seeger, R. Coutrez and W.J.G. Beynon. Van der Pol, the strong supporter of science, died in October. Oort (1960) published a special public lecture given at WARC-1959.

The main burden at the WARC fell on Charles Seeger (secretary of URSI Sub-Commission Ve). John Findlay (1988) describes the support of The Netherlands and the United Kingdom, and the opposition of the United States (Sullivan 1959, Lear 1959, Finney 1959). Seeger tells of later charges against him of "un-American activities;" being the brother of folk singer Pete Seeger didn't help his position.

Although falling considerably below the hopes of radio astronomers, WARC-1959 laid a good basis for radio astronomy, notably the exclusive "passive" band at 1400–1427 MHz for 21 cm hydrogen line observations. In 1959 there was little commercial or military demand for frequencies above 1000 MHz, and allocations (in footnote form) of reasonable bandwidth were possible. There was less success with the request for allocations of a 2.5 percent band every octave in frequency for continuum observations.

The ITU recognizes radio services, such as broadcasting, fixed networks, radionavigation, radar, mobile communications, amateur use, and meteorological aids. A major victory at the World Conference in 1959 was to have the ITU accept the idea of a "passive service." Until that point, a service had to transmit radio waves; those who only received signals had been a totally different group, outside the ambit of the ITU.

The WARC agreed that the needs of radio astronomy and space science would be included in a "Space" WARC to be held in 1963.

In his report to URSI, Findlay (1960) recommended that Sub-Commission Ve be dissolved, to be replaced by a group representing URSI, IAU, and COSPAR. But Sub-Commission Ve continued until the 1966 URSI General Assembly in Munich.

As well as Commission V in URSI, radio astronomy had also found a place as Commission 40 in the International Astronomical Union. This Commission Pour les Observations Radioelectriques did not meet at the Rome IAU General Assembly in 1952, but its president sent a report to the IAU Transactions (Woolley 1952). By the 1955 General Assembly of IAU in Dublin, Commission 40 had its first meeting, with the title Commission de Radio-Astronomie. Appendix I of the Proceedings (Pawsey 1955) addressed the general requirements for frequency allocations (proposed by Laffineur, Hagen, de Voogt, Ryle, and Vitkevitch).

The last IAU General Assembly before WARC-1959 was the 1958 meeting in Moscow, which re-affirmed the broad requirements for protection of radio astronomy. At the following IAU General Assembly (Berkeley 1961) J-F. Denisse (1961) reported that *CCIR (Los Angeles 1958) a enterine dans ses grandes lignes les demandes d'allocation de frequences formulees par l'UAI à Moscou* [which translates to English as, "CCIR (Los Angeles 1958) ratified in the main proposals for frequency allocations formulated by the IAU in Moscow"].

2. THE FORMATION OF IUCAF: 1960 to 1971

By 1960 two significant developments occurred: the formation of COSPAR, and the proposal from the MIT Lincoln Laboratory for Project WEST FORD (the Needles Project, funded by the U.S. Air Force). They proposed a reflective belt of dipoles in a circular orbit around the Earth at a height of 3000 kilometers. The orbital scatter concept was proposed by W.E. Morrow (of Lincoln Lab) in 1958 in collaboration with H. Meyer (of Thompson Ramo Wooldridge Inc.).

2.1 IUCAF: London 1960

Project WEST FORD presented a great threat to radio astronomy, optical astronomy, and space research, and called for a combined response by URSI, IAU, and COSPAR. An ICSU Inter-Union Commission was set up by URSI's Executive Committee on September 2, 1960, encouraged by Lloyd Berkner. The members initially appointed were J.A. Ratcliffe (Chair), R. Emberson, L.G.H. Huxley (on behalf of COSPAR), V. Ilyin and H. Sterky.

At their first meeting in London on September 5, 1960, Ratcliffe suggested that coordination of activities with Commission V of URSI would be achieved if J.W. Findlay were to be co-opted to the committee as its secretary. Findlay attended the second meeting on September 8, 1960, presented the report of Sub-Commission Ve (Findlay 1960) and outlined the work that had been done in getting protected frequencies for radio astronomy during 1958 and 1959.

IUCAF's Terms of Reference were published by Smith-Rose (1960). By October 1960 the membership had evolved to R. Emberson, V. Ilyin, J.A. Ratcliffe, H. Sterky, and R.L. Smith-Rose (secretary-general) for URSI; J-F. Denisse (chairman), V.V. Vitkevitch, J.H. Oort, and A. Unsold for IAU; and J.P. Hagen, L.G.H. Huxley, A.P. Mitra, and H.C. van de Hulst for COSPAR.

Under chairmen J-F. Denisse (1960–64) and Graham Smith (1964–1975) 13 meetings of a large IUCAF group were held all over Europe, plus a meeting in Washington D.C. in 1967.

Many articles pointed out the dangerous pollution of space around the Earth by the Needles proposal, and the likely effects on radio astronomy, optical astronomy, and space research (see Findlay 1962 and Lovell & Ryle 1962). Four papers appeared in *Astronomical Journal* in April 1961. A.E. Lilley (1961) analyzed the radio properties of an orbiting scattering medium. After considering Project WEST FORD and anticipating new communications and navigation systems employing powerful transmitters in earth satellites, Lilley's article concluded:

> The pursuit of basic science and the progress of space radio technology represent needs of man which must be advanced. For the impending interference a simple solution exists: allocation of clear frequency bands for basic science. This action is imperative and must ultimately rest on national and international agreements.

Figure 1 The six chairmen of IUCAF. *Top*: J-F. Denisse (1960–1964), F.G. Smith (1964–1975). *Middle*: J.P. Hagen (1975–1981), J.W. Findlay (1981–1987). *Bottom*: B.J. Robinson (1987–1995), W.A. Baan (1995–1999). In March 1999 Klaus Ruf became chairman.

IAU Involvement Denisse & Seeger (1961) gave an account of the formation of IUCAF to the 1961 IAU General Assembly in Berkeley. IAU Commission 40 formed a working group (Lilley, Seeger, Smith, Haddock, McClain, Molanchov, Gold, and Blum) to study the conditions under which the experimental belt of dipoles could be observed by radio astronomy observatories.

At the 1961 General Assembly the majority of members of IAU Commission 40 were completely opposed to Project WEST FORD. The members of Commission 40 were not convinced that the belt of dipoles would have a limited lifetime. This point of view was expressed in IAU resolutions proposed by Bondi, Christiansen & Gold (1961), and by Seeger & Pawsey (1961).

Resolution 1 said:

> Commission 40 views with concern the increasing contamination of the space around the Earth by radiating and scattering objects. It feels that no group has the right to change the Earth's environment in any significant way without full international study and agreement.

Resolution 2 said:

> Such projects should not be undertaken unless sufficient safeguards have been obtained against harmful interference with astronomy. Nevertheless, Commission 40 views with the utmost concern the possibility that the belt of dipoles proposed in Project WEST FORD might be permanent, and is completely opposed to such an experiment until this question is clearly settled in published papers and time has been given for their study. Whatever the limitations of present radio astronomical equipment, the Commission is inflexibly opposed to any steps that might permanently compromise future development in radio and optical astronomy.

The IAU executive committee announced a resolution (Denisse 1961):

> *Maintaining* that no group has the right to change the Earth's environment in any significant way without full international study and agreement; the International Astronomical Union *gives* clear warning of the grave moral and material consequences which could stem from a disregard of the future of astronomical progress, and *appeals* to all Governments concerned with launching space experiments which could possibly affect astronomical research to consult with the International Astronomical Union before undertaking such experiments and to refrain from launching until it is established beyond doubt that no damage will be done to astronomical research.

See also the reports of Smith-Rose (1962), Lovell & Ryle (1962), and Blackwell & Wilson (1962). The threat posed by WEST FORD was publicized in *New Scientist*, February 2, 1962, and updated in March 1, 1962.

Project West Ford Launches An abortive WEST FORD launch on October 21, 1961 (Morrow 1962) deployed on an Air Force Atlas-Agena B rocket carried

75 pounds of fine copper dipole fibers embedded in naphthalene. The package was expected to release the fibers slowly to form a belt at an altitude of 2000 miles. There was no sign that the dipole fibers dispersed from the package. The fibers remained clustered in five or six small clumps that were tracked by the MIT Millstone Hill UHF radar for several months.

A second WEST FORD launch scheduled for 1962 was delayed until May 12, 1963 (Morrow 1963). The 1963 launch put into polar orbit 400 million dipoles, more than half of which failed to separate into individual reflectors, remaining loosely tangled in small clusters or chains (Smith-Rose 1964c). The WEST FORD Project expired after these attempts.

The threats posed by Project WEST FORD to optical and radio astronomy and to space research brought home forcefully to scientists that strong international action was required to prevent major changes to the Earth's environment, particularly if these might be long lasting changes. It was a propitious time for the birth of IUCAF as an ICSU Inter-Union Committee to coordinate scientific needs for frequency allocations for scientific research. The threat of Project WEST FORD came from a military project. Later in this review we will discuss other military projects, or civil projects such as the Geostationary Meteorological Satellite, or commercial projects such as IRIDIUM.

The Committee on Radio Frequencies In the spring of 1961 the U.S. National Academy of Sciences and the National Research Council established the Committee on Radio Frequencies (CORF) "to serve as the United States counterpart to IUCAF" (Dellinger 1961). CORF first met on May 3, 1961 and was to play a major part in preparations for future WARCs.

3. SPACE WARC OF 1963

After the formation of IUCAF in 1960, the next ITU World Conference was the Space WARC of 1963. At the IUCAF meeting at the Royal Society London in October 1961 Smith-Rose (1961) reported that IUCAF "is now formally recognized as an active participant in the work of CCIR and the ITU." The meeting confirmed a recommendation, already provisionally forwarded to ITU, seeking the inclusion of radio astronomy and space science in the agenda of the 1963 Extraordinary Administrative Radio Conference to be held in Geneva from October 7 through November 8, 1963.

In 1959 Charles H. Townes and his group at Columbia University succeeded in establishing the laboratory frequencies of two radio spectral lines for the hydroxyl molecule. In the fall of 1963 Sander Weinreb applied his new correlation receiver technique on the MIT Millstone Hill 84-foot radio telescope. Weinreb, Barrett, Meeks, and Henry (1963) found absorption on October 15 at 1667.46 MHz in the direction of Cassiopeia A. The observers then found absorption at 1665.34 MHz, at 5/9ths of the strength at 1667 MHz. Interstellar OH molecules to be sure (Robinson 1965).

During the 1963 ITU conference the U.S. delegation was able to announce the discovery of the main OH lines at 1665 and 1667 MHz. A secondary allocation was made to radio astronomy in the 1664.4–1668.4 MHz band (shared with meteorological aids and meteorological satellites). Footnote F 353A described the discovery of the main OH lines and mentioned the intent to remove meteorological aids (meteorological balloons) from the 1664.4–1668.4 MHz band. The IUCAF Report on WARC-1963 appeared in *Nature* (Smith-Rose 1964a, 1964b).

In April 1964 Australian radio astronomers observing Sgr A discovered two "satellite" lines of hydroxyl at 1612.201 and 1720.559 MHz (Gardner et al 1964). At the time these satellite lines of OH were believed to be of lesser astrophysical importance than the main lines reported at WARC-1963. The satellite OH lines were later given a footnote allocation by WARC-1971.

4. MOLECULES IN SPACE: Millimeter-Wave Radio Astronomy

Great excitement greeted the discovery of the OH lines in 1963 and 1964. Theoreticians such as Bates & Massey (1947) had worked out the theory of two-body recombination, and we were comfortable to find OH lines at radio wavelengths to parallel CH and CN in optical spectra. Bates showed that the rate of three-body recombination in the interstellar medium was negligible.

Townes and his collaborators ignored Bates and found ammonia NH_3 at 23.7 GHz (Cheung et al 1968), and H_2O at 22.235 GHz (Cheung et al 1969). Formaldehyde H_2CO was found at 4.829 GHz (Palmer & Zuckerman 1969). Then came the 1970 discovery of CO at 115.271 GHz with the Kitt Peak 11-meter precision radio telescope (Wilson et al 1970). Over the next six years there was an avalanche of discoveries of molecular spectral lines (see the reviews by Rank et al 1971, Zuckerman & Palmer 1974, and Robinson 1976).

The flood of molecular line discoveries was timely in the preparations for the second "Space WARC" in 1971, discussed in the next section. The richness of the millimeter-wave spectrum in molecular lines also gave radio astronomers entrée to the top end of the radio spectrum, which other services had not begun to use. By 1972 there were 137 observed molecular lines (between 0.8 GHz and 173 GHz); most were assigned to 35 terrestrially identified molecules; eight of the lines were unidentified or produced by molecules that are transient or rare under laboratory conditions.

5. WARC-SPACE TELECOMMUNICATIONS 1971

Smith-Rose wrote to the ITU on March 4, 1969, requesting participation by IUCAF in the 1971 WARC on Space Telecommunications. The requirements of radio astronomy and space research were reviewed by Smith (1970) and in IUCAF Document 142 (Smith-Rose 1969). Improvements sought in the existing allocations were *(a)* at least part of each frequency band be made an exclusive radio astronomy allocation which is the same throughout the world; *(b)* improve and complete the series of bands throughout the spectrum; *(c)* provide wider bandwidths for observations in the bands at 2690–2700 MHz and above, and thus achieve higher sensitivities.

WARC-ST 1971 was held in Geneva June 7 to July 17 1971 and allocated frequencies up to 275 GHz. A detailed report (Horner 1971) listed the decisions affecting radio astronomy at WARC-ST 1971. The report's appendices give complete details of the allocations made to radio astronomy over the range of frequencies from 21.85 MHz to 240 GHz. (For a corresponding list of the frequencies allocated to space research see Horner 1972.)

A downward extension of the hydrogen band to 1400 MHz on a shared basis was made to cater for redshifts in emissions from more remote sources. Protection of the main OH lines was improved considerably by deleting meteorological satellite allocations in the band 1660–1670 MHz.

The 1971 ITU allocations covered these spectral lines above 1.4 GHz:

(a) By table allocation: Hydroxyl (1665 & 1667 MHz), Ammonia (23.7 GHz), Hydrogen Cyanide (86.3 & 88.6 GHz).

(b) By footnote allocation: Hydroxyl (1611.5–1612.5 and 1720–1721 MHz), Formaldehyde (4.829 GHz), Excited Hydrogen (5.763 GHz), Formaldehyde (14.489 MHz), Water Vapor (22.235 GHz), Excited Hydrogen (36.466 GHz), and Carbon Monoxide (115.271 GHz).

Recommendation Spa2-8 from WARC-ST 1971 was entitled "Relating to the Protection of Radio Astronomy Observations on the Shielded Area of the Moon." Arthur C. Clarke (1961) had suggested that lunar area as a location for radio astronomy. From 1974 CCIR studied the need to keep the radio spectrum quiet on the far side of the moon (CCIR Report 539-1 and CCIR Recommendation 479-1). When commercial pressures pollute the whole terrestrial radio spectrum and the sky contains swarms of radiating satellites, the far side of the moon would become the last oasis for radio astronomy.

6. INTERFERENCE FROM SATELLITES AND AIRCRAFT

6.1 Broadcasting Satellites

The allocation of the 2670–2690 MHz band to radio astronomy at WARC 1971 (on a shared basis) was nullified by new allocations for the Broadcasting Satellite Service in the 2500–2690 MHz band. Out-of-band energy from broadcasting satellites up to 2690 MHz would also cause interference to radio astronomy in the 2690–2700 MHz band, which was allocated exclusively to radio astronomy. The Broadcasting Satellite Service was unable to guarantee that such interference would be kept below limits recommended by the CCIR; nor would they agree to a guard band of several megahertz. During the Project WEST FORD planning, a television picture had been sent in April 1962 from California to Massachusetts using the ECHO-1 space balloon (Morrow 1963). This had shown the feasibility of radiating TV from a satellite in geostationary orbit to a whole hemisphere of the Earth.

The chairman of IUCAF (Smith 1972a) wrote to all radio astronomy observatories about the very serious threat to observations using one of the most important frequency bands allocated to radio astronomy. The WARC 1971 allocation to broadcasting satellites was taken despite strong objections by IUCAF and others representing the interests of radio astronomers.

Smith (1972b) discussed how the launch of the sixth Applications Technology Satellite (ATS-F) could threaten radio observations of the universe. The ATS-F, broadcasting at 2670 MHz, was a precursor to satellites that would broadcast television programs direct to domestic receivers. Smith voiced the concern of radio astronomers over the ATS-F affair because of the cavalier treatment of the international regulations governing allocation of radio spectrum frequencies. The broadcasting signal would spill into the exclusive radio astronomy band at a level 30 times the permissible level in the CCIR report to the ITU for the 1971 WARC. Smith quoted a radio astronomer who was deeply involved in the protection of the RA bands as saying the ATS-F was "a sad and ironic example of the rapidly growing misuses of the radio spectrum." NASA agreed to insert a filter in ATS-F to protect the radio astronomy band (Howard 1973). The satellite was launched as ATS-6 in May 1974.

6.2 Satellite Beacons

Not all radio astronomers were aware of the great danger of emissions from spacecraft in or near ITU allocated radio astronomy bands. They had not taken in Graham Smith's dire warnings on 22nd January 1972. On 22nd March 1972, Paul Wild wrote to Bernie Mills: "During the past year I have been in touch with American authorities investigating the possibility of their providing a radio astronomy calibration service consisting of one or more radio beacons located in a geostationary satellite.... Our specific interest is for an aid to calibrate the phase and amplitude of the Culgoora radioheliograph, and we need three frequencies, 160, 80, and 43.5 MHz. I am now writing to other radio astronomers in Australia to find out whether there are requirements for other frequencies and, if so, how important the requirement is judged to be."

On March 23 Bernie Mills replied:

> I was very disturbed to receive your letter about the proposed beacons located in a geostationary satellite. This could be a very dangerous precedent which, although it might provide certain groups immediately with easier calibration procedures, could also lead eventually to a host of special purpose beacons polluting the whole frequency spectrum. Radio astronomers should be opposing this concept, not supporting it.

Mills contacted IUCAF. Graham Smith wrote to Paul Wild:

> I hear from Bernie Mills that you are exploring the idea of using calibrating radio transmitters in satellites. My initial reaction is much the same as his.

> ...I think perhaps the correct approach would be to make a case for such transmitters and ask for a discussion in IUCAF or in Commission V of the URSI General Assembly.
>
> We are planning to have a meeting of IUCAF during the Warsaw (URSI) assembly, and I would be happy for this and any other matters to be discussed if you wish. At the moment the main topic seems to be the problem of broadcasting satellites.... I would be embarrassed about trying to stop one transmitter and encourage another during the same meeting of IUCAF.

Martin Ryle joined in on March 29:

> I must say that I am appalled at the idea. I do not see how you could operate such a system.
>
> ...I agree with Bernie that radio astronomers have a good record as conservationists, and that to intentionally produce radiation in radio astronomy bands and from a satellite is about as serious a pollution as one could devise!
>
> ...I do not see the point of it; Nature has provided us with a large number of sources—many of which are of very small angular size. Special observations (both optical and radio) can now locate these to better than $1''$ arc. ...they are there all night, every night—and you don't have to seek permission from the world's radio astronomers before you have a brief period of calibration....
>
> The I.T.U. has recognized radio astronomy and has provided me with the right to observe on internationally agreed bands. I intend to fight to preserve that right!

On March 20, 1972 Frank Kerr wrote to Bill Howard at NRAO, who had circulated a letter to CORF members. John Hagen was chairman of CORF, due to meet on May 5, 1972:

> The Paul Wild–John Hagen proposal to instrument a NASA satellite with calibration beacons operating in radio astronomy bands is an attractive one.... The possibilities are certainly worth exploring. At the same time, we must be quite sure that the system contains a completely foolproof off-switch for each frequency!

The exchanges of letters continued. Mills wrote to Graham Smith on April 18, 1972:

> Needless to say, I am thoroughly opposed to the concept of radio beacons in radio astronomy bands and, among Australian radioastronomers, I know that my feelings are shared by Christiansen, Bolton and Robinson so that there seems little prospect of Paul Wild's proposal making much headway at this end. The danger would seem to be with the U.S. radioastronomers.

Ultimately, world opposition to the beacon proposal led to the wires to the beacons being cut while the NASA satellite was on the launch pad.

6.3 Aircraft Landing System at 5 GHz

The second Space WARC in 1971 not only allowed broadcasting satellites adjacent to the 2700 MHz radio astronomy band. That WARC had also made an allocation for aeronautical radionavigation from 5.0 to 5.25 GHz. Immediately below 5 GHz was another very important radio astronomy band. WARC-1971 had allocated 4.7 to 4.99 GHz as a band for radio astronomy, shared with fixed and mobile services; 4.99 to 5.00 GHz was a primary radio astronomy allocation.

The International Civil Aviation Organization (in Montreal) announced in 1972 an international competition to design a 5 GHz microwave landing system (MLS), for use at airports, to replace the old Instrument Landing System operating at 200 MHz. Radio astronomers were concerned that out-of-band emissions from MLS would cause harmful interference to their observations below 5 GHz.

Paul Wild had just become Chief of the CSIRO Radiophysics Laboratory in Sydney. The winds of economic rationalism were blowing. So Paul entered CSIRO in the ICAO competition, and was ultimately successful. High-caliber engineers such as Brian Cooper and John Brooks were taken from the radio astronomy program at CSIRO to design and test the MLS system.

Fortunately, by 1998 only a handful of MLS systems have been installed at airports. One reason is that air traffic has not grown as ICAO projected, and airports are not required to handle very high arrival and departure rates. Also, as each year passed, the possibility of using the very accurate GPS and GLONASS satellite navigation systems became a better alternative to MLS. The accuracy of differential GPS is excellent.

6.4 The GMS Meteorological Satellite

A bright spot in the 1970s was successful negotiations with Japan on the out-of-band transmissions from the Geostationary Meteorological Satellite. Effective filtering of wide unwanted sidebands was shown to be feasible. Each day television viewers in Asia and the South Pacific see a satellite image of the Earth. They can see a typhoon approaching Guam, or a cyclone bearing down on Fiji. Great armchair viewing.

The GMS Meteorological Satellite was to be constructed and launched by Hughes Aircraft Company and operated by Japan. Its central frequency was to be 1681.5 MHz. That is of no concern to Radio Astronomers. But the modulation on the image beamed down to Asia and the South Pacific would have extended sidelobes that would cause harmful interference to radio astronomy observations in the OH band 1660 to 1670 MHz.

At the time the GMS satellite was being designed, IUCAF correspondent Brian Robinson contacted the Japanese Meteorological Agency and encouraged them to build a filter to protect the OH band. Yasushi Horikawa (National Space Development Agency of Japan) and Seiji Yoshimoto (Nippon Electric Company) discussed the problem. When Hughes Aircraft Company engineers said that such a filter could not be built, the Japanese decided to design and build it. Vacuum chamber tests at the end of 1975 showed that the radio astronomy specifications could be met.

When the satellite was launched in 1977, Horikawa and Yoshimoto came to the Parkes 64-meter radio telescope. Observations made then of the GMS satellite are reproduced in Robinson & Whiteoak (1979). In the radio astronomy band the filter gave 60 decibels attenuation of the GMS emissions. The radio astronomers were delighted. The Japanese engineers were delighted. A win-win situation.

6.5 SSU Series Satellites—Interference at 1420 MHz

In April 1976 the U.S. launched three "SSU" satellites, which produced strong interference to Canadian radio astronomy observations at 1420 MHz (Argyle et al 1977). The signals were a billion times greater than the hydrogen emissions being studied. The Canadians protested to the U.S. that the satellites were transmitting, in contravention of ITU allocations, close to the band reserved solely for radio astronomical observations of neutral hydrogen.

Investigations showed that the satellites stored large amounts of data during their 107.5 minute orbit, and dumped the data (when controlled from the ground) near Alaska, the Pacific Northwest and Midwestern United States. Argyle et al said that the signals are "in contravention of the International Telecommunications Union's Table of Frequency Allocations. ...Therefore, we urge scientists in the United States who are concerned with the orderly management of the electromagnetic spectrum to press their government to limit the use of bands near radio astronomy allocations to ground-based services." This was a salutary warning to radio astronomers as the preparations for the next WARC (in 1979) got under way.

7. DOWNSIZING OF IUCAF IN 1973

Smith-Rose (1972) reported to the 1972 URSI General Assembly in Warsaw that "IAU, URSI, and COSPAR have agreed to reduce the number of their representatives on IUCAF to two and it is intended to have an annual meeting beginning in 1973."

This decision, for financial reasons, removed many influential radio astronomers from IUCAF and cut down on the number of meetings. At the 1972 General Assembly in Warsaw, URSI appointed J.W. Findlay and J.P. Hagen as its IUCAF representatives. In 1973 IAU appointed Graham Smith and Richard Wielebinski as its representatives. COSPAR representatives were Fred Horner and M.M. Thue.

With the Cold War mentality of the time, there was no Soviet representation on IUCAF after 1972. The exclusion of Eastern Europe was odd, given that the Communist Bloc had a significant vote at ITU conferences on frequency allocations. Graham Smith became Astronomer Royal in 1975, and his IAU slot on IUCAF was filled by Gart Westerhout. John Hagen became chairman of IUCAF at the Paris meeting in October 1975.

7.1 IUCAF Correspondents

At the IUCAF meeting in Konstanz (26 May 1973) an important decision was made to appoint IUCAF Correspondents in 27 countries (Minnis 1973). This was a recognition that, when frequency allocations were considered by the ITU at WARCs, only national administrations have a vote. Bodies like IUCAF are recognized at WARCs; they can speak and lobby, but not vote. The IUCAF Correspondents would be kept informed of IUCAF policy, and communicate the needs of radio astronomy and Space Science to their national spectrum administrators. The 1973 reorganization reduced the effectiveness of IUCAF, but gave it a coordinating role to the network of Correspondents. A list of the Correspondents was published by Minnis (1974).

7.2 WARC 1979: A WARC General

The ITU proposed a general WARC for 1979, to consider all frequencies up to 285 GHz. It would be like WARC 1959—considering all frequencies, all services. No holds barred.

IUCAF began discussing WARC-1979 at the Bonn meeting July 29, 1974. IUCAF proposed discussing with its correspondents "questions concerned with, for example, the usefulness of existing frequency allocations, the need for changes in the bandwidths available at present, and requirements for allocations to meet new scientific needs. The increasing use of artificial earth satellites which emit radio frequency energy for various purposes represents a growing interference problem, especially when wide-band emissions spread into adjacent bands and cause interference. IUCAF is actively engaged in studying current problems of this kind and the potential sources of interference in satellites to be launched in the future."

A meeting of IUCAF to discuss preparations for WARC 1979 was held in 1976 at the IAU General Assembly in Grenoble. John Findlay conducted the meeting (Findlay 1976, Horner 1976).

7.3 IAU Priorities for Molecular Lines

The members of IUCAF (Hagen, Findlay, Westerhout, Wielebinski, Horner and Thue) were not conversant with the flood of discoveries of molecular spectral lines. By 1976, 220 molecular lines were known in the radio spectrum.

IAU Commission 40 set up a Working Group, chaired by Brian Robinson, to determine the priorities to be given to the molecular lines during IUCAF's preparations for WARC 1979. By 1979 around 600 molecular lines were known between 0.8 and 346 GHz. An IAU list of 30 key spectral line frequencies was agreed to at the 1979 IAU General Assembly in Montreal (Robinson & Whiteoak 1979, Westerhout 1979). The IAU submitted the list of key spectral lines to CCIR in Geneva. The listing was agreed to by the relevant study group and incorporated in CCIR Recommendation No. 314. As previously noted, CCIR Recommendations carry great weight in the deliberations at a WARC.

The IAU Working Group continued to evaluate the key spectral lines at every IAU General Assembly for the next 21 years. Each time the revisions were adopted as Recommendations by CCIR (or, as it is now known, ITU-R).

7.4 The CORF "Green Book"

The U.S. National Academy of Sciences Committee on Radio Frequencies (CORF) produced the "Green Book," the key document for the radio astronomy case at WARC 1979 (and the CCIR Special Preparatory Meeting of 1978).

The chairman of CORF was Bernard F. Burke. Members were Alan H. Barrett, Otis Brown, William C. Erickson, John P. Hagen (chairman of IUCAF), David C. Hogg, Hein Hvatum, Frank J. Kerr, Nancy G. Roman, and, serving as secretary, Richard Y. Dow. The hard work was done by the Sub-Committee on radio astronomy: William C. Erickson (chairman), Donald C. Backer, Alan H. Barrett, Barry G. Clark, Thomas A. Clark, Michael M. Davis, Andrea K. Dupree, Alan T. Moffat, and A. Richard Thompson. The Green Book was distributed to all IUCAF members and correspondents. In most cases they interacted with their national telecommunication administration to have the radio astronomy position inserted into their country's submission to WARC 1979. Those "input document" submissions to the ITU in Geneva were numerous and weighty (in kilograms). Reviewing the whole spectrum was a monumental task.

7.5 WARC-1979 in Geneva

A summary of what the radio astronomers sought from WARC 1979 was published by Robinson & Whiteoak (1979). WARC 1979 ran from September 24 to December 6, finishing a week late. IUCAF delegates who were unattached to any national delegation and could express individual views on behalf of the scientific community were John Hagen, Marcus Price, and John Findlay. Some members of IUCAF, IUCAF correspondents, or other scientists were members of national delegations: Bajaja (Argentina), Block (ESA), Doherty (Canada), Dubinski (USSR), Grahl (West Germany), Horner (U.K.), Kahlmann (Netherlands), Okoye (Nigeria), Pankonin (USA), Pezzani (France), Schilizzi (Netherlands), Swarup (India), Thue (France) and Whiteoak (Australia).

Two meetings of radio astronomers were called by IUCAF in the early stages of the WARC, but most work was carried out by personal contacts. For some it was instantaneous training in diplomacy. A report on WARC-1979 was published by Fred Horner (1980). An extensive report was also published by Bodson et al (1981). Horner's report says of radio astronomy and space research:

> The overall outcome is widely considered to be satisfactory for these sciences; indeed the general opinion is that they have received very favorable treatment, but some provisos will be mentioned below.
> There was a strong and cohesive radio astronomy group... five (radioastronomers) were involved for various periods as representatives of IUCAF.

... In the case of radioastronomy there was a significant increase in the number of (frequency) table allocations.... In space research... there are considerable gains in allocations for sensing of the Earth's surface and atmosphere by both radiometric and radar techniques.

WARC 1979 allocated radio astronomy 16 bands between 322 to 328.6 MHz and 105 to 116 GHz. Allocations "by footnote" were made to 18 bands between 140 GHz and 348 GHz. The WARC also approved a Dutch proposal for Article 36 in the Radio Regulations, setting out the basic operation and needs of radio astronomy. There were also sections on "radio astronomy in the Shielded Zone of the Moon" and "The Search for Extraterrestrial Emissions" plus new definitions of interference and "unwanted emissions."

At WARC 1979 India had proposed that the band 322–328.6 MHz be allocated to radio astronomy. The band contains the important spin-flip transition of Deuterium. The cosmological importance of Deuterium was, and remains, high. The NATO countries and the USA supported this 327 MHz allocation: they knew that the Soviet Union had an extensive radar network around the Middle East at 327 MHz. On May 1, 1960, it had tracked Gary Powers' U2 spy plane from Pakistan to Sverdlovsk, 1200 miles inside Soviet territory. A re-allocation of the 322–328.6 MHz band to radio astronomy was voted through at WARC 1979.

7.6 Recommendation 66 of WARC-1979

Within ITU activities, "Recommendations" are the highest level of directive. Also the World Administrative Conferences are the highest level of authority. The ITU Secretariat and organs of the ITU like CCIR must take note of recommendations—especially if they come from a WARC.

When WARCs allocate frequency bands to different users ("services") it is hard to avoid incompatible neighbors. We have already seen the problems of allocating to satellite broadcasting a band that is next to an important radio astronomy band (2700 MHz); or an aircraft landing system next to another important radio astronomy band (5000 MHz); or the GMS meteorological satellite close to the OH band (1660–1670 MHz). Similar problems arose during WARC-1979.

The hypersensitivity of radio telescopes to low-level "unwanted," spurious and out-of-band emissions leaves them particularly exposed to these emissions. Throughout the "Radio Regulations"—the allocations made by WARCs—there are "footnotes." Every footnote relating to radio astronomy says: "Emissions from space or airborne stations can be particularly serious sources of interference to the radio astronomy Service." So WARC-1979 issued "Recommendation 66" as a directive to CCIR to carry out studies of the Maximum Permitted Levels of Spurious Emissions. Rec. 66 directed the CCIR to carry out these studies of space services transmissions "as a matter of urgency."

CCIR did absolutely nothing. The vested interests of the satellite service operators made sure that nothing happened! We will see that, thirteen years later, WARC-1992 repeated the instruction to CCIR in the form of Recommendation 66-2. We'll see how committees set up by CCIR (now renamed as ITU-R) sabotaged the process, and that Recommendation 66 from the ITU has been issued yet again by WRC-1997.

8. ORBITAL AND MOBILE WARCS: 1983–1988

From 1976, IUCAF meetings were held every three years, at the URSI General Assemblies in Helsinki (1978), Washington (1981), Florence (1984) and Tel Aviv (1987). After the great effort made at WARC-1979 and the very satisfactory outcome, radio astronomers looked forward to a 20-year period of creative discovery to the outer fringes of the universe. The next WARC-General was expected to be in 1999—far away! New radio telescopes had been built: the Very Large Array was commissioned in January 1980, and the opportunities for new observations were tantalizing. The Australia Telescope was funded in 1982.

But a series of crucial WARCs was being held, dealing with allocations to mobile and satellite services. These were WARC-1983 (MOBILE 1), WARC-1985 (ORBITAL 1), WARC-1987 (MOBILE 2) and WARC-1988 (ORBITAL 2). Unfortunately there was little IUCAF involvement at these WARCs.

Plans for the Broadcasting Satellite Service in the 12 GHz band (and its feeder links) were produced by the WARCs in 1983 and 1988. WARC-1988 looked at a broadcasting satellite allocation (HDTV) at 23 GHz—adjacent to the important 22 GHz transition of water vapor where H_2O masers were being found all over our Galaxy and in distant galaxies. At WARC-1987 allocation was made to a radiodetermination satellite in the 1612 MHz OH band. But a key footnote was added saying that "Harmful interference shall not be caused to stations of the radio astronomy Service using the band 1610.6–1613.8 MHz by stations of the radiodetermination-satellite service."

8.1 IUCAF Stand on the VEGA Mission in 1984

In 1984 IUCAF took a strong stand against the Soviet VEGA Mission (Findlay 1984a). In the build-up to Comet Halley's appearance in 1986, a Soviet VENERA spacecraft was to drop off probes at Venus in June 1995 and continue on a trajectory to intercept Comet Halley. The problem was that the VEGA project managers planned to transit radio signals at 1667.8 MHz, falling within the 1660–1670 MHz band where radio astronomy has a primary, world-wide allocation. The managers had chosen 1667.8 MHz because they knew that radio telescopes in many countries were equipped with extremely sensitive receivers in this band. They saw these radio telescopes as a readily available network of tracking stations for their mission.

CORF, chaired by David Staelin, expressed great concern about the frequency chosen by Vega. IUCAF discussed the Soviet plan during the 1984 General Assembly of URSI in Florence. Radio astronomers could not sanction the use of transmitters within a passive band. It was just like the satellite-beacon proposals of 1972, and undermined the strong IUCAF objections to broadcasting satellites adjacent to the 2700 MHz radio astronomy band. The URSI Council met in Florence and resolved "to urge the Member Committees of COSPAR, IAU and URSI to work with IUCAF as a consultative body when planning any active radio frequency usage in future scientific missions which may cause interference to passive observations" (Findlay 1984a).

Also URSI "resolves, in view of the specific danger of interference to radio astronomy from space-based radio transmissions, to urge all those concerned in the design of experiments requiring radio transmissions from space to consult with IUCAF at the planning stage to ensure that the protection of sensitive passive radio observations which has been acquired through wide cooperation and with great effort is not jeopardized in the future."

The linking of the Vega mission to the comet encounter in March 1986 made the mission time-critical and the IUCAF meeting in August/September 1984 was too late to alter this frequency. The Vega mission was launched in December 1984. Most radio astronomy observatories boycotted the mission, depriving the Soviet managers of the convenient world-wide tracking network they had hoped to exploit. Radio observatories wished to search for hydroxyl molecules in Comet Halley and its tail, and were concerned that the in-band transmissions from the Vega satellite would compromise the observations.

A proposal emanating from Jodrell Bank (initiated by Ray Norris and John Ponsonby) was to transmit a strong beam of radiation at Comet Halley at 1.6 GHz to investigate the pumping of the OH masers. A French student at Jodrell Bank, Laurent Pagani, did an excellent thesis study and came up with a detailed proposal for sweeping a radio beam in frequency across the line profile of the cometary maser and studying the time it took the comet to recover. It was an elegant proposal to do a laboratory-type measurement on a cosmic maser source for the first time. Eric Gerard had the job of taking this idea to the International Halley Watch Committee (meeting in the U.S.) where it was opposed unanimously—everyone was afraid the comet might not recover at all.

While passing Venus, the Vega mission had 1667.8 MHz transmitters on two balloons and another on the spacecraft. The balloons drifted in the upper atmosphere winds of Venus and, from differential VLBI between the spacecraft (a known orbit) and the balloons, quite a lot was learned about the atmosphere and winds. The Vega VLBI project had Bob Preston from JPL as its principal investigator, and the Jodrell Bank telescope took part.

8.2 Spread-Spectrum Modulation

From 1983 to 1988, at the MOBILE and ORBITAL WARCs, communication engineers had seized the opportunities they saw in using transmitters on satellites to illuminate the earth. They had also embraced "spread-spectrum modulation." Originally developed by the military in the 1950s, spread-spectrum modulation appeared to be a panacea for everything (Pickholtz et al 1982). The high-speed modulation produced extended unwanted/spurious/out-of-band emissions that interfered seriously with those listening in nearby or adjacent bands. The energy from the spread-spectrum sidebands fell off slowly with the frequency separation x from the carrier frequency—only as the square of ($\sin x/x$).

8.3 WARC-1988 ORB 2

Attempts were made by IUCAF to make up lost ground at ITU WARC ORB 88. The sole IUCAF representative at the 1988 WARC was B.H. Grahl. Ten years on, IUCAF is still sorting out how to share bands allocated to satellite and mobile users by the WARCs of 1987 and 1988.

8.4 GPS: The Global Positioning Satellites

The U.S. Air Force designed and launched a family of navigation satellites designed to give an accurate position of a receiver on or near the earth. The motto over the entrance to GPS Headquarters said: "Two bombs in the one hole." The U.S. Air Force demonstrated that feat in a raid on Baghdad in January 1991.

The GPS emissions had extended sidebands, which caused harmful interference in the OH bands 1610.5–1613.8 MHz and 1660–1670 MHz. Protests from radio astronomers were heeded. The space shuttle Challenger disaster of January 28, 1986 took three GPS satellites into the Atlantic Ocean. After the Challenger disaster no GPS satellites were sent into orbit until 1989.

IUCAF made a formal protest to IFRB in Geneva in February 1990. Canada also made a formal protest to the U.S. Government in March 1990 about harmful interference, and the International Frequency Registration Board in Geneva was informed. Sweden also complained. The U.S. Department of Commerce confirmed that the interference originated on older "Block I" GPS satellites and that the system was being upgraded. Canada also complained to the USSR about interference from GLONASS satellites (see the next section).

By 1993 the sideband emissions of a new series of "Block II" GPS satellites were all filtered to remove the harmful interference with OH observations. The spurious signal levels were 38 decibels lower than those of their predecessors. Spectra of GPS emissions before and after filtering are shown by Ponsonby (1991).

One benefit of GPS emissions was that they made effective use of spread spectrum technology. All carriers had the same frequency (1574.4 MHz). The spread-spectrum modulation for each satellite was different, but there was no mutual interference.

8.5 Global Navigation Satellite System—GLONASS

A navigation system with satellites launched by the USSR Space Forces from 1983, GLONASS was similar to the U.S. Air Force GPS. The important differences were *(a)* each satellite emitted on a different frequency, with the carrier frequencies spread over the range of 1597–1617 MHz; *(b)* the level of the interference was incredibly high because the range of carrier frequencies stepped right through the radio astronomy allocation of 1610.6 to 1613.8 MHz; *(c)* the satellites were "frequency mobile"—no filtering seemed possible, whereas filtering had been possible with the GPS satellite emissions.

9. IUCAF ACTIVITY FROM 1980 TO 1987

The list of radio spectral lines of the greatest astrophysical significance was extended at the 1982 IAU General Assembly in Patras (Robinson 1982) and conveyed to CCIR. Fifteen lines were added between 279 and 691 GHz on the expectation that a future WARC would consider this very high frequency range.

Fred Horner, a COSPAR representative on IUCAF, had become chairman of CCIR Study Group 2 (dealing with radio astronomy) in 1980. At the 1987 IUCAF meeting in Tel Aviv, Fred commented that IUCAF had not provided any advice to CCIR in the previous three years. What had IUCAF been doing?

The minutes of the 1987 Tel Aviv meeting reveal despondence. Regarding the IUCAF correspondents system, the minutes record: "It only worked once, before the 1979 WARC. It's pretty useless otherwise." IUCAF funds were low, and John Findlay asked, "Should we quit?"

At the Tel Aviv meeting Hans Kahlmann announced European proposals to ITU for a major WARC in 1992. It became clear that IUCAF had a lot of work to do, that it needed to have more members and that more funds were required from URSI and IAU. John Findlay's term as chairman of IUCAF was to end in 1989, just when the bulk of the preparatory work for the major WARC would be needed. So discussions began in Tel Aviv about replacing John as chairman in 1988. A ballot of IUCAF members was held in March 1988, and Brian Robinson was elected chairman.

New members were appointed to IUCAF, and membership increased to ten: B.H. Grahl, H.C. Kahlmann, R.M. Price, and B.J. Robinson for URSI; B.A. Doubinsky, N. Kaifu, V.L. Pankonin, and G. Swarup for IAU; F. Horner and S. Hieber for COSPAR.

It was timely to have a Soviet member of IUCAF, after the long gap since the reorganization of IUCAF in 1972. The network of IUCAF correspondents was revived, with correspondents in 35 countries. Eastern Europe is well represented, but there is only one correspondent from Africa. URSI and IAU increased their annual funding of IUCAF, and additional funds were obtained from ICSU for the extensive preparations for WARC-1992.

9.1 The Foundation of CRAF

At the 1987 meeting of IUCAF, Hans Kahlmann announced the foundation of a European parallel to CORF and link to IUCAF. The moves to create a EuroCORF began in 1987 at a meeting in Paris. H.C. Kahlmann was the first chairman, and T.A.Th. Spoelstra was the secretary. The timing suggests that this move was a European preparation for WARC-1992. Another factor was the need for concerted European action against the strong interference from Soviet GLONASS satellites.

Those involved in the foundation of CRAF were Kahlmann, Schilizzi, and Spoelstra from the Netherlands; Paul Burgess, Cohen, and Ponsonby from the United Kingdom; Daigne, Kovalevsky, Pezzani, and Pick from France; Ellder from Sweden; Gorgolewski from Poland; Grahl and Ruf from Germany; and Tomassetti from Italy. The IUCAF (Findlay, Horner), CORF (Price), NSF (Gergeley) and ESO/SFCG (Block) contributed to the effort.

In March 1988 the European Science Foundation (ESF) accepted in principle a proposal that CRAF become an ESF Committee with the name "ESF Committee on Radio Astronomy Frequencies." This paralleled the link between CORF and the U.S. National Academy of Sciences. One represented Europe, the other the United States. Internationally, the position was coordinated by IUCAF.

Two members of CRAF (Kahlmann and Cohen) became full members of IUCAF to reinforce the collaboration. Other members of CRAF became correspondents of IUCAF. CRAF met frequently and has continued to meet two or three times a year, while CORF in the U.S. meets but once a year.

Through its national members, CRAF hoped to influence the WARC-1992 positions of individual countries (given that only countries affiliated with the ITU have the right to vote at WARCs). Hans Kahlmann retired as chairman of CRAF in 1995, and Jim Cohen was appointed chairman. Titus Spoelstra continued as secretary; in 1997 he became a full-time CRAF frequency manager for European radio astronomy.

10. ITU WORLD ADMINISTRATIVE RADIO CONFERENCE 1992

The 1989 Plenipotentiary Conference of ITU in Nice resolved to hold a special World Administrative Conference in Spain in 1992 (WARC-1992). Such a wide-ranging WARC had not been expected until 1999 (in keeping with the established pattern: 1959, 1979, 1999, and so on). But the increasing pace of technological change forced the ITU to hold the conference earlier, to embrace satellite mobile communications, digital radio broadcasting, high definition TV and other technologies not envisaged at WARC-1979. The conference agenda for WARC-1992 was drawn up in June 1990.

10.1 IUCAF Preparations for WARC-1992

The re-igniting of IUCAF to prepare for a wide-ranging WARC began at the 1988 IAU General Assembly in Baltimore. After the IAU assembly, IAU Colloquium 112 discussed "Light Pollution. Radio Interference and Space Debris" (Crawford 1991).

Strenuous activities were needed for the input to CCIR Study Group 2 in 1989. More work was done at an IUCAF meeting during the 1990 URSI General Assembly in Prague. That general assembly was followed by key meetings in the Washington area of CCIR Working Parties and a meeting of SFCG (Robinson

1990). A second CCIR Working Party meeting was held in Geneva in February 1991. These meetings kept the members of IUCAF very busy.

In March 1991 the CCIR held a major conference in Geneva to establish "The Technical and Operational Bases for the World Administrative Radio Conference 1992." Represented by Robinson and Kahlmann, IUCAF played a major part in this meeting, which would generate technical decisions that would circumscribe the political decisions to be faced at WARC-1992 (Robinson 1991).

continued to be a busy year for IUCAF. It participated in the SFCG meeting in October to coordinate the preparations for WARC-1992 on space research and space exploration. Then discussions about GLONASS were held with the Russian Space Forces in Moscow in October. IUCAF took part in CCIR meetings in Geneva during October 1991, converting existing CCIR reports into recommendations that would carry legal force at WARC-1992. Finally, in negotiations with Motorola Corporation, IUCAF and Motorola tried with little success to find a *modus vivendi* between vital astronomical observations of OH (related to star formation and to the late stages of stellar evolution) and a proposed mobile telephone system linked by a swarm of low-flying satellites—fundamental knowledge versus big bucks.

A summary of the status of radio astronomy prior to WARC-1992 was published by Thompson, Gergely, and Vanden Bout (1991).

10.2 WARC-1992

IUCAF was heavily involved in WARC-1992 itself, from February 3 to March 4, 1992 (Robinson 1992). Hans Kahlmann, Berndt Grahl, Brian Robinson, Govind Swarup, Dick Thompson, and Boris Doubinsky comprised the IUCAF delegation, supported by IUCAF correspondents Tom Gergely, Ramesh Sinha, and Rob Roger. At the WARC the status of space research and radio astronomy was significantly enhanced between 137 and 3000 MHz and above 13.5 GHz. Delegates from 125 countries clearly recognized the importance of scientific use of the radio spectrum in the face of increasing pressure from telecommunications, broadcasting, and navigation interests, particularly when the proposed commercial transmissions are from satellites.

IUCAF delegates became immersed in the projected characteristics of digital radio broadcasting, high definition TV, and the "big LEOs" and "little LEOs"— mobile communications facilitated by low-earth-orbit satellites. WARC-1992 made allocations for radio astronomy up to 285 GHz; these were mainly CO bands.

Despite the efforts of CRAF, there were moves in 1990 through 1991 to bind the 28 European countries to a single point of view. This solid block of votes was seen as a counter to the U.S.-Canada-Mexico cast-iron position, or the Russia-Belorussia-Ukraine cast-iron position. At WARC-1992 the position of the 28 European countries discounted the recommendations of CRAF and ran contrary to the needs of radio astronomy. However, the U.K. voted independently on some key radio astronomy issues. The rock-solid position of the European bloc (now 27 countries) held until 1 a.m. on the morning of the last day of the WARC.

Recommendation 66, the WARC-1979 directive to CCIR to examine spurious and unwanted emissions, was renewed by the 1992 WARC.

10.3 Illegality of GLONASS Transmissions

Prior to WARC-1992, radio astronomy use of the most important 1612 MHz satellite line of OH had been authorized by Footnote 352K in the Radio Regulations, first inserted by the "Space WARC" of 1971. The footnote warned that "Emissions from space and airborne stations can be particularly serious sources of interference to the radio astronomy service." But later WARC-1979 inserted Footnote 352A, reserving the whole band 1610–1626.5 MHz on a worldwide basis for "airborne electronic aids to air navigation and any directly associated ground-based or satellite-borne facilities."

In April 1983 the Soviet government used the regular ITU coordination procedures to advise of its plan to operate its Global Navigation Satellite System (GLONASS). Only one country responded to the announcement of this satellite system, and the response was submitted after the 45-day cut-off for objections. The U.S. objection, in August 1983, expressed its concern for the impact on radio astronomy, and requested more information on the GLONASS signal character. The USSR did not respond.

No radio astronomer outside the U.S. noticed the announcement of the USSR system. Then the horrific interference produced by GLONASS satellites near 1612 MHz closed down radio astronomy observations of the OH line. Two "footnote" ITU allocations had equal legal status, and radio astronomers could only lament their oversight. By January 1984 Vernon Pankonin identified that eight GLONASS satellites were operational. Epherimedes of the satellites were difficult to obtain and the interference was unpredictable (Crane 1985).

The first change to the GLONASS situation occurred in May 1991, when the Russian government approached the ITU seeking improved status for GLONASS in the 1597–1617 MHz band. This time IUCAF was immediately aware of the request to ITU, and in June 1991 ten countries organized to oppose the Russian request to ITU, which then lapsed. IUCAF argued that GLONASS should contain its emissions to the 1559–1610 MHz band, which had been allocated by WARC-1979 to radio navigation satellites and aeronautical radionavigation.

IUCAF members Brian Robinson and Boris Doubinsky confronted the Russian Space Forces at a meeting in Moscow in October 1991, under the umbrella of Academician V.A. Kotelnikov. After much negotiation, the Russians agreed "to consider all possibilities to decrease the interference to the radio astronomy in mentioned frequency band in the process of upgrading of the GLONASS system." They undertook to eliminate the out-of-band emissions of the GLONASS system "beginning from 1994 in process of replacement of the Space apparatuses with new ones equipped by improved filters."

At WARC-1992 the situation changed dramatically. While the conference was going through a very moribund state, locked in some stalemate between Europe and the rest of the world, an Australian proposal to enhance the status of radio astronomy at 1612 MHz was approved. During the crucial vote, the Russian, Belorussian, and Ukrainian delegates at the WARC were deep in discussing something else, had taken off their earphones and were not listening to the simultaneous translation of the presentations. With that vote, radio astronomy gained full primary status in the band. Services with secondary or (lower still) footnote status could not interfere with a primary service: GLONASS had only footnote status.

In June 1992 Brian Robinson, Boris Doubinsky, and Willem Baan confronted the Russian Space Forces again, at a second meeting in Moscow. All proposals from IUCAF were greeted with a stony "*niet*." So it was arranged that a delegation from the Space Forces and the Institute of Space Device Engineering would visit the Jodrell Bank telescope in November 1992 to witness the interference for themselves. During the tests at Jodrell Bank the Space Forces switched off or retuned nine of the 13 GLONASS satellites and observed the effect on the interference.

At a further meeting in Moscow (June 1993), the Space Forces recognized the illegality of their transmissions and agreed to move their operating frequencies away from the 1610.6–1613.8 MHz radio astronomy band. A provisional agreement was signed, and finalized November 4, 1993. As a result, all GLONASS satellite transmissions will be out of the band by the year 2007.

Was this a hollow victory for radio astronomy? Did the emptying of the radio astronomy band by GLONASS satellites open the door to allow Motorola IRIDIUM satellites to move in? Out of the frying pan...

10.4 IRIDIUM Satellites

The fight with Motorola Corporation regarding its IRIDIUM satellites began in 1991 at CCIR meetings in Geneva, where the spurious and unwanted emissions from the satellites were ill defined but recognized as threatening. (The danger posed by IRIDIUM is discussed by Abbott 1996 and Ponsonby 1996.) The name IRIDIUM came from the original plan to have 77 low-flying satellites: 77 electrons orbit the iridium nucleus. Later Motorola reduced the proposal to 66 satellites (but the company has not yet renamed the system DYSPROSIUM).

"Negotiations" with Motorola Corporation to resolve the dilemma continued. Some of the participating scientists were surprised by the dictatorial style of the would-be satellite operators. The Russian Space Forces were more pliable.

The U.S. National Radio Astronomy Observatory signed a Memorandum of Understanding with IRIDIUM in June 1994. All other observatories reacted with horror, and adopted a strong defensive position behind the IUCAF banner. Later, in March 1998, Arecibo Observatory signed an MOU with IRIDIUM on much better terms than those accepted by NRAO.

Outside the U.S. the negotiations with Motorola are tougher. Australia has signed an agreement with Motorola allowing use of IRIDIUM mobiles under the condition that "the licensee must not cause harmful interference to Australian radioastronomy services." The agreement is to be reviewed after five years, at which time "IRIDIUM will be expected to demonstrate... that it has reached agreement with the Australian Radio Astronomy Service on the steps necessary to prevent harmful interference to Australian radioastronomy services." At the time of writing, negotiations with Motorola about IRIDIUM continue in Canada and India, as well as with a bloc of European countries.

10.5 Broadcasting Satellites

After WARC-1971 Graham Smith had alerted radio astronomers to the threat from broadcasting satellites adjacent to the RA band at 2700 MHz. Later, WARC-1988 authorized broadcasting satellites near 12 GHz.

On May 9, 1991, Masaki Morimoto alerted IUCAF to Japanese plans to license a broadcasting satellite at 22.6 GHz, using a bandwidth of 120 MHz. IUCAF sprang into action. The 22.21–22.5 GHz band is a primary allocation to space research, radio astronomy and earth exploration. The band contains an important transition of water vapor. The Japanese plan would obviously generate harmful band-edge interference to these passive services.

The Japanese proposal had to go through ITU registration procedures (so-called "Article 14" procedures). IUCAF organized protests by eight countries in the Asia-Pacific region, and the ITU registration failed to gain approval—a close shave for the passive services. More steps to protect the water-vapor band were taken at WARC-1992.

11. ADVERSE ENVIRONMENTAL IMPACTS ON ASTRONOMY

In July 1992 an IAU/ICSU/UNESCO Exposition was held at UNESCO Headquarters in Paris to publicize adverse environmental impacts on astronomy. The target audiences were government representatives, science writers and the public. The case for radio astronomy was made by Marcus Price, member of IUCAF and chairman of CORF. The exposition produced an excellent book: *The Vanishing Universe—Adverse Environmental Impacts on Astronomy* (McNally 1994). The book is aimed at scientists and the scientifically curious public.

The 1992 Exposition was a follow-up to IAU Colloquium 112 (Crawford 1991) which addressed light pollution, radio interference and space debris. At the 1992 Exposition, UNESCO was exploring the possibility of designating a few selected observatories as World Heritage sites. Nothing has yet come of this.

12. WORLD RADIO CONFERENCES 1993, 1995, 1997

From 1993, ITU World Radio Conferences (now called WRCs) are held every two years, with preparatory meetings every two months. How can the members of IUCAF cope with the flood of proposals, with the mountains of paper? Can they combine that responsibility with productive research activities in radio astronomy?

There is a contrast with the situation in the 1980s, when there were WARCs in 1983, 1985, 1987, and 1988 with little or no IUCAF participation. IUCAF has been very active in the 1993, 1995 and 1997 WRCs, and is busy preparing for the issues to be discussed at WRC 2000. The WRC agenda items relating to passive scientific use of the spectrum can be found at the IUCAF Web site, http://www.nfra.nl/iucaf.

12.1 WRC-1993

There was an unexpected German proposal at WRC-1993 that was very threatening to radio astronomy. The proposal had come up between meetings of CRAF, so European astronomers had not discussed it. At the last moment, IUCAF heard about it; Robinson and Baan rushed to Geneva for the opening of the WRC.

The German proposal was to throw open all the allocated radio astronomy bands for discussion at WRC-1997. Apparently the motive was to propose that radio astronomers share their allocations with other services "in view of new technology and new observation methods." In exchange, radio astronomers could ask for access to other bands where interesting molecular lines were to be found. This suggestion was extremely naïve: radio astronomy would have been eaten alive at WRC-1997 by the greedy satellite operators.

The German proposal was not on the agenda of the other 27 European countries. When the motion was introduced at WRC-1993, it was immediately opposed by Australia and Canada, with the U.S. ready to join in. No European country spoke for or against the proposal, and it was excised from the WRC-1997 agenda. A very close shave.

The unplanned visit of Robinson and Baan to WRC-1993 had two positive benefits: *(a)* the Agreement between GLONASS and IUCAF, signed in Moscow on November 4, 1993 was distributed by the ITU to all WRC delegates in English, French, and Spanish—the three formal languages of the ITU; *(b)* IUCAF was alerted to the inadequate response of ITU-R to Recommendation 66 (WARC-1979 and WARC-1992). (See the section on WRC-1997.)

12.2 Belgian MLMS LEO Satellite

On December 3, 1993 the Belgian Administration announced through ITU the launching of a satellite in 1995 for a "micro low-earth orbit (LEO) message service" called MLMS, orbiting 800 km above the Earth. This launch was to be followed in mid-1995 by a second satellite, and two more satellites at a later date. The Belgians proposed a 400.15–401.0 MHz downlink using direct-sequence spread-spectrum modulation. They also planned an FSK telemetry downlink at 136.0–138.0 MHz.

Radio astronomers panicked, fearing harmful interference at 406.1–410 MHz and 322–328.6 MHz. But information from Paul Delogne and a paper by Delogne & van Himbeeck (1995) showed that out-of-band radiation can be carefully controlled in well-designed spread-spectrum systems. They showed that the power-spectral-density of spread-spectrum signals can be perfectly confined in a nominal bandwidth of 1.5 times the "chip rate." The techniques described are also "rather cheap."

Scary experiences with widespread out-of-band radiation from earlier spread-spectrum systems will diminish, hopefully. Cures for those less-sophisticated systems were expensive, involving radio-frequency filters on satellites.

12.3 WRC-1995

Nine radio astronomers from five countries attended WRC-1995 (Ananthakrishnan, Baan, Doubinsky, Gergely, Gorgolevsky, Roger, Ruf, Sinha, and Thompson). A delegation of this size is needed to deal with the multiple committees and working groups that process the work of the WRC. About 15 million pages of text were produced during the WRC. In established WRC tradition, many important issues remained stand-offs until very late in the conference, until the clock forced compromises to be forged.

The main topics of WRC-1995 were new allocations for the mobile satellite services and issues regarding their feeder links. Although radio astronomy was not directly on the agenda, new footnotes were inserted in the Radio Regulations to urge protection of radio astronomy bands from mobile satellite downlinks and feeder links. These footnotes specified protection of bands below 1 GHz, protection of the 4.99–5.0 GHz band from harmonic radiation, and protection of the 6.668 GHz methanol line and the 15.35–15.4 GHz band.

12.4 WRC-1997

WRC-1997 had around 1700 delegates from 140 countries. 15 radio astronomers attended, including 7 IUCAF representatives (Willem Baan, Jim Cohen, Boris Doubinsky, Tom Kuiper, Dave Morris, Masatoshi Ohishi, and Anders Winnberg). The mid-size IUCAF delegation was vocal during the WRC, playing a major role in the areas where radio astronomy needed to be protected. The technological/industrial pressures at the WRC can be seen from the size of the largest delegations: the United States (99), Japan (81), France (74), and the United Kingdom (47). IUCAF received considerable support from the delegations of Canada, India, France, The Netherlands, Sweden, the United Kingdom and the United States.

As at all WRCs, national delegates arrived in Geneva with established positions for the agenda items. The work of the WRC is to merge these into internationally acceptable conclusions. Some of the agenda items requiring IUCAF involvement were:

Earth-Looking Radar (94 GHZ)   To explore the planet and profile clouds—the clash of astronomers wanting to look out and earth resources scientists wanting to look in. There had been no Earth resources allocation since WARC-1979. WRC-1997 allocated the band 94.0–94.1 GHz for use by a very limited number of cloud-profiling radar satellites. There is a need to reduce satellite transmissions when the satellites are overhead major radio astronomy sites operating at 3-mm wavelength.

Recommendation 66 (From WARC-1979 and WARC-1992) For several years an ITU-R task group had aimed to produce a set of protection levels from out-of-band and spurious signals that could be approved by WRC-1997. The task group could agree only on a compromise of "design objective" levels, which were two or three orders of magnitude too high to protect radio astronomy. WRC-1997 recognized the inability of the task group to suggest adequate protection for radio astronomy. Task Group 1/5 has therefore been set up to report to a future WRC. There has been very little progress in the 20 years since Recommendation 66 was first introduced at WARC-1979.

13. WRC 2000 AND BEYOND

The agenda for WRC-2000 contains 15 items of concern to IUCAF. The threats from allocations in and adjacent to radio astronomy bands continue to roll up, while ITU-R action on Recommendation 66 stalls. Some of the proposals for the 2000 WRC are:

High-Altitude Platforms There are proposals for 200 high-altitude balloon platforms, known as skystations, positioned 20 to 80 km above major cities and operating at frequencies adjacent to radio astronomy bands at 42.5–43.5 GHz and 48.94–49.09 GHz. Beam-forming antennas proposed for the balloon platforms will inherently generate unwanted emissions. Suppression of these emissions requires the development of new filtering techniques.

Mobile Satellite Services There will be consideration of the sharing of the 1610.6–1613.8 MHz and 1660–1660.5 MHz bands with radio astronomy. Another mobile satellite service is proposed in the band 405–406 MHz, which is likely to interfere with radio astronomy observations in the prime band 406.1–410 MHz. Another proposal is for fixed satellite transmissions at 15.43–15.63 GHz, adjacent to radio astronomy observations at 15.35–15.4 GHz.

Inmarsat An old problem will be reexamined at WRC-2000: an extension band for INMARSAT around 1600 MHz. The IUCAF Report to ICSU in 1992 discussed "sharing problems in the 1660 to 1660.5 MHz band and out-of-band interference in the 1660 to 1670 MHz band from aircraft communicating with INMARSAT satellites." The 1992 report said: "Discussions with INMARSAT on these problems have been unproductive." Nothing has changed.

Bands Above 71 GHZ There will be a major examination of the many radio astronomy allocations between 71 GHz and 285 GHz. Most of these were allocated at WARC-1979, soon after the rush of discoveries of millimeter-wave molecular lines. An IUCAF working group considered which of these bands are of high priority, and whether a new approach to sharing with active services might give radio astronomers access to other molecular lines that were not known in 1979. There would be protection zones around millimeter-wave observatories. Satellite downlinks and aeronautical operations need to be located adjacent to each other at the edge of atmospheric spectral windows. Allocations for the Earth Exploration Service (passive) will also be on the agenda at WRC-2000.

13.1 Prospects for the 2000s

The exciting prospects for radio astronomy in the next century are described by Kellerman (1997). Many imaginative and expensive facilities are planned, and the best sites for them have been carefully chosen.

But radio astronomers face hazards arising from digital radio broadcasting, high-definition TV, new mobile services using satellites, high-altitude platforms, and the information-age high-density data systems in the millimeter-wave spectral region (Baan 1996, Roth 1997). Already on the agenda for a later WRC are "Little LEO" mobile satellite "feeder links" adjacent to both edges of the hydrogen line band 1400–1427 MHz, a prime passive band. A future WRC will also need to allocate bands above 285 GHz.

Other advances in technology are just around the corner. How will radio astronomy cope with the flood? Do we have to go to the far side of the moon to find a radio quiet zone? How "quiet" can we keep the far side of the moon? (Morimoto 1993).

URSI has recommended (1996) that electromagnetic environmental impact statements be required before satellite transmission systems are authorized.

14. CONCLUSION

Galileo found in 1609 that exploration of the cosmos was opposed by the College of Cardinals. Now, on the threshold of the year 2000, the opposition comes from multinational corporations.

IUCAF members had to evolve from being starry-eyed astronomers as they encountered a world of politics, lobbying, entertainment, threats, espionage and bribery. On one occasion, an offer (in Geneva) of two million dollars in cash "to shut up" proved no match for dedication to the joys and excitement of twentieth-century astrophysics.

There is an insatiable demand for use of the radio spectrum. Private enterprise seeks the largest possible return on its investment dollar. At a 1995 meeting in Tel Aviv the meeting organizer said, "the communication people see the business, and will not be impressed by the importance of the science"—a case of greed versus curiosity. Fortunately, the spectrum is a renewable resource. As more TV and data links transfer to optical fibers, and as the demand for more bandwidth pushes data links into the infrared, perhaps pressure on the radio spectrum will ease. Also, far better use of coding systems and better data compression could lead to much more efficient use of the spectrum.

There is another approach: the national park system model, which limits the relentless exploitation of land through the establishment of large oases of land that are protected from development. Could the passive bands required by radio astronomy, space research, and earth exploration be kept in a pristine state as "national parks" in the spectrum, under the protection of UNESCO? The question was raised at the 1992 Exposition at UNESCO Headquarters (McNally 1994) and discussed further at the 1994 IAU General Assembly and 1996 URSI General Assembly. This issue needs to be followed up with ICSU and UNESCO. A good opportunity to discuss "national parks" was IAU Symposium 196, held in Vienna in July 1999 just before the third United Nations meeting, UNISPACE 3, on the peaceful uses of outer space.

We trust that UNESCO will support this proposal.

ACKNOWLEDGMENTS

In 1993 Sir Francis Graham Graham-Smith suggested the writing of a forty-year history of IUCAF, and has been most supportive. I have also received much help from Jim Cohen, Fred Horner, Willem Baan, and Emile Blum, to name but a few.

Visit the Annual Reviews home page at http://www.AnnualReviews.org

LITERATURE CITED

Abbott A. 1996. *Nature* 380:569
Argyle E, Costain CH, Dewdney PE, Galt JA, Landecker T, Roger RS. 1977. *Science* 195: 932–33
Baan WA. 1996. *ICSU Sci. Int.* 62:5–17
Bates DR, Massey HSW. 1947. *Proc. R. Soc.* A(192):1–16
Blackwell DE, Wilson R. 1962. *Q. J. R. Astron. Soc.* 3:109–14
Bodson D, Gould RG, Hagn GH, Utlaut WF. 1981. *IEEE Trans.* EMC-23:165–333
Bondi H, Christiansen WN, Gold T. 1961. *Trans. IAU* XI (B):354–55
Cheung AC, Rank DM, Townes CH, Thornton DD, Welch WJ. 1968. *Phys. Rev. Lett.* 21: 1701–5
Cheung AC, Rank DM, Townes CH, Thornton DD, Welch WJ. 1969. *Nature* 221:626–28
Clarke AC. 1961. *Harper's* 223:56–62
Crane PC. 1985. *NRAO News* 24:13
Crawford DL, ed. 1991. *Light Pollution, Radio Interference and Space Debris, Astron Soc. Pacific Conf. Ser.* 17:1–331
Dellinger JH. 1961. *URSI Info. Bull.* 128:78–79

Delogne P, van Himbeeck C. 1995. *Radio Sci. Bull.* 275:23–29
Denisse JF. 1961. *Trans. IAU* XI (B):83–87
Denisse JF, Seeger CL. 1961. *Trans. IAU* XI (B):348
Denisse JF, Lilley AE. 1961. *Trans. IAU* XI (B):349–50
Ewen HI, Purcell EM. 1951. *Nature* 168:356
Findlay JW. 1960. *URSI Proc.* XII (V):20–31
Findlay JW. 1962. *URSI Info. Bull.* 130:1–4
Findlay JW. 1976. *Trans. IAU* XVI (B):264–266
Findlay JW. 1984a. *URSI Info. Bull.* 230:10–11
Findlay JW. 1984b. *URSI Info. Bull.* 230:34–35
Findlay JW. 1988. *URSI Info. Bull.* 246:14–19
Finney J. 1959. *New York Times*, CIX (Oct.17):47
Gardner FF, Robinson BJ, Bolton JG, van Damme KJ. 1964. *Phys. Rev. Lett.* 13:3–5
Haddock FT. 1957. *URSI Proc.* XI:146–47
Horner F. 1971. *URSI Info. Bull.* 181:9–22
Horner F. 1972. *URSI Info. Bull.* 183:31–32
Horner F. 1976. *URSI Info. Bull.* 200:17–18
Horner F. 1980. *URSI Info. Bull.* 212:14–16
Howard WE. 1973. *Trans IAU* XV(B):166–67

Jansky CM. 1979. *Cosmic Search* 1(4)
Kellerman KI. 1997. *Sky Telesc.* 93(2):26–33
Kirby RC. 1978. *Telecomm. J.* 45:267–75
Kirby RC. 1989. *Telecomm. J.* 56:741–44
Kirby RC, Struzak RG. 1994. See McNally 1994, pp. 85–93
Lear J. 1959. *Sat. Rev.* 42(Oct 3):47–49
Lilley AE. 1961. *Astron. J.* 66:116–18
Lovell B, Ryle M. 1962. *Q. J. R. Astron. Soc.* 3:100–8
McNally D. 1994. *The Vanishing Universe*, Cambridge: Cambridge Univ. Press
Minnis CM. 1973. *URSI Info. Bull.* 188:16
Minnis CM. 1974. *URSI Info. Bull.* 192:11–14
Morimoto M. 1993. *Modern Radio Science*, pp. 226–29. Oxford: Oxford Univ. Press
Morrow WE. 1962. *URSI Info. Bull.* 130:5–12
Morrow WE. 1963. *URSI Info. Bull.* 138:89–99
Oort JH. 1960. *Am. Sci.* 48:160–78
Palmer P, Zuckerman B. 1969. *Ap. J. Lett.* 156:L147
Pawsey JL. 1955. *Trans. IAU* IX:563–89
Pickholz RL, Schilling DL, Milstein LB. 1982. *IEEE Trans. Commun.* 30:855–84
Piddington JH, Minnett HC. 1949. *Aust. J. Sci. Res. A* 2:63
Ponsonby J. 1991. *J. Navig.* 44:392–98
Ponsonby J. 1996. *Nature* 381:550
Rank B, Townes CH, Welch WJ. 1971. *Science* 174:1083–1107
Robinson BJ. 1965. *Sci. Am.* 211(7):26–33
Robinson BJ. 1976. *Proc. Astron. Soc. Aust.* 3:12–19
Robinson BJ. 1982. *Trans. IAU* XVIII (B):273–78
Robinson BJ. 1990. *URSI Info. Bull.* 255:103–13
Robinson BJ. 1991. *URSI Info. Bull.* 257:49–52
Robinson BJ. 1992. *URSI Info. Bull.*, 261:60–67
Robinson BJ, Whiteoak JB. 1979. *Proc. Astron. Soc. Aust.* 3:396–400
Roth J. 1997. *Sky Telesc.* 93(4):40–44
Seeger CL, Pawsey JL. 1961. *Trans. IAU* XI (B):354–55
Shain CA. 1951. *Aust. J. Sci. Res. A* 4:258–67
Smith FG. 1960. *Radio Astronomy*, p. 19. Harmondsworth, UK: Pelican
Smith FG. 1970. *Nature* 228:419
Smith FG. 1972a. *URSI Info. Bull.* 182:26–27
Smith FG. 1972b. *Nature* 239:61–62
Smith-Rose RL. 1956. *URSI Info. Bull.* 100:64–65
Smith-Rose RL. 1960. *URSI Info Bull.* 123:130–31
Smith-Rose RL. 1961. *URSI Info. Bull.* 128:76–80
Smith-Rose RL. 1962. *J. IEE.* 8:323–24
Smith-Rose RL. 1964a. *Nature* 203:7–11
Smith-Rose RL. 1964b. *URSI Info. Bull.* 142:39–52
Smith-Rose RL. 1964c. *URSI Info. Bull.* 147:41–42
Smith-Rose RL. 1969. *URSI Info. Bull.* 171:21–27
Smith-Rose RL. 1972. *URSI Proc.* XVI:84–85
Stumpers FLHM. 1989. *URSI Info. Bull.* 251:14–16
Sullivan W. 1959. *New York Times* CIX(Sept. 20):27
Thompson AR, Gergely T, Vanden Bout P. 1991. *Phys. Today* 44(11):41–49
Weinreb S, Barrett AH, Meeks ML, Henry JC. 1963. *Nature* 200:829–31
Westerhout G. 1979. *Trans. IAU* XVII(B):245–47
Wilson RW, Jefferts KB, Penzias AA. 1970. *Astrophys. J. Lett.* 161:L43
Woolley RvdR. 1952. *Trans IAU* VIII:610–12
Zuckerman B, Palmer P. 1974. *Annu. Rev. Astron. Astrophys.* 12:279–313

REFERENCE FRAMES IN ASTRONOMY

K. J. Johnston
U.S. Naval Observatory, 3450 Massachusetts Ave. NW, Washington, D.C. 20392-5420; e-mail: kjj@astro.usno.navy.mil

Chr. de Vegt
Hamburger Sternwarte, Gojenbergsweg 112, 21029 Hamburg, Federal Republic of Germany

Key Words astrometry, reference frames, reference systems

■ **Abstract** Advances in wide-angle astrometric measurements of three to four orders of magnitude in the last thirty years have resulted in a redefinition of the fundamental astronomical reference frame. This new frame, the International Celestial Reference Frame (ICRF), is based on the radio positions of 212 compact extragalactic radio sources. The ICRF defines the direction of the axes of the International Celestial Reference System (ICRS) with a precision of approximately 20 μas. At optical wavelengths, the Hipparcos catalog is the realization of this frame. The precision with which the ICRF is now determined requires that the ICRS models for precession, nutation, and others, be revised. Increases in the precision of measurements from astrometric space missions will further improve the celestial reference frame and may require its redefinition within the next ten years. These improvements will again challenge the models for the celestial reference system.

1. INTRODUCTION

Reference frames used in astronomy are defined by the positions of objects on the celestial sphere, usually specified only by direction. Astrometric advances provided by Very Long Baseline Interferometry (VLBI), radar ranging to the inner planets, laser ranging to the moon, and the European Space Agency (ESA) Hipparcos Space Astrometry mission have improved the determination of the position of celestial objects by three to four orders of magnitude. Figure 1 shows these advances in the context of achievements in astrometry since 150 B.C. Increasing accuracy dictates that relativistic effects (which are of order 10^{-8}, and are at the milliarcsecond or mas level for angles of order of a radian) be taken into account. Further, the distances of some of these objects (i.e. extragalactic radio sources) are so large that kinematic motions of these objects do not contribute to apparent temporal positional changes greater than a few microarcseconds (μas). According to Mach's principle, there are no fixed (inertial) reference points for the frame because all

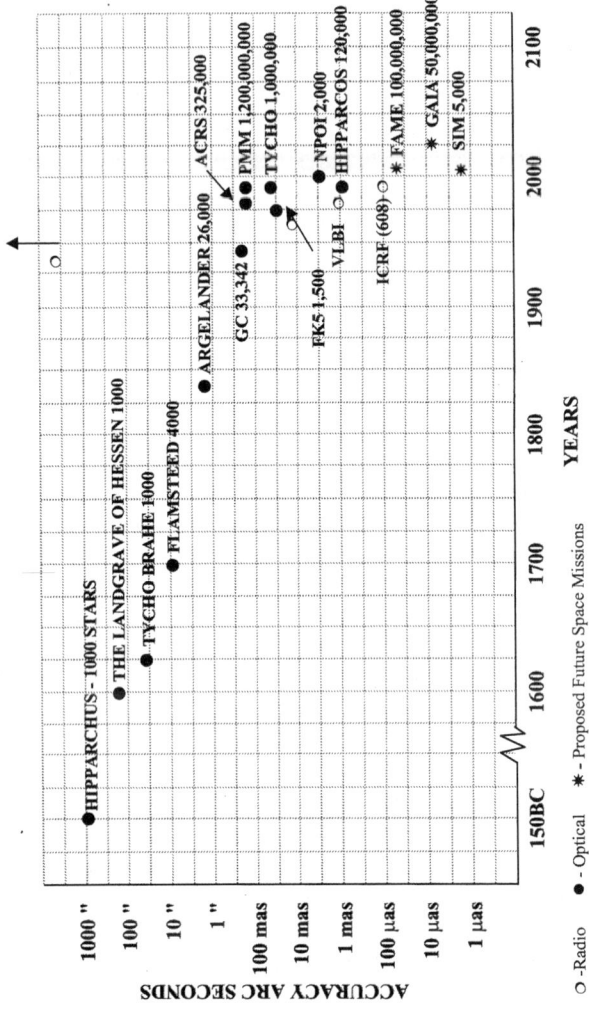

Figure 1 The increase in accuracy of astrometric measurements. Note the revolutionary increase in accuracy during 1970 to 1990.

objects in the universe interact with one another. The positions of these objects then define a quasi-inertial reference frame. Previously the positions of nearby bright stars had been employed in this capacity. The system's zero point was based upon the motions of the Earth and thus was a dynamical/kinematic reference system. The system defined now by the extragalactic radio sources is a purely kinematic system.

At the XXIII IAU General Assembly in 1997, the new ICRS and its realization, the ICRF, were adopted to replace the traditional optical fundamental reference system as realized by the FK5 reference frame. The ICRF is based exclusively on the directions to selected strong compact extragalactic radio sources determined by VLBI techniques. The IAU has further resolved that the realization of the radio reference frame in the optical domain shall be primarily the high-precision stellar net of the 118,218 stars constructed from the ESA Hipparcos Space Astrometry mission. With this development, the optical and radio regions of the spectrum are now related at the mas level.

Previous Annual Reviews articles associated with this topic have addressed radio astrometry (Counselman 1978) and astrometry (van Altena 1983, Monet 1988, Kovalevsky 1998). Other reviews in the field of astrometry and reference frames may be found in Sovers et al (1998); Vondrak & Capitaine (1998); Kovalevsky et al (1989); IAU Colloquium 127, *Reference Systems*, edited by Hughes, Smith, & Kaplan (1991); IAU Symposium 166, *Astronomical and Astrophysical Objectives of Sub-millarcsecond Optical Astrometry*, edited by Hog & Seidelmann (1995); IAU Symposium 172, *Dynamics, Ephemerides and Astrometry of the Solar System* edited by Ferraz-Mello et al (1995); IAU Colloquium 165, *Dynamics and Astrometry of Natural and Artificial Celestial Bodies*, edited by Wytrzyszczak et al (1996); and the International Earth Rotation Service (IERS) Conventions (IERS 1996), http://maia.usno.navy.mil/conventions.html.

A basic overall reference is the Explanatory Supplement to the Astronomical Almanac (Seidelmann 1992) and references therein. Radio astronomy and astrometry techniques are dealt with in Meeks (1976), Thompson et al (1986), and Zensus et al (1995). The Hipparcos mission is described in a 17-volume set (ESA 1997).

With the refinement of radio interferometric techniques for astrometry, especially VLBI, and the success of the Hipparcos mission, research efforts have intensified during the past 30 years. This review presents the background for the adoption of the ICRF, its realization at optical and radio frequencies, its limitations, and future outlook for increased precision. The references mentioned here deal in great detail with reference systems and reference frames as well as optical and radio astrometry, which are the basis for the adoption of the ICRF.

2. PRINCIPLES OF CELESTIAL REFERENCE SYSTEMS AND REFERENCE FRAMES

The reference system defines the coordinate system by specifying the direction of the axes, and specifying the zero points. The reference frame is defined by the positions of objects along with their coordinates in the system and is the practical realization of the reference system. In principle, two fixed points can define the reference frame, allowing the reference system to define its principal plane, pole, and zero point. In practice, many fixed points are used to define the reference frame globally because the definition of the axes of the system improves with the number of points defined. Also many fiducial points are needed to improve the ability to access the reference system anywhere on the celestial sphere.

In view of the high precision achieved by modern astrometric measuring methods from ground and space, and considering that those in the future will approach the microarcsecond level, a clear, rigorous specification of terms is required so that inconsistencies are not introduced into the definitions of the reference system and reference frame. Following the terminology introduced by Kovalevsky & Mueller (1981) (see also Kovalevsky et al 1989 for a detailed discussion) the purpose of a reference frame is to provide the means to materialize a reference system so that it can be used for a quantitative description of the positions and motions of celestial bodies.

A reference system is the underlying theoretical concept for the construction of the corresponding frame. An ideal celestial reference system, then, would be an inertial (i.e. nonrotating) system in the Newtonian definition which, however, is applicable only locally in General Relativity. The actual construction of a reference system requires the identification of a physical system to which the ideal reference frame definition is applied. Such a choice is naturally not unique, but the following two definitions are suitable.

In an ideal dynamical reference system, the equations of motion of a celestial body do not contain any rotational or acceleration terms: an inertial system in the Newtonian definition of a universal reference system and applicable only locally in General Relativity, where the transport of a coordinate system from one point to another is a complex transformation involving the exact knowledge of the mass distribution everywhere in that part of the space involved. (For a detailed discussion see Kovalevsky et al 1989, Seidelmann 1992, and Soffel 1989). Because of the local restriction of a given ideal reference system, the term "quasi-inertial system" is used.

In an ideal kinematic reference system, it is assumed that the universe does not rotate. Thus a suitable class of extragalactic objects, such as quasars or remote galaxies, do not display any group rotation. The ideal kinematic reference system is essentially based on the kinematic properties of very distant objects, which are assumed to be random. In Newtonian mechanics these different reference system definitions are equivalent. In General Relativity the local character of any reference system must be taken into account.

The actual modeling of this ideal reference system concept will depend on the numerical values of a number of parameters, which, as a result of observations, are not known exactly and must be adopted. Therefore the model is only an approximation of the ideal situation and is called a conventional reference system.

For centuries the primary astronomical reference systems and frames were restricted to the optical spectral region based on the continuous observations of a small number of bright fundamental stars and members of the solar system. These data resulted in a series of fundamental star catalogs, the latest being the Fifth Fundamental Catalog (FK5) (Fricke et al 1988). In the radio domain, the now mature and highly precise VLBI observing technique has been used to define and maintain a celestial (and make a contribution to the terrestrial) coordinate system with sub-mas precision which now supersedes the optical systems precision by at least one order of magnitude.

Following the recommendations of the IAU working group on Reference Systems (Hughes et al 1991, Proc. IAU Coll. 127), in 1991 the IAU decided that the future IAU conventional celestial reference system should be based on a set of distant extragalactic objects with no global rotation (Bergeron 1992). A list of suitable candidate objects (mainly QSOs, AGNs, and compact extragalactic radio sources) to define the associated new conventional reference frame was adopted (Appenzeller 1994). The XXIII General Assembly 1997 (Bergeron 1997) approved the total concept, resolving that (*a*) from January 1, 1998 the IAU celestial reference system shall be the International Celestial Reference System (ICRS) as specified by the 1991 IAU resolutions and as defined by the International Earth Rotation Service (IERS) (see Arias et al 1995, IERS 1997); (*b*) the corresponding fundamental celestial reference frame shall be the International Celestial Reference Frame (ICRF) constructed by the IAU Working Group on Reference Frames (WGRF); (*c*) the Hipparcos Catalog shall be the primary realization of the ICRF at optical wavelengths; (*d*) IERS should take appropriate measures, in conjunction with WGRF, to maintain the ICRF and its ties to the reference frames at other wavelengths.

According to the IAU resolutions, the origin of the ICRS coordinate axes is at the barycenter of the Solar System. To achieve this condition, all observations are modeled within the framework of General Relativity. The directions of the axes are fixed with respect to the extragalactic sources. To maintain continuity with the FK5 System, the principal plane of the new system is kept close to the mean equator at J2000.0 and the direction of the new conventional pole is held consistent within the errors of the FK5 system. The origin of right ascension of the new reference system will be close to the dynamical equinox at J2000.0. With reference to the definitions discussed here, the ICRS is the realization of a conventional kinematic quasi-inertial reference system. The corresponding frame (Section 6) is materialized by a catalog of extragalactic source positions.

The conventional dynamical reference system is based on the theory of the motions of the bodies of the solar system and is constructed in such a way that there are no rotational terms in the equations of motion. Thus the corresponding dynamical reference frame (Section 4) is based on the specific choice of solar-system ephemerides, the origin of the frame, and the adopted set of fundamental constants (masses of planets and auxiliary constants).

The fundamental reference system of observational optical astronomy has been based for the last 100 years on a net of selected bright stars, the so-called fundamental stars, covering the whole sphere somewhat uniformly. The term "fundamental" refers to the adopted observational procedures—that is, "fundamental observations"—that allow the construction of a net of star positions at a specified epoch without reference to previous determinations. The materialization of the system is achieved by a subsequent series of fundamental catalogs, containing the positions and proper motions of these fundamental stars together with a set of fundamental constants, in particular an adopted precession constant and nutation theory.

This conventional celestial reference system (CCRS) is modeled by these basic structures: (*a*) the origin of the CCRS (barycenter of the solar system), (*b*) the fundamental plane (celestial equator), and (*c*) the zero point of the fundamental plane (vernal equinox, intersection of mean equator and ecliptic). The actual procedures for establishing a fundamental star system have been modified continuously over time depending on available instrumentation and accuracy requirements. However, two basic operations were used throughout: measuring the positions of stars relative to one another, and determining the position of the pole and the plane of the equator and the equinox with respect to these stars by observing the apparent motions of the Sun and selected planets relative to them. Until recently, the precision with which these measurements were made over large angles was 0.1 arcseconds with a transit circle. The celestial reference system must have a reference epoch and specify all the necessary procedures and constants required to transform the frame from the reference epoch to any other date. These procedures involve precession, nutation, space motion, parallax, aberration, and light deflection, as well as those local to the observer: earth orientation parameters, refraction, and time.

3. THE FK5 REFERENCE FRAME

The CCRS was materialized by a consecutive sequence of fundamental catalogs, the most important being the FK-series of Fundamental Catalogs of the Astronomische Rechen–Institut in Germany. These catalogs were a compendium based on catalogs containing fundamental observations, meaning that the positions were on an instrumental system, the pole was determined independently, and the zero point of right ascension was adjusted to a dynamical system via observation of the sun and Solar System objects. These fundamental catalogs contained three major elements: positions, proper motions, and an adopted value for precession in order to define the fundamental coordinate system at its initial epoch, which then could be projected to other epochs and equinoxes.

Newcomb (1905a) published a catalog of 1257 stars with positions reduced to epochs 1875 and 1900 using a precession constant he derived from fundamental stellar proper motions (Newcomb 1905b). Boss (1937) published the General Catalog (GC), a catalog containing 33,342 stars from 238 catalogs obtained from observations as early as 1777.

The Dritter Fundamental-Katalog des Berliner Astronomischen Jahrbuchs (FK3) (Kopff 1937) based on the FC and NFK catalogs originated by Auwers (1879), contains stars with long and good observational histories, which limited the number of entries in the FK3 (Part I) to 925 stars.

The FK3 and FK4 catalogs were in use during most of the twentieth century. The FK3 is a realization of the dynamical reference system based on Newcomb's value of precession and a dynamical theory of motion in the solar system. The FK4 (Fricke & Kopff 1963) is essentially an improvement of FK3 (improved star positions and proper motions removing regional errors only). Thus the system of the FK3 was still represented. (For details, see Woolard & Clemence 1966, Mueller 1969, Eichhorn 1974).

Because Newcomb's determination of precession is also based on the analysis of the proper motion of fundamental stars and a model of local solar motion, all fundamental catalogs contain aspects of a kinematic system.

With the last catalog in this series, the FK5 (Fricke et al 1988), major changes and improvements were introduced. The FK5 system uses the IAU 1976 value of precession (Lieske et al 1977) and nutation (Seidelmann 1982), a new determination of the equinox and equator (Fricke 1982), a precessional correction determined from FK4 proper motion assuming a kinematic model of parallactic motion and galactic rotation (Fricke 1981) and sidereal time correction (Aoki et al 1982). The equinox of the accepted dynamical ephemerides (DE200/LE200) was made to agree with the catalog equinox of the FK5 (Standish 1982).

This implies that the FK5 system is not a pure dynamical system but partly kinematic (see also Kovalevsky et al 1989). The FK5 at epoch approximates an inertial reference frame related to the dynamical equinox as right ascension zero point.

The FK5 contains 1535 primary (FK5 I, Fricke et al 1988) and 3117 additional bright fundamental stars (FK5 II, Extension, Fricke et al 1991). The mean precision of the FK5 catalog positions and proper motions at average mean epoch 1950 is ± 0.02 arcsecond and ± 0.8 mas/yr, respectively. The quality of the FK5 frame is time dependent and is decreasing continuously by propagation of systematic and random proper motion errors, which introduce regional distortions at different epochs. A detailed comparison of the FK5 frame with the Hipparcos frame (Section 8) at epoch 1991.25 has shown large regional distortions up to 150 mas (Mignard & Froeschle 1997). However, the inherent high quality of the basic FK5 I proper motions due to their large time basis will be used together with the Hipparcos data in order to obtain improved proper motions in particular for unresolved astrometric binaries (FK6 project, Wielen et al 1997).

4. THE DYNAMICAL REFERENCE FRAME

The Dynamical Reference Frame is defined by the motions of objects in the solar system. The standard system makes use of the equator of the Earth and equinox defined by the intersection of the mean plane of the equator and ecliptic. This reference frame is defined by the process of calculating the ephemerides of solar system bodies using the equations of motion in the chosen frame, adjusting these calculated positions to the frame of observation and observed positions of these bodies. In this way these computational ephemerides are then consistent with the observational data. When sufficiently accurate observational data are available, this fitting procedure can provide the constants necessary for the ephemerides.

Range measurements and spacecraft observations are independent of any external frame but have a strong dependence on observing sites and an accurate basis for time. These observations refer the position of the planet or moon with respect to the orbit of the Earth. Other measurements such as the optical positions of the planets are dependent on the stellar reference frame. The accuracies of the range and spacecraft measurements are by far superior to those of the optical. For example, radar ranging measurements to the planets have accuracies of 2 km to 100 m, which correspond to a precision of 10^{-8} to 10^{-9}, while the optical transit circle observations are precise at the 10^{-6} level. Jet Propulsion Laboratory's ephemerides before DE130 were oriented by the optical observations, which were on the FK4 reference frame.

Beginning with the DE200 ephemeris, JPL attempted to put the ephemerides onto the mean equator and equinox of J2000 (Standish 1982), but this was not strictly possible because of the uncertainty in the definition of the ecliptic (Standish 1981), the accuracy of the optical observations of the planets were poor, and the precession of the equator from mean epoch to J2000 was inaccurate by $0.3''$/century. Although these ephemerides were on the accepted celestial system in use, they did not combine observations of the inner planets and moon—whose positions were determined very accurately by ranging and spacecraft measurement—with those of the outer planets, which are dominated by measurements in the accepted celestial reference frame. This resulted, effectively, in two frames: one for the inner planets and one for the outer planets.

5. THE RADIO REFERENCE FRAME

With the evolution of radio interferometry from 1950 to 1985, two major developments occurred. Intense, nonthermal, compact radio sources were discovered (Allen et al 1962a,b, Palmer 1962), and later identified as the extragalactic source known as quasars (Matthews & Sandage 1963, Schmidt 1963, Oke 1963). Interferometric techniques were developed for locating the positions and determining the spatial sizes of discrete radio sources (Bolton & Stanley 1948, Ryle & Smith 1948), resulting in the association of the radio source Cygnus A with a distant galaxy (Baade & Minkowski 1954). The accuracy of the position was ten seconds of arc in right ascension and forty seconds in declination. Radio interferometry evolved into linear arrays such as the Cambridge Array (Ryle 1972) and the Green Bank Interferometer (Hogg et al 1969) which was the prototype for the Very Large Array (Hjellming & Bignell 1982). The antennas in these arrays were connected via cables. In the 1960s Very Long Baseline Interferometry (VLBI), using independent local oscillators to link telescopes that were not physically connected via cable or radio link, was demonstrated (Carr et al 1965, Bare et al 1967, Broten et al 1967, Moran et al 1967).

The positions of compact radio sources were determined to less than 1″ by Wade (1970) and Brosche et al (1973), and refined to 0″.02 (Wade & Johnston 1977) using the Green Bank Interferometer. This was paralleled by accuracies achieved via VLBI: 1″ (Cohen & Shaffer 1971) and 0″.02 (Clark et al 1976). At this point the accuracy of radio positions surpassed that of optical transit circles.

In 1978 at the IAU Symposium 182, *Time and the Earth's Rotation*, an IAU Working Group was formed to promote the comparison and evaluation of techniques for the determination of Earth rotation and high-precision data for scientific analysis. The efforts of the IAU Working Group later became known as Project MERIT ("MERIT" is an acronym for "Monitor Earth Rotation and Intercompare Techniques"). During the 1980s Project MERIT carried out many campaigns evaluating laser ranging and VLBI techniques, and adopted standards for obtaining the highest precision of measurements (Melbourne et al 1983). Project MERIT's efforts led to the establishment of the International Earth Rotation Service (IERS) on January 1, 1988. The IERS replaced the Bureau International de l'Heure and the International Polar Motion Service, and serves as a general coordinating agency for the development of reference frames and the relationship of geodetic and astronomical methods for precise measurements. The IERS adopted conventions for the celestial, dynamical, and terrestrial reference systems and frames, as well as numerical standards (precession, masses of the planets, etc), and models for nutation, tropospheric refractivity, geopotential, site displacement, etc. These conventions are updated as often as necessary and were most recently published in the IERS Technical Note 21 (1996).

By 1978, the promise of improved accuracy was about to be fulfilled. At IAU Colloquium 48, *Modern Astrometry*, a working group was established under IAU Commission 24 to select candidate radio sources with optical counterparts for a reference frame. The working group identified 234 candidate sources, the majority of which had positional accuracies of 10 mas. The positions of these sources were obtained by a weighted average of eight catalogs, seven of which were obtained with interferometers connected with cables (Argue et al 1984).

In the 1980s great advances were made using VLBI techniques. Observations were standardized. Two frequencies, 2.3 (S-band) and 8.4 GHz (X-band) generally using fourteen channels of two MHz bandwidth each were used. Six S-band channels spanned about 85 MHz while eight X-band channels spanned about 360 MHz. The basic observable is group delay (Rogers 1970). The two frequencies allow for accurate calibration of the frequency-dependent propagation delay in the ionosphere. Catalogs of positional accuracy reached a few milliarcseconds (Fanselow et al 1984), and survey observations extended coverage to the Southern Hemisphere (Morabito et al 1986a,b) with accuracies of 300 mas.

The VLBI technique made a major advance with the introduction of the Mark III VLBI system (Rogers et al 1983, Clark et al 1985). Catalogs using Mark III technology achieved an accuracy of less than 0.5 mas for the best positions (Ma et al 1986, Robertson et al 1986); Mark II measurement by Sovers et al (1988) achieved the same level of accuracy.

To obtain these accuracies, certain astronomical constants such as precession and nutation were solved from the data. The MERIT standards were brought forth to update the 1976 IAU standards (Melbourne et al 1983). Two different software analysis systems evolved for the reduction of VLBI data: MASTERFIT (Sovers & Fanselow 1987), developed by the Jet Propulsion Laboratory, and CALC /SOLVE/GLOBL, developed jointly by the Goddard Space Flight Center, the Harvard-Smithsonian Center for Astrophysics, and the National Geodetic Survey (Robertson 1975, Ma 1978, Gordon 1985, Ma et al 1986). Both sets of software were found to deviate in theoretical delays by 1.5 to 50 picoseconds, not including the comparison of the different tropospheric and relativistic models used by each (Sovers and Ma 1985).

Various catalogs of radio source positions were produced by JPL, GSFC, NGS, and USNO/NRL in the late 1980s. These catalogs were combined by the IERS to obtain a reference frame. By 1990, it was concluded that the accuracy of the source positions was better than 1 mas with regional deformations between catalogs of a few mas (Arias et al 1991). Later, the precision of source coordinates was found to be a function of the number of observations with an estimate of 0.2 mas for positions based on 100 observations (Arias et al 1995).

In 1986, a program to establish a global radio reference frame of 400 sources was undertaken (Johnston et al 1988). The resulting reference frame would be radio/optical: all the radio sources would have optical counterparts whose positions would also be measured. The program proposed to use all Mark III data available in a consistent solution (Johnston et al 1991), and resulted in a number of catalogs between 1991 and 1995. The first was a catalog of 182 sources with a positional accuracy of one mas (Ma et al 1990). The right ascension zero point was defined by the FK5 based optical positions of 28 quasars. These sources were all north of -30 degrees declination. This first program was followed by campaigns to increase the density of sources in the northern hemisphere and to add an equal grid of sources in the southern hemisphere (Russell et al 1991, 1992, 1994; Fey et al 1992, 1994; Reynolds et al 1994). These campaigns resulted in a catalog consisting of a total of 403 sources with 208 in the northern hemisphere and 195 in the southern hemisphere.

These data, together with all available dual-frequency bandwidth synthesis Mark III VLBI data from the geodetic and Earth orientation programs, were used to solve for catalog positions from first principles in a single solution (Johnston et al 1995). This data set consisted of all data (1,015,292 pairs of group delay and phase delay rate observations) collected between 1979 and 1993. The majority of the data were from the geodetic programs, Earth orientation programs, and source surveys.

The astrometric campaign contributed only 23,000 observations but filled out the reference frame especially in the southern hemisphere. It resulted in a catalog of 436 sources with positional accuracies better than three mas in each coordinate with the accuracies of the majority of sources smaller than one mas. These 436 sources defined the reference frame. The sources were divided into two classes. Class 1 sources (163 in northern hemisphere, 48 in southern hemisphere) have positional weighted rms accuracies of less than one mas, while class 2 sources (98 in the northern hemisphere, 127 in the southern hemisphere) have weighted rms accuracies of less than three mas. Also measured were an additional 124 objects that needed further observation to improve their positions or were unsuitable for the reference frame because of complex source structure.

6. THE INTERNATIONAL CELESTIAL REFERENCE FRAME

As discussed in Section 2, in 1991 the XXI General Assembly of the IAU passed a resolution stating that the celestial reference system would be realized by a celestial reference frame defined by the precise coordinates of extragalactic radio sources. A working group on reference frames was established to create this catalog, which was accomplished in 1998. It includes primary sources, which define the frame, and secondary sources that later may be added to—or replace—the defining sources. The criteria for the defining sources were sufficient data (observation for more than two years, with more than twenty observations) showing lack of position variation (differences less that 0.5 mas or 3σ in either coordinate); have submilliarcsecond positional formal errors; a structure index (Fey & Charlot 1997b) at X Band (if available) of one or two and show no significant apparent proper motion.

This work has been detailed by Ma et al (1998). A single solution of 1.6 million pairs of group delay and phase delay rates data obtained between August 1979 and July 1995 was made. A frame based on the positions of 212 defining extragalactic radio sources distributed over the entire sky has been established. The positional accuracy of these sources is less than one mas. The positions are at a frequency of 8.4 GHz, and were obtained from 1.6 million pairs of group and phase delay observations obtained by dual frequency 2.3 and 8.4 GHz VLBI observations. Figure 2 shows the distribution of these sources on an Aitoff Equal Area

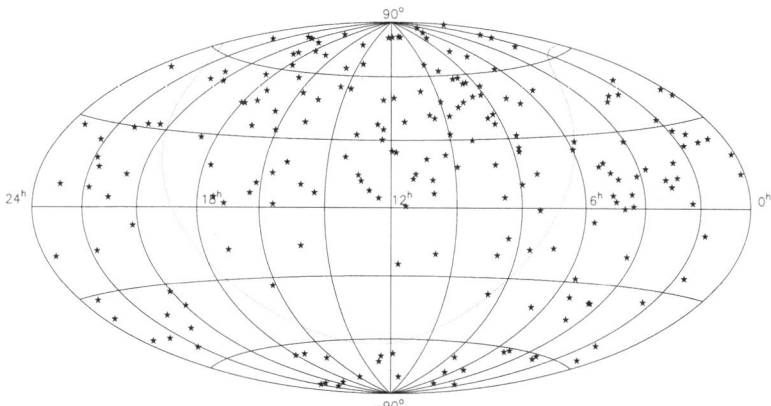

Figure 2 The distribution of ICRF defining sources. There is a lack of sources south of the equator due to lack of observations. A large number (more than 100) of observations are needed to obtain formal positional accuracy at the 100 μas level.

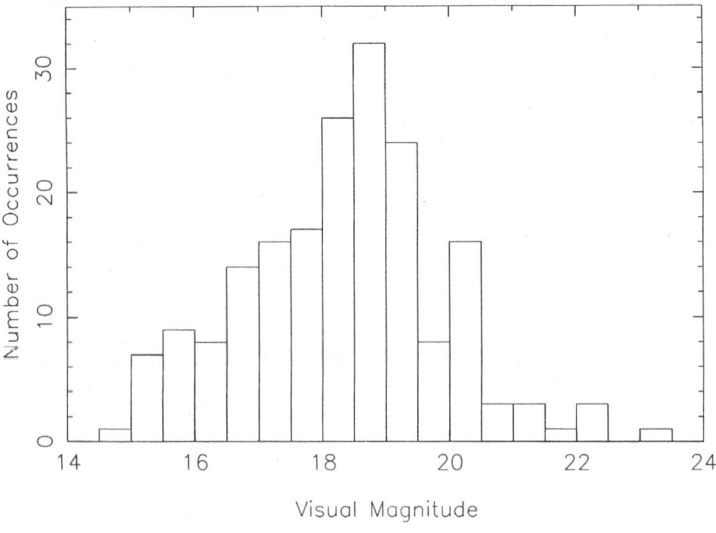

Figure 3 The visual magnitude of the optical counterparts of the ICRF sources. Note that they are much fainter than the Hipparcos stars and peak at about the 18th visual magnitude. Thus bright stars with radio counterparts were used to link the ICRF and Hipparcos frames.

Projection. Only 22 percent of the defining sources are in the southern hemisphere, a fact attributable to a lack of VLBI observations: most radio observatories are in the northern hemisphere.

The accuracy of the positions of these sources is estimated by RSS, the formal error of the solution multiplied by 1.5 with a systematic error estimated to be 0.25 mas. As of January 1, 1998, the IAU adopted this frame as the fundamental reference frame. The optical magnitudes of the defining sources are presented in Figure 3. The majority of the sources have optical counterparts at the 18th to 19th visual magnitude. Through future observations, an additional 294 candidate sources may become defining sources. There are also 102 "other" sources whose positions may show variations with time, or whose positions are less well known.

7. MAINTENANCE AND LIMITATIONS OF THE ICRF

7.1 Limitations Due to Source Structure

Extragalactic radio sources display structure on spatial scales from hundreds to one mas. The spatial structure is usually more compact at higher radio frequencies. The mechanism giving rise to these sources is believed to be the existence of massive

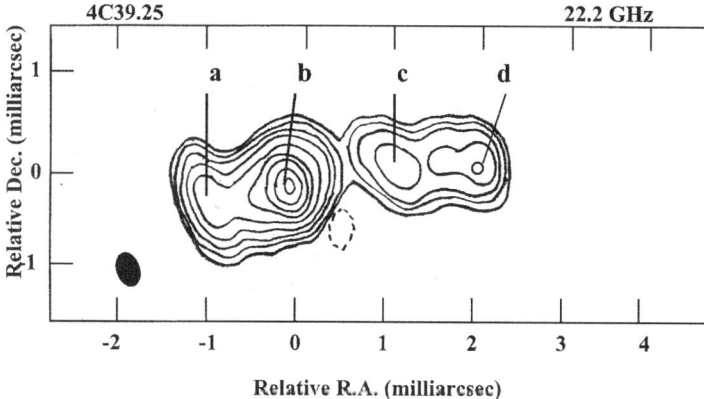

Figure 4 The distribution of flux density at 22 GHz for the extragalactic radio source 4C39.25 (Guirado et al 1995). The restoring beam is shown as a filled ellipse. Note that 4C39.25 is distributed over 3 mas in the east-west direction.

black holes in the center of galaxies. The radio emission takes the form of jets. There is a wide variety of source structures. These sources are all variable on timescales of weeks to years at radio wavelengths. Recent observations have found timescales of hours for some of these objects.

Figure 4 presents typical source structure for one of these compact objects (4C39.25), obtained at a frequency of 22 GHz (taken from Guirado et al 1995). Note that the source has at least four components distributed over about four mas in right ascension. The effect of variations in this structure is illustrated in Figure 5. A position for this source has been measured over a 15-year period using all geodetic Mark III VLBI observations. The solutions for right ascension and declination show variations due to changes in source structure with time. This source has an apparent proper motion of 13.6 μas/yr in right ascension and 6.8 μas/yr in declination (Fey et al 1997a), consistent with variations in components a and b identified in Figure 4. In the period 1980 through 1982 components b and c dominate the position. In the period 1982 through 1985 component a dominates, and in 1985 through 1997 component b dominates (Fey et al 1997a). Also note that the quality of the positions markedly improved after 1985. This improvement was probably due to the introduction of the Mark III system and improved observing techniques for geodetic measurements.

Maps of sources will also have different appearances when mapped with different arrays of antennas. Maps of the source VR422201 (BL Lac) at 8.4 GHz are displayed in Figure 6. The top figure was obtained using the VLBA, and the bottom figure was made from the VLBA data with the addition of geodetic antennas. Note that the structure appears more compact when mapped with the VLBA plus geodetic antennas, which could result in a source position offset of a few tenths of a mas in declination.

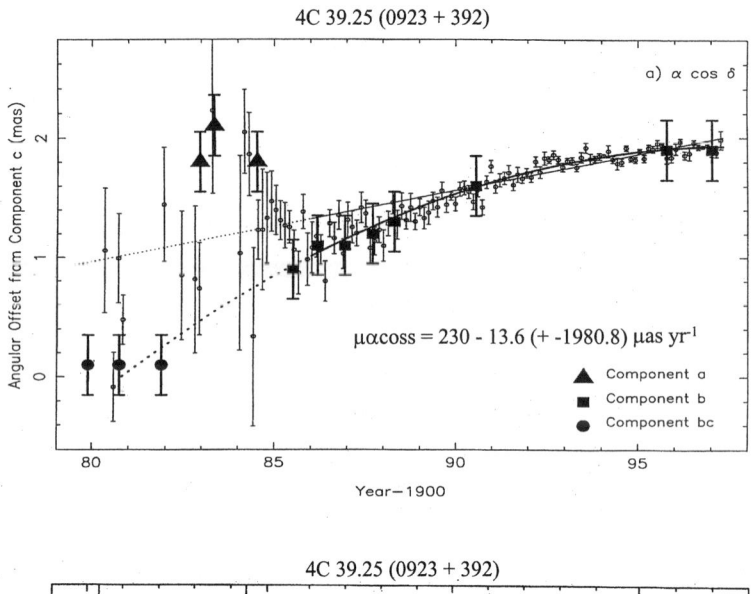
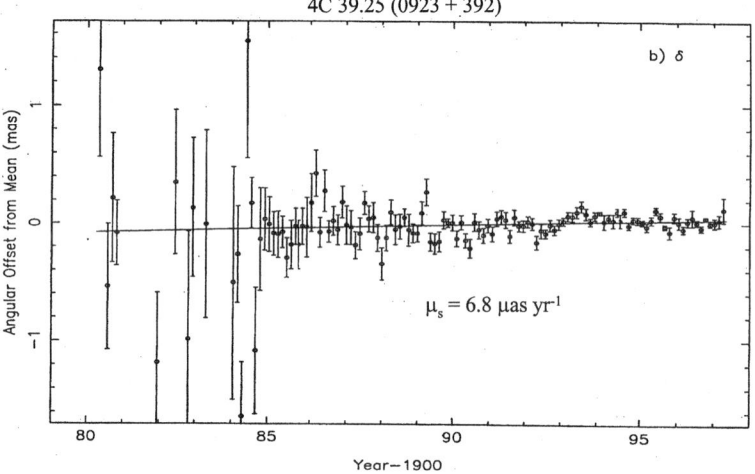

Figure 5 The position of 4C39.25 as measured with the geodetic database over the time period 1979–1997. There is significant apparent motion in right ascension, while the declination position is quite stable. Note that the quality of the data increased markedly after 1985. The motion in right ascension is attributed to variations in the flux density of the components.

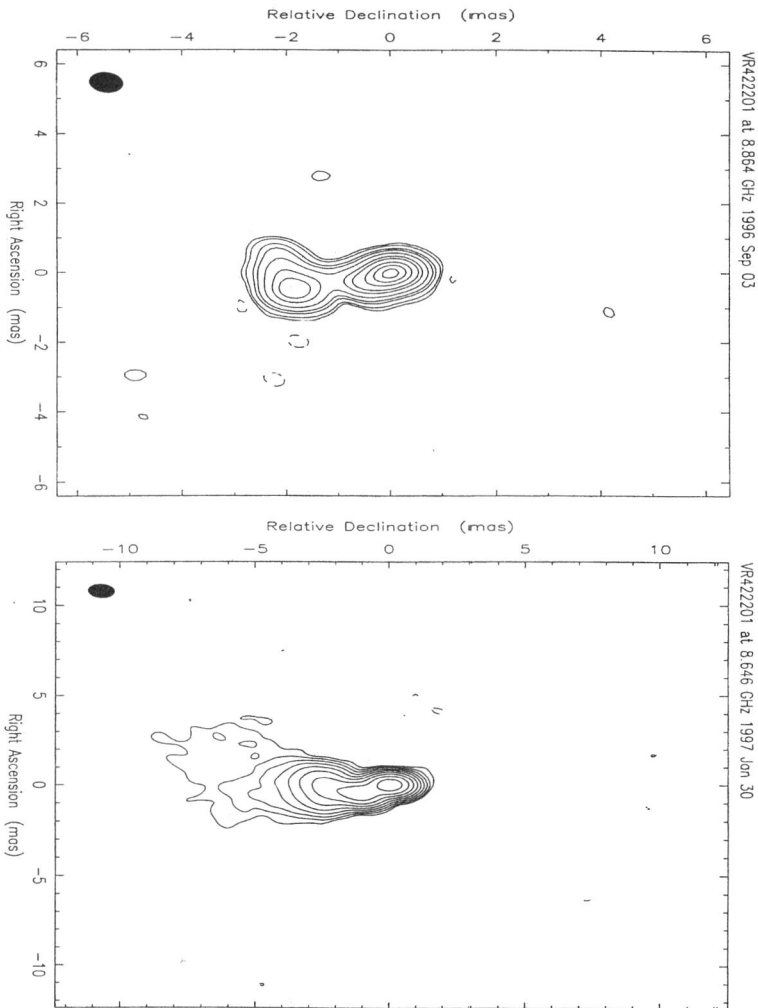

Figure 6 Maps showing the distribution of radio emission from the sources and VRO 42201 as mapped using the VLBA array (*top*) and the VLBA and geodetic antennas (*bottom*). The restoring beams are shown by filled ellipses. The emission is quite complex and appears slightly different as mapped by the different arrays.

The effect of source structure on position can be as large as tens of mas. From an investigation of source structure corrections to source positions, Fey & Charlot (1997b) found a correlation between compactness of the sources and their formal positional uncertainties, indicating that more extended sources have larger positional uncertainties. They define a structure index to estimate the astrometric quality of the sources. An index of one at 8.4 GHz is very good, two at 8.4 GHz is good; three is marginal and should be used with caution and four should not be used.

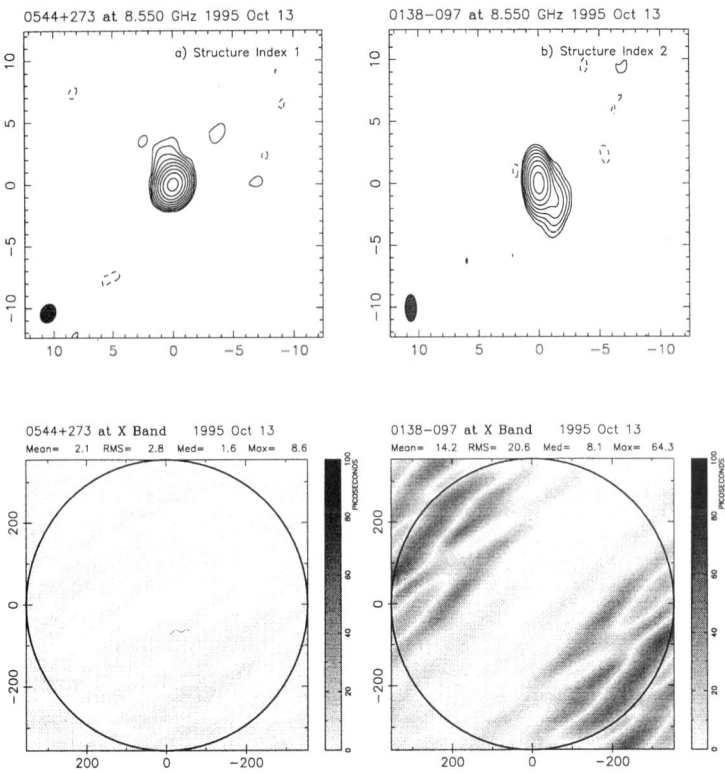

Figure 7 Structure contribution to measured interferometric delay. *Top* displays the 8.4 GHz radio emission. *Bottom* shows the structure of the induced bandwidth synthesis delay. The structure index is one for the left source and rises at intervals of one with each source.

Figure 7 shows the effect of structure on delay. In order to obtain good astrometry, they also recommend that the source have an index of one at 2.3 GHz also. Maps of the radio sources making up the ICRF may be found in Fey et al (1996 & 1997b) and at http://www.usno.navy.mil. Figures 4, 6, and 7 show the complexity in spatial structure from a sample of extragalactic radio sources at 8.4 GHz. The sources in the ICRF are class 1 and 2.

Because the structure of these sources is variable in time, it is wise to measure their positions for apparent motions. Only those sources observed frequently over the past 25 years have sufficient history to yield excellent positions. A subset of these sources making up the ICRF has a large number of observations—these are the sources that were observed frequently on the geodetic programs. The southern hemisphere sources especially suffer from a lack of observations. In the future, some of the defining sources may become unsuitable. A maintenance program to obtain a large number of observations of the defining and candidate sources must be undertaken.

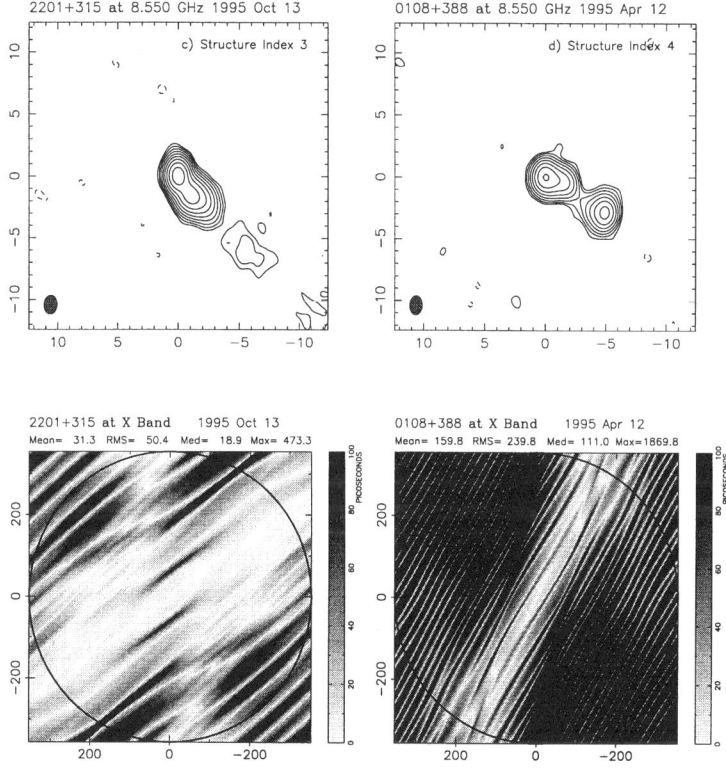

Figure 7 (*Continued*)

The geodetic programs will be a good source of a large number of observations of a subset of these sources. Astrometric maintenance observations will have to be made of a large number of sources using the available resources. This effort will require time on the VLBA, the EVN, MERLIN, and southern hemisphere telescopes in Australia, South Africa, and Brazil. Positional variation in the sources may be modeled if the variations in source structure are known. Charlot (1994) has shown that modeling the source structure in the complex source 3C273 significantly improves its positional stability.

Due to the lack of telescopes and land area in the southern hemisphere, it will be difficult to obtain observations of the source structure south of declination -20 degrees. The Highly Advanced Laboratory for Communication and Astronomy (HALCA) space mission may help this situation by obtaining observations for southern-hemisphere sources of a large number of spatial frequencies. Another method of monitoring these sources for changes in structure would be to monitor their flux density. There are no plans to do so in an organized way at this time.

Ma & Shaffer (1991) have investigated using the geodetic database of the time varying positions for quasars. Ma & Shaffer give limits of less than 50 μas/yr, while

Eubanks et al (1995) find many sources with motions of order 30 μas/yr. Many of these motions have not been confirmed (Fey et al 1997a). As already shown, apparent motions in the source 4C39.25 are attributable to variations in the spatial emission from the source (Fey et al 1997a). For well-behaved sources the positions are estimated to be repeatable at the level of 0.2 to 0.3 mas (Fey et al 1997a).

An estimate of the best stability in position of a compact radio source may be obtained from the numerous experiments made on close source pairs such as 1038 + 52 A and B. For this pair of sources separated by 33 arcseconds, observations spanning a decade at wavelengths of 3.6 and 13 cm find changes in the angular separation of nearly 0.2 mas (Rioja et al 1997). These changes can be accounted for by motion of the reference feature in quasar B, which is a component of the jet displaying superluminal motion. The position stability of component A is estimated to be better than 10 μas.

7.2 Limitations Due to the Atmosphere, Scaling Factors and Models

The propagation of signals in the troposphere causes one of the largest sources of measurement error. The error varies as a function of elevation and azimuth. Mapping functions that give the elevation dependence of tropospheric delay have reduced the systematic and random errors (Davis et al 1985, Herring 1992, Niell 1996). The effect of these azimuthal gradients is to cause a north-south asymmetry due to the greater troposphere thickness near the equator. Estimating these gradients reduces the radio source declinations by 0.5 mas at the equator (MacMillian & Ma 1997). In the future, mapping the troposphere via GPS satellites may give detailed models of the troposphere that will allow the delay path lengths in the atmosphere to be estimated with greater precision than at present from the VLBI data.

It is difficult to estimate the systematic errors introduced by the software used in the data reduction models. Comparison of different software routines such as GSFC's CALC and JPL's MODEST has recently shown discrepancies at the 5 μas level (Ma et al 1998). Previous analysis of geodetic data has shown that the formal errors in station coordinates should be multiplied by 1.5 (Ryan et al 1993). Ma et al (1998) have applied this to the errors in the positions of the defining sources. In addition, they have added 0.25 mas in quadrature in reporting the errors of the defining sources. Typical formal errors for sources in the ICRF are on the order of 0.1 mas. In the future a more complete understanding of the random and systematic errors should allow further refinement in the positions of quasars to levels well below 0.1 mas. The level at which variable source structure effects contribute significantly to these errors remains to be seen. For many sources, it is a significant effect at levels of a mas.

7.3 Maintenance

The IAU resolution adopting the ICRF also requested the IERS to provide its maintenance. The IERS VLBI coordinating center has been designated to carry out this task. It was also agreed at the IERS Directing Board meeting in San Francisco in 1998 that the ICRF be maintained by the VLBI coordinating center of the IERS. The level of observations for geodetic and earth orientation programs should increase over the next few years with the inception of NASA's Continuous Observation for Rotation of the Earth (CORE) program. The Mark IV correlator should come into operation in 1999, allowing larger arrays and more frequent observations needed by the CORE program. The U.S. Naval Observatory/NASA GSFC plan future observing programs to maintain the ICRF with yearly observations of the program sources. MERLIN also plans to contribute to this effort (Morrison et al 1997). The number of sources available for the reference frame will certainly increase. A VLBI survey of the nearly 2000 flat-spectrum sources from the Jodrell Bank-VLA Astrometric Survey is under way (Peck & Beasley 1998).

8. THE HIPPARCOS OPTICAL REFERENCE FRAME

The extremely successful Hipparcos mission (1989–1993) established a global astrometric catalog that is not affected by atmospheric refraction and turbulence and is independent of the Earth orientation parameters. However, the construction of the consistent instrumental Hipparcos reference frame, the so-called sphere reconstruction problem, has a rank deficiency of 6. To align the Hipparcos system with the ICRF, two small rotations were determined, fixing the final orientation of the Hipparcos net at epoch 1991.25 and removing a global rotation from the proper motion. Because Hipparcos was not able to observe directly the optical counterparts of the defining sources of the ICRF, link procedures were adopted to achieve *a posteriori* adjustment (Lindegren & Kovalevsky 1995).

The Hipparcos catalog contains 118,218 stars with typical precisions of 1.5 mas of the five astrometric parameters: position, parallax, and proper motion for the majority of the stars. Figure 8 shows the distribution in magnitudes of the catalog stars and the percentage of completeness as compared to the expected global galactic star counts in each magnitude interval. The Hipparcos catalog provides, for the first time at optical wavelengths, a global reference frame that is not affected by zonal and magnitude-dependent errors found in ground-based fundamental catalogs. However, the average star density of 2.7 stars/deg^2 and the relatively bright magnitude of the catalog stars are not suitable for the direct adjustment of fainter objects to the ICRF in the small arcminute-sized fields of modern large telescopes. Hipparcos provides a high precision, first order net that can be extended by a secondary and much denser stellar net of fainter stars. This is currently achieved by new astrograph programs (Zacharias et al 1997) and eventually will be provided by future space missions.

In contrast to the extragalactic ICRF sources, the Hipparcos catalog stars display proper motions due to the angular component of the stars' space motion within the gravity field of our galaxy. As a consequence, the accuracies of the Hipparcos positions are strongly time dependent due to the proper motion error propagation.

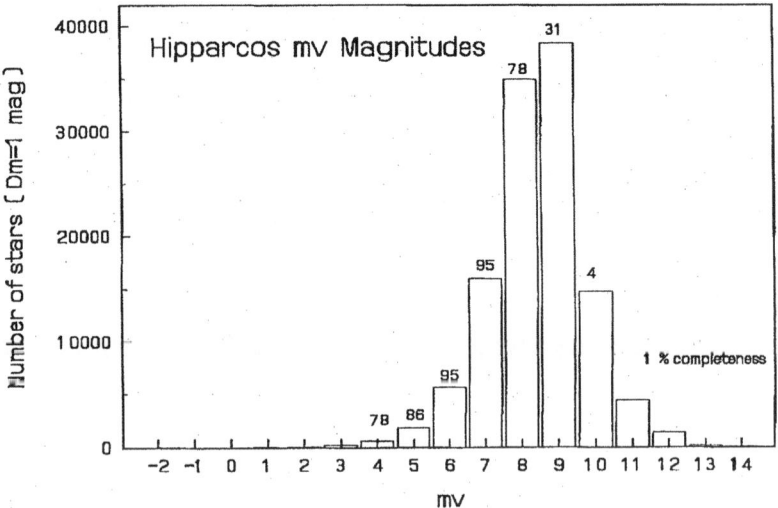

Figure 8 Number of Hipparcos stars versus visual magnitude. The completeness of the survey for each magnitude is shown at the top of the column. Note that Hipparcos is nearly complete to the eighth visual magnitude.

An additional smaller systematic contribution is due to the time-dependent error of the Hipparcos zero point adjustment (Figure 9). To preserve the high quality of the Hipparcos net at central epoch 1991.25 over an extended period, future space astrometry missions are indispensable. As an example, the proposed Deutsches Interferometer für Vielkanalphotometrie und Astrometrie (DIVA) (Roser et al 1997)

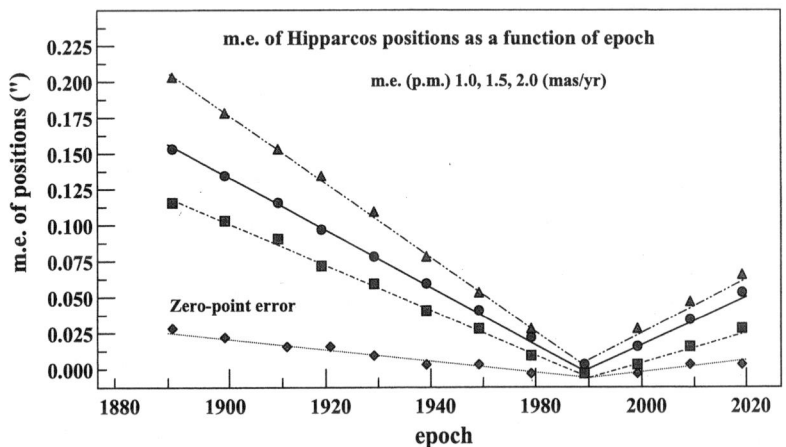

Figure 9 The accuracy of the Hipparcos positions as a function of time. The accuracy of the positions degrades with time because of imperfect knowledge of the proper motion of the stars.

and Full-Sky Astrometric Mapping Explorer (FAME) (Johnston et al 1998) missions would limit the Hipparcos catalog error propagation to a few mas for many decades. The present Hipparcos extragalactic link can be improved further by re-observation and inclusion of additional radio stars with the VLBA, EVN, and MERLIN to extend the proper motion time basis.

9. LINKING THE HIPPARCOS OPTICAL REFERENCE FRAME TO THE ICRF

The Hipparcos mission's high-precision relative global stellar net of positions and proper motions had to be linked to the ICRF with similar precision. Because all suitable optical counterparts of the ICRF sources were too faint to be measured directly by Hipparcos (with the exception of 3C273B, but only with reduced accuracy), a direct link to the ICRF will be made only by future space missions that extend the magnitude range beyond that of Hipparcos.

The Hipparcos Extragalactic Link Working Group has used a variety of methods (Kovalevsky et al 1997), in particular VLBI, MERLIN, and VLA observations of selected radio stars that are optically bright Hipparcos catalog stars, and the direct measurement of ICRF optical counterparts. The most precise contribution has been obtained from long-term VLBI observations of 12 radio stars, which were tied to the ICRF using the closest ICRF source as the primary reference. From these long-term measurements the five astrometric parameters of these stars were determined to sub-mas precision (Lestrade et al 1995). A weighted least-squares solution using all of the methods was used to adjust the Hipparcos catalog to the ICRF with uncertainties of 0.6 mas and 0.25 mas/yr at epoch 1991.25 in the position and proper motion rotation parameters (Kovalevsky et al 1997).

10. LINKING THE DYNAMICAL REFERENCE FRAME TO THE ICRF

The advent of VLBI observations of spacecraft at planets and satellites has put the positions directly on the radio frame at accuracies of 1 to 3 mas. Further CCD observations of the planets using Hipparcos reference stars can also place the positions of planets and satellites at accuracies of 30 mas (Stone & Dahn 1994). The major problem alluded to earlier is the transformation or alignment of the two different types of data onto a common reference frame, which is accomplished by taking VLBI measurements and solving for the rotation of the ephemeris that gives the best fit to the data. From a joint analysis of VLBI and LLR observations, a tie between the JPL planetary ephemerides and the IERS radio catalogs has been determined (Folkner et al 1994). The frame tie between the Hipparcos realization of the ICRF and the ICRF may be used to relate optical observations of the planets. Thus the present JPL ephemerides of the planets, DE405 and of the moon, LE405 (Standish et al 1995), use the reference frame of the ICRF.

11. REFERENCE FRAMES AT OTHER SPECTRAL RANGES

The ICRF is defined at a radio frequency of 8.4 GHz. Care must be taken when extending this frame to other radio wavelengths. Most likely the nearby ICRF defining source will show little or no structure at higher radio frequencies, but at lower frequencies where the emission is self absorbed its position may change by the order of a mas or more. Other spectral ranges may be linked to the ICRF by use of link objects, which display emission at 8.4 GHz and the spectral range. The Hipparcos realization of the ICRF also allows the frame to be transferred to other spectral ranges such as the IR through the IR counterparts of Hipparcos stars. The radio and optical regions of the spectrum both have frame ties to the IR. For example, Menten et al (1997) have determined the position of Sgr A* at IR wavelengths to 30 mas via the radio emission from SiO masers arising from the innermost parts of circumstellar envelopes of giant and supergiant stars that appear as compact IR sources.

12. FUTURE IMPROVEMENTS IN REFERENCE FRAMES: Space and Ground Systems

Hipparcos has demonstrated the feasibility of mas precision global astrometry from space. Follow-up missions aiming at an extension to much fainter limiting magnitudes and increased accuracy have been proposed. It should be noted that all these astrometry missions, including Hipparcos, primarily are directed to the solution of astrophysical problems—for example, galactic kinematics and the galactic distance scale from Cepheids and RR Lyras stars. To cover a substantial galactic volume and to obtain a complete sample of different stellar groups, therefore, the star selection may not be always optimal with regard to reference frame requirements.

In the future, reference frames at optical wavelengths will reach improved accuracies. Three proposed space missions, FAME (Johnston et al 1998), Global Astrometric Interferometer for Astrophysics (GAIA) (Gilmore et al 1998), and Space Interferometry Mission (SIM) (Shao 1998) propose to achieve global accuracies of 50, 4, and 4 μas respectively. These missions will extend the accuracy of the optical reference frame by a factor of 20 to 250.

FAME and GAIA are survey missions that will determine the positions of 40 to 500 million stars. The brighter stars will have positions at global accuracies of 50 and 4 μas respectively. To achieve this, the parallax and proper motions of these stars along with their orbital motions, if they are in multiple systems, must be measured to these accuracies. SIM (Shao 1998) is an integrating steerable space interferometer. The other missions are more or less based on modifications of Hipparcos' one-dimensional scanning principle with two superimposed fields of view separated by a constant basic angle of 60 to 100 degrees. Optical and detector technologies have improved significantly and all of these space missions will reach comparable or much higher accuracies than Hipparcos has reached, increasing the number of stars and magnitude range by orders of magnitude. A direct link to the ICRF optical counterparts can be made by all these missions, with SIM projected to reach the highest accuracy.

Improvement in the accuracy of wide-angle ground-based astrometric observations will be achieved by optical/IR interferometry. Hummel et al (1994) summarizes the results obtained with the Mount Wilson Interferometer. The Navy Prototype Optical Interferometer located near Flagstaff, Arizona is nearing completion. The astrometric accuracy of the instrument over wide angles should approach one mas (Johnston et al 1997). This instrument will be capable of measuring the positions of approximately 2000 bright Hipparcos stars and will be capable of maintaining the accuracy of the Hipparcos Frame in the northern hemisphere. The Keck Interferometer (Colavita et al 1998) may contribute to wide-angle astrometry although its primary function is 2-micron narrow-angle astrometry.

The premier accuracy of the ICRF will be challenged if FAME, GAIA, or SIM reach their goals. Frequent VLBI observations of the defining sources may allow their source structures to be modeled. Observations at radio frequencies higher than 8.4 GHz may also reduce this effect because the structure may be dominated by a central compact source. Observations are under way at two cm with the VLBA (Eubanks, personal communication 1998). The models for the reduction of VLBI data in CALC/SOLVE and MODEST may be improved and further understood so that a factor of 1.5 does not need to be applied to the errors and systematic effects may become better understood. The atmospheric delay residuals pose the principal problem to improving the accuracy of VLBI astrometry. Figure 10 shows the formal errors in the ICRF solution. Future space VLBI missions with several elements in space can eliminate this source of error. A radio array in space will be very expensive. It would appear, however, that without a radio array in space the accuracy of VLBI astrometry over wide angles may be limited to more than 10 μas. Lensing may impose a limit of 10 μas as well.

13. NEED FOR THE DEFINITION OF ASTRONOMICAL CONSTANTS AND TIME FOR THE REFERENCE SYSTEM

The International Celestial Reference System was defined by the XXIII General Assembly of the IAU to take effect 1 January 1998. It is in accordance with the 1991 IAU recommendations that the origin be located at the solar system barycenter via modeling of VLBI observations in the framework of General Relativity. The pole is in the direction defined by the conventional IAU models for precession (Lieske et al 1977) and nutation (Seidelmann 1982) and the origin of right ascension is defined by fixing the right ascension of 3C273's (Hazard et al 1971) FK5 value transferred to J2000.0. The Hipparcos catalog contains all of the FK5 fundamental stars. Thus the location of the FK5 pole and origin of right ascension are related to an accuracy of a few mas. The coordinates of the mean pole at J2000 have been found from analysis of a long series of VLBI observations using a state-of-the-art

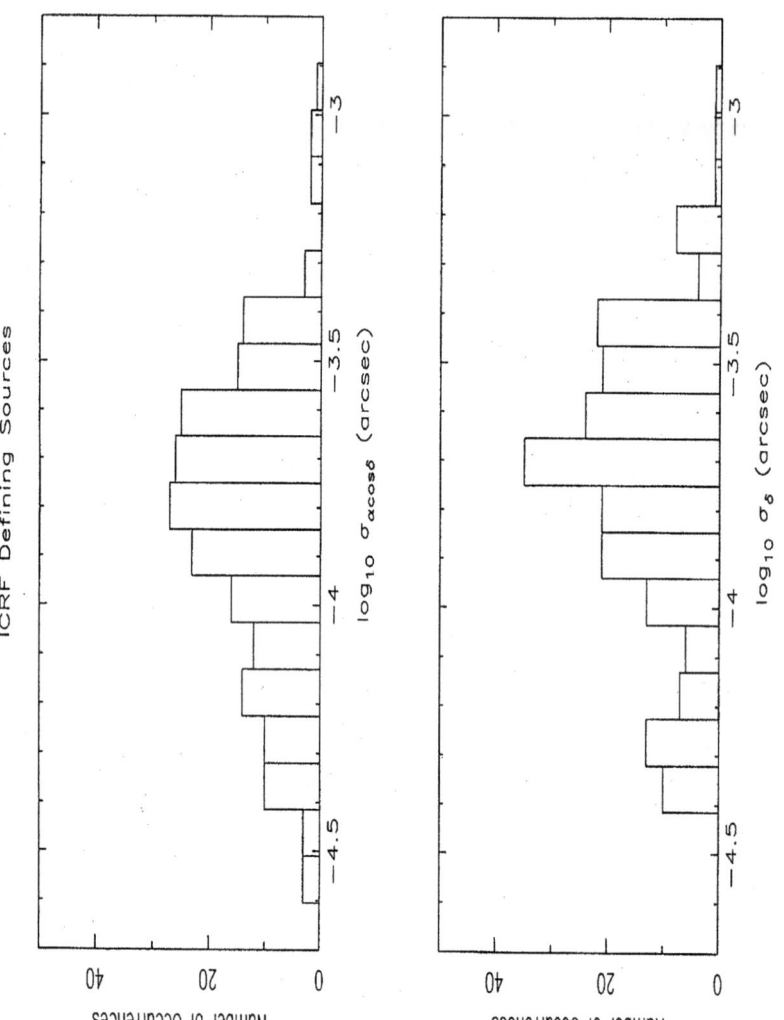

Figure 10 The formal accuracy of the 212 defining sources. If the systematic errors were understood, the accuracy of the VLBI positions would be less than 100 μas. Future improvements may lead to accuracies of less than 10 μas.

precession model to 17.3 ± 0.2 mas in the direction 12 h and 5.1 ± 0.2 mas in the direction of 18 h (IERS 1997).

McCarthy & Luzum (1991) documented that the models for precession (Lieske et al 1997) and the IAU 1980 nutation theory (Seidelmann 1982) do not describe precisely the position of the pole. Ma et al (1998), in determining the positions of the ICRF sources, solve for an offset between the J2000 pole and the Celestial Ephemeris Pole of the ICRF in the form of celestial pole offsets angles in ψ (longitude) and ε (obliquity). The IERS conventions give improved values for these and other phenomena as well as constants such as the mass of the planets and gravitational constant. The present status of IAU standards is given in Fukushima (1997). There is no physical theory to adequately describe many of the phenomena at the level of precision needed. The IERS conventional model for nutation is precise at the 0.1 mas level. Therefore a numerical fit to the data may have to suffice to describe these phenomenon at the $10 - 1$ μas level. New constants for precession and other effects, as well as models for nutation and other phenomena, must be adopted by the IAU. The adoption of the nonrotating origin (Guinot 1979, Captaine et al 1986) would simplify future definitions of precession and models for nutation.

There is also the problem with the definition of time. Special Relativity dictates that there is no independent order of events that is universal at all locations. Therefore the coordinate frame must be looked upon as a four-dimensional space. The position of the Earth-moon barycenter is looked upon as a quasi-inertial space because it contains a "weak" gravitational field in which the effects of General Relativity will be small. Transformations from an Earth-centered geocentric coordinate time frame to the barycentric coordinate time are given in the IERS Conventions.

14. CONCLUSIONS

The ICRF has been established using the positions of 212 extragalactic radio sources whose positions are accurate to one mas with the majority being less than 0.5 mas. The Hipparcos frame is the realization of this frame at optical wavelengths. This has come about as a result of the accuracies achieved by VLBI radio astrometry and the characteristics of the emitting sources. The ICRF establishes the ICRS. However, the definition of the ICRS needs to be refined by updating astronomical constants and models in time, precession, etc to take advantage of this increase in accuracy. This should be accomplished at the next IAU meeting in 2000. Further increases in measurement precision to 4 μas from space missions at optical wavelengths may dictate a redefinition of the ICRF, accompanied by improvement in constants and models for the ICRS.

ACKNOWLEDGMENTS

The authors would like to thank PK Seidelmann, DD McCarthy and A Fey for critical reading of the manuscript, A Fey for assistance with the figures, and L Treadway for preparing the manuscript.

Visit the Annual Reviews home page at http://www.AnnualReviews.org

LITERATURE CITED

Aoki S, Guinot B, Kaplan GH, Kinoshita H, McCarthy DD, et al. 1982. *Astron. Astrophys.* 105:359–61

Allen LR, Anderson B, Conway RG, Palmer HP, Reddish VC, et al. 1962a. *MNRAS* 125: 477–99

Allen LR, Hanburry-Brown R, Palmer HP. 1962b. *MNRAS* 125:57–74

Appenzeller I. 1994. *Highlights of Astronomy: XXII IAU Gen. Assem.* Vol 10. Dordrecht: Kluwer

Argue AN, de Vegt C, Elsmore B, Fanselow J, Harrington R, et al. 1984. *Astron. Astrophys.* 130:191–99

Arias EF, Charlot P, Feissel M, Lestrade JF. 1995. *Astron. Astrophys.* 303:604–08

Arias F, Lestrade JF, Boucher C, Feissel M. 1991. *Astrophys. Space Sci.* 177:187–92

Auwers A. 1879. *Pub. Astr. Ges.* p. 14

Baade W, Minkowski R. 1954. *Ap. J.* 119:206–14

Bare C, Clark BG, Kellermann KI, Cohen MH, Jauncey DL. 1967. *Science.* 157:189–91

Bergeron J, ed. 1992. *Highlights of Astronomy. XXI IAU Gen. Assem.*, Vol. 9. Dordrecht: Kluwer

Bergeron J, ed. 1997. *Highlights of Astronomy. XXIII IAU Gen. Assem.*, Vol. 10. Dordrecht: Kluwer

Bolton JG, Stanley GJ. 1948. *Aust. J. Sci. Res. Ser. A.* 1:58–69

Bross B. 1937. *General Catalog of 33,342 Stars for the Epoch 1950.* Wash. DC: Carnegie Inst. Wash.

Brosche P, Wade CM, Hjellming RM. 1973. *Ap. J.* 183:805–18

Broten NW, Legg TH, Locke JL, McLeish CW, Richards RS et al. 1967. *Nature.* 215: 38

Capitaine N, Guinot B, Souchay J. 1986. *Celest. Mech.* 39:283–307

Carr TD, May J, Olson CN, Walls GF. 1965. *IEEE NEREM Rec.* 7:222

Charlot P. 1994. *Proc. VLBI Technology-Progress and Future Observational Possibilities, Kyoto, Japan* (September 1993), ed. T Sasso, S Manabe, O Kameya, M Inoue, p. 287. Tokyo: Terra Sci. Publ. Co.

Clark TA, Corey BE, Davis JL, Elgered G, Herring TA, et al. 1985. *IEEE Trans. Geosci. Remote Sens.* GE-23:438

Clark TA, Hutton LK, Marandino GE, Counselman CC, Robertson DS, et al. 1976. *Astron. J.* 84:599–603

Cohen MH, Shaffer DB. 1971. *Astron. J. Let.* 76:91–100

Colavita MM, Boden AF, Crawford SL, Meinel AB, Shao M. 1998. *Proc. SPIE 3350.* pp. 776–84

Counselman CC III. 1978. *Annu. Rev. Astron. Astrophys.* 14:197–214

Davis JL, Herring TA, Shapiro II, Rogers AEE, Elgered G. 1985. *Radio Sci.* 20:1593–1607

Eichhorn H. 1974. *Astronomy of Star Positions: A Critical Investigation of Star Catalogues, the Methods of their Construction, and their Purpose.* New York: Ungar

Eubanks TM, Matsakis DN, Josties FJ, Archinal BA, et al. 1995. *Proc IAU Symp. 166*, ed. E. Hog, PK Seidelmann. Dordrecht: Kluwer Academic

European Space Agency. 1997. *The Hipparcos and Tycho Catalogs.* sci. coord. MAC Perryman, ESA SP-1200

Fanselow JK, Sovers OJ, Thomas JB, Purcell GH, Cohen E et al. 1984. *Astron. J.* 89:987–98

Ferraz-Mello S, Morando B, Arlot J-E, eds. 1995. *IAU Symp. No. 172, Dynamics, Ephemerides and Astrometry of the Solar System.* Dordrecht: Kluwer

Fey AL, Charlot P. 1997b. *Ap. J. Suppl.* 111:95–142

Fey AL, Clegg AW, Fomalont EB. 1996. *Ap. J. Suppl.* 105:299–330

Fey AL, Eubanks TM, Kingham KA. 1997a. *Astron. J.* 114:2284–91

Fey AL, et al. 1992. *Astron. J.* 104:891–96

Fey AL, Russell JL, de Vegt C, Zacharias N, Johnston KJ, et al. 1994. *Astron. J.* 107:385–91

Folkner WM, Charlot P, Finger MH, Williams JG, Sovers OJ, et al. 1994. *Astron. Astrophys.* 287:279–89

Fricke W. 1981. Definition and practical realization of the reference frame in the FK5—the role of planetary dynamics and stellar kinematics in the definition. In *Reference Coordinate Systems for Earth Dynamics: Proceedings of the 56th Colloquium of the IAU, Warsaw, Poland*, ed. EM Gapschkin, B Kolaczek, *Astrophys. Space Sci. Lib. Proc.* Dordrecht: Kluwer

Fricke W. 1982. *Astron. Astrophys.* 107:L13–16

Fricke W, Kopff A. 1963. *FK4 Veroffentl. Astron. Rechen-Institut. Heidelberg. No 10.* Germany: G. Braun, Karlsruhe

Fricke W, Schwan H, Corbin T. 1991. *FK5 II Extension. Veroffentl. Astron. Rechen-Institut. Heidelberg. Vol 33.* Germany: G. Braun, Karlsruhe

Fricke W, Schwan H, Lederle T. 1988. *FK5 I Veroffentl. Astron. Rechen-Institut. Heidelberg. Vol 32.* Germany: G Braun, Karlsruhe

Fukushima T. 1997. *IAU Symp. 172, Dynamics, Ephemerides and Astrometry of the Solar System.* pp. 461–68

Gilmore G, Perryman M, Lindegren L, Favata F, Hoeg E. 1998. *Proc. SPIE 3350.* pp. 541–50

Gordon D. 1985. *NASA Intern. Memo.* Greenbelt, MD: Goddard Space Flight Center

Guinot B. 1979. *IAU Symposium No. 82, Time and the Earth's Rotation*, ed. DD McCarthy, JD Pilkington, pp. 7–18. Dordrecht: D. Reidel Publ. Co.

Guirado JC, et al. 1995. *Astron. J.* 110:2586–96

Hazard C, Sutton J, Argue AN, Kenworthy CM, Morrison LV, Murray CA. 1971. *Nature.* 233:89–91

Herring TA. 1992. In *Symposium on Refraction of Transatmospheric Signals in Geodesy*, ed. JC De Munck, TA Spoelstra, p. 157. Delft: Netherlands Geod. Comm.

Hjellming RM, Bignell RC, 1982. *Science.* 216:1279–85

Hog E, Seidelmann PK, eds. 1995. *IAU Symp. No. 166, Astronomical and Astrophysical Objectives of Sub-Milliarcsecond Astrometry.* Dordrecht: Kluwer

Hogg DE, Macdonald GH, Conway RG, Wade CM. 1969. *Astron. J.* 74:1206–13

Hughes JA, Smith CA, Kaplan GH, eds. 1991. *Proc. IAU Coll. 127, Reference Systems.* Wash. DC: USNO

Hummel CA, Mozurkewich D, Elias N, Quirrenbach A, Buscher DF, et al. 1994. *Astron. J.* 108:326–36

International Earth Rotation Service. 1996. *IERS Conv. (1996), IERS Tech. Note 21*, ed. DD McCarthy. Paris: Obs. Paris

International Earth Rotation Service. 1997. *IERS Annu. Rep. (1996).* Paris: Obs. Paris

International Earth Rotation Service. 1997. *Tech. Note 23.* Paris: Obs. Paris

Johnston KJ, Fey AL, Zacharias N, Russell JL, Ma C, et al. 1995. *Astron. J.* 110:880–915

Johnston KJ, Hutter DF, Benson JA, Elias N, Armstrong JT, et al. 1997. *Proc. IAU Symp.189, Fundamental Stellar Properties.* ed. T Bedding, A Booth, J Davis, pp. 39–42. Dordrecht: Kluwer

Johnston KJ, Russell J, de Vegt C, Hughes J, Jauncey DL, et al. 1988. *IAU Symp. 129, The Impact of VLBI on Astrophysics and Geophysics*, ed. M Reid, J Moran, pp. 317–18. Dordrecht: Reidel

Johnston KJ, Russell JL, de Vegt C, Zacharias N, Hindsley R, et al. 1991. *Proc of IAU Coll. No. 127*, ed. J Hughes, CA Smith, GH Kaplan, pp. 123–29. Wash. DC: USNO

Johnston KJ, Seidelmann PK, Reasonberg RD, Phillips JD, Horner S, et al. 1998. *Pap. USNO*, Wash. DC: USNO

Kopff A. 1937. *Dritter Fundamental-Katalog Part I Veroffentl. Astron. Rechen-Institut. zu Berlin-Dahlem No 54.* Berlin: Ferd. Dummlers

Kovalevsky J. 1998. *Annu. Rev. Astron. Astrophys.* 36:99–129

Kovalevsky J, Lindegren L, Perryman MAC, Hemenway PD, Johnston KJ, et al. 1997. *Astron. Astrophys.* 323:620–33

Kovalevsky J, Mueller II. 1981. *Proc. IAU Coll. 56*

Kovalevsky J, Mueller II, Kolaczek B. 1989. *Reference Frames in Astronomy and Geophysics*. APSC. Lib. 154. Dordrecht: Kluwer

Lestrade JF, Jones DL, Preston RA, Phillips RB, Titus MA. 1995. *Astron. Astrophys.* 304:182–88

Lieske JH, Lederle T, Fricke W, Morando B. 1977. *Astron. Astrophys.* 58:1–16

Lindegren L, Kovalevsky J. 1995. *Astron. Astrophys.* 304:189–201

Ma C. 1978. *PhD dissertation. NASA TM 79582*. Greenbelt, MD: Goddard Space Flight Center

Ma C, Arias EF, Eubanks TM, Fey AL, Gontier AM, et al. 1998. *Astron. J.* 116:516–46

Ma C, Clark TA, Ryan JW, Herring TA, Shapiro II, et al. 1986. *Astron. J.* 92:1020–29

Ma C, Shaffer DB, de Vegt C, Johnston KJ, Russell JL. 1990. *Astron. J.* 99:1284–98

Ma C, Shaffer DB. 1991. *IAU Coll. 127*, ed. JA Hughes, CA Smith, GH Kaplan, pp. 135–44. Wash. DC: USNO

MacMillan DS, Ma C. 1997. *Geophys. Res. Lett.* 24:453–56

Matthews TA, Sandage A. 1963. *Ap. J.* 138:30–56

McCarthy DD, Luzum BJ. 1991. *Astron. J.* 102:1889–95

Meeks ML. 1976. Ed. *Astrophysics: Radio Observations, Vol. 12, Part C, Methods of Experimental Physics*, pp. 531–69. New York: Academic Press

Melbourne W, Anderle R, Feissel M, King R, McCarthy D, et al. 1983. *USNO Circ. 167*. Wash. DC: USNO

Menten KM, Reid MJ, Eckart A, Genzel R. 1997. *Ap. J. Lett.* 475:111–14

Mignard F, Froeschle M. 1997. *Proc. ESA Symp. Hipparcos-Venice 1997*. ESA SP-402. pp. 57–60

Monet DG. 1988. *Annu. Rev. Astron. Astrophys.* 26:413–40

Morabito DD, Niell AE, Preston RA, Linfield RP, Wehrle AE, Faulkner J. 1986a. *Astron. J.* 91:1038–50

Morabito DD, Preston RA, Linfield RP, Slade MA, Jauncey DL. 1986b. *Astron. J.* 92:546–51

Moran JM, Crowther PP, Burke BF, Barrett AH, Rogers AEE, et al. 1967. *Science* 157:676–77

Morrison LV, Garrington ST, Argyle RW, Davis RJ. 1997. *Proc. ESA Symp. Hipparcos-Venice 1997*. ESA SP-402. pp. 143–45

Mueller II. 1969. *Spherical and Practical Astronomy*. New York: F Ungar Publ.

Newcomb S. 1905a. *Astronomical Papers of the American Ephemeris and Nautical Almanac. Vol 8, Part 2*. Wash. DC: Nautical Almanac Office, USNO

Newcomb S. 1905b. *Astronomical Papers of the American Ephemeris and Nautical Almanac. Vol 8, Part 1*. Wash. DC: Nautical Almanac Office, USNO

Niell AE. 1996. *J. Geophys. Res.* 101(B2):3227–46

Oke JB. 1963. *Nature* 197:1040–41

Palmer HP. 1962. *IAU Symposium No. 15, Problems of Extragalactic Research*, ed. GC McVittie, pp. 315–25. New York: Macmillan

Peck AB, Beasley AJ. 1998. *IAU Coll. 164*, ed. JA Zensus, GB Taylor, JM Wrobel. ASP Conf. Ser. Vol. 144:155–56

Reynolds JE, Jauncey DL, Russell JL, King EA, McCulloch PM, et al. 1994. *Astron. J.* 108:725–30

Rioja MJ, Marcaide JM, Elodegui P, Shapiro II. 1997. *Astron. Astrophys.* 325:383–90

Robertson DS. 1975. *Geodetic and Astrometric Measurements with Very Long Baseline Interferometry*. PhD thesis. MIT

Robertson DS, Fallon FW, Carter WE. 1986. *Astron. J.* 91:1456–62

Rogers AEE. 1970. *Radio Sci.* 5:1239–47

Rogers AEE, Cappallo RJ, Hinteregger HF, Levine JL, Nesman EF, et al. 1983. *Science* 219:51–53

Roser S, Bastian U, de Boer KS, Hog E, Roser

HP, et al. 1997. *Proc. ESA Symp. Hipparcos-Venice 1997*. ESA-SP 402. pp. 777–82
Russell JL, Johnston KJ, Ma C, Shaffer D, de Vegt C. 1991. *Astron. J.* 101:2266–73
Russell JL, et al. 1992. *Astron. J.* 103:2090–98
Russell JL, et al. 1994. *Astron. J.* 107:379–84
Ryan JW, Ma C, Caprette DS. 1993. *NASA Tech. Mem. 104572*. Greenbelt, MD: NASA
Ryle M, Smith FG. 1948. *Nature* 162:462–63
Ryle M. 1972. *Nature* 239:435–38
Schmidt M. 1963. *Nature* 197:1040
Seidelmann PK. 1982. *Celest. Mech.* 27:79–106
Seidelmann PK. 1992. Ed. *Explanatory Supplement to the Astronomical Almanac*. Mill Valley, CA: Univ. Sci. Books
Shao M. 1998. *Proc. SPIE 3350*. pp. 536–40
Soffel MH. 1989. *Relativity in Astrometry, Celestial Mechanics, and Geodesy*. Berlin: Springer-Verlag
Sovers OJ, Edwards CD, Jacobs CS, Lanyi KM, Liewer KM, Treuhaft RN. 1988. *Astron. J.* 95:1647–58
Sovers OJ, Fanselow JL. 1987. *JPL Publ. 83–39 Rev. 3*. Pasadena: Jet Propulsion Laboratory
Sovers OJ, Fanselow JL, Jacobs CS. 1998. *Rev. Mod. Phys.* 70:1393–1454
Sovers OJ, Ma C. 1985. *EOS*. 66:858 (Abstr.)
Standish EM. 1981. *Astron. Astrophys.* 101:L17–L18
Standish EM. 1982. *Astron. Astrophys.* 114:297–302
Standish EM, Newhall XX, Williams JG, Folkner WF. 1995. *JPL IOM*. 314:10–127
Stone RC, Dahn CC. 1995. *IAU Symp. 166, Astronomical and Astrophysical Objectives of Sub-milliarcsecond Astrometry*, ed. E Hog, PK Seidelmann, P Dordrecht. pp. 3–8. Dordrecht: Kluwer
Thompson AR, Moran JM, Swenson GW Jr. 1986. *Interferometry and Synthesis in Radio Astronomy*. New York: Wiley
van Altena WF. 1983. *Annu. Rev. Astron. Astrophys.* 21:131–64
Vondrak J, Capitaine N. 1998. *Journees 1997, Reference Systems and Frames in the Space Era, Present and Future Astrometric Programs*. Paris: Obs. Paris
Wade CM. 1970. *Ap. J.* 162:381–90
Wade CM, Johnston KJ. 1977. *Astron. J.* 82:791–95
Wielen, et al. 1997. *Proc. ESA Symp. Hipparcos-Venice 1997*. ESA SP-402. pp. 727–32
Woolard EW, Clemence GM. 1966. *Spherical Astronomy*. New York: Academic Press
Wytrzyszczak IM, Lieske JH, Feldman RA, eds. 1996. Ed. *IAU Coll. No. 165, Dynamics and Astrometry of Natural and Artificial Celestial Bodies*. Dordrecht: Kluwer
Zacharias N, de Vegt C, Murray CA. 1997. *Proc. ESA Symp. Hipparcos-Venice 1997*. ESA SP-402. pp. 177–80
Zensus JA, Diamond PJ, Napier PJ. 1995. *Very Long Baseline Interferometry and the VLBA. ASP Conf. Proc. 82*. San Francisco: ASP

PROBING THE UNIVERSE WITH WEAK LENSING

Yannick Mellier[1,2]

[1]*Institut d'Astrophysique de Paris, 98 bis Boulevard Arago, 75014 Paris, France.*
[2]*Observatoire de Paris, DEMIRM, 61 avenue de l'Observatoire, 75014 Paris, France; E-mail: mellier@iap.fr*

Key Words cosmology, gravitational lensing, dark matter, clusters of galaxies, evolution of galaxies

■ **Abstract** Gravitational lenses can provide crucial information on the geometry of the Universe, on the cosmological scenario of formation of its structures as well as on the history of its components with look-back time. In this review, I focus on the most recent results obtained during the last five years from the analysis of the weak lensing regime. The potential of weak lensing as a probe of dark matter and the study of the coupling between light and mass on scales of clusters of galaxies, large-scale structures and galaxies is discussed first. Then I present the impact of weak lensing for the study of distant galaxies and of the population of lensed sources as a function of redshift. Finally, I discuss the potential of weak lensing to constrain the cosmological parameters, either from pure geometrical effects observed in peculiar lenses, or from the coupling of weak lensing with the CMB.

1. INTRODUCTION

Matter intervening along the light paths of photons causes a displacement and a distortion of ray bundles. The properties and the interpretation of this effect depend on the projected mass density integrated along the line of sight and on the cosmological angular distances to the observer, the lens and the source.

The sensitivity to mass density implies that gravitational lensing effects can probe the mass of deflectors, without regard to their dynamical stage and the nature of the deflecting matter. This is therefore a unique tool to probe the dark matter distribution in gravitational systems as well as to study the dynamical evolution of structures with redshift. The dependence on the various angular distances involved in the lens configuration means that the deviation angle depends on the cosmological parameters, H_o, Ω and λ, so that the analysis of gravitational lensing can potentially provide a diagnosis on cosmography. Of course, the sensitivity to cosmological parameters is not unique to gravitational lensing, and many other astrophysical phenomena depend on them. However, owing to magnification, image multiplicity and deflection angle produced by lensing, it is possible to use the lensing effect as a bonus when compared with other experiments: image magnification permits observation of the high-redshift universe, study of the evolution of galaxies with look-back time and comparison with theoretical cosmological scenarios. Image multiplicity probes different light paths taken by photons emitted by one source. By computing time delays of the same transient event observed in each individual image, one can measure H_o. Finally, for high-redshift sources the deflecting angle depends on the geometry of the universe and provides a unique tool for measuring the cosmological parameters.

The interest in gravitational lensing for cosmology started very early, after Zwicky's discovery (Zwicky 1933) of the apparent contradiction between the visible mass of the Coma cluster and its virial mass, which could not be explained without recognizing that it is dominated by *unseen mass*. This surprising statement could not be confirmed without an independent mass estimator, which could probe the total mass directly, without using the light distribution or critical assumptions on the dynamical stage of the cluster components. Four years later, Zwicky (1937) envisioned that *extragalactic nebulae* could be efficient gravitational lenses and provide an invaluable tool for weighting the gravitational systems of the Universe.

The other works that raised interest in lensing for cosmology are more contemporary. Refsdal (1964) first emphasized that time delays in multiple images could be used to measure H_o, and the very first considerations of light propagation and deformation of ray bundles in inhomogeneous universes were discussed initially by Sachs (1961) and Zel'dovich (1964) and later by Gunn (1967). From an observational point of view, the discoveries of the first multiply imaged quasar (Walsh et al 1979) and the first distorted galaxies (Soucail et al 1987, Lynds & Petrosian 1986) were major steps that boosted theoretical and observational investigations of gravitational lenses.

Most of the cosmological interest in gravitational lenses has already been reviewed by Blandford & Narayan (1992), Schneider et al (1992) and Refsdal & Surdej (1994). Fort & Mellier (1994) presented the first review which focused particularly on the use of arc(let)s in cosmology, and the interest in the use of lensed galaxies to probe the deep universe has been recently reviewed by Ellis (1997). With the amazing observational and theoretical developments in the field, in particular in weak lensing, it seems timely to review all these results and to address the new and future issues in the area.

During the last five years, thanks to the seminal work on mass reconstruction from weak lensing analyses (Tyson et al 1990, Kaiser & Squires 1993), mass reconstruction algorithms have provided new and robust tools for studying the mass distribution of gravitational systems and have permitted the establishment of a link between theoretical investigations of weak lensing and the observations of weakly distorted galaxies. In particular, there have been impressive developments in cosmological diagnoses from the analysis of weak lensing induced by large-scale structures. Theoretical and numerical studies demonstrate that the statistical analysis of gravitational lensing will provide valuable insights on the mass distribution as well as on the cosmological parameters. With the coming of new wide field surveys with subarcsecond seeing [such as Megacam at the Canada-France-Hawaii Telescope (CFHT) or the VLT-Survey-Telescope (VST) at Paranal] or very wide field shallow surveys [such as the VLA-Faint Images of the Radio Sky at Twenty-Centimeters (FIRST) survey or the Sloan Digital Sky Survey (SDSS)], weak-lensing analysis should probe the power spectrum of the projected mass density, from arcminutes up to degree scales. Visible weak lensing surveys should also be capable of providing a projected mass map of the universe, just as the Automated Plate Machine (APM) survey provides the visible light distribution (Maddox et al 1990). From the observational point of view, the outstanding images coming from Hubble Space Telescope (HST) had a considerable positive impact on our intuitions about the potential usefulness of gravitational distortion. The wonderful shear pattern around lensing cluster A2218 is visual proof that weak lensing works and that it directly reveals the mass distribution. One of the most spectacular uses of HST images for lensing was done by Kneib et al (1996), also in A2218. The superb HST images allowed them to demonstrate, from the morphology of only one arclet, and without the need of a spectroscopic redshift, that it must be a lensed image associated with the same source as the giant arc. The similarity of the morphologies of the giant arc and the counter-image is so impressive that it cannot be questioned that they are images of the same source. In parallel, the Keck telescope, which is currently detecting the most distant galaxies, reveals the obvious importance of *giant gravitational telescopes*. Finally, the impressive results obtained by the Submillimetre Common-User Bolometer Array (SCUBA) in the submillimeter wavebands have shown that the joint use of a submillimeter instrument with magnification of high-redshift galaxies is an ideal tool for studying the evolution and content of distant galaxies.

In the following I review most of these recent works and discuss their impact for cosmology. Although this review focuses on weak lensing, the distinction between arclets and the weak lensing regime is somewhat arbitrary, and both are relevant for our purpose. Furthermore, because some of the results cannot be discussed without referring to strong lensing, I often include new results from arcs and multiple image studies. Section 2 recalls the basic equations useful in gravitational lensing which help in the understanding of this review. The definitions for strong lensing cases are not presented again, and I will refer to the review by Fort & Mellier (1994) for all these aspects. In Section 3 I focus on the mass distribution in clusters of galaxies from arc(let)s or mass reconstruction from weak lensing inversion. I also address the issues concerning the measurement of weak shear because it appears to be a major challenge for observers. Section 4 presents weak lensing induced by large-scale structures, and Section 5 presents weak lensing induced by foreground galaxies on the background sources (the so-called galaxy-galaxy lensing analysis). I then move toward the high-redshift universe in Section 6. Sections 7 and 8 are devoted to cosmological parameters and weak lensing on the cosmological microwave background (CMB), respectively. Conclusions and future prospects are discussed in the last section.

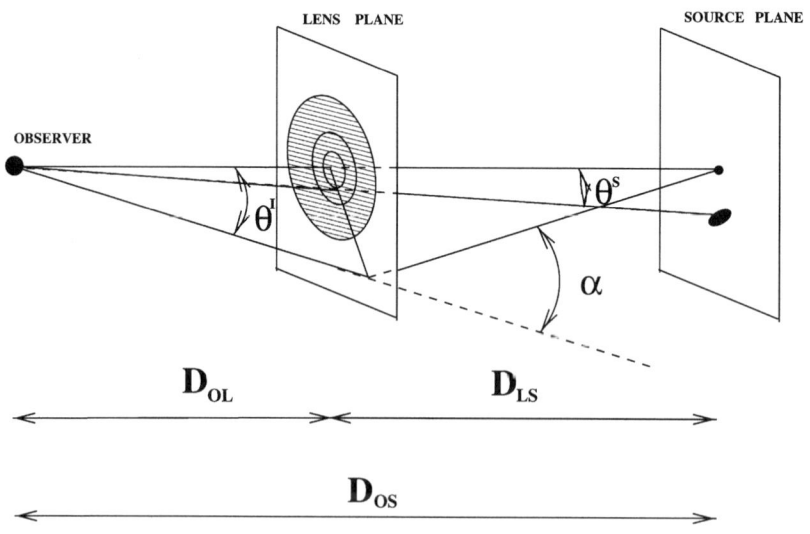

Figure 1 Description of a lensing configuration.

2. DEFINITIONS

2.1 Lensing Equations

In this preliminary section, I do not discuss at length the theoretical basis of the gravitational lens effect because all the details can be found in the comprehensive textbook written by Schneider et al (1992). I focus on concepts and basic equations of the gravitational lensing theory, in the thin lens approximation and for small deviation angles, which are necessary for this review.

The apparent angular position of a lensed image, θ^I (in this review, bold symbols denote vectors), can be expressed as a function of the (unlensed) angular position of the source, θ^S, and the deflection angle, $\alpha(\theta^I)$ as follows (see Figure 1):

$$\theta^I = \theta^S + \frac{D_{LS}}{D_{OS}} \alpha(\theta^I). \tag{1}$$

$\alpha(\theta^I)$. depends on the projected mass density of the lens, $\Sigma(\theta^I)$, and the cosmological parameters through the angular-diameter distances from the lens L to the source S, D_{LS}, from the observer o to the source, D_{OS}, and from the observer to the lens, D_{OL}:

$$\alpha(\theta^I) = \frac{4\pi G}{c^2} D_{OL} \frac{1}{\pi} \int \Sigma(\theta') \frac{\theta^I - \theta'}{|\theta^I - \theta'|^2} d^2\theta', \tag{2}$$

where G is the gravitational constant and c is the speed of light. $\Sigma(\theta^I)$ can be expressed as a function of the Poisson equation, and the strength of the lens is characterized by the ratio of the projected mass density of the lens to its critical projected mass density Σ_{crit} (see Fort & Mellier 1994):

$$\frac{\Sigma(\theta^I)}{\Sigma_{crit}} = \frac{4\pi G}{c^2} \frac{D_{LS} D_{OL}}{D_{OS}} \Sigma(\theta^I) = \frac{1}{2}\Delta\varphi(\theta^I), \qquad (3)$$

where Δ is the 2-dimension Laplacian and φ is the dimensionless gravitational potential projected along the line of sight which is related to the projected gravitational potential Φ as follows:

$$\varphi = \frac{2}{c^2} \frac{D_{LS} D_{OL}}{D_{OS}} \Phi. \qquad (4)$$

From the differentiation of Equation (1), we can express the deformation of an infinitesimal ray bundle as a function of the Jacobian

$$\frac{d\theta^S}{d\theta^I} = A(\theta^I), \qquad (5)$$

where $A(\theta^I)$ is the *magnification matrix*:

$$A = \begin{pmatrix} 1 - \partial_{11}\varphi & \partial_{12}\varphi \\ \partial_{12}\varphi & 1 - \partial_{22}\varphi \end{pmatrix}. \qquad (6)$$

It can be written as a function of two parameters (similar to the magnification and the astigmatism terms in classical optics), the *convergence*, κ, and the *shear* components γ_1 and γ_2 of the complex shear $\gamma = \gamma_1 + i\gamma_2$:

$$A = \begin{pmatrix} 1 - \kappa - \gamma_1 & \gamma_2 \\ \gamma_2 & 1 - \kappa + \gamma_1 \end{pmatrix}. \qquad (7)$$

The isotropic component of the magnification, $\kappa = 1/2\Delta\varphi(\theta^I)$, is directly related to the projected mass density, and the two components γ_1 and γ_2 describe an anisotropic deformation produced by the tidal gravitational field. The eigenvalues of the magnification matrix are $1 - \kappa \pm |\gamma|$, where $|\gamma| = \sqrt{\gamma_1^2 + \gamma_2^2}$. They provide the elongation and the orientation produced on the images of lensed sources. The magnification of an image is:

$$\mu = \frac{1}{det(A)} = \frac{1}{(1-\kappa)^2 + |\gamma|^2}. \qquad (8)$$

The points of the image plane where $det(A) = 0$ are called the critical lines. The corresponding points of the source plane are called the caustic lines and produce infinite magnification (see Schneider et al 1992; Blandford & Narayan 1992; Fort & Mellier 1994 for more detailed descriptions of caustic and critical lines). The strong lensing cases correspond to configurations where sources are close to the caustic lines. These lenses have $\Sigma(\theta^I)/\Sigma_{crit} \geq 1$ and the convergence and shear are strong enough to produce giant arcs and multiple images for suitably positioned sources (Figures 2 and 3). The weak lensing regime,

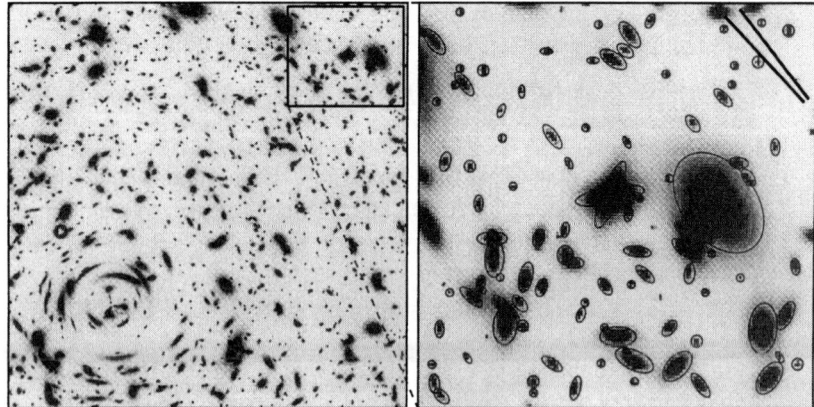

Figure 2 Illustration of the two lensing regimes. The left panel is a simulation of a cluster of galaxies at redshift 0.15, modeled by an isothermal sphere with a velocity dispersion of 1300 kms^{-1}. The lensed population has an average redshift of one. In the innermost region (*bottom left part of the panel*), tangential and radial arcs are clearly identified. As the radial distance of lensed galaxies increases, the shear decreases, and far from the cluster center, the ellipticity produced by the shear is lower than the intrinsinc ellipticity of the galaxies. The lensing signal must be averaged over a large number of galaxies in order to be measured accurately. The zoom on the right panel shows the images of the galaxies in the weak lensing regime. The contours show their shape as determined from their second moments. The average orientation of these galaxies is given by the solid lines at the top right. The lower line is the true orientation of the shear produced by the cluster at that position and the upper line is the orientation computed from 92 galaxies of the zoomed area. The difference between the two orientations is random noise due to the intrinsinc ellipticity and orientation distributions of the galaxies.

which is the main topic of this review, corresponds to lensing configurations where $\kappa \ll 1$ and $|\gamma| \ll 1$. In this regime, the magnification and the distortion of background galaxies are so small that they cannot be detected on individual objects. In that case, it is necessary to analyze statistically the distortion of the lensed population.

2.2 Relation with Observable Quantities

Let us assume that, to first approximation, faint galaxies can be described as ellipses. Their shape can be expressed as a function of their weighted second moments which fully define the properties of an ellipse,

$$M_{ij} = \frac{\int S(\boldsymbol{\theta}) \left(\theta_i - \theta_i^C\right)\left(\theta_j - \theta_j^C\right) d^2\theta}{\int S(\boldsymbol{\theta}) 2\theta}, \quad (9)$$

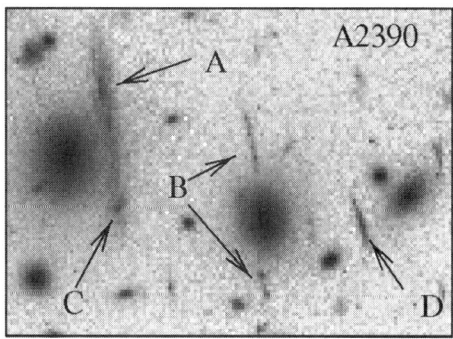
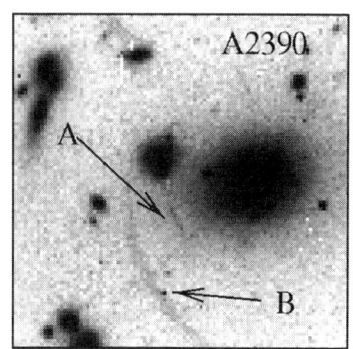
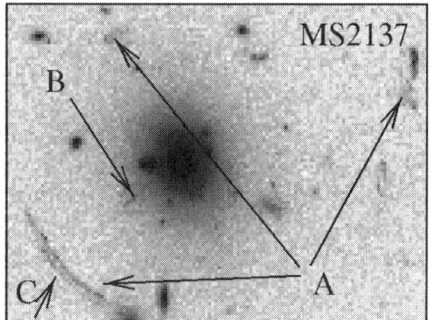
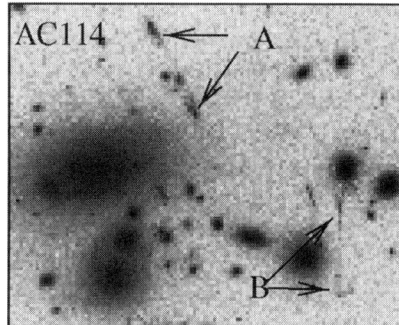
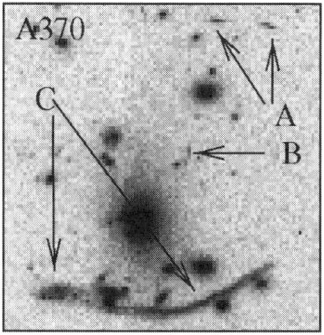
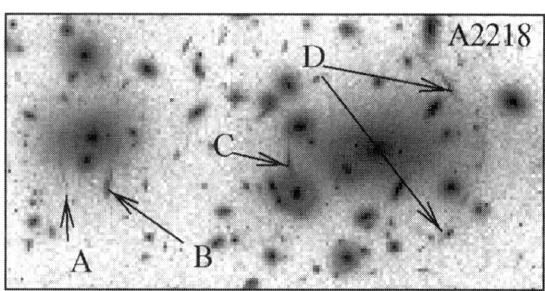

Figure 3 A panel of lensing clusters observed with HST. The arc(let)s and multiple lensed images are indicated by a letter. In A2390 (*top left*), the straight arc is made of two different galaxies corresponding to images A and C. The pairing of some images is obvious, like B in A2390 (*top-left*), A in AC114, or A in A370. Image B in MS2137 and B in A370 are radial arcs. A in MS2137 is a triple image from an almost ideal configuration of a fold caustic.

where the subscripts ij denote the axes (1, 2) of coordinates θ in the source and the image planes, $S(\theta)$ is the surface brightness of the source and θ^C is the center of the source.

Since the surface brightness of the source is conserved through the gravitational lensing effect (Etherington 1933), it is easy to show that, if one assumes that the magnification matrix is constant across the image (lensed source), the relation between the shape of the source, M^S and the lensed image, M^I is

$$M^I = A^{-1} M^S A^{-1}. \tag{10}$$

Therefore, to first approximation, the gravitational lensing effect on a circular source changes its size (magnification) and transforms it into an ellipse (distortion) with axis ratio given by the ratio of the two eigenvalues of the magnification matrix. The shape of the lensed galaxies can then provide information about these quantities. The approximation that the magnification matrix is constant over the image area is always valid in the weak-lensing regime, because the spatial scale variation of the magnification is much larger than the typical size of the lensed galaxies (a few arcseconds). This is not the case when the magnification tends to infinity, but this case is beyond the scope of this review (see Schneider et al 1992 and Fort & Mellier 1994).

The relation between the lens quantities described in Section 2.1 and the shape parameters of lensed galaxies is not immediately apparent. Although γ_1 and γ_2 describe the anisotropic distortion of the magnification, they are not directly related to observables (except in the weak-shear regime). It is preferable to use the *reduced complex shear*, **g**, and the complex polarization (or distortion), δ, which is an observable,

$$\mathbf{g} = \frac{\gamma}{(1-\kappa)}; \quad \delta = \frac{2g}{1+|\mathbf{g}|^2} = \frac{2\gamma(1-\kappa)}{(1-\kappa)^2+|\gamma|^2}, \tag{11}$$

because δ can be expressed in terms of the observed major and minor axes a^I and b^I of the image, I, produced by a circular source S:

$$\frac{a^2-b^2}{a^2+b^2} = |\delta|. \tag{12}$$

In this case, the two components of the complex polarization are easily expressed with the second moments:

$$\delta_1 = \frac{M_{11}-M_{22}}{Tr(M)}; \quad \delta_2 = \frac{2M_{12}}{Tr(M)}, \tag{13}$$

where $Tr(M)$ is the trace of the magnification matrix. For non-circular sources, from Equations (8) and (11) it is possible to relate the ellipticity of the image ϵ^I to the ellipticity of the lensed source, ϵ^S. In the general case, it depends on the sign of $Det(A)$ (that is the position of the source with respect to the caustic lines) which expresses whether images are radially or tangentially elongated. In most cases of interest, $Det(A) > 0$ (the external regions, where the weak lensing regime applies) and:

$$\epsilon^I = \frac{1+b^I/a^I}{1+b^I/a^I} e^{2i\vartheta} = \frac{\epsilon^S+\mathbf{g}}{1-\mathbf{g}^* \epsilon^S} \tag{14}$$

(Seitz & Schneider 1996), but when $Det(A) < 0$:

$$\epsilon^I = \frac{1 + \epsilon^{S*}\mathbf{g}}{\epsilon^{S*} + \mathbf{g}^*}. \tag{15}$$

Equations 14 and 15 summarize most of the cases that will be discussed in this review.

3. MASS DISTRIBUTION IN CLUSTERS OF GALAXIES

3.1 Mass Reconstruction with Arclets

Recent Developments The use of arc(let)s and multiple images has already been discussed in detail in Fort & Mellier (1994). In the meantime, with the refurbishment of the HST, spectacular images of arc(let)s and multiply imaged galaxies have permitted enormous progress in this field. It turns out that giant arcs are no longer the strongest constraints on cluster mass distribution, because similar and even better information can be obtained with spatially resolved HST images of arclets.

The usual mass reconstruction technique with arc(let)s present in the innermost regions (close to the critical lines where arcs are located) is based on the assumption that the cluster mass density is smoothly distributed and can be expressed analytically, possibly with addition of some substructures, and on the hypothesis that the observed arc(let)s correspond to rather generic lens configurations, such as folds, cups or lisp caustics. These assumptions have already provided some convincing results with the use of ground-based images, in particular with predictions of the position of additional images (counter images) associated with arc(let)s (Hammer & Rigaud 1989; Mellier et al 1993; Kneib et al 1993, 1995). In all cases, it was found that the core radius of the dark matter distribution is small ($<50h_{100}^{-1}$ kpc) and that its geometry is compatible with the faint extended envelopes of light surrounding the giant cluster galaxies. The new investigations using the detailed morphology of the numerous arc(let)s visible in the HST images (see Figure 3) have provided more refined constraints on the dark matter distribution on the 100 kpc-scale (Hammer et al 1997, Gioia et al 1998, Kneib el al 1996, Tyson et al 1998). They confirm the trends inferred from previous ground-based data.

One critical issue concerning the approach described above is the possible sensitivity of the result to the analytical mass profile used for the modeling. Because none of the mass distribution has an unrealistic shape, the large-scale global property is expected to be well described. However, a direct comparison of the detailed mass distribution with theoretical expectations seen in simulations is difficult. Furthermore, the redshift distribution inferred from the lensing inversion (see section 5) can be strongly affected by the properties of the analytical model. AbdelSalam et al (1998a,b) have recently proposed a non-parametric mass reconstruction algorithm which helps to overcome the limitation of analytical modeling. The technique uses arc(let)s with known redshifts as strong constraints to recover a pixelised mass map of the lensing-cluster. The pixel-mass reconstruction uses the smoothed projected light distribution of the galaxy distribution which is then pixelised exactly like the projected mass map. A fit of a pixelised Mass-to-Light ratio (M/L) permits one to relate the projected light distribution to the projected mass distribution for each pixel. The results found for A370 and A2218 (for A2218, weak and strong lensing features are used) are similar to those obtained otherwise, but this approach appears to be a very interesting alternative that permits a complete lens modeling based only on arc(let)s properties. Dye & Taylor (1998) attempted to generalize this approach in order to compute the convergence and the shear, in the weak lensing regime.

The X-ray/Lensing Mass Discrepancy The use of X-ray and optical images (ground-based or from the HST) of arc(let)s reveals that the X-ray peaks are located at the center of the most massive clumps of dark matter (Kneib et al 1995, Pierre et al 1996, AbdelSalam et al 1998a, Gioia et al 1998, Hammer et al 1997, Kneib el al 1996, Ota et al 1998, Kneib et al in preparation). On the other hand, the apparent contradiction between the mass estimated from X-ray data and the lensing mass ($M_{lensing} \approx 2\text{--}3\, M_X$), initially raised by Miralda-Escudé & Babul (1995), is not totally clear. The puzzling results obtained on several clusters, sometimes on the same cluster but analyzed by different groups, have not yet provided conclusive statements about the mass density profile and the X-ray versus dark matter dynamics. Böhringer et al (1998) find an excellent agreement between X-ray and lensing masses in A2390 which confirms the view claimed by Pierre et al (1996); Gioia et al (1998) show that the disagreement reaches a factor of 2 at least in MS0440+0204; Schindler et al (1997) find a factor of 2-3 discrepancy for the massive cluster RX 1347.4-1145, but Sahu et al (1998) claim that the disagreement is marginal and may not exist; Ota et al (1998) and Wu & Fang (1997) agree that there are important discrepancies in A370, Cl0500-24 and Cl2244-02.

As yet, there are no definitive interpretations of these contradictory results. It could be that the modeling of the gravitational mass from the X-ray distribution is not as simple. By comparing the geometry of the X-ray isophotes of A2218 to the mass isodensity contours of the reconstruction, Kneib et al (1995) found significant discrepancies in the innermost parts. The numerous substructures visible in the X-ray image have orientations which do not follow the projected mass density. They interpret these features as shocks produced by the infalling X-ray gas, which implies that the current description of the dynamical stage of the inner X-ray gas is oversimplified (see Markevitch 1997 and Girardi et al 1997 for similar views). Recent ASCA observations of three lensing-clusters corroborate the view that substructures are the major source of uncertainties (Ota et al 1998).

To study this possibility in more detail, Smail et al (1997a) and Allen (1998) have performed a detailed comparison between the lensing mass and X-ray mass for a significant number of lensing clusters. Both works conclude that the substructures have a significant impact on the estimate of X-ray mass. More remarkably, the X-ray clusters where cooling flows are present do not show a significant discrepancy with X-ray mass, whereas other X-ray clusters do (Allen 1998). This confirms that the discrepancy is certainly due to wrong assumptions on the physical state of the gas. These two studies provide strong presumptions that we are now close to understanding the origin of the X-ray and lensing discrepancy.

An alternative has been suggested by Navarro et al (1997) who proposed that the analytical models currently used for modeling mass distributions may be inappropriate. Instead, they argue that the universal profile of the mass distribution produced in numerical simulations of hierarchical clustering may reconcile the lensing and X-ray masses. This is an attractive possibility because the universal profile is a natural outcome from the simulations that do not use external prescriptions. However, Bartelmann (1996) emphasized that the caustics produced by the universal profile predict that radial arcs should be thicker than observed in MS2137-23 (Fort et al 1992; Mellier et al 1994; Hammer et al 1997) and in A370 (Smail et al 1995b), unless the sources are very thin (≈ 0.6 arcsecond for MS2137-23). This is not a strong argument against the universal profile because this is possible in view of the shapes of some faint galaxies observed with HST that some distant galaxies are indeed very thin. However, it is surprising that no radial arcs produced by "thick galaxies" have been detected so far. Even a selection bias would probably favor the observation of large sources rather than small, thin and hardly visible ones. Evans & Wilkinson (1998) have explored the range of slopes of cusp-like mass profiles which would produce radial arcs with thicknesses similar to those observed. As found by Bartelmann, they too found that the universal profile does not work well, but that a more singular mass profile could be satisfactory. They do not mention, however, whether these new profiles are compatible with the numerical simulations of Navarro et al (1997).

Probing the Clumpiness of Clusters HST images have also revealed the clumpiness of the cluster mass distribution on small scales. Although most of the HST images of lensing-clusters show arc(let)s with a coherent polarization on scales of 100 kpc, numerous perturbations are visible on scales of about 10 kpc. The long-range pattern is disrupted around most of the bright galaxies and shows saddle-shaped configurations as expected for clumpy mass distributions. In some extreme cases, giant arcs appear as broken filaments, probably disrupted by the halos around the brightest galaxies. With so much detail, one can therefore make a full mass reconstruction which takes into account all these clumps and possibly constrains the mass of individual cluster galaxies. For giant arcs, this has already been stressed by Kassiola et al (1992a), Mellier et al (1993), Kneib et al (1993), Dressler et al (1994), Wallington et al (1995) and Kneib et al (1996). They used the breaks (or the absence of breaks) in arcs to put upper limits to the masses of a few cluster galaxies which are superimposed on the arcs. The masses found for these cluster galaxies range between 10^{10} $M_\odot$ and 2×10^{11} $M_\odot$, with typical mass-to-light ratios between 5 and 15.

With the details visible on the HST images of arclets in A2218, AC114 or A2390, the sample of halos which can be constrained by this method is much larger and can provide more significant results. The number of details also permits use of more sophisticated methods of investigation. The most recent procedure uses the galaxy-galaxy lensing analysis. This technique is described in Section 5, but because the clumpiness of dark matter in clusters is strongly related to the halos of cluster galaxies, I present the use of galaxy-galaxy lensing for cluster galaxies in this section.

The simplest strategy is to start with an analytical potential that reproduces the general features of the shear pattern of HST images, and in a second step, to include analytical halos around the brightest cluster members in the model. In practice, additional mass components are put in the model in order to interpret the arc(let)s which cannot be easily explained by the simple mass distribution. Some guesses are made in order to pair unexplained multiple images. The colors of the arc(let)s as well as their morphology help to make these associations. This approach was proposed by Natarajan & Kneib (1997), and Natarajan et al (1998). The detailed study done in AC114 by Natarajan et al indicates that about 10% of the dark matter is associated with halos of cluster galaxies. These halos have truncation radii smaller than field galaxies ($r_t \approx 15$ kpc) with a general trend of S0-galaxies to be even more truncated than the other galaxies. If this result is confirmed, it would be a direct evidence that truncation by tidal stripping is very efficient in rich clusters of galaxies. This result is somewhat contradictory with the absence of a clear decrease of rotation curves of spiral galaxies in nearby clusters (Amram et al 1993) which is interpreted as a proof that massive halos of galaxies still exist in cluster galaxies. However, it could be explained if the spirals that have been analyzed only appear to be in the cluster center because of projection effects, but actually are not located in the very dense region of the clusters where stripping is efficient.

Geiger & Schneider (1998, 1999) used a maximum likelihood analysis that explores simultaneously the distortions induced by the cluster as a whole and by its individual galaxies. They applied this analysis to the HST data of Cl0939+4713 and reached conclusions similar to those of Natarajan et al (1998). Several issues limit the reliability of their analyses and of the other methods as well (Geiger & Schneider 1998). First, depending on the slope of the mass profile of the cluster, the contributions of the cluster mass density and of the cluster galaxies may be difficult to separate. Second, it is necessary to have a realistic model for the redshift distribution of the background and foreground galaxies. Finally, the mass sheet degeneracy (see Section 3.2) is also an additional source of uncertainties. Regarding these limitations, Geiger & Schneider discuss the capability of the galaxy-galaxy lensing in clusters to provide valuable constraints on the galactic halos from the data they have in hand. Indeed, some of the issues they raised can be solved, such as the redshift distribution of the lensed galaxies. It would be interesting to take a more detailed look at how the analysis could be improved with more and better data.

3.2 Mass Reconstruction from Weak Lensing

A powerful and complementary way to recover the mass distribution of lenses has been proposed by Kaiser & Squires (1993). It is based on the distribution of weakly lensed galaxies rather than the use of giant arcs. In 1988, Fort et al (1988) obtained at CFHT deep sub-arcsecond images of the lensing-cluster A370 and observed the first weakly distorted galaxies ever detected. The galaxy number density of their observation was about 30 arcmin^{-2}, mostly composed of background sources, far beyond the cluster. These galaxies lensed by the cluster show a correlated distribution of ellipticity/orientation which maps the projected mass density. The first attempt to use this distribution of arclets as a probe of dark matter was made by Tyson et al (1990), but the theoretical ground and a rigorous inversion technique was first proposed by Kaiser & Squires.

By combining Equations 2, 4, and 5, one can express the complex shear as a function of the convergence, κ (see Seitz & Schneider 1996 and references therein):

$$\gamma(\boldsymbol{\theta}) = \frac{1}{\pi} \int \mathcal{D}(\boldsymbol{\theta} - \boldsymbol{\theta}') \kappa(\boldsymbol{\theta}') d^2\theta', \qquad (16)$$

where

$$\mathcal{D}(\boldsymbol{\theta} - \boldsymbol{\theta}') = \frac{(\theta_2 - \theta_2')^2 - (\theta_1 - \theta_1')^2 + 2i(\theta_1 - \theta_1')(\theta_2 - \theta_2')}{|(\boldsymbol{\theta} - \boldsymbol{\theta}')|^4}. \qquad (17)$$

This equation can be inverted in order to express the projected mass density, or equivalently κ, as function of the shear:

$$\kappa(\boldsymbol{\theta}) = \frac{1}{\pi} \int \Re[\mathcal{D}^*(\boldsymbol{\theta} - \boldsymbol{\theta}') \gamma(\boldsymbol{\theta}')] d^2\theta' + \kappa_0, \qquad (18)$$

where $\Re$ denotes the real part. From Equation 14 we can express the shear as a function of the complex ellipticity. Hence, if the background ellipticity distribution is randomly distributed, then $\langle |\epsilon^S| \rangle = 0$ and

$$\langle |\epsilon^I| \rangle = |\mathbf{g}| = \frac{|\gamma|}{1 - \kappa}, \qquad (19)$$

(Schramm & Kayser 1995). In the most extreme case, when $\kappa \ll 1$ (the linear regime discussed initially by Kaiser & Squires 1993), $\langle |\epsilon^I| \rangle \approx |\gamma|$, and therefore, the projected mass density can be recovered directly from the measurement of the ellipticities of the lensed galaxies.

The first cluster mass reconstructions using the Kaiser & Squires linear inversion have been done by Smail (1993) and Fahlman et al (1994). Fahlman et al estimated the total mass within a circular radius using the *Aperture densitometry* technique (or the "ζ-statistics"), which consists of computing the difference between the mean projected mass densities within a radius r_1 and within an annulus $(r_2 - r_1)$ (Fahlman et al 1994, Kaiser 1995) as function of the *tangential shear*, $\gamma_t = \gamma_1 \cos(2\vartheta) + \gamma_2 \sin(2\vartheta)$ (see Equation 14), averaged in the ring:

$$\zeta(r_1, r_2) = \langle \kappa(r_1) \rangle - \langle \kappa(r_1, r_2) \rangle = \frac{2}{1 - r_1^2/r_2^2} \int_{r_1}^{r_2} \langle \gamma_t \rangle d\ln r. \qquad (20)$$

TABLE 1 Main results obtained from weak lensing analyses of lensing-clusters. The scale is the typical radial distance with respect to the cluster center. The last cluster has two values for the M/L ratio. This corresponds to two extreme redshifts assumed for the lensed population, either $z = 3$ or $z = 1.5$. For this case, the two values given for the velocity dispersion are those inferred when $z = 3$ or $z = 1.5$ are used.

Cluster	z	σ_{obs} (kms^{-1})	σ_{wl} (kms^{-1})	M/L (h_{100})	Scale (h_{100}^{-1} Mpc)	Tel.	Ref.
A2218	0.17	1370	—	310	0.1	HST	Smail et al (1997)
A1689	0.18	2400	1200–1500	—	0.5	CTIO	Tyson et al (1990)
		—	—	400	1.0	CTIO	Tyson & Fischer (1995)
A2163	0.20	1680	740–1000	300	0.5	CFHT	Squires et al (1997)
A2390	0.23	1090	≈1000	320	0.5	CFHT	Squires et al (1996b)
Cl1455+22	0.26	≈700	—	1080	0.4	WHT	Smail et al (1995)
AC118	0.31	1950	—	370	0.15	HST	Smail et al (1997)
Cl1358+62	0.33	910	780	180	0.75	HST	Hoekstra et al (1998)
MS1224+20	0.33	770	—	≈800	1.0	CFHT	Fahlman et al (1994)
Q0957+56	0.36	715	—	—	0.5	CFHT	Fischer et al (1997)
Cl0024+17	0.39	1250	—	150	0.15	HST	Smail et al (1997)
			1300	≈900	1.5	CFHT	Bonnet et al (1994)
Cl0939+47	0.41	1080	—	120	0.2	HST	Smail et al (1997)
		—	—	≈250	0.2	HST	Seitz et al (1996)
Cl0302+17	0.42	1080	—	80	0.2	HST	Smail et al (1997)
RXJ1347−11	0.45	—	1500	400	1.0	CTIO	Fischer & Tyson (1997)
3C295	0.46	1670	1100–1500	—	0.5	CFHT	Tyson et al (1990)
		—	—	330	0.2	HST	Smail et al (1997)
Cl0412-65	0.51	—	—	70	0.2	HST	Smail et al (1997)
Cl1601+43	0.54	1170	—	190	0.2	HST	Smail et al (1997)
Cl0016+16	0.55	1700	—	180	0.2	HST	Smail et al (1997)
			740	740	0.6	WHT	Smail et al (1993)
Cl0054-27	0.56	—	—	400	0.2	HST	Smail et al (1997)
MS1137+60	0.78	859[1]	—	270	0.5	Keck	Clowe et al (1998)
RXJ1716+67	0.81	1522[2]	—	190	0.5	Keck	Clowe et al (1998)
MS1054-03	0.83	1360[3]	1100–2200	350–1600	0.5	UH2.2	Luppino & Kaiser (1997)

[1]Gioia, private communication.
[2]Gioia et al (1998).
[3]Donahue et al (1998).

This quite robust mass estimator minimizes the contamination by foreground and cluster galaxies and permits a simple check that the signal is produced by shear, simply by changing γ_1 in γ_2 and γ_2 in $-\gamma_1$ which should cancel out the true shear signal.

The mass maps inferred from their images coincide with the light distribution from the galaxies. But, the impressive M/L found in the lensing cluster MS1224+20 by Fahlman et al (see Table 1) led to a surprisingly high value of Ω (close to 2!). This result is somewhat questionable and is probably due to the various sources of errors, possibly in the correction of PSF anisotropy (see Section 3.3). Furthermore, the two-dimensional mass reconstructions presented in the earliest papers looked noisy, probably because of boundary effects from the intrinsically non-local reconstruction, the geometry of the finite-size charge coupled device (CCD), and the reconstruction algorithm that can have terrible effects on the inversion. These problems have been discussed in several papers (Schneider & Seitz 1995, Seitz & Schneider 1995a, Schneider 1995, Seitz & Schneider 1996). In particular, Kaiser (1995) and Seitz & Schneider (1996) generalized the inversion to the non-linear regime, by solving the integral equation obtained from Equation 18 by replacing γ by $(1 - \kappa)\mathbf{g}$, or similarly by using the fact that both κ and γ depend on second derivatives of the projected gravitational potential φ (Kaiser 1995) which permits one to recover the mass density by this alternative relation:

$$\nabla \log(1 - \kappa) = \frac{1}{1 - |\mathbf{g}|^2} \begin{pmatrix} 1 + g_1 & g_2 \\ g_2 & 1 - g_1 \end{pmatrix} \begin{pmatrix} \partial_1 g_1 + \partial_2 g_2 \\ \partial_1 g_2 - \partial_2 g_1 \end{pmatrix}. \quad (21)$$

Both Equations 18 and 21 express the same relation between κ and γ and can be used to reconstruct the projected mass density. The improvements that have been proposed and discussed in detail by Seitz & Schneider (1995a), Kaiser (1995), Schneider (1995), Bartelmann (1995c), Squires & Kaiser (1996), Seitz & Schneider (1996, 1997) and Lombardi & Bertin (1998a,b) lead to reliable mass reconstructions from lensing inversion, and comparison with simulated clusters proves that it can now be considered a robust technique (see Figure 4, in particular the comparison between Figure 4a, and Figures 4g and 4h). However, the recovered mass distribution is not unique because the addition of a lens plane with constant mass density will not change the distortion of the galaxies (see Equation 18). Furthermore, the inversion only uses the ellipticity of the galaxies without regard to their dimension, so that changing $(1 - \kappa)$ in $\lambda(1 - \kappa)$ and γ in $\lambda \gamma$ keeps $\mathbf{g}$ invariant. This so-called *mass sheet degeneracy* initially reported by Gorenstein et al (1988), has been pointed out by Schneider & Seitz (1995) as a fundamental limitation of the lensing inversion.

The degeneracy could in principle be broken if the magnification can be measured independently, because it is not invariant under the linear transformation mentioned above, but instead it is reduced by a factor of $1/\lambda^2$. Broadhurst et al (1995) proposed measuring the magnification directly by using the magnification bias which changes the galaxy number-counts (see Section 3.4), whereas Bartelmann & Narayan (1995) explored their *lens parallax method* which compares the angular sizes of lensed galaxies with an unlensed sample. The lens-parallax method requires a sample of unlensed population having the same surface brightness distribution. However, in the case of ground-based observations, for the faintest (i.e. smallest galaxies), the convolution of the signal by the seeing disk can significantly affect the measurement of their surface brightness. Therefore, the method requires a careful handling of small-sized objects. A more promising approach is the use of wide field cameras with a typical field of view much larger than clusters of galaxies. In that case κ should vanish at the boundaries of the field, so that the degeneracy could in principle be broken.

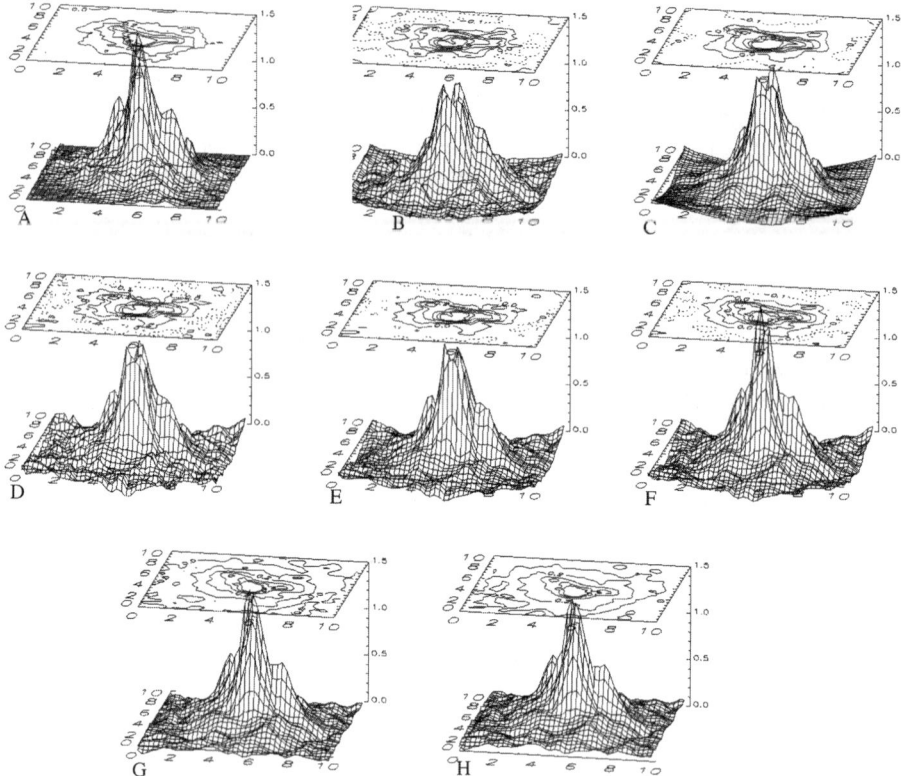

Figure 4 Examples of different algorithms for the mass reconstruction of clusters. Image A (*top left*) is the original simulated lensing cluster. Panel B shows an original Kaiser & Squires (1993) mass reconstruction, assuming that all background sources are circular. Panel C shows the same reconstruction as panel B, but with another smoothing method, which was proposed by Seitz & Schneider (1995a), in order to smooth the distortion distribution. Panel D shows the same result but the sources have an ellipticity distribution. Panel E shows the same reconstruction as panel D, but uses an adaptive smoothing scale. In panel F the linear and non-linear weak lensing regimes are now used in the inversion. Panel G shows the same reconstruction algorithm as F, with an additional extrapolation of the distortion field outside the field. The last panel shows the mass reconstruction with the constraint that the minimal mass density at any point is zero (from Seitz & Schneider 1995a).

An attractive alternative has been suggested by Bartelmann et al (1996) who proposed a maximum likelihood reconstruction algorithm. The advantage is that this is a *local* approach because it fits the projected potential at each grid-point. The observables are constrained by the second-derivatives of the potential, using a least-χ^2 which simultaneously computes both the magnification and the distortion, which are compared to the ellipticity and the sizes of the galaxies. Squires & Kaiser (1996) and Bridle et al (1998) have investigated similar maximum-likelihood techniques with different regularizations, though they fit the projected mass density rather than the potential. It seems however more attractive to use the deflection potential rather than the projected mass distribution in order to avoid incomplete knowledge of the contribution to the projected mass density of the matter outside the observed area (Seitz et al 1998).

The comparison done by Squires & Kaiser (1996) between the *direct reconstructions*, like the Kaiser & Squires (1993) approach, and the *inversion methods*, like the maximum likelihood reconstructions, did not lead to conclusive results, though the maximum likelihood inversion looks somewhat better. It is worth pointing out that one of the advantages of the maximum likelihood inversion is that it eases the addition of some observational constraints, such as strong lensing features (Bartelmann et al 1996; Seitz et al 1998). More recently, Seitz et al (1998) proposed an improved entropy-regularized maximum-likelihood inversion in which they no longer smooth the data, but instead use the ellipticity of each individual galaxy.

Since 1990, many clusters have been investigated using the weak lensing inversion, using either ground-based or HST data that are summarized in Table 1, but the comparison of these results is not straightforward because of the different observing conditions which produced each set of data and the different mass reconstruction algorithms used by each author. Nevertheless, all these studies show that on scales of about 1 Mpc, the geometry of mass distributions, the X-ray distribution and the galaxy distribution are similar (see Figure 5), although the ratio of each component compared with the others may vary with radius. The inferred M/L ratio lies between $100h_{100}$ to $1600h_{100}$, with a median value of approximately $300h_{100}$, with a trend to increase with radius. Contrary to the strong lensing cases, there is no evidence of discrepancies between the X-ray mass and the weak lensing mass. It is worth noting that the strong-lensing mass and the weak-lensing mass estimates are consistent in the region where the amplitudes of two regimes are very close. This is an indication that the description of the X-ray gas, and its coupling with the dark matter on the scales corresponding to strong-lensing studies, is oversimplified, whereas on larger scales, described by weak-lensing analysis, the detailed description of the gas has no strong impact.

The large range of M/L could partly be a result of one of the issues of the mass reconstruction from weak lensing. As shown in Equation 1, the deviation angle depends on the ratios of the three angular-diameter distances, which vary with the redshift assumed for the sources. For low-z lenses, the dependence with redshift of the background galaxies is not considerable, so the calibration of the mass can

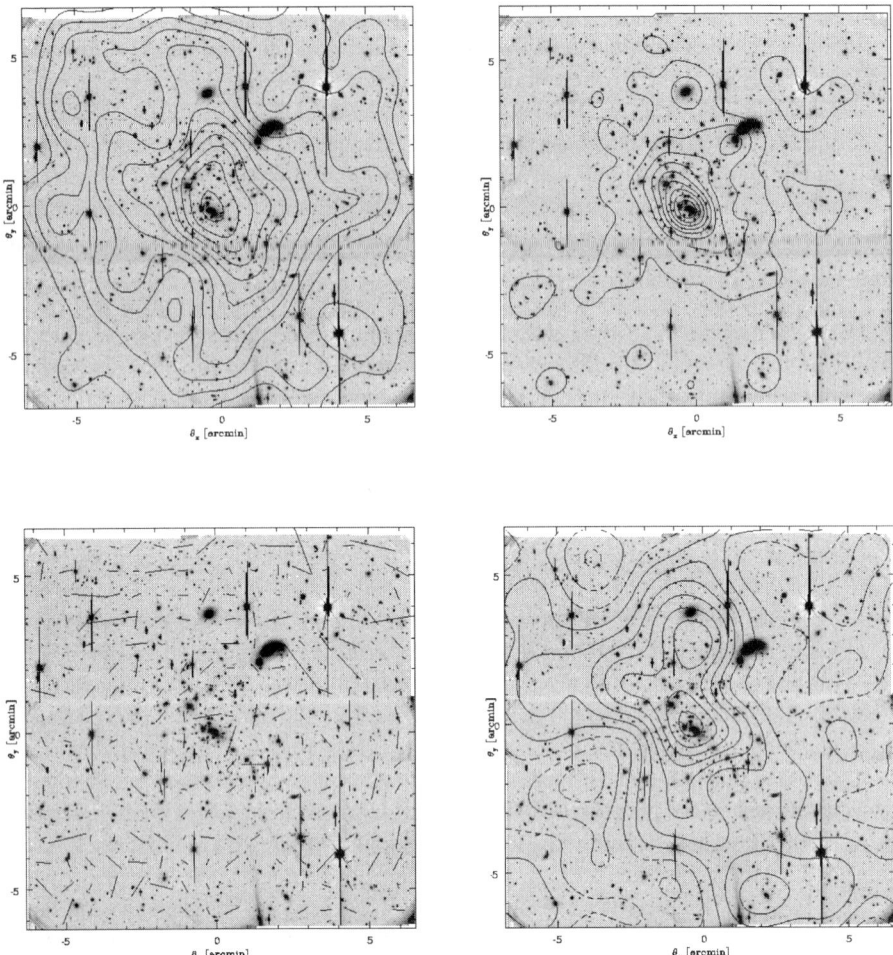

Figure 5 Weak lensing analysis and mass reconstruction of A2218 (from Squires et al 1996a). The images have been obtained at CFHT in I-band. The *top-left* panel shows the smoothed galaxy number density and the *top-right* shows the smoothed light distribution. The *bottom-left* is the shear map. The length of each line is proportional to the amplitude of the shear. From this shear map, the mass reconstruction of the Kaiser & Squires (1993) algorithm produces the mass map on the *bottom-right*. The correlation between the light and the matter distribution is clear.

be provided with a reasonable confidence level. However, distant clusters are highly sensitive to the redshift of the sources, and it becomes very difficult to scale the total mass without this information, even though the shape of the projected mass density is reconstructed correctly. The case of high-redshift clusters is more complicated. For a low-z cluster (say $z < 0.4$), it is not necessary to go extremely deep since the background galaxies are between $z = 0.4$ to 1. Thus, spectroscopic surveys can provide the redshift distribution with a good accuracy. In contrast, the background sources lensed by high-z clusters are beyond $z = 1$ and therefore are dominated by very faint galaxies ($I > 22.5$) which cannot be observed easily by spectroscopy.

The masses inferred from the strong lensing and weak lensing reconstruction put valuable constraints on the median M/L of lensing clusters. From the investigation of about 20 clusters, the median M/L is lower than $400h_{100}$. This implies that weak lensing analyses predict $\Omega < 0.3$ with a high significance level. Even if the uncertainties are large and if the weak lensing inversion needs to be improved, the HST data, in particular for clusters with giant arcs and many arc(let)s with known redshift, imply that the mass of clusters of galaxies cannot be reconciled with an Einstein-de Sitter (EdS) universe. The constraints on Ω are in good agreement with other observations (see the recent discussion by Krauss 1998).

Another strong statement results from the mass reconstruction obtained by Luppino & Kaiser (1997) and Clowe et al (1998) or the detection of giant arcs in very distant clusters (Deltorn et al 1997): massive clusters do exist at redshift ≈ 1! Though the total mass and the M/L cannot be given with a high accuracy, it cannot change the conclusion, unless unknown important systematics have been neglected. Therefore, we now have the first direct observational evidences that high mass-density peaks have generated massive clusters of galaxies at redshift 1. These results are extremely promising and are corroborated by other weak lensing studies around radiosources and quasars (see Section 4.2). Indeed, because these studies question the standard Cold Dark Matter (CDM) model and instead favor low-density universes, we can certainly expect fantastic developments from the investigation of high-redshift clusters with weak lensing during the coming years.

The impressive and spectacular results obtained by weak-lensing have demonstrated the power of this technique. However, although the results seem reliable in the cluster center (say within 500 kpc), there are still uncertainties outward, where the shear becomes very small. A critical issue of the lensing inversion is the reliability of the mass reconstruction and how it degrades when the shear decreases. Kaiser (1995) emphasized that Equation 20 can be used as a check of the mass reconstruction since the curl of $\nabla \log(1 - \kappa)$ should be zero only if the shear is recovered properly. Van Waerbeke (1999) has recently proposed an elegant way to estimate the accuracy of the mass reconstruction from the noise properties of the reconstruction. Nevertheless, the comparison of results using different algorithms and the stability and the reproducibility of each inversion have still to be done in order to demonstrate that weak lensing analysis produces reliable results. This is important for the future, when ground-based observations of very large fields will be performed. Mellier et al (1997) have compared the shear maps obtained in A1942 by using the Bonnet-Mellier and the autocorrelation function (ACF) methods (see next section). Though the two maps are similar, discrepancies are visible at the periphery, but no quantitative estimates of the similarity of both maps are given. An important step was taken by Van Waerbeke et al (in preparation) who analyzed A1942 using different shear measurements, different mass reconstruction algorithms and different data, obtained with two CCD cameras mounted at CFHT. Three sets of data have been used, all of them having a total exposure time of 4 hours and a seeing of $0.7''$. The results show impressive similarities even in the details of the mass reconstruction, down to a shear amplitude of 2%. This is the first work that demonstrates that results are stable and are not produced by artifacts, even at a very low shear amplitude. The main concerns are the capabilities of instruments, and the image analysis algorithms to measure very weak shear. This critical issue deserves a detailed discussion, which is the subject of the next section.

3.3 Measuring Weak Shear

In addition to the technical problem of the mass reconstruction algorithm and the redshift distribution of the sources, weak lensing is also sensitive to the accuracy of the measurement of ellipticities of lensed galaxies. The atmosphere has dramatic effects; in particular, the seeing circularizes the innermost part of galaxies which affects the measurements of shapes of faint galaxies (see for example the simulations by Bartelmann 1995c). These issues have been investigated in detail by Bonnet & Mellier (1995), Mould et al (1994), Kaiser et al (1995) and Van Waerbeke et al (1997). Atmospheric effects (seeing, atmospheric refraction, atmospheric dispersion), telescope handling (flexures of the telescopes, bad guiding) and optical distortions are extrinsic problems that can bias the measurements, though in principle they can be corrected in various ways. The atmospheric dispersion can be minimized by using an I-band filter and by observing clusters close to zenith, which also minimizes the flexures. Optical distortions can be corrected either analytically, if the optics are known perfectly, or by using the stars located in the fields. On the other hand, the ellipticity distribution of the galaxies is an intrinsic source of noise.

The extrinsic and intrinsic noises compete together: the circularization by the seeing is important only for the faint galaxies because their typical size is of the same order as the seeing disk. So, in principle it should be better to use only large (bright) galaxies, though they are not as numerous as the small (faint) galaxies. On the other hand, the noise produced by the intrinsic ellipticity distribution of the galaxies is minimized by averaging the shape of a large number of galaxies. The typical scale on which galaxies can be averaged is defined by the spatial resolution of the reconstructed mass map. For intermediate and high-redshift clusters of galaxies the typical angular scale is a few arcminutes, so that galaxies must be averaged on less than one arcminute in order to map the projected mass-density with a good sampling. If we assume an ellipticity dispersion, σ_ε of about 0.3, as it is suggested by nearby surveys and the distribution of galaxies in the Hubble Deep Field (HDF), then we can measure an ellipticity, ϵ, produced by a gravitational shear of $|\gamma| = 10\%$ if the number of averaged galaxies, N is:

$$N > \left(\frac{\sigma_\varepsilon}{\epsilon}\right)^2 \approx 10. \tag{22}$$

It is therefore quite easy to measure gravitational shear of 10% on an arcminute scale. However, going down to 1% would require about 900 galaxies which is not feasible on such an angular scale, unless many fields of 1 arcminute are averaged. An alternative is to go deeper in order to increase the galaxy number density. However, this is not sufficient to increase the accuracy of the results, because most of the faint galaxies have an unknown redshift distribution which prevents scaling the mass properly, and also because it is difficult to correct them from PSF.

The procedure to recover the shear field from the ellipticities of individual galaxies has several solutions. Bonnet & Mellier (1995) compute the second moment of galaxies (see Equation 9) within a circular annulus and average the signal on a given area (a *superpixel*) by using only the faint galaxies that dominate their deep observations. The size of the inner radius is constant and close to the seeing disk, which minimizes the effect of the circularization of the innermost isophotes on the measurement of the ellipse. The outer radius is also constant and has been optimized by using simulations of galaxies in order to get the highest signal-to-noise ratio on each second moment. The drawback of this approach is that the second moment of this annulus is no longer a direct measurement of the shape of the galaxy and it must be calibrated by simulations for each observing condition. The anisotropy of the PSF is corrected on the averaged signal, assuming that it is dominated by optical defects and that it behaves like a stretching of the image. This assumption is not valid for individual galaxies, but for each superpixel, deep observations average so many galaxies that it is possible to assume that the resulting signal reflects the one produced by an ideal galaxy having the same profile on each superpixel. In this case, Bonnet & Mellier have shown from simulations that the correction works very well down to a shear amplitude of 3% (see Bonnet et al 1994, and Schneider et al 1998a).

Kaiser et al (1995) (see also Luppino & Kaiser 1997, Hoekstra et al 1998 for further developments) compute the second moment within a variable aperture which depends on the size of individual galaxies; however, instead of an annulus they use a Gaussian filter, and introduce a more rigorous correction of the PSF anisotropy. Because they do not make selection from the size of the galaxies to measure the shear, it is clear that the largest galaxies require less correction. Therefore their correction depends on the total area of each individual galaxy. Assuming that the anisotropy of the PSF is small, Kaiser et al introduced a smearing correction, defined by a linear smear polarizability which expresses the (small) shift of polarization of galaxies induced by an anisotropic PSF. To first order, this shift can be expressed analytically and provides the correction from the shapes of the stars visible in the field, by dividing the smear polarizability by the observed polarization of stars. The efficiency of this method has been tested by using HST images which were degraded to the corresponding PSF anisotropy expected on ground-based images. They proved that the correction works very well. However, the method only works for bright galaxies. For fainter samples, they calibrated the polarization-shear relation by artificially lensing HST images and then by degrading them by the PSF observed in their data. However, it seems more preferable to calibrate the PSF anisotropy directly from the images. Luppino & Kaiser (1997) calibrated the anisotropic correction only from the observations of the stars in their fields, without auxiliary data.

Mould et al (1994) proposed a different procedure: they computed the second moment within a limiting isophote rather than a finite aperture and corrected linearly from the PSF anisotropy, assuming, as did Kaiser et al, that the correction is inversely proportional to the area of the source.

Van Waerbeke et al (1997) proposed an original alternative that fully exploits the signal down to the noise level on each CCD image and reduces the error of the second moment of galaxies. Instead of using individual objects, they compute the local auto-correlation function (ACF), $\xi(\theta)$, of pixels, averaged on a given area:

$$\xi(\theta) = \langle (I(\theta) - \langle I \rangle)(I(\theta + \theta') - \langle I \rangle) \rangle, \qquad (23)$$

where $\langle I \rangle$ is the mean surface brightness of the area. The (unlensed) ACF of the averaged source population, ξ^S, is an isotropic quantity, which by definition is centered, and therefore does not depend on the detection procedure and on the computation algorithm of its centroid. When a lensing signal is present, the shape of the ACF of the lensed population, $\xi^I(\theta) = \xi^S(A\theta)$ is no longer isotropic. For example, in the weak lensing regime:

$$\xi^I(\theta) = \xi^S(|\theta|) - |\theta| \, \partial_\theta \xi^S(|\theta|) \, (I - A), \qquad (24)$$

where I is the identity matrix. The ACF is now composed of the unlensed isotropic component plus an anisotropic term which depends on the magnification matrix, A, which stretches it like a real object. The second moment of the ACF can also be expressed as a function of the distortion and the magnification (Van Waerbeke et al 1997). Because all the pixels of the image are used, the ACF uses the full information of the image; in particular, the flux coming from extremely faint objects, for which the measurement of a centroid and the second moments are not measurable precisely, is also taken into account. For that reason, in principle, the ACF can work on images that reach the confusion limit (Van Waerbeke et al 1997, Réfrégier & Brown 1998). It turns out that in practice it is better to use the ACF around detected galaxies than on the total image because correlated noises, such as electronics cross-talk or shift-and-add residuals, may generate spurious coherent signals (Van Waerbeke & Mellier 1997). This method looks ideal for the optimal extraction of weak lensing signals because the signal-to-noise ratio of the ACF is always high enough and spreads over sufficient pixels to avoid the need for circular filtering, such as faint galaxies, and to provide an accurate estimate of its shape parameters.

The reliability of the relation between measured ellipticities and shear, and of the mass reconstruction obtained from observations, has been checked by Kaiser & Squires (1993) and Bonnet & Mellier (1995), and by the independent simulations of Bartelmann (1995c) and Wilson et al (1996a). Despite careful studies to check that images are corrected accurately from circularization by seeing and PSF anisotropy (in particular the spatial variation of the PSF in the field can be modeled), there is still a lot of work to do in this area. For instance, the weighting functions proposed to measure the ellipticities are based on intuition but no complete investigation has been done so far to find the optimal one. Moreover, for each of these procedures, it is assumed that the PSF anisotropy is unidirectional. This may not be true, in particular when instruments with poor optical design are used. In that case, the correction becomes non-trivial, and paradoxically this could appear on the image with the best image quality because details of the PSF are no longer smeared by circularization of the seeing. This domain is certainly at its infancy and much work and many new ideas should appear during the next years, mainly because the shear produced by large-scale structures is expected to be very small.

3.4 Mass Profile from the Magnification Bias

Parallel to the mass reconstruction using weak shear measurements, one can use the direct measurement of the magnification from the local modification of the galaxy number density. This "magnification bias" expresses the simultaneous effects of the gravitational magnification, which increases the flux received from any lensed galaxies and permits the detection of galaxies enhanced by the amplification, but also magnifies by the same amount the area of the projected lensed sky and thus decreases the apparent galaxy number density. The total amplitude of the magnification bias depends on the slope of the galaxy counts as a function of magnitude and on the magnification factor of the lens. For a circular lens, the radial galaxy number density of background galaxies writes:

$$N(<m, r) = N_0(<m)\,\mu(r)^{2.5\alpha-1} \approx N_0\,(1+2\kappa)^{2.5\alpha-1} \quad \text{if } \kappa \text{ and } |\gamma| \ll 1, \quad (25)$$

where $\mu(r)$ is the magnification, $N_0(< m)$ is the intrinsic (unlensed) number density, obtained from galaxy counts in a nearby empty field, and α is the intrinsic count slope:

$$\alpha = \frac{\mathrm{d}\log N(<m)}{\mathrm{d}m}. \quad (26)$$

A radial magnification bias $N(<m, r)$ shows up only when the slope $\alpha \neq 0.4$; otherwise, the increasing number of magnified sources is exactly compensated by the apparent field dilatation. For slopes larger than 0.4 the magnification bias increases the galaxy number density, whereas for slopes smaller than 0.4 the radial

density will show a depletion. Hence, no change in the galaxy number density can be observed for $B(<26)$ galaxies, since the slope is almost this critical value (Tyson 1988). However, it can be detected in the $B > 26$, $R > 24$ or $I > 24$ bands when the slopes are close to 0.3 (Smail et al 1995c).

The change of the galaxy number density can be used as a direct measurement of the magnification and can be included in the maximum likelihood inversion as a direct observable in order to break the mass sheet degeneracy (see Section 3.2). Alternatively, it can also be used to model the lens itself. In the case of a singular isothermal sphere, the magnification can be expressed as a function of the velocity dispersion of the lens, σ, and the radial distance $\theta = r/r_E$, where $r_E = 4\pi\sigma^2/c^2 \, D_{LS}/D_{OS}$:

$$\mu(r) = \frac{4\pi\sigma^2}{c^2} \frac{D_{LS}}{D_{OS}} \frac{\theta}{\theta - 1}. \tag{27}$$

Reconstruction of cluster mass distribution using magnification bias was initially explored by Broadhurst et al (1995), and has been used by Taylor et al (1998) in A1689, and by Fort et al (1997) in Cl0024+1654 (see also a generalization by Van Kampen 1998). The masses found are consistent with those inferred from gravitational weak shear or from strong lensing.

This magnification bias is an attractive alternative to the weak shear because it is only based on the galaxy counts and does not require outstanding seeing to measure ellipticities and orientations of galaxies. However, it is more sensitive to shot noise, which unfortunately increases when the number density decreases in the depletion area. Furthermore, it also depends on the galaxy clustering of the background sources which can have large fluctuations from one cluster to another. Indeed, in the weak lensing regime, assuming that $\kappa \approx |\gamma|$, the ratio of the signal-to-noise ratios of the shear and the depletion, $R_{Sh/Dep}$, clearly favors the shear analysis:

$$R_{Sh/Dep} = \frac{|\gamma|}{\sigma_\varepsilon} \frac{1}{\kappa |(5\alpha - 2)|} \approx 3, \tag{28}$$

for a dispersion of the intrinsic ellipticity distribution of the galaxies, $\sigma_\varepsilon = 0.3$, and $\alpha = 0.2$. Despite these limitations, this is a simple way to check the consistency of the mass reconstruction. Its great merit is that it is not sensitive to systematic effects, such as the weak shear measurement which depends on the correction from the PSF of the observed ellipticities.

4. LARGE-SCALE STRUCTURES AND COSMIC SHEAR

The idea that mass condensations and the geometry of the Universe can alter light bundles and distort the images of distant galaxies was emphasized by Kristian & Sachs (1966), and later by Gunn (1967) and Blandford & Jaroszyński (1981), who first gave a quantitative estimate of the amplitude of this effect. Kristian

(1967) looked at this effect on photographic plates of six clusters of galaxies using the Palomar Telescope, but found nothing significant. Valdes et al (1983) were the first to attempt to measure a coherent alignment of distant galaxies generated by large-scale structures. They used about 40,000 randomly selected field galaxies with J magnitudes between 22.5 and 23.5, but, like Kristian, they did not find any conclusive signal. These negative measurements were not definitely interpreted as important cosmological constraints on the curvature and the mass distribution in our Universe, but instead as a result of technical limitations related to the poor image quality of the photographic data. Indeed, the recent weak lensing analysis produced by a supercluster candidate done by Kaiser et al (1998) seems to show that large-scale structures produce gravitational shear which is already detectable. Numerical simulations by Schneider & Weiss (1988) using point-mass models, or by Babul & Lee (1991) using a smooth mass distribution, showed that both the ellipticity distribution and the apparent luminosity function of distant galaxies could be modified, in particular if the fraction of small-scale structures such as clusters of galaxies is important (Webster 1985). Therefore, two different effects produced by the cosmological distribution of structures in the universe are expected: a change of the galaxy number count correlated with the mass distribution, namely a magnification bias; and a change of the ellipticity distribution, namely a shear pattern, correlated with the mass distribution as well. Because the expectation values strongly depend on the fraction of non-linear systems and the redshift distribution of the galaxies, it is clear that the analysis of weak lensing effects by large-scale structures is an interesting test of cosmological scenarios.

4.1 Theoretical Expectations

The theoretical investigations of the effect of the large-scale mass distribution on the distribution of ellipticity/orientation of distant galaxies are somewhat simplified by the low density contrast of structures. Beyond 10 Mpc scales, $\delta\rho/\rho \approx 1$ and linear perturbation theory can be applied. On these scales, lenses are no longer considered individually but they are now viewed as a random population which has a cumulative lensing effect on the distant sources. Blandford (1990), Blandford et al (1991) and Miralda-Escudé (1991) first investigated the statistical distribution of distortions induced by large-scale structures in an EdS universe. They computed the two-point polarization (or shear) correlation function and established how the rms value of the polarization depends on the power spectrum of density fluctuations. Kaiser (1992) extended these works and showed how the angular power spectrum of the distortion is related to the three-dimension mass density power spectrum, without assumptions on the nature of fluctuation. These works were generalized later to any arbitrary value of Ω by Villumsen (1996) and Bar-Kana (1996). All these studies concluded that the expected rms amplitude of the distortion is of about one percent, with a typical correlation length of a degree. Therefore it should be measurable with present-day telescopes.

These promising predictions convinced many groups to start investigating more thoroughly how weak lensing maps obtained from wide field imaging surveys could constrain cosmological scenarios. To go into further detail, it is necessary to generalize the previous works to any cosmology and to describe in detail observables and physical quantities that could be valuable to the constraint of cosmological models. Indeed, the investigations of weak lensing by large-scale structures require theoretical and statistical tools that are not different from those currently used for catalogues of galaxies or cosmological microwave background (CMB)-maps. In this respect, the perturbation theory, which has already been demonstrated to work, describes the properties of large-scale structures very well (see Bouchet 1996 and references therein), and seems to be an ideal approach for such large scales. In addition, the use of similar statistical estimators for catalogues of galaxies seems perfectly suited. Bernardeau et al (1997) used the perturbation theory to explore the sensitivity of the second and third moments of the gravitational convergence κ (rather than the distortion whose third moment should be zero), to cosmological scenarios, and to cosmological parameters, including λ-universes. The small angle deviation approximation implies that the distortion of the ray bundle can be computed on the unperturbed geodesic (Born approximation). In the linear regime, if lens-coupling is neglected (see Section 4.4.3), the cumulative effect of structures along the line of sight generates a convergence in the direction θ,

$$\kappa(\theta) = \frac{3}{2}\Omega_0 \int_0^{z_s} n(z_s)\,dz_s \int_0^{\chi(s)} \frac{D_0(z,z_s)D_0(z)}{D_0(z_s)} \delta(\chi,\theta)\,(1+z(\chi))\,d\chi, \quad (29)$$

where χ is the radial distance, D_0 is the angular diameter distance, $n(z_s)$ is the redshift distribution of the sources, and

$$\delta = \int \delta_k D_+(z) e^{i\mathbf{k}\cdot\mathbf{x}}\,d^3k \quad (30)$$

is the mass density contrast, which depends on the evolution of the growing modes with redshift, $D_+(z)$. It is related to the power spectrum as usual: $\langle \delta_\mathbf{k} \delta_{\mathbf{k}'} \rangle = P(k)\delta_{Dirac}(\mathbf{k}+\mathbf{k}')$. It is worth noting that κ depends explicitly on Ω_0 and not only δ because the amplitude of the convergence depends on the projected mass, not only on the projected mass density contrast.

The dependence of the angular power spectrum of the distortion as a function of (Ω, λ), of the power spectrum of density fluctuations and of the redshift of sources has been investigated in detail in the linear regime by Bernardeau et al (1997) and Kaiser (1998). Bernardeau et al (1997) and Nakamura (1997) computed also the dependence of the skewness of the convergence on cosmological parameters, arguing that it is the first moment which directly probes non-linear structures. From perturbation theory and assuming Gaussian fluctuations, the variance, $\langle \kappa(\theta)^2 \rangle$, and the skewness, $s_3 = \langle \kappa(\theta)^3 \rangle / \langle \kappa(\theta)^2 \rangle^2$, have the following dependencies with the cosmological quantities:

$$\langle \kappa(\theta)^2 \rangle^{1/2} \approx 10^{-2} \, \sigma_8 \, \Omega_0^{0.75} \, z_s^{0.8} \left(\frac{\theta}{1°}\right)^{-(n+2)/2}, \tag{31}$$

and

$$s_3(\theta) \approx -42 \, \Omega_0^{-0.8} \, z_s^{-1.35}, \tag{32}$$

for a fixed source redshift z_s, where n is the spectral index of the power spectrum of density fluctuations and σ_8 is the normalization of the power spectrum (the rms mass density fluctuation within a sphere of $8h_{100}^{-1}$ Mpc). Hence, since the skewness does not depend on σ_8, the amplitude of fluctuations and Ω_0 can be recovered independently using $\langle \kappa(\theta)^2 \rangle$ and s_3. The slope of the projected power spectrum can, in principle, be recovered from the complete reconstruction of the projected mass density, using weak-lensing inversion as discussed in Section 3.

Jain & Seljak (1997), generalizing the early work by Miralda-Escudé (1991), have analyzed the effects of non-linear evolution on $\langle \kappa(\theta)^2 \rangle$ and s_3 using the fully non-linear evolution of the power spectrum (Peacock & Dodds 1996). They found formal relations similar to those found by Bernardeau et al. However, on a scale below 10 arcminutes, $\langle \kappa(\theta)^2 \rangle$ increases more steeply than the theoretical expectations of the linear theory and is 2 or 3 times higher on scales below 10 arcminutes. These predictions are strengthened by numerical simulations (Jain et al 1998). Therefore, a shear amplitude of about 2–5% is predicted on these scales which should be observed easily with ground-based telescopes (Figure 6). Schneider et al (1998a) recently claimed that they have detected this small-scale *cosmic-shear* signal.

The previous studies are based on the measurements of ellipticities of individual galaxies in order to recover the stretching produced by linear and non-linear structures. Like the mass reconstruction of clusters, it demands high-quality images and an accurate correction of systematics down to a percent level. An alternative to this strategy has been investigated by Villumsen (1996) who looked at the effect of the magnification bias on the two-point galaxy correlation function. Because the magnification may change the galaxy number density as a function of the slope of the galaxy number counts, it similarly modifies the apparent clustering of the galaxies. From Equations 23 and 25, the two-point correlation function averaged over the directions θ is changed by the magnification of the sources and, in the weak lensing regime, its contribution writes (Kaiser 1992, Villumsen 1996, Moessner & Jain 1998):

$$\omega(\theta) = \frac{\langle [N_0(m)(5\alpha - 2)\kappa(\theta + \theta')][N_0(m)(5\alpha - 2)\kappa(\theta)] \rangle}{N_0(m)^2}, \tag{33}$$

that is,

$$\omega(\theta) = (5\alpha - 2)^2 \langle \kappa(\theta + \theta')\kappa(\theta) \rangle. \tag{34}$$

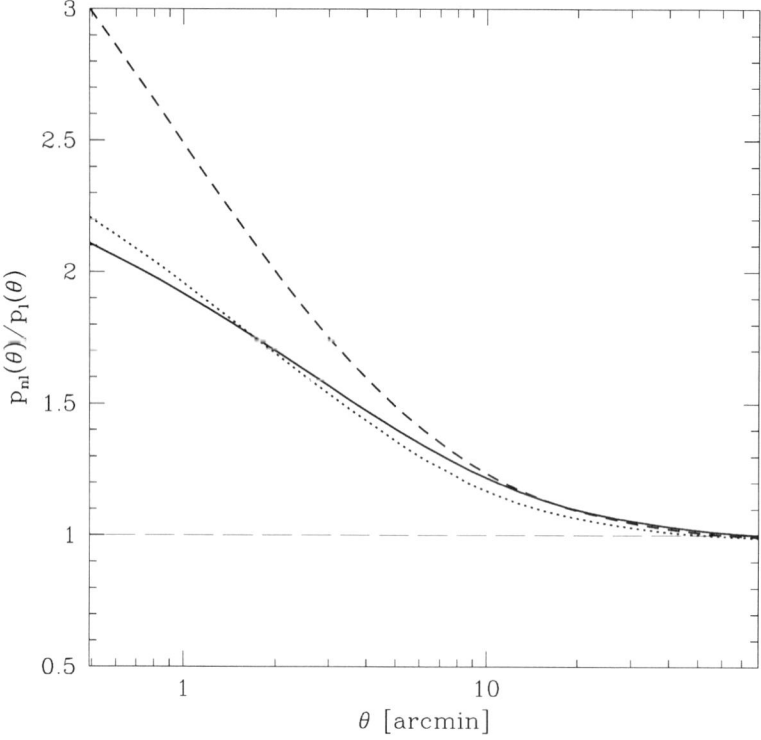

Figure 6 Ratio of the amplitude of the polarization predicted by the non-linear and linear evolution of the power spectrum as a function of angular scale (from Jain & Seljak 1997). The normalization is $\sigma_8 = 1$. The plot shows the expectations for three cosmologies: $\Omega = 1$ (*solid line*), $\Omega = 0.3$ (*dashed line*) and $\Omega = 0.3$, $\lambda = 0.7$ (*dotted line*). The difference between the two regimes becomes significant below the $10'$ scale.

The galaxy two-point correlation function is therefore sensitive to the correlation function of the convergence and to the slope of galaxy counts. If the unlensed two-point correlation function is known, it is then possible to compute the local correlation function of the convergence from the local two-point correlation function of the galaxies.

Detailed investigations of the capability of this technique have been discussed by Moessner et al (1998) who looked at the effect of non-linear clustering on small scales and for different cosmologies. They raised the point that the correlation function can be also affected by the evolution of galaxies which also modifies the two-point correlation function of galaxies, but in an unknown way. Moessner & Jain (1998) proposed a way to disentangle these two effects by using the cross-correlation of two galaxy samples having different redshift distributions that do not overlap. This minimizes the effect of intrinsic galaxy clustering, but it requires the knowledge of the biasing which can also depend on the redshift. Therefore, the magnification bias method needs auxiliary input that can constrain the biasing independently.

4.2 Measuring the Biasing

Because gravitational lensing is directly sensitive to the total mass responsible for the deflection, it provides a potentially important tool for measuring the biasing factor, as it has been demonstrated by the recent weak lensing analysis of the supercluster MS0302+17 (Kaiser et al 1998). In particular, a well-known effect of the magnification bias is the generation of correlations between foreground and background luminous systems observed in catalogues. The matter associated with the foreground systems can amplify the flux received from background objects, which results in an apparent correlation between the number density or the luminosity of backgound objects and the number density of foregrounds (Canizares 1981). The correlation has been first detected by Fugmann (1990) and confirmed later by Bartelmann & Schneider (1993a,b; 1994) who found correlations of the galaxy number density with the radio-sources of the 1-Jy catalogues, the IRAS catalog, and the X-ray galaxies. Further independent analyses of such associations showed also evidences of magnification bias (Bartelmann et al 1994, Rodrigues-Williams & Hogan 1994, Seitz & Schneider 1995b, Wu & Han 1995, Benítez & Martínez-González 1997, Williams & Irwin 1998). The statistical basis of these associations is surprisingly robust and difficult to explain physically without invoking lensing effects.

Bartelmann et al (1994) argued that the correlation can be interpreted has a magnification bias only if it is produced by condensations of matter, like clusters of galaxies (Bartelmann & Schneider 1993a). In order to check this hypothesis, Fort et al (1996) have attempted to detect weak shear around some selected bright quasars which could be good lensing candidates. They found strong evidence of weak shear around half of their sample with, in addition, a clear detection of galaxy overdensities in the neighborhood of each quasar. The Fort et al sample has been re-analyzed recently by Schneider et al (1998a) and the detection has been strengthened on a more reliable statistical basis, with a firm confirmation for one of the quasars from the HST images obtained by Bower & Smail (1997). In addition, some other observations have also detected gravitational shear around bright radiosources: Bonnet et al (1993) found a shear signal around the double imaged quasar Q2345+007, which was later confirmed to be associated with a distant cluster (Mellier et al 1994; Fischer et al 1994; Pelléo et al 1996). Similar detection of a cluster has recently been reported around the Cloverleaf (Kneib et al 1998) and around 3C324, which also clearly shows a shear pattern from the HST images (Smail & Dickinson 1995). The sample is nevertheless too small to provide a significant direct evidence that magnification bias is detected in quasar catalogues. Indeed, it would be important to pursue this program using a large sample of bright quasars or radiosources. The field of view does need to be large, so the HST with the Advanced Camera could be a perfect instrument for this project. The impact of such magnification bias could be important for our understanding of the evolution scheme of quasars and galaxies, because it changes the apparent luminosity functions of these samples (Schneider 1992, Zhu & Wu 1997).

If these correlations are caused by magnification bias, a quantitative value of the biasing can in principle be estimated, for instance from the angular foreground-background correlation function (Kaiser 1992), where foreground could be galaxies or dark matter. Bartelmann (1995a) expressed the angular quasar-galaxy correlation, ξ_{QG}, as a function of the biasing factor, b, and the magnification-mass density contrast cross-correlation function, $\xi_{\mu\delta}$, in the weak lensing regime:

$$\xi_{QG}(\theta) = (2.5\alpha - 1)\, b\, \xi_{\mu\delta}(\theta), \tag{35}$$

where α is the slope of the background galaxy counts. However, since $\xi_{\mu\delta}$ depends on the power spectrum of the projected mass density contrast, the determination of the biasing factor is possible only if one independently obtains the amplitude of the power spectrum. Generalizations to non-linear evolution of the power spectrum and to any cosmology were explored by Dolag & Bartelmann (1997) and Sanz et al (1997). As expected, the non-linear condensations increase the correlation on small scales by a very large amount; that, however, still strongly depends on the amplitude of the power spectrum.

With the observation of weak lensing induced by large-scale structures, it becomes possible to observe directly the correlation between the background galaxies and the projected mass density of foreground structures instead of using the light distribution emitted by the foreground galaxies. Following the earlier development by Kaiser (1992), Schneider (1998) has computed the cross-correlation between the projected mass, M, and the galaxy number density of foreground galaxies, N, on a given scale θ, $\langle MN \rangle_\theta$, as a function of the biasing factor of the foreground structures, b. Like previous studies, it is simply proportional to b, but it also depends on σ_8 and the slope of the power spectrum, and as such it is not trivial to estimate it without ambiguity. A more detailed investigation of the interest of $\langle MN \rangle_\theta$ has been done by Van Waerbeke (1998a) who computed the ratio, $R(\theta)$ of the density-shear correlation over the two-point galaxy correlation function for a narrow redshift distribution of foregrounds and a narrow range in scale (Van Waerbeke 1998a,b):

$$R_\theta = \frac{3}{2}\frac{\Omega}{b}\frac{g(w_f) f_K(w_f) N_f(w_f)}{a(w_f)\, \int N_f^2(w)\, dw}, \tag{36}$$

where a is the expansion factor, w_f is the comoving distance of the foreground, f_K is the comoving angular diameter distance, N_f is the redshift distribution of the foreground galaxies, and

$$g(w) = \int_w^{w_f} N_b(w') \frac{f_K(w' - w)}{f_K(w')}, \tag{37}$$

$N_b(w')$ being the redshift distribution of the background galaxies. Using this ratio, he investigated the scale dependence of the biasing, including the non-linear spectrum and different cosmologies, and discussed how it can be used to analyze the evolution of the biasing with redshift, if two different populations of foregrounds are observed. The ratio $R_\theta/R_{\theta'}$ permits one to compare the biasing on two different scales. This quantity does not depend on Ω, on the power spectrum or on the smoothing scale, so it is a direct estimate of the evolution of biasing with scale. Van Waerbeke predicts that a variation of 20% of the bias on scales between 1' and 10' will be detectable on a survey covering 25 square degrees. This ratio is therefore a promising estimator of the scale dependence of the biasing.

The analysis discussed above indicates the potential interest of lensing to the study of the evolution of bias with scale and redshift of foreground, with the direct use of the correlation between the ellipticity distribution of background sources and the projected mass density inferred from mass reconstruction. It thus permits an accurate and direct study of the biasing and its evolution with lookback time and possibly with scale. However, it is worth stressing that Van Waerbeke (like Bartelmann 1995a and others) assumed a linear biasing. In the future it will be important to explore in detail the generalization to a non-linear bias (Van Waerbeke, private communication) as described by Dekel & Lahav (1998).

4.3 Strategies for Weak Lensing Surveys

The rather optimistic predictions from theoretical work reported in Section 4.1 are convincing enough to investigate in great detail how weak lensing surveys can be designed in order to maximize the signal to noise ratio of statistical and physical quantities, as well as to minimize the time spent for the survey. The definition of a best strategy addresses many issues: size, shape, topology of the survey, best statistical estimators, optimal analysis of the catalogues and best extraction of the signal from the raw data. This last point (measurement of ellipticities, correction of the PSF) has already been discussed in Section 3.3, so I will focus on the other aspects. Indeed, weak lensing surveys have rather generic constraints which are common to any survey in cosmology (see Szapudi & Colombi 1996, Kaiser 1998, Colombi et al 1998 and references therein). The main sources of noise are the cosmic variance, the intrinsic ellipticity distribution of the lensed galaxies and possibly the noise propagation during the mass reconstruction from lensing inversion. In addition, at such low shear level, the correction of the PSF, as well as the removals of systematics coming from optical and atmospheric degradation of the images, or from the method used to measure the shear from galaxy ellipticity, are crucial steps.

The impact of the size, the shape and the deepness of weak lensing surveys has been addressed by Blandford et al (1991) and Kaiser (1992) and has been investigated in more detail by Kaiser (1998), Kamionkowski et al (1998), Van Waerbeke et al (1999), and Jain et al (1998). Depending on the scientific goals and the nature of the data, medium-size deep, very-wide shallow and very-small ultra-deep surveys have been proposed.

The most detailed investigation done so far has been conducted by Van Waerbeke et al (1999). From a sample of 60 simulations per set of parameters they computed the expected signal to noise ratio of the variance and the skewness of the convergence, and they attempted to reconstruct the projected mass density from the lensing signal in order to recover the projected power spectrum for each model on $5° \times 5°$ and $10° \times 10°$ noisy maps. They included various cosmologies, mass density power spectra, sampling strategies and redshifts of sources. This work is a preparation of the subarcsecond seeing survey which should be conducted at CFHT with the new wide field Megacam camera (which will cover one square degree, Boulade et al 1998). These simulations show that the projected mass density maps can be recovered with an impressive accuracy and demonstrate that such a survey will be able to recover the projected power spectrum of mass density fluctuations, from 2.5 arcminutes up to 2 degrees with a signal to noise ranging from 10 to 3. The power spectrum on small scales is dominated by the shot noise due to the intrinsic ellipticity distribution of the background galaxies, but it follows the statistics predicted by Kaiser (1998) which permits its easy removal. The best strategy suggested by these simulations is a shallow survey with a typical galaxy number density of 30 arcmin^{-2} over 10×10 square degrees. According to Van Waerbeke et al (1999) it will permit one to separate $\Omega = 0.3$ from $\Omega = 1$ universes at a 6σ confidence level Figure 7). Kaiser (1998) recommends a wide field as well, but suggests a sparse sampling on a very large scale rather than a compact topology. This alternative has not been investigated yet using simulations. In particular, it would be useful to have quantitative estimates of the scale beyond which the survey should switch from a compact to a sparse sampling of the sky.

Alternatively, Stebbins (1996) and Kamionkowski et al (1998) have explored the possibility of using shallower surveys, which sample the sky with only a few galaxies per arcminute, but cover about half the sky. Contrary to the Megacam-like surveys which aim at mapping the shear field and building a map of the projected mass density, these shallow surveys only aim at measuring the correlation of ellipticities on very large scales. Stebbins (1996, 1999) used the linear theory to compute the angular power spectrum of the shear inferred from a tensor spherical harmonic expansion of the shear pattern of an all-sky survey. He argues that the SDSS could provide reliable information on the projected angular power spectrum of the shear. The expected signal is very small, because most of the galaxies will be at a redshift lower than 0.2. The seeing of the site and the sampling of the images could have minor impact on the signal because these nearby galaxies should be much larger than the seeing disk. However, a simulation of the systematics which include the quality of the telescope and the instruments, as well as an estimate of the possible systematics produced by the drift-scanning are now necessary in order to have a clear quantitative estimate of the expected signal to noise ratio.

The use of the VLA-FIRST radio survey for weak lensing proposed by Kamionkowski et al (1998) looks promising. FIRST covers about the same area

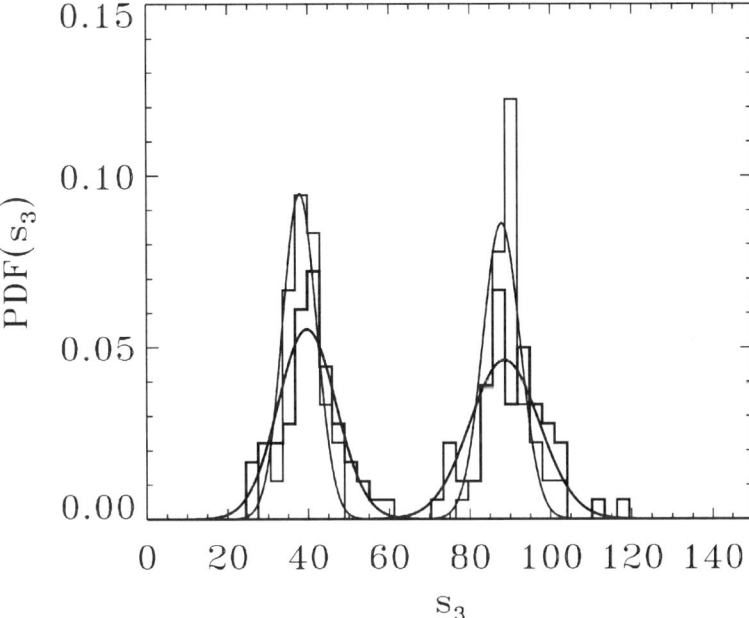

Figure 7 Histograms of the values of the skewness of the convergence for an $\Omega = 1$ (*right*) and $\Omega = 0.3$ (*left*) universe and for a 5×5 (*thick lines*) and 10×10 (*thin lines*) square degrees survey. This simulates the possible surveys to be done with MEGACAM at CFHT (from Van Waerbeke et al 1998).

as the SDSS but the radiosource sample has a much broader redshift distribution and a median redshift beyond $z = 0.2$. However, with less than 100 sources per square degree, the sampling is rather poor. Kamionkowski et al predict an rms ellipticity of about 3% on 6 arcminutes, 1% on 20 arcminutes and on one degree scale, with a signal to noise ratio larger than 5, in good agreement with Bernardeau et al (1997) and Jain & Seljak (1997). Preliminary results from this survey should soon be available.

On small scales, first attempts for measuring cosmic shear have already been made by several groups. Mould et al (1994) used a very deep image obtained at the Palomar Telescope to measure cosmic shear on a scale of 5 arcminutes. They do not find a significant signal (however, see JV Villumsen 1995, unpublished results) and argued that the cosmological signal of their field is below 4%. A similar attempt has been made by Fahlman et al (1994) on a 15 arcmin field. The images were obtained at CFHT in subarcsecond seeing conditions ($\approx 0.5''$, against $0.9''$ for Mould et al), but significantly less deep than the Mould et al observations. Fahlman et al did not detect any significant signal, though owing to the excellent seeing, they expected it to be three times more sensitive. On the other hand, a weak lensing signal has already been detected on a 2 arcmin scale around radiosources and quasars (see Section 4.2), which shows that cosmic shear on a small scale is detectable. Schneider et al (1998a) argued that the shear detected around radiosources is cosmic shear, though the sample is probably biased toward non-linear structures because the fields are preferentially selected around bright quasars. This is probable because the motivation of the Fort et al (1996) observations (from which the Fort et al and Schneider et al sample is based) was the verification of the Bartelmann et al (1994) hypothesis that the quasar-galaxy association results from a magnification bias of bright quasar samples by large-scale structures. Therefore, if the magnification bias really works then the cosmological significance of the Schneider et al (1998a) interpretation is difficult to quantify. Conversely, if the quasar-galaxy associations are not produced by such magnification, then Schneider et al measured the cosmic shear for the first time.

Several low-angular scale surveys are under way to provide significantly better statistics. These surveys have been summarized by Seitz et al (1999). A promising strategy consists of using the parallel Space Telescope Imaging Spectrometer (STIS) observations of HST as randomly selected fields. For each of those outstanding images, it is possible to measure the average distortion with a very high confidence level. Preliminary analyses by Seitz et al (1999) show that the PSF is incredibly stable and that the first observations lead to an rms ellipticity of 3% on one arcminute scale, which is very encouraging. However, for the moment it is difficult to interpret the rms shear in terms of constraints for cosmological models because the STIS parallel observations are done without filter, so there is no color information on the lensed galaxies to constrain their average redshift.

In addition to the definition of the survey strategy, the scientific return is sensitive to the technique used to analyze the catalogues. In this respect, the best procedure depends on the noise properties on the survey. In theory, if the noise follows Gaussian statistics, the Wiener filtering should provide a minimum variance estimator of cosmological quantities (Seljak 1997a). However, it is not obvious that this is the best approach, in particular on small scales where non-linear features deviate significantly from gaussianity. For instance, Kruse & Schneider (1999) have explored a strategy for an optimal extraction of the high density peaks present in surveys. It uses the aperture mass densitometry for cosmic shear measurement proposed by Schneider et al (1998b) which permits the detection of peaks of the projected mass density as function of the tangential shear, γ_t (Kaiser 1995). This simple and rather robust approach which focuses on the most contrasted systems should provide in a simple way some constraints of the number density of high peaks, and therefore of the cosmological scenarios. Schneider et al (1998b) argued that the use of a compensated filter is an optimal procedure to measure κ from γ without mass reconstruction because in contrast to the top hat filter, it selects a sharp scale range that cancels the additional noise produced by the adjacent wave numbers. The merit of the two filterings has also been investigated in detail in the simulations of Van Waerbeke et al (1999). They found that the best choice is unclear: the compensated filter is better for the variance, but is worse for the skewness (see Figure 8, color).

4.4 Critical Issues

The critical issues already discussed for the weak lensing and mass reconstruction of clusters of galaxies are valid and even more critical for large-scale structures. This will not be discussed again here, but we should mention that systematic effects are a real concern which could be an ultimate limitation of the weak lensing capabilities, at least on ground-based telescopes.

Redshift of Sources Equations 27 and 28 show a strong dependence of the variance and the skewness of the convergence on the redshift distribution of the lensed sources. As far as weak lensing is concerned, from the investigations of the effect of redshift distribution of sources on cluster mass reconstruction, it seems that only the averaged redshift of the galaxies and the width of their distribution are needed (Seitz & Schneider 1997), even with a bad precision (say, $\Delta z_s \approx 0.5$). This requirement is therefore not severe and can be obtained using photometric redshifts. The work by Connolly et al (1995) demonstrates that galaxies brighter than $I = 22.5$ can easily be calibrated using spectroscopic redshifts and that photometric information in 4 different filters constrain the redshift of these galaxies with an accuracy of about $\Delta z_s \approx 0.05$. The future surveys with 10-meter class telescopes will calibrate photometric redshift up to $I = 23.5$ and even fainter, by using Lyman-break galaxies (Steidel et al 1998) and near-infrared photometry. This is deep enough for shallow wide field surveys so one can be reasonably confident that the redshift distribution of the lensed galaxies will not be a major problem. Conversely, this is another argument in favor of shallow rather than deep surveys which would reach limiting magnitudes beyond the capabilities of spectrographs. Indeed, as shown by Van Waerbeke et al (1999) the signal to noise ratio of the variance and the skewness does not strongly depend on the redshift of the sources, so it is useless to reach very faint magnitudes.

Source Clustering Due to the intrinsic clustering of the galaxies, the redshift distribution can be broad enough to mix together the population of lensed galaxies and the galaxies associated with the lensing structures. In particular, an extended massive structure at high redshift can play simultaneously the role of a lens and a reservoir of lensed galaxies. The average redshift distribution of the sources can therefore be biased by the galaxies located within the massive structure, which can bias as well the estimated value of the convergence in a similar way. Indeed, the variance of the convergence is not affected by this clustering (Bernardeau 1998a). However, the skewness is much more affected, mainly because the overlapping acts exactly as a non-linear evolution of the projected density. Bernardeau (1998a) shows that most of these effects are negligible on scales beyond 10 arcminutes. It would also be interesting to investigate more deeply how the source clustering may contribute to spurious signals on small scales.

The apparent change of the two-point correlation function by magnification bias can also change the local redshift distribution of lensed sources. This effect, though mentioned by Bernardeau (1998a), has not been investigated in detail.

Lens Coupling When ray bundles cross two lenses by accident, the cumulative convergence is given by the product of the magnification matrix, and not simply the sum of the two convergences. The resulting convergence contains additional coupling terms that must be estimated. Fortunately, in the weak lensing regime, this coupling appears to be negligible. It does not change the value of the variance, and the skewness is only weakly modified (Bernardeau et al 1997, Schneider et al 1997).

Validity of the Born Approximation The effects of mass density fluctuations caused by large-scale structures on the deformation of the ray bundles are computed assuming that the Born approximation is valid, that is, the deformation can be computed along the unperturbed geodesic. In the case of linear perturbations, this assumption is valid at the lowest order. As discussed by Bernardeau et al (1997), the correction on the skewness should be at the percent level. However, this is less obvious once lens couplings are taken into account. The validity of the Born approximation has not been tested in detail, so far. This certainly deserves inspection using high-resolution numerical simulations. The simulations done by Jain & al (1998) or Wambsganss et al (1998) could provide valuable information on this issue.

Intrinsic Correlated Polarization of Galaxies It is possible that the intrinsic orientations of galaxies are not randomly distributed but have a coherent alignment correlated to the geometry of the large-scale structures in which they are embedded (Bingelli 1982). If so, the coherent alignment produced by weak lensing will be contaminated by the intrinsic alignment of the galaxies, and a mass reconstruction based on the shear pattern will be impossible. Such alignments could appear during the formation of large-scale structures or could result from dynamical evolution of galaxies within deep potential wells, such as superclusters or clusters of galaxies (see Coutts 1996, Garrido et al 1993 and references therein). However, the most recent numerical simulations do not show such correlations. Many attempts have been made to search for signatures of these intrinsic coherent patterns. So far, no convincing observations of nearby structures have demonstrated that there are large-scale coherent alignments. This possibility has to be investigated thoroughly, in particular in nearby large-scale structures where a coherent alignment from gravitational lensing effect is negligible. It would be valuable to have more quantitative estimates of possible trends for alignments from future very high-resolution numerical simulations of structure formation.

5. GALAXY-GALAXY LENSING IN FIELD GALAXIES

Evidence that galaxies have dark halos comes from the kinematical and dynamical studies of galaxies. However, the geometry of the halos and the amount and distribution of dark matter are unknown and in practice difficult to probe. Gravitational lensing could provide valuable insight in this field: because it works on all scales, in principle the halos of galactic dark matter should be observed from their gravitational lensing effects on background galaxies. The first Einstein rings and the other galaxy-scale lensing candidates have provided unique opportunities to measure the mass-to-light ratios and to probe the mass profiles of a few galaxies (Kochanek 1991). In the case of rings, the mass of the lensing galaxies can be very well constrained (see, for instance, Kochanek 1995), so the properties of the halos inferred from modeling are reliable. However, Einstein rings are rare lensing events, so the sample is still very small.

A more promising approach consists of a statistical study of the deformation of distant galaxies by foreground galactic halos. The galaxy-galaxy lensing analysis uses the correlation between the position of foreground galaxies and the position-alignment of their angular-nearest neighbors among the background population. If the alignment is assumed to be produced by the gravitational shear of the foreground halos, then it is possible to probe the mass of the halos, if the redshift distributions of the foregrounds and the backgrounds are known. A statistical analysis is then possible, if one assumes that all the foreground galaxies have similar halos, which can be scaled using the Tully-Fisher relation and the photometric data (galaxy luminosity). The expected gravitational distortion is very weak: for foregrounds at redshift $\langle z_l \rangle = 0.1$, backgrounds at $\langle z_s \rangle = 0.5$, and typical halos with velocity dispersion of 200 kms^{-1} and radius of 100 kpc, $|\gamma| \approx 1\%$ at about 20 kpc from the center. But if the observations go to very faint magnitudes there is a huge number of background lensed galaxies, so that the weakness of the signal is compensated by the large statistics. It is worth pointing out that the signal-to-noise ratio depends on the number of galaxy pairs, so either very wide field shallow surveys or ultra-deep imaging can be used for the statistics (although they will not probe similar angular scales).

Tyson et al (1984) made the first attempt, using approximately 50,000 background and 11,000 foreground galaxies obtained from photographic plates. The 3σ upper limit of the circular velocity that they found was 160 km.sec^{-1}, with a maximum cutoff radius below $50h_{100}^{-1}$ kpc. These values are significantly smaller than theoretical expectations from rotation curves and dynamical analyses of galaxies. However, despite a careful examination of possible systematics (Tyson et al 1984, Tyson 1985), there are two limitations to Tyson et al's pioneering work. First, as emphasized by Kovner & Milgrom (1987), the assumption that background galaxies are at infinite distances has considerable impact on the constraints on the circular velocity and the cutoff radius. If one includes a corrective factor which takes into account the distances of the sources, the upper limit for the circular velocity is considerably higher (330 kms^{-1} for a L_* galaxy) and no constraints can be put on the cutoff radius (Kovner & Milgrom 1987). Second, the image quality of the photographic plate is poor and may also affect the measurement of weak distortions.

The first attempt to use deep CCD subarcsecond images was made by Brainerd et al (1996) using about 5,000 galaxies. The distortion was compared with simulations, based on analytical profiles for the dark matter halos, and the Tully-Fisher

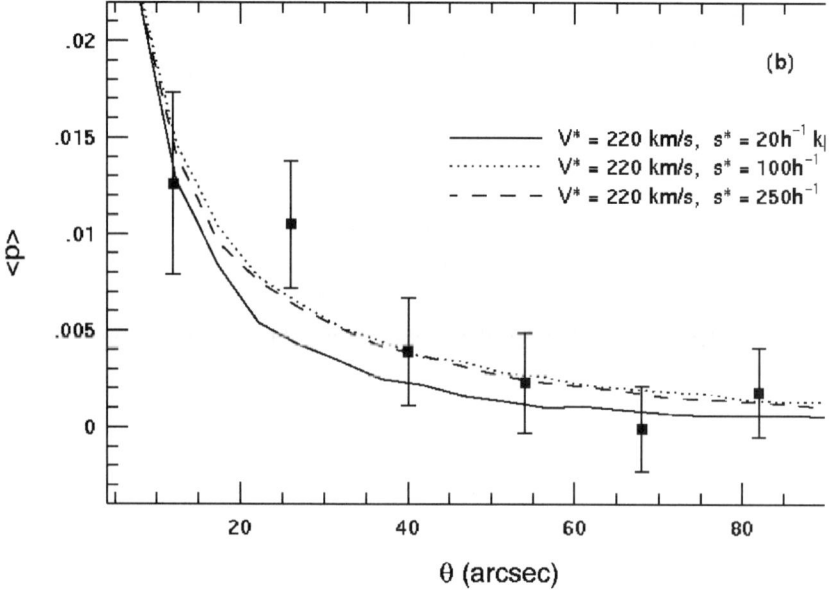

Figure 9 Angular variation of polarization produced by weak lensing of foreground galaxies on the background (lensed) sources in the Brainerd et al (1996) sample. The lines show theoretical expectations for three models of halos having different velocity dispersion and scales.

relation, in order to relate mass models to observations. After careful investigations of systematics, they detected a significant polarization of about 1%, averaged over a separation between $5''$ and $34''$ (see Figure 9). They concluded that halos smaller than $10h^{-1}$ kpc are excluded at a 2σ level, but the data are compatible with halos larger than $100h^{-1}$ kpc and circular velocities of 200 kms^{-1}. Similar works using 23,000 galaxies have been done recently using very deep images obtained with MOCAM at CFHT (Cuillandre et al 1996) with seeing below $0.7''$. They found remarkably similar results as Brainerd et al for the polarization and its evolution with radius (Erben et al, in preparation).

The HST data look perfectly suited for this kind of program which demands high image quality and the observation of many field galaxies. Griffiths et al (1996) used the Medium Deep Survey (MDS) and measured the distortion produced by foreground elliptical and spiral galaxies. They found results similar to those of Brainerd et al (1996) but with a signal more significant for foreground elliptical than spiral galaxies. The comparison with shear signals expected from various analytical models seems to rule out de Vaucouleurs' law as mass density profile of ellipticals. Ebbels et al (in preparation) are now extending the MDS work to a larger sample, trying to simulate more carefully the selection effects. Dell'Antonio & Tyson (1996) and Hudson et al (1998) analyzed the galaxy-galaxy lensing signal in the HDF. As compared with the ground-based images or the MDS, the field is small but the depth permits the use of many background galaxies even on a scale smaller than 5 arcseconds. Furthermore, the UBRI data of the HDF permit the inference of accurate photometric redshifts for the complete sample of galaxies. Dell'Antonio & Tyson compared the lensing signal with predictions from an analytical model for the halo. They found a significant distortion of about 7% at $2''$ from the halo center which corresponds to halos with typical circular velocities of less than 200 km.sec^{-1}. The results obtained by Hudson et al (1998) are consistent with those of Dell'Antonio & Tyson (1996) and Brainerd et al (1996). However, their maximum-likelihood analysis more accurately takes into account the collective effects of large-sized halos (Schneider & Rix 1997). In contrast to previous studies, Hudson et al made careful corrections of images from the PSF and scaled the magnitude inferred from the analytical models of the halos using the Tully-Fischer relation.

Galaxy-galaxy lensing is potentially a very valuable tool for studying the dynamics of galaxies, complementary of standard methods using photometry and spectroscopic data. Because the foreground galaxies have an average redshift of approximately 0.1, galaxy-galaxy lensing also offers an opportunity to look at the dynamical evolution of galaxies by comparing local galaxies at redshift zero to intermediate-redshift galaxies (Hudson et al 1998). However, there are still some limitations due to the rather small number of galaxies used in each sample. As shown from the simulations by Schneider & Rix, it is rather easy to constrain the velocity dispersion of the halos, but more difficult to put limits on their physical scale (see Figure 10). There are also uncertainties coming from the models of halos (which are assumed to be spherical) and the additional noise produced by cosmic shear which can contaminate the galaxy-galaxy signal. Although these issues should be analyzed in more detail in the future, dramatic changes in the results are not expected (Schneider & Rix 1997). In particular, the weak lensing induced by large-scale structures should indeed be canceled by the averaging procedure of galaxy-galaxy lensing.

6. GRAVITATIONAL TELESCOPE AND HIGH-Z UNIVERSE

6.1 Redshift Distribution of Galaxies Beyond B = 25

Gravitational lensing magnifies part of the distant universe and permits exploration of the redshift distribution of faint galaxies as well as the morphology and the contents of very distant galaxies. As discussed in Section 4.4, information on the distances of the sources is relevant for the weak lensing inversion, because the mass reconstruction uses a grid of faint distant sources whose redshift distribution is basically unknown. In particular, this hampers the mass estimates of high-redshift lensing clusters which are very sensitive to the redshifts of the background

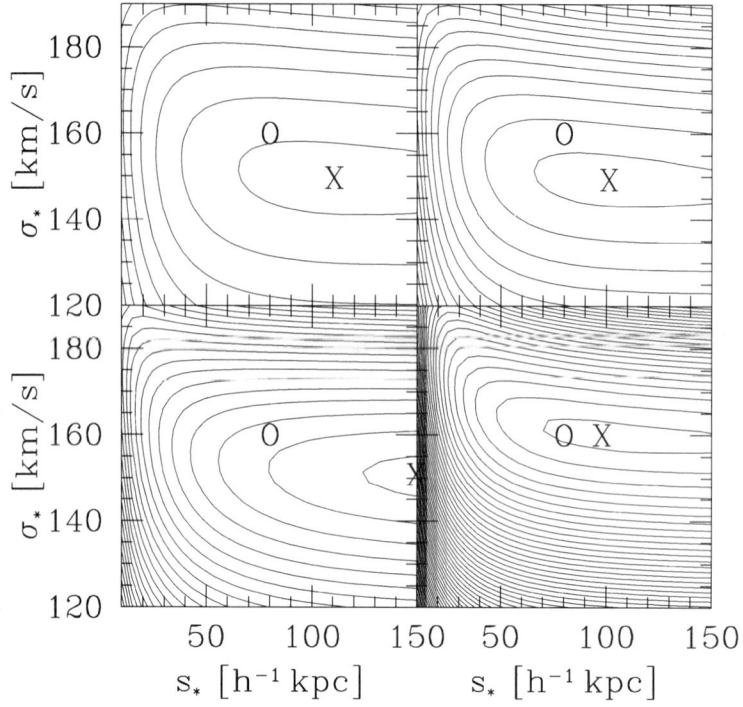

Figure 10 Simulations of galaxy-galaxy lensing done by Schneider & Rix (1997) for four samples of galaxies: 795 (*top left*), 1165 (*top right*), 2169 (*bottom left*) and 3137 (*bottom right*). The O symbol indicates the input parameters of the model and the X shows the maximum of the likelihood function. The likelihood isocontours shows that the velocity dispersion is recovered easily, in contrast to the scale which requires many more galaxies.

sources (Luppino & Kaiser 1997). Unfortunately, beyond $B = 25$, even giant optical telescopes are too small for spectroscopy and the redshift of a complete sample of $B > 25$ galaxies cannot be secured in a reasonable amount of observing time. The possibility of using photometric redshifts is very promising, but observations as well as tests of the reliability of the method for the faintest galaxies are still underway and require careful control. Moreover, because it is hopeless to calibrate the photometric redshifts of the faint samples with spectroscopic data, a cross-check of the predictions of photometric redshift and "lensing-redshift" is important.

Spectroscopic Surveys of Arclets Spectroscopic surveys of redshifts of arc(let)s are long and difficult tasks but are definitely indispensable for lensing studies. They permit computation of the angular distances D_{OL}, D_{LS} and D_{OS} and therefore obtain the absolute scaling of the projected mass density. These redshifts also directly

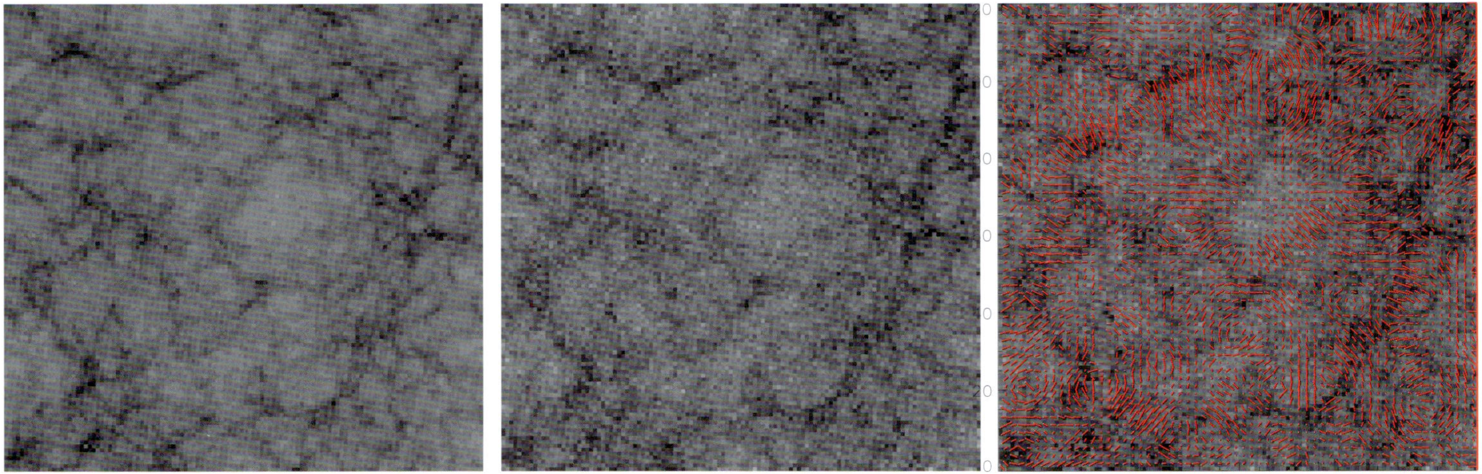

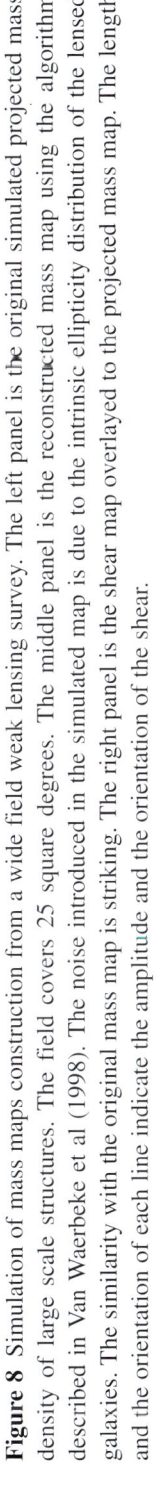

Figure 8 Simulation of mass maps construction from a wide field weak lensing survey. The left panel is the original simulated projected mass density of large scale structures. The field covers 25 square degrees. The middle panel is the reconstructed mass map using the algorithm described in Van Waerbeke et al (1998). The noise introduced in the simulated map is due to the intrinsic ellipticity distribution of the lensed galaxies. The similarity with the original mass map is striking. The right panel is the shear map overlayed to the projected mass map. The length and the orientation of each line indicate the amplitude and the orientation of the shear.

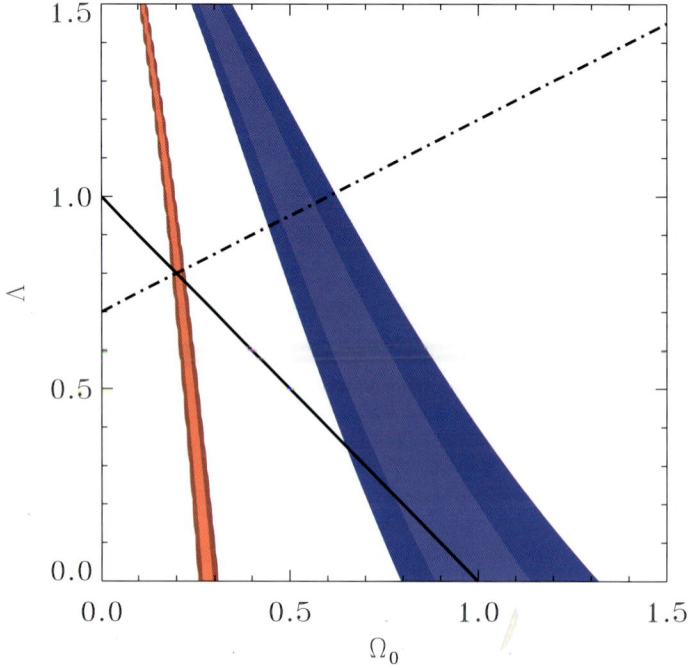

Figure 13 Constraints on (Ω, λ) from a weak lensing survey covering 10×10 square-degrees. The bright and dark regions refer to 1σ and 2σ level. The left strip is for an $\Omega = 0.3$ universe, and the right band for an $\Omega = 1$ universe. The solid and dot-dashed lines correspond to a zero-curvature universe and to a fixed deceleration parameter, respectively (from Van Waerbeke et al 1999).

probe the positions of critical lines which eventually constrain the local mass distribution for some detailed models (Kneib et al 1993, 1996; Natarajan et al 1998; Kneib et al, in preparation). The development of the lensing inversion technique (see Section 6.1.2) also requires spectroscopic confirmations of its predictions to demonstrate that this is a reliable and efficient method. Last but not least, it is in principle also possible to obtain information on the cosmological parameters if one could have enough redshifts to constrain both the mass distribution of the lens and the geometry of the Universe.

Spectroscopic surveys of the "brightest" arclets in many clusters are progressing well. Some extremely distant galaxies have been discovered, such as the arc(let)s in Cl13587+6245 at $z = 4.92$ (Franx et al 1997, see Figure 11), in A1689 at $z = 4.88$ (Frye & Broadhurst, private communication), in A2390 at $z = 4.04$ (Frye & Broadhurst 1998), in Cl0939+4713, where three $z > 3$ arcs have been detected (Trager et al 1997), or the hyperluminous galaxy in A370 at $z = 2.8$ (Ivison et al 1998). Extensive spectroscopic follow-up is also under way in A2390 (Bézecourt & Soucail 1997, Frye et al 1998) and in A2218 (Ebbels et al 1996, 1998). The spectroscopic observation by Frye et al (1998) is a good example of what could be expected from redshifts of arc(let)s: from their Keck observations,

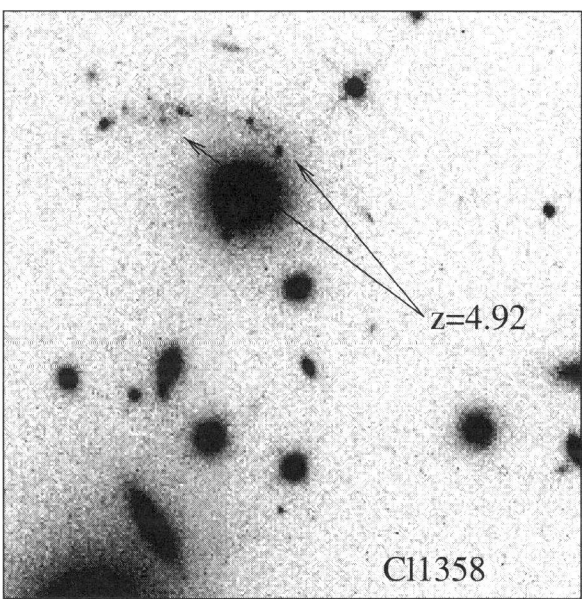

Figure 11 The giant arc detected in Cl1358+6245 is the most distant arc ever observed ($z = 4.92$). The strong magnification permits the detection of some inner structures in a lensed galaxy at $z \approx 5$. Many bright spots are visible (Franx et al 1997). The brightest spot on the left is also visible in the Soifer et al (1998) J-band observations and seems to be a dense core.

they show that the straight arc in A2390 is actually composed of two lensed galaxies aligned along the same direction, one at $z = 0.931$ (reported earlier by Pelló et al 1991) and the other one at $z = 1.033$. These observations confirm the early conclusions from multicolor photometry (Smail et al 1993) as well as from theoretical considerations (Kassiola et al 1992b) that this straight arc should be composed of two galaxies.

About 50 redshifts of arc(let)s have been measured so far. However, from this sample it is difficult to infer valuable information on the redshift distribution of $B > 25$ galaxies or to constrain evolution models of galaxies because it is biased in an unknown way. Since most of these objects are very faint, only arclets showing bright spots on HST images revealing star-forming regions are generally selected. These features, which increase the probability of detecting emission lines, optimize the chance to get reliable redshifts but generate a sample of arclets where star forming galaxies are preferentialy selected. Furthermore, owing to the steep slope of galaxy counts beyond $B = 25$ the magnification bias favors observations of blue galaxies rather than red ones. So, even if the spectroscopy of arclets is crucial for the lens modeling and eventually to obtain the spectral energy distribution of high-redshift galaxies, the spectroscopic sample of arc(let)s must be handled carefully and a detailed analysis of the selection function is needed prior to statistical studies. In the meantime, it is important to focus on getting a few spectra of extremely high-redshift galaxies ($z > 5$) that could be observable thanks to high magnification.

Redshift Distribution from Lensing Inversion If it is possible to recover the lensing potential with good accuracy, the lensing equation can be inverted in order to send the lensed image back to its source plane. The shape of the source can in principle be predicted for any redshift beyond the lens position. The basic principle of the lensing-inversion approach was initially discussed by Kochanek (1990) and refined later by Kneib et al (1994, 1996). If the shape of the galaxies sent back beyond the lens plane is parameterized by the quantity $\tau = (a^2 - b^2)/2ab \; e^{2i\theta}$, then it is easy to show that in the weak lensing regime, the complex quantity τ_I and its projection on a y-axis, τ_y, writes (Kneib et al 1994, 1996):

$$\tau_I = \tau + \tau_S; \quad \tau_{S_y} = \tau_{I_y}, \tag{38}$$

where the subscripts I and S refer to the image and the source, respectively. Therefore, τ_y is an invariant. The conditional probability of a source to be at redshift z, given the shape and the position of the image, and for a given mass model is:

$$p(z \mid model) = \frac{p(\tau_{S_x}; \tau_{S_y})}{p(\tau_{S_y})}. \tag{39}$$

For a simple distribution of the shape of the sources, it turns out that this probability is maximum at the redshift where the deformation of the source is minimized. Therefore, the lensing-inversion predicts that the most probable redshift, for a given model, is where the unlensed galaxies have a minimum ellipticity. This intuitive assumption proposed by Kneib et al (1994) was established on an observational basis by Kneib et al (1996) using the HST-MDF galaxies as a fair sample of unlensed sources. The obvious interest of this method is that it does not depend on the magnitude of the arclet but on its position and its shape in the image plane. Potentially, it provides the redshift of any arclet up to the limiting magnitude of the observations.

The lensing inversion was first applied on A370 (Kochanek 1990, Kneib et al 1994), from the lens modeling of the giant arc and some multiple images. Though the (unlensed) magnitude-redshift diagram found for these arclets shows a good continuity with the faint spectroscopic surveys (Mellier 1997), some of the predicted redshifts are uncertain. In fact, as shown by Fort & Mellier (1994, Figure 12), the X-ray isophotes and the arclet positions do not follow the expectations of the lens modeling of the eastern region. This is an indication that although the modeling is excellent in the cluster center, the mass distribution does not have a simple geometry beyond the giant arc and therefore the lens model in this region is uncertain. Similar complex substructures could exist on scales below the resolution of the mass maps and could also produce wrong redshift estimates. Furthermore, the lensing inversion is also sensitive to the accuracy of the shape measurements of each arclet, which in the case of very faint objects could be an important source of uncertainty.

There are two solutions to solve these issues: first, it is highly preferable to use HST images instead of ground based images. The recent spectroscopic confirmations by Ebbels et al (1996, 1998) of most of the redshifts predicted from the lensing-inversion in A2218 from the HST data (Kneib et al 1996) are wonderful demonstrations of the capabilities of such a technique when used with superb images. Second, it is important to focus on lensing-clusters with simple geometry in order to lower the uncertainties on the lens modeling. In this respect, though A370 and A2218 are rather well modeled, they are not the simplest, and clusters such as MS0440, A1689 or MS2137-23 appear to be better candidates.

Probing Source Redshifts Using Various Lens Planes A more natural and simple way to infer the redshift distribution of the faint galaxies is to look for arc(let)s or weak lensing signals through a set of different lensing clusters having increasing redshifts. The ratio of lensed versus unlensed faint galaxies and the amplitude of the shear patterns as a function of redshift directly probe the spatial distribution of the galaxies along the line of sight. This idea was tentatively explored by Smail et al (1994) who analyzed the lensing signal in three lensing clusters at redshifts 0.26, 0.55 and 0.89. They found that most $I < 25$ objects cannot be low-z dwarf galaxies and concluded that a large fraction of $I = 25$ galaxies are beyond $z = 0.55$. The absence of any significant lensing signal in the most distant cluster led to inconclusive results on the fraction of these galaxies that could be at very large redshift. Fortunately, the distant clusters observed by Luppino & Kaiser (1996) and Clowe et al (1998) provided considerable insights about the high-redshift tail of faint galaxies. The detection of weak lensing in three $z > 0.75$ clusters put strong constraints on their samples of $23.5 < I < 25.5$ galaxies, which must be dominated by a $z > 1$ population. These three lensing clusters strengthen the conclusions obtained from the shear detected around Q2345+007 (Bonnet et al 1993) which is also produced by a high-redshift cluster (Mellier et al 1994, Fischer et al 1994, Pelló et al 1996).

The use of distant clusters is promising because it is a direct consequence of the detection of weak lensing signals, regardless of the accuracy of the mass reconstruction. The shape of the redshift distribution of the galaxies can be inferred if many clusters at different redshifts map the lensing signal. Up to now, the number of clusters is still low, but it will continuously increase during the coming years. However, it is worth noting that the derived shape of the redshift distribution also depends on the accuracy of the lens modeling, as well as on the dynamical evolution of clusters with look-back time. At high redshift it is possible that the lensing signal decreases rapidly, not only because of the absence of background sources, but also because clusters of galaxies are no longer dense and massive enough to produce gravitational distortion. It will be important to disentangle these two different processes.

The Distribution of Faint Galaxies From the Magnification Bias When the slope of the galaxy number count is lower than 0.3, a sharp decrease in the galaxy number density is expected close to the critical radius corresponding to the redshift of the background sources (see Equation 24). For a broad redshift distribution, the cumulative effect of each individual redshift results in a shallow depletion area which spreads over two limiting radii corresponding to the smallest and the largest critical lines of the populations dominating the redshift distribution. Therefore, the shape and the width of depletion curves reveal the redshift distribution of the background sources, and their analysis should provide valuable constraints on the distant galaxies. Similar to the lensing inversion, this is a statistical method that also needs a very good modeling of the lens; but in contrast to it, it does not need information on the shapes of arclets, so the "depletion-redshift" could be a more relevant approach for very faint objects.

This method was first used by Fort et al (1997) in the cluster Cl0024+1654 to study the faint distant galaxy population in the extreme magnitude ranges $B = 26.5$–28 and $I = 25$–26.5. The (unlensed) galaxy number counts were first calibrated using CFHT blank fields and checked from comparison with the HST-HDF counts. In this magnitude range, the slopes are close to 0.2, so these populations can produce a highly contrasted depletion area, so they are best suited for this project. In this cluster, the lower boundary of the depletion is sharp and the growth curve toward the upper radius extends up to 60 arcseconds from the cluster center, as expected if the high-redshift tail is a significant fraction of the lensed galaxies. Fort et al concluded that $60\% \pm 10\%$ of the B-selected galaxies are between $z = 0.9$ and $z = 1.1$ while most of the remaining $40\% \pm 10\%$ galaxies appear to be broadly distributed around a redshift of $z = 3$. The I selected population shows a similar bimodal distribution, but spreads up to a larger redshift range with about 20% above $z > 4$.

There is no spectroscopic confirmation yet that the double-shape redshift distribution predicted by Fort et al is real. Indeed, there are still uncertainties related to this method: in the particular case of Cl0024+1654, the redshift of the arc used to scale the mass was assumed to be close to 1. We know from recent Keck spectroscopic observation that this arc is at redshift 1.66 (Broadhurst et al 1999), so the innermost calibration of the sources has to be rescaled. The method is also sensitive to the lens modeling of the projected mass density. In the case of Cl0024+1654, it is rather well constrained from the subarcsecond ultra-deep images of Fort et al, and the predicted velocity dispersion is very close to the measured values from the galaxy radial velocities (Dressler & Gunn 1992).

This approach was recently generalized by Bézecourt et al (1998), in order to predict the number counts of arc(let)s with a magnification larger than a lower limit. The prediction first needs a model for galaxy evolution that reproduces some typical features of field galaxies, such as the observed galaxy counts and redshift distributions of spectroscopic surveys. The best final model can then be used to predict the number of arc(let)s brighter than a lower surface brightness limit and with magnification larger than a lower limit, for any lensing cluster whose mass distribution can be modeled properly. Bézecourt et al (1998, 1999) used this technique to build the best model capable of producing the number of arc(let)s observed in A370 and A2218. They also predicted a bimodal redshift distribution, but their expected number of giant arcs is overestimated by a factor of two. This inconsistency is still difficult to interpret. Indeed, the method should be handled carefully, because it both depends on the modeling of the lens and on the modeling of galaxy evolution. In particular, fitting of counts and redshift distributions of galaxies produces models that have degeneracies (Charlot 1999) that are potential limitations. Nevertheless, like the Fort et al (1997) method, the Bézecourt et al generalization is an interesting and original idea which can certainly be improved in the future using much better data. As compared with the lensing-inversion, it does not depend on the shapes of distant galaxies, and just needs very deep counts [the Bézecourt et al (1998, 1999) method does require shapes of lensed galaxies, but they only use reasonably magnified arc(let)s, so it is not a difficulty]. In the "depletion-redshift" technique this is a great advantage for the faintest (most distant?) objects, because, as the HDF images show, many of them have bright spots but do not show regular morphologies, so the lensing-inversion procedure could be inefficient for these galaxies. These very first attempts must be pursued on many lensing clusters in order to provide reliable results on the redshift distribution of the faintest galaxies.

The redshift distribution obtained by these various techniques is summarized in Figure 12. The distribution is broad and the comparison with Figure 9 of Fort & Mellier (1994) shows that the median redshift and the width of the distribution increases continuously. The median redshift distribution of giant arcs was close to 0.4 four years ago and has increased up to 0.7 but with a more pronounced high-redshift tail. A significant fraction of the new redshifts exceeds 1.5, with a visible trend toward very high redshifts ($z > 2.5$). The median redshift obtained from lensing inversions is close to 0.7, a good correlation with other methods. It

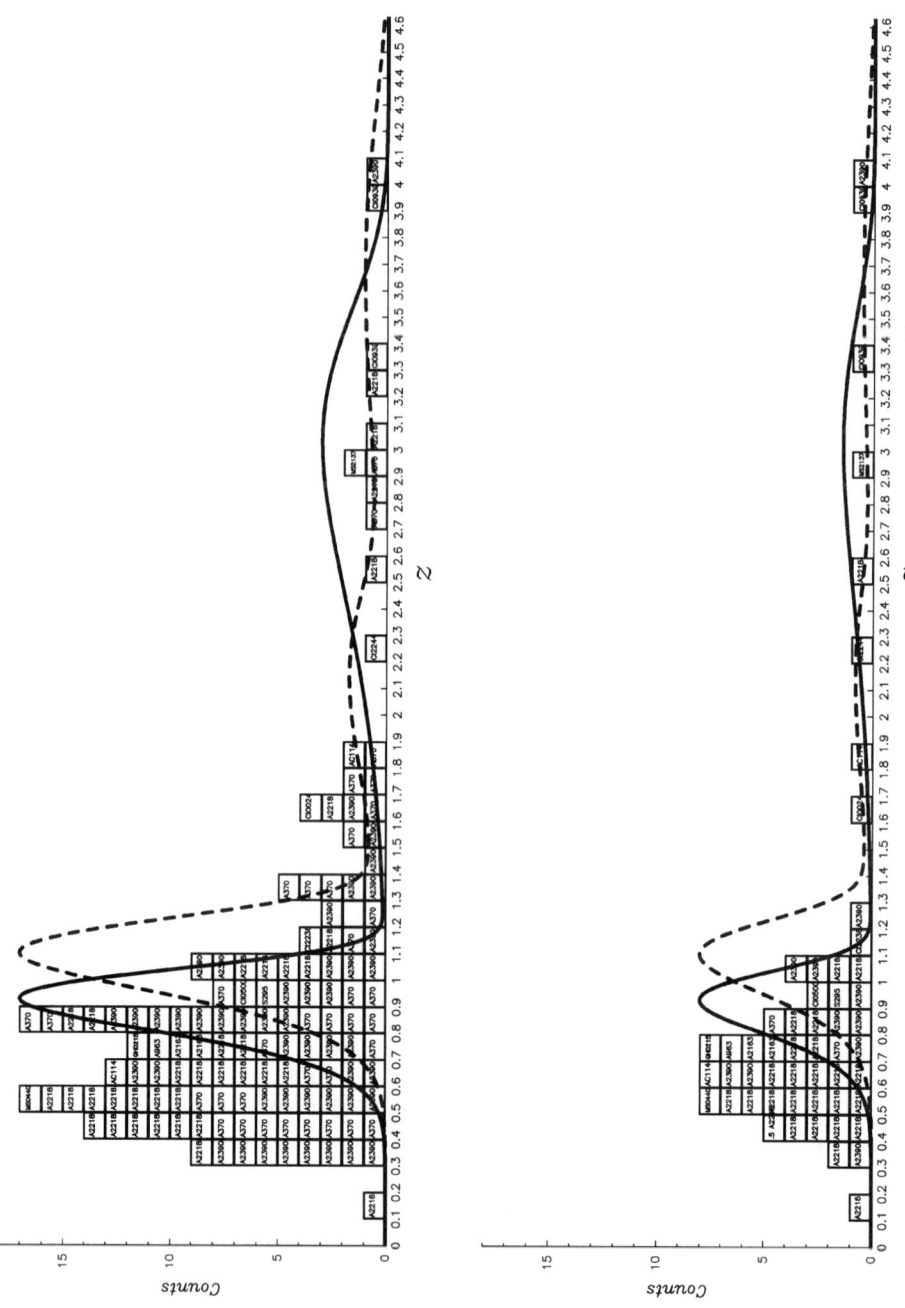

is surprising to see that the redshift distribution obtained by the depletion curves predict two peaks which seem to be visible in the redshift distribution of arc(let)s as well, both on the spectroscopic and lensing inversion samples. Owing to the somewhat different selection criteria used for these two samples, a resemblance was not really expected, and it may be an indication that selection biases in the spectroscopic sample of arc(let)s are not critical.

Because the observations of lensed galaxies simultaneously use magnification, color selection, shape selection (elongation) and relative position with respect to the cluster center (deviation angle), it is possible to jointly select drop-out galaxies with the radial-distance criterion of an elongated object in order to select extremely distant galaxies. Because this method seems to be efficient, it will certainly be applied to select samples of $z > 5$ galaxies. This will probably be a main goal for the future.

6.2 Spectral Content of Arc(let)s

The strong magnification of giant arcs also permits one to study the content and star formation rates of high-redshift galaxies. Preliminary studies started with Mellier et al (1991) and Smail et al (1993) who explored the spectral content of some arcs. These samples do not show spectacular starburst galaxies and seem to be compatible with a continuous star formation rate. The HST images confirm that many of these galaxies have bright spots with ongoing star formation. The star formation rates inferred from new optical spectra of arcs in A2390 (Bézecourt & Soucail 1996), in A2218 (Ebbels et al 1996), in Cl1358+6245 (Franx et al 1997) or in Cl0939 (Trager et al 1997) range from 5 to 20 $M_\odot$/yr and are consistent with other observations (Bechtold et al 1998), but none of the rates computed for arclets are corrected from dust extinction.

It is only very recently that the material of magnified arcs has been studied in detail. Trager et al (1997) made the first attempt to estimate the metallicity of the arclets at $z > 3$ detected in Cl0939 with the Keck telescope, and found that they are metal-poor systems, having $Z < 0.1\ Z_\odot$. The very first CO observations at IRAM of the giant arc in A370 [Casoli et al 1996: CO(J=2-1) detected] and at Nobeyama Observatory in MS1512-cB58 [Nakanishi et al 1997: CO(J=3-2) undetected] have demonstrated that (sub)millimeter observations are feasible thanks to the magnification and can provide useful diagnoses on the molecular and gas content of galaxies at high redshifts. If, as suggested by the cosmic infrared background (Puget et al 1996; see also Guiderdoni 1998 and references therein), a significant fraction of the UV emission of distant galaxies is released in the

←

Figure 12 Redshift distribution inferred from lensing techniques. The bottom panel is the spectroscopic sample of arc(let)s compiled from the literature. The top panel includes the spectroscopic sample and the redshift predictions of arclets from lensing inversion. On both histograms, the plots of the redshift distributions from the depletion curves in B (*solid line*) and I (*dashed line*) prediction by Fort et al (1997) are shown.

submillimeter range, the observations of lensed galaxies in the submillimeter and millimeter bands could be a major step in our understanding of the history of star formation in galaxies. Blain (1997) emphasized that the joint submillimeter flux-density/redshift relation and the steep slope counts make the observations of lensed distant galaxies in this waveband an optimum strategy, so a large number of bright (magnified) sources are expected. Both SCUBA (at 450 μm and 850 μm) and IRAM (at 1.3 mm) can therefore benefit from magnification of distant lensed galaxies. The large field of view and the wavelength range of SCUBA at JCMT seem perfectly suited for this program. Smail et al (1997b) are carrying out a long-term program of observations of lensing clusters with this instrument. They detected sources in A370 and Cl2244-02 with a success rate which implies that the number density of these galaxies is about 3 times higher than expected from the 60 μm IRAS count. Their observations of a new sample of 7 lensing clusters (Smail et al 1999) show that the energy emitted by these galaxies is much higher than the expectations from nearby galaxies. Most of these galaxies are at redshift larger than 1, and probably less than 5.5. The star formation activity of high-redshift galaxies is therefore important, and for those which have an optical counterpart, their morphology reveals signs of merging processes. Therefore, the star formation activity seems frequently triggered by interactions (Smail et al 1999), which corroborates the recent ISOCAM observations in some giant arcs, as in A2390 (Lémonon et al 1998). The star formation rates measured from the various fluxes have a very broad range, between 50 to 1000 $M_\odot$/yr, but they are difficult to estimate accurately, in particular for the hyperluminous system in A370 (Ivison et al 1998) because AGNs could contribute significantly to the flux.

The submillimeter observation is certainly one of the most promising tools for the future. The magnification and the shape of the continuum produced by dust make the "submillimeter gravitational telescope" perfect for studying the high-redshift Universe and the star formation history of galaxies.

6.3 Morphology of Highly Magnified Galaxies

Although the coupling of gravitational telescopes with the high resolution images from HST is superb for probing the intrinsic morphology of arc(let)s, no significant results have been raised from recent studies. Many HST images of arcs are made of two parallel elongated arcs, as in A2390 (Kneib et al, in preparation) or in MS2137-23 (Hammer et al 1997; see arcs A and C in MS2137-23 in Figure 3), which supports the idea that these distant galaxies are interacting systems, as it was reported in the previous section. Other images show bright spots which are interpreted as star forming regions, like Cl2244-02 (Hammer & Rigaud 1989), A2390 (Pelló et al, in preparation) or Cl0024+1654 (Colley et al 1996). In the case of the giant arc in Cl1358+62 at $z = 4.92$, the comparion of the visible and the near-infrared observations obtained with the HST and the Keck telescopes (Soifer et al 1998) reveals that one of its bright spots already contains half the stellar mass of the galaxy. It shows that at this redshift, galaxies may already have dense cores.

Furthermore, the visible and near infrared imaging and spectrospic data show also that reddening produced by dust is important, even at that redshift.

The detailed description of a $z \approx 5$-galaxy given by Soifer et al strengthens the usefulness of image reconstruction techniques of high-redshift lensed galaxies, as the one initially proposed by Kochanek et al (1989). Some attempts have already been made in Cl0024+1654 (Wallington et al 1995, Colley et al 1996) or in MS2137-23 (Hammer et al 1997). They succeed in reproducing a single image in the source plane, but the details of the morphology are still uncertain and do not yet provide valuable information on distant galaxies. Therefore, it is still premature to present quantitative results on the distant galaxies, first because this sample is poor and incomplete, and second because there are too many uncertainties in the source reconstruction.

7. COSMOLOGICAL PARAMETERS

The potential of gravitational lensing for cosmography, already discussed by Blandford & Narayan (1992) and Fort & Mellier (1994) for the particular cases of arc(let)s, is clear but also challenging. The increasing evidence that the whole set of observations are compatible with a non-zero cosmological constant, for instance, motivated many new studies devoted to the constraints on λ, following the early suggestions by Paczynski & Gorski (1981). One of the most promising and reliable approaches explored by recent theoretical studies has been raised by Bernardeau et al (1997) and Van Waerbeke et al (1999). These works clearly demonstrate that the determination of the cosmological parameters (Ω_0, λ) using future surveys of weak lensing induced by large-scale structures should provide Ω with a high accuracy (see Figure 7 and Figure 13). Since this was already pointed out in Section 4.1, I do not discuss it in more detail here. In the following sections, I focus on new and more speculative investigations, which are still difficult to implement but seem promising in the future.

7.1 Constraints from Cluster Reconstruction

Following the early suggestion by Paczynski & Gorski (1981) for multiply imaged quasars, Breimer & Sanders (1992) (see also Fort & Mellier 1994, Link & Pierce 1998) emphasized that the ratio of angular diameter distances of arc(let)s having different redshifts does not depend on the Hubble constant and therefore can constrain (Ω, λ). This ratio still depends on the mass distribution within the two critical lines corresponding to redshifts z_1 and z_2, so it is worth noting that it is sensitive to the modeling of the lens. It is only in the case of an isothermal sphere model that the radial positions of the critical lines θ, where arcs at a given redshift are formed, only depend on the angular distances $D(z_s, \Omega_0, \lambda)$:

$$\left(\frac{\theta_1}{\theta_2}\right) = \left(\frac{D_{LS}(z_1, \Omega_0, \lambda)}{D_{OS}(z_1, \Omega_0, \lambda)}\right) \left(\frac{D_{OS}(z_2, \Omega_0, \lambda)}{D_{LS}(z_2, \Omega_0, \lambda)}\right). \tag{40}$$

Because cluster potentials are by far more complex than isothermal spheres, in practice the method works only for very specific cases, such as clusters with regular morphology, and if auxiliary independent data, such as high-quality X-ray images or additional multiple images, help to constrain the lens model. So far, no case has been found where the modeling of two (or more) arc systems at very different redshifts is sufficiently reliable. However, the joint HST images and spectroscopic redshifts obtained with new giant telescopes should provide such perfect configurations in the near future. A1689 or MS0440 seem like good examples of such candidates because both show many arc(let)s and have a regular shape.

A similar approach has been proposed by Hamana et al (1997), using the arc cB58 observed in the lensing cluster MS1512.4+3647. Assuming that the dark matter distribution is sufficiently constrained by the ROSAT and ASCA data, the magnification and number of multiple images of cB58 only depend on the cosmology. One should therefore use the detection of counter-image to cB58 to constrain the domain (Ω_0, λ) which cannot produce a counter-arc. This point was discussed by Seitz et al (1997) who argue that in practice it cannot work because it depends too much on the modeling of the lensing cluster. The variation of the lensing strength as a function of cosmology is small, lower than 0.5% between an EdS universe and an $\Omega = 0.3$, $\lambda = 0$ universe. Furthermore, the use of independent X-ray data to model the dark matter demands a very good understanding of the physics of the hot gas for each individual cluster considered.

More recently, Lombardi & Bertin (1999) have proposed the use of weak lensing inversion to recover simultaneously the cluster mass distribution and the geometry of the universe. The method assumes that the redshifts of the lensed galaxies are known. In that case, for a given cosmology, it is possible to compute the shear at a given angular position which is produced on a lensed galaxy located in a narrow redshift range, from the observed ellipticities of the galaxies at that angular position. Conversely, if the shear is known, then it is possible to infer the best set of cosmological parameters which reproduce the observed ellipticities of the galaxies. Therefore, it is possible to iterate a procedure, starting from an arbitrary guess for the set (Ω, λ), which at the final step will simultaneously procure the best mass inversion with the most probable (Ω, λ). The key point is the assumption that the redshift of each individual source is known. The method should provide significant results if at least a dozen of clusters with different redshift are reconstructed using this iterative procedure (Lombardi & Bertin 1999). Indeed, this inversion is demanding in telescope time since a very good knowledge of the redshifts of many lensed sources is necessary; but otherwise the method seems promising.

7.2 Statistics of Arc(let)s

Up to now, several tens of multiply imaged distant galaxies have already been detected in clusters of galaxies, and this number will probably increase by a large factor within the next few years. Since the fraction of rich clusters (and therefore lensing-clusters) strongly depends on the cosmological scenario, we expect the number of arc(let)s to depend on cosmological parameters (Wilson et al 1996b).

It is well known that present-day statistical studies of clusters of galaxies are limited by the few samples of cluster catalogs with well understood selection function. Wu & Mao (1996) analyzed the statistics of arcs in the EMSS cluster sample (Gioia et al 1990) and show that the fraction is twice the one expected for an EdS Universe, but compatible with a flat $\lambda = 0.7$ model (see also Cooray 1999). Unfortunately, the geometry of the mass distribution and the substructures present in the lensing cluster increase the shear contribution to the magnification and change dramatically the expected number of arcs (Bartelmann et al 1995, Bartelmann 1995b, Hattori et al 1997, Bartelmann et al 1998). The importance of accurate simulations of clusters is clearly obvious from the recent studies. Bartelmann et al (1998) find a totally opposite result to that of Wu & Mao, and conclude that an open model ($\Omega = 0.3$ and $\lambda = 0$) is preferred to any flat models to reproduce the number of arcs observed. The other models, including the $\lambda = 0.7$ model, fail by about of factor of ten. This is a clear demonstration that the use of statistics of arc(let)s to constrain the cosmological scenario is very sensitive to the assumptions. The constraint on λ from the fraction of arc(let)s is therefore rather weak and hopeless for the moment.

7.3 Magnification Bias

The depletion of the galaxy number density as a function of radial distance from the cluster center can potentially provide information on the cosmological constant. The reason for this is ultimately the same as for giant arcs—namely, the ratios of angular distances which strongly depend on the cosmological constant. Therefore, if the redshift distribution of the sources and the mass distribution of the lensing cluster are known, the shape of the depletion curve—in particular, its extension at a large radius—is constrained by λ.

Fort et al (1997) have used this property in order to constrain the cosmological constant. They used ultra-deep images of the lensing clusters Cl0024+1654 and A370 which permit a good signal to noise ratio of the depletion curves. These clusters have giant arcs with known redshift so the mass at a given critical line can be scaled. The method provides jointly the redshift of the sources and the cosmological parameters. Fort et al (1997) found that the location of this high redshift critical line rather favors a flat cosmology with λ greater than 0.6.

It is remarkable that from these two clusters only the method predicts a value of λ compatible with other independent approaches (see White 1998 and references therein). Since it needs a good model for the lens, this method has still many uncertainties and can be significantly improved with a large sample of arc clusters, in particular by using a maximum likelihood analysis applied to the probabilities of reproducing their observed local shears and convergences. A strong improvement can come from the new possibility of using the redshift distribution found independently. This should be possible using photometric redshifts. Even more promising, Gautret et al (1998) proposed to use triplets of neighbouring arclets at a different redshift. Because they are close together, the positions of these arclets are independent of the mass profile but only depend on λ. This in principle breaks the degeneracy and solves this problem.

All the methods described above do not yet provide convincing results on λ mainly because they use simultaneously different quantities which are degenerate without external information: mass distribution of the lensing-cluster, redshift distribution of the sources, cosmological parameters, and evolution scenarios of clusters and of sources. The approach using statistics of arc(let)s looks promising but demands very good simulations and a good understanding of selection functions of cluster samples which are used for comparison with observations. The method using lens modeling requires very good lens models and information on the redshift distribution of galaxies, in particular for the most distant ones, since they contain the population which depends the most on λ. This approach can use the redshift distribution obtained from photometric redshifts, and should focus on regular lensing clusters containing giant arcs with known redshift. As emphasized by Fort et al (1997), Lombardi & Bertin (1999) and Gautret et al (1998), significant results cannot be expected until many clusters have been investigated. This should be done within the next few years, in particular using 10-meter class telescopes. However, it is remarkable that the Fort et al limit corresponds to the value given by Im et al (1997) from the measurement of strong lensing produced by elliptical galaxies, and to the upper limit given by Kochanek (1996) from the statistics of lensed quasars.

8. LENSING THE CMB

The measurement of CMB fluctuations is a major goal for cosmology in the next decade (see White et al 1994, and references therein). The shape of the power spectrum over the whole spectral range contains a huge amount of information which permits one to constrain the cosmological scenario with an incredibly high accuracy. However, the reliability of the interpretation of the features visible on the spectrum requires a complete and detailed understanding of all the physical mechanisms responsible for its final shape. Gravitational lensing induced by foreground systems along the line of sight may play a role, so it is important to predict in advance whether it can modify the signal from the CMB and, if so, what the expected amplitudes of the effects are.

Because surface brightness is conserved by the gravitational lensing effect (Etherington 1933), only fluctuations of the CMB temperature maps can be affected by lensing. However, even for strong lenses, no significant modification of the power spectrum is expected on large scales (Blanchard & Schneider 1987) and therefore there is no hope of detecting positive signals of the coupling between CMB and gravitational lensing in the Cosmological Background Explorer (COBE) maps. Nevertheless, since the COBE-Differential Microvawe Radiometer (DMR) experiment has demonstrated that fluctuations exist (Smooth et al 1992), the study of the lensing effect on smaller scales than COBE resolution is potentially interesting and has some advantages with respect to weak lensing on distant galaxies. First, contrary to lensed galaxies, the redshift of the source, namely the last scattering surface, is well known and spreads over a very small redshift range. Second, with the on-going and the coming of high-resolution ground-based and balloon observations as well as the two survey satellites MAP and Planck-Surveyor, observation of low-amplitude temperature distortions on small scales will become possible and will permit investigation of possible lensing effects.

Early theoretical expectations from Blanchard & Schneider (1987) or Cole & Efstathiou (1989) show that the *shape* of the small-scale temperature fluctuations can be modified by lensing effects; in particular they can redistribute the power in the power spectrum. In contrast, the *amplitude* of the temperature anisotropy on medium and small scales has been a matter of debates during the last decade (see the review by Blandford & Narayan 1992, and more recently Fukushige & Makino 1994; Fukushige et al 1994; Cayón et al 1993a,b, 1994). The conclusions of these works showed strong discrepancies, depending on the assumptions used to explore the deflection of photons and to model inhomogenous universes. Furthermore, the expectation values also depend on the cosmological models. Indeed, the most recent critical studies show that the effect of lensing on large scales is negligible (Seljak 1996, Martínez-González et al 1997). In particular, the non-linear evolution of the power spectrum does not significantly increase the amplitude on these scales. On the other hand, the gravitational lensing effect reduces the power spectrum on small scales, and eventually can smooth out acoustic peaks on scales below $l \approx 2000$ (Seljak 1996). Martínez-González et al (1997) obtained conclusions similar to those of Seljak—that is, the contribution of lensing is small but not negligible and should be taken into account in the detailed analysis of future CMB maps. Furthermore, the transfer of power from large to small scales induces an important increase of power in the damping tail, which results in a decrease of very small scale amplitudes at a smaller rate than expected without lensing (Metcalf & Silk 1997, 1998). According to Metcalf & Silk (1997), 30% of the additional power at $l = 3000$ comes from $l < 1000$ scales, and 8% from $l < 500$ in the case of a $\sigma_8 = 0.6$, $h = 0.6$ model.

Zaldarriaga & Seljak (1998) pointed out that gravitational lensing not only smoothes the temperature anisotropy, but can also change the polarization. The polarization spectra are more sensitive to gravitational lensing effects than the power spectrum of the temperature because the acoustic oscillations of polarization spectra have sharper oscillations and can be smoothed out more efficiently by lensing than temperature fluctuations. The effect is small but can reach amplitudes of approximately 10% for $l < 1000$ scales. More remarkably, because of the coupling between E-type and B-type polarizations (Seljak 1997b), gravitational lensing can generate low amplitude B-type polarization, even if none is predicted from primary fluctuations (for instance, for scalar perturbations).

Because the signal is weak and only concerns the small scales, temperature and polarization fluctuations induced by gravitational lensing will be difficult to measure with high accuracy and seems a hopeless task before the Planck-Surveyor mission. It is therefore valuable to explore alternatives which could provide better or complementary information which couples lensing and CMB. An interesting idea is to analyze the non-Gaussian features induced by the displacement fields generated by gravitational lensing on the CMB maps (Bernardeau 1997, 1998b). As for weak lensing on distant galaxies, the CMB temperature map can be sheared and magnified. The resulting distortion patterns are direct signatures of the coupling between the CMB and the foreground lenses. Bernardeau (1998b) argued that the distortion map produced by lensing can be decoupled from other fluctuation patterns because it generates similar magnification and deformation on close temperature patches, which therefore can be correlated. He also explored the consequences of the non-Gaussian signal on the four-point correlation function. Unfortunately, the signal is very small, and it is even not clear on which scale the signal is highest, in particular because the non-linear evolution of the power spectrum was not considered by Bernardeau. The weakness of the signal and the fact that the four-point correlation function could be contaminated by other non-Gaussian features are strong limitations of Bernardeau's suggestion. Therefore, Bernardeau (1998b) preferred to focus on the modification of the ellipticity distribution function of the temperature patches induced by lensing. From his simulated lensed maps, a clear change of the topology of the temperature maps is visible: the shapes of the structures are modified and their contours look sharper than for the unlensed maps. However, the signal is still marginally detectable on a $10° \times 10°$ map, even with Planck-Surveyor.

From these investigations it is clear that weak lensing on the CMB has small effects on the spectrum of the temperature and polarization power spectra and on the non-Gaussianity of the CMB temperature maps. However, with typical amplitude of 1% to 10% percent they can be detected with future missions, so they must be taken into account for detailed investigations of the CMB anisotropy on small scales. This is an important prediction since the detection of gravitational lensing perturbations of the CMB will be possible with Planck-Surveyor. Its high sensitivity and spatial resolution are sufficient to permit one to break the geometrical degeneracy expected from linear theory, and to disentangle different (Ω, λ) common models (Metcalf & Silk 1998; Stompor & Efstathiou 1999). It is worth noting that these analyses can be used jointly with the weak lensing maps of large-scale structures on background galaxies which will also provide (Ω, λ) with a very good accuracy.

9. FUTURE PROSPECTS

Although the use of weak lensing analysis and its applications in cosmology made spectacular progress during the last five years, most of the astrophysical questions addressed by Fort & Mellier (1994) in their conclusions are still pending.

However, it seems that we are progressing quickly in the right direction, even if some of these problems are complex and should be envisioned in a long-term perspective.

The HST images have dominated most of the results, in particular in the modeling of clusters of galaxies. Thanks to the formidable work devoted to mass reconstruction, the projected mass density of clusters of galaxies are now robust and reliable. It is now important to couple strong and weak lensing features (Seitz et al 1998, AbdelSalam et al 1998b, Dye & Taylor 1998, Van Kampen 1998) in order to build consistent models for clusters. It is worth noting that for many of the issues discussed in this review, it was emphasized that precise and reliable mass reconstructions of clusters of galaxies are crucial and determine the reliability of many scientific outcomes. In this respect, it is important to keep in mind that the redshift distribution of the sources is indispensable and that the new giant telescopes will be the best tools for this purpose.

From the sample of clusters already analyzed, there are converging results that $\Omega < 0.3$ with a high confidence level. However, complete cluster samples are necessary for deeper investigations of cluster properties. They should come out rapidly from weak lensing studies of ROSAT samples (Rosati 1999). Indeed, we now have a very good understanding of the mass distribution of each component (dark matter, hot gas, galaxies) in clusters of galaxies and we are close to understanding the discrepancy between the lensing mass and the X-ray mass of clusters. During the next five years, one can reasonably expect significant improvements in our knowledge of the dynamics of clusters of galaxies by jointly using weak lensing reconstruction, from HST and giant telescopes images, and a full description of the hot gas, from AXAF and XMM observations.

In contrast, the investigation of galaxy halos from galaxy-galaxy lensing is still in its infancy and the preliminary results presented in this review must be confirmed. A new generation of instruments will contribute to the development of this hot topic. Below 10 arcseconds down to 2 arcseconds, "wide field" HST observations with the new Advanced Camera devoted to deep galaxy-galaxy lensing studies appear to be a promising approach. Beyond this scale, the high image quality of telescopes like Keck, GEMINI, Subaru, Magellan, the VLT or CFHT will be decisive in obtaining valuable constraints from galaxy-galaxy lensing analysis between 10 and 60 arcseconds.

Parallel to these studies, we can now envision fully exploiting some of the most valuable information that gravitational lensing can provide, namely the relation(s) between light and mass distributions in the universe. Theoretical studies have demonstrated that in the near future the weak lensing analysis coupled with the study of the galaxy distribution will allow us to understand the evolution of the biasing factor with scale and redshift. However, it is important to explore the case of non-linear and stochastic biasing in order to understand which parameters can be reasonably constrained. The possible existence of large dark halos around galaxies or in clusters of galaxies is also an unknown but fascinating topic. In this respect, the dark cluster candidates discussed by Hattori et al (1997) and Erben et al (1999), or the remarkable distortion field detected by Bonnet et al (1994) in the periphery of Cl0024+1654, which does not seem to be correlated to luminous matter, deserves more careful investigation.

The study of the contents and the past history of galaxies made formidable progress as well. It is clear that jointly using the magnification of cluster-lenses with the unprecedented image quality of HST, or with the wide field and high sensitivity of SCUBA, results in highly competitive tools. In the future, continuous developments are expected, but the observation of extremely high redshift galaxies which could not be observed without magnification is certainly a major goal, in particular in the submillimeter wavebands. As demonstrated by the recent study of Soifer et al (1998), the coming of optical and near-infrared spectroscopic capabilities on the giant telescopes will permit one to study in detail their spectral energy distribution and the kinematics of their stellar and gas components. The magnification permits the viewing of a huge amount of detail on the images of these galaxies, and we can envision probing small details of these galaxies from image reconstruction "à la Kochanek" (Kochanek et al 1989). Unfortunately, though theoretical tools have been developed in order to recover the morphology of these lensed galaxies, the quality of optical and submillimeter data are not good enough to produce reliable details of the sources from inversion. This is probably a goal for the Large Southern Array (LSA) which will have much better sensitivity and resolution.

With new instruments, such as Megacam at CFHT (Boulade et al 1998) or the VST at the European Southern Observatory (Arnaboldi et al 1999) in Paranal, we are now entering the era of wide field subarcsecond imaging surveys which will produce the first shear-limited samples or, similarly, the first mass selected catalogues of gravitational condensations (Reblinsky & Bartelmann 1999). Their designs are optimized to investigate weak lensing induced by large-scale structures in order to produce the first mass maps of the universe. They will permit recovering of the detailed spectrum of the projected power spectrum of mass density fluctuations as well as measuring (Ω, λ) with an accuracy greater than 10% (see Figure 13, color). There are still some theoretical issues (see Sections 3.3 and 4.4) that must be addressed in detail from both the theoretical and simulation points of view. For most of them, there are no crucial conceptual difficulties, so they should be fixed rapidly. On the other hand, the control of systematics which can affect weak lensing measurement, as well as the correction for an anisotropic PSF, could be critical and should be considered seriously in the future for very low shear amplitudes ($<1\%$). Nevertheless, these cameras, as well as the very wide field surveys of the VLA-FIRST and the SDSS, should provide a major breakthrough in weak lensing applications for cosmology.

In the longer term, after the crucial results expected for mass maps with wide-field CCD cameras, the New Generation Space Telescope (NGST) and Planck Surveyor could be ultimate steps in this area. The potential interest of NGST for weak lensing has been summarized by Schneider & Kneib (1998), who argued that low-mass clusters and groups of galaxies as well as very distant clusters should be detectable with this telescope. In parallel, as reported by Stompor & Efstathiou (1998) and Metcalf & Silk (1998), weak lensing on the CMB should be able to break the geometrical degeneracy and ultimately provide the (Ω, λ) parameters. This coupling between observations of CMB fluctuations and weak lensing analyses emerges as a consecration illustrating the major roles played by these two complementary approaches to present-day cosmology.

ACKNOWLEDGMENTS

I am particularly grateful to B. Fort for his advices, and his friendly and continuous encouragements during the long period of preparation of the review. I would like to thank first F Bernardeau, F Casoli, S Charlot, R Ellis, P Schneider and L Van Waerbeke for their careful reading and useful comments of the manuscript as well as for their strong support during its writing. I thank all the other close collaborators who participate to our gravitational lensing projects, and the colleagues with whom we had many fruitful discussions, namely M Bartelmann, H Bonnet, T Broadhurst, J-C Cuillandre, T Erben, H Hoekstra, B Jain, N Kaiser, J-P Kneib, C Kochanek, J-F Le Borgne, P-Y Longaretti, G Luppino, D Narasimha, R Pelló, M Pierre, C Seitz, S Seitz, U Seljak, I Smail, G Soucail and G Squires. I thank I Gioia for providing me data prior to publication, and T Brainerd, B Jain, P Schneider and G Squires, for giving me their authorization to publish figures of their papers in this review. I thank especially M Dantel-Fort for her assistance and also for the crucial work she does in order to have all these data set ready for our scientific objectives. Part of this work was supported by the Programme National de Cosmologie which is funded by the Centre National de la Recherche Scientifique, the Commissariat à l'Énergie Atomique and the Centre National d'Études Spatiales, under the responsibility of the Institut National des Sciences de l'Univers and the Indo-French Centre for the Promotion of Advanced Research IFCPAR grant 1410-2.

Visit the Annual Reviews home page at http://www.AnnualReviews.org

LITERATURE CITED

AbdelSalam HM, Saha P, Williams LLR. 1998a. *MNRAS* 294:734–46
AbdelSalam HM, Saha P, Williams LLR. 1998b. *Astron. J.* 116:1541–52
Allen SW. 1998. *MNRAS* 296:392–406
Amram P, Sullivan WT, Balkowski C, Marcelin M, Cayatte V. 1993. *Ap. J. Lett.* 403:L59–62
Arnaboldi M, Capaccioli M, Mancini D, Rafanelli P, Sedmak G, Scaramella R, Vettolani GP. 1999. In *Wide Field Surveys in Cosmology*, ed. S Colombi, Y Mellier, B Raban. Paris: Frontières
Babul A, Lee MH. 1991. *MNRAS* 250:407–13

Bar-Kana R. 1996. *Ap. J.* 486:17–27
Bartelmann M. 1995a. *Astron. Astrophys.* 298:661–71
Bartelmann M. 1995b. *Astron. Astrophys.* 299:11–16
Bartelmann M. 1995c. *Astron. Astrophys.* 303:643–55
Bartelmann M. 1996. *Astron. Astrophys.* 313:697–702
Bartelmann M, Narayan R. 1995. *Ap. J.* 451:60–75
Bartelmann M, Narayan R, Seitz S, Schneider P. 1996. *Ap. J. Lett.* 464:L115–18

Bartelmann M, Schneider P. 1993a. *Astron. Astrophys.* 268:1–13
Bartelmann M, Schneider P. 1993b. *Astron. Astrophys.* 271:421–24
Bartelmann M, Schneider P. 1994. *Astron. Astrophys.* 284:1–11
Bartelmann M, Schneider P, Hasinger G. 1994. *Astron. Astrophys.* 290:399–411
Bartelmann M, Steinmetz M, Weiss J. 1995. *Astron. Astrophys.* 297:1–12
Bartelmann M, Huss A, Colberg JM, Jenkins A, Pearce FR. 1998. *Astron. Astrophys.* 330:1–9
Bechtold J, Elston R, Yee HKC, Ellingson E, Cutri RM. 1998. In *The Young Universe*, ed. S D'Odorico, A Fontana, E Giallongo, pp. 241–48. PASP Conf. Series. Vol. 146
Benítez N, Martínez-González E. 1997. *Ap. J.* 477:27–35
Bernardeau F. 1997. *Astron. Astrophys.* 324:15–26
Bernardeau F. 1998a. *Astron. Astrophys.* 338:375–82
Bernardeau F. 1998b. *Astron Astrophys.* 338:767–76
Bernardeau F, Van Waerbeke L, Mellier Y. 1997. *Astron. Astrophys.* 322:1–18
Bézecourt J, Soucail G. 1997. *Astron. Astrophys.* 317:661–69
Bézecourt J, Pellé R, Soucail G. 1998. *Astron. Astrophys.* 330:399–411
Bezecourt J, Kneib J-P, Soucail G, Ebbels TMD. 1999. *Astron. Astrophys.* 347:21–29
Bingelli B. 1992. *Astron. Astrophys.* 107:338–49
Blain A. 1997. *MNRAS* 290:553–65
Blanchard A, Schneider J. 1987. *Astron. Astrophys.* 184:1–6
Blandford RD. 1990. *Q. Jl. R. Astron. Soc.* 31:305–31
Blandford RD, Jaroszyński M. 1981. *Ap. J.* 246:1–12
Blandford RD, Saust AB, Brainerd TG, Villumsen JV. 1991. *MNRAS* 251:600–27
Blandford RD, Narayan R. 1992. *Annu. Rev. Astron. Astrophys.* 30:311–58
Bonnet H, Fort B, Kneib J-P, Mellier Y, Soucail G. 1993. *Astron. Astrophys. Lett.* 280:L7–10
Bonnet H, Mellier Y, Fort B. 1994. *Ap. J. Lett.* 427:L83–86
Bonnet H, Mellier Y. 1995. *Astron. Astrophys.* 303:331–44
Böhringer H, Tanaka Y, Mushotzky RF, Ikebe Y, Hattori M. 1998. *Astron. Astrophys.* 334:789–98
Boulade O, Vigroux L, Charlot X, de Kat J, Borgeaud P, Roussé JY, Mellier Y, Gigan P, Crampton D. 1998. *Astronomical Telescopes and Instrumentation.* SPIE Vol. 3355
Bouchet F. 1996. Preprint astro-ph/9603013
Bower RG, Smail I. 1997. *MNRAS* 290:292–302
Brainerd TG, Blandford RD, Smail I. 1996. *Ap. J.* 466:623–37
Breimer TG, Sanders RH. 1992. *MNRAS* 257:97–104
Bridle SL, Hobson MP, Lasenby AN, Saunders R. 1998. *MNRAS* 299:895–903
Broadhurst T. 1995. Preprint astro-ph/9511150
Broadhurst T, Taylor AN, Peacock J. 1995. *Ap. J.* 438:49–61
Canizares CR, 1981. *Nature* 291:620–24
Casoli F, Encrenaz P, Fort B, Boissé P, Mellier Y. 1996. *Astron. Astrophys. Lett.* 306:L41–44
Cayón L, Martínez-González E, Sanz JL. 1993a. *Ap. J.* 403:471–75
Cayón L, Martínez-González E, Sanz JL. 1993b. *Ap. J.* 413:10–13
Cayón L, Martínez-González E, Sanz JL. 1994. *Astron. Astrophys.* 284:719–23
Charlot S. 1999. In *The Next Generation Space Telescope: Science Drivers and Technological Challenges*, ed. P Benvenuti et al. ESA SP-429
Clowe D, Luppino GA, Kaiser N, Henry JP, Gioia IM. 1998. *Ap. J. Lett.* 497:L61–64
Cole S, Efstathiou G. 1989. *MNRAS* 239:195–200
Colombi S, Szapudi I, Szalay A. 1998. *MNRAS* 296:253–74
Colley WN, Tyson JA, Turner E. 1996. *Ap. J. Lett.* 461:L81–86
Connolly AJ, Csabai I, Szalay A, Koo DC, Kron RG, Munn JA. 1995. *Astron. J.* 110:2655–64

Cooray AR. 1999. *Astron. Astrophys.* 341:653–61
Coutts A. 1996. *MNRAS* 278:87–94
Cuillandre J-C, Mellier Y, Dupin J-P, Tilloles P, Murowinski R, Crampton D, Woof R, Luppino GA. 1996. *Pub. Astron. Soc. Japan* 108:1120–28
Dekel A, Lahav O. 1998. Preprint astro-ph/9806193
Dell'Antonio IP, Tyson JA. 1996. *Ap. J. Lett.* 473:L17–20
Deltorn J-M, Le Fèvre O, Crampton D, Dickinson M. 1997. *Ap. J. Lett.* 483:L21–24
Dolag K, Bartelmann M. 1997. *MNRAS* 291:446–54
Donahue M, Voit M, Gioia IM, Hughes J, Stocke J. 1998. *Ap. J.* 502:550–57
Dressler A, Gunn JE. 1992. *Ap. J. Supp.* 78:1–60
Dressler A, Oemler A, Sparks WB, Lucas RA. 1994. *Ap. J. Lett.* 435:L23–26
Dye S, Taylor A. 1998. *MNRAS* 300:L23–28
Ebbels T, Le Borgne J-F, Pelló R, Ellis RS, Kneib J-P, Smail I, Sanahuja B. 1996. *MNRAS* 281:75–81
Ebbels T, Le Borgne J-F, Pelló R, Ellis RS, Kneib J-P, Smail I, Sanahuja B. 1998. *MNRAS* 295:75–91
Ellis RS. 1997. *Annu. Rev. Astron. Astrophys.* 35:389–444
Erben T, Van Waerbeke L, Mellier Y, Schneider P, Cuillandre J-C, Castander F, Dantel-Fort M. 1999. Preprint astro-ph/9907134
Etherington IMM. 1933. *Phil. Mag.* 15:761–75
Evans NW, Wilkinson Ml. 1998. *MNRAS* 296:800–12
Fahlman G, Kaiser N, Squires G, Woods D. 1994. *Ap. J.* 437:56–62
Fischer P, Tyson JA. 1997. *Astron. J.* 114:14–24
Fischer P, Tyson JA, Bernstein GM, Guhathakurta P. 1994. *Ap. J. Lett.* 431:L71–74
Fischer P, Bernstein G, Rhee G, Tyson JA. 1997. *Astron. J.* 113:521–30
Fort B, Prieur J-L, Mathez G, Mellier Y, Soucail G. 1988. *Astron. Astrophys. Lett.* 200:L17–20
Fort B, Le Fèvre O, Hammer F, Cailloux M. 1992. *Ap. J. Lett.* 399:L125–29
Fort B, Mellier Y. 1994. *Astron. Astrophys. Rev.* 5:239–92
Fort B, Mellier Y, Dantel-Fort M, Bonnet H, Kneib J-P. 1996. *Astron. Astrophys.* 310:705–14
Fort B, Mellier Y, Dantel-Fort M. 1997. *Astron. Astrophys.* 321:353–62
Franx M, Illingworth GD, Kelson DD, Van Dokkum PG, Tran K-V. 1997. *Ap. J. Lett.* 486:L75–78
Frye BL, Broadhurst TJ. 1998. *Ap. J. Lett.* 499:L115 18
Frye BL, Broadhurst TJ, Spinrad H, Bunker A. 1998. In *The Young Universe*, ed. S D'Odorico, A Fontana, E Giallongo, pp. 182–85. PASP Conf. Series. Vol. 146
Fugmann W. 1990. *Astron. Astrophys.* 240:11–21
Fukushige T, Makino J. 1994. *Ap. J. Lett.* 436:L111–14
Fukushige T, Makino J, Ebisuzaki T. 1994. *Ap. J. Lett.* 436:L107–10
Garrido JL, Battaner E, Sánchez-Saavedra ML, Florido E. 1993. *Astron. Astrophys.* 271:84–88
Gautret L, Fort B, Mellier Y. 1998. Preprint astro-ph/9812388
Geiger B, Schneider P. 1998. *MNRAS* 295:497–510
Geiger B, Schneider P. 1999. *MNRAS* 300:118–30
Gioia IM, Shaya EJ, Le Fèvre O, Falco EE, Luppino G, Hammer F. 1998. *Ap. J.* 497:573–86
Gioia IM, Maccacaro T, Schild RE, Wolter A, Stocke JT, Morris SL, Henry JP. 1990. *Ap. J. Suppl.* 72:567–619
Gioia IM, Henry JP, Mullis CR, Ebeling H, Wolter A. 1999. *Astron. J.* 117:2608–16
Girardi M, Fadda D, Escalera E, Giuricin G, Mardirossian F, Mezzetti M. 1997. *Ap. J.* 490:56–62
Gorenstein MV, Falco EE, Shapiro II. 1988. *Ap. J.* 327:693–711
Griffiths RE, Casertano S, Im M, Ratnatunga KU. 1996. *MNRAS* 282:1159–64

Guiderdoni B, 1998. In *The Young Universe*, ed. S D'Odorico, A Fontana, E Giallongo, pp. 283–88. PASP Conf. Series. Vol. 146
Gunn J. E. 1967. *Ap. J.* 150:737–53
Hamana T, Hattori M, Ebeling H, Henry P, Futumase T, Shioya Y. 1997. *Ap. J.* 484:574–80
Hammer F, Rigaud F. 1989. *Astron. Astrophys.* 226:45–56
Hammer F, Teyssandier P, Shaya EJ, Gioia IM, Luppino GA. 1997. *Ap. J.* 491:477–82
Hattori M, Ibeke Y, Asaoka I, Takeshima T, Böhringer H, Mihara T, Neumann DM, Schindler S, Tsuru T, Tamura T. 1997. *Nature* 388:146–48
Hattori M, Watanabe K, Yamashita K. 1997. *Astron. Astrophys.* 319:764–80
Hoekstra H, Franx M, Kuijken K, Squires G. 1998. *Ap. J.* 504:636–60
Hudson MJ, Gwyn S, Dahle H, Kaiser N. 1998. *Ap. J.* 503:531–42
Im M, Griffiths RE, Ratnatunga KU. 1997. *Ap. J.* 475:457–61
Ivison RJ, Smail I, Le Borgne J-F, Blain AW, Kneib J-P, Bézecourt J, Kerr TH, Davies JK. 1998. *MNRAS* 298:583–93
Jain B, Seljak U. 1997. *Ap. J.* 484:560–73
Jain B, Seljak U, White S. 1998. In *Fundamental Parameters in Cosmology*, ed. J Tran Thanh Van, Y Giraud-Héraud, F Bouchet, T Damour, Y Millier. Les Arcs: Frontières
Kaiser N. 1992. *Ap. J.* 388:272–86
Kaiser N. 1995. *Ap. J. Lett.* 439:L1–3
Kaiser N. 1998. *Ap. J.* 498:26–42
Kaiser N, Squires G. 1993. *Ap. J.* 404:441–50
Kaiser N, Squires G, Broadhurst T. 1995. *Ap. J.* 449:460–75
Kaiser N, Wilson G, Luppino G, Kofman L, Gioia I, Metzger M, Dahle H. 1998. Preprint astro-ph/9809268
Kamionkowski M, Babul A, Cress CM, Réfrégier A. 1998. *MNRAS* 301:1064–72
Kassiola A, Kovner I, Fort B. 1992a. *Ap. J.* 400:41–57
Kassiola A, Kovner I, Blandford RD. 1992b. *Ap. J.* 396:10–19
Kneib J-P, Mellier Y, Fort B, Mathez G. 1993. *Astron. Astrophys.* 273:367–76
Kneib J-P, Mathez G, Fort B, Mellier Y, Soucail G, Longaretti P-Y. 1994. *Astron. Astrophys.* 286:701–17
Kneib J-P, Mellier Y, Pelló R, Miralda-Escudé J, Le Borgne J-F, Böhringer H, Picat J-P. 1995. *Astron. Astrophys.* 303:27–40
Kneib J-P, Ellis RS, Smail I, Couch W, Sharples RM. 1996. *Ap. J.* 471:643–56
Kneib J-P, Alloin D, Mellier Y, Guilloteau S, Barvainis R, Antonucci R. 1998. *Astron. Astrophys.* 329:827–39
Kochanek CS, Blandford RD, Lawrence CR, Narayan R. 1989. *MNRAS* 238:43–56
Kochanek CS. 1990. *MNRAS* 247:135–51
Kochanek CS. 1991. *Ap. J.* 373:354–68
Kochanek CS. 1995. *Ap. J.* 445:559–77
Kochanek CS. 1996. *Ap. J.* 466:638–59
Kovner I, Milgrom M. 1987. *Ap. J. Lett.* 321:L113–15
Krauss LM. 1998. Preprint astro-ph/9807376
Kristian J, Sachs RK. 1966. *Ap. J.* 143:379–86
Kristian J. 1967. *Ap. J.* 147:864–67
Kruse G, Schneider P. 1999. *MNRAS* 302:821–29
Lémonon L, Pierre M, Cesarsky C, Elbaz D, Pelló R, Soucail G, Vigroux L. 1998. *Astron. Astrophys. Lett.* 334:L21–25
Link R, Pierce M. 1998. *Ap. J.* 502:63–74
Lombardi M, Bertin G. 1998a. *Astron. Astrophys.* 330:791–800
Lombardi M, Bertin G. 1998b. *Astron. Astrophys.* 335:1–11
Lombardi M, Bertin G. 1999. *Astron. Astrophys.* 342:337–52
Luppino G, Kaiser N. 1997. *Ap. J.* 475:20–28
Lynds R, Petrosian V. 1986. *BAAS* 18:1014
Maddox SJ, Efstathiou G, Sutherland WJ, Loveday J. 1990. *MNRAS* 243:692–712
Markevitch M. 1997. *Ap. J. Lett.* 483:L17–20
Martínez-González E, Sanz JL, Cayón L. 1997. *Ap. J.* 484:1–6
Mellier Y. 1997. In *The Hubble Space Telescope and the High Redshift Universe*, ed. NR Tanvir, A Aragón-Salamanca, JV Wall. pp. 237–48. Cambridge: World Scientific
Mellier Y, Fort B, Soucail G, Mathez G, Cailloux M. 1991. *Ap. J.* 380:334–43

Mellier Y, Fort B, Kneib J-P. 1993. *Ap. J.* 407:33–45

Mellier Y, Dantel-Fort M, Fort B, Bonnet H. 1994. *Astron. Astrophys. Lett.* 289:L15–18

Mellier Y, Van Waerbeke L, Bernardeau F, Fort B. 1997. In *Neutrinos, Dark Matter and the Universe*, ed. T Stolarcyk, J Tran Thanh Van, F Vannucci, pp. 191–204. Paris: Frontières

Metcalf B, Silk J. 1997. *Ap. J.* 489:1–6

Metcalf B, Silk J. 1998. *Ap. J. Lett.* 492:L1–4

Miralda-Escudé J. 1991. *Ap. J.* 380:1–8

Miralda-Escudé J, Babul A. 1995. *Ap. J.* 449:18–27

Moessner R, Jain B. 1998. *MNRAS Lett.* 294:L18–24

Moessner R, Jain B, Villumsen J. 1998. *MNRAS* 294:291–98

Mould J, Blandford R, Villumsen J, Brainerd T, Smail I, Small T, Kells W. 1994. *MNRAS* 271:31–38

Nakamura TT 1997. *Publ. Astron. Soc. Japan* 49:151–57

Nakanishi K, Ohta K, Takeuchi TT, Akiyama M, Yamada T, Shioya Y. 1997. *Publ. Astron. Soc. Japan* 49:535–38

Navarro JF, Frenk CS, White SDM. 1997. *Ap. J.* 490:493–508

Natarajan P, Kneib J-P. 1997. *MNRAS* 287:833–47

Natarajan P, Kneib J-P, Smail I, Ellis RS. 1998. *Ap. J.* 499:600–7

Ota N, Mitsuda K, Fukazawa Y. 1998. *Ap. J.* 495:170–78

Paczynski B, Gorski K. 1981. *Ap. J. Lett.* 248:L101–4

Peacock JA, Dodds S. 1996. *MNRAS Lett.* 280:L19–26

Pelló R, Sanahuja B, Le Borgne J-F, Soucail G, Mellier Y. 1991. *Ap. J.* 366:405–11

Pelló R, Miralles J-M, Le Borgne J-F, Picat J-P, Soucail G, Bruzual G. 1996. *Astron. Astrophys.* 314:73–86

Pierre M, Le Borgne J-F, Soucail G, Kneib J-P. 1996. *Astron. Astrophys.* 311:413–24

Puget J-L, Abergel A, Boulanger F, Bernard J-P, Burton WB, Désert F-X, Hartmann D. 1996. *Astron. Astrophys.* 308:L5–8

Reblinksy K, Bartelmann M. 1999. *Astron. Astrophys.* 345:1–6

Réfrégier A, Brown ST. 1998. Preprint astro-ph/9803279

Refsdal S. 1964. *MNRAS* 128:307–10

Refsdal S, Surdej J. 1994. *Rep. Prog. Phys.* 56:117–85

Rodrigues-Williams LL, Hogan CJ. 1994. *Astron. J.* 107:451–60

Rosati P. 1999. In *Wide Field Surveys in Cosmology*, ed. S Colombi, Y Mellier, B Raban. pp. 219–29. Paris: Frontières

Sachs RK. 1961. *Proc. R. Soc. London* A264:309

Sahu KC, Shaw RA, Kaiser ME, Baum SA, Fergusson HC, Hayes JJE, Gull TR, Hill RJ, Hutchings JB, Kimble RA, Plait P, Woodgate BE, et al. 1998. *Ap. J. Lett.* 492:L125–29

Sanz JL, Martínez-González E, Benítez N. 1997. *MNRAS* 291:418–24

Saraniti DW, Petrosian V, Lynds R. 1996. *Ap. J.* 458:57–66

Schindler S, Hattori M, Neumann DM, Böhringer H. 1997. *Astron. Astrophys.* 317:646–55

Schneider P. 1992. *Astron. Astrophys.* 254:14–24

Schneider P. 1995. *Astron. Astrophys.* 302:639–48

Schneider P. 1998. *Ap. J.* 498:43–47

Schneider P, Weiss A. 1988. *Ap. J.* 327:526–43

Schneider P, Ehlers J, Falco EE. 1992. In *Gravitational Lenses*. New York: Springer

Schneider P, Rix HW. 1997. *Ap. J.* 474:25–36

Schneider P, Seitz C. 1995. *Astron. Astrophys.* 294:411–31

Schneider P, Kneib J-P. 1999. In *The Next Generation Space Telescope: Science Drivers and Technological Challenges*, ed. P Benvenuti et al. ESA SP-429

Schneider P, Van Waerbeke L, Mellier Y, Jain B, Seitz S, Fort B. 1998a. *Astron. Astrophys.* 333:767–78

Schneider P, Van Waerbeke L, Jain B, Kruse G. 1998b. *MNRAS* 296:873–92

Schramm T, Kayser R. 1995. *Astron. Astrophys.* 299:1–10

Seitz C, Kneib J-P, Schneider P, Seitz S. 1996. *Astron. Astrophys.* 314:707–20

Seitz C, Schneider P. 1995a. *Astron. Astrophys.* 297:287–99

Seitz C, Schneider P. 1997. *Astron. Astrophys.* 318:687–99

Seitz S, Collodel L, Pirzkal N, Erben T, Freudling W, Schneider P, Fosbury R, White SDM. 1999. In *Wide Field Surveys in Cosmology*, ed. S Colombi, Y Mellier, B Raban, pp. 203–8. Paris: Frontières

Scitz S, Saglia RP, Bender R, Hopp U, Belloni P, Ziegler B. 1998b. *MNRAS* 298:945–65

Seitz S, Schneider P. 1995b. *Astron. Astrophys.* 302:9–20

Seitz S, Schneider P. 1996. *Astron. Astrophys.* 305:383–401

Seitz S, Schneider P, Bartelmann M. 1998a. *Astron. Astrophys.* 337:325–37

Seljak U. 1996. *Ap. J.* 463:1–7

Seljak U. 1998. *Ap. J.* 506:64–79

Seljak U. 1997. *Ap. J.* 482:6–16

Smail I. 1993. *Gravitational Lensing by Rich Clusters of Galaxies*. PhD Thesis, University of Durham

Smail I, Ellis RS, Aragón-Salamanca A, Soucail G, Mellier Y, Giraud E. 1993. *MNRAS* 263:628–40

Smail I, Ellis RS, Fitchett M. 1994. *MNRAS* 270:245–70

Smail I, Ellis RS, Fitchett M, Edge AC. 1995a. *MNRAS* 273:277–94

Smail I, Couch W, Ellis RS, Sharples RM. 1995b. *Ap. J.* 440:501–9

Smail I, Dickinson M. 1995. *MNRAS* 455:L99–102

Smail I, Hogg DW, Yan L, Cohen JG. 1995c. *Ap. J. Lett.* 449:L105–8

Smail I, Ellis RS, Dressler A, Couch WJ, Oemler A, Sharples R, Butcher H. 1997a. *Ap. J.* 479:70–81

Smail I, Ivison RJ, Blain AW. 1997b. *Ap. J. Lett.* 490:L5–8

Smail I, Ivison RJ, Blain AW, Kneib J-P. 1999. *MNRAS* 302:632–48

Smooth GF, Bennett CL, Kogut A, Wright EL, Aymon J, Boggess NW, Cheng ES, De Amici G, Gulkis S, Hauser MG, Hinshaw G, Jackson DD, Janssen M, Kaita E, Kelsall T, Keegstra P, Lineweaver G, Loewenstein K, Lubin P, Mather J, Meyer SS, Moseley SH, Murdock T, Rokke L, Silverberg RF, Tenorio L, Weiss R, Wilkinson DT, et al. 1992. *Ap. J. Lett.* 396:L1–5

Soifer BT, Neugebauer G, Franx M, Matthews K, Illingworth GD. 1998. *Ap. J.* 501:L171–74

Soucail G, Fort B, Mellier Y, Picat J-P. 1987. *Astron. Astrophys. Lett.* 172:L14–17

Soucail G, Mellier Y, Fort B, Mathez G, Cailloux M. 1998. *Astron. Astrophys. Lett.* 191:L19–22

Squires G, Kaiser N, Babul A, Fahlman G, Woods D, Neumann DM, Böhringer H. 1996a. *Ap. J.* 461:572–86

Squires G, Kaiser N, Fahlman G, Babul A, Woods D. 1996b. *Ap. J.* 469:73–77

Squires G, Kaiser N. 1996. *Ap. J.* 473:65–80

Squires G, Neumann DM, Kaiser N, Arnaud M, Babul A, Böhringer H, Fahlman G, Woods D. 1997. *Ap. J.* 482:648–58

Stebbins A. 1996. Preprint astro-ph/9609149

Stebbins A. 1999. In *Wide Field Surveys in Cosmology*, ed. S Colombi, Y Mellier, B Raban, pp. 197–201. Paris: Frontières

Steidel C, Adelberger K, Giavalisco M, Dickinson M, Pettini M, Kellogg M. 1998. In *The Young Universe*, ed. S D'Odorico, A Fontana, E Giallongo, pp. 428–35. PASP Conf. Series. Vol. 146

Stompor R, Efstathiou G. 1999. *MNRAS* 302:735–47

Szapudi I, Colombi S. 1996. *Ap. J.* 470:131–48

Taylor AN, Dye S, Broadhurst TJ, Benítez N, Van Kampen E. 1998. *Ap. J.* 501:539

Tomita K, Watanabe K. 1990. *Prog. Theo. Phys.* 83:467–90

Trager SC, Faber SM, Dressler A, Oemler A. 1997. *Ap. J.* 485:92–99

Tyson JA. 1985. *Nature* 316:799–800

Tyson JA, Valdes F, Jarvis JF, Mills AP. 1984. *Ap. J. Lett.* 281:L59–62
Tyson JA. 1988. *Astron. J.* 96:1–23
Tyson JA, Valdes F, Wenk RA. 1990. *Ap. J. Lett.* 349:L1–4
Tyson JA, Fischer P. 1995. *Ap. J. Lett.* 446:L55–58
Tyson JA, Kochanski GP, Dell'Antonio IP. 1998. *Ap. J. Lett.* 498:L107–10
Valdes F, Tyson JA, Jarvis JF. 1983. *Ap. J.* 271:431–41
Van Kampen E. 1998. *MNRAS* 301:389–404
Van Waerbeke L. 1998a. *Astron. Astrophys.* 334:1–10
Van Waerbeke L. 1998b. In *Wide Field Surveys in Cosmology.* pp. 189–93. Paris: Frontières
Van Waerbeke L. 1999. *MNRAS.* In press
Van Waerbeke L, Mellier Y. 1997. In *Proceedings of the XXXIst Rencontres de Moriond.* Les Arcs: Frontières
Van Waerbeke L, Mellier Y, Schneider P, Fort B, Mathez G. 1997. *Astron. Astrophys.* 317:303–17
Van Waerbeke L, Bernardeau F, Mellier Y. 1999. *Astron. Astrophys.* 342:15–33
Villumsen JV. 1996. *MNRAS* 281:369–83
Wambsganss J, Cen R, Ostriker JP. 1998. *Ap. J.* 494:29–46
Wallington S, Kochanek CS, Koo DC. 1995. *Ap. J.* 441:58–69
Walsh D, Carswell RF, Weymann RJ. 1979. *Nature* 279:381
Webster R. 1985. *MNRAS* 213:871–88
White M. 1998. Preprint astro-ph/9802295
White M, Scott D, Silk J. 1994. *Annu. Rev. Astron. Astrophys.* 32:319–70
Williams LLR, Irwin M. 1998. *MNRAS* 298:378–386
Wilson G, Cole S, Frenk CS. 1996a. *MNRAS* 280:199–218
Wilson G, Cole S, Frenk CS. 1996b. *MNRAS* 282:501–10
Wu X-P, Mao S. 1996. *Ap. J.* 463:404–8
Wu X-P, Fang L-Z. 1996. *Ap. J.* 461:L5–8
Wu X-P, Fang L-Z. 1997. *Ap. J.* 483:62–67
Zaldarriaga M, Seljak U. 1998. Preprint astro-ph/9803150
Zel'dovich YB. 1964. *Sov. Astron.* 8:13–16
Zhu Z-H, Wu X-P. 1997. *Astron. Astrophys.* 324:483
Zwicky F. 1933. *Helv. Phys. Acta* 6:10
Zwicky F. 1937. *Phys. Rev.* 51:290

THE HR DIAGRAM AND THE GALACTIC DISTANCE SCALE AFTER HIPPARCOS

I. Neill Reid

Dept. of Physics and Astronomy, University of Pennsylvania, Philadelphia, Pennsylvania 19104; e-mail: inr@herschel.physics.upenn.edu

Key Words Stars: subdwarfs, parallaxes, RR Lyraes; Globular clusters: distances; Galactic structure

■ **Abstract** The completion and publication of the Hipparcos astrometric catalogue has revitalized studies in many fundamental areas of Galactic structure and stellar evolution. This article reviews the impact of the new parallax results on our understanding of the location of the main-sequence as a function of abundance, of the luminosity calibration of primary distance indicators and of the Galactic distance scale. Many of these issues remain to be resolved.

1. INTRODUCTION

Calibration of the extragalactic distance scale rests on our ability to determine accurate locations in the observational Hertzsprung-Russell diagram for specific types of stars at particular stages in their evolution. Whether one uses Cepheids, RR Lyraes, planetary nebulae or type I supernovae as links in the distance chain, the underlying assumptions are that in matching similar types of stars in different stellar systems, one is comparing like with like; and that if the calibrators are not identical, systemic variations can be traced to systematic trends with changes in physical characteristics. Given that those criteria are satisfied, in principle all distance-scale investigations can determine *relative* distances between different external galaxies. However, determining absolute distances requires absolute luminosities for individual calibrators. Thus, the accuracy with which one can derive parameters such as the Hubble constant, H_0, depends on the accuracy with which one can measure distances within the Galaxy so as to calibrate the various distance indicators.

Trigonometric parallax measurements offer the only method of directly measuring distances to almost all single stars or multiple star systems. The application of CCD detectors to astrometry (Monet 1988) has shown that sub-milliarcsecond (mas) precision can be achieved from the ground. However, the need for adequate reference stars within the small field of view allowed by current CCDs limits observations to faint apparent magnitudes (V ≥ 14), and hence to nearby stars of low intrinsic luminosities. Moreover, the accuracy of the final parallax is dependent on the transformation from relative to absolute reference frames.

The ESA Hipparcos satellite was designed to address this issue. Full details of the scope of this mission and the subsequent analysis are given by van Leeuwen (1997) and Kovalevsky (1998). Two 13-cm diameter optical telescopes were used to image 1 square-degree regions of the sky, separated by 58°, onto the same focal plane. A photodiode timed stellar transits across a reference grid as the satellite rotated on its axis. Ground-based parallax observations are limited to small angular fields of view, and reference stars share the same parallactic motion as the target—hence the necessity for correction from relative to absolute systems. Hipparcos measured angular separations of stars with significantly different parallax factors, allowing direct determination of absolute parallaxes in the final astrometric solution.

Hipparcos was a targeted mission, rather than a sky survey, with programme objects limited generally to surface densities of eight per square degree. Photon-counting statistics rendered impracticable observations of sources fainter than 13th magnitude. The final catalogue includes positions, proper motions and parallax measurements, as well as BV photometry, for 118,000 stars, including nearly every star brighter than 7th visual magnitude, but only 45,000 stars with $9 < V < 11$ and 4,000 stars between 11th and 13th magnitude.

Those data have formed the basis for a large number of investigations over the last two years (160 refereed publications as of May, 1998). The present article does not aim at a comprehensive review of all issues raised in those papers. Rather, our intention is to consider the impact of the Hipparcos results on investigations of the distance scale. This leads us to focus on two issues: the location of the main-sequence as a function of age and chemical abundance, based partly on observations of nearby stars and partly on distance estimates to clusters; and direct distance measurements of primary distance indicators. We consider how those re-calibrations affect distance estimates to the Magellanic Clouds and M31, but carry extragalactic distance scale arguments no further.

2. ASTROMETRIC ACCURACY AND SYSTEMATIC BIASES

2.1 Measurement Precision

The formal precision of the astrometry of an individual star depends primarily on the number of observations and the signal-to-noise of each observation. Scans were made with the satellite rotational axis precessing at 6.4 revolutions yr^{-1} at an angle of 43° to the direction to the Sun. As a result, stars at ecliptic latitudes $\beta > 47°$ (particularly those at $\beta \sim 47°$) were observed more frequently than those at lower latitudes, with a corresponding ~50% increase in accuracy (ESA 1997).

Figure 1 illustrates the typical distribution of formal uncertainties ($\sigma_{\mu(\alpha)}$, $\sigma_{\mu(\delta)}$, σ_π) as a function of apparent magnitude.

Data reduction was undertaken independently by two consortia, NDAC and FAST, with the final catalogue consisting of the merged astrometric parameters. Both undertook extensive tests and comparisons to verify the accuracy of the satellite astrometry, particularly the reliability of the absolute zeropoints in μ and π. Those tests are discussed by Arenou et al (1995), Lindegren et al (1995) and Lindegren (1995) for the initial 30-month analysis, and in the first and third volumes of the Hipparcos catalogue (ESA 1997). In brief, the positional data show clear evidence for distortions in the FK5 ground-based frame at the 60–100 milliarcsecond level, while a comparison of the proper motions against ground-based data, which have longer time baselines and comparable accuracy, confirms random uncertainties of $\epsilon_H(\mu) \sim 1\text{--}2$ mas. Similarly, comparisons of the Hipparcos parallax data against the most accurate available ground-based measurements (by the US Naval Observatory) generally confirm the 1–2 mas quoted uncertainties of individual observations, as does Lindegren's (1995) analysis of the distribution of negative parallaxes.

Determining the reliability of the absolute zeropoints, notably in parallax, is a more complex issue. Coordinates are on the J2000 system (epoch 1991.25), referenced to the International Celestial Reference System (ICRS) via secondary standards (no extragalactic reference sources were observable by Hipparcos). The zeropoint for the proper-motion system is tied to the extragalactic frame through reference sources with absolute motions, either directly from VLBI or indirectly, from photographic surveys (see Johnson 1999, this volume). The zeropoint of the Hipparcos parallaxes, π_H, was tested by matching against independent parallax estimators (e.g. photometric parallax), primarily for distant stars where $\sigma_\pi(other) \ll \sigma_\pi(H)$. In particular, Hipparcos data for 46 Magellanic Cloud stars ($\pi \approx 0.02$ mas) give $\bar{\pi} = -0.16 \pm 0.26$ mas and $\sigma_\pi = 1.72 \pm 0.18$ mas. Arenou et al conclude that the overall results are consistent with a *global* zeropoint error of <0.1 mas in π_H. However, this does not exclude larger deviations on smaller angular scales, as discussed further below.

These tests of the parallax zeropoint rest primarily on relatively faint stars, since few stars brighter than 7th magnitude have high-accuracy ground-based parallaxes. Harris et al (1997) and Gatewood et al (1998) present comparisons based on 23 and 63 stars respectively. A weighted mean of their results (Hanson, priv. comm.) gives $\pi_H - \pi_{other} = 0.30 \pm 0.24$ mas, where the uncertainty is the standard error of the mean. While this is consistent with the absolute zeropoint defined at fainter magnitudes, there is no stringent test for the presence of any magnitude-dependent systematics.

2.2 Systematic Biases in Parallax Determination

Lutz-Kelker Bias The measured parallax provides the best (trigonometric) estimate of the distance to a particular star system. If, however, one considers that

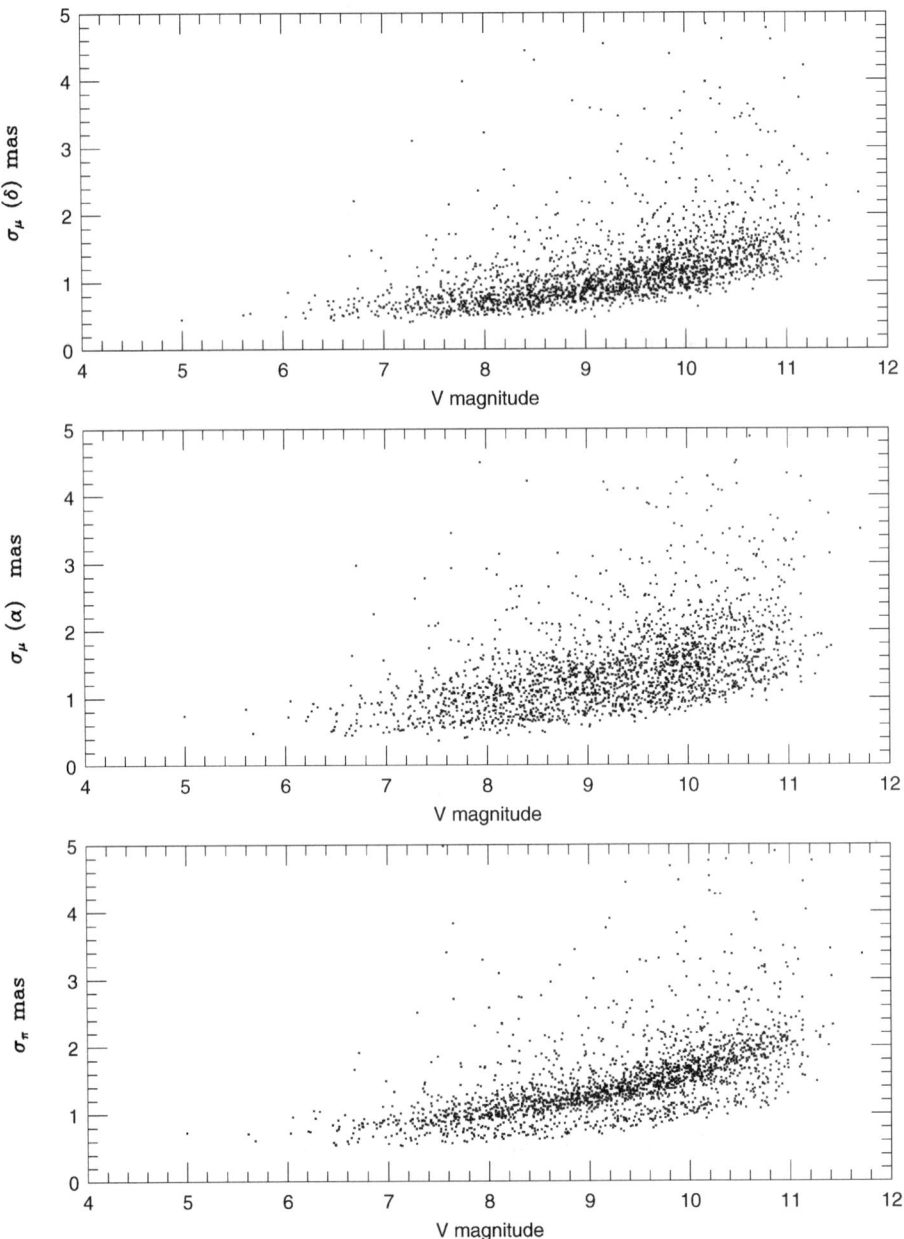

Figure 1 Formal uncertainties in Hipparcos astrometry as a function of apparent magnitude—the sample consists of stars drawn from the Lowell Observatory proper motion survey.

same object in a statistical sense—as a representative of the disk main sequence, for example—then one must allow for possible systematics introduced by the sample-selection criteria adopted. This potential for bias in statistical analysis of parallax-selected datasets was noted early this century by Russell and is also discussed by Trumpler & Weaver (1953). Lutz & Kelker (1973) provide the first quantitative description of this effect.

Under the standard assumption that the measured parallax, π_o, is an unbiased estimate of the true parallax, π, the conditional distribution $C_o(\pi_o \mid \pi)$ has a mean value π. However, we are aiming to determine π given π_o, and are therefore concerned with the conditional distribution $C(\pi \mid \pi_o)$. In that case,

$$C(\pi \mid \pi_o) \sim C_o(\pi_o \mid \pi) * N_D(\pi), \quad (1)$$

where $N_D(\pi)$ is the true parallax distribution of stars in dataset D. Note that $N_D(\pi)$ is often *not* identical with the parallax distribution of the parent Galactic stellar population, since D is usually pre-selected based on a combination of magnitude, color and proper motion. The dependence on $N_D(\pi)$ leads to systematic bias in the mean parallax if one selected a sub-sample, S, from dataset D on the basis of π_o.

Lutz-Kelker corrections

$$\Delta M_{LK} = \langle M_{true} - M_o \rangle = \left\langle 5 \log \frac{\pi}{\pi_o} \right\rangle \quad (2)$$

provide a means of correcting this bias in the absolute magnitude plane. The bias is not present if there is no selection based on π_o, as in the methods of mean parallaxes (e.g. averaging open cluster parallaxes) or reduced parallaxes (cf. Feast & Catchpole's (1997) Cepheid analysis and Feast (2000)). The LK bias, however, is always present when one averages absolute magnitudes, since the latter cannot be calculated if π_o is negative.

Lutz & Kelker computed ΔM_{LK} for the specific case of a uniform distribution, $N_D(\pi) \propto \pi^{-4}$, limited to $\frac{\sigma_\pi}{\pi} < 0.175$. Hanson (1979) showed that common observational selection effects (magnitude limits, proper motion limits) lead to less steep power-law density distributions, and used series expansion to derive general solutions for $N_D(\pi) \propto \pi^{-n}$. This results in both smaller LK corrections, and reliable corrections for higher values of $\frac{\sigma_\pi}{\pi}$ (see also Koen 1992).

With an apparent magnitude limit of $V \sim 12$, the Hipparcos dataset is magnitude-limited for stars with $M_{true} > 5.0$. In that case, the power-law exponent n of the density distribution and the resultant Lutz-Kelker bias are reduced significantly (Lutz 1983), since imposing a magnitude limit modifies $N_D(\pi)$ without direct reference to π_o. The situation is analogous to observations toward an opaque molecular cloud at π_{MC}: no stars have $\pi < \pi_{MC}$, reducing the number of stars with smaller π, and the probability that, for any given star, $\pi < \pi_o$. This emphasises that there is no unique ΔM_{LK} for a given star: the correction depends on context. If one modifies $N_D(\pi)$ based on parallax-independent criteria before selecting the π-defined subsample, one also modifies ΔM_{LK}.

Effects on Small Angular Scales The surface density of Hipparcos targets was set to match the limits of the observing mode. As a result, the same objects were generally observed on successive scans of a particular region. This leads to correlations in the derived astrometric data over scales of up to ≈2 degrees (twice the field of view). One notable effect is that the precision of the mean parallax of stars with separations within this régime is $\frac{\sigma_\pi}{n^{0.35}}$ rather than $\frac{\sigma_\pi}{n^{0.5}}$ (Lindegren 1988).

These correlations *may* lead to systematic bias in π_H at those angular scales. A comparison between the results from NDAC and FAST (Hipparcos catalogue, ESA 1997; vol. 3, ch. 16 and 17) shows that differences of up to ∼2 mas can occur within 2 × 2 degree regions near the ecliptic, corresponding to potential offsets of ≈1 mas in the final, merged catalogue. This result does not contradict the finding that the *global* zeropoint in π_H is reliable at the 0.1 mas level, but it serves a cautionary note for studies which require sub-milliarcsecond accuracy over 1–2 degree scalelengths.

Temporal Effects in Parallax Measurement In simple terms, the aim of the Hipparcos data reduction is to solve the following equations of condition for each target:

$$X = X_0 + \mu_X(T - T_0) + F_X\pi \qquad (3)$$

$$Y = Y_0 + \mu_Y(T - T_0) + F_Y\pi, \qquad (4)$$

where (X, Y) are the observations, (X_0, Y_0) the co-ordinates, (μ_X, μ_Y) the proper motions and (F_X, F_Y) the parallax factors. If observations are not well-distributed in time, the unknowns are correlated. Hipparcos observations span the period January, 1990 to March, 1993, but no data were obtained between September and November, 1992, due to the loss of a gyro and subsequent reconfiguration of the control system. As a result, January to March scans are available at four epochs, but September to November at only two. This imbalance leads to correlation coefficients, ρ, of up to ±0.6 (Hipparcos catalogue, ESA 1997; vol. 1, ch. 3.2), particularly in $\rho(\pi, \alpha)$. Ground-based parallax analyses almost always achieve $\rho < 0.1$ since one of the usual criteria is equal numbers of morning and evening observations. The existence of such high correlations does not guarantee bias, but the data are clearly vulnerable to such effects.

3. DISTANCES TO OPEN CLUSTERS

The typical Galactic open cluster is populated by from a few tens to a few hundred stars, the product of a single star-forming region, spanning a narrow range in age and in chemical composition. Several fiducial clusters fall within the range of Hipparcos (Mermilliod et al 1997, van Leeuwen et al 1997). Since open clusters have a small line-of-sight depth and since cluster membership can be decided based on photometric, proper motion or radial velocity criteria, individual parallax measurements can be combined without incurring significant Lutz-Kelker bias, although other systematic sources of error may be present. In particular, the small angular size of the more distant clusters renders the astrometry susceptible to the small-scale systematic problems outlined in Section 2. Indeed, Lindegren (1988) identified observations of open cluster stars as a powerful diagnostics of the extent of biases on small angular scales.

3.1 The Hyades

The metal-rich Hyades is near enough to permit ground-based trigonometric and, in resolved binaries, orbital parallax measurements, as well as statistical convergent-point analysis of the proper motions and radial velocities. The most recent latter study, by Schwan (1991), derived a mean modulus of 3.40 ± 0.04 mag. (47.9 ± 0.9 pc.) based on 145 stars with proper motion measurements from the FK5/FK4Sup, N30 and PPM catalogues. Perryman et al (1998) have analyzed the Hipparcos astrometry, complemented by additional radial velocity data. Conventional convergent-point analyses derive the cluster distance from the equation

$$\bar{d} = \sum_i^N \left(\frac{V_S \sin \lambda_i}{\kappa \mu_i} \right) \bigg/ N, \qquad (5)$$

where $\kappa = 4.74$, V_S is the cluster space velocity, μ_i the individual proper motions toward the convergent point and λ_i the angular distance between each star and the convergent point. As Perryman et al point out, the exact location of the last parameter depends on the extent to which one allows for random cluster motions ($\sigma_{cl} \sim 0.2$ to 0.4 kms^{-1}) in defining the membership list. The Hipparcos data provide full positional and velocity information, and allow iterative membership analysis. Following that approach, Perryman et al derive a distance of 46.34 ± 0.27 pc. ((m-M) = 3.33) to the center of mass defined by 134 stars within a radius of 10 parsecs of the cluster center.

The main source of the 3.4% discrepancy between the Hipparcos result and Schwan's analysis is the value adopted for the cluster space motion: Schwan's datum is $V_S = 46.6$ kms^{-1}, based on Detweiler et al (1984), while Perryman et al derive $V_S = 45.72$ kms^{-1} from CORAVEL data, a 2% difference. Most of the remaining discrepancy rests with the proper motions, and those differences are also responsible for the larger mean distance derived by Torres et al (1997) based on precise orbital parallax determination. Thus, the new Hyades distance determination is fully consistent with previous observations.

3.2 The Pleiades and Other Clusters

The same consistency with previous work is not apparent in initial analyses of the distance to the Pleiades cluster, which offered potentially the most significant result of the Hipparcos mission. The consensus of pre-Hipparcos studies located the Pleiades at a distance of $\sim$130 pc., $(m-M)_0 = 5.6$ mag. However, both van Leeuwen & Hansen-Ruiz (1997) and Mermilliod et al (1997) derive a mean parallax from Hipparcos observations of $\bar{\pi} = 8.6 \pm 0.24$ mas, or $r = 116 \pm 3$ pc. and $(m-M)_0 = 5.33 \pm 0.05$ mag., placing the (solar-abundance) Pleiades main-sequence $\sim$0.2 magnitudes fainter than the relation defined by nearby field stars ($\langle[Fe/H]\rangle \sim -0.15$ dex). Such a result can be accommodated within standard stellar evolution models, but only at the expense of invoking substantial anomalies, such as a helium abundance $Y_{MS} \sim 0.35$, and casting severe doubt on the global utility of distance determination by main-sequence fitting.

Pinsonneault et al (1998) re-examine this issue and show that plotting parallax against the $\rho(\pi, \alpha)$ correlation (see Section 2) for individual Pleiads reveals a clear trend, with stars having higher $\rho(\pi, \alpha)$ also having larger π (their Figure 18). As noted in Section 2, observations with high $\rho(\pi, \alpha)$ correlation indices *may* also have biased parallax determinations. Since the relevant Pleiads are also predominantly bright stars near the cluster center, carrying most of the weight in deriving $\bar{\pi}$, the small angular-scale correlation inherent in Hipparcos astrometry leads one to expect correlated bias. Given that Pleiads with $\rho(\pi, \alpha) < 0.2$ give $\bar{\pi} = 7.49 \pm 0.50$ mas, $(m-M)_0 = 5.63 \pm 0.14$ mag., and that there is an absence of any [Fe/H] ~ 0 field stars which reproduce the van Leeuwen/Mermilliod main-sequence (Soderblom et al 1998), it is reasonable to reject the initial Pleiades distance estimates as likely to be biased. The current best estimate of the distance of the Pleiades remains at $(m-M)_0 = 5.6$ magnitudes, or 132 parsecs (although see van Leeuwen 1999, for a dissenting viewpoint).

Mermilliod et al (1997) and Robichon et al (1997) estimate distances to other nearby clusters, including Coma, Praesepe, IC 2602 and α Persei. Most are significantly more distant than the Pleiades and, in general, the Hipparcos distance moduli are consistent with previous determinations. An exception is the sparse, nearby cluster Coma Berenices, where Pinsonneault et al find a discrepancy of $\sim$0.2 magnitudes with respect to isochrone-fitting, and trends in $(\pi, \rho(\pi, \alpha))$ similar to those in the Pleiades.

In summary, Hipparcos astrometry does not provide the major challenge to stellar evolution theory suggested by the preliminary analyses. Systematic biases, correlated over scales of <1 degree, limit the reliability of results derived for most open clusters. To echo Pinsonneault et al, it seems more prudent to regard these comparisons as tests of small-scale zeropoint errors, rather than as measurements of cluster distances.

4. THE HR DIAGRAM DEFINED BY FIELD STARS

The technique of distance determination through main-sequence fitting rests on the hypothesis that the absolute magnitude (luminosity) of normal, single stars of a given color (effective temperature) is a monotonic function of metallicity:

the Vogt-Russell theorem. Hipparcos provides the first opportunity for a thorough empirical test of that hypothesis.

4.1 The Abundance Scale

In principle, fine analysis of high-resolution optical spectra, coupled with accurate temperature determination, provides the best basis for measuring stellar metallicity. However, it is only recently that such studies have been extended to significant numbers of even moderately faint stars. As a result, most investigations of abundance distributions are based on a variety of lower-resolution spectroscopic or photometric techniques, each anchored to the high-resolution scale through a limited (and non-unique) set of standards. The latter step can lead to systematic zeropoint offsets between different studies, particularly given the relatively recent revision in the accepted value of the solar iron abundance (Biémont et al 1991).

Advances in atmosphere modelling are opening the door for the first quantitative abundance analyses of late-type (K, M) dwarfs, notably using metal hydride bandstrengths (Allard et al 1997, Gizis 1997). However, most investigations continue to concentrate on F, G and early-K stars on the upper main-sequence. Sandage & Eggen (1959) originally devised δ(U-B), the ultraviolet excess index, as a means of using Johnson broadband photometry to measure relative line-blanketing shortward of 4000 Å. Intermediate-band systems, such as Strömgren or Geneva photometry, have refined that measurement to some extent, while spectroscopic calibrations center on Fe, CH, CN and Mg features in the 4000–5300 Å range (Rose 1991, Carney et al 1994, Jones et al 1996).

Two of the most extensive abundance catalogues are Schuster et al's (1993) Strömgren-based dataset and Carney et al's (1994: CLLA) spectroscopic calibration. Schuster & Nissen (1989: SN) derive relations, tailored for main-sequence stars, between the (b-y), c_1 and m_1 indices and [Fe/H], while the CLLA calibration is based on high dispersion, but low signal-to-noise, spectra covering $\sim$50 Å centered on the Mgb triplet. Abundances are derived by cross-correlating the latter spectra against observations of standard stars. Jones et al (1996) adopt a similar approach, basing their calibration on the CLLA system, while Flynn & Morell (1997) have devised an abundance index for G and K dwarfs which combines Geneva photometry with a Cousins (R-I)-based temperature scale.

All of these calibrations (save the last) are tied to standards with high-resolution abundance analyses which predate the revision in [Fe/H]$_\odot$. This leads to an offset between those abundance scales and metallicities defined in the more recent high-resolution studies by Axer et al (1994: AFG) and Gratton et al (1997a: GCC), while there are further systematic differences between the individual datasets. Reid (1998) finds

$$[Fe/H]_{GCC} \sim [Fe/H]_{AFG} \sim [Fe/H]_{SN} + 0.15 \sim [Fe/H]_{CLLA} + 0.3,$$

with dispersions of $\sigma \sim 0.2$ dex. The offsets in the latter two scales become less pronounced at near-solar abundances. There is also evidence for a color term (temperature scale difference) between the Schuster & Nissen and Carney et al calibrations, while Axer et al clearly underestimate log g for a number of Hipparcos stars (see also Nissen et al 1997). Clementini et al (1999) generally confirm these offsets,

$$[Fe/H]_{GCC} = [Fe/H]_{SN} + 0.102 \pm 0.012$$

$$[Fe/H]_{GCC} = (0.935 \pm 0.032)[Fe/H]_{CLLA} + 0.181 \pm 0.173,$$

while comparison with data from Ryan & Norris (1991) gives

$$[Fe/H]_{GCC} = [Fe/H]_{RN} + 0.40 \pm 0.04.$$

Since the Jones et al scale is based on CLLA standards, one expects similar systematic errors in their abundances.

4.2 The HR Diagram of the Local Disk

Stars within the immediate Solar Neighborhood can be used to probe the general properties of the disk, since the overall velocity dispersion permits stars born at Galactic radii from 3 to 13 kpc to migrate through the Solar Radius. However, one of the early results from the Hipparcos survey was an indication that a significant fraction of the stars included in the most recent incarnation of the Nearby Star catalogue (CNS3: Jahreiss & Gliese 1991) are not, in fact, within the nominal distance limits. Analysis of the final catalogue confirms that ∼40% of the CNS3 stars observed by Hipparcos lie beyond 25 parsecs. This has obvious implications for statistical investigations which require unbiased, volume-limited samples. For example, Figure 2 plots Hipparcos data for 106 nominally-single stars from the "G-dwarf" sample which are used by Wyse & Gilmore (1995) to analyze the disk abundance distribution. Approximately 10 percent of the stars lie well above the main-sequence, and are likely to be evolved subgiants rather than dwarfs, while a further 5–10 stars are probably previously unrecognized binaries. To date, no comparable study has been undertaken based on a volume-limited sample drawn from the Hipparcos catalogue.

While the Wyse/Gilmore stars do not constitute a volume-limited sample, these observations can be used to examine the variation in main-sequence location with changing abundance. Wyse & Gilmore use Strömgren photometry (from Schuster & Nissen 1989) to estimate metallicities. Dividing the stars into four subsets,

Figure 2 Hipparcos-based (M_V, (B-V)) data for nominally-single nearby stars from Wyse & Gilmore's analysis of the local abundance distribution. The dashed lines are 1- and 5-Gyr isochrones from Bertelli et al (1994) for [Fe/H] = 0.0 and −0.4 dex; the solid line is the mean relation described by local stars (Reid & Murray 1992).

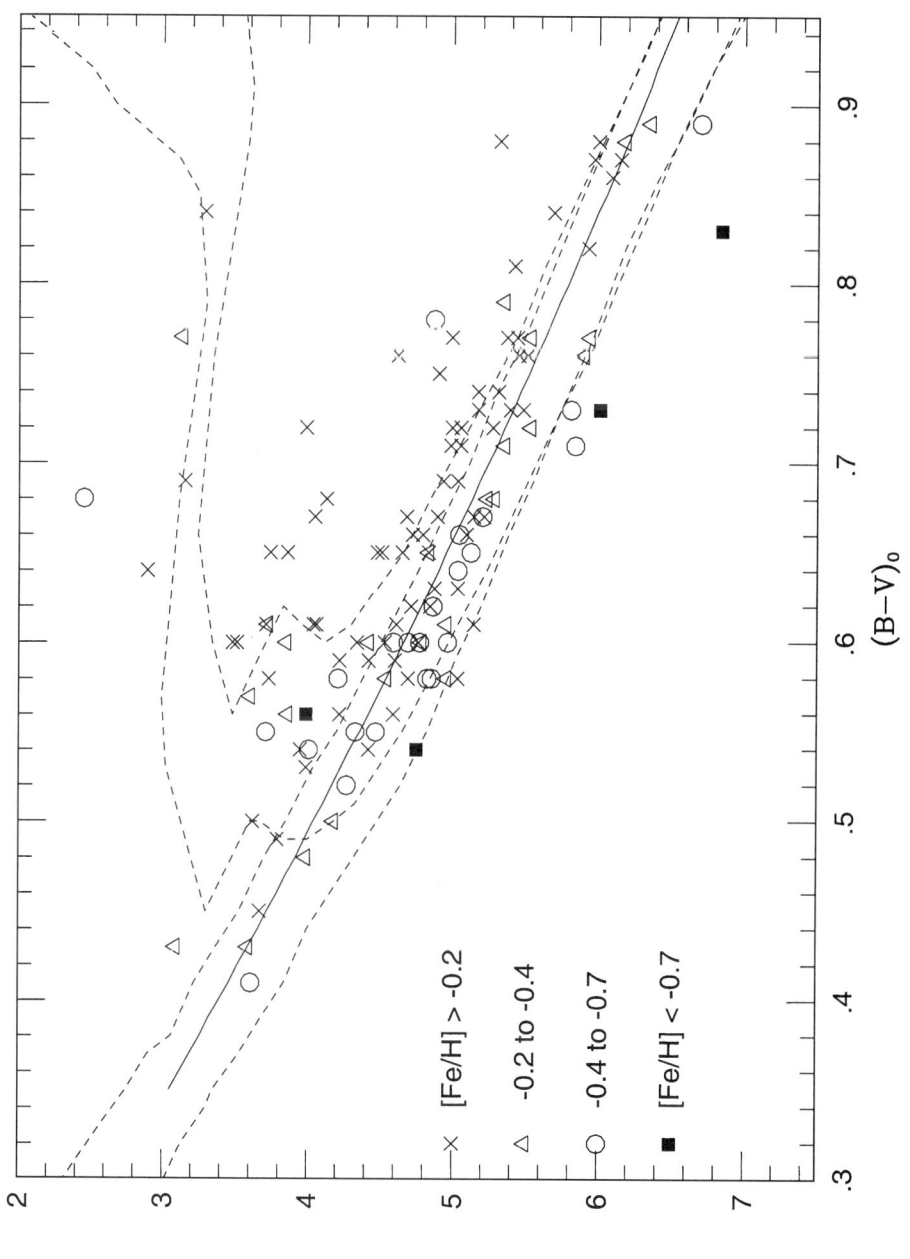

Figure 2 compares the (M_V, (B-V)) distribution against 1-Gyr and 5-Gyr isochrones for solar abundance and [Fe/H] $= -0.4$ dex as computed by Bertelli et al (1994). Ignoring evolved stars, the relative ranking is consistent with expectations, with M_V generally becoming fainter with decreasing [Fe/H]$_{SN}$ at a given color. However, there is a systematic offset with respect to the isochrones, in the sense that the lower-abundance stars are brighter than expected given the Strömgren metallicity estimates. This is consistent with the abundance-scale comparisons given in the previous section.

The data plotted in Figure 2 are consistent with the hypothesis of a monotonic change in the location of the main-sequence with varying metallicity. Moreover, the comparison with the isochrones suggests that most disk stars have abundances within ± 0.2 dex of the solar value (see also Figure 4 of Pinsonneault et al 1998). Jimenez et al (1998) find a comparable systematic offset between the lower-abundance ($-0.6 <$ [Fe/H] < -0.4) stars in the Flynn & Morell sample and their [Fe/H] $= -0.5$ isochrones. The implication is that previous studies, such as Wyse & Gilmore's analysis, overestimate the extent of the metal-poor tail to the disk abundance distribution, an inference consistent with Reid's (1998) analysis of Hipparcos data for Lowell proper-motion stars. Definitive results await more extensive observations, particularly detailed abundance determinations, of an Hipparcos-based volume-limited sample.

4.3 The HR Diagram for Halo Subdwarfs

The low space density of the local halo coupled with the apparent magnitude limits of the Hipparcos survey lead to the final catalogue including only a few tens of metal-poor subdwarfs or subgiants with parallaxes of even moderate accuracy. As a result, those data offer only weak constraints on theoretical calculations. Cayrel et al (1997) compare the observed (M_{bol}, log T_{eff}) distributions for stars with $-1.8 <$ [Fe/H] < -1.2, with temperatures taken from Alonso et al (1996), against (as-yet unpublished) isochrones computed by Lebreton and Vandenberg[1] They find a systematic offset of 0.01 in log T_{eff}, in the sense that the models are too hot. Adjusting the isochrones, Cayrel et al deduce an age of ~ 14 Gyrs for the field halo, although it should be noted that that result rests on the location of two subgiants.

Figure 3 matches D'Antona et al's (1997) theoretical isochrones for 12 Gyr-old populations against data for subdwarfs from the AFG and GCC analyses, restricting the latter to stars with no known binary companion having $\Delta m < 4$ mag., with $\frac{\sigma_\pi}{\pi} < 0.15$ and with [Fe/H] < -0.7. As with the disk stars, the overall trend matches theoretical expectations. At metallicities exceeding [Fe/H] $= -1.7$ there is reasonable agreement between the observations and the predicted isochrones, a circumstance also noted by Chaboyer et al (1998). There is a suggestion that

---

[1] As discussed further in Section 5, the D'Antona et al (1997) models predict a similar (M_{bol}, log T_{eff}) relation for main-sequence stars with $M_V > M_V(TO)_1$.

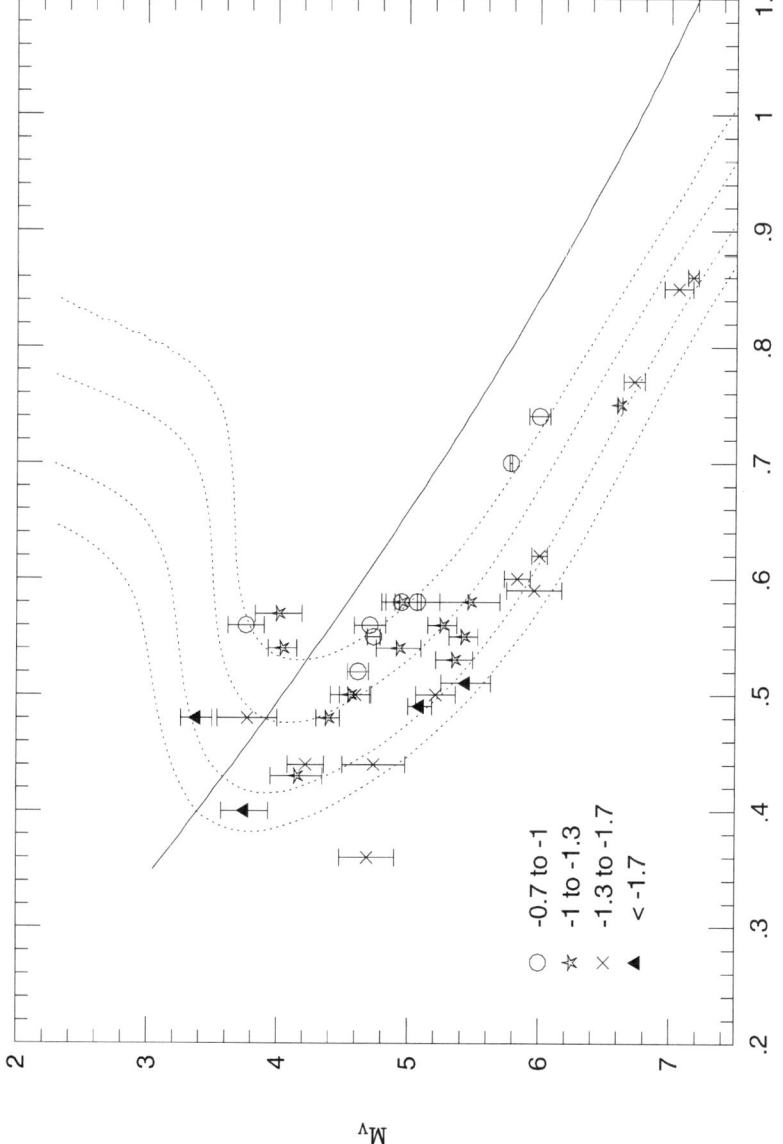

Figure 3 (M_V, (B-V)) data for single subdwarfs with abundance determinations by Gratton et al (1997) or Axer et al (1994) compared against 12-Gyr isochrones predicted by D'Antona et al (1997) for [Fe/H] = -0.7, -1.0, -1.5 and -2.0.

the lowest-abundance subdwarfs are more luminous (redder?) than expected, but this may reflect small-number statistics or an inherent selection effect rather than intrinsic stellar properties.

These data can also be used to estimate $\frac{\Delta Y}{\Delta Z}$, the proportional increase in helium and heavy element abundance due to stellar nucleosynthesis—an important quantity in determining the primordial helium abundance, Y_P. Helium abundance cannot be measured directly for stars on the lower main-sequence, but can be inferred by matching the observed distribution in the (M_{bol}, log T_{eff}) plane against theoretical isochrones. Based on pre-Hipparcos data for nearby disk G-dwarfs, Fernandes et al (1996) deduce that $\frac{\Delta Y}{\Delta Z} > 2$. Pagel and Portinari (1998) extend this analysis by matching Hipparcos data for stars in the abundance range[2] $-3 < $ [Fe/H] $ < 0$ against isochrones computed for $\frac{\Delta Y}{\Delta Z} = 0, 3, 6$. Limiting analysis to stars with $M_V > 5.75$, Pagel & Portinari deduce a best-fit value of $\frac{\Delta Y}{\Delta Z} = 3 \pm 2$, with the large uncertainty reflecting primarily the small sample size.

5. DISTANCES TO GLOBULAR CLUSTERS

While globular clusters lie well beyond the scope of present parallax measurement, high-accuracy astrometry which Hipparcos provides for metal-poor stars allow re-examination of the cluster distance scale through indirect techniques, notably main-sequence fitting. In particular, new data for field subdwarfs offer the chance to retrace the observer's route to globular cluster distances (Sandage 1970; 1986). A pre-requisite for the successful application of those techniques is a consistent metallicity scale for both clusters and field stars.

5.1 Cluster Abundances

The most frequently referenced source of cluster metallicities remains the extensive study by Zinn & West (1984). Their calibration is based on large-aperture broad- and narrow-band photometry, with individual magnitudes combined to give reddening-free color indices; and on integrated low-resolution spectrophotometry, isolating the Ca II K-line, G-band and Mgb features. Those measurements are scaled and averaged to give a composite estimate of the Q_{39} index, measuring the line-blanketing in the 3820–4010 Å region. That index was calibrated itself in terms of [Fe/H] using Cohen's (1983) analysis of high-resolution, image-tube spectra of giants in sixteen clusters.

Cohen's observations were, of necessity, limited to stars near the tip of the red-giant branch. Advances in detector technology over the last decade have allowed high signal-to-noise, high-resolution spectroscopy of larger samples of cluster

---

[2]Note that the stellar abundances are drawn either from Carney et al (1994) or the compilation by Cayrel de Strobel et al (1997), and can differ by 0.2–0.3 dex with more recent analyzes.

TABLE 1 Globular cluster abundances

Cluster	[Fe/H]$_C$	[Fe/H]$_{ZW}$	[Fe/H]$_{LT}$	[O/Fe]$_{LT}$	[Fe/H]$_{CG}$
M71	−0.7	−0.58	−0.79	0.2	−0.70
M5	−1.4	−1.40	−1.17	0.35/−0.25	−1.11
M3	−1.8	−1.66	−1.47	0.3/−0.2	−1.34
M13	−1.6	−1.65	−1.49	0.3/−0.15/−0.6	−1.39
M10		−1.60	−1.52	0.3	−1.41
M92	−2.35	−2.24	−2.25	0.3/−0.05	−2.16
M15	−2.20	−2.17	−2.40	0.2	−2.12

giants, notably by the Lick/Texas group (Sneden et al 1997 and refs within: hereinafter LT) and by Gratton and co-workers (Carretta & Gratton 1997, and refs within: CG97). Both of the latter groups tie their analyses to the revised value of the solar iron abundance, adopting $\log \epsilon(Fe) = 7.52$, ∼0.15 dex lower than the previously-accepted value. The LT analysis, aimed primarily at determining [O/Fe] ratios, spans stars in seven clusters (Table 1), with homogeneous observations, and temperatures derived from either (B-V) [M15, M92] or (V-K) colors, using MARCS model atmospheres (Gustafsson & Bell 1979) as a reference.

Carretta & Gratton analyze both their own CASPEC echelle spectra and equivalent-width data from the literature, including the LT dataset. Temperatures are based primarily on (V-K) colors, while CG97 adopt the Kurucz (1992) model atmosphere grid as their reference. The resulting abundances are ∼0.1 dex more metal-rich than previous analyzes, including the LT determinations. CG97 ascribe this systematic discrepancy to the different model atmospheres used in the different studies (although one should note that JR King et al (1998) derive a *lower* abundance in their re-analysis of LT's M92 data).

Both LT and CG97 compute true iron abundance, [Fe/H], based on equivalent widths of Fe I and Fe II lines. As with field subdwarfs, population II globulars are expected to have non-solar elemental abundance ratios, with enhanced oxygen and α-elements. The LT observations, however, reveal more complex behavior among the cluster giants than in the field, with [O/Fe] values ranging from +0.3 dex to <−0.4 dex in some clusters (notably M13). [α/Fe] ratios are typically +0.2 to +0.3, typical of the halo. The oxygen abundance anomalies are coupled with an observed anti-correlation of Na-O and correlation of Na-N, and are probably explicable in terms of deep-mixing nucleosynthesis (Kraft et al 1993), but possible inconstancy in the total CNO abundance may reflect primordial variations. The latter would represent a potential complication for MS-fitting. If [O/Fe] = +0.3 is adopted as the standard value for [Fe/H] < −1, then following Salaris et al (1993),

$$\log Z = [M/H] - 1.66 = [Fe/H] - 1.45. \qquad (6)$$

All of these analyses center on stars on the giant branch, whereas MS-fitting demands (by definition) matching cluster subdwarfs against field subdwarfs of comparable abundance. While there is no *a priori* reason to distrust the existing cluster [Fe/H] data, the variations in [O/Fe] emphasise the complications in dealing with evolved stars. Spectroscopy of main-sequence cluster members would allow abundance calibration using the same criteria as used in field-subdwarf calibration. Such observations are within the scope of the new generation of 8-meter class telescopes and are under way (J Cohen, priv. comm.), but no studies have been completed. JR King et al (1998) have obtained the first high-resolution spectra of six subgiant stars near the M92 turnoff. Their analysis of those data, based on the Kurucz (1992) atmospheres and with temperatures estimated from (B-V) colors, imply [Fe/H] ~ -2.5, significantly lower than either the LT or CG97 analyses, but matching Peterson et al's (1990) earlier analysis of two M92 red giants. One should note, however, that they derive a higher reddening ($E_{B-V} = 0.07$) than any other study, suggesting possible systematic errors in the colors and the temperature scale. Observations of a larger sample of stars, preferably with accurate infrared colors, in this and other clusters should be given high priority.

5.2 Cluster Distances

Empirical main-sequence fitting, the observer's route to cluster distances, is simple in principle. Observations of nearby stars with known trigonometric parallax are used to define the location of the main-sequence in the absolute magnitude/color plane; photometry of cluster stars allows one to determine a fiducial cluster main-sequence in the apparent magnitude/color plane; and the offset between the two sequences is the distance modulus of the cluster. In practice, this idealized concept is vulnerable to a number of complications.

1. Photometry: the field-star and cluster observations must both be of high accuracy and tied to a consistent photometric system. Crowding is clearly a problem in analyzing cluster data, although the high spatial resolution offered by the Hubble Space Telescope (HST) addresses that issue. Most field subdwarfs have photometry in only the Johnston B and V passbands, and individual (B-V) measurements drawn from the literature can differ by up to 0.04 magnitudes. Since the slope of the main-sequence is $\frac{\partial M_V}{\partial (B-V)} \approx 5$, systematic uncertainties at that level lead to errors of up to 0.2 magnitudes in (m-M). Accurate, homogeneous UBVRIJHK photometry for both field and cluster subdwarfs is eminently desirable.

2. Abundances: the metallicity of the calibrating stars should be either close to that of the individual cluster under study, or the stellar parameters at least adjusted reliably to match the appropriate abundance. Variations in the [O/Fe] and [α/Fe] ratios from the nominal values of $\sim +0.2/+0.3$ amongst field subdwarfs (e.g. HD 134439 & 134440: King 1997) further complicate this issue. The most recent stellar models (e.g. D'Antona et al

1997) indicate $\frac{\partial(B-V)}{\partial[\text{Fe/H}]} \sim 0.14$ at $[\text{Fe/H}] = -1$ and $\frac{\partial(B-V)}{\partial[\text{Fe/H}]} \sim 0.05$ at $[\text{Fe/H}] = -2$ for $M_V > 5$.

3. Binary and multiple stars: unrecognized duplicity can lead not only to erroneous absolute magnitude estimates for calibrating subdwarfs, but can distort the position of the fiducial (magnitude, color) relation for the lower main-sequence in binary-rich clusters, such as NGC 288 (Bolte 1992).

4. Reddening: most clusters are subject to foreground Galactic obscuration, as are a few calibrating subdwarfs. Qualitatively, if the cluster reddening is *overestimated* (cluster sequence too blue), the distance modulus is also *overestimated*. On the other hand, if the reddening of the calibrating subdwarfs is *overestimated* (reference sequence too blue), the cluster distance is *underestimated*.

5. Distribution in (M_V, color): even with the addition of Hipparcos data, few subdwarfs are available over the full range of $M_V/(B-V)$ and $[\text{Fe/H}]$. Typically, the distance to a given cluster is estimated using only five to ten subdwarf calibrators, so decisions on whether to include individual field stars can influence the derived (m-M) by ~ 0.1 magnitude.

6. Evolution off the main-sequence: theoretical isochrones show that at ages of from 10 to 16 Gyrs, metal-poor ($[\text{Fe/H}] < -1$) stars within 1.5 magnitudes of the turnoff have luminosities which lie above the zero-age main-sequence, although the offset is less than 0.1 magnitude for $\Delta M < 1$ magnitude. This reduces further the potential number of calibrators. One should note, however, that evolution will bias MS-fitting to larger distances only if the calibrating field turnoff stars are significantly older (i.e. more luminous) than the cluster.

The Pre-Hipparcos Cluster Distance Scale The sparse numbers of local subdwarfs with accurate ground-based parallax measurements discouraged purely empirical MS-fitting analyses, and most pre-Hipparcos analyses either match model isochrones directly against cluster data (the theoretician's route to distances), or adopted a semi-empirical approach. In some studies (e.g. Richer & Fahlman 1986) isochrones are used to adjust subdwarf colors to those appropriate to the cluster abundance, and the MS-fitting distance derived using the resultant monometallicity sequence; alternatively, the subdwarfs calibrate zeropoint offsets in the isochrones, and the latter are adjusted and matched to the cluster color-magnitude data. In one of the more influential studies, Bolte & Hogan (1995) adopted the latter approach in deriving the distance and age of M92, using data for HD 103095 (Groombridge 1830), to derive an offset of 0.015 magnitude in the (B-V) colors predicted by Bergbusch & Vandenberg's (1992—BV92) models. Clearly, this single-point adjustment assumes that the models have no systematic errors as a function of luminosity and/or effective temperature (i.e. the isochrones have the correct shape in the color-magnitude plane).

Bolte & Hogan's results were not unchallenged. D'Antona et al (1997—DCM) derive a distance modulus of 14.80 ± 0.1 to M92, based on weighted MS-fitting against eight nearby subdwarfs with accurate ground-based parallaxes, rather than relying on only Groombridge 1830. Distance estimates to the extreme metal-poor clusters M30 and M68 exceed previous analyses by a similar factor.

RR Lyrae stars provide an alternative to MS-fitting in cluster distance determination, and we discuss their luminosity calibration in Section 6.3. The other main options are astrometric distances and, recently, white dwarf sequence fitting. In the former case, measurement of the proper-motion dispersion of cluster members is combined with observations of the radial velocity dispersion and a dynamical model to derive a distance estimate. This method has the advantage of being reddening-independent, but requires relative astrometry of extremely high precision. Rees (1997) has summarized results for eight clusters, and typical uncertainties are from ± 0.2 to ± 0.4 mag. in $(m-M)_0$. As yet, those results offer only weak constraints on the distance scale.

Accurate photometry for white dwarfs in globular clusters has become possible only with the existence of HST. Deep imaging has led to the detection of degenerate sequences in at least three clusters: NGC 6752 (Renzini et al 1996), NGC 6397 (Cool et al 1996) and M4 (Richer et al 1997). Renzini et al use their observations to estimate a distance modulus of 13.05 to NGC 6752, choosing as calibrators nearby disk white dwarfs with spectroscopic masses of $\sim 0.5\ M_\odot$, close to the expected value for halo white dwarfs. The latter mass estimates, however, are based on white dwarf models with thin-hydrogen atmospheres: thick-hydrogen atmosphere models provide a better match to masses derived using gravitational redshifts (Reid 1996a), and imply higher masses for these disk calibrators. Lower-mass white dwarfs have larger radii, and therefore higher luminosities at a given effective temperature, so Renzini et al are likely to have underestimated $(m-M)_0$ in their analysis. Concordant with this hypothesis, Richer et al note an offset of 0.1 magnitude between their MS-fitting distance estimate to M4 (1.73 kpc—in good agreement with the astrometric estimate) and the WD-fitting result. Finally, Cool et al do not use their data to estimate directly a distance to NGC 6397, but inspection of their Figure 6 shows a systematic offset of $\sim +0.3$ magnitudes between the $0.5\ M_\odot$ theoretical white dwarf sequence and their data for the adopted $(m-M)_0 = 11.7$ magnitudes.

Table 2 presents distance estimates, derived through these various techniques, for a representative sample of clusters. The "standard" pre-Hipparcos distance scale is characterized using the values listed in Harris' (1996) compilation. In each case we list both the true distance moduli and the reddening, E_{B-V}, (from Harris). The latter values are derived using a variety of techniques, and as a self-consistent reference we list reddenings derived from Schlegel et al's (1998) 100 μm COBE/IRAS all-sky analysis. In general, increasing the adopted cluster reddening by ΔE_{B-V} leads to an increase of $2 \times \Delta E_{B-V}$ in $(m-M)_0$. In addition, we list the apparent magnitude (uncorrected for reddening) at the main-sequence turnoff.

Those values are taken from the fiducial cluster sequences referenced in Reid (1997, 1998).

Post-Hipparcos Cluster Distance Determinations   Even with the completion of the Hipparcos catalogue, there are few lower main-sequence subdwarfs which are unambiguously single, have reliable abundance determinations and accurate parallaxes. Extreme metal-poor clusters, such as M92 and M15, are ill-served with calibrators for distance measurement. Table 2 lists distance modulus determinations from the several post-Hipparcos MS-fitting analyses. Apart from the choice of abundance calibration and foreground reddening, each study employs a different set of subdwarf calibrators and uses different color-corrections and residual-minimisation techniques in MS-fitting. Uncertainties in $(m-M)_0$ are at least ± 0.1 magnitude.

The largest discrepancies between the pre- and post-Hipparcos distance estimates listed in Table 2 lie among the extreme metal-poor clusters, notably M92, where Harris' distance modulus is 0.1 magnitude shorter than even Bolte & Hogan's result. Reid's (1997) initial Hipparcos analysis produces the largest distances, but, apart from being based on the Carney et al (1994) abundance scale, Pont et al (1998) point out that several metal-poor calibrators used in that survey are binaries.[3] In their M92 analysis, Pont et al apply statistical corrections to adjust M_V for known binary subdwarfs, a procedure criticised by both Chaboyer et al (1998) and Reid (1998); excluding those stars increases their derived modulus to $(m-M)_0 = 14.74$, in close agreement with other recent studies.

NGC 6397 represents a special case. The pre-Hipparcos distance estimate rests on one subdwarf, HD 64090, whose parallax is reduced by 20% by Hipparcos astrometry. Lying at low latitude, this cluster is subject to significant field-star contamination, but deep HST observations span a sufficient baseline in time to allow proper-motion identification of members, and the resultant color-magnitude diagram extends to luminosities approaching the hydrogen-burning limit (IR King et al 1998). Exploiting these results, Reid & Gizis (1998) have used nearby M-subdwarfs with accurate ground-based parallax data to undertake MS-fitting on the lower main-sequence, and derive a distance modulus of 12.12 ± 0.15.

Given a distance to NGC 6397, that cluster can be used as a template to estimate distances to more metal-poor systems by extending the differential upper-MS matching technique devised by Vandenberg et al (1990). Theoretical isochrones can be used to make explicit allowance for intrinsic abundance-dependent variations in the color and luminosity at the turnoff, corrections analagous to those required in constructing a mono-metallicity sequence of subdwarf calibrators. This requires that the models are accurate in a differential sense over a limited range in luminosity and effective temperature. The resultant distance estimates (Table 2—from Reid

---

[3]Note that Reid (1997) derives $(m-M)_0 = 14.80$ for M92 based on pre-Hipparcos data, in good agreement with D'Antona et al, and $(m-M)_0 = 14.82$ by direct calibration against the metal-poor F-type subdwarf, HD 19445.

TABLE 2 Main-sequence fitting distance determinations

Cluster	E_{B-V}(IR) V_{TO}	$[Fe/H]_{CG}$	Harris E_{B-V}	Rees/WD E_{B-V}	DCM E_{B-V}	R97 E_{B-V}	G97 E_{B-V}	CDKK[1] E_{B-V}	R98
47 Tuc	0.03 17.65	−0.70	13.17 0.05				13.47 0.055		13.56 0.04
M71	0.34 18.00	−0.70	12.92 0.25						13.19 0.28
NGC 288	0.015 19.02	−1.07	14.55 0.03				14.85 0.03		15.00 0.01
M5	0.045 18.48	−1.11	14.32 0.03	14.4 ± .4		14.45 0.03	14.51 0.035	14.42 0.03	14.52 0.02
NGC 362	0.035 18.80	−1.15	14.60 0.05				14.88 0.055		14.79[2] 0.04
M4[3]	0.495 16.90	−1.19	11.41 0.36	11.2 ± .2					
M3	0.015 19.14	−1.34	15.01 0.01	14.9 ± .6					15.10[4] 0.01
M13	0.02 18.50	−1.39	14.22 0.02	14.06 ± .23		14.48 0.02	14.41 0.02		14.45 0.02
NGC 6752	0.06 17.30	−1.41	12.96 0.04	13.05^{WD} 0.04		13.17 0.02	13.24 0.035	13.20 0.04	13.16 0.04
NGC 6397	0.18 16.60	−1.82	11.75 0.18						12.24 0.18

M30	0.06	−1.91	14.46		14.80	14.95	14.84		14.85[5]
	18.60		0.03		0.04	0.05	0.04		0.04
M68[6]	0.07	−1.99	15.01		15.10	15.24	15.15		15.15[5]
	19.05		0.04		0.06	0.05	0.04		0.06
M15	0.10	−2.12	15.03			15.38			15.23[5]
	19.40		0.09			0.09			0.09
M92[7]	0.02	−2.16	14.54	14.76 ± .3	14.80	14.93	14.74	14.76	14.78[5]
	18.60		0.02		0.02	0.02	0.02	0.02	0.02

Globular cluster distance moduli and reddening:
Column 2 lists E_{B-V} deduced from 100 μm maps (Schlegel et al 1998) and the apparent magnitude (uncorrected for reddening) at the turnoff; column 3 gives the abundance on the Carretta & Gratton (1997—CG97) scale. Succeeding columns list derived distance moduli and adopted reddenings from individual investigations: column 4, from the pre-Hipparcos compilation by Harris (1996); column 5, astrometric distances from Rees (1996) and WD-fitting distance for NGC 6752 (Renzini et al 1996); column 6, from D'Antona, Caloi & Mazzitelli (1997—DCM); column 7, Reid (1997—R97); column 7, Gratton et al (1997b—G97); column 9, Chaboyer et al (1998—CDKK); and column 10, data from Reid (1998) and Reid & Gizis (1998—RG98).
The G97, CGKK, R98 and RG98 analyses are tied to the CG97 abundance scale and use subdwarf metallicities from GCC and AFG. The remaining studies adopt the Zinn & West cluster abundances and, in most cases, CLLA's subdwarf metallicities.

Notes:
[1] Main-sequence fitting distance estimates only.
[2] Relative distance modulus (Vandenberg et al 1990) with respect to M5.
[3] R = 3.8 is assumed for M4.
[4] Based on matching the turnoff region against M13 (Reid & Gizis 1998).
[5] Relative distance moduli with respect to NGC 6397 (Reid & Gizis 1998).
[6] MS-fitting results adjusted to match photometry from Walker (1994).
[7] Pont et al (1998) derive $(m-M)_0 = 14.67$ for $E_{B-V} = 0.02$ for M92.

& Gizis 1998) are in good agreement with more direct analyses. Overall, MS-fitting based on Hipparcos subdwarf data points to a modest increase of 0.1 to 0.2 magnitudes in the distance moduli, or 5 to 7% in distance, of the majority of globular clusters.

5.3 Globular Cluster Ages

Much of the interest in globular cluster distances centers on their use as Galactic age indicators. The luminosity at the main-sequence turnoff remains the most effective absolute-age determinant, although there are observational difficulties in defining the parameter. The main-sequence is vertical near the turnoff, so $M_V(TO)$ can be defined with a typical accuracy of only $+0.1$ magnitude, equivalent to an uncertainty of ~ 1.5 Gyrs in age. Matching M92 observations against the available stellar models, Bolte & Hogan set a firm lower bound of 16 ± 2 Gyrs on the age of the Galaxy, identifying a strong conflict between stellar evolution-based and (short distance-scale) cosmological estimates of the age of the Universe (see also Chaboyer et al 1996). As in all classical drama, hubris stepped forth on cue, not only in the guise of a longer distance scale, but also in the form of refined physics of stellar interiors.

Main-sequence fitting based on Hipparcos-calibrated subdwarfs implies $M_V(TO)$ values that are brighter by 0.1 to 0.15 magnitudes, corresponding to an age reduction of 1.5 to 2 Gyrs. However, revisions in the ($M_V(TO)$, [Fe/H], age) relation lead to significant changes in the age calibration. The last decade has seen substantial improvement in defining many of the physical parameters underlying stellar interior theory, prompted primarily by theoretical inconsistencies revealed through helioseismological observations. In particular, while most model isochrones predating 1994 are based on the equation of state derived by either Eggleton et al (1973) or Straniero (1988), and opacities from the Los Alamos calculations (Huebner et al 1977), recent calculations employ the new generation of OPAL opacity data (Rogers & Iglesias 1992; Seaton et al 1994; Iglesias & Rogers 1996), and the corresponding equations of state (Rogers et al 1996). The first extensive set of models to incorporate these new data were computed by Mazzitelli, D'Antona & Caloi (1995—MDC; see also DCM), and those calculations predict significantly lower turnoff luminosities for a given age and abundance: there is a ~ 0.2 magnitude (~ 3 Gyr) offset in $M_V(TO)$ between the DCM calibration and the pre-OPAL Straniero & Chieffi (1991) models (Figure 4a).[4] Matched against the DCM calibration (Figure 4b), the Hipparcos-based distance scale implies cluster ages of 11 to 13 Gyrs.

Figure 4a shows that the BV92 models predict turnoff absolute magnitudes within 0.1 magnitude of the OPAL-based DCM calculations. This is due partially to the highly enhanced [O/Fe] ratios adopted in the former models—both DCM

---

[4]Calculations by Cassisi et al (1998), incorporating further modifications in the input physics, predict turnoff luminosities lower by $\Delta \log L \sim 0.05$ than even the MDC calibration.

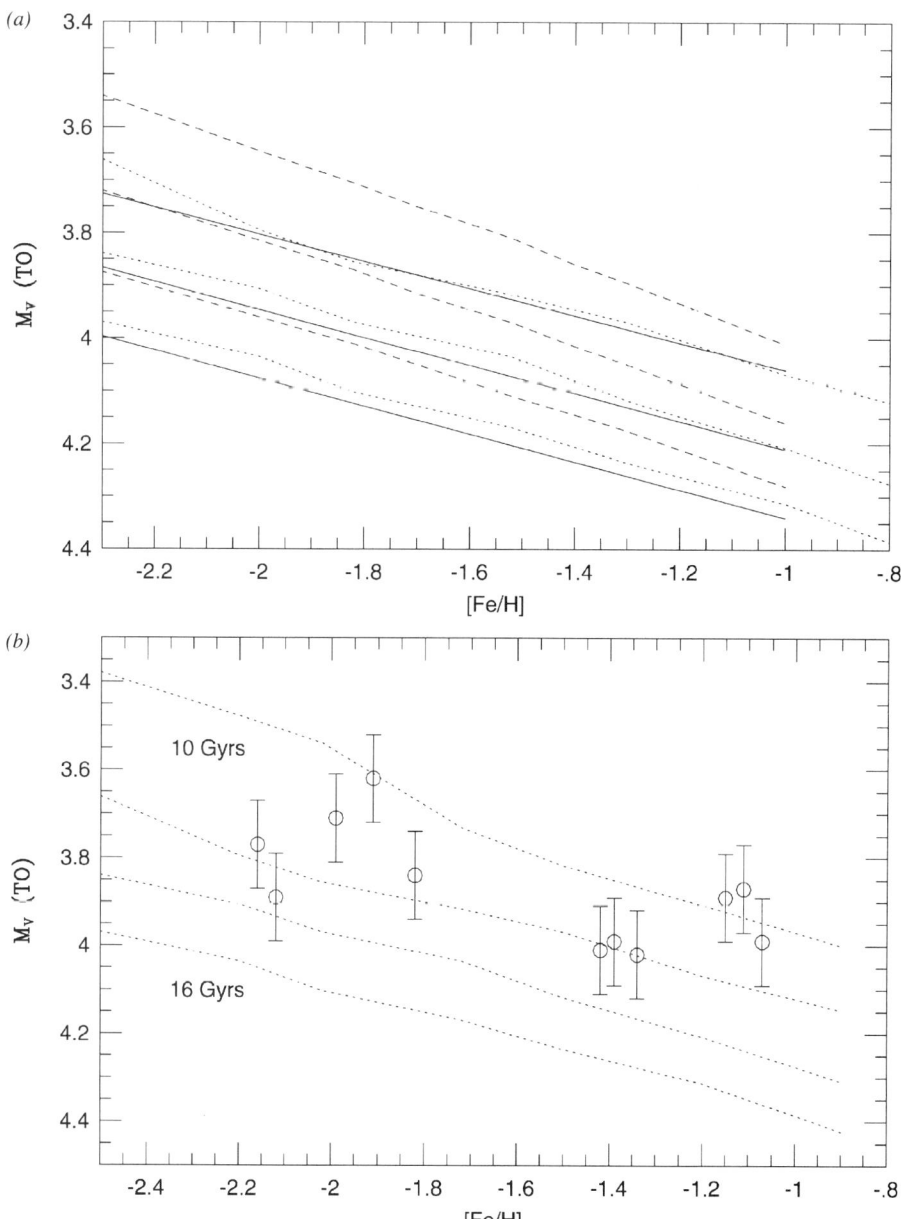

Figure 4 (*a*) The predicted relation between $M_V(TO)$ and [Fe/H] for 12, 14 and 16 Gyrs given by Bergbusch & Vandenberg (*solid lines*), Straniero & Chieffi (*dashed lines*) and D'Antona, Caloi & Mazzitelli (*dotted*); (*b*) $M_V(TO)$ deduced for clusters in Table 2 (except the highly reddened M4) compared to the DCM predictions for ages of 10, 12, 14 and 16 Gyrs. The models have $[\alpha/Fe] = 0$ and are therefore offset by -0.2 dex.

and Straniero & Chieffi employ scaled-solar abundances. Given this concordance in M_V(TO), one might ask why Bolte & Hogan deduced an age of 16 Gyrs for M92. The answer lies with the second parameter of the color-magnitude diagram. Figure 5 compares 12-Gyr isochrones from BV92 and DCM for [Fe/H] = -2.26 and -2.03. respectively. The isochrones in (M_{bol}, log T_{eff}) are almost identical on the lower main-sequence, but diverge near the turnoff, with Bergbusch & Vandenberg predicting higher temperatures. However, there is a significant, luminosity-dependent color offset between the isochrones in the (M_V, (B-V)) observational plane. These discrepancies reflect several factors: different effective-temperature/color transformations; different bolometric corrections; different treatment of convection; and inclusion or exclusion of helium sedimentation. In the last context, Straniero et al (1997) demonstrate that allowing for helium diffusion in their OPAL-based models moves the turnoff redward by up to 0.06 magnitudes in (B-V). Matching their models directly against Walker's (1994) M68 data, they estimate $(m-M)_0 = 15.10$ (for $E_{B-V} = 0.04$ cf. Table 2) and an age of $\sim$13 Gyrs.

Clearly, even if one aims to use M_V(TO) as the age indicator, there is a strong tendency to select an isochrone which matches the observed color of the turnoff—particularly given the observational ambiguity in locating M_V(TO). Figure 5 shows that that choice drives one to choosing an older age if the BV92 isochrones are taken as the reference set than if one matches against the DCM calculations: the former predict an intrinsically bluer turnoff for a given age and composition. In general, the agreement in age estimates derived from M_V(TO) and full-CMD comparisons against OPAL-based stellar models favors cluster ages in the range 11 to 13 Gyrs.

5.4 Summary

The changes in the cluster distance scale resulting from the addition of accurate Hipparcos parallax data for halo subdwarfs are relatively minor: no more than 0.15 magnitudes, or $\sim$7% in distance. Nonetheless, when combined with the recent revisions in the physics of stellar interiors, the net result is a reduction of 30 to 40% in the likely ages of even the most metal-poor globular clusters. A final resolution of the age question must always rest with the stellar models, but the availability of accurate parallaxes for a larger number of unambiguously single subdwarfs, particularly at [Fe/H] < -1.5, would go a long way toward setting distance estimates of a firm foundation. This requires sub-milliarcsecond astrometry to at least 15th magnitude, and probably the next astrometric space mission. In the

Figure 5 (M_V, T_{eff}) and (M_V (B-V)) isochrones predicted for 12-Gyr old metal-poor systems by Bergbusch & Vandenberg (BV92) and D'Antona et al (DCM). The former models adopt [O/Fe] = $+0.4$ while the latter have scaled-solar abundance ratios. To allow for this we compare the BV92 [Fe/H] = -2.26 model against the [Fe/H] = -2.03 DCM isochrone.

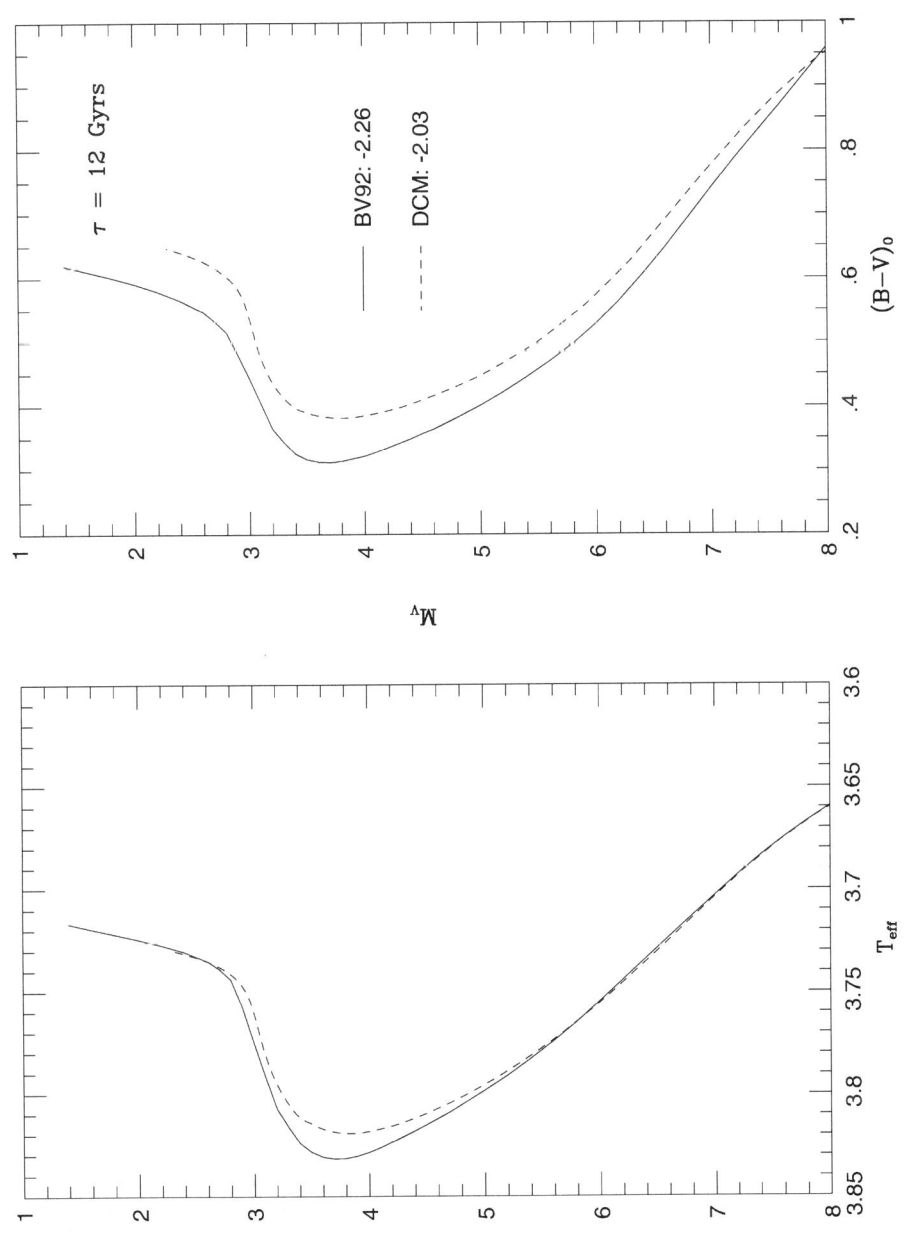

interim, more accurate photometric and spectroscopic data for both subdwarfs and a larger number of globular clusters (including CCD photometry of stars *above* the turnoff) would not be amiss.

6. DISTANCE INDICATORS

6.1 Cepheid Variables

The Cepheid period-luminosity (P-L) relation constitutes the backbone of the extragalactic distance scale. Descendants of intermediate-mass stars, Cepheids have typical ages of $\sim 10^8$ years and therefore are found predominantly near the parent star-forming region. As a result, only a handful of variables lie within 1 kpc. of the Sun, and most Cepheids are subject to considerable foreground reddening. However, the continued spatial association with coeval lower-mass stars offers a potential means of distance determination, albeit for a limited number of stars. Until recently, the primary calibrators of the Galactic P-L relation were thirteen Cepheids in clusters and associations (Sandage & Tamman 1968), whose distances were estimated using main-sequence fitting. More recent calibrations of the same kind are by Feast & Walker (1987) and Laney & Stobie (1994) (see later). Theoretical models (e.g. Chiosi et al 1993) predict a small metallicity dependence in M_{bol}, which is amplified by differential blanketing at optical wavelengths. Metal-rich Cepheids are cooler and brighter in V and I. The amplitude of variation is currently under debate (Sasselov et al 1997; Kochanek 1997; Sandage, Bell & Tripicco 1999), but since M31 Cepheids are near solar abundance and LMC Cepheids have [O/H] $\sim -0.3(\delta M_V \sim -0.05)$, we consider this effect no further.

Hipparcos Astrometric Analyses Observations of over 200 Galactic Cepheids are included in the final Hipparcos catalogue, with most lying at distances exceeding 1 kiloparsec. As a result, few variables have parallaxes measured to an accuracy of better than 2σ. Lynga & Lindergren (1998) estimate parallaxes for two Cepheids associated with open clusters, but with accuracies of no better than 40%. However, since the Hipparcos Cepheids were not selected for observation based on parallax, the sample as a whole is amenable to statistical analysis. Feast & Catchpole (1997) have used the method of reduced parallaxes to complete the most rigorous such study to date.

Given parallax data for an unbiased sample of stars at effectively the same distance (e.g. open cluster members), $\langle \pi \rangle = \Sigma \frac{\pi_i}{N}$. Consider an unbiased (as regards π) sample of stars spanning a range of distances with identical, but unknown, M_V. The apparent magnitude can be used to scale each parallax measurement to a single distance, e.g. $\pi'_i = \pi 10^{-0.2 \times V_0}$, and $\langle \pi' \rangle = \Sigma \frac{\pi'_i}{N}$ gives the appropriate zeropoint. Cepheids do not have identical magnitudes, but follow the P-L relation

$$\langle M_V \rangle = \delta \log P + \rho,$$

which can be used as a scaling relation. Feast and Catchpole adopt $\delta = -2.81$ from Magellanic Cloud Cepheid data, and solve for ρ using

$$10^{0.2\rho} = 0.01\pi \, 10^{0.2(\langle V_0 \rangle - \delta \log P)}.$$

The Cepheid period-color relation is used to estimate foreground reddening, and hence $\langle V_0 \rangle$, for each star. Since Cepheids follow a period-luminosity-color (PLC) relation, the P-L relation has intrinsic scatter. However, the slope of the color dependence in (B-V) ($\sim$2.5) is close to the ratio between total and selective reddening ($\sim$3.15). Thus, using the period-color relation to estimate E_{B-V} compensates for a substantial proportion of the intrinsic scatter in the P-L relation. Feast and Catchpole derive $\rho = -1.43 \pm 0.1$.

Feast & Whitelock (1998) follow similar precepts to derive the zeropoint, ρ_2, of the PLC relation as $\rho_2 = -2.38 \pm 0.1$. This calibration allows the derivation of Galactic rotation constants from proper motion data, and Feast and Whitelock are in good agreement with radial velocity analyses. Inverting the process, they note that Pont et al's (1997) Galactic rotation study implies $\rho_2 = -2.42 \pm 0.13$.

Madore & Freedman (1998) and Oudmaijer et al (1997) have criticized the Feast/Catchpole analysis and derive zeropoints fainter by $\sim$0.1 magnitude. However, both calibrations are based on variables with higher-accuracy Hipparcos parallaxes; that is, the samples are parallax-selected, and therefore liable to bias. Oudmaijer et al apply estimated Lutz-Kelker corrections to individual stars, but the substantial values of $\frac{\sigma_\pi}{\pi}$ mean that those corrections are sensitive to the assumed spatial distribution. In contrast, the individual weights in the Feast/Catchpole reduced parallax analysis are determined primarily by the apparent magnitude, rather than π, giving an estimate of ρ which is effectively unbiased by parallax uncertainties (Feast, Pont & Whitelock 1998).

The highest-weight Cepheids in the Feast/Catchpole analysis have near naked-eye apparent magnitudes, significantly brighter than the average among stars used to test the Hipparcos absolute zeropoints, and the possibility of low-amplitude magnitude terms has not yet been eliminated fully. However, Sandage & Tamman (1998) point out that Feast & Catchpole's P-L calibration is no more than 0.1 magnitudes brighter than previous Galactic calibrations, notably their 1968 cluster-Cepheid calibration (with the cluster distances given by Kraft's (1961) main-sequence fitting). Those discrepancies lie at the 1σ level, even without allowing for possible unrecognized systematics in the Hipparcos analysis.

Pulsational Parallax Analyses The concept of using Cepheid radial pulsations to estimate distances originated with Baade (1926) and was elaborated further by Wesselink (1946). In the simplest terms, the difference in apparent magnitude at phases with identical colors (temperatures) is interpreted as a difference in angular size; integrating over the velocity curve gives the corresponding change in linear diameter; and the distance follows from the ratio of those quantities. Analyzing the velocity curve requires correcting the observed velocities, integrations over the full stellar disk, to give pulsational velocities. Recent studies are based primarily on variations on the surface-brightness method introduced by Barnes & Evans (1976), who derived a linear relation between visual surface brightness, S_V, and Johnson (V-R) color. The technique has been extended to near-infrared colors (cf. Di Benedetto 1997), with the zeropoints for these relations set through observations of non-variable stars, generally late-type giants, whose diameters can be measured directly using interferometric techniques (Fouqué & Gieren 1997).

The most recent analyses are by Di Benedetto (1997), based on an $(S_V, (V-K)_0)$ calibration, and Gieren et al (1997, 1998), who use both $(S_V, (V-K)_0)$ and $(S_K, (J-K)_0)$ relations. Gieren et al (1996 and refs within) have also applied an $(S_V, (V-R)_0)$ to analysis of individual Cepheids associated with Galactic open clusters and associations. Comparing the derived distances, Di Benedetto finds excellent agreement between his Baade-Wesselink estimates and main-sequence fitting distances (from Feast & Walker 1987, and Laney & Stobie 1994) for cluster Cepheids: 25 variables give

$$\Delta\mu_0 = (m-M)_0^{DB} - (m-M)_0^{ZAMS} = -0.05 \pm 0.23.$$

All of these studies are based on a Hyades distance modulus of 3.27, and hence the zeropoint should be made brighter by 0.06 magnitudes to match the Hipparcos analysis (Perryman et al 1998). Comparison between the Di Benedetto and Gieren et al analyses (17 stars in common) gives

$$\Delta\mu_0 = (m-M)_0^{DB} - (m-M)_0^{GFG} = 0.01 \pm 0.21.$$

Many of these Cepheids have substantial foreground reddening: ten of the 34 stars in Gieren et al's sample have $E_{B-V} > 0.5$ mag., while only three have $E_{B-V} < 0.25$ magnitudes. Thus, the deduced M_V is subject to uncertainties in both E_{B-V} and in R, the ratio of total to selective absorption.[5] In general, the slope (δ) adopted for the period-luminosity relation is taken from observations of LMC variables. Gieren et al derive $\delta = -2.76$, $\rho = -1.29$, a relation less steep than the Laney & Stobie LMC calibration, and 0.16 fainter than Feast & Catchpole's relation at $\log P = 0.5$, and 0.2 magnitudes fainter at $\log P = 1.5$. Di Benedetto finds an intermediate zeropoint (again adjusted to $(m-M)_{Hyades} = 3.33$) of $\rho = -1.41$ from his analysis of Galactic Cepheids.

6.2 Red Clump Stars

Population II horizontal branch stars are at the core-helium/H-shell burning stage of evolution. Their higher-abundance counterparts from the "red clump" evident on the giant branch in intermediate-age disk open clusters, such as NGC 2420 and 2243, and in the field population of the Galactic Bulge and the Magellanic Clouds. Hipparcos provides the first accurate definition of the Solar Neighborhood population, with over 1000 evolved stars having both $\frac{\sigma_\pi}{\pi} < 10\%$ and BVI photometry.

---

[5]These considerations are of lesser importance in near-infrared wavelength. Extensive JHK observations of both Galactic and Magellanic Cloud Cepheids are being undertaken by SE Persson and collaborators.

Analyzing that sample, Paczyński and Stanek (1998) derive $\langle M_I \rangle = -0.185 \pm 0.016$, with the majority of the clump stars lying in the range $0.85 < (V\text{-}I)_C < 1.1$, $-0.4 < M_I < 0.1$.

Paczyński & Stanek propose using these stars as distance indicators. Neither the Solar Neighborhood stars nor the Bulge population ($1.0 < (V\text{-}I)_C < 1.4$) show evidence for significant trends in $\langle M_I \rangle$ with color. Bolstered by evolutionary models (Jimenez et al 1998), and despite the near-solar abundance derived by McWilliam & Rich (1994), the difference in mean colors is interpreted as a metallicity effect. The invariance in $\langle M_I \rangle$ then implies at most a weak dependence on abundance. Building on that inference, Udalski et al (1998) extend analysis to the Magellanic Clouds.

The lynchpin of this technique is the assumption that, over the age range ~2–10 Gyrs, $\langle M_I \rangle$ is invariant with metallicity. However, Cole (1998), Girardi et al (1998) and Beaulieu & Sackett (1998) show that evolutionary models predict that red clump stars spanning a range of $\delta I \sim 0.5$ magnitudes, depending on age and abundance. Cole, in particular, points out that the small dispersion in M_I within individual systems can be understood as reflecting internal homogeneity, rather than global uniformity. Moreover, contrary to expectations, Paczyński (1998) finds that individual Bulge stars show no correlation between (V-I) and [Fe/H] (assessed from Washington photometry). Finally, one should note that none of the stellar models used in these analyses are based on the new OPAL opacities, rendering the relative zeropoints uncertain. Given these complications, and the lack of consistency with other distance indicators, it is prudent to defer full acceptance of red clump stars as distance indicators until a reliable independent calibration has been obtained.

6.3 RR Lyraes and Blue Horizontal Branch Stars

Although less luminous than Cepheids, RR Lyraes are identifiable in most Local Group galaxies, providing an independent test of the distance scale. The most common, and most easily identified, are the Bailey RRab-type fundamental-mode pulsators, with asymmetric, large-amplitude ($\Delta V \sim 1$ mag) light curves. First-overtone RRc-type variables have near-sinusoidal light curves, $\Delta V \sim 0.4$ mag., while a growing number of double-mode RRd-type stars and a few second-overtone RRe-type variables are being identified, primarily through high-accuracy CCD photometric monitoring of globular clusters. Discrepancies persist, however, among the various techniques used to calibrate the mean luminosity of both variable and non-variable horizontal branch stars, centering particularly on the metallicity dependence (see Smith 1995). While only a subset of those analyses rely directly on astrometry, we review recent results from each method to set the Hipparcos-based results in context.

Direct Trigonometric Parallax Measurements Like Cepheids, RR Lyraes have short evolutionary lifetimes ($\tau \sim 10^8$ years) and the local space density is correspondingly low ($\sim 2 \times 10^{-8}$ stars pc^{-3}). Indeed, the eponymous star of this class (RR Lyrae) is the only star with an Hipparcos parallax determined to a precision better than 15% ($\pi = 4.38 \pm 0.59$ mas), implying an uncorrected absolute magnitude of $M_V = 0.82^{+0.28}_{-0.31}$ at [Fe/H] $= -1.37$. Both de Boer et al (1997) and Gratton (1998) have analyzed Hipparcos astrometry of larger samples of field horizontal branch stars. de Boer et al note that the eight blue horizontal branch (BHB) stars in their sample have individual absolute magnitudes in agreement with Dorman's (1992) model predictions, although the inferred masses are only ~ 0.4 M$_\odot$, significantly lower than expected. Gratton estimates a mean absolute magnitude for his larger sample of 22 stars using a variation of the reduced parallax method, with the M5 horizontal branch taken as a template of the absolute magnitude distribution. He derives a mean absolute magnitude of $\langle M_V \rangle = 0.69 \pm 0.1$ for $\langle [Fe/H] \rangle = -1.4$. Finally, Tsujimoto et al (1998) have analyzed data for 125 RR Lyraes ($-2.5 <$ [Fe/H] < 0.06) using the maximum likelihood technique proposed by Smith (1988), and derive

$$M_V(RR) = (0.59 \pm 0.37) + (0.20 \pm 0.63)([Fe/H] + 1.60). \quad (7)$$

Statistical Parallaxes The mean absolute magnitude of a kinematically distinct group of stars can be estimated using statistical methods if one has radial velocity and proper motion data (and photometry) for a representative sample, and field RR Lyraes are well suited to this type of analysis. Most studies are based substantially on stars drawn from the Shanghai observatory catalogue (Wan et al 1980), and almost all use the maximum likelihood method outlined by Murray (1983). As Table 3 shows, the results show a remarkable consistency, with the most recent

TABLE 3 RR Lyrae statistical parallax analyses

Source	$\langle M_V \rangle$	n_*	$\langle [Fe/H] \rangle$	U	V	W	σ_U	σ_V	σ_W
Hawley et al[a]	0.79 ± 0.21	65	-0.7	5	-120	-14	128	120	78
	0.73 ± 0.18	65	1.5	-21	-184	-4	166	114	91
Strugnell et al[b]	0.90 ± 0.21	64	-0.68	4	-110	-17	124	109	70
	0.85 ± 0.19	75	-1.43	-2	-185	-3	161	108	94
Layden et al	0.94 ± 0.30[c]	51	-0.76	6	-45	-16	52	48	29
	0.71 ± 0.12[d]	162	-1.61	9	-210	-12	168	102	97
Fernley et al	0.77 ± 0.15	84	-1.66						
	0.69 ± 0.21	60	-0.85						
Tsujimoto et al	0.69 ± 0.10	99	-1.58	-12	-200	1	160	104	86
Gould & Popowski	0.77 ± 0.13[e]	147	-1.60	6	15	1	171	99	90

[a]Reddenings based on Sturch's (1966) method.
[b]Reddenings based on Burstein & Heiles (1982) HI maps.
[c]Disk-3 sample.
[d]Halo-3 sample.
[e]Subset of Halo-3 sample.

analyses agreeing to within 0.15 magnitudes with the mean absolute magnitudes derived at the outset by Strugnell et al (1986), Barnes & Hawley (1986) and Hawley et al (1986).

Layden et al (1996) supplement the Shanghai catalogue with data for field RR Lyraes included in the Lick Northern Proper Motion sample, almost doubling the number of metal-poor stars in the sample. Their analysis also has the benefit of improved radial velocity determinations for many stars. Using Monte Carlo simulations to test the robustness of their solution, they find no suggestion of systematic bias due to inherent sample properties.

Fernley et al (1998) and Tsujimoto et al (1998) have undertaken similar analyses based on Hipparcos data. While the proper motions are typically a factor of two more accurate than in the ground-based catalogue, there is no evidence for any systematic differences. Thus, the derived mean absolute magnitudes are little changed. Gould and Popowski (1998) have also examined critically the photometry, radial velocities and reddening determinations, and find no indication of potential systematic biases in $\langle M_V \rangle$. Similar results follow if they constrain the kinematics using data for spectroscopically selected field halo stars from Beers & Sommer-Larsen (1995). In summary, statistical parallax analysis of field RR Lyraes consistently leads to faint absolute magnitudes, and provides no empirical evidence for the trend in $\langle M_V \rangle$ with [Fe/H] which is predicted theoretically and derived empirically by other analysis methods, as discussed in the following sections.

Theoretical Analyses Luminosities for both RR Lyraes and non-variable HB stars can be estimated directly from theoretical evolutionary models. Building on the fundamental calculations by Sweigart & Gross (1976), extensive grids of "standard" models (solar CNO/Fe, no core rotation) were computed by Lee & Demarque (1990—LD90) and Castellani et al (1991—CCP), while Dorman (1992) extended calculations to include enhanced CNO abundances. In broad terms, the location of a zero-age horizontal branch (ZAHB) star depends on the mass of the helium core, M_C ($\approx 0.50\ M_\odot$ for globulars); the mass of the residual envelope, which is predominantly hydrogen, but with a helium fraction enriched through dredge-up to 0.01-0.02 above Y_{MS}; and the metallicity. Increasing M_C leads to higher luminosities, while the effective temperature is highest (and log L lowest) at the minimum envelope mass: that is, BHB stars are lower mass than RHB stars at given abundance, Z.

Increasing Z for a given total mass leads to lower temperatures, with the effect becoming more pronounced with increasing envelope mass. Hence, while the LD90 models predict only a modest change in the effective temperature range spanned by [Fe/H] $= -2.25$ and -1.25 ZAHB models ($4.4 > \log T_{\it eff} > 3.75$ versus $4.32 > \log T_{\it eff} > 3.72$), the mass distribution is markedly less uniform, with the higher mass stars concentrated at low temperatures. This is the first-order explanation for the change in HB morphology with abundance, and there are three important consequences for the expected properties of RR Lyrae stars. First, Bono & Stellingwerf (1994) have shown that the location of the instability strip is relatively insensitive to abundance, with (to within $\sim$150 K) overtone pulsation stable for 7200 > T_{eff} > 6500 K and fundamental pulsation viable over the range 6900 > T_{eff} > 5900.[6] Combined with evolutionary models, the prediction is that RR Lyraes in intermediate-abundance clusters (such as M5) are lower-mass than their counterparts in metal-poor systems. In particular, the LD90 models predict 0.65–0.70 $M_\odot$ and 0.75–0.90 $M_\odot$ respectively.

These predictions can be tested using data for RRd variables in globulars. Analyzing double-mode Cepheids, Petersen (1973) demonstrated that, since the pulsation constants depend only weakly on luminosity, T_{eff} and composition, the overtone/fundamental period ratio, P_1/P_0, is determined primarily by the mass and radius (or $\langle \rho \rangle$). Hence, stars of a given mass occupy a narrow locus in the (P_1/P_0, P_0) plane—the Petersen diagram. The same technique can be applied to RRd stars, and initial results indicated masses $\sim$0.1 $M_\odot$ lower than the evolutionary predictions: $\sim$0.55 $M_\odot$ at [Fe/H] $\sim$ −1.3 (M3) and $\sim$0.65 $M_\odot$ at [Fe/H] $\sim$ −2.1. However, both Kovacs et al (1991) and Cox (1991) showed that adopting Iglesias & Rogers' (1992) opacities, rather than the Los Alamos calculations used previously, led to increases of $\sim$0.1 $M_\odot$ in the mass derived from linear pulsational analysis. Full nonlinear calculations by Bono et al (1996) give even higher mass estimates, with $\sim$0.70 $M_\odot$ for M3 variables and $\sim$0.82 $M_\odot$ for M15 and M68 stars. These results are therefore in good agreement with direct evolutionary calculations.

The second consequence is that as RR Lyrae stars evolve, the predominant direction of evolution is expected to be a function of abundance. In M5-like systems, the majority of variables should evolve from low to high temperature, while the reverse holds for M15-like systems. Recent observations of variable star period-changes (Silbermann & Smith 1995; Reid 1996b) tend to support this hypothesis. In particular, Clement et al (1997) have shown that the M3 star V79 has evolved from a fundamental pulsator to a mixed-mode RRd variable, while the strength of the overtone pulsations in V68, an RRD star, have increased. Both results imply blueward evolution. In contrast, there is evidence for more dominant fundamental-mode oscillations in the M15 RRd variable V30, implying redward evolution.

The third consequence concerns the average luminosity of RR Lyraes, which is expected to exceed the level of the ZAHB. In most clusters the offset is $\sim$0.1 magnitudes, but Lee et al (1990) emphasize that in clusters with an extremely blue horizontal branch, those stars which reach the instability strip have evolved significantly, and can be brighter than M_V(ZAHB) by $\sim$0.3 magnitudes. This is only a serious consideration for clusters where Lee's HB-type parameter exceeds 0.9 (M13, NGC 6397, etc).

Quantitative predictions of the luminosity of the ZAHB have been influenced significantly by the inclusion of the revised OPAL opacities (Iglesias & Rogers 1996) and equation of state (Rogers et al 1996). Mazzitelli et al (1995)

---

[6]Note the overlap in temperature, considered further in discussion of the Oosterhoff dichotomy.

demonstrated that the revised physical parameters predict an increase over the LD90 and CCP models of 0.01 to 0.015 $M_\odot$ in the core mass at the tip of the red giant branch. Cassisi et al (1998) further extend the analysis by including revised He-burning (α) rates and updated neutrino energy-loss calculations. They derive slightly higher core masses than even MDC, and HB lifetimes $\sim$23% shorter than the CCP models ($\approx$80 Myrs vs. 100 Myrs). The latter result has important repercussions, beyond the scope of the current article, for estimating Y_{MS}. Including the effects of helium diffusion, luminosities on the ZAHB at log $T_{\mathit{eff}} = 3.85$ (i.e. the center of the instability strip) are predicted as $\log(L) = 1.68$ at [M/H] $= -1.3$ and $\log(L) = 1.74$ at [M/H] $= -2.0$. Given bolometric corrections of 0.03 mag. (Flower 1996) and $M_{bol}(\odot) = 4.70$ (see above), these imply M_V(ZAHB) $= 0.53$ and 0.38 mag, with $\langle M_V(RR) \rangle$ brighter by up to 0.1 magnitude.

The higher HB luminosities predicted by these most recent models are also in accord with earlier spectroscopic analysis of BHB stars. Stellar atmosphere models can be used to match spectral line profiles to derive (T_{eff}, log(g)) and, given a distance estimate, the mass. The average mass derived in previous studies of metal-poor clusters (M15: Moehler et al 1995; NGC 6397: de Boer et al 1995) is no more than 0.5 $M_\odot$, significantly below theoretical expectations. However, this discrepancy is mitigated substantially if one adopts the longer distance scale implied by either the Hipparcos MS-fitting, the OPAL-based ZAHB calibration (Heber et al 1997), the brighter absolute magnitudes advocated by Walker (1992), Sandage (1993b) and McNamara (1997), and the models calculated by Caloi et al (1997) and Cassisi et al (1998).

Baade-Wesselink and Pulsational Analyses Early studies, based primarily on BV photometry, derived inconsistent results, largely because shocks in the RR Lyrae atmospheres (higher density than Cepheids) invalidate the assumption of equal surface brightness at equal color. Later analyses center on longer wavelength colors, notably (V-I) and (V-K), which are less susceptible to these effects, and use more sophisticated model atmospheres. Even so, most are limited to well-behaved phases of the light-curve. Fernley et al (1989) have also combined this technique with the infrared flux method, directly determining M_{bol} at all phases.

Results from field star studies are summarized by Cacciari et al (1992), Skillen et al (1993) and McNamara (1997). Direct analyses of cluster variables (e.g. Cohen 1992; Storm et al 1994) give results consistent with the field star data, but with typical uncertainties of ± 0.15 to 0.2 magnitudes. Skillen et al's final calibration, based on combined infrared flux/Baade-Wesselink analysis of 29 stars, gives a mean relation of

$$\langle M_V \rangle = (0.21 \pm 0.05)[\text{Fe/H}] + (1.04 \pm 0.10). \tag{8}$$

However, McNamara has re-analyzed these same stars using more recent Kurucz model atmospheres, and finds a systematic offset of 200–300 K between temperatures derived from optical photometry (including (V-I) and (V-R)) and from (V-K) data. Adopting the optical scale, McNamara derives a steeper, more luminous calibration,

$$\langle M_V \rangle = (0.29 \pm 0.05)[Fe/H] + (0.98 \pm 0.04). \tag{9}$$

Further systematic uncertainties remain in analyzing the spectroscopic data. The observed radial velocity is an integration over the stellar disk, and therefore underestimates the true pulsational velocity by a factor, p. Fernley (1994) argues that that factor is underestimated in standard analyses; hence the change in radius, linearly dependent on p, is also underestimated; and the zeropoint should be brighter by 0.07 magnitudes. Adjusting McNamara's relation, this implies $\langle M_V \rangle = 0.58 \pm 0.1$ for M5 variables and 0.29 ± 0.1 in M15. On the other hand, Bono et al (1994) point out that velocity gradients and changes in the opacity within the atmosphere during the pulsational cycle can lead to cross-correlation techniques *overestimating* the true velocity.

Another absolute-magnitude calibration method tied to the pulsational characteristics of RR Lyraes is Fourier decomposition of the light curves. Simon & Clement (1993) have applied this technique to RRc variables, concentrating on the Fourier term, ϕ_{31}. Matched against hydrodynamic models, their results indicate a steep gradient and bright zeropoint, $M_V \approx 0.36$ [Fe/H] + 0.96, close to McNamara's recent Baade-Wesselink results. In contrast, Kovacs & Jurcsik (1996) use similar techniques to analyze RRab cluster variables, deriving $M_V \approx 0.19$[Fe/H] + 1.04.

Period Shifts and the Oosterhoff Dichotomy Following Bailey's discovery of the first cluster variables, globular clusters became the targets of many photographic studies during the early 20th century. Collecting the results from those studies, Oosterhoff (1939, 1944) computed the mean period for the RRab-types in each cluster, and identified an apparent bimodality in that distribution, with peaks at $\langle P_{ab} \rangle \sim 0^d.55$ (type I) and $0^d.65$ (type II). Arp (1955) first noted that the division between type I and II was also a division by abundance, with type II systems limited to the most metal-poor clusters, such as M15 and M30. Finally, in a detailed study of the RR Lyraes in several clusters, notably M3 and M15, the archetypical type I and II clusters, Sandage (1981) presented evidence for both systematic cluster-to-cluster offsets in the mean period-amplitude and period-temperature (but not amplitude-temperature) relation, and star-to-star offsets at fixed temperature. Interpreting and understanding these observational results is crucial to determining the RR Lyrae (M_V/[Fe/H]) relation.

Most interpretations of these effects are based on the pulsation equation for RR Lyraes. The fundamental equation, $P\sqrt{\langle \rho \rangle} = Q$, where P is the period, ρ the density and Q the pulsation constant, holds for homologous systems. Substituting for ρ allows one to write this equation in terms of mass, luminosity and effective temperature. In the case of RR Lyraes, as with all stars, differences in chemical composition mean that the stars are not homologous, but van Albada & Baker

(1971), from analysis of stellar models, showed that a very similar relation holds

$$\log P_{ab} = -0.34\,\mathrm{M}_{bol} - 0.68 \log \frac{M}{M_\odot} - 3.48 \log \mathrm{T}_{\it eff} + const. \qquad (10)$$

Kovacs et al (1991: see also Fernley 1993) repeat this analysis, using models based on the revised Livermore opacities (Rogers & Iglesias 1992), and find that these coefficients vary by up to $\approx 10\%$ over the abundance range spanned by the Galactic halo. Irrespective of those variations, this equation indicates that a difference in period at the same effective temperature implies either a difference in mass or a difference in luminosity.

In qualitative terms, the Oosterhoff dichotomy is generally agreed to result from the change in morphology of the horizontal branch with decreasing abundance in old ($\gtrsim 10$ Gyr) stellar systems (Catelan 1992; Sandage 1993a and refs within). In metal-rich clusters, such as 47 Tuc, the horizontal branch lies redward of the instability strip, and there are no "traditional" RR Lyraes. Decreasing abundance leads to a migration to higher temperatures, until, at abundances of between ~ -1.7 and -1.9 dex, the horizontal branch lies blueward of the instability strip. At lower abundances, the trend reverses, and the instability strip is well populated in clusters such as M15 and M68, although less so in M92. The physical origin of the Oosterhoff phenomenon, however, remains a subject of debate, with two main explanations proposed: evolutionary hysteresis and an increased luminosity with decreasing abundance.

van Albada & Baker (1973) originated the hysteresis hypothesis. As discussed above, evolutionary models predict (and observations tend to confirm) that metal-poor RR Lyraes originate from the blue HB and evolve from high to low temperatures, while the opposite holds for intermediate abundance systems. Hence Oo I HB stars evolve initially from long-period to short-period as RRab variables before becoming higher-temperature RRc-types. In contrast, type II variables evolve from RRc-type to fundamental-mode pulsators. van Albada and Baker proposed the existence of an "either-or" temperature zone, where both fundamental and first overtone pulsations were stable. Under this scenario, Oo II variables switch to fundamental pulsation at a lower temperature than their type I counterparts, leading to a longer average period $\bar{P}_{ab}$ and a higher fraction of c-type stars in metal-poor systems.

The alternative model (Sandage 1958, 1982) is that HB luminosity (hence $\bar{P}_{ab}$) increases as [M/H] decreases. Analyzing both cluster data and Lub's (1987) observations of field stars, Sandage (1993a) argues that this is a continuous relation, and that the apparent dichotomy amongst cluster variables arises solely from the change in HB morphology with [M/H]. A similar trend is evident among RRc cluster variables, although Simon & Clement (1993) identify temperature differences as the likely source of $\delta \langle P_c \rangle$ between M5 and M68. Sandage, however, also finds systematic differences in period when comparing *individual* stars at the same temperature in clusters of different abundances. The last is the Sandage period-shift effect (SPSE). Given a gradient of $\frac{\partial \log(P_{ab})}{\partial [\mathrm{Fe/H}]} = -0.12$ for stars at the blue fundamental edge (BFE), combined with abundance-dependent relations for mass, T_{eff} (BFE) ($\frac{\partial T}{\partial [Fe/H]} = 0.012$) and bolometric correction, Sandage (1993b) deduces a mean relation for the horizontal branch of $\langle M_V \rangle \propto 0.30[Fe/H]$.

Both hysteresis and luminosity variation likely play a role in the Oosterhoff dichotomy. As noted above, Bono et al (1995) predict that the fundamental and overtone pulsation zones overlap in temperature, while RRc variables are more than twice as common in type II than type I clusters (45% vs 20%). Hysteresis, however, cannot account for the SPSE. Lee et al's (1990) proposed explanation of higher luminosities due to evolution above the ZAHB is not viable in Oo II clusters with both BHB and RHB stars (M15, M68). Caputo & de Santis (1992) have challenged both the cluster reddenings adopted by Sandage and his temperature scale for RRab variables. Catelan (1992), however, has shown that uncertainties in the former are insufficient to account fully for the SPSE, while Fernley (1993) finds excellent agreement between the latter scale and his (V-K)-based temperature calibration.

Catelan and Fernley both derive shallower gradients than Sandage's period-abundance relation, $\frac{\partial P}{\partial [Fe/H]} \sim -0.06$ to -0.09. However, a shallower period-shift relation does not necessarily imply a shallower gradient $\frac{\partial \langle M_V \rangle}{\partial [Fe/H]}$. Fernley uses his near-infrared SPSE analysis and the theoretical pulsation relations to derive

$$\langle M_V \rangle = 0.19[Fe/H] + 0.84. \tag{11}$$

That analysis, however, is based on a value of $M_{bol}(\odot) = 4.75$ and assumes little variation in mass between Oo I and Oo II RR Lyraes. If one adopts the most recent RRd mass calibration (0.65 and 0.80 $M_\odot$ respectively) and $M_{bol}(\odot) = 4.70$, then Fernley's near-infrared data imply $\langle M_V(RRab) \rangle \sim 0.45$ for M15 and ~ 0.8 for M5, i.e. $\langle M_V \rangle \propto 0.35[Fe/H]$.

Main-Sequence Fitting Calibration of Cluster Variables The RR Lyrae ($\langle M_V \rangle$, [Fe/H]) relation can be estimated directly from globular cluster color-magnitude diagrams given a distance scale calibration, such as the Hipparcos-based results tabulated in the previous section. The main limitation rests with the availability of high-accuracy CCD photometry, since most studies concentrate on low-luminosity main-sequence stars, rather than evolved stars. Time-resolved data are particularly sparse. To date, the clusters with sufficient observations to defined $\langle V_0(RRab) \rangle$ are the Oo I cluster M5 (Reid 1996b) and the three Oo II clusters M15 (Silbermann & Smith 1995), M68 (Walker 1994) and M92 (Carney et al 1992). The corresponding values of $\langle M_V(RRab) \rangle$, given the distance moduli and reddening listed in the final column of Table 2, are 0.5, 0.32, 0.32 and 0.26—implying $M_V \propto 0.2[Fe/H]$ and a bright zeropoint.

Summary The various techniques outlined above can be divided into two categories based on the deduced RR Lyrae absolute magnitude calibration: direct trigonometric parallax and statistical parallax analyzes favor $\langle M_V \rangle \sim 0.75 \pm 0.15$, with little or no variation with abundance; other methods favor a steeper

(M_V, [Fe/H]) relation and, usually, a brighter zeropoint (Figure 6). Deciding between these alternatives is not straightforward. The local RR Lyrae sample is dominated by relatively metal-rich stars ([Fe/H] > -1.4), but a brighter zeropoint in M_V also implies higher velocity dispersions, at odds with other tracers of the halo population. Gould & Popowski (1998) suggest that the problem lies in a mismatch between the cluster and field-subdwarf metallicity scales, with the clusters significantly lower abundance than the CG97 giant-based scale. However, reducing the M15 distance modulus by 0.4 magnitudes to accommodate $\langle M_V \rangle$(RR) ~ 0.8 requires ajusting the (B-V) colors of the field subdwarfs by -0.07 mag, placing those stars well blueward of even the most extreme metal-poor isochrones.

Demanding that the field and cluster calibrations agree assumes that the two sets of variables have identical absolute magnitude distributions. Cluster red giants exhibit abundance anomalies which have yet to be detected in halo giants in the field (Shetrone 1997). Sweigart (1997) points out that if those anomalies stem from deep mixing, the atmospheric helium content can also be increased, leading to higher luminosities and hotter temperatures on the horizontal branch. Catelan (1998), however, suggests that the good agreement between the period-temperature distributions of field and cluster RR Lyraes argues against significant luminosity differences. Currently, the majority of analysis techniques favor the steep (M_V, [Fe/H]) relation derived originally by Sandage (1982).

6.4 Mira Variables

Observations of long-period asymptotic giant-branch variables (Miras) in the Magellanic Clouds have established that, like Cepheids, those stars follow a period-luminosity relation, best defined in the 2.2 μm K-band (Feast et al 1989). Hipparcos obtained astrometry for 180 Galactic Miras, with the highest-precision measurement having an accuracy of 10.5% (R Car). van Leeuwen et al (1997) follow the precepts outlined by Feast & Catchpole in the latter's Cepheid study in analyzing a sub-sample of 16 Miras, pre-selected using parallax-independent criteria. Fixing the slope of the period-luminosity relation using LMC variables, the preferred relation $((\frac{1}{\sigma_\pi})^2$ solution) is

$$M_K = -3.47 \log P + 0.88 \pm 0.18.$$

7. THE LOCAL DISTANCE SCALE

Revised calibrations for distance indicators inevitably lead to revised distance estimates. We summarize those results in the context of other recent analyses, concentrating on Cepheids and horizontal branch stars. More extensive reviews of measurements of R_0 and distance estimates to the Magellanic Clouds are provided by Reid (1993) and Walker (1998) respectively. Given the continuing uncertainties, as far as possible we avoid the temptation to rationalize discrepancies and arrive at preferred values.

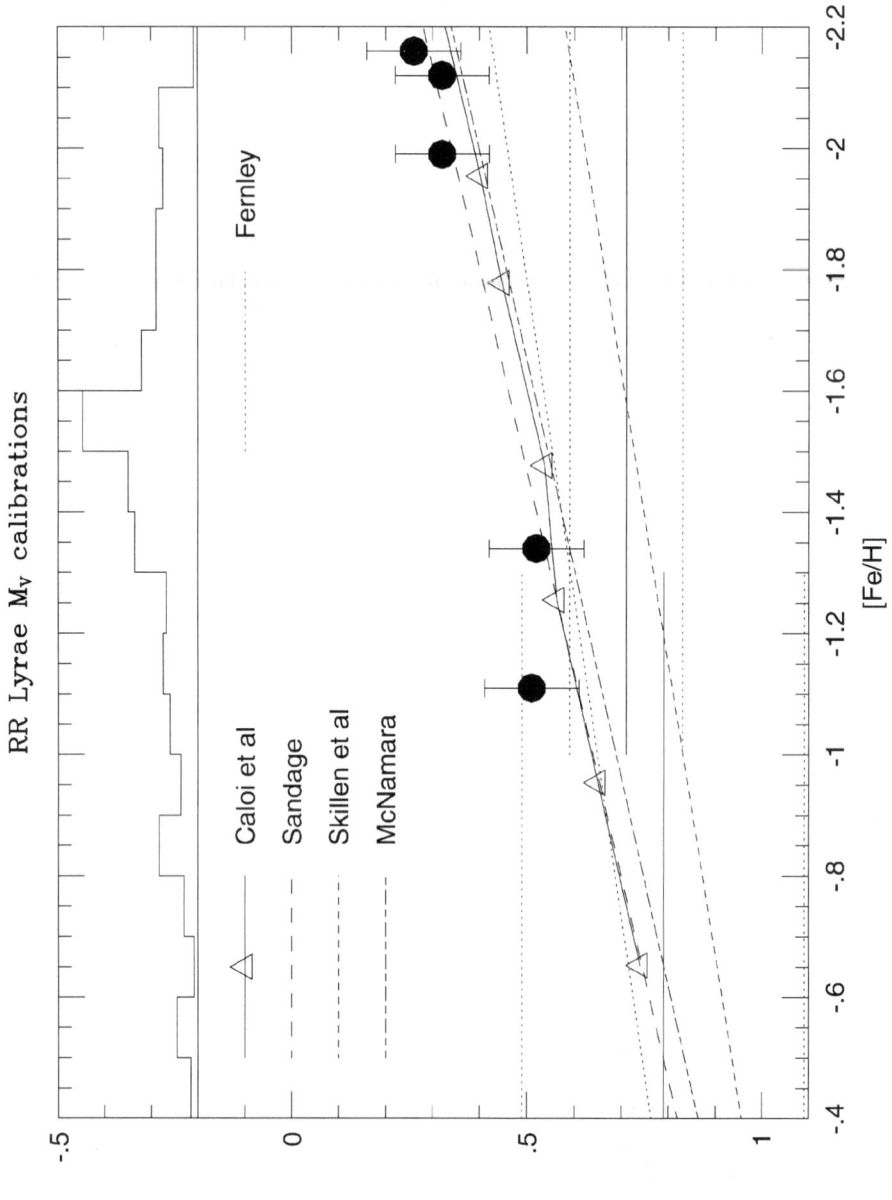

7.1 The Distance to the Galactic Center

Rotation constants: The Solar Radius can be derived from the Oort constants, $A = -\frac{1}{2}R_0(\frac{d\Omega}{dR})_0$ and $B = -\Omega_0 - \frac{1}{2}R_0(\frac{d\Omega}{dR})_0$, where Ω_0 is the local angular rotation velocity, $\frac{\Theta_0}{R_0}$. Feast and Whitelock (1998) combine Hipparcos Cepheid proper-motion data with Pont et al's (1997) radial velocities to derive $\Omega_0 = 27.19 \pm 0.87$ kms^{-1} kpc^{-1} and $R_0 = 8.5 \pm 0.5$ kpc. They derive similar results if they scale Metzger et al's (1998) Cepheid distance scale to match the Hipparcos P-L relation zeropoint; Metzger et al's own determination is 7.66 ± 0.54 kpc.

Feast and Whitelock's analysis implies $\Theta_0 \sim 230$ kms^{-1}. Olling & Merrifield (1998), however, have modeled the local rotation curve, and find that variations in the surface density lead to non-linear behavior in the Oort functions, A(R) and B(R). They argue that those variations account for the range of values for A & B derived in studies sampling sources at different heliocentric distances. Their best-fit solution has a significantly lower local velocity of rotation, $\Theta_0 = 184 \pm 8$ kms^{-1} and $R_0 = 7.1 \pm 0.4$ kpc.

Geometric determinations: Proper-motion and radial velocity data for H$_2$O masers allow the derivation of an expansion parallax. Analysis of the Sgr B2 (North) complex, ≈300 pc. from the Galactic center, leads to a value of $R_0 = 7.1 \pm 1.5$ kpc, while averaging several other studies gives $R_0 = 7.2 \pm 0.7$ kpc (Reid 1993). Alternatively, the measured proper motion of Sgr A*($\mu_l = -6.55 \pm 0.17$ mas yr^{-1}, $\mu_b = -0.48 \pm 0.06$ mas yr^{-1}; Backer, 1996, interpreted as reflex Solar motion, can be used to constrain $\frac{\Theta_0}{R_0}$. Adopting $V_\odot = 12$ kms^{-1} relative to the Local Standard of Rest, then if $170 < \Theta_0 < 240$ kms^{-1}, R_0 is constrained to lie between 5.9 and 8.1 kiloparsecs.

Red clump stars: The OGLE and MACHO microlensing surveys have fueled extensive studies of the color-magnitude distribution of both the Magellanic Cloud and Bulge stellar populations. Paczyński & Stanek (1998) derive a mean magnitude for $\langle I_0 \rangle = 14.32$ for red clump stars in the lower-reddening ($A_V < 1.5$) regions of Baade's window. Locally, the red clump stars observed by Hipparcos have $\langle M_I \rangle = -0.18$, which Paczyński and Stanek correct to $\langle M_I \rangle = -0.28$. As noted in the previous section, the local clump stars are bluer than their Bulge counterparts ($\langle (V-I)_0 \rangle \sim 1.0$ versus $\langle (V-I)_0 \rangle \sim 1.25$), but if one assumes that the stellar populations are compatible, then this implies a distance modulus of (m-M)$_0 = 14.60$ magnitudes to the centroid of the Baade's window population. Adjusting that estimate to the Galactic center, Paczyński & Stanek derive (m-M)$_0 = 14.62$, or $R_0 = 8.4 \pm 0.4$ kpc.

Figure 6 A summary of the various RR Lyrae (M_V, [Fe/H]) relations. The five solid points mark $\langle M_V \rangle$ for M5, M13, M15, M68 and M92 variables, using the distance moduli given in the final column of Table 2. The dotted horizontal lines mark the 1σ limits deduced in the Layden et al statistical-parallax analysis, while the upper histogram plots the abundance distribution of field RR Lyraes.

RR Lyraes: The Galactic bulge variables have an average abundance of [Fe/H] ~ -1 (Walker & Terndrup 1991), comparable with the globular cluster M5. Both cluster and Bulge RR Lyraes describe well-defined relations in the (logP, K) plane, with an offset in zeropoint in the sense that the Bulge stars are fainter (Reid 1998). If both sets of RR Lyraes have identical absolute magnitude distributions, then a distance modulus of 14.5 to M5 implies $(m-M)_0 = 14.8$ to the centroid of the Bulge variables, or $R_0 = 9.1 \pm 0.4$ kiloparsecs.

7.2 The Distance to the Large Magellanic Cloud

SN 1987A: The fortuitous existence of a gaseous ring surrounding the progenitor of SN 1987A permits a geometric estimate of its distances. As originally discussed by Panagia et al (1991), the time delay between the onset of fluorescence in forbidden lines of C and N, and the maximum luminosity in those lines (observed by the IUE satellite) provides a measure of the linear dimensions of the ring, while imaging by HST allows measurement of the angular dimensions. Gould & Uza (1998) employ sophisticated light-curve analysis to determine an upper limit to the distance modulus of $(m-M)_0 = 18.44$ (for an elliptical ring, $\frac{b}{a} \sim 0.95$). However, Panagia (1998) points out that their analysis omits a crucial factor: the SN 1987A ring is expanding; hence the diameter measured by HST in 1990 overestimates the angular size at maximum fluorescence in 1988. Taking the expansion into account, Panagia derives a distance modulus of 18.57 ± 0.1, or a distance of 51.7 ± 2.3 kiloparsecs.

Cepheids and Miras: Laney & Stobie's (1995) pre-Hipparcos Galactic calibration (scaled to a Hyades modulus of 3.33 magnitudes) gives a true distance modulus of 18.58 to the LMC. In comparison, Feast & Catchpole use their Galactic calibration to estimate $(m-M)_0 = 18.70$. As both Sandage & Tamman (1998) and Feast & Whitelock (1998) emphasize, these estimates depend on both the assumed foreground reddening toward the LMC and metallicity corrections. Madore & Freedman (1991) adopt $E_{V-I} = 0.10$ for LMC Cepheids lacking direct measurement, while Feast & Whitelock argue for $E_{V-I} = 0.075$. This translates to $\delta(m-M)_0 \sim 0.1$ mag for identical Galactic calibrations. Gieren et al (1998) adopt the same reddening corrections as Feast & Catchpole, and no metallicity corrections (Feast & Catchpole adopt +0.042 mag), and use their Galactic calibration to derive a shorter distance modulus of $(m-M)_0 = 18.54$ magnitudes, while scaling Di Benedetto's calibration gives $(m-M)_0 = 18.64$ (Walker 1998). Finally, van Leeuwen et al's preferred calibration of the Mira period-luminosity relation leads to an estimate of $(m-M)_0 = 18.54$ magnitudes. Averaging these three estimates gives $(m-M)_0 = 18.57$, matching the pre-Hipparcos Cepheid value.

Horizontal branch and red clump stars: Population differences between the Galaxy and the LMC complicate the use of these stars, particularly the red clump. The LMC clusters with the highest RR Lyrae frequency (such as NGC 1466) have abundances of [Fe/H] ~ -1.8, at which abundance the horizontal branch lies exclusively blueward of the instability strip in the (older) Galactic clusters.

Theoretical models predict relatively small luminosity differences, but one must assume that a similar (M_V, [Fe/H]) relation holds for both systems. Given $\langle V_0 \rangle = 18.98$ for the LMC cluster variables (Walker 1994), the Galactic RR Lyrae statistical parallax calibration ($M_V = 0.77 \pm 0.15$) implies $(m-M)_0 \sim 18.2$, compatible with the value derived from red clump stars (Udalski et al 1998). On the other hand, the steeper (M_V, [Fe/H]) relation favored by the preponderance of Galactic cluster calibrations implies $\langle M_V \rangle \sim 0.35$ to 0.4 at [Fe/H] $= -1.8$, and an LMC modulus of $(m-M)_0 \sim 18.60$. The latter is in better accord with estimates based on other distance indicators, notably SN1987A.

7.3 Distances to the SMC and M31

The SMC The relative distance modulus between the LMC and the SMC can be determined through observations of both Cepheids and RR Lyraes, making due allowance for the significant depth of the latter system over much of its area. RR Lyraes in NGC 121 ([Fe/H] ~ -1.9) have $\langle V_0 \rangle = 19.46 \pm 0.07$ (Walker & Mack 1988), implying $\delta(m-M)_0 = 0.48 \pm 0.07$ relative to the average of the LMC clusters. Udalski's (1998) analysis of field RR Lyraes leads to a slightly higher relative distance, $\delta(m-M)_0 = 0.52 \pm 0.05$, although reddening and possible abundance differences are a complication in this comparison. Analyses of LMC and SMC Cepheids (Laney & Stobie 1995, di Benedetto 1997), favor $\delta(m-M)_0 = 0.42 \pm 0.04$. Thus, the available estimates span only a limited range in $\delta(m-M)_0$.

M31 Distance estimates for M31 are based primarily on observations of fundamental distance indicators, with the high spatial resolution provided by HST rendering globular cluster HB stars accessible. As with the SMC, direct determinations based on Galactic calibrations can be complemented by measuring the relative distance modulus with respect to the LMC.

Cepheids: Freedman & Madore (1990) derive a relative distance modulus of $\langle \delta(m-M)_0^{M31-LMC} \rangle = 5.89$ mag using observations of variables in Baade's fields I, III and IV, while Kochanek (1997) derives $5.8 < \delta(m-M)_0 < 5.97$ for a range of metallicity relations. One should note, however, that Freedman & Madore derive $\delta(m-M)_0 = 6.10$ mag. from data for ~ 50 Cepheids in Baade's field IV, the outermost field, which has both negligible foreground reddening and a metallicity ($\frac{Z}{Z_\odot} \sim 0.3$) close to that of the LMC. Comparative results for RR Lyraes are also consistent with a longer modulus.

RR Lyraes: Pritchet & Van den Bergh (1987) obtained the first photometry of RR Lyraes in the halo of M31, deriving $\langle B \rangle = 25.68 \pm 0.06$. Fusi-Pecci et al (1996) and Holland et al (1997) have taken advantage of HST to determine color-magnitude diagrams for ten M31 globulars ranging from [Fe/H] $= -0.2$ to -2.3 dex. Assuming a linear relation, $V_0(HB) = \alpha[Fe/H] + \beta$, a least-squares fit to the mean HB magnitudes gives $\alpha = 0.13 \pm 0.07$, $\beta = 25.38 \pm 0.09$, slightly shallower than deduced for Galactic cluster variables. Considering individual clusters, the three lowest-abundance systems are G219 ([Fe/H] = −2.28), G351 (−1.95) and G302 (−1.85), with $V_0(HB) = 25.10, 25.16$ and 25.13 respectively. Adopting $V_0(HB) = 25.13$ at [Fe/H] = −2.1 gives $\delta(m-M)_0^{M31-LMC} = 6.15$ to 6.25 for $0 < \alpha < 0.35$, or $(m-M)_0^{M31} = 24.75$ to 24.85 for $(m-M)_0^{LMC} = 18.6$. Direct comparison with data for the Galactic clusters M68 and M15 gives $\delta(m-M)_0^{M31-M68} = 9.70$ ($(m-M)_0^{M31} = 24.81$ if the M68 modulus is taken from the final column of Table 2) and $\delta(m-M)_0^{M31-M15} = 9.59$ ($(m-M)_0^{M31} = 24.82$) respectively. Similarly, one can average Fusi-Pecci et al's result for G1 ([Fe/H] = −1.33, $V_0(HB) = 25.15$) and G105 (−1.25, 25.30) to estimate $V_0(HB) = 25.22$ at [Fe/H] = −1.3. Matching against M5 gives a relative distance modulus of 10.22 to 10.29 ($0 < \alpha < 0.35$) and $(m-M)_0^{M31} = 24.72$ to 24.79, consistent to within 0.1 magnitude with the estimates based on the metal-poor RR Lyrae.

Red clump stars: Stanek & Garnavich (1998) have used HST observations to measure $\langle I_0 \rangle = 24.27$ for red clump stars in three fields in M31. Coupled with their Hipparcos-based Galactic calibration, they derive $(m-M)_0 = 24.47 \pm 0.05$, a value identical with Freedman & Madore's (1990) result. However, the latter value is based on $(m-M)_0^{LMC} = 18.50$ and $\delta(m-M)_0^{M31-LMC} = 5.89$ mag, while Udalski et al (1998) derive $\langle I_0 \rangle = 17.85$ for LMC red clump stars, giving $\delta(m-M)_0^{M31-LMC} = 6.42$ mag and $(m-M)_0^{LMC} = 18.04$. These inconsistencies underline current uncertainties in using red clump stars as distance indicators.

8. CONCLUSIONS AND PROGNOSIS

From the preceding discussion it should be clear that the Hipparcos results do not, in and of themselves, provide definitive resolution of any of the major questions associated with the determination of the Galactic distance scale. This is not surprising, since the main achievement lies in extending greatly the sample of stars with accurate astrometry, rather than pushing that accuracy beyond the best ground-based limits. Nonetheless, these data have catalysed a general reappraisal of several issues of fundamental importance and have underlined the potential of space-based astrometry. Several satellite missions are under consideration for launch over the next decade to capitalize on this potential.

Among the acronym-rich proposed future projects are DIVA, GAIA and SIM. DIVA (Röser et al 1997) is proposed for launch as early as 2002, and marks an extension of the Hipparcos project, with twin-beamed Fizeau interferometers used to determine positions to 0.5 mas for all stars with V < 10.5 and a limited sample extending to 15th magnitude. GAIA (Lindegren & Perryman 1996) is an interferometric satellite observatory, proposed for ESA's Horizon 2000+ plan. The aims include observations of astronomical targets to 16th visual magnitude, including all sources brighter than 15th magnitude (a total of 50 million objects), with an accuracy of 10 μarcsec at V = 15.

Finally, and most ambitiously, the Space Interferometry Mission (SIM), a 5- to 10-year mission scheduled for launch in 2006, aims to achieve a precision of

1 μarcsec in relative measurements over small (<5°) angles, and an overall accuracy of 4 μarcsec in absolute astrometry. The latter capability will permit trigonometric parallax measurements of not only all primary distance-scale calibrators, but also direct distance determinations for stars in the nearer globular clusters. If achieved, those observations should finally settle the remaining discrepancies in at least the local distance scale.

Bob Hanson provided helpful comments and insight on systematic problems associated with parallax measurements.

Visit the Annual Reviews home page at http://www.AnnualReviews.org

LITERATURE CITED

Allard F, Hauschildt PH, Alexander DR, Starrfield S. 1997. *Annu. Rev. Astron. Astrophys.* 35:137–77

Alonso A, Arribas S, Martínez-Roger C. 1996. *Astron. Astrophys. Suppl.* 117:227–54

Arenou F, Lindegren L, Froeschle M, Gómez AE, Turon C, Perryman MAC, Wielen R. 1995. *Astron. Astrophys.* 304:52–60

Arp HC. 1955. *Astron. J.* 60:317–37

Axer M, Fuhrmann K, Gehren T. 1994. *Astron. Astrophys.* 291:895–909

Baade W. 1926. *Astr. Nacht.* 228:359–62

Backer DC. 1996. In *IAU Symposium 169*, eds. L Blitz & P Teuben, pp. 23–29. Reidel: Dordrecht, The Netherlands

Barnes TG, Evans DS. 1976. *MNRAS* 174:489–502

Barnes TG, Hawley SL. 1986. *Ap. J.* 307:L9–13

Beaulieu JP, Sackett PD. 1998. *Astron. J.* 116:209–19

Beers TC, Sommer-Larsen J. 1995. *Ap. J. Suppl.* 96:175–221

Bergbusch P, Vandenberg DA. 1992. *Ap. J. Suppl.* 81:163–220

Bertelli G, Bressan A, Chiosi D, Fagotto F, Nasi E. 1994. *Astron. Astrophys. Suppl.* 106:275–302

Biémont E, Baudoux M, Kurucz RL, Ansbacher W, Pinnington EH. 1991. *Astron. Astrophys.* 249:539–44

de Boer KS, Schmidt JHK, Heber U. 1995. *Astron. Astrophys.* 303:95–106

de Boer KS, Tucholke HJ, Schmidt JHK. 1997. *Astron. Astrophys.* 317:L23–26

Bolte M. 1992. *Ap. J. Suppl.* 82:145–65

Bolte M, Hogan CJ. 1995. *Nature* 376:399–402

Bono G, Caputo F, Stellingwerf RF. 1994. *Ap. J.* 432:L51–54

Bono G, Caputo F, Marconi M. 1995. *Astron. J.* 1110:2365–68

Bono G, Caputo F, Castellani V, Marconi M. 1996. *Ap. J.* 471:L33–36

Bono G, Stellingwerf RF. 1994. *Ap. J. Suppl.* 93:233–69

Burstein D, Heiles C. 1982. *Astron. J.* 87:1165–89

Cacciari C, Clementini G, Fernley JA. 1992. *Ap. J.* 396:219

Caloi V, D'Antona F, Mazzitelli I. 1997. *Astron. Astrophys.* 320:823–30

Caputo F, De Santis R. 1992. *Astron. J.* 104:253–61

Carney BW, Latham DW, Laird JB, Aguilar LA. 1994. *Astron. J.* 107:2240–89

Carney BW, Storm J, Trammell SR, Jones RV. 1992. *Publ. Astron. Soc. Pac.* 104:44–56

Carretta E, Gratton RG. 1997. *Astron. Astrophys. Suppl.* 121:95–112

Cassisi S, Castellani V, Degl'Innocenti S, Weiss A. 1998. *Astron. Astrophys. Suppl.* 129:267–79

Castellani V, Chieffi A, Pulone L. 1991. *Ap. J. Suppl.* 76:911–77

Catelan M. 1992. *Astron. Astrophys.* 261:457–71

Catelan M. 1998. *Ap. J.* 495:L81–84
Cayrel de Strobel G, Soubiran C, Friel ED, Ralite N, Francois P. 1997. *Astron. Astrophys. Suppl.* 124:299–305
Cayrel R, Lebreton Y, Perrin M-N, Turon C. 1997. In *Hipparcos Venice '97*, ed. B Battrick (ESA), pp. 219–23
Chaboyer B, Demarque P, Kernan PJ, Krauss LM. 1996. *Science* 271:957–61
Chaboyer B, Demarque P, Kernan PJ, Krauss LW. 1998. *Ap. J.* 494:96–110
Chiosi C, Wood PR, Capitanio N. 1993. *Ap. J. Suppl.* 86:541–98
Clement CM, Hilditch RW, Kaluzny J, Rucinski SM. 1997. *Ap. J.* 480:L55–58
Clementini G, Gratton RG, Carretta E, Sneden C. 1999. *MNRAS* 302:22–36
Cohen JG. 1983. *Ap. J.* 270:654–65
Cohen JG. 1992. *Ap. J.* 400:528–34
Cole AA, 1998. *Ap. J.* 500:L137–40
Cool AM, Piotto G, King IR. 1996. *Ap. J.* 468:655–62
Cox AN. 1991. *Ap. J.* 381:L71–74
D'Antona F, Caloi V, Mazzitelli I. 1997. *Ap. J.* 477:519–34
Detweiler HL, Yoss KM, Radick RR, Becker SA. 1984. *Astron. J.* 89:1038
di Benedetto GP. 1997. *Ap. J.* 486:60–74
Dorman B. 1992. *Ap. J. Suppl.* 81:221–50
Eggleton PP, Faulkner J, Flannery BP. 1973. *Astron. Astrophys.* 23:325–30
ESA. 1997. The Hipparcos and Tycho Catalogues, ESA SP-1200 (Noordwijk: ESA)
Feast MW, Walker AR. 1987. *Annu. Rev. Astron. Astrophys.* 25:345–75
Feast MW, Glass IS, Whitelock PA, Catchpole RM. 1989. *MNRAS* 241:375–92
Feast MW, Catchpole RM. 1997. *MNRAS* 286:L1–5
Feast MW, Whitelock PA. 1998. *MNRAS* 291:683–93
Feast MW, Pont F, Whitelock P. 1998. *MNRAS* 298:L43–44
Feast MW. 2000. *Publ. Astron. Soc. Pac.* In press
Fernandes J, Lebreton Y, Baglin A. 1996. *Astron. Astrophys.* 311:127–34

Fernley J. 1993. *Astron. Astrophys.* 268:591–606
Fernley J. 1994. *Astron. Astrophys.* 284:L16–18
Fernley JA, Lynas-Grey AE, Skillen I, Jameson RF, Marang F, Kilkenny D, Longmore AJ. 1989. *MNRAS* 236:447–73
Fernley JA, Barnes TG, Skillen I, Hawley SL, Hanley CJ, Evans DW, Solano E, Garrido R. 1998. *Astron. Astrophys.* 330:515–20
Flower PJ. 1996. *Ap. J.* 469:355–65
Flynn C, Morell O. 1997. *MNRAS* 286:617–25
Fouqué P, Gieren WP. 1997. *Astron. Astrophys.* 320:799–810
Freedman WF, Madore BF. 1990. *Ap. J.* 365:186–94
Fusi-Pecci F, Buonnanno R, Cacciari C, Corsi CE, et al. 1996. *Astron. J.* 112:1461–71
Gatewood G, De Jonge JK, Persinger T. 1998. *Astron. J.* 116:1501–3
Gieren WP, Mermilliod J-C, Matthews JM, Welch DL. 1996. *Astron. J.* 111:2059–65
Gieren WP, Fouqué P, Gómez M. 1997. *Ap. J.* 488:74–88
Gieren WP, Fouqué P, Gómez M. 1998. *Ap. J.* 496:17–30
Girardi L, Groenewegen MAT, Weiss A, Salaris M. 1998. *MNRAS* 301:149–60
Gizis JE. 1997. *Astron. J.* 113:806–22
Gould A, Popowski P. 1998. *Ap. J.* 508:844–53
Gould A, Uza O. 1998. *Ap. J.* 494:118–24
Gratton RG. 1998. *MNRAS* 296:739–45
Gratton RG, Carretta E, Castelli F. 1997a. *Astron. Astrophys.* 314:191–203
Gratton RG, Fusi-Pecci F, Carretta E, Clementini G, Corsi CE, Lattanzi M. 1997b. *Ap. J.* 491:749–71
Gustafsson B, Bell RA. 1979. *Astron. Astrophys.* 74:313–52
Hanson RB. 1979. *MNRAS* 186:875–96
Harris HC, Dahn CC, Monet D. 1997. In *Hipparcos Venice '97*, ed. B Battrick (ESA), pp. 105–10
Harris WE. 1996. *Astron. J.* 112:1487–88
Hawley SL, Jeffreys WH, Barnes TG, Wan L. 1986. *Ap. J.* 302:626–31
Heber U, Moehler S, Reid IN. 1997. In

Hipparcos Venice '97, ed. B Battrick (ESA), pp. 461–64
Holland S, Fahlman GG, Richer HB. 1997. *Astron. J.* 114:1488–1507
Huebner WF, Merts AL, Magee NH, Argo MF. 1977. *Los Alamos Sci. Lab. Rept.* (LA-6760-M)
Iglesias CA, Rogers FJ. 1992. *Ap. J.* 443:460–63
Iglesias CA, Rogers FJ. 1996. *Ap. J.* 464:943–53
Jahreiss H, Gliese W. 1991. Preliminary version of the third Nearby Star Catalogue
Jimenez R, Flynn C, Kotoneva E. 1998. *MNRAS* 219:515–19
Jones JB, Gilmore GF, Wyse RFG. 1996. *MNRAS* 278:146–82
King IR, Anderson J, Cool AM, Piotto G. 1998. *Ap. J.* 492:L37–40
King JR. 1997. *Astron. J.* 113:2302–11
King JR, Stephens A, Boesgaard AM, Deliyannis CP. 1998. *Ap. J.* 115:666–84
Kochanek CS. 1997. *Ap. J.* 491:13–28
Koen C. 1992. *MNRAS* 256:65–68
Kovacs G, Buchler JR, Marom A. 1991. *Astron. Astrophys.* 252:L27–30
Kovacs G, Jurcsik J. 1996. *Ap. J.* 466:L17–20
Kovalevsky J. 1998. *Annu. Rev. Astron. Astrophys.* 36:99–129
Kraft RP. 1961. *Ap. J.* 134:616–32
Kraft RP, Sneden C, Langer GE, Shetrone MD. 1993. *Astron J.* 106:1490–1507
Kurucz RL. 1992. In *The Stellar Populations of Galaxies*, IAU NSymposium 149, ed. B Barbuy & A Renzini, pp. 225–28. Luwer: Dordrecht
Laney CD, Stobie RS. 1994. *MNRAS* 266:441–54
Laney CD, Stobie RS. 1995. In *Astrophysical Applications of Stellar Pulsation*, ed. RS Stobie & PA Whitelock, ASP Conf. Ser. 83: 254–58
Layden AC, Hanson RB, Hawley SL, Klemola AR, Hanley CJ. 1996. *Astron. J.* 112:2110–31
Lee Y-W, Demarque P. 1990. *Ap. J. Suppl.* 73:709–46
Lee Y-W, Demarque P, Zinn R. 1990. *Ap. J.* 350:155–72
Lindegren L. 1988. In *Scientific Aspects of the Input Catalogue Preparation II*, Proc. Colloq. Sitges (Catalonia), ed. J Torra & C Turon, pp. 179–85
Lindegren L. 1995. *Astron. Astrophys.* 304:61–68
Lindegren L, Perryman MAC. 1996. *Astron. Astrophys. Suppl.* 116:579–95
Lindegren L, Röser S, Schrijver H, Lattanzi MG, van Leeuwen F, Perryman MAC, Bernacca PL, Falin JL, Froeschle M, Kovalevsky J, Lenhardt H, Mignard F. 1995. *Astron. Astrophys.* 304:44–51
Lub J. 1987. In *Stellar Pulsations*, pp. 218–29. Springer-Verlag: New York & Berlin
Lutz TE, Kelker DH. 1973. *Publ. Astron. Soc. Pac.* 85:573–78
Lutz TE. 1983. In *IAU Coll. 76. The Nearby Stars and the Stellar Luminosity Function*, ed. AGD Phillips & AR Upgren, pp. 41–50. Schenectady: L Davis Press
Lynga G, Lindegren L. 1998. *New Astron.* 3:121–23
McNamara DH. 1997. *Publ. Astron. Soc. Pac.* 109:857–67
McWilliam A, Rich RM. 1994. *Ap. J. Suppl.* 91:749–91
Madore BF, Freedman WL. 1991. *Publ. Astron. Soc. Pac.* 103:933–57
Madore BF, Freedman WL. 1998. *Ap. J.* 492:110–15
Mazzitelli I, D'Antona F, Caloi V. 1995. *Astron. Astrophys.* 302:382–400
Mermilliod J-C, Robichon N, Arenou F, Lebreton Y. 1997. In *Hipparcos Venice '97*, ed. B Battrick (ESA), pp. 643–50
Metzger MR, Caldwell JA, Schechter PL. 1998. *Astron. J.* 15:635–47
Moehler S, Heber U, de Boer KS. 1995. *Astron. Astrophys.* 294:65–79
Monet DG. 1988. *Annu. Rev. Astron. Astrophys.* 26:413–40
Murray CA. 1983. In *Vectorial Astrometry*, p. 297. Bristol: Adam Hilger
Nissen PE, Hog E, Schuster WJ. 1997. In

Hipparcos Venice '97, ed. B Battrick (ESA), pp. 225–30
Olling RP, Merrifield MR. 1998. *MNRAS* 297:943–52
Oosterhoff P. 1939. *Observatory* 62:104–9
Oosterhoff P. 1944. *Bull. Astr. Inst. Netherlands* 10:55–58
Oudmaijer RD, Groenewegen MAT, Schrijver H. 1997. *MNRAS* 294:L41–46
Paczyński B. 1998. *Acta Astron.* 48:405–12
Paczyński B, Stanek K. 1998. *Ap. J.* 494:L219–22
Pagel BEJ, Portinari L. 1998. *MNRAS* 298:747–52
Panagia N, Gilmozzi R, Machetto F, Adorf HM, Kirshner RP. 1991. *Ap. J.* 380:L23–26
Panagia N. 1998. *Mem. Soc. Astron. Ital.* 69:225–35
Perryman MAC, Brown AGA, Lebreton Y, Gómez A, Turon C, Cayrel de Strobel G, Mermilliod J-C, Robichon N, Kovalevsky J, Crifo 1998. *Astron. Astrophys.* 331:81–120
Petersen JO. 1973. *Astron. Astrophys.* 27:89–93
Peterson RC, Kurucz RL, Carney BW. 1990. *Ap. J.* 350:173–85
Pinsonneault MH, Stauffer J, Soderblom DR, King JR, Hanson RB. 1998. *Ap. J.* 504:170–91
Pont F, Queloz D, Bratschi P, Mayor M. 1997. *Astron. Astrophys.* 318:416–28
Pont F, Mayor M, Turon C, Vandenberg DA. 1998. *Astron. Astrophys.* 329:87–100
Pritchet CJ, van den Bergh S. 1987. *Ap. J.* 316:517–29
Rees RF. 1997. In *Proper Motions and Galactic Astronomy*. ASP Conf. Ser. vol. 127, ed. RM Humphreys, pp. 109–15
Reid IN, Murray CA. 1992. *Astron. J.* 103:514–28
Reid IN. 1996a. *Astron. J.* 111:2000–16
Reid IN. 1996b. *MNRAS* 278:367–94
Reid IN. 1997. *Astron. J.* 114:161–80
Reid IN. 1998. *Astron. J.* 115:204–28
Reid IN, Gizis JE. 1998. *Astron. J.* 116:2929–35
Reid MJ. 1993. *Annu. Rev. Astron. Astrophys.* 31:345–72
Renzini A, Bragaglia A, Ferraro FR, Gilmozzi R, Ortolani S, et al. 1996. *Ap. J.* 465:L23–26
Richer HB, Fahlmann GG. 1986. *Ap. J.* 304:273–83
Richer HB, Fahlmann GG, Ibata RA, Pryor C, Bell RA, et al. 1997. *Ap. J.* 484:741–60
Robichon N, Arenou F, Turon C, Mermilliod J-C, Lebreton Y. 1997. In *Hipparcos Venice '97*, ed. B Battrick (ESA), 567–70
Rogers FJ, Iglesias CA. 1992. *Ap. J. Suppl.* 79:507–68
Rogers FJ, Swenson FJ, Iglesias CA. 1996. *Ap. J.* 456:902–8
Rose JA. 1991. *Astron. J.* 101:937–49
Röser S, et al. 1997. In *Hipparcos Venice '97*, ed. B Battrick & MAC Perryman (ESA), pp. 777–82
Ryan SG, Norris JE. 1991. *Astron. J.* 101:1835–64
Salaris M, Chieffi A, Straniero O. 1993. *Ap. J.* 414:580–600
Sandage A. 1958. In *Vatican Conf. Stellar Populations*, ed. D O'Connell. *Specola Vaticana* 5:41–50
Sandage A, Eggen OJ. 1959. *MNRAS* 119:278–91
Sandage A. 1970. *Ap. J.* 162:841–70
Sandage A. 1981. *Ap. J.* 248:161–76
Sandage A. 1982. *Ap. J.* 252:574–81
Sandage A. 1986. *Annu. Rev. Astron. Astrophys.* 24:421–58
Sandage A. 1993a. *Astron. J.* 106:687–701
Sandage A. 1993b. *Astron. J.* 106:702–18
Sandage A, Bell RA, Tripicco MJ. 1999. *Ap. J.* 552
Sandage A, Tammann GA. 1968. *Ap. J.* 151:531–45
Sandage A, Tammann GA. 1998. *MNRAS* 293:L23–26
Sasselov DD, Beaulieu FP, Renault C, Grison P, Ferlet R, et al. 1997. *Astron. Astrophys.* 324:471–82
Schlegel DJ, Finkbeiner DP, Davis M. 1998. *Ap. J.* 500:525–53
Schuster WJ, Nissen PE. 1989. *Astron. Astrophys.* 221:65–77
Schuster WJ, Parrao L, Contreras Martinez

ME. 1993. *Astron. Astrophys. Suppl.* 97:951–83

Schwan H. 1991. *Astron. Astrophys.* 243:386–400

Seaton MJ, Yan Y, Mihalas D, Pradhan AK. 1994. *MNRAS* 266:805–28

Shetrone MD. 1997. *Astron. J.* 112:1517–35

Silbermann N, Smith HA. 1995. *Astron. J.* 110:704–28

Simon NR, Clement CM. 1993. *Ap. J.* 410:526–33

Skillen I, Fernley JA, Stobie RS, Jameson RF. 1993. *MNRAS* 265:301–15

Smith H. 1988. *Astron. Astrophys.* 198:365–69

Smith HA. 1995. *RR Lyrae Stars.* Cambridge, UK: Cambridge University Press

Sneden C, Kraft RP, Shetrone MD, Smith GH, Langer GE, Prosser CF. 1997. *Astron. J.* 114:1964–81

Soderblom DR, King JR, Hanson RB, Jones BF, Fischer D, Stauffer J, Pinsonneault MH. 1998. *Ap. J.* 504:192–99

Stanek KZ, Garnavich PM. 1998. *Ap. J.* 503:L131–34

Storm JS, Carney BW, Latham DW. 1994. *Astron. Astrophys.* 290:443–57

Straniero O. 1988. *Astron. Astrophys. Suppl.* 76:157–84

Straniero O, Chieffi A. 1991. *Ap. J. Suppl.* 76:525–616

Straniero O, Chieffi A, Limongi M. 1997. *Ap. J.* 490:425–36

Strugnell P, Reid N, Murray CA. 1986. *MNRAS* 220:413–27

Sturch CR. 1966. *Ap. J.* 143:774–808

Sweigart AV, Gross PG. 1976. *Ap. J. Suppl.* 32:367–98

Sweigart AV. 1997. *Ap. J.* 474:L23–26

Torres G, Stefanik RP, Latham DW. 1997. *Ap. J.* 485:167–81

Tsujimoto T, Miyamoto M, Yoshii Y. 1998. *Ap. J.* 492: L79–82

Trumpler RJ, Weaver HF. 1953. *Statistical Astronomy.* Berkeley: Berkeley University Press

Udalski A, Szymański M, Kubiak M, Pietrzyński G, Woźniak P, Żebruń LK. 1998. *Acta Astron.* 48:1–17

van Albada TS, Baker N. 1971. *Ap. J.* 169:311–26

van Albada TS, Baker N. 1973. *Ap. J.* 185:477–98

van Leeuwen F. 1997. *Sp. Sci. Rev.* 81:201–409

van Leeuwen F, Feast MW, Whitelock PA, Yudin B. 1997. *MNRAS* 287:955–60

van Leeuwen F, Hansen-Ruiz CS. 1997. In *Hipparcos Venice '97*, ed. B Battrick (ESA), pp. 689–92

vaan Leeuwen F. 1999. *Astron. Astrophys.* 341:L71–74

Vandenberg DA, Bolte M, Stetson PB. 1990. *Astron. J.* 100:445–68

Walker AR. 1992. *Ap. J.* 390:L81–84

Walker AR. 1994. *Astron. J.* 108:555–84

Walker AR. 1998. In *Post-Hipparcos Cosmic Candles*, ed. F Caputo & A Heck, Kluwer Acad. Publ.: Dordrecht

Walker AR, Mack P. 1988. *Astron. J.* 96:872–76

Walker AR, Terndrup DW. 1991. *Ap. J.* 378:119–26

Wan L, Mao Y-Q, Ji D. 1980. *Ann Shanghai Obs. Acad. Sinica* 2:1

Wesselink AJ. 1946. *Bull. Astron. Inst. Netherlands* 10:91–98

Wyse RFG, Gilmore G. 1995. *Astron. J.* 110:2771–87

Zinn R, West MJ. 1984. *Ap. J. Suppl.* 55:45–66

NUCLEOSYNTHESIS IN ASYMPTOTIC GIANT BRANCH STARS: Relevance for Galactic Enrichment and Solar System Formation

M. Busso,[1] R. Gallino,[2] and G. J. Wasserburg[3]

[1]Osservatorio Astronomico di Torino, 10025 Pino Torinese, Italy,
e-mail: busso@to.astro.it; [2]Dipartimento di Fisica Generale, Universita' di Torino, Via P
Giuria 1, 10125 Torino, Italy, e-mail: gallino@ph.unito.it; [3]Lunatic Asylum, Division of
Geological and Planetary Sciences, California Institute of Technology, Pasadena
California 91125, e-mail: isotopes@gps.caltech.edu

Key Words stellar evolution, supernovae, extinct radioactivities, s-process, r-process, red giants, low mass stars

■ **Abstract** We present a review of nucleosynthesis in AGB stars outlining the development of theoretical models and their relationship to observations. We focus on the new high resolution codes with improved opacities, which recently succeeded in accounting for the third dredge-up. This opens the possibility of understanding low luminosity C stars (enriched in s-elements) as the normal outcome of AGB evolution, characterized by production of ^{12}C and neutron-rich nuclei in the He intershell and by mass loss from strong stellar winds. Neutron captures in AGB stars are driven by two reactions: $^{13}C(\alpha,n)^{16}O$, which provides the bulk of the neutron flux at low neutron densities ($N_n \leq 10^7$ n/cm^3), and $^{22}Ne(\alpha,n)^{25}Mg$, which is mildly activated at higher temperatures and mainly affects the production of s-nuclei depending on reaction branchings. The first reaction is now known to occur in the radiative interpulse phase, immediately below the region previously homogenized by third dredge-up. The second reaction occurs during the convective thermal pulses. The resulting nucleosynthesis phenomena are rather complex and rule out any analytical approximation (exponential distribution of neutron fluences). Nucleosynthesis in AGB stars, modeled at different metallicities, account for several observational constraints, coming from a wide spectrum of sources: evolved red giants rich in s-elements, unevolved stars at different metallicities, presolar grains recovered from meteorites, and the abundances of s-process isotopes in the solar system. In particular, a good reproduction of the solar system main component is obtained as a result of Galactic chemical evolution that mixes the outputs of AGB stars of different stellar generations, born with different metallicities and producing different patterns of s-process nuclei. The main solar s-process pattern is thus not considered to be the result of a standard archetypal s-process occurring in all stars. Concerning the ^{13}C neutron source, its synthesis requires penetration of small amounts of protons below the convective envelope, where they are captured by the abundant ^{12}C forming a ^{13}C-rich *pocket*. This penetration cannot be modeled in current evolutionary codes, but is treated as a free parameter. Future hydrodynamical studies of time dependent mixing will be required to attack this problem. Evidence of other insufficiencies in the current mixing algorithms is common throughout the evolution of low and intermediate mass stars, as is shown by the inadequacy of stellar models in reproducing the observations of CNO isotopes in red giants and in circumstellar dust grains. These observations require some circulation of matter between the bottom of convective envelopes and regions close to the H-burning shell (cool bottom processing). AGB stars are also discussed in the light of their possible contribution to the inventory of short-lived radioactivities that were found to be alive in the early solar system. We show that the pollution of the protosolar nebula by a close-by AGB star may account for concordant abundances of ^{26}Al, ^{41}Ca, ^{60}Fe, and ^{107}Pd. The AGB star must have undergone a very small neutron exposure, and be of small initial mass ($M \leq 1.5\ M_\odot$). There is a shortage of ^{26}Al in such models, that however remains within the large uncertainties of crucial reaction rates. The net ^{26}Al production problem requires further investigation.

1. INTRODUCTION

This paper presents an updated view of nucleosynthesis in asymptotic giant branch (AGB) stars and of the implications for solar system composition and formation, as well as for galactic chemical history. This area of study is of growing interest and complexity owing to the confluence of results in nuclear physics, astronomical observations, laboratory studies of meteorites, and theoretical studies of AGB evolution. AGB stars show various degrees of photospheric enhancements of C and of elements produced by slow neutron captures (the s process), in particular those elements belonging to the Zr and to the Ba abundance peaks. Among these red stars are the carbon stars that, because of their peculiar spectrum, were classified by Secchi (1868) as of "class four."

The definition of the s process as the neutron addition mechanism running along the valley of β stability dates back to Burbidge et al (1957) and to Cameron (1957). It occurs when the neutron density is so small that most unstable isotopes have time to decay before capturing a neutron. Since the above cited work by Burbidge et al, an extensive body of work has been dedicated to the phenomenological treatment of the s process and of its signatures in the solar system abundance distribution (Clayton et al 1961, Seeger et al 1965, Clayton & Ward 1974, Käppeler et al 1982). For an account of this research see Käppeler et al (1989) and Meyer (1994). Studying the occurrence of these processes in stars then required the development of detailed stellar models for low- to intermediate-mass stars ($1 \leq M/M_\odot \leq 10$). In this work a fundamental role was played by Icko Iben and coworkers (Iben 1975, 1976, 1982, 1983; Truran & Iben 1977; Iben & Truran 1978, Iben & Renzini 1982a,b; Hollowell & Iben 1988, 1989, 1990). For extensive reviews see Iben & Renzini 1983, Sackmann & Boothroyd 1991, Iben 1991, and Wallerstein et al 1997.

Several observational and experimental results have contributed, in the last 15 years, to an increased interest in AGB stars. In particular, developments in high-resolution spectroscopy of evolved red giants have yielded a more precise understanding of their photospheric composition, where freshly synthesized materials dredged up from the interior are observed. Fundamental work on s-process–enriched AGB stars of classes MS and S was done by Smith & Lambert (1985, 1986). Important results were obtained also through the extensive search for Tc in evolved red giants (Dominy & Wallerstein 1986, Wallerstein & Dominy 1988, Little et al 1987, Little-Marenin 1989), after the original discovery of this unstable nuclide by Merrill (1952) in S stars. A more difficult but relevant issue was the attempt to analyze the composition of C stars (cf Kilston 1985, Utsumi 1985). The above studies made clear the correlation of s-process abundances with C enhancement (see e.g. Lambert 1985). Recently s-process abundances have been traced for all of the zoo of AGB stars and their descendants, from MS and S stars (Smith & Lambert 1990) to various classes of Ba stars (Luck & Bond 1991, North et al 1994) to the metal poor CH giants (Vanture 1992).

In addition the observational work at long wavelengths, using improved infrared (IR) and radio techniques, has greatly increased our knowledge of circumstellar environments. From ground-based IR observations, the signatures of circumstellar grains were recognized [e.g. SiC (see Cohen 1984, Little-Marenin 1986, Martin & Rogers 1987)]. Infrared studies received an enormous impetus from data collected by the Infrared Astronomical Satellite (IRAS), which yielded a comprehensive picture of the general evolution of dusty envelopes (see e.g. van der Veen & Habing 1988, Willems & de Jong 1988, Chan & Kwok 1988, Groenewegen & de Jong 1999). The results from the Infrared Space Observatory (ISO) mission have recently completed and extended this view (see e.g. Blommaert et al 1998, Aoki et al 1998a, 1998b; Voors et al 1998; Yamamura et al 1998; Waters et al 1998). Radio observations of molecular lines then provided information on the isotopic composition of such environments (Knapp & Chang 1985, Jura et al 1988, Nyman et al 1993, Kahane et al 1992, Bujarrabal et al 1994a,b).

In the dusty envelopes, newly synthesized species are trapped in molecules, and some condense in dust grains before being ultimately injected into the interstellar medium (ISM). Several types of such presolar dust grains (SiC, graphite, corundum, etc) have been recovered from a wide variety of meteorites (see e.g. Huss & Lewis 1995). A large fraction of these grains were identified as circumstellar condensates formed around AGB stars, as inferred from clear nucleosynthetic signatures (see e.g. Anders & Zinner 1993, Ott & Begemann 1990, Zinner 1997, Hoppe & Ott 1997). Some graphite grains include crystals of different compounds in a regular relationship to the host grain (Bernatowicz et al 1991). This shows that dust can form in dense stellar winds, often governed by efficient grain growth in chemical equilibrium (Sharp & Wasserburg 1995; Bernatowicz et al 1996; Lodders & Fegley 1995, 1997; Glassgold 1996).

Finally, the existence of short-lived radioactive nuclei (especially ^{26}Al and ^{41}Ca), in condensates of the early solar system (ESS) indicates injection from a nearby star and rapid solar system formation. This has stimulated a search for a plausible stellar source (a supernova, Wolf-Rayet, or AGB star) that might have polluted the ESS with radioactivity and perhaps triggered the collapse of the proto-sun in a dense molecular cloud (see e.g. Truran & Cameron 1978; Cameron 1993;

Wasserburg et al 1994, 1995; Cameron et al 1997; Vanhala & Cameron 1998). Hydrodynamical models of triggered collapse and injection from such an event were presented (cf Foster & Boss 1998, Vanhala & Cameron 1998). An alternative hypothesis considers a local high-energy particle bombardment from the early sun during late stages of accretion (Lee et al 1999, Clayton & Jin 1995, Glassgold et al 1998, Shu et al 1996). This mechanism may have played a role in generating some short-lived nuclei in the ESS, although there are distinct conflicts in explaining the observations (Ramaty et al 1996, Lee et al 1998). The interpretations presented in this review are based on the assumption that the observed abundances of short-lived nuclei are global solar system properties reflecting the initial composition of the protosolar cloud (SC). This approach would be subject to major changes if the early sun had played an important role in providing them.

All of the above considerations call for an interpretation based on stellar nucleosynthesis. A comprehensive review dealing with the most recent developments in AGB models and with their relevance for galactic enrichment and pollution of the ESS was therefore needed, and this need, in part, motivates this work. After a short outline of stellar evolution before the AGB phase and of previous studies on AGB stars (Sections 2 and 3), we give a rather extended account of recent calculations both for AGB evolution and for neutron capture processes (Sections 4 and 5). We show how predicted abundances of s-process nuclei are strongly affected by the initial metallicity and are characterized by a consistent star-to-star scatter, in which the solar s-process distribution is only a particular sample of a continuously changing mixture, established by galactic chemical evolution (Section 6). Concerning the problem of solar system formation, we focus our attention on the possibility that a close encounter with an AGB star may have been the source of some of the extinct radioactivity in the ESS (Section 7). For reasons of space we do not address the important constraints coming from the observed isotopic composition of circumstellar dust grains preserved in meteorites. Recent reviews on this subject can be found in Anders & Zinner 1993, Zinner 1997, and Gallino et al 1997. The final conclusions are drawn in Section 8.

2. STELLAR EVOLUTION PRIOR TO THE ASYMPTOTIC GIANT BRANCH PHASE

For clarity, we discuss the phases of stellar evolution that are of relevance here, making use of a schematic view of the track followed by the stellar representative point in the H-R diagram. We adopt models of 1-$M_\odot$ and 5-$M_\odot$ stars as examples (see Figure 1a,b). Core hydrogen burning starts on the zero age main sequence (ZAMS; the diagonal line in Figure 1a,b) and goes on until H is exhausted in the core over a mass fraction close to 10% (Iben & Renzini 1984). Then the He core

Figure 1 a. Schematic evolution in the H-R diagram of a 1-$M_\odot$ stellar model and solar metallicity. All of the major evolutionary phases discussed in the text are indicated. b. Same as a, but for a stellar model of 5 $M_\odot$ with solar metallicity.

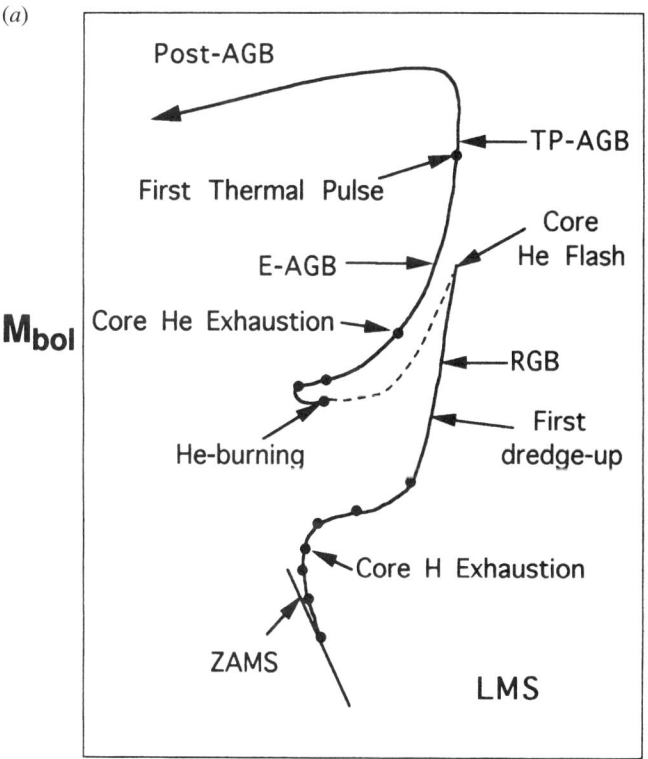

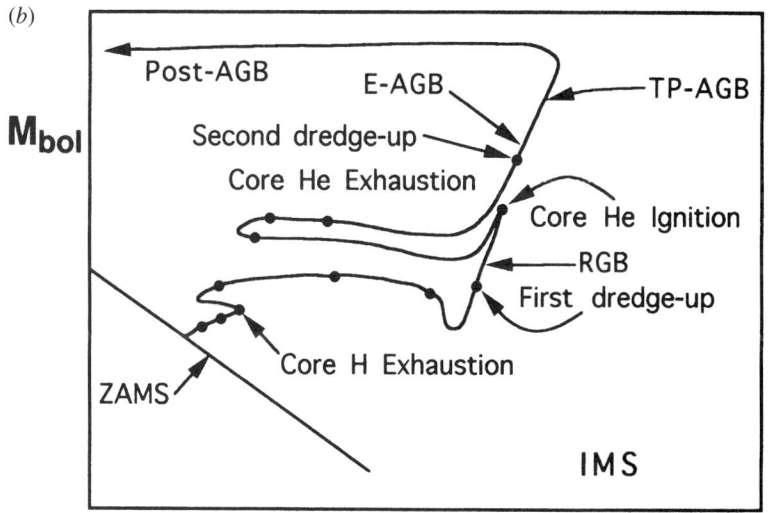

shrinks, while the stellar radius increases to carry out the energy produced by the H-burning shell (Sandage & Schwarzschild 1952, Hoyle & Schwarzschild 1955). As a consequence of envelope expansion, the stellar representative point in the H-R diagram moves to the red and to increasing luminosity, into the region occupied by stars with fully convective envelope structures, and then climbs a track called the red giant branch (RGB; see Figure 1a,b). The details of such an evolution have been the object of many dedicated studies over the last 20 years (e.g. Renzini 1984, Chiosi et al 1992).

While the envelope expands outward, convection penetrates into regions that, during the main sequence, had already experienced partial C-N processing and whose abundances of light elements had therefore been changed by proton captures (Figure 2). This occurs after the star reaches the red giant stage and ascends for

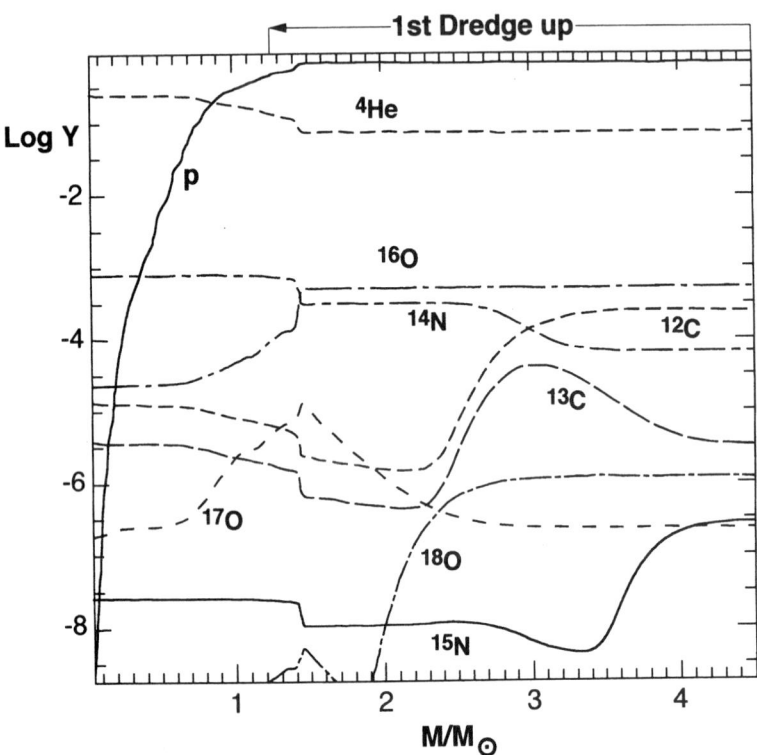

Figure 2 Distribution of the abundances of light nuclei in the inner regions of a 5-$M_\odot$ star with $Z = Z_\odot$ after core hydrogen burning. The region subsequently mixed by the first dredge-up is indicated by the *arrow*. Before first dredge-up, the original composition is unchanged down to ~4.0 $M_\odot$; below that region there are increases in ^{13}C and ^{14}N and sharp decreases in ^{12}C and ^{15}N. After first dredge-up, the envelope composition is then changed to the average value mixed to the depth indicated.

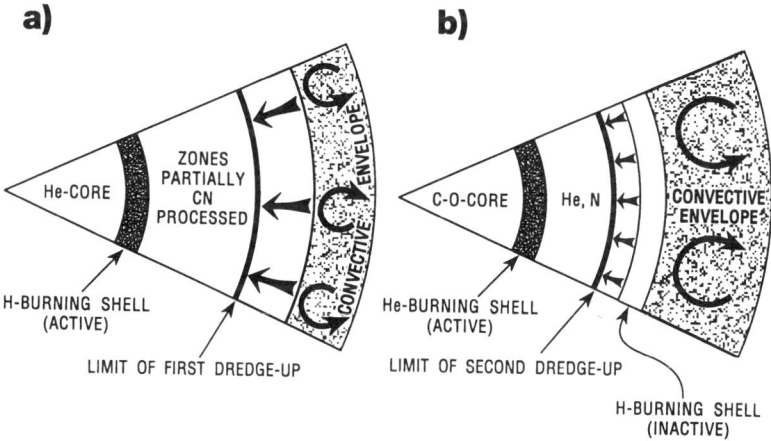

Figure 3 *a*. Sketch of the internal structure of a star at the moment of first dredge-up. The energy is produced by the H-burning shell above the He core. The first dredge-up extends downward to incorporate regions where partial CN processing occurred during the main sequence phase and are thus incorporated into the convective envelope (see Figure 2). *b*. Second dredge-up for intermediate mass stars. Here the energy is produced by the He-burning shell during the time when H burning is extinguished in the H shell. Convection penetrates from the envelope to below the former position of the H shell and hence mixes with the envelope products of H burning.

the first time along the RGB. The ensuing process, which mixes the pre-existing layered stellar structure, homogenizing about 80% of the mass (see Figure 3*a*), is called the first dredge-up. In the H-R diagram the position of stars undergoing the first dredge-up is indicated in Figure 1*a* and *b*. As a consequence, a sharp chemical discontinuity forms between the homogenized envelope and the radiative interior. The mixing modifies the surface composition: models predict a drop of ∼30% in ^{12}C and an increase by a factor of 2–3 in ^{13}C. The N and O isotopes are also affected: ^{14}N increases by a factor of 3–4, ^{15}N is decreased by a factor of ∼2, ^{17}O is enhanced by up to 1 order of magnitude, and ^{18}O is slightly depleted (∼30%) (see Dearborn 1992, El Eid 1994, Lattanzio & Boothroyd 1997). Later in their evolution, for all stars below ∼2.5 $M_\odot$ the H-burning shell reaches the discontinuity in molecular weight and erases it. This does not occur in more massive stars, because the remaining evolutionary time is too short.

Observations showed that model predictions for the first dredge-up are in rough agreement with spectroscopic abundances of C and O isotopes in red giant stars of mass above ≈2.5 $M_\odot$ (Lambert & Ries 1981, Luck & Lambert 1982, Kraft et al 1993, Kraft 1994). Below this mass limit, photospheric CNO abundances of star clusters reveal remarkable discrepancies (e.g. Gilroy 1989, Gilroy & Brown 1991, Charbonnel 1994). Figure 4 shows a comparison of carbon and oxygen isotopic ratios predicted by stellar models with observations. The low ^{12}C/^{13}C ratio found

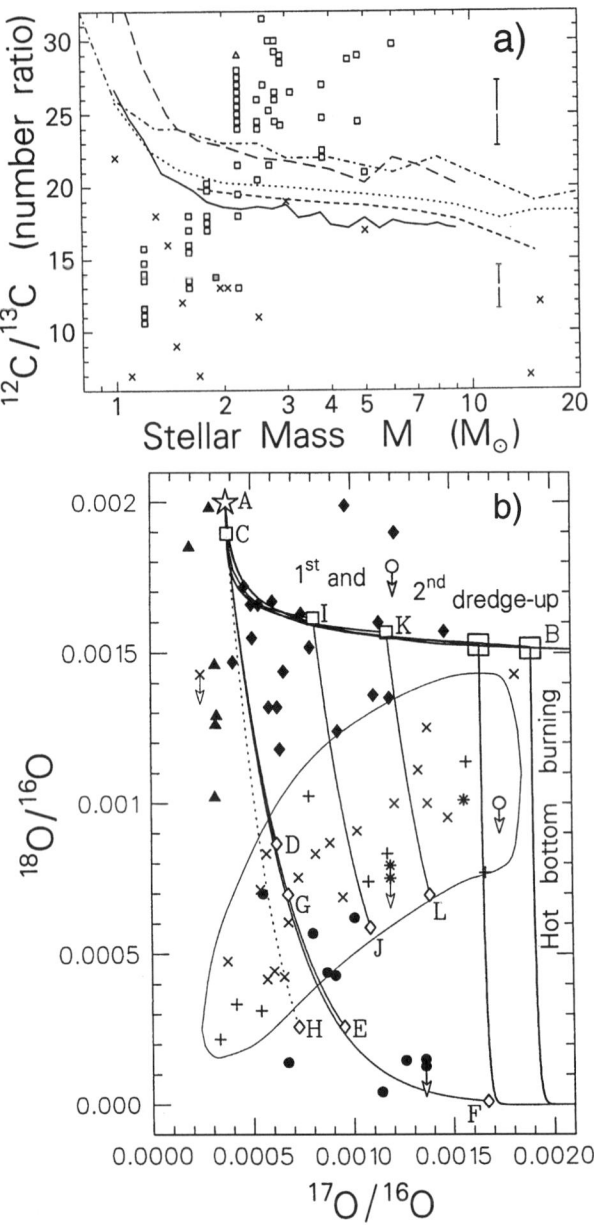

for low-mass stars approaching CNO equilibrium is in gross disagreement with the standard models (see Figure 4). It has also been noted (cf Pilachowski et al 1997) that giants with $M_{Bol} \gtrsim 0.5$ mag consistently have a sharp drop in $^{12}C/^{13}C$. These observations appear to require an extra mixing process from the bottom of stellar envelopes to near the H shell to produce the extra ^{13}C and to decrease the $^{18}O/^{16}O$ ratio to the level observed in circumstellar Al_2O_3 grains and in stellar envelopes. Its nature is still much debated; it has been ascribed to turbulent diffusion, shear instabilities, or semiconvection and has sometimes been associated with rotation (Sweigart & Mengel 1979, Zahn 1992, Charbonnel 1995, Denissenkov & Weiss 1996, Langer et al 1999). In general, the fact that, below 2.5 $M_\odot$, the discontinuity in molecular weight left by first dredge-up is subsequently erased by the advancing H shell may allow the mixing of material at the bottom of the convective envelope with regions close to the H burning shell, where partial nuclear burning modifies the chemical composition. This process was called cool bottom processing (CBP; Boothroyd et al 1994, 1995). It appears to be independent of the details of transport so long as the extra mixing gets sufficiently close in temperature to that of the H shell. It was shown, from consideration of both stellar observations and data on oxygen isotopes in circumstellar grains (cf Huss et al 1994, Nittler et al 1994, Choi et al 1998) that many observational constraints can be accounted for by CBP (Boothroyd, Sackmann & Wasserburg 1995, Wasserburg et al 1995).

For stars below a critical mass M_{HeF} [M_{HeF} is the mass at helium flash (see Alongi et al 1991, Bertelli et al 1985)], the conditions to ignite He are reached while the stellar interior is strongly electron degenerate. M_{HeF} spans the range from 1.7 to 2.5 $M_\odot$, for different choices of the mixing algorithms (Shaviv & Salpeter 1973, Maeder 1975, Sweigart & Gross 1978, Langer 1986, Renzini 1987, Chiosi et al 1992). When He is ignited in degenerate conditions, a thermal runaway (core He flash) develops (Hartwick et al 1968, Buzzoni et al 1983, El Eid 1994). After a number of flashing episodes, which burn some He into C, the electron degeneracy is removed without modification of the envelope composition (Renzini & Fusi Pecci 1988). The start of core He burning, occurring at the so-called red giant tip (Figure 1a,b), forces the representative point of the star to move off the RGB, while the H-burning shell is extinguished. A low-mass star of solar metallicity

←

Figure 4 a. Comparison of the $^{12}C/^{13}C$ ratios observed in red giant stars with model predictions of first dredge-up, as a function of stellar mass. Note that stars with $M_\odot \leq 2\ M_\odot$ have $^{12}C/^{13}C$ far (10-fold) below the five theoretical curves shown. Stars above this mass have $^{12}C/^{13}C$ typically within $\sim$20% of the theoretical curves shown. b. Comparison between observations on oxygen in Al_2O_3 grains formed as circumstellar condensates (*blackened symbols*) and stellar observations (+). Note that the stellar observations have large errors (not shown). The trajectory A–B corresponds to first and second dredge-up for standard stellar models for $Z = Z_\odot$. Trajectories A, D, J, L, and F correspond to cool-bottom-processing conveyor belt calculations (after Wasserburg et al 1995) and data sources cited therein, particularly Nittler et al (1994).

moves slightly toward the blue and to lower luminosities in the H-R diagram, to the clump region, where He burning proceeds quietly up to central He exhaustion (Figure 1a). Stars of low mass and metallicity, like those in globular clusters, in this phase populate a wider and bluer track called the horizontal branch (Faulkner 1966, Hartwick et al 1968, Castellani et al 1969). For initial masses above M_{HeF}, electron degeneracy at core He ignition is avoided, and core He burning occurs quietly in an extended blue loop of the H-R diagram (as illustrated in Figure 1b). The rest of the evolution is similar to that of lower masses.

At core He exhaustion, the star, whose mass has been reduced by stellar winds by up to 10% (Renzini 1977, Iben & Renzini 1984), becomes powered by He burning in a shell and partly by the release of potential energy from the gravitationally contracting C-O core. The large energy output pushes the representative point upward in the H-R diagram along a track that, for low-mass stars (see Figure 1a), asymptotically approaches the former RGB and is therefore known as the AGB. During the early phases (E-AGB; see Figure 1a,b), for all stars more massive than about 3.5 $M_\odot$, the energy output from the He shell forces the star to expand and cool so that the H shell remains inactive. While on the E-AGB, all stars more massive than $\sim$3.5 $M_\odot$ [hereafter referred to as intermediate-mass stars (IMS), while those below this limit are called low-mass stars (LMS)] experience a second mixing episode, called second dredge-up, whose position in the H-R diagram is shown in Figure 1b. This is driven by the expansion of the envelope, which is forced to radiate the energy produced by He burning shell and core contraction. The base of the convective envelope penetrates beyond the H-He discontinuity (Figure 3b), mixing H-burning ashes (essentially ^{4}He and ^{14}N) with the envelope (Becker & Iben 1980, Becker 1981). Note that the H-processed material dredged into the envelope is also ^{15}N poor, so that the ratio ^{14}N/^{15}N undergoes a large increase (six- to sevenfold), and that smaller changes affect ^{12}C, whereas O isotopes are almost unchanged (see Lattanzio & Boothroyd 1997 for details).

3. ASYMPTOTIC GIANT BRANCH STARS AND THE SLOW NEUTRON CAPTURE PROCESS: General Background

3.1 Stellar Evolution Along the Asymptotic Giant Branch

When the E-AGB phase is terminated, the H shell is reignited, and from then on it dominates energy production, whereas the He shell is almost inactive, ($L_{He}/L_H \leq 10^{-3}$; see Iben & Renzini 1983). This behavior is interrupted at regular intervals by thermal instabilities of the He shell [thermal pulses (TP)]. The phenomenon was discovered by Schwarzschild & Härm (1965) in LMS and then confirmed by Weigert (1966) in IMS. The structure of the star in this stage is schematically represented in Figures 3b and 5. It is characterized by a degenerate C-O core, by two shells (of H and He) burning alternatively, and by an extended convective

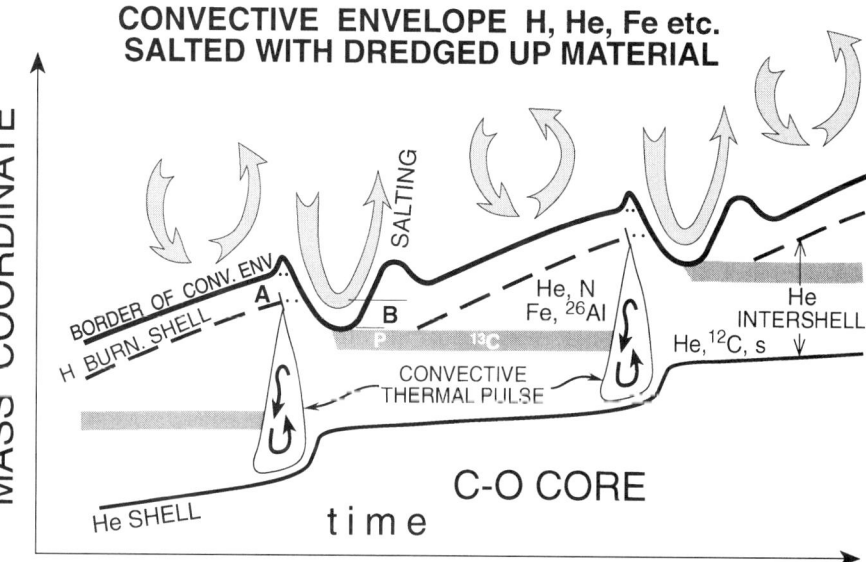

Figure 5 Illustration of the structure of a thermal pulse-asymptotic giant branch star over time, showing the border of the convective envelope, the H-burning shell, and the He-burning shell. The region between the H and He shells is the He intershell. *Horizontal gray bars* represent zones where protons are assumed to be ingested to make ^{13}C. In the earlier models, ^{13}C was not allowed to burn until the region was engulfed in a convective pulse. In the newer models, ^{13}C is naturally burned under radiative conditions in the gray area before ingestion because of the progressive heating of the region. The slow neutron capture (s) products are then engulfed by the thermal pulse, and further processing occurs owing to neutrons from the ^{22}Ne(α,n)^{25}Mg source. Region A between the H shell and the border of the convective zone and region B in the He intershell are mixed into the convective envelope during TDU, and these regions salt the envelope with freshly synthesized material. The remaining part of the He intershell region below B is also enriched in s-process nuclides and is partly mixed over subsequent cycles. Note that the convective thermal pulse does not reach the H-burning shell, as found by Iben (1977).

envelope that reaches dimensions of a few AU (Wallerstein & Knapp 1998). The two shells are separated by a thin layer in radiative equilibrium that we call the He intershell (see Figure 5). Initially, $M_H - M_{He} \simeq 1 \times 10^{-2}\ M_\odot$–$2 \times 10^{-2}\ M_\odot$ for LMS, and a factor of 5–10 less for IMS. As shell H burning proceeds while the He shell is inactive, the mass of the He intershell $M_H - M_{He}$ increases (owing to sinking of newly formed He) and attains higher densities and temperatures. This results in a dramatic increase of the He-burning rate for a short period of time, that is, in the development of a TP. The radiative state of the He intershell is thereby interrupted, and the shell then becomes almost completely convective, so that nucleosynthesis products manufactured by He burning at its bottom are carried close to the H-He interface (see Figure 5). Then the star readjusts its structure, expanding to radiate the energy produced by the temporarily effective He burning. This expansion pushes the H-He interface to low temperatures, so that H burning is shut off, whereas the convective burning in the intershell ceases. The process is repeated many times (from $\sim$10 to 100s of cycles) before the envelope is completely eroded by mass loss. This evolutionary phase is usually referred to as the TP-AGB stage. It occupies the zone of the H-R diagram shown in Figure 1a and b. The general integrated properties of stars in this phase were early studied by Paczyński (1970, 1971, 1975) and expressed through analytical relations between key stellar parameters (luminosity, interpulse period, etc.) and the mass of the H-exhausted core (M_H). The TP-AGB is characterized by the stellar structure shown in Figure 6

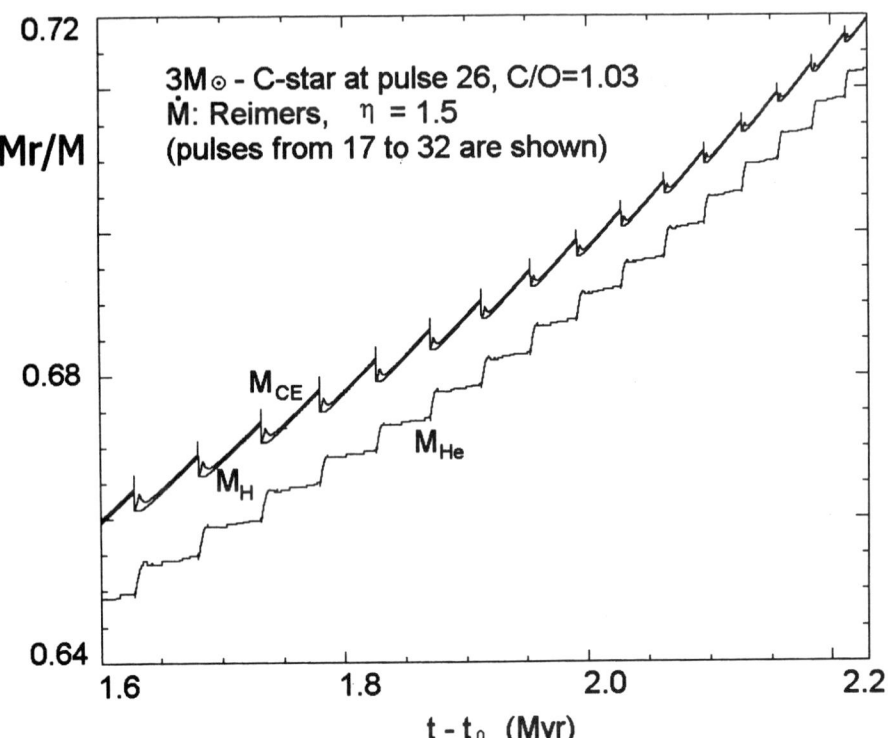

Figure 6 Plot of the internal structure of a thermal-pulse–asymptotic giant branch star as a function of time, for a 3-$M_\odot$ model with $Z = 0.02$ (Straniero et al 1997). The positions in mass of the H-burning shell (M_H), of the He-burning shell (M_{He}), and of the bottom border of the convective envelope (M_{CE}) are shown. Convective pulses (not shown) occupy almost the whole intershell region during the sudden advancement in mass of the He shell. Periodic penetration of the envelope into the He intershell (third dredge-up) is clearly visible. This model reaches the C star phase (C/O $\geq$ 1) at the 26th pulse.

as a function of time, for a 3-$M_\odot$ model star with solar metallicity. A more complex evolution is followed by AGBs with masses in the range of 8–10 $M_\odot$; here the C-O core becomes only partially degenerate, and both He- and C-thermal pulses occur (Ritossa et al 1996, Iben et al 1997, García-Berro et al 1997).

A final phase of enhanced mass loss [superwind (see Iben & Renzini 1983)] by which the star ejects the remaining envelope mass and produces a planetary nebula, is believed to terminate the AGB evolution. As a consequence, in the post-AGB phase the stellar core (i.e. the planetary nebula nucleus) evolves first to the blue (Figure 1a,b) and then toward lower luminosities, along the cooling sequence of white dwarfs. Models for the post-AGB phases were presented by Schönberner (1979, 1983), Iben (1984), Blöcker (1995a), and Blöcker & Schönberner (1996).

Since the first models were computed for TP-AGB stars, a critical problem has been that of understanding how freshly synthesized nuclei are mixed from the He intershell with the convective envelope, where they can be observed at the surface. Iben (1977) pointed out that during a thermal instability an entropy barrier exists between the He intershell and the envelope, preventing direct penetration of the convective pulse into the H shell. The excess entropy is later transferred outward, quenching the convective instability and causing the already mentioned sudden expansion and cooling of the envelope, whereas the inner border of envelope convection suddenly retreats. Then the system relaxes, the stellar structure shrinks again, and the bottom border of the envelope convection penetrates below the H-He discontinuity, that is, below the former position of the H shell, now inactive. This results in a mixing process called third dredge-up (TDU), which carries some fraction of H- and He-burning products to the surface. From the structural point of view, the TDU is very similar to the second dredge-up in Figure 3b; however, its occurrence is much faster and is expected to repeat many times.

Modeling TDU was always very difficult; it was related to the choice of the opacity tables and, in the framework of the mixing-length theory, to the value of α_P (the ratio of the mixing length l and the pressure scale height H_P). Using opacities that considered only atomic lines, the value of α_P that allowed a reproduction of the present luminosity and structure of the sun was $\alpha_P \sim 1$–1.5. Adopting this value, negative results on TDU models (Boothroyd & Sackmann 1988a,b,c,d) and occasional successes (Iben 1975, Wood 1981) were both reported. Successes were sometimes favored by various specific inputs, for example, a low metallicity. TDU was found more easily in models for population-II stars or in IMS with massive envelopes (Iben 1975, Wood 1981). In most works, sufficiently high C/O ratios to account for the formation of C stars could not be achieved or were obtained only for luminosities much higher than observed and/or by applying extra-mixing algorithms (Vassiliadis & Wood 1993). Only Lattanzio (1986, 1989) and subsequently Straniero et al (1995), using LMS models with $\alpha_P \simeq 1.5$ and the Schwarzschild criterion for convection, obtained C stars of low enough luminosity and high enough metallicity to agree with observations. The problems in explaining self-consistently TDU and the long computational times required by this work caused some authors to develop semianalytical models of the TP-AGB phases, in which TDU and other model parameters were chosen to be constrained by observations (see Marigo et al 1996 and Wagenhuber & Groenewegen 1998 for recent work in this field). The reader is referred to Wood (1981) and Iben & Renzini (1983) for the first models with third dredge-up and to Straniero et al (1995, 1997) for an updated view.

3.2 Observational Evidence of Asymptotic Giant Branch Nucleosynthesis

The relevance of the evolutionary phases briefly outlined above for the nucleosynthesis of heavy elements was demonstrated observationally long before any stellar model could address the problem. Indeed, after Merrill (1952) discovered that the chemically peculiar S stars, enriched in elements heavier than iron, contain the unstable isotope ^{99}Tc ($\bar{\tau} = 2 \times 10^5$ years) in their spectra, it was clear that ongoing nucleosynthesis occurred in situ in their interior and that the products were mixed to the surface. The fact that Tc is widespread in S stars and also in the more evolved C stars was subsequently confirmed by many workers on a quantitative basis (Smith & Wallerstein 1983, Dominy & Wallerstein 1986, Little et al 1987, Wallerstein & Dominy 1988, Kipper 1991). Coupling of high-resolution spectroscopic observations with sophisticated stellar atmosphere models allowed the determination of heavy-element abundances in AGB stars (see Gustaffson 1989 for a discussion). In particular, Smith & Lambert (1985, 1986, 1990) and Plez et al (1992) revealed that MS and S stars show an increased concentration of s-process elements. Despite large observational uncertainties, this was recognized to apply also to C stars, characterized by a photospheric C/O ratio above unity (Utsumi 1985, Kilston 1985, Olofsson et al 1993a,b, Busso et al 1995b; see also Figure 7). Direct information on AGB nucleosynthesis can also be derived spectroscopically from stars belonging to the post-AGB phase and evolving to the blue (see Figure 1a,b) after envelope ejection (Gonzalez & Wallerstein 1992, Waelkens et al 1991, Decin et al 1998). Since the pioneering work of McClure et al (1980) and McClure (1984), another source of information has come from the observation of surface abundances for the binary relatives of AGB stars, that is, for the various classes of binary sources whose enhanced concentrations of n-rich elements are caused by mass transfer in a binary system (Pilachowski 1998, Wallerstein et al 1997). These involve the classical Ba stars (Luck & Bond 1991) and their dwarf star equivalents (North et al 1994), as well as the low-metallicity CH giants (Vanture 1992). All of these works lead to the conclusion that, despite a large intrinsic scatter, an anticorrelation exists between the abundance of heavy n-rich elements and the initial metal content of the star, at least for metallicities typical of the galactic disk (see Figure 8 and Busso et al 1995b for a discussion). A summary of abundance "anomalies" of many elements was recently given by Pilachowski (1998).

From extended surveys of C stars and of MS and S giants in the Magellanic Clouds, an estimate of the luminosities (hence masses) could be derived. By such techniques, Blanco et al (1980) and Mould & Reid (1987) showed that C stars

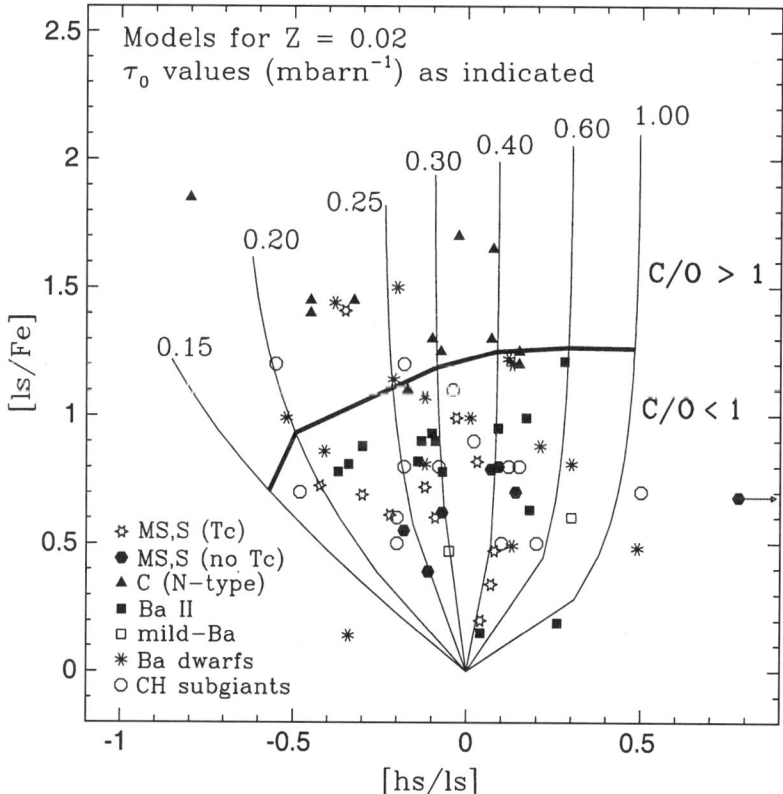

Figure 7 Observations of the logarithmic ratios [ls/Fe] of light s elements (Y, Zr) with respect to the logarithmic ratios between heavy (Ba, La, Nd, Sm) and light (Y, Zr) slow neutron capture (s) elements. Symbols refer to different types of s-enriched stars. Stars with the higher s-element enrichments are C-rich (adapted from Busso et al 1995b). Model curves were calculated for the previous generation of s-process models with convective burning of ^{13}C, for different values of the mean neutron exposure τ_0 ($\tau_0 = 0.28$ mbarn^{-1} corresponds to the solar s-process main component for the phenomenological model). Each curve represents the evolution of the envelope composition, for increasing dredge-up of processed material. Models with radiative burning of ^{13}C cover similar sequences when the amount of ^{13}C burnt is varied (upward or downward) within a factor of 3 compared with the standard (*ST*) choice of Gallino et al (1998).

are generally much fainter than the most luminous AGB stars, which appear to be O rich. This fact was confirmed more recently (see Frogel et al 1990, van Loon et al 1998, Wallerstein & Knapp 1998). In particular, Frogel et al (1990) showed how the masses of C stars in Magellanic Cloud clusters do not generally exceed 2.7 $M_\odot$. On the other side, Smith & Lambert (1989) and Smith et al (1995) showed that the brightest, Li-rich AGB objects are invariably of class S, that is, O rich. For a recent review of C star observations and evolution, see Wallerstein & Knapp

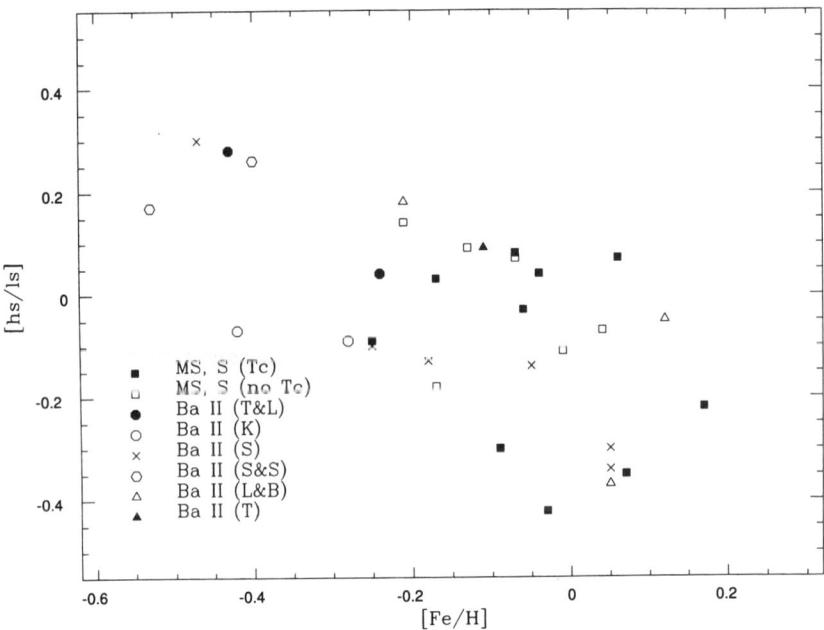

Figure 8 Observations of the logarithmic abundance ratios [hs/ls] as a function of [Fe/H] for slow neutron capture-enriched stars in the galactic disk (symbols are as in Figure 7). On average, and despite considerable scatter, the neutron exposure (which grows with the ratio [hs/ls]) appears to increase toward low metallicities. (Adapted from Busso et al 1995b).

(1998). The above studies provided the observational basis for neutron capture nucleosynthesis studies in AGB stars, in particular for discriminating between possible neutron sources. In summary, direct observations contain compelling evidence that AGB stars are the main astrophysical site for the s process and provide abundant constraints on its occurrence (e.g. its neutron exposure, correlation with ^{12}C production, inferred masses of the parent stars, etc). It is therefore not surprising that red giants in the TP-AGB phase were suggested as the site for s processes as early as in the 1960s (Sanders 1967).

3.3 The Phenomenological Approach to the Slow Neutron Capture Process

Long before the physics of AGB stars became known in some detail, the mathematical tools for the analysis of the nucleosynthesis mechanisms controlling s processes had been described by Clayton et al (1961) and by Seeger et al (1965), who pursued the original idea by Burbidge et al (1957) and built a large part of what later became known as the phenomenological approach to the s process.

They showed how considerable insight could be obtained with analytical methods. They noticed that, far from magic nuclei, the product $\sigma(A)N_s(A)$ of neutron capture cross-sections times s-process abundances is smoothly variable, so that the rule that $\sigma(A)N_s(A)$ is constant can be applied locally. They also recognized that, to reproduce the s-element abundances in the solar system, a single neutron fluence is insufficient. Defining the time-integrated neutron flux, or neutron exposure τ, as $\tau = \int N_n v_T \, dt$, where v_T is the thermal velocity and N_n the neutron density, Seeger et al (1965) pointed out that a series of physical processes, providing a number of different values of τ, were necessary to bypass the bottlenecks introduced in the neutron capture path by the very small cross-sections of the neutron magic nuclei ($N = 50, 82, 126$). Both a limited series of relatively large neutron exposures and a decreasing distribution of them (e.g. like an exponential form) were shown to provide a satisfactory fit to the solar system composition (see Clayton 1968). The exponential distribution soon became very popular, because it allowed a simple analytical formulation (Clayton & Ward 1974). If $\rho(\tau)d\tau$ is the number of iron nuclei in the solar system that experienced a neutron exposure between τ and $\tau + d\tau$, one has $\rho(\tau) = (fN_{56}^{\odot}/\tau_0)e^{-\tau/\tau_0}$, where $f(\equiv N_{56}^{capt}/N_{56}^{\odot})$ is the fractional number of iron seeds that captured neutrons and τ_0 is the mean neutron exposure. The mean neutron exposure required to reproduce the main s-process component in the solar system is $\tau_0 = 0.28$ mbarn^{-1} (Käppeler et al 1989, and Gallino et al 1993). For the evolution of a particular star, the value of τ_0 depends on the metallicity, the mass, and the size of the ^{13}C pocket; hence τ_0 for a model star may be quite different from the value for the solar fit. It was also realized that it is unlikely to obtain all the required neutron fluences in a single astrophysical site, so that three different components of the s process, for the build-up of nuclei in different atomic-mass regions, were suggested: (a) the weak component, necessary for producing s isotopes from Fe to Sr; (b) the main component, for s nuclei from Sr to Pb; and (c) a strong component, devised to provide at least 50% of ^{208}Pb. The observations discussed in the previous subsection make clear that the main component is associated with AGB stars. Each of the three components was supposed to be characterized by its own exponential distribution of exposures $\rho(\tau)$.

After the first TP-AGB models became available, Ulrich (1973) showed that, whatever the neutron source is (with the only assumption being that it burns inside the convective intershell), the recurrent mechanism of the partially overlapping TPs could easily produce an exponential distribution of neutron exposures. Indeed, the s-processed material is forced to undergo repeated n-irradiations, each followed by dilution with fresh Fe seeds. It was subsequently assumed that an exponential distribution was really required, at least for the main component, which fact, however, is not true (see Section 5).

The analyses by Seeger et al (1965) and Clayton & Ward (1974) were followed by a long series of phenomenological studies aimed at deriving more stringent constraints on the stellar models. This approach (also known as the "classical" analysis) was rapidly developed into a technique sophisticated enough to account

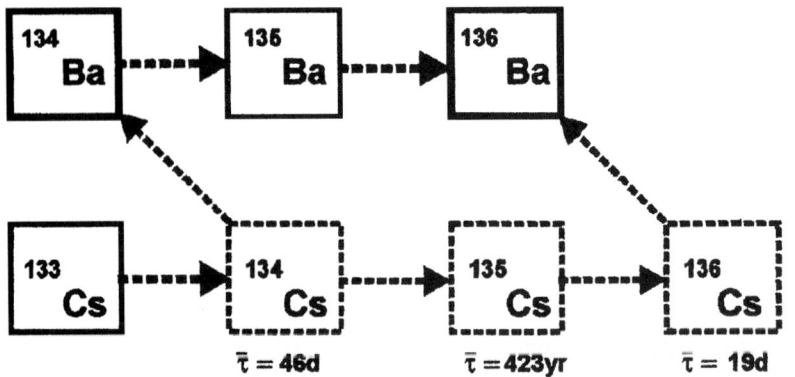

Figure 9 Neutron capture reactions branching along the slow neutron capture path at ^{134}Cs and ^{136}Cs. Here the relatively long-lived ^{135}Cs may be considered as a stable isotope, as shown for simplicity. The flow to ^{134}Ba and to ^{135}Ba depends on the branch at ^{134}Cs. Lifetimes of unstable nuclei are from Takahashi & Yokoi (1987), for $T = 3 \times 10^8$ K.

for reaction branchings along the s path. Even at the low neutron densities characterizing the s process, the competition between capture and decay still had to be considered for a number of crucial unstable isotopes (e.g. ^{79}Se, ^{85}Kr, ^{148}Pm, and ^{151}Sm). The existence of isomeric states in a number of nuclei, such as ^{85}Kr, pointed in the same direction (Ward 1977). An analytical formulation of reaction branchings for a time-dependent neutron flow was given by Ward et al (1976). The branching analysis became a useful tool for deriving information on the physical parameters at the production site (average neutron density, temperature, and electron density). As an example consider Figure 9, showing the (simplified) s flow through cesium and barium. Neutron captures on ^{133}Cs feed ^{134}Cs; this nucleus is unstable against β^- decay to ^{134}Ba, but its mean life is appreciable, from 0.97 years down to 46 d in the range $T = 1 \times 10^8$–3×10^8 K (see Takahashi & Yokoi 1987). The estimated neutron capture cross-section for ^{134}Cs is large (about 1 barn at 30 KeV; see Holmes et al 1976). Hence, ^{134}Cs becomes a branching point of the reaction chain, which can produce not only ^{134}Ba but also the relatively long-lived ^{135}Cs. In particular, this means that ^{134}Ba is partly bybassed, whereas ^{136}Ba is unbranched (as the two channels join on it).

For each branching, a branching ratio f_β can be defined comparing the rates for β^- decay and neutron capture, so that $f_\beta = \lambda_\beta/(\lambda_\beta + \lambda_n)$, where $\lambda_n = N_n \langle \sigma \rangle_b v_T$. Here $\langle \sigma \rangle_b$ is the Maxwellian averaged (n,γ) cross-section of the nucleus at the branching point (^{134}Cs in the above example). Note that $\lambda_\beta(T)$ can be dependent on temperature and thus on the stellar model. N_n is the neutron density and v_T is the thermal velocity. The operation of a branch sometimes controls the abundance ratio of two s-only isotopes (i.e. nuclei that are shielded against the r process, as for the pair ^{134}Ba and ^{136}Ba). Using the solar system s-process abundances, assuming constant neutron density, and looking for branchings at which the decay rate is independent of temperature, one can then derive N_n from the formula $N_n = (1 - f_\beta)\lambda_\beta/(f_\beta v_T \langle\sigma\rangle_b)$, by choosing f_β so that the abundance ratio of the two s only nuclei involved is solar. The temperature can subsequently be derived by branchings in which λ_β strongly depends on T, as for ^{134}Cs shown in Figure 9. It was shown that, in some cases, this procedure can also yield information on the time dependence of the neutron flux (Ward & Newman 1978). By considering several different branchings together, the classical analysis sought to derive a self-consistent set of parameters characterizing the s process. This became a guideline for stellar models, which had the task of discovering the astrophysical sites where the conditions derived a priori by the classical analysis could be fulfilled. Abundance distributions expected with an exponential form of $\rho(\tau)$ and different values of the mean neutron exposure and of the neutron density were published by Malaney (1987) and were then extensively used in the literature as approximations of what real stars should do.

The method briefly described above was continuously updated over the past two decades, to take into account progress in neutron capture cross-section measurements along the s path. After the pioneering work of the Oak Ridge group led by Macklin (see e.g. Winters & Macklin 1982 and references therein), a fundamental role in precise measurements of the key cross-sections was played by the Karlsruhe group of Franz Käppeler and coworkers (see e.g. Käppeler et al 1989, Bao & Käppeler 1987, Beer et al 1992). They sustained the stellar s-process studies with a continuous flow of high-precision cross-section data, without which the recent improvements in the field would have been impossible. The level of accuracy reached in cross-section measurements (typically of the order of a few percent) has finally demonstrated that the phenomenological approach, based on an exponential distribution of exposures, can no longer be seen as an acceptable approximation of the s process. Indeed, using the new, high-precision cross-sections, this treatment leads to large discrepancies with the solar σN_s curve for isotopes of elements like Sn, Ba, and Nd (see Käppeler 1997). Hence, the new results from updated stellar models described later (Section 5) find their correspondence in the acknowledgement that the classical analysis of s processes, after its many important contributions in the past, is now superseded. A more complete account of the phenomenological approach in various stages of its development, up to the most recent works with refined cross-section results, can be found in Käppeler et al (1982), Mathews & Ward (1985), Käppeler et al (1989), Wheeler et al (1989), Meyer (1994), and Käppeler (1997, 1999). This approach will, in any case, continue to be a key guide to nuclear astrophysical processes and problems.

3.4 The First Models of Nucleosynthesis in Asymptotic Giant Branch Stars

Since the works of Schwarzschild & Härm (1967) and Sanders (1967), AGB stars have been recognized as the most promising sites for slow neutron capture processes, and the success of Ulrich's model (1973) gave new support to this idea. Computing detailed nucleosynthesis results for LMS and IMS thus became a very important task of stellar astrophysics. This could be done owing to improvements obtained over the years in the estimates (either theoretical or experimental) of reaction rates for charged-particle interactions in stellar interiors. This work has been carried out in laboratories dedicated to nuclear physics. For many years this field was led by William Fowler and his coworkers (e.g. see Fowler et al 1967, 1975, Caughlan et al 1985, Caughlan & Fowler 1988, Kavanagh 1982). Several other groups made very important contributions, which unfortunately cannot be sufficiently detailed here (see reviews in Barnes 1982, Champagne & Wiescher 1992, Rolfs et al 1987, Rolfs & Rodney 1988). Reports on recent experimental and theoretical nuclear physics problems pertinent to this area of research can be found in the volumes of the *Nuclei in the Cosmos* series (e.g. see the issues edited by Käppeler & Wisshak 1993, Busso et al 1995a, Görres et al 1997, Prantzos 1999).

In the He-burning conditions typical of the AGB phases, two neutron sources can in principle be at play: the $^{13}C(\alpha,n)^{16}O$ reaction, originally introduced by Cameron (1954, 1957) and by Greenstein (1954), and the $^{22}Ne(\alpha,n)^{25}Mg$ reaction, suggested by Cameron (1960). ^{22}Ne is naturally produced in the He intershell by conversion of the original CNO nuclei first to ^{14}N in the H-burning shell and then to ^{22}Ne at the very beginning of a thermal pulse, through the reaction chain $^{14}N(\alpha,\gamma)^{18}F(\beta^+\upsilon)^{18}O(\alpha,\gamma)^{22}Ne$. Formation of a consistent amount of ^{13}C instead needs some mixing process to bring protons into the He intershell. As mentioned (Section 3a), the direct engulfment of protons from the H shell (that had been suggested by Schwarzschild & Härm 1967) was found to be impossible (Iben 1977). Alternatively, Iben (1975) showed that in IMS the reaction $^{22}Ne(\alpha,n)^{25}Mg$ is naturally activated, as the temperature at the bottom of the convective intershell (T_b) reaches values of $T_b \geq 3.5 \times 10^8$ K. In this scenario, the same repeated generation of n exposures envisaged by Ulrich (1973) would result, but with the $^{13}C(\alpha,n)^{16}O$ reaction replaced by the $^{22}Ne(\alpha,n)^{25}Mg$ one.

The nucleosynthesis induced by ^{22}Ne burning was analyzed by Truran & Iben (1977), who stated that an asymptotic distribution of s-process abundances rather similar to the solar pattern for $80 \leq A \leq 210$ could be achieved. Their result was, however, strictly related to the status of knowledge of cross-sections and solar abundances available at that time and was subsequently questioned (Mathews et al 1986). The combined information coming from detailed calculations of stellar models and from phenomenological analyses of the s process soon showed that the $^{22}Ne(\alpha,n)^{25}Mg$ source in IMS suffers from serious problems. Indeed, the peak neutron density reached in thermal pulses through ^{22}Ne burning is rather high (above 10^{11} cm^{-3} for $T \geq 3.5 \times 10^8$ K). This implies an n-irradiation intermediate between the s and the r process (Despain 1980), leading to a reaction network that becomes rather complex, with many branchings open. This inevitably leads to a nonsolar distribution of s isotopes and especially to excesses in the neutron-rich nuclides affected by s branchings, like ^{86}Kr, ^{87}Rb, and ^{96}Zr. The possibility was envisaged that the decline of the n density after its peak value could overcome this difficulty, because the abundances of several isotopes on the s-path freeze out along the n-density tail (Cosner et al 1980). However, the final drop in neutron density is too rapid, and it was also shown that the neutron exposure that can be achieved with the ^{22}Ne source is insufficient to produce enough heavy s-isotopes to account for the solar system main component (Howard et al 1986, Mathews et al 1986, Busso et al 1988).

4. EVOLUTION AND NUCLEOSYNTHESIS IN LOW-MASS ASYMPTOTIC GIANT BRANCH STARS

4.1 The First Models for Low Mass Star Nucleosynthesis

The stellar models so far discussed and the works on neutron captures by the ^{22}Ne neutron source were increasingly difficult to apply in explaining the observed characteristics of C stars. The main problems can be summarized as follows: (*a*) the luminosities obtained by the models were far in excess of the observations for C stars; (*b*) there had been great difficulty in obtaining TDU; and (*c*) there was no well-developed technique to treat mass loss in a physically reliable way, to account for both the low- and the high-mass loss phases observed in AGB stars (this latter problem remains a fundamental issue today). Another difficulty was that stars more massive than 5–6 $M_\odot$ were found to experience partial H burning at the base of the convective envelope [hot bottom burning (HBB); see Sugimoto 1971, Scalo & Ulrich 1973, Iben 1973, Renzini & Voli 1981, Blöcker & Schönberner 1991, Sackmann & Boothroyd 1992, Vassiliadis & Wood 1993, D'Antona & Mazzittelli 1996]. Proton captures in HBB partially convert carbon to ^{14}N. This prevents the formation of a C star unless third dredge-up (TDU) can continue to operate after HBB ceases. For example, recent calculations by Frost et al (1998) found that heavy-mass loss switches HBB off before the end of TDU. In this case, the last pulses can bring the C/O ratio above unity. In any case, the formation of a C star is at least delayed and may produce ^{12}C/^{13}C ratios lower than those observed in normal, C- (N-type) stars. Furthermore, a series of dedicated spectroscopic observations of s-enriched stars did not find any sign of the expected excesses in ^{25}Mg, which would result by the activation of the ^{22}Ne$(\alpha,n)^{25}$Mg source. Indeed Clegg et al (1979), Smith & Lambert (1986), and McWilliam & Lambert (1988) found that the isotopic mix of Mg deduced for such cool stars by the MgH molecular lines is essentially solar. In addition, current estimates for the masses of C stars suggested that they were of low mass (Frogel et al 1990). For these reasons, while the evolution of massive AGB stars remained a fascinating and open research field, it was gradually recognized that the major role in s processing and C production had to be played by LMS.

The maximum temperature achieved in LMS at the bottom of TPs does not exceed $T_b = 3 \times 10^8$ K, hence ^{22}Ne is only marginally consumed. This fact pushed some authors to reanalyze the conditions for the activation of the alternative

$^{13}C(\alpha,n)^{16}O$ source that had been previously abandoned. The main difficulty concerning this reaction is that of finding some mixing mechanism (other than the direct engulfment of a small amount of hydrogen from the envelope into a convective pulse), to bring protons into the intershell region, where they are subsequently captured by the abundant ^{12}C at H reignition, producing ^{13}C.

Iben & Renzini (1982a,b) suggested that a suitable mechanism may operate in LMS of low metallicity. Immediately after the occurrence of each thermal instability, in the cooling and expansion phase, the C-rich material is pushed into a low-temperature regime, and the local opacity strongly increases owing to partial C recombination. It was proposed that this might allow the formation of a semiconvective layer in which a small amount of hydrogen is dredged down into carbon-enriched zones. The entropy barrier problem discussed by Iben (1977) would thus be overcome. When the region heats up again, the reactions $^{12}C(p,\gamma)^{13}N(\beta^+v)^{13}C(p,\gamma)^{14}N$ consume all the hydrogen, producing a ^{13}C (and ^{14}N) pocket of a few 10^{-4} $M_\odot$. Iben & Renzini (1982a,b) suggested that this pocket is engulfed by the next convective pulse, where ^{13}C nuclei easily suffer α captures, releasing neutrons. It was immediately clear that the effectiveness of the above mechanism would critically depend on the complex physics of convective boundaries and on the treatment of opacities in the recombination zone. However, Boothroyd & Sackmann (1988a) did not find semiconvection to be important in their LMS models, and Iben (1983) demonstrated that it is ineffective in population-I stars. Nevertheless, for sufficiently low metallicity ($Z = 0.001$), Hollowell & Iben (1988) confirmed the possibility of formation of a consistent ^{13}C pocket, through a local time-dependent treatment of semiconvection.

Other processes may be at play in bringing protons below the inner border of the convective envelope. In fact the TDU forces the H-rich and the ^{12}C-rich layers to establish a contact, leaving a sharp H/C discontinuity; this is likey to be smoothed by some sort of mixing at the interface, e.g. by chemical diffusion during the interpulse phase (as considered by Iben 1982, but the effectiveness of diffusive mixing is difficult to quantify) or by rotational shear (Langer et al 1999). In the future, modeling of the proton mixing below the envelope should become the object of hydrodynamical simulations, similar to those performed for subphotospheric solar layers (see e.g. Nordlund & Stein 1995); these should be run with inputs suitable for AGB conditions. Recently, diffusive approximations of the dynamical processes playing a role at the base of the convective envelope have started to provide encouraging results, in which a ^{13}C pocket appears to be produced (Herwig et al 1997, 1998). Fully hydrodynamical calculations have also been started, and these offer promising possibilities (Singh et al 1998). As a matter of fact, the occurrence of hydrogen penetration at the top of the ^{12}C-rich intershell appears plausible, but the mass involved and the resulting ^{13}C and ^{14}N profiles at H reignition have still to be treated as relatively free parameters, constrained only by the rates of proton capture on carbon isotopes. A dedicated effort on hydrodynamical grounds constitutes a major challenge for AGB calculations and s-process nucleosynthesis studies.

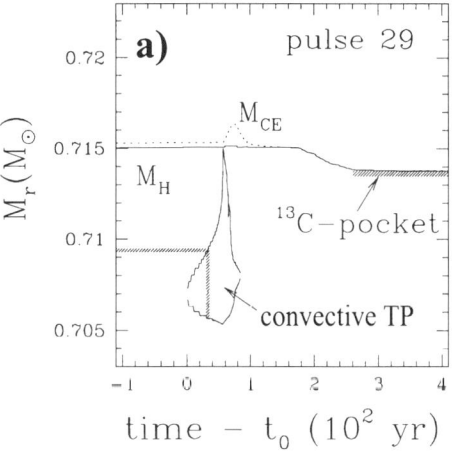 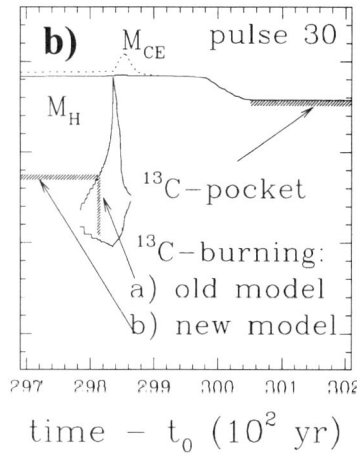

Figure 10 Two successive thermal pulses for the 3-$M_\odot$ model with $Z = Z_\odot$ are shown in their relative positions as calculated from the stellar model. The *shaded zone* is the ^{13}C pocket, in which protons are captured by ^{12}C. *a*. Ingestion and burning of ^{13}C in a pulse, based on the older models, where ^{13}C(α,n)^{16}O is first burned convectively, producing the major neutron exposure, followed by a small exposure from the ^{22}Ne(α,n)^{25}Mg neutron source in the pulse. *b*. The newer model in which ^{13}C burns in the thin radiative layer where it is produced, releasing neutrons locally. After ingestion into the convective intershell region, this is then followed by a second small neutron exposure from the marginal activation of the ^{22}Ne source.

A more indirect, but fruitful, approach was to study the implications of an assumed ^{13}C pocket on nucleosynthesis results, by comparing the computed abundances with observations. The first models presented with ^{13}C as the dominant neutron source were based on the hypothesis (Iben & Renzini 1982a,b, Malaney 1986) that the ^{13}C pocket is formed and remains inactive until the occurrence of the next instability; there, it was assumed to be engulfed and burned in convective conditions. Figure 10 illustrates the development of two subsequent pulses (the 29th and 30th) for a 3-$M_\odot$ star model in which TDU was successfully achieved. The area marked "convective thermal pulse" represents the region swept by the pulse. Figure 10*a* shows the position of the ^{13}C pocket before the 29th pulse, its ingestion into the 29th pulse, and the position of the ^{13}C pocket after the pulse. The vertical line represents the ingestion and mixing of the ^{13}C pocket into the convective pulse where it is burned following the standard old model. Figure 10*b* shows the position of the ^{13}C pocket after pulse 29 and the position after pulse 30. In the new model, the ^{13}C pocket is consumed under radiative, not convective conditions (i.e., along the line marked ^{13}C pocket). The products are then mixed into the convective thermal pulse after ^{13}C consumption and processed further by neutrons from the ^{22}Ne source (see Section 5). When convection in the He intershell reaches its maximum extent and the bottom temperature increases to $T_b \sim 3 \times 10^8$ K, the

^{22}Ne source is marginally activated, providing a second small neutron burst of relatively high peak N_n. This neutron burst was recognized to account for several details of the solar s-process abundance distribution, mainly for nuclei dependent on branchings, for which the phenomenological approach suggested a relatively high temperature (Käppeler et al 1989, 1990).

Based on this approach and the idea that the two neutron sources could be activated at different times in the same TP, a number of works were dedicated to understanding the conditions under which the solar system main component could be matched in low-mass TP-AGB stars (Gallino et al 1988; Hollowell & Iben 1989, 1990; Gallino 1989). These studies were first aimed at fitting the solar system main s-process component inside a single, archetypal AGB star. From these results, in the early 1990s, a more physical approach began to be followed, in which the main component was considered as the natural outcome of different s-process distributions averaged by galactic evolution (for an early attempt see Iben & Truran 1978). The exploration of a wide range of stellar conditions for providing a range of s-process distributions was also supported by new observational constraints, coming from the growing database on abundances for evolved and unevolved stars at various metallicities (see e.g. Gratton 1996 and references therein) and from the s-process isotopic signatures found in presolar dust grains recovered from meteorites (see e.g. Zinner 1997).

A detailed comparison between the stellar conditions suitable for ^{13}C burning and the results from the phenomenological approach was presented by Käppeler et al (1990). This comparison showed for the first time a remarkable general agreement, in sharp contrast with the many problems previously encountered with the ^{22}Ne source. The need for averaging stellar contributions from different generations of stars was clearly stated, but a best-fit solution was also found to be possible inside individual AGB models slightly less metal rich than the Sun ($Z_\odot/3$). Despite this success, the idea of ^{13}C burning in the convective pulses was also shown to have some drawbacks. In general, it provided values of $N_n = 4 \times 10^8$–10×10^8 n cm^{-3}, higher than previously expected, unless very slow ingestion rates were assumed. Moreover, Bazan & Lattanzio (1993) pointed out that there could be a risk of producing excessive energy by ^{13}C burning, thus modifying the pulse structure in a direction that would increase N_n up to values incompatible with the s process.

4.2 Recent Improvements: the Formation of Low-Mass C Stars

The field has recently gone through a phase of rapid change. In part this has been caused by the advent of new opacity tables, including contributions from molecules (Kurucz 1991, Alexander & Ferguson 1994, Rogers & Iglesias 1992, Sharp 1992), with large effects on the energy transfer in stellar envelopes. As discussed in detail by Sackmann & Boothroyd (1992), the match of the solar structure with these increased opacities required values for the parameter α_P of $\sim$2–2.2. When applied to the AGB phase, this choice has major consequences. In particular, owing to the inverse dependence of the stellar radius on α_P, new stellar models provided more compact envelopes, which made TDU occur more easily. The new opacities stimulated a number of new calculations of AGB models by various groups (Vassiliadis & Wood 1993, Straniero et al 1995, Forestini & Charbonnel 1997, Frost et al 1998). However, despite these improvements, a general consensus on TDU was not achieved. Some of the new models found third dredge-up, but this was not established as a self-consistent process agreed on by all researchers. The complexities of the AGB structure, involving extreme contrasts in local material properties, the use of the mixing-length theory for describing convective transport, and the short duration of the interpulse phases available for mixing, continue to make it difficult to address the problem (Frost & Lattanzio 1996, Lattanzio & Boothroyd 1997).

Straniero et al (1995, 1997) emphasized the necessity of adopting a computational scheme with high resolution in time and mass. With opacities based only on atomic lines and using the Schwarzschild criterion for convection, they were able to obtain a self-consistent model for a 3-$M_\odot$ star exhibiting regular dredge-up above $M_H = 0.63\ M_\odot$ and evolving into a C star (Straniero et al 1995). After introducing the new opacity tables requiring the increased value of $\alpha_P = 2.1$, using their Frascati Raphson Newton Evolutionary Code (FRANEC), these workers found that TDU occurred even more efficiently. It was reported (Straniero et al 1997) that, for models of solar metallicity, formation of a C star occurs for $M \geq 1.5\ M_\odot$, if no mass loss is included, and for $M \geq 2\ M_\odot$ if mass loss is taken into account through the Reimers' formula (1975): the values of the free parameter η were 0.7 for $M = 2\ M_\odot$ and either 1.5 or 3 for $M = 3\ M_\odot$. TDU was found to start after a limited number of pulses (when the core mass reached $M_H \simeq 0.60\ M_\odot$) and to vanish while TPs were still active, when the envelope mass was decreased below a critical limit ($\sim 0.5\ M_\odot$). It was stressed in these works that the time and mass resolution must be high and tuned so that variations of the physical quantities through each mass layer and through each time step remain very small and under control. These findings seem to express necessary, but not sufficient, conditions to model TDU through the Schwarzschild criterion alone. The results of other groups, using high-resolution prescriptions in different codes, sometimes confirm the above findings (Frost & Lattanzio 1996, case C), and sometimes do not (Herwig et al 1998, Mowlavi et al 1998).

Apart from these problems, there is now general consensus on the fact that, for sufficiently low initial masses (below ~ 1.3–$1.5\ M_\odot$), stellar winds should prevent the star from suffering more than a few pulses (and hence a few TDU episodes) before evolving to the blue and eventually becoming a white dwarf. They would thus not contribute much to the galactic synthesis of carbon and s nuclei. In this respect, we note that AGB stars with few, if any, TDU episodes are very common. Because the initial mass function favors the lower-mass stars, we actually expect that the majority of AGBs will be of low mass and hence will not mix significant amounts of C and neutron-rich nuclei into the envelope. Observational evidence confirms this conclusion and shows that several AGBs, showing IR excesses caused by dust (including the O-rich Mira prototype o Cet, $M = 1.2\ M_\odot$), show Tc in their atmosphere, with no other indications of s-process activity (Dominy & Wallerstein 1986, Little et al 1987).

A crucial parameter for the understanding of the relevant mass range for s processing is mass loss, because AGB stars are believed to be responsible for $\sim$50% of the mass return to the galaxy from evolved stars (Wallerstein & Knapp 1998). Mass loss is currently supposed to be driven by radiation pressure on dust grains, which are consequently ejected supersonically into the ISM, dragging gas with them (Sedlmayr & Dominik 1995). On this very important aspect of AGB evolution, stellar models have so far been essentially incapable of providing useful insights. Indeed, current laws used in modeling the mass loss rate by stellar winds are simple parameterized fits to the average mass loss rates observed [e.g. $\sim 10^{-6}\ M_\odot$/year for C stars (Olofsson et al 1993a,b, Loup et al 1993)]. But fits to average observations obviously do not sufficiently weight the objects with the strongest winds, which predominantly contribute to the return of mass to the galaxy, in the form of both gas and dust particles (Jura 1997). They also cannot account for the extreme variability shown by mass loss rates (Jura & Kleinmann 1989), modulated on a short time scale ($\sim$1 year) by stellar pulsations (Le Bertre 1998), on an intermediate time scale ($\sim$100 years) by unknown envelope instabilities (Sahai et al 1998), and on a longer time scale by the TPs (Waters et al 1994, Izumiura et al 1996). These variable mass loss phases also result in observations of detached circumstellar shells (Sahai et al 1998). A final phase of fast mass loss or "superwind" (Iben & Renzini 1983) is also to be expected in the preplanetary nebula stage. Attempts to introduce these enhanced mass loss rates in the models have been performed (Vassiliadis & Wood 1993, Blöcker 1995b, Schröder et al 1998), but the real situation is much too complex to be followed in a physically realistic way. The problem of the inadequate treatment of mass loss is relevant, especially because stellar winds affect the final evolution of an AGB star and the envelope composition. Strong stellar winds limit the number of pulses that a star experiences and, hence, the integrated amount of material dredged up into the envelope. Important improvements in this field are now coming from hydrodynamical models of AGB stellar winds, made by different groups (Hron et al 1997, Arndt et al 1997).

5. RADIATIVE ^{13}C BURNING AND THE NEW SLOW-NEUTRON-CAPTURE PROCESS

The works by Straniero et al (1995, 1997) represent an important step in studies of AGB evolution and nucleosynthesis, owing to the recognition that the neutron release by ^{13}C burning actually occurs in the interpulse phases, under radiative and not convective conditions. This result has been verified by various authors (Lattanzio 1995, Mowlavi et al 1996) and could have been obtained previously, if it were not for the common procedure used in past stellar models to exclude proton captures during He burning. Recognizing that neutrons are released by the ^{13}C source in radiative conditions makes a profound difference for all phenomenological treatments based on an exponential distribution of neutron exposures. The *s*-process abundances in AGB stars now appear to be controlled by how the hydrogen profile penetrates into the ^{12}C-rich intershell; the corresponding neutron exposure is better described as a superposition of a few single irradiations (Clayton et al 1961) than as an exponential distribution. Owing to the importance of this change, we shall discuss recent results in some detail, focusing on *s*-process calculations performed by Gallino et al (1998), using the new stellar models computed by Straniero et al (1997).

In the above works it was assumed that penetration of a small amount of hydrogen below the formal envelope border occurs at each TDU episode. TDU itself was found to start after 11 pulses for a 2-$M_\odot$ model and after 9 pulses for a 3-$M_\odot$ Model. At H reignition the temperature in the He intershell quickly reaches values of ~ 0.8–0.9×10^8 K, so that the time scale $\tau_{13,\alpha}$ for α captures on ^{13}C becomes much smaller than the available time before the next pulse (several 10^4 years for LMS). Hence, any ^{13}C left behind by partial-mixing mechanisms will burn in radiative conditions in the tiny zone ($\sim 1/20$ of the intershell) where it was produced, that is, the ^{13}C pocket; and only the reaction products will be ingested into the convective pulse. This avoids the risk of structural changes from the energy release by the ^{13}C$(\alpha,n)^{16}$O reaction (Bazan & Lattanzio 1993). Indeed, the H shell is active in the star and dominates any other energy source. The rate for the ^{13}C$(\alpha,n)^{16}$O reaction has recently been revised on experimental grounds by Denker et al (1995), whose results were used by Gallino et al (1998). In the temperature range of interest, this rate is up to a factor of 2 higher than estimated by Caughlan & Fowler (1988).

Figure 10 illustrates the new situation for ^{13}C burning, as compared with the previous scenario. In the present point of view, the ^{13}C pocket is the region where *s* processing occurs under radiative conditions. The hydrogen profile adopted for these zones by Gallino et al (1998) was similar to that found by Hollowell & Iben (1988), but was restricted to hydrogen mass fractions below $X_H \sim 0.0015$, that is, to number ratios $N(H)/N(^{12}C) < 0.1$. (A higher H abundance would be accompanied by a correspondingly higher abundance of ^{14}N, which is the dominant neutron absorber in these zones.) Due to the gradient in H and to the radiative conditions that maintain the layered structure, the concentration of ^{13}C which is locally formed when protons are captured by the ambient ^{12}C is dependent on the position. The same is true for the neutron density and the neutron exposure characterizing the *s* process that ensues when this ^{13}C undergoes (α,n) captures. For the sake of simplicity Gallino et al (1998) modeled these continuously changing conditions using three layers. The temporal behavior of N_n (*dashed line*) in the central layer of their simulation during ^{13}C consumption is shown in Figure 11a. It reaches rather low values, at most 10^7 n cm^{-3}. Consequently, only a small number of reaction branchings along the *s* path are open during radiative ^{13}C burning. In contrast, the neutron exposure $\Delta\tau$ in the same zone is quite large ($\Delta\tau \sim 0.40$ mbarn^{-1}). The *s* processing in each layer is highly effective; enhancement factors for the

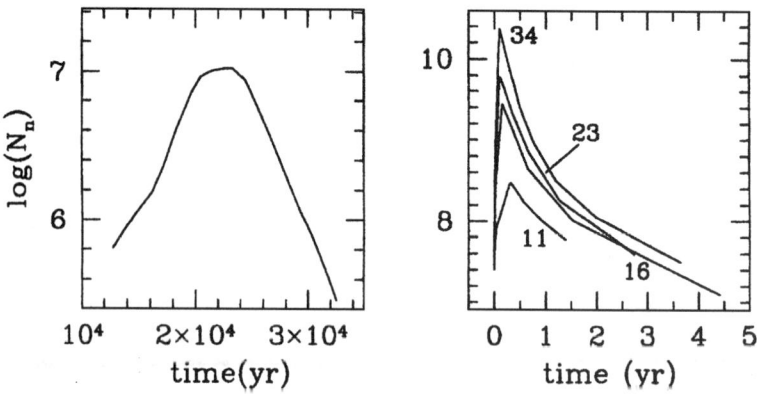

Figure 11 Neutron density during slow neutron capture processing in thermal-pulsing asymptotic giant branch stars. *a*. The situation in the central layer of the radiative zone where ^{13}C burns in the interpulse period, according to the schematic model of Gallino et al (1998). *b*. Neutron density from activation of the ^{22}Ne$(\alpha,n)^{25}$Mg neutron source in different pulses. The bottom temperature increases with pulse number (11, 16, etc), as does the neutron density.

initial composition of *s*-only isotopes range from 1×10^3 to 5×10^4. The next convective instability of the He shell mixes the *s*-process nuclei manufactured in the ^{13}C pocket over the intershell zone, so that the material over which the next neutron capture episode operates is a mixture of original Fe nuclei, of light elements (some coming from previous shell H burning), and of *s*-processed material left behind in the intershell at the quenching of the previous pulse. Note that the mass spanned by convective instabilities and their mutual overlap factor *r* decrease with pulse number. As the H shell advances in mass into the envelope, another reservoir of ^{13}C becomes available, as nuclei present in the H-burning ashes are added to the He-intershell. This occurs above the ^{13}C pocket, in regions that experience a lower temperature and are rich in ^{14}N. They can survive the interpulse phase and be engulfed by the next pulse. However, the ^{13}C abundance established here by H-burning equilibrium is small and so is its contribution to *s*-process synthesis.

During late pulses, an additional neutron flux is released by the ^{22}Ne source, whose strength varies with pulse number owing to a slight increase of the maximum T_b in the convective shell. The average N_n caused by the ^{22}Ne source is shown in Figure 11*b* for different thermal pulses of a 3-$M_\odot$ model with $Z = Z_\odot$. The neutron density reaches a high value of a few times 10^{10} n/cm^3. In Gallino et al (1998), the rates of the ^{22}Ne$(\alpha,n)^{25}$Mg reaction and of the competing channel ^{22}Ne$(\alpha,\gamma)^{26}$Mg were taken from recent experimental measurements (Käppeler et al 1994, Denker et al 1995), after excluding a possible resonance at 633 keV. In the temperature range of interest, these rates are up to a factor of 3 higher than the Caughlan & Fowler (1988) estimates. The consequences of this second neutron irradiation on the final abundances of *s*-branched nuclei are important. They result from the high peak N_n and from the temperature dependence of the decay rate of many unstable nuclei along the s path. Note that the s-processed material retains a memory of all the previous high-temperature phases. An example of the effect of the second small ^{22}Ne neutron burst is the production of s-only ^{164}Er, which is possible because ^{163}Dy, a normally stable nuclide, becomes unstable in He shell conditions (Takahashi & Yokoi 1987). Other examples are given by the branching-dependent s-only nuclei ^{134}Ba and ^{152}Gd. They are actually highly overproduced during the major ^{13}C neutron exposure at low neutron density, but subsequently suffer an important depletion during the short ^{22}Ne neutron peak and are eventually restored when the neutron density falls below $\sim 10^8$ n/cm^3 (Käppeler et al 1990). A last example is the depletion of the pair ^{86,87}Sr (which are sensitive to the branching at ^{85}Kr), relative to the unbranched ^{88}Sr (which in contrast is mostly fed by the major neutron exposure released during ^{13}C consumption). Isotopes on the neutron-rich side of s branchings generally receive a low contribution (or are partially destroyed) by the ^{13}C neutron source, because of the low neutron density, whereas they can be fed by the ^{22}Ne source when the neutron density reaches a high peak value. This is particularly true near the neutron magic zones. Characteristic isotopes are ^{86}Kr, ^{87}Rb, ^{96}Zr, and also ^{60}Fe, whose production is controlled by the branching at unstable ^{59}Fe.

Based on the above analysis, s-process nucleosynthesis in an AGB star can be summarized as occurring in different phases: i) penetration of a small amount of protons into the top layers of the cool He intershell (to form a proton pocket), ii) formation of a ^{13}C pocket at H reignition, iii) release of neutrons by the ^{13}C$(\alpha,n)^{16}$O reaction when the region is subsequently compressed and heated to $T = (0.8–0.9) \times 10^8$ K [here s processing takes place locally under radiative conditions (s-enhanced pocket)], iv) ingestion into the convective thermal pulse, where the s-enhanced pocket is mixed with H-burning ashes from below the H shell (Fe seeds, ^{14}N) and with s-processed material already from the previous pulses, v) exposure to a small neutron irradiation at high N_n by the ^{22}Ne source over the mixed material in the pulse, vi) occurrence of the TDU episode after the quenching of the thermal instability, so that part of the s-processed and ^{12}C-rich material is mixed into the envelope, vii) repetition of the above cycle until the TP phase is over.

6. METALLICITY EFFECTS AND THE CHEMICAL EVOLUTION OF THE GALAXY

6.1 Model Predictions at Different Metallicities

To illustrate the dependence of s-process predictions on metallicity, we consider as an example the results obtained with the thin ^{13}C pocket of Gallino et al (1998). This pocket was used as a standard (ST) choice by Gallino et al (1997) [$M(^{13}$C$) = 3 \times 10^{-6}$ $M_\odot$ burnt per pulse]. We recall here that both smaller and larger values are expected in MS, S, and C stars, with a considerable star-to-star scatter, as required by the observed spreads in the s-process enhancements of chemically peculiar stars (Smith & Lambert 1990, Luck & Bond 1991, Busso et al 1995b). On galactic time scales this spread of s-process abundance distribution is melded by the general process of mixing into the ISM. Hence, only the average abundance of ^{13}C burnt at each metallicity is really relevant for studies of galactic enrichment, whereas the detailed structure of the ^{13}C pocket plays a role in explaining the abundances of individual stars.

For a given choice of the ^{13}C pocket, the efficiency of s processing turns out to be strongly dependent on metallicity. Under the assumptions discussed above, the results are shown in Figure 12, in terms of the logarithm of the yields of crucial

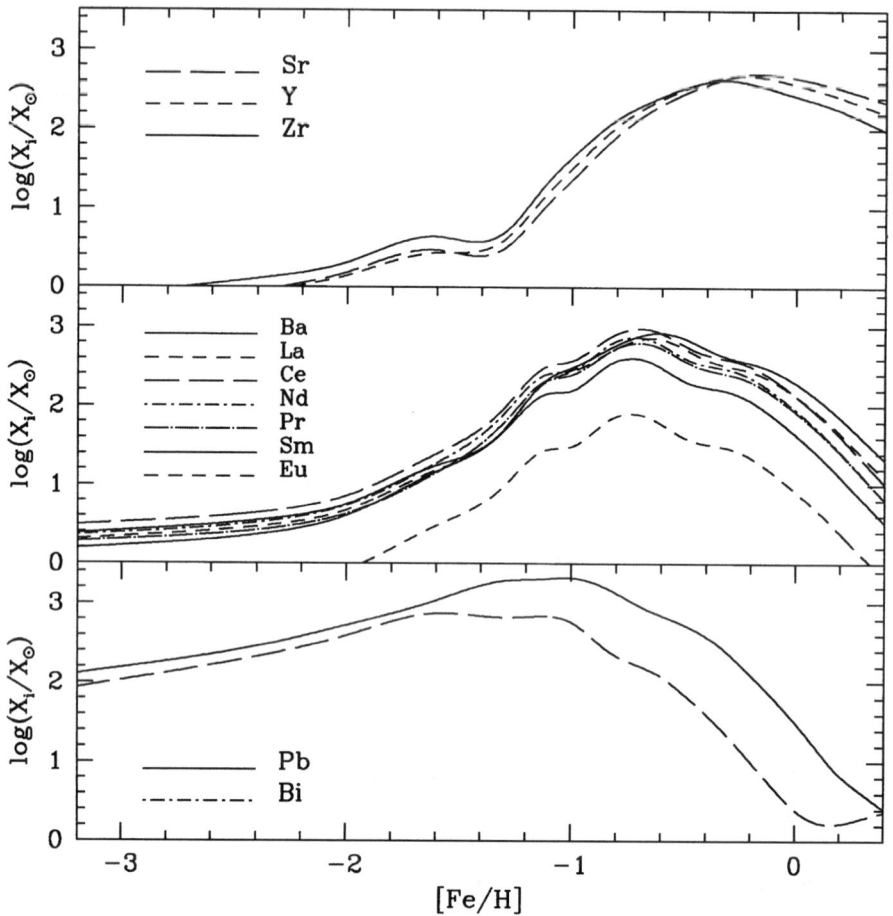

Figure 12 Logarithm of the enhancement factors relative to solar abundances as a function of metallicity for a 2-$M_\odot$ model asymptotic giant branch star with the standard (ST) ^{13}C pocket discussed in the text. ^{13}C is burned under radiative conditions. It can be seen that Pb (dominantly ^{208}Pb) and Bi are the main slow neutron capture-process products in extremely metal-poor stars and that Ba and the rare-earth elements are enhanced for intermediate metallicities, whereas at near solar metallicity Zr, Y, and Sr are enhanced. This has clear implications for galactic chemical evolution.

elements. Here the adopted stellar models are for $M = 2\,M_\odot$, but the trend shown has a general application, at least in the mass range $1.5 \leq M/M_\odot \leq 3\text{--}4$. One should notice that, in computing models for metal-poor stars, the initial abundance ratios are not simply a scaling of solar abundances but reflect the observed enhancement of α-rich elements (e.g. O, Ne, Mg, etc) in population-II stars (e.g. see Gratton & Ortolani 1986, Sneden et al 1991, Spite & Spite 1991, Edvardsson et al 1993, Israelian et al 1998). In the calculations, the assumption was made that, below [Fe/H] $= -1$, oxygen is enhanced by 0.5 dex and other α elements are enhanced by 0.3 dex. For galactic-disk metallicities up to solar value, the enhancements are scaled linearly between these extreme values and zero. As is clear from Figure 12, at very low metallicity all of the s-element yields decrease for decreasing Z, reflecting the secondary nature of neutron captures that require seeds, mainly ^{56}Fe, from the ISM. ^{208}Pb (which reaches its maximum at [Fe/H] ~ -1) is the most abundantly produced nucleus in such metal-poor stars. This fact is sufficient to explain the galactic abundance of this isotope (Gallino et al 1998), without requiring any ad hoc strong component for the s process (see also Section 3). At intermediate metallicities, ([Fe/H] ~ -0.8, as typical of early galactic-disk stars), the Ba peak elements become the dominant products of neutron captures in AGB models. At still higher metallicities, the Zr peak elements become progressively more important. These nuclei are also efficiently contributed by more massive AGBs (IMS), which are instead never effective in producing much Ba (see the discussion of Section 4 and Busso et al 1988).

Some of the effects introduced by the primary nature of the ^{13}C neutron source on s-process abundances were recognized by Clayton (1988), who noticed that, although neutron captures are secondary in nature, a primary neutron source roughly implies that, for metallicities not far from solar, the neutron exposure behaves as $1/Z$. The higher the neutron exposure τ is, the more the distribution of nuclei is peaked toward heavy species. Mathews et al (1992) also pointed out that an important role in controlling the dependence of the neutron flux on metallicity is played by the light neutron absorbers, especially by ^{16}O. These workers simply scaled the solar main component distribution with [Fe/H]. However, not using suitable nucleosynthesis models, their analysis could not identify the strong non-linearity in the s-process trend versus metallicity (see Figure 12). In stellar models the trend of the neutron exposure as a function of Z is actually more complex, and the dependence becomes $\sim Z^{-0.6}$ down to $Z = Z_\odot/10$ (Gallino et al 1999).

For the same choice of the ^{13}C pocket, Figure 13 shows the distribution of the enhancement factors of s-only isotopes (and of isotopes mostly fed by the s process) in the He intershell material cumulatively dredged up into the envelope, for metallicities $Z = (1/10, 1/3, 1/2, 1) \times Z_\odot$. Here an initial mass $M = 2\,M_\odot$ was chosen, but the trend is generally applicable. Enhancements are computed for the initial composition; solar abundances are taken from Anders & Grevesse (1989) and Grevesse et al (1996). As can be seen from the plot, at metallicities $Z = (1/3\text{--}1/2) \times Z_\odot$, the distribution for the s-only nuclei of $A \geq 85$ (*heavy dots*) is remarkably flat. This guarantees that such models are capable of mimicking the

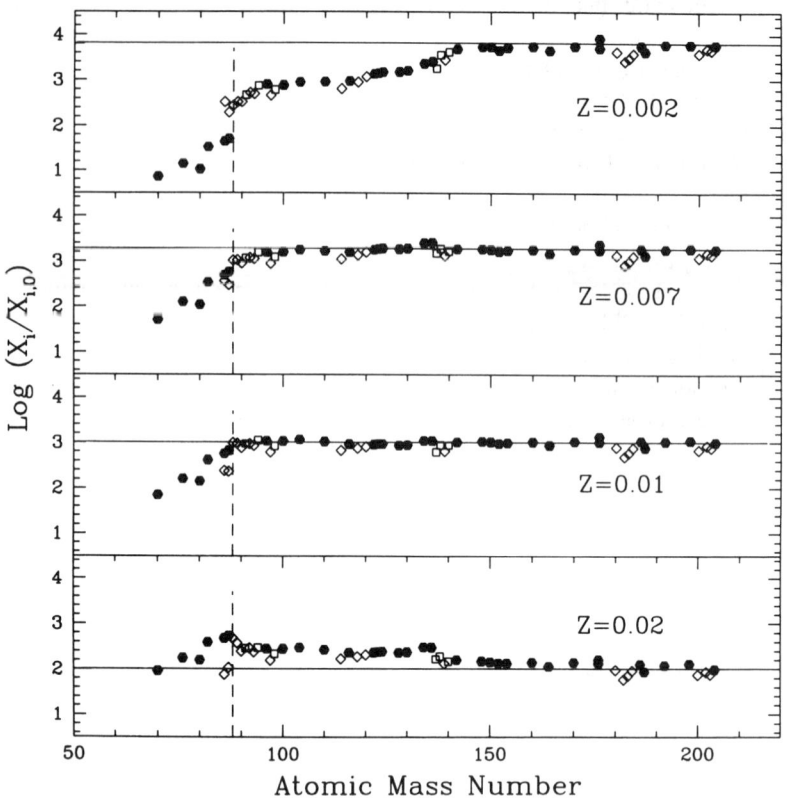

Figure 13 Distributions of enhancement factors of neutron-rich slow neutron capture (s)-process isotopes compared with the initial composition for different Z values in the material cumulatively mixed to the surface of thermal-pulse–asymptotic giant branch stars. The *horizontal line* through Bi is drawn as a guide to the eye. Models are for a 2-$M_\odot$ star, run at different metallicities, with the same standard (*ST*) choice of the ^{13}C pocket. *Heavy dots* represent s-only nuclei; *open squares* are nuclei that are at least 80% s process; *open diamonds* are nuclei with s contributions between 60% and 80%. For decreasing Z, progressively heavier nuclei are favored. In the range of $Z = (1/2–1/3) \times Z_\odot$ the distribution is relatively flat, and $Z = 0.01$ is a good approximation of the solar system main component. Note that the results are dependent on the ^{13}C pocket amount and inversely proportional to Z. For example, a case with $Z = 0.01$ and a ^{13}C pocket 1/2 of the ST value would be indistinguishable from that of the bottom panel.

solar main component; in particular the case at $Z = Z_\odot/2$ reproduces it in remarkable detail, although, as previously stated, the main component is actually the result of the galactic average of many s-process events. Note that each distribution shown in Figure 13 can also be obtained in a different way, when both the metallicity and the amount of ^{13}C burnt are scaled by the same factor (upward or downward), so that the changes mutually compensate. For example, if one wishes to consider an increase of the ^{13}C pocket by a factor of 2 for $Z = 0.02$, then the results will be very close to the curve shown for $Z = 0.01$. It is through the interplay of the complex phenomena briefly outlined above that neutron captures in low-mass AGB stars ultimately produce s-process distributions that are suitable to account for the observational constraints at various metallicities, as we outline in the next subsections.

6.2 Comparison with Observations: Evolved Stars

The above models and their dependence on metallicity allow a consistent interpretation of the abundances observed in AGB stars of various metallicities. A first constraint concerns the neutron density. High-resolution spectroscopic observations showed that AGB stars of the galactic disk that are s-process enhanced have a low Rb/Sr abundance ratio. This ratio is sensitive to the neutron density and increases with it (Lambert et al 1995). Values of the Rb/Sr ratio as low as those measured can be reproduced only if the neutron density is below $\sim 10^7$ n cm^{-3}, as is indeed the case for radiative ^{13}C burning (Figure 14).

Important observational constraints derive from the abundances of elements near the major s-process peaks observed in AGB stars or their progeny and from their variation as a function of stellar metallicity. To explain such data, Busso & Gallino (1997) constructed sequences of envelope compositions for AGB stars by following the mixing into the envelope of the s elements produced in the intershell zone. The adopted models are the same as described above. Those authors compared their model abundances with spectroscopic observations of evolved stars. At galactic-disk metallicities (see Busso et al 1995b), the observed counterparts of the model stars are s-process– and carbon-enriched AGBs, both intrinsic and extrinsic, that is, both single objects of classes MS-S-C and giant or dwarf Ba stars. The Ba stars were most probably produced by the so-called wind accretion mass exchange in a binary system, by an AGB primary star that subsequently evolved to a white dwarf (e.g. see Jorissen & Mayor 1992). Among metal-poor stars, s-process–enriched atmospheres are generally shown by binary CH and Ba stars, although an increasing role is played by newly available observations of post-AGB B-A-F supergiants (Decin et al 1998).

Comparisons with observations are shown by Figures 15 and 16. In particular, Figure 15a and b illustrates the degree to which the abundances of specific AGB stars can be reproduced. Here the observed compositions of a Ba star of the galactic disk (HR774) and of a CH subgiant of the galactic halo (CPD $-62°6195$) are modeled by mixing to the surface products of the He intershell, using stellar evolutionary computations of the appropriate metallicities ([Fe/H] $= -0.3$ and

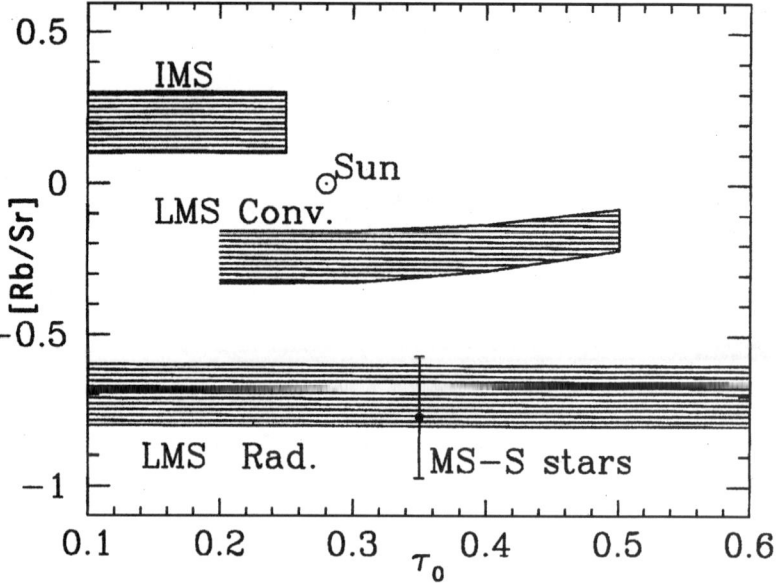

Figure 14 The average abundance ratio Rb/Sr, as deduced by Lambert et al (1995) from measurements in MS and S stars. The shaded regions cover the predictions we derive from s-process models in IMS and in LMS. In the first case, neutrons are produced by ^{22}Ne burning and in the second by ^{13}C burning. Convective (*Conv.*) and radiative (*Rad.*) ^{13}C burnings are shown, and it is clear that the convective model prediction is far above the observed value, whereas the radiative model is consistent with the data on MS stars.

−1, respectively). A less detailed, but more comprehensive, comparison is shown in Figure 16, in terms of the [hs/ls] abundance ratios. This quantity (Luck & Bond 1991) represents the mean logarithmic abundance ratio between the elements from Ba to Sm (hs) and those from Sr to Zr (ls), respectively. The abundances of various classes of observed AGB, post-AGB, or mass-transfer stars are indicated by different symbols. Model curves represent the computed surface compositions near the end of the AGB phase (typical dilution between envelope and He shell material ∼20) for some representative cases. They correspond to the ST choice of the ^{13}C pocket discussed above (*heavy line*) and to choices in which the ^{13}C content is scaled upward by a factor of 2.5 and downward by factors of 2 and 4. The lowest curve represents the case of no ^{13}C pocket. The *horizontal, dash-dotted line* shows by comparison the maximum [hs/ls] attained by previous models, in which ^{13}C was supposed to burn in convective pulses, producing exponential distributions of neutron exposures (Gallino et al 1993, Busso et al 1995b). It is clear that the new models are compatible with several observations at low Z, especially for some CH giants, whereas the old convective burning models are not in accord with such observations (Vanture 1992).

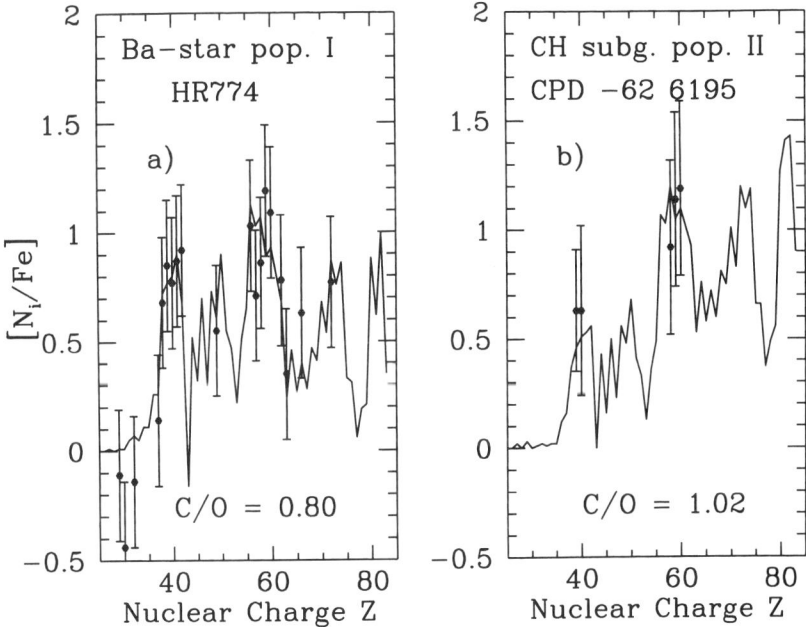

Figure 15 The slow neutron capture (s)-element abundances of (a) a classical Ba star of the galactic disk ($Z = Z_\odot/3$) and (b) a metal-poor CH subgiant ($Z = Z_\odot/10$). Observations are from (a) Smith (1984) and Tomkin & Lambert (1983) and (b) Luck & Bond (1991). Curves represent model surface compositions in models of the appropriate metallicity, using a Reimers' mass loss ($\eta = 1.5$), an initial mass $M = 3\,M_\odot$, and the standard (ST) choice of the ^{13}C pocket. Mixing to the envelope is performed during post-processing, using the TDU extensions found in a 3-$M_\odot$ model of $Z = Z_\odot/3$ and stopping when the C/O ratio is matched. As is seen from the figures, this procedure automatically produces a good fit to the s-process abundances, without adjusting other parameters.

6.3 Comparison with Observations: Unevolved Stars

It has been known for almost two decades that elements dominantly produced by the s process, like Sr, Ba, or La, apparently behave like primary products in field dwarfs or first-ascent red giants of population I. This means that [X_i/Fe] remains roughly constant as a function of [Fe/H], within the (large) observational uncertainties. The growing sets of data on metal-poor stars then showed that those abundance ratios remain consistently high even at lower metallicities, for stars belonging to the galactic halo (e.g. see Ryan et al 1991, Gratton & Sneden 1994, McWilliam et al 1995, McWilliam 1995, 1998). The interpretation of such a behavior is not trivial and involves consideration not only of the s-process, but also of the r-process contribution to each nuclide.

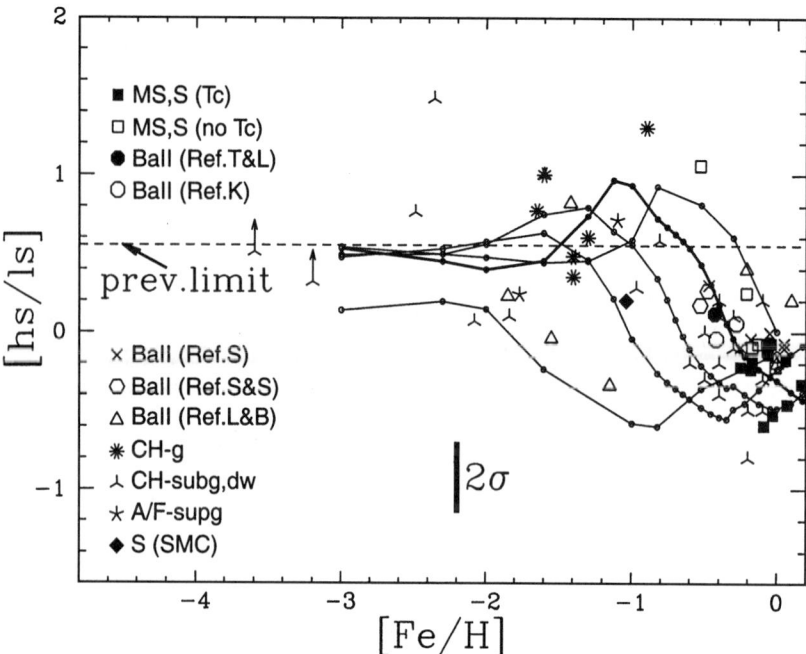

Figure 16 The logarithmic ratio [hs/ls] between heavy (Ba, La, Nd, Sm) and light slow neutron capture elements (Y, Zr) as a function of [Fe/H]. Observations are from Smith & Lambert (1990), Luck & Bond (1981, 1984, 1985, 1991), Vanture (1992), Plez et al (1993), Luck et al (1990), Waelkens et al (1991), and Sneden & Parthasarathy (1983). Each model curve represents surface compositions obtained with a particular choice for the ^{13}C pocket and various initial metallicities. The *curve in bold* is for the standard (ST) case. Other curves are for choices of the ^{13}C pocket with extensions scaled by 0.25, 0.5, and 2.5 compared with the ST case. The lowest curve is for no ^{13}C pocket. The dispersion in the observations may indicate a real spread in the magnitude of the ^{13}C pocket. These models appear to account for the observations of CH giants. The *horizontal dashed line* shows the maximum [hs/ls] with previous (convective) ^{13}C-burning models with an exponential distribution of neutron exposures.

We discuss the above trend by using Ba as an example (Figure 17*a,b*), following results obtained by Travaglio et al (1998). These workers used the stellar yields shown in Figures 12 and 13 and computed the chemical enrichment of the galaxy in neutron-rich elements, adopting a previously established chemical-evolution model (Ferrini et al 1992) that explained the galactic evolution of major light elements. Models that consider only the *s*-process contribution to Ba as coming from the integrated production of AGB stars throughout the galactic life yield the continuous line shown in Figure 17*a*. A comparison with observations of unevolved stars shows a sharp disagreement for low metallicities. However, for stars of the galactic disk, the model curve roughly reproduces the observed data.

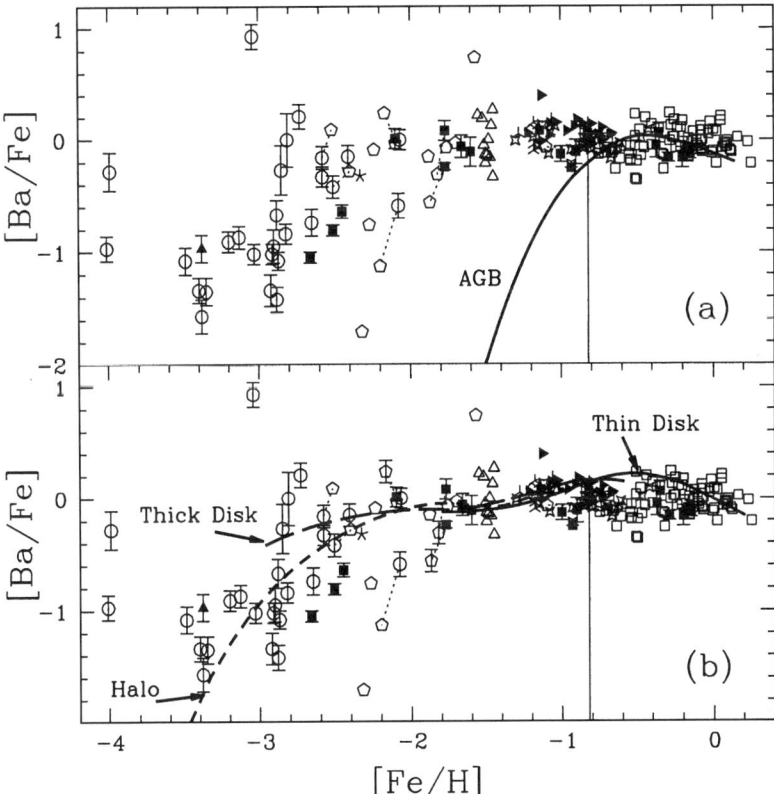

Figure 17 Observed abundances of [Ba/Fe] in unevolved stars as a function of [Fe/H]. *a.* The curve for galactic enrichment of purely slow neutron capture (*s*)-process Ba, as produced by asymptotic giant branch stars of different metallicities calculated by using the galactic chemical-evolution model of Travaglio et al (1998). The curve shows a clear disagreement with the data for low-metallicity stars but provides a reasonable reproduction of Ba abundances in the galactic disk. The *s* contribution to Ba dominates after the first Gyr of galactic evolution (corresponding to the vertical line in the two panels). *b.* Results after adding the *r*-process contribution, determined to be 20% of the solar values. Different model curves in panel *b* represent the halo population (*short dashed line*), thick disk (*long dashed line*), and thin disk of the galaxy (*continuous line*). In panel *a* these curves are indistinguishable (see Travaglio et al 1998, for details).

Here the lower and upper limits of AGB stellar masses were taken to be 2 and 4 $M_\odot$, respectively, and the ST choice for the ^{13}C pocket was made. The results were found to remain qualitatively the same when the upper mass limit of AGB stars is increased, because more massive AGBs do not contribute much to Ba. Similarly, the curves are about the same when a reasonable spread in *s*-process efficiencies at each metallicity is taken into account. The *s*-process fraction of Ba at $t = t_\odot$ is calculated to be 80%. Travaglio et al (1999) estimated the *r* contribution to Ba by subtracting the s contribution shown in Figure 17a: for the sun, at $Z = Z_\odot$, $(1 - N_s/N)_{Ba} = 0.20$.

From Figure 17a, one can see that the s-process contribution begins to dominate the galactic enrichment for [Fe/H] $\simeq -1$, corresponding to a galactic age of ~1 Gyr. At lower [Fe/H] values, the s process is clearly incapable of explaining the observed data. This is linked both to the long time scales of LMS evolution and to the efficiency of s-element production in AGB stars of different metallicities, as summarized in Figures 12 and 13. Indeed, at low Z values the Ba peak nuclei do not receive strong contributions from AGB stars, and it is conceivable that they are dominated by their r-process components, as early suggested by Truran (1981). It is in fact known from observations that the heavy r-process nuclei are produced by short-lived massive stars, presumably in supernova explosions that also produce ^{16}O (cf Cowan et al 1996).

Travaglio et al (1999) made the hypothesis that the r-process yield of Ba comes from moderately massive supernovae of type II (8–10 $M_\odot$). Hence, they assumed that the yield of Ba relative to O is independent of metallicity for the producing supernovae. This means that, at low Z, where the O/Fe ratio is higher than in the sun (by a factor of 3–4), the Ba_r/O ratio increases by the same factor. In the above hypotheses, Travaglio et al (1999) reconstructed the galactic history of Ba, including both r and s contributions. The results are shown in Figure 17b. Here the model curves for the three major components of the galaxy are shown (halo, *short dashed line*; thick disk, *long dashed line*; thin disk, *continuous line*). The curves extend through the intervals of metallicity (in part mutually overlapping) over which star formation is active in the three zones. As can be seen from the plot, the model curves are compatible with the bulk of observed data. The galactic evolution model used for the halo phase implies a well-homogenized situation and cannot explain the large scatter of spectroscopic data. This spread has been commented upon by various authors (Griffin et al 1982; Gilroy et al 1988, Ryan et al 1991; McWilliam et al 1995, McWilliam 1998). As more and more data is collected, the scatter appears to be largely intrinsic in nature, a clear indication of an inhomogeneous evolution of the halo, with incomplete mixing of stellar ejecta over the extended interstellar medium. This role can now be studied using numerical simulation techniques, either as a stochastic analysis with Monte Carlo simulation (see McWilliam & Searle 1998) or with N-body smooth particle hydrodynamics (SPH) (Raiteri et al 1999) to create models of the chemical and dynamical evolution of the Galaxy to account for the incomplete mixing. A similar effort has been made by Ishimaru & Wanajo (1999). There is a wide scatter that is not easily explained. Raiteri et al (1999) show that part of the scatter at low metallicity can be caused by observations of stars from different galactic subsystems. Later, when star formation becomes limited to the thin disk, the s-process component to Ba from AGB stars becomes dominant. These results are in support of the suggestions by Truran (1981) that at low Z the abundance of neutron-rich nuclei is caused by the r process. Important progress in this respect has been obtained through observations in very metal-poor stars (Gilroy et al 1988; Sneden et al 1996, 1998; Norris et al

1997). What is relevant in the present context is that the r component to Ba was deduced as a difference, after the s contribution from AGB stars was modeled.

From the examples discussed above we can say that the present status of s-process nucleosynthesis in AGB stars appears to offer an understanding of the behavior of n-capture nuclei in stars of the galactic disk. It also serves to infer the expected r-process contributions and hence to aid in a reconstruction of the whole galactic history of the nucleosynthesis of neutron-rich isotopes.

7. SOURCES OF SHORT-LIVED NUCLEI IN THE EARLY SOLAR SYSTEM

From studies of meteorites that formed early in solar system history, which were based on long-lived (mean life $\bar{\tau} \geq 10^9$ years) radioactive parents and their daughter products, it has been established that the age of the solar system is about 4.556×10^9 years (cf Bahcall et al 1995). The problem of the galactic time scale and the abundances of the actinides was formulated by Burbidge et al (1957) and Fowler & Hoyle (1960). In addition to the long-lived actinides, many of the meteorites also contain evidence that some radioactive nuclei with relatively short lifetimes were present at the time of their formation. The extent to which a radioactive nucleus will survive until the present time is dependent on its mean lifetime. The shortest-lived species that is preserved today from the beginning of the solar system, with a small but significant abundance, is ^{235}U ($\bar{\tau} = 1.015 \times 10^9$ years). For an isotope with $\bar{\tau} = 2 \times 10^8$ years, the initial abundance would be decreased today by a factor of 10^{10}. We thus consider the current inventory of all original radioactivities in the solar system with $\bar{\tau} \leq 1 \times 10^8$ years to be extinct (this does not relate to those few nuclei produced later by cosmic rays or local nuclear interactions or swept up from the interstellar medium). Considering long-term production in the history of the galaxy (over a time T) prior to solar system formation, the abundance of a stable nuclide (S) is $N_S(T) = P_S \langle p \rangle T$. Here $P_S \langle p \rangle$ is the average production rate, assumed to be the product of a time-invariant stellar production factor P_S and of a time-dependent scaling factor $p(t)$. A short-lived radioactive nuclide R produced in the same process that is in steady state in the ISM between production and decay will have the abundance $N_R(T) \approx P_R p(T) \bar{\tau}_R$. Here $p(T)$ is the value of the scaling factor close to the formation of the solar system. Then $N_R(T)/N_S(T) \approx P_R p(T) \bar{\tau}_R / P_S \langle p \rangle T$. If material is separated from the ISM for a period Δ without further addition of freshly nucleosynthesized nuclei, then its abundance will be decreased by $\exp(-\Delta/\bar{\tau}_R)$. For nuclides with $\bar{\tau} \leq 10^6$ years (e.g. ^{41}Ca, ^{26}Al), isolation times of a few Myr are sufficent to greatly diminish their abundance (see Schramm & Wasserburg 1970 for a full treatment).

Consider a reference state in which solar materials first began to form from the interstellar medium, during which a diverse set of radioactive nuclei was present. Their existence in the early solar system cannot be determined directly by measuring them today but depends on showing that their decay products are present and that these decay products are quantitatively correlated with the chemical properties of the parent element, not with the daughter element. For a sample that has been preserved as an isolated system from its initial state (0) until today, the relationship governing an extinct radioactive parent nucleus R that decays to a daughter D (a different element) is $N_D^{today} = N_D^0 + N_R^0$. Here N_D^0 is the initial number of D nuclei. Consider now a stable isotope I (index) of the same element as the daughter and another stable isotope S (stable) of the same element as the unstable parent R. The isotope S serves as a surrogate of R of the same element. One has $(N_D/N_I)^{today} = (N_D/N_I)^0 + (N_R/N_S)^0 (N_S/N_I)^{today}$. Here $N_I^{today} = N_I^0$ and $N_S^{today} = N_S^0$. N_D^0/N_I^0 is the initial isotopic ratio before decay of the parent nucleus R. The above also applies to the bulk solar ratio today, using solar abundance ratios. Thus for ^{26}Al (which decays to ^{26}Mg) we use the index isotope ^{24}Mg, and for ^{107}Pd (which decays to ^{107}Ag) we use the index isotope ^{109}Ag. In these examples, the ratio N_S^{today}/N_I^{today} corresponds to ^{27}Al/^{24}Mg and to ^{108}Pd/^{109}Ag, respectively, whereas $(N_R/N_S)^0$ would correspond to $(^{26}$Al/^{27}Al$)^0$ or $(^{107}$Pd/^{108}Pd$)^0$. This exhibits the relationship that the isotope ratio N_D^{today}/N_I^{today} linearly correlates with the ratios of N_S^{today}/N_I^{today}. A sample with none of the parent element (neither R nor S) would show the initial value for the ratio of the daughter nuclide to I. To see observable effects in N_D/N_I today, the samples studied must have been formed early with a large range in values of N_S/N_I and must have been preserved until today as isolated systems. Evidence for the presence of the radioactive nuclides can thus be found only by isotopic shifts relative to the solar value, which requires that there have been very large chemical fractionates of the element represented by S as compared with the element represented by I at the time of formation (Figure 18). By chemical fractionation we mean that a phase (crystal or liquid) is formed from some uniform bulk material but has major differences in proportions of some elements (large variations in N_S/N_I for a sample) as compared with the bulk material (e.g. the crystallization of a salt from a liquid or the condensation of Al_2O_3 or SiC crystals from a gas of stellar composition). For simplicity we assume that the solar nebula was once reasonably well mixed. For a complete presentation of the formalism, see Lee et al (1977) and Wasserburg (1987).

There is now clear evidence for the presence of many radioactive nuclei in the ESS (see Table 1 and Figure 18). The first short-lived nuclide discovered was ^{129}I by JH Reynolds (1960). It was a crucial discovery. The discovery of ^{26}Al in the early solar system (Lee et al 1976, 1977) juxtaposed the presence of two radioactive nuclides (^{129}I and ^{26}Al) with grossly disparate lifetimes. It was to take over three decades to obtain a fuller assessment of the early solar system inventory. We have indicated our assessment of the degree of certainty with which the presence of these nuclei has been established and their abundance relative to a stable or long-lived isotope of the same element (N_R/N_S) in the ESS at 4.55×10^9 years ago. References cited here are usually for recent publications covering the observations. A recent review of the evidence and interpretation of the short-lived nuclei is given by Podosek & Nichols (1997). More extensive literature references can be found in these citations. We indicate the nucleosynthetic processes associated with these nuclei. An extensive summary of the ^{26}Al as observed in the galaxy from gamma spectroscopy is given by Prantzos & Diehl (1996). Abundances of r-process nuclei in the ISM have been calculated for constant production rates derived by r-process systematics, assuming a timescale $T = 10^{10}$ years before isolation of the solar system. For ^{60}Fe, ^{53}Mn, ^{41}Ca, and ^{26}Al, we have used estimates of supernova type-II (SNII) yields from Woosley & Weaver (1995) and an SNII rate of 0.03 SN year^{-1} for the galaxy (see also Wasserburg et al 1996, 1998). Table 1 makes the implicit assumption of a unique and time-invariant r-process site, associated with SN. Both r and p processes are currently believed to be associated with supernovae (Cowan et al 1996). The simple case for uniform production (hereafter UP) made in Table 1 is an approximation representing a stochastic input of discrete events. If we consider a model with discrete events, then there is a distinct granularity in the abundances caused by the most recent sources. A late-stage addition close to the time of formation of the solar system will enhance the short-lived nuclei. Note that the isotopic ratios inferred from meteoritic studies for the various species are from a multiplicity of objects found in meteorites or from different bulk meteorites that formed under different conditions at different times. This is particularly important for nuclei with $\bar{\tau} < 10^7$ years, (i.e. ^{41}Ca, ^{26}Al, ^{60}Fe, and ^{53}Mn). Their abundance critically depends both on the initial inventory at some reference time (0) near the formation of the solar system and the time of formation of the meteorite or meteorite subsystem that was studied. These times are not, in general, known (Figure 19). Aspects of the theoretical and observational estimates of various short-lived nuclei and the accretionary lifetime of the solar nebula are given by Podosek & Cassen (1994). It is evident from studies of meterorites that these objects contain a reliable record of the first several million years of solar system history and also of individual stellar contributions to the early sun (cf Anders & Zinner 1993).

If we consider the abundances of these nuclei at the time of formation of the solar system, we must recognize that they came from various stellar sources. It is actually well established that some of the materials in the ISM from which the solar system formed are assemblages of debris from different stars born in different molecular clouds at different times (Figure 20). For radioactive nuclei with mean lives sufficiently long compared with the local astration rate, the abundance will reflect the longer-term galactic inventory in the ISM. In contrast, for those radioactive nuclei with short mean lives, it may be required that their abundance reflects injection of freshly synthesized material from stellar sources immediately before the solar nebula formed (Figure 20). The earliest objects formed within the solar system by melting, cooling, and crystallization are considered to be the calcium-aluminum–rich inclusions (CAIs). In contrast, the chondrites are assemblages of early and later formed solar system debris and usually contain in their matrix a very small amount of unprocessed presolar grains (cf Huss & Lewis 1995). They thus contain a variety of different objects (e.g. CAIs, chondrules, matrix, and metal) formed at different times and each with its own previous history. The other objects (e.g. iron meteorites and eucrites) are the result of planetary melting that took place after the formation of CAIs and, presumably (but not certainly), after chondrites were assembled. Most meteorites have been subjected to some heating and chemical changes after their "initial" formation (metamorphism). We note

TABLE 1 Mean life times and abundances of short-lived nuclides, steady-state production and early solar system inventory

Radioactive Isotope (R)	Reference Isotope (S)	Process[1]	Mean Life $\bar{\tau}_R$ (Myr)	$(N_R/N_S)_{UP}$[2]	$(N_R/N_S)_{ESS}$[2]	Epoch[3]	Q[4]	Ref.
^{244}Pu	^{232}Th	r	115	6×10^{-3}	3×10^{-3}	CAI	EX	a)
^{247}Cm	^{232}Th	r	23	1.1×10^{-3}	$<6 \times 10^{-4}$	CAI	UB	b)
^{182}Hf	^{180}Hf	s, r	13	4.5×10^{-4}	2.0×10^{-4}	PD	EX	c)
^{146}Sm	^{144}Sm	p	148	1.5×10^{-2}	1.0×10^{-2}	PD	EX	d)
^{135}Cs	^{133}Cs	r, s	2.9	2.1×10^{-4}	1.6×10^{-4}?	CAI	H	e)
^{129}I	^{127}I	r	23	5×10^{-3}	1.0×10^{-4}	CAI	EX	f)
^{107}Pd	^{108}Pd	r, s	9.4	6.2×10^{-4}	2.0×10^{-5}	PD	EX	g)
^{60}Fe	^{56}Fe	eq, exp, s	2.2	2.6×10^{-8}	$\sim 10^{-8}$	PD	LD	h)
^{53}Mn	^{55}Mn	p, exp	5.3	1.0×10^{-4}	$\sim 6 \times 10^{-5}$	CAI	LD	i)
					4.7×10^{-6}	PD	EX	j)
					2.3×10^{-6}	PM		
^{41}Ca	^{40}Ca	s	0.15	2×10^{-8}	1.5×10^{-8}	CAI	EX	k)
^{36}Cl	^{35}Cl	s	0.43	3.8×10^{-7}			H	l)
^{26}Al	^{27}Al	p	1.03	$\sim 10^{-7}$	5×10^{-5}	CAI	EX	m)

[1]eq: equilibrium processes; exp: explosive nucleosynthesis; s: s-process; r: r-process; p: p-process.

[2]$(N_R/N_S)_{UP}$ is the ratio of the number of radioactive isotopes R to the number of reference isotopes S as a result of uniform production (UP). $(N_R/N_S)_{ESS}$ is the ratio of the number of radioactive isotopes R to the number of reference isotopes S in the early solar system (ESS).

[3]Epoch corresponds to the event at which major parent-daughter fractionation took place: CAI–contemporaneous with Calcium–Aluminum-rich inclusions which contained ^{26}Al with $(^{26}Al/^{27}Al)^0 = 5 \times 10^{-5}$; PD: planetary differentiate reflecting melting and crystallization processes on a planet; PM: planetary metamorphism. Some nuclides have two entries depending on the epoch.

[4]Q corresponds to the quality and reliability of inferred early solar system ratios at the time of the corresponding epoch. EX = excellent widespread quantitative correlations; LD = Limited data, qualitative correlations; H = hint, no strong evidence; UB = upper bound.

TABLE REFERENCES

a) Rowe & Kuroda 1965. Alexander et al 1971. Hudson et al 1989.
b) Chen & Wasserburg 1981a,b.
c) Harper & Jacobsen 1994, 1996. Lee & Halliday 1995, 1996, 1997, 1998, 1999.
d) Lugmair & Marti 1977. Jacobsen & Wasserburg 1984. Prinzhofer et al 1989. Stewart et al 1996.
e) McCulloch & Wasserburg 1978.
f) Reynolds 1960. Jeffrey & Reynolds 1961. Brazzle et al 1999.
g) Kelly & Wasserburg 1978. Chen et al 1996.
h) Shukolyukov & Lugmair 1993a,b
i) Birck & Allègre 1985. Rotaru et al 1992.
j) Lugmair & Shukolyukov 1998. Hutcheon et al 1998.
k) Srinivasan et al 1994. Srinivasan et al 1996. Sahijpal et al 1998.
l) Murty et al 1997.
m) Lee et al 1976, 1977. MacPherson et al 1995. Russell et al 1996.

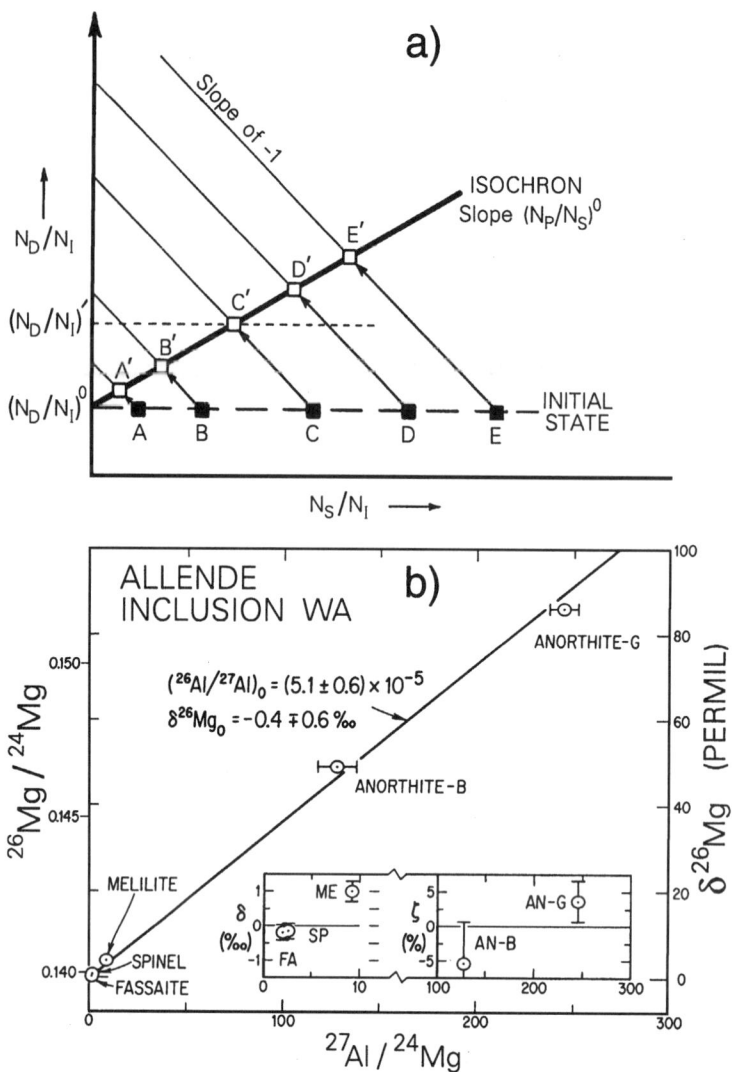

Figure 18 (*a,b*) *a*. Evolution diagram of N_D/N_I for a series of samples with different values of N_S/N_I but with the same initial value $(N_D/N_I)^0$. The evolution through time for sample E follows the line of slope -1 to the point E' when the parent nucleus P has completely decayed; all such samples will lie on the isochron with slope $(N_P/N_S)^0$ (after Wasserburg 1987); *b*. Plot of ^{26}Mg/^{24}Mg vs ^{27}Al/^{24}Mg, showing the correlation of excess ^{26}Mg with ^{27}Al. Data are from various phases with different ^{27}Al/^{24}Mg from the same calcium–aluminum-rich inclusion from the Allende meteorite. This demonstrates the presence of ^{26}Al at an abundance of ^{26}Al/^{27}Al $= 5 \times 10^{-5}$ at the time of crystallization. Inset shows the deviations of the data from the line in parts per thousand (Lee et al 1977).

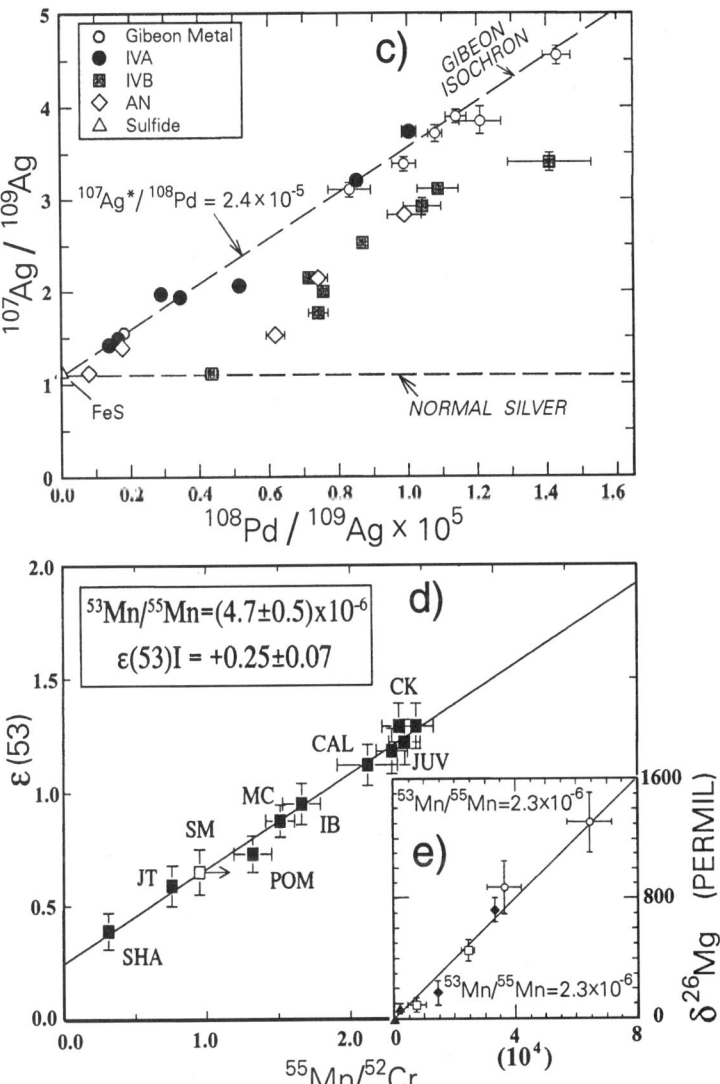

Figure 18 (*c–e*) *c*. Correlation of ^{107}Ag/^{109}Ag with ^{108}Pd/^{109}Pd in different iron meteorites, demonstrating the presence of ^{107}Pd in these planetary segregates at the time they formed. A suite of different samples from the meteorite Gibeon defines the reference isochron (after Chen & Wasserburg 1996). Some iron meteorites appear contemporaneous with Gibeon, and others appear more recent. *d*. Correlation of ^{53}Cr/^{52}Cr vs ^{55}Mn/^{52}Cr, showing the presence of ^{53}Mn in different eucritic meteorites (basaltic rocks) at the time the melts formed on their parent planet (Lugmair & Shukolyukov 1998). Here $\epsilon_{53} \equiv [(^{53}Cr/^{54}Cr)_{sample}/(^{53}Cr/^{54}Cr)_\odot - 1] \times 10^4$. *e*. Isochron for chondrules from a CV meteorite (same general class as Allende) also demonstrating the presence of ^{53}Mn but showing that large Fe and very large Mn enrichments caused by planetary metamorphism in the olivine, occurring 5×10^6 years after the eucrites were crystallized (Hutcheon et al 1998). This illustrates some of the basic problems of establishing a truly refined, self-consistent chronology.

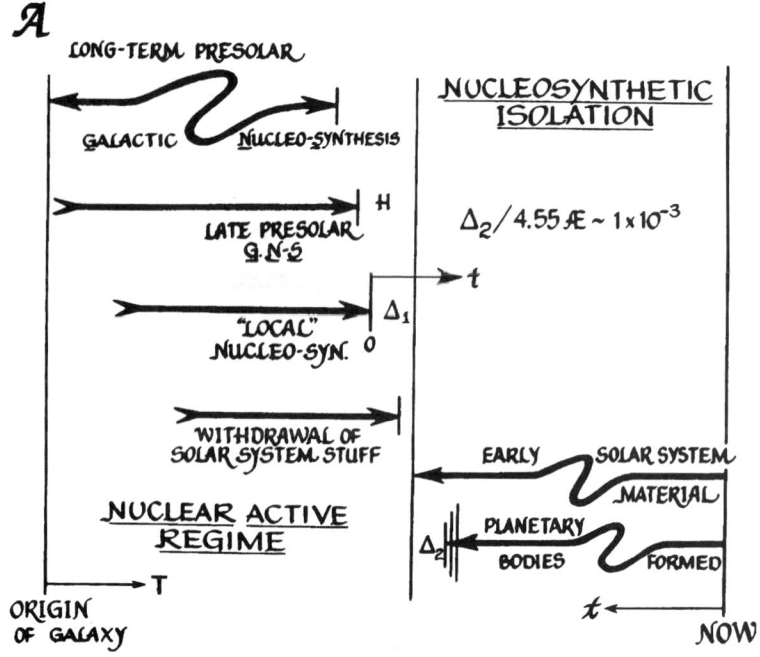

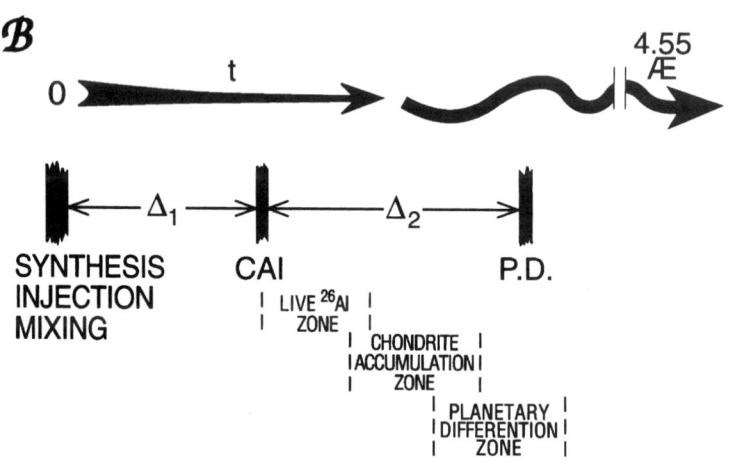

that ^{26}Al is an extremely potent heat source that would cause major melting and metamorphism even on small planetesimals that formed early (cf LaTourrette & Wasserburg 1998). In comparing the observations with stellar-source models, it is imperative that possible differences in formation times are taken into account.

Table 1 and Figure 21 (see also Wasserburg et al 1996) show that ^{244}Pu (a pure r-process nuclide with a relatively long $\bar{\tau}$) is fully compatible with long-term uniform production (UP), much like all the longer-lived radioactivities. Hence, ^{244}Pu does not require a special stellar source close to the time of formation of the solar system. The only other information on transuranics is the upper bound on ^{247}Cm ($\bar{\tau} = 22.5$ Myr), which should be present at the value (^{247}Cm/^{235}U) $\sim 10^{-3}$ in the ESS. We note that determining ^{247}Cm from the variations in ^{235}U/^{238}U also requires estimates on the Cm/U fractionation between phases. This factor is not well known. In the above hypothesis, all of the other nuclides would have substantial inventories in the ISM owing to long-term production. As long recognized by AGW Cameron (e.g. see Cameron 1993, Cameron et al 1993), this is the case for ^{129}I, a pure r-process nuclide that cannot be produced in an AGB source. We not that the UP ratio of ^{129}I/^{127}I in Table 1 is far above the value observed; also ^{107}Pd($s + r$) is overproduced. The pure p-process (or γ,n) nuclide ^{146}Sm is instead compatible with long-term production. The same is true for ^{53}Mn, which also may be associated with supernovae and cannot be produced in AGB stars. ^{41}Ca and ^{26}Al are underproduced by UP. The presence of the r-nucleus ^{182}Hf can be accounted for by UP, but this, together with the high ^{129}I production, requires that at least two r-process mechanisms are active, one producing the actinides and ^{182}Hf, and the other explaining the lighter ^{129}I and ^{107}Pd. This last process must be associated with rarer supernovae, to reconcile the high production with observations. The assignment of ^{182}Hf to long-term production requires that the replenishment of heavy r-process nuclei in molecular clouds must have a time scale of $\sim 10^7$ years (Wasserburg et al 1996).

Figure 19 *a.* Schematic diagram showing the time scale and events of nucleosynthesis contributing to the solar nebula. The time from the origin of the galaxy to the last significant addition of nucleosynthetic material to the solar system is T. The time from the formation of the solar system and its state of isolation to the present is 4.55×10^9 years; this time is indicated in the figure as 4.55 AE, with 1 AE $= 1 \times 10^9$ years. The period T covers long-term and late-galactic nucleosynthesis (GNS). The first is responsible for the general inventory of nuclei. Late GNS events (i.e. supernovae) at time H before the formation of the solar system are considered to be responsible for ^{182}Hf and ^{53}Mn, whereas some local late-stage stellar source (supernovae, asymptotic giant branch star, Wolf Rayet) is responsible for injection of ^{26}Al, ^{41}Ca, ^{107}Pd, and ^{60}Fe (after Wasserburg 1987). *b.* Schematic diagram showing the time scale and schedule of events after local late-stage injection. Δ_1 is the interval between the last injection of freshly synthesized stellar materials and formation of calcium–aluminum-rich inclusions in the solar system. Note the range in time over which ^{26}Al is alive (up to $\sim 4 \times 10^6$ years). Here Δ_2 is a time of planetary melting, differentiation, and crystallization.

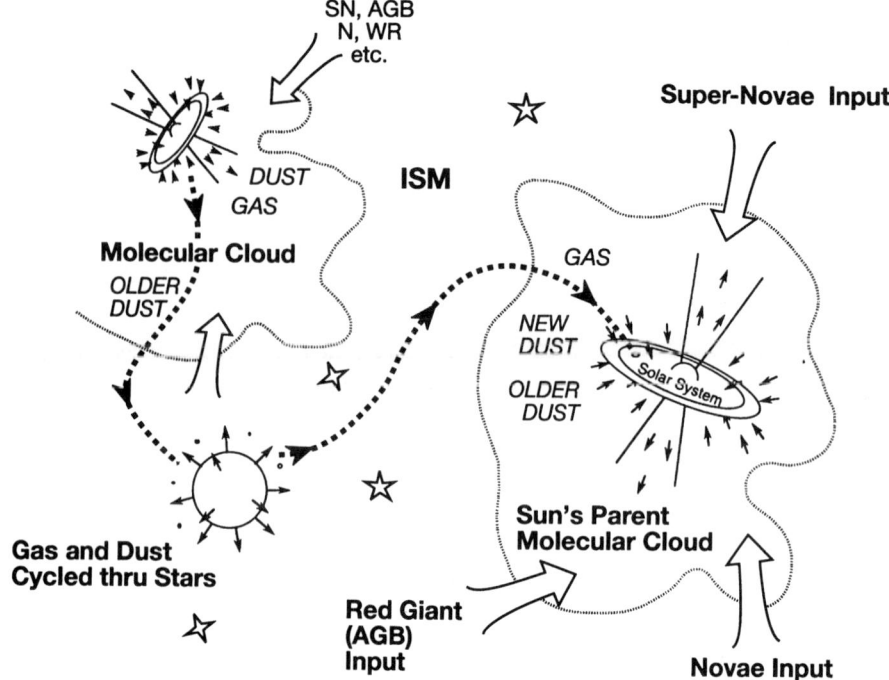

Figure 20 Injections of material from different stellar sources into a sequence of molecular clouds. The parent cloud of the protosun contains gas and preserved dust grains from several generations of stars formed in different molecular clouds, as well as the local inventory of short-, intermediate-, and long-lived nuclei produced over long time scales (after Huss GR, Nichols RH Jr, Wasserburg 1996, unpublished data).

In summary, some radioactive nuclei (^{238,235}U, ^{232}Th, ^{244}Pu, ^{247}Cm(?), ^{146}Sm, ^{129}I, ^{107}Pd, ^{60}Fe, ^{53}Mn, and ^{182}Hf) can be understood in terms of UP from supernovae sources. However, some isotopes (^{129}I and ^{107}Pd) are overproduced. If UP is assumed for a single r-process–type site for both nuclei, then the solar system inventory of ^{129}I and ^{107}Pd would require a withdrawal of the ISM material from

---

Figure 21 *a.* Graph of the log of the measured ratios (M) of the actinides (r-process) relative to ^{232}Th and of ^{146}Sm (p-process) relative to ^{232}Th in the early solar system as a function of their mean lifetimes. $\bar{\tau}_i$ is the reciprocal of their decay constant. The straight line is a reference line corresponding to unit production ratios for all species. UP are the calculated values using best estimates of the relative production rates (P_i) for $T = 10$ AE (1 AE $= 10^9$ years). *SSP* corresponds to the steady-state case using (P_i) values. The ^{247}Cm is an upper bound (Wasserburg et al 1996). *b.* Same as *a* but for mean lives $< 2 \times 10^8$ years; these include the r-process nuclei ^{182}Hf, ^{129}I, and ^{107}Pd, as well as ^{26}Al. Note that ^{129}I and ^{107}Pd for UP are far above the measured points. The upper right corner corresponds to the normalization point for ^{232}Th.

AGB NUCLEOSYNTHESIS 287

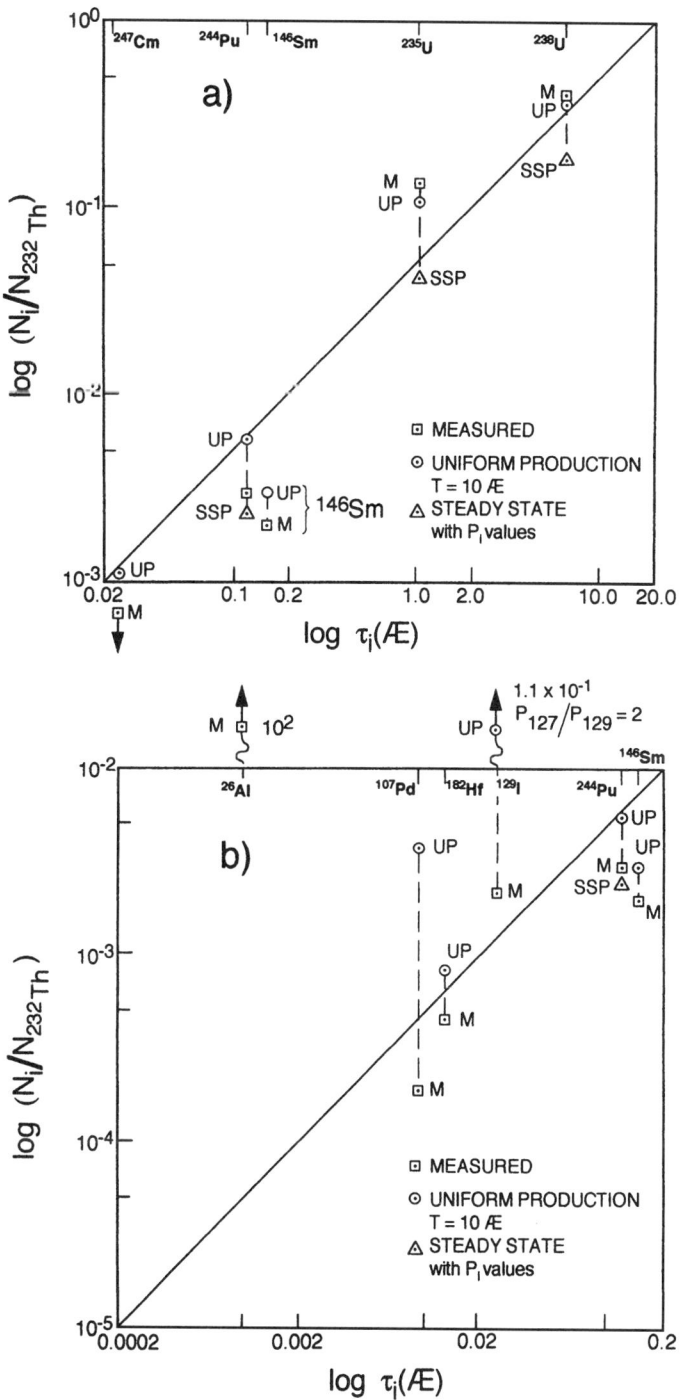

further fresh addition for $\sim 10^8$ years, which has long been recognized as a requirement of $^{129}\text{I}/^{127}\text{I} = 10^{-4}$ (cf Wasserburg et al 1960, Schramm & Wasserburg 1970). This is in conflict with ^{182}Hf and ^{53}Mn data. We follow the approach of Wasserburg et al (1996) for the model of two sites producing r-process nuclei at different rates. Other non–r-process nuclei that can be produced in supernovae (e.g. ^{26}Al) are inadequately contributed by UP. There are thus three distinct issues: how can the observations on nuclides produced solely in supernovae be reconciled with theoretical models; what is the role that an AGB star may play in providing some of the nuclei; and how can rapid addition of freshly synthesized nuclei be achieved in a physically reasonable scenario?

7.1 The Asymptotic Giant Branch Contamination Model

It is assumed that a mass of stellar ejecta is well mixed with some part of a molecular cloud and that the solar nebula was formed from this material. We assume that both the stellar ejecta and cloud each have their own uniform composition. The composition of the ejecta is determined by the stellar model and by the time scale between production ($t = 0$) and the formation of an object from the solar mix ($t = \Delta$). For reference, we take the formation of CAIs to be Δ_1 and of later objects to be $\Delta_1 + \Delta_2$ (Figure 19). If the ejection and mixing were instantaneous, then, in the mixture (mix), the ratio of a short-lived radioactive nuclide R to the net number of stable nuclei S (of the same element) is given by $(N_R^0/N_S)^{mix} \equiv \alpha_{R,S} = N_R^{AGB}/(N_S^0 + N_S^{AGB})$. Here N_S^0 and N_S^{AGB} are the numbers of S nuclei per gram in the molecular protosolar cloud (SC) and the AGB ejecta, respectively. If the contribution of the stable nuclide S from the star to the cloud is small compared with the amount in the protosolar cloud SC, we obtain $\alpha_{R,S} \approx (N_R/N_S)^{AGB} q_S^{AGB} M_{AGB}/q_S^{SC} M_{SC}$. Here q_S^{SC} and q_S^{AGB} are the numbers of S nuclei per gram in the cloud and in the AGB ejecta, respectively. It is evident that the ratios N_R/N_S in the envelope of the star and the yields (q'S) of stable nuclei S in the star's ejecta compared with that in the ambient ISM are the governing factors. A self-consistent scenario would require that the dilution factor M_{AGB}/M_{SC} be the same for all isotope pairs. If we consider only nuclei produced in the He shell (or the H shell), then M_{AGB} should simply refer to the mass of He-intershell material (M_{He}) or the mass of H-shell material (M_H) cumulatively dredged up into the envelope, and the values of the other ratios are computed for the same stellar layers (e.g., q_S^{He} and N_R^{He}/N_S^{He}). If an object forms at time Δ after the production event, then the $\alpha_{R,S}$ becomes $\alpha_{R,S}(\Delta) = \alpha_{R,S}(0) e^{-\Delta/\bar{\tau}_R}$. The term $q_S^{AGB}/q_S^{SC} (\equiv q_S/q_S^0)$ is >1 if S is produced in the star and <1 if S is destroyed. We note that, for a given stellar model, the values of q_S/q_S^0 may range over several orders of magnitude for different species S; the same is true for the term $\exp(-\Delta/\bar{\tau}_R)$ for different species R. To compare the data on short-lived nuclei in the ESS to any AGB model, we must first establish a dilution factor for one pair of nuclei and then compare the results for all others. Because the formation time Δ is critical, we have chosen the pair with well-defined relative production characteristics and with a lifetime that is sufficiently long so that an uncertainty in Δ will not seriously alter the estimate of the dilution factor. For this reason, the pair ^{107}Pd ($\bar{\tau} = 9 \times 10^6$ years) and ^{108}Pd were chosen for $\Delta_1 + \Delta_2 = 5 \times 10^6$ years (see Wasserburg et al 1994, 1995).

7.2 Asymptotic Giant Branch Sources of Short-Lived Nuclei

Results of Previous Models Estimating the possible contributions from different stellar sources is dependent on the existence of adequate stellar models. Detailed stellar models of *s*-process nucleosynthesis have been developed, based on a prescription for the ^{13}C neutron source and activation of the ^{22}Ne neutron source as determined by the internal temperature structure. Although there are serious concerns with the actual mechanisms governing the formation of the ^{13}C neutron source, the general AGB results appear to be consistent with observations and have provided a good guide for their interpretation (see Sections 5 and 6). The *s*-process calculations are based on stellar models that follow the evolution of a star through the TP-AGB phase. The nucleosynthetic results are, of course, dependent on the specific nuclear-reaction rates. The stellar structure, instead, is not governed by the minor energetic contributions from neutron production and *s* processing. It follows that the fundamental, independent parameters determining the production of short-lived nuclei for the TP-AGB phase are initial stellar mass and composition (metallicity Z), mass loss rates, and the magnitude of the ^{13}C pocket that is assumed. Note that AGB sources have well-mixed envelopes, which is not the case for supernovae. There have been some efforts at evaluating the yields of a large number of radioactive products in a self-consistent fashion for AGB stars, which lead to clear predictions that appear to be reasonably reliable.

Previous estimates of some *s*-process species were made by using a simplified steady-flow approximation for the neutron exposure and a constant neutron density (cf Cameron 1993). Isotopic ratios for a single element for steady-state flow patterns provided an excellent guide. There are also significant contributions from non–*s*-process isotopes in AGB stars. Forestini et al (1991) evaluated the production mechanism of ^{26}Al in the H shell and showed that this could contribute significantly to the ^{26}Al inventory. However, no self-consistent yields for different isotopes were obtained by this approach, nor were the destruction mechanisms properly considered. If one is to consider late contributions of an AGB star to the protosolar nebula, then it is clear that a comprehensive analysis is required that follows AGB evolution and takes the net production (including destruction) from H burning and He burning in a star. One has also to consider the dilution effects of mixing nuclear-processed matter to the surface through TDU and the final losses to the ISM through stellar winds.

A first effort in this direction (Wasserburg et al 1994) was the calculation of AGB yields for $Z = Z_\odot = 0.02$ and for masses from 1.5 to 3 $M_\odot$ from the models then available (see Section 4). A series of yields were obtained for a wide variety of isotopes. Results for the production ratios were not significantly sensitive to the stellar mass, but the yields of the different isotopes depended critically on the extent of neutron exposure and neutron density. Those nuclei produced by an AGB star could be reasonably estimated, and others were clearly excluded. For ^{26}Al, a time scale $\Delta_1 \approx 10^6$ years was found. The nuclide ^{26}Al is produced in the H-burning shell, but it was also found to be extensively destroyed in the He shell. In the H shell, the main production process of ^{26}Al is through ^{25}Mg(p,γ)^{26}Al. The rate for the competing reaction ^{26}Al(p,γ)^{27}Si is still uncertain (Arnould et al 1995) and constitutes the main problem for estimating the ^{26}Al production. The ^{26}Al nuclei surviving p captures sink into the He zone with H-burning ashes. Here they are efficiently destroyed by ^{26}Al(n,p)^{26}Mg and by ^{26}Al(n,α)^{23}Na reactions, because the total neutron capture cross section of ^{26}Al is 385 mbarn in the temperature conditions of a convective pulse. Somewhat surprisingly, it was found that AGB models could contribute ^{60}Fe. Production of ^{60}Fe requires substantial neutron densities and occurs during the high-neutron-density peak in the convective He zone, owing to the ^{22}Ne neutron source. ^{60}Fe is fed through the minor (n,γ) channel on ^{59}Fe ($\bar{\tau} = 65.1$ d), which is not affected by the temperature in the He zone (Takahashi & Yokoi 1987). Once produced, ^{60}Fe does not suffer appreciable destruction by neutron captures because of its low cross-section ($\sigma_{60} \simeq 3.2$ mbarn). Agreement was obtained for ^{60}Fe/^{56}Fe, assuming $\Delta_1 + \Delta_2 \leq 5 \times 10^6$ years. Upon the discovery of ^{41}Ca ($\bar{\tau} = 0.15 \times 10^6$ years), it was found that this datum could also be matched by the same AGB source if $\Delta_1 \approx (5 \times 10^5)$–$(7 \times 10^5)$ years (Wasserburg et al 1995). As in the case of ^{60}Fe, ^{41}Ca is produced by ^{40}Ca(n,γ)^{41}Ca when the ^{22}Ne source is activated. Because ^{40}Ca has no precursor that is fed by neutron captures, production of ^{41}Ca depends on the initial ^{40}Ca abundance (i.e. Z). ^{41}Ca has a high neutron capture cross-section (560 mbarn) undergoing the reactions ^{41}Ca(n,α) ^{38}Ar and ^{41}Ca(n,p) ^{41}K. Hence we have the ratio ^{41}Ca/^{40}Ca $\approx \sigma_{40}/\sigma_{41} \approx 10^{-2}$ (Cameron 1993). Note that the mean life of ^{41}Ca for e^- capture is longer in stellar conditions than in the laboratory (by 1 order of magnitude, see Fuller et al 1985). This means that ^{41}Ca is virtually stable in the He shell and through most of the envelope, so that its decay before ejection is small. The results for ^{129}I and ^{53}Mn showed that these nuclei could not be produced in an AGB source, in agreement with earlier calculations by Cameron (1993). For ^{129}I, the very low yield is caused by the low neutron density and the branching at ^{128}I. For ^{53}Mn, there are no channels feeding this nuclide during the neutron exposure. A good quantitative agreement was obtained between some of the observed isotopic ratios and those for a late AGB injection. This accord was obtained for the nuclides ^{107}Pd, ^{26}Al, ^{60}Fe, and ^{41}Ca by matching the dilution factor M_{AGB}/M_{SC} and the neutron exposure (τ_0).

From the above results it was suggested that a late-stage injection from an AGB star of ~ 3 $M_\odot$ into the protosolar cloud could provide the inventory of ^{107}Pd, ^{26}Al, ^{60}Fe, and ^{41}Ca in the early solar system with a dilution factor of $M_{He}/M_{SC} = 1.46 \times 10^{-4}$ for a low neutron exposure ($\tau_0 = 0.03$ mbarn^{-1}). The corresponding ratio of the mass of the AGB envelope to the contaminated solar cloud was $M_{AGB}/M_{SC} \sim 10^{-2}$. The choice of 3-$M_\odot$ models was based on obtaining adequate ^{60}Fe production. This scenario required that the time between

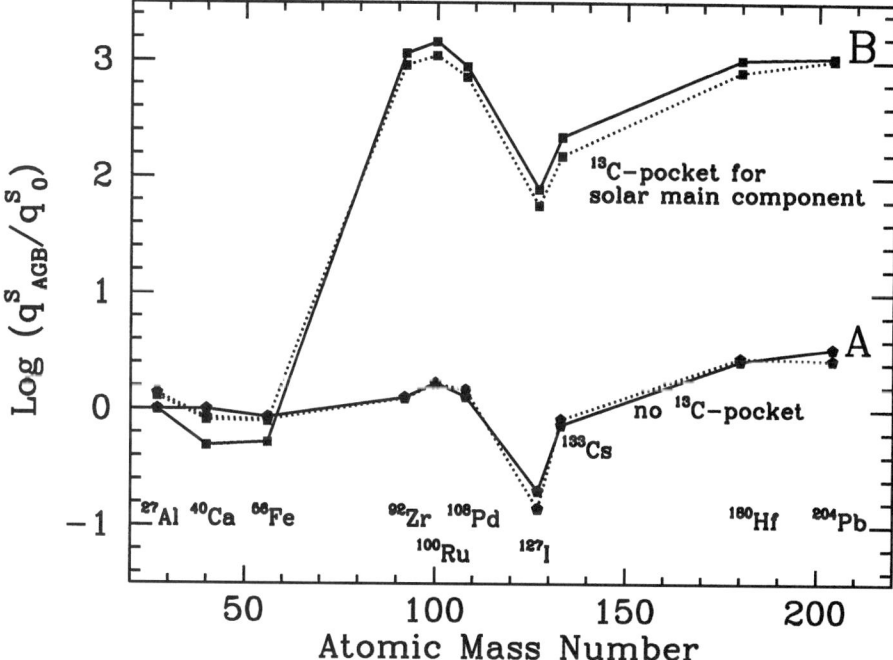

Figure 22 Logarithm of the enrichment factors of a slow neutron capture s-process nuclide S from a model asymptotic giant branch (AGB) source relative to its original abundance as a function of atomic mass number for selected nuclides. *Full curves (a)* are for earlier models with $\tau_0 = 0.03$ and 0.28 mbarn^{-1}, respectively. *Dotted curves (b)* are for new models, either without a carbon pocket or with the choice standard (ST) for it discussed in the text. Note that AGB models with $\log(q_{AGB}^s/q_o^s) \approx 1)$ appear to provide ^{26}Al, ^{41}Ca, ^{60}Fe, and ^{107}Pd, whereas all models with ^{13}C pockets overproduce the heavy nuclei compared with the light ones so that no concordant solution is possible.

production and injection and the formation of the earliest solar system condensates be very small (0.5–0.7 × 10^6 years). This time interval places strong dynamical constraints on the injection, mixing, and collapse of the protosolar cloud. The basic requirement on the model AGB star was that the neutron exposure would have to be low ($\tau_0 = 0.03$ mbarn^{-1}) as compared with $\tau_0 = 0.28$ mbarn^{-1}, which is the value used to obtain a good reproduction of the solar main s-process component in this class of models. The low value of τ_0 was necessary owing to the dependence of the yields on the neutron exposure. From Figure 22 it can be seen that stellar models with high values of τ_0 produce large amounts of ^{107}Pd and low amounts of ^{26}Al, ^{60}Fe, and ^{41}Ca. The low value of τ_0 corresponds to the minimum neutron exposure generated by the ^{22}Ne source (plus a marginal contribution from the low ^{13}C abundance in the H shell ashes); no ^{13}C pocket was assumed by Wasserburg et al (1995).

The AGB models also provided explicit predictions of other radioactive species. The prediction of ^{182}Hf/^{180}Hf was of particular interest. Lee & Halliday (1995) and Harper & Jacobsen (1994) discovered ^{182}W deficiencies in iron meteorites. It was further demonstrated by Lee & Halliday (1996) that there was a widespread occurrence of the ^{182}W isotopic effects. The observed value of ^{182}Hf/^{180}Hf was a factor of 100 greater than the predicted value from the AGB model. This was true even though ~43% of ^{182}W is produced by the s process caused by the branching at ^{181}Hf. The high (^{182}Hf/^{180}Hf)$_{ESS}$ has been explained (Wasserburg et al 1996) by showing that the ^{182}Hf abundance, as well as ^{53}Mn, is a consequence of the UP model (see Table 1). Thus, like the actinides, ^{182}Hf and ^{53}Mn are considered to be present in the inventory of the precursor ISM owing to long-time-scale production by supernovae. The ambient ISM is not a reasonable source for ^{41}Ca, because its abundance would require a time scale of $< 3 \times 10^5$ years for Δ. However, as noted earlier, this model requires that the usual assumption of a single r process must be rejected. The question of the diversity of r-processes is now the subject of vigorous investigation of the abundances of elements in very low metallicity ([Fe/H] < -2) stars. If the hypothesis of two distinct r-process sources for the two r-abundance peaks (and the inferred time scales) is valid, then this could be used to identify the first generation of stars with r nuclei in the Galaxy. It is necessary to have relative quantitative abundances of r-elements at both of the r-process peaks to test this model (Sneden et al 1998, Crawford et al 1998, Cowan et al 1999). This problem is both technically and observationally very demanding. The possibility of multiple r-processes is a fundamental one. If the hypothesis adopted here is not valid, there are major problems with regard to our understanding of the short- and intermediate-lived nuclei found in the early solar system. In addition to the optical observations, there are proposed studies using both gamma-rays and X-rays to study r-process sites (Diehl & Timmes 1998, Qian et al 1998).

New Generation Models We now consider a new set of calculations of yields for short-lived nuclei based on the advances in stellar models outlined in Section 5, using recently revised reaction rates. In these calculations, the abundances in the various stellar layers (He intershell, H shell, and envelope) were followed in detail, based on the prescriptions given by the new stellar models for processes like dredge-up and mass loss. This allowed us to obtain a precise estimate of the relative mass contributions of the reservoirs producing n-rich radioactive nuclei (e.g. ^{60}Fe and ^{107}Pd) and the regions producing ^{26}Al. As for the contribution to ^{26}Al that comes from the H shell, it has a rather complex fate, a large fraction of it being cycled and partially destroyed in the He shell. For this reason, in the new model we present results on the production of ^{26}Al split into two parts by whether it is injected into the AGB winds either directly from the H-burning reservoir (either H shell or HBB) or after partial depletion in the He shell. Concerning ^{26}Al, note that there is a very large experimental uncertainty in the reaction rate for ^{26}Al(p,γ)^{27}Mg (Arnould et al 1995). On the whole, any difference between the model and the ESS values of less than a factor of 2 should not be considered critical. The results are now discussed in terms of the following parameters: (a) the efficiency of the ^{22}Ne source, which is essentially given by the stellar model, (b) the magnitude of the ^{13}C pocket, and (c) the initial mass and metallicity. The effectiveness of ^{22}Ne burning is controlled by the maximum bottom temperature T_b, and is dependent on the stellar mass and Z. As discussed in Sections 3, 4, and 5, in low-mass stars $T_b \sim 3 \times 10^8$ K, and the ^{22}Ne source is only marginally activated; in more massive AGBs, T_b increases up to $\sim 3.5 \times 10^8$ K, and the ^{22}Ne source is very effective. As for the ^{13}C source, in most cases discussed here, we did not introduce a ^{13}C pocket, because Wasserburg et al (1994, 1995) found only models with very low s-process exposures to be consistent with the record of extinct radioactivities in the ESS. This was confirmed with calculations of some new generation models (see Figure 22). In the calculations listed in Table 2, the neutron sources were within the convective He pulse and consisted of the ^{22}Ne source and the limited ^{13}C ingested into the pulse from H-burning ashes.

Table 2 shows the results for a model of 1.5 $M_\odot$. The value of d_0 corresponds to the dilution factor calculated as in Wasserburg et al (1994), using the net ^{107}Pd produced in the He intershell region (in this case, the ^{107}Pd produced both during the radiative burning in the ^{13}C pocket and the contribution from the intershell convective pulse). The ratio M_E/M_{He} is the ratio of the mass of the convective envelope to the net contribution of mass from the He intershell after repeated dredge-up when the star reaches C/O about unity. M_H/M_{He} is the ratio of the masses of the contributions from the H-shell to those from the He-intershell. The ^{26}Al/^{27}Al ratio in the H-shell and the ^{26}Al/^{27}Al in the He-intershell are listed in Table 2. The values calculated for the early solar system at different times represent the total from these 2 sources (H-shell + He-intershell). These results were obtained by starting from the model by Straniero et al (1997), in which TDU was found, but mass loss was not included. For the present calculations, the envelope evolution was monitored through a post-processing phase in which we introduced a mass loss with the Reimer's (1975) formula using $\eta = 0.5$. (For all the other stellar models discussed here, TDU was self-consistently produced in the stellar models, and mass loss was included from the beginning of the calculation.) These results (Table 2) give production ratios and q_S/q_S^0 values that are very similar to those of the previous-generation models for $\tau_0 = 0.03$ mbarn^{-1}. As in that case, the abundances of the short-lived nuclei are also in good agreement with the observed ESS values. There is an excess of ^{41}Ca (5.7-fold) with $\Delta_1 = 0.5$ Myr, which could readily be accommodated if $\Delta_1 = 0.7$ Myr. ^{60}Fe production is increased but remains, as before, dependent on $\Delta_1 + \Delta_2$ and the corresponding value of (^{60}Fe/^{56}Fe)$_{PD}$, where the subscript PD is for planetary differentiate, as well as on the uncertainty in the neutron capture cross-section of its precursor ^{59}Fe (of up to a factor of 2) and on the ^{22}Ne(α,n)^{25}Mg rate. We also found that, for a 1.5-$M_\odot$ model and $Z = Z_\odot/2$, the results are essentially unchanged. There is thus no requirement that a polluting AGB star be initially of solar composition.

Models at 3 $M_\odot$ were explored for the case of no ^{13}C pocket. The results are again roughly commensurate with those found for 1.5 $M_\odot$. However, the ^{60}Fe is

TABLE 2 Case 1.5 $M_\odot$; no ^{13}C pocket; $Z = 0.02$; Age of PD(^{107}Pd) = 5 Myr

Radioactive isotope (R)	Reference isotope (S)	N_R^{He}/N_S^{He}	q_S^{He}	Isotope ratios (Δ_1 = 0.5 Myr)	Isotope ratios (Δ_1 = 1 Myr)	Isotope ratios ($\Delta_2 + \Delta_2$ = 5 Myr) [age of PD]	Measured ratio
^{107}Pd	^{108}Pd	0.167	1.56	3.23×10^{-5}	3.06×10^{-5}	2.00×10^{-5} ($d_0 = 1.31 \times 10^{-4}$)[c]	$(2 \times 10^{-5})_{PD}$
^{60}Fe	^{56}Fe	1.86×10^{-4}	0.85	1.65×10^{-8}	1.31×10^{-8}	2.13×10^{-9}	$(4 \times 10^{-9})_{CK}$
^{41}Ca	^{40}Ca	1.20×10^{-2}	0.88	4.93×10^{-8}	1.76×10^{-9}	~ 0	$(1.5 \times 10^{-8})_{CAI}$
^{182}Hf	^{180}Hf	0.011	2.56	3.55×10^{-6}	3.36×10^{-6}	2.43×10^{-6}	$(3.4 \times 10^{-4})_{PD}$
^{36}Cl	^{35}Cl	0.073	0.84	2.5×10^{-6}	7.85×10^{-7}	~ 0	—
^{26}Al(*)	^{27}Al	0.6	1.0				
^{26}Al(**)	^{27}Al	0.077	0.94	1.60×10^{-5}	9.84×10^{-6}	2.02×10^{-7}	$(5 \times 10^{-5})_{CAI}$

[a]For abbreviations, see footnotes to Table 1.
[b](*)Al from the H-shell (**)Al from the He-shell. M_E/M_{He} = 48; M_H/M_{He} = 0.21. The value of the isotopic ratios in the mixtures at different times was calculated with the sum of the contributions from the H-shell and the He-intershell.
[c]d_0 is the dilution factor derived from ^{107}Pd.

enhanced by a factor of 5.5 owing to the higher T_b in the pulses. The difference in the ^{60}Fe yields compared with the earlier model is principally due to the somewhat higher temperatures obtained for the convective pulse region in the new model and to the increased rate for ^{22}Ne$(\alpha,n)^{25}$Mg reaction (Denker et al 1995). ^{26}Al is down by a factor of 5 owing to the narrowing of the H shell with increased stellar mass, so that production of ^{26}Al appears to be more difficult in this case. ^{41}Ca/^{40}Ca is high by a factor of 5.5 and can be readily accommodated by taking $\Delta_1 = 0.7$ Myr. Any ^{13}C pocket introduced would grossly overproduce ^{107}Pd, and no match would be obtained for the other observations.

Figure 23 shows the N_R/N_S ratios in the envelope for ^{26}Al and ^{107}Pd for the models of a 1.5- and of a 3-$M_\odot$. The ratios are plotted together with the C/O ratio as a function of fractional envelope mass lost owing to stellar winds. The vertical line indicates the point at which a C star (C/O = 1) is formed. The ^{26}Al/^{27}Al ratios achieved in the envelope match typical values measured in circumstellar SiC grains recovered from meteorites (see e.g. Zinner 1997).

We also explored models of higher AGB masses that were experiencing HBB, because this is known to give a potentially high contribution to ^{26}Al (Lattanzio et al 1996, Frost et al 1998). We computed HBB for a 7-$M_\odot$ star and $Z_\odot$ through a post-processing code, using the stellar parameters from the complete stellar model. At this mass, we considered two cases—one without and one with a ^{13}C pocket. In the latter case, we used a ^{13}C pocket that was found necessary to fit the observed abundances of some post-AGB stars that are probably descendants of IMS (e.g. see Decin et al 1998). The ^{13}C pocket was introduced in this massive AGB (Table 3) to compensate for the high T_b at the bottom of the pulses, which induces a high neutron density and, hence, a high ^{60}Fe abundance. Neutrons from the ^{13}C$(\alpha,n)^{16}$O reaction, indeed, increase the abundance of ^{107}Pd at low n density without affecting ^{60}Fe. The values of N_R^{He}/N_S^{He} for all species are very similar to what was found in lower stellar masses; however, q_S^{He}/q_S^0 is high for ^{108}Pd and $(q_S^{He}/q_S^{SC} = 103)$ when a ^{13}C pocket (ST case) is included. ^{26}Al is produced in high abundance by HBB. Here the ^{26}Al in Table 3 is produced in the envelope by HBB. The ratio of the mass of the envelope (M_E) to the mass of the cumulatively dredged-up material from the He-intershell (M_{He}) is high. Due to the fairly high neutron exposure in the He-intershell, all of the ^{26}Al initially present in the H-burning ashes is completely consumed in the case of the 5 M_v (see Table 3). ^{60}Fe is especially high, as shown in Table 3, but could possibly be accomodated with a large $\Delta_1 + \Delta_2$ (not shown). In both cases (with and without a ^{13}C pocket), a 7-$M_\odot$ AGB source is excluded if we use the underproduction of ^{41}Ca as a criterion.

In summary, both the new-generation results and the earlier calculations require a very small ^{13}C pocket if an AGB is to roughly explain the observed abundances of ^{41}Ca, ^{26}Al, ^{60}Fe, and ^{107}Pd in the ESS. There is a clear overall self-consistency in the results that have been obtained with the new model and the earlier calculations, and we consider that they are robust. The basis for rejecting any of the AGB models must rest on the very low relative yields of some nuclei with ^{107}Pd/^{108}Pd as the primary reference. The ratio and yields of ^{107}Pd/^{108}Pd are by far the least

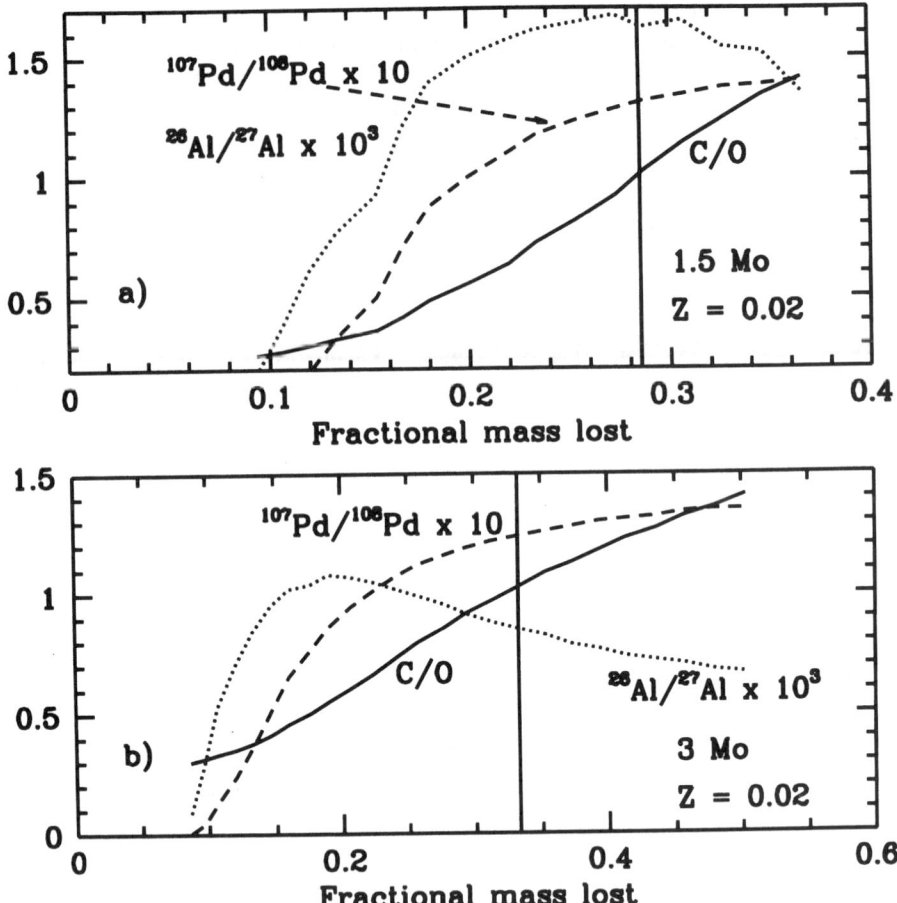

Figure 23 Ratios N_D/N_R for ^{26}Al and ^{107}Pd in the envelope of (*a*) the 1.5-$M_\odot$ asymptotic giant branch (AGB) model and (*b*) the 3-$M_\odot$ AGB model of solar metallicity. No ^{13}C pocket was assumed. The *abscissa* represents the mass cumulatively lost through stellar winds, expressed as a fraction of the initial mass. Also shown is the C/O ratio. The *vertical line* indicates when C/O = 1. In the subsequent evolution (to the right of the line) the star is carbon-rich, i.e. when a C star develops.

susceptible to any possible changes due to the well-defined production rates. In addition, we note that ^{107}Pd is abundant in the UP model. The high abundance of ^{107}Pd in the UP model is diminished by decay (over $\sim 10^8$ years) after the multiple *r*-process source proposed by Wasserburg et al (1996). As a result, ^{107}Pd from the ISM was taken to be negligible at the time of solar system formation. If, however, some of the ^{107}Pd in the ESS came from UP, then this would require an even lower production of ^{107}Pd from an AGB source. It therefore follows that, if the source

TABLE 3 Case: 7 $M_\odot$; ^{13}C pocket with M(^{13}C) = $2 \times 10^{-7} M_\odot$; Z = 0.02; Age of PD(^{107}Pd) = 5 Myr

Radioactive isotope (R)	Reference isotope (S)	N_R^{He}/N_S^{He}	q_S^{He}	Isotope ratios ($\Delta_1 = 0.5$ Myr)	Isotope ratios ($\Delta_1 = 1$ Myr)	Isotope ratios ($\Delta_1 + \Delta_2 = 5$ Myr) [Age of PD]	Measured ratio
^{107}Pd	^{108}Pd	0.168	103	3.23×10^{-5}	3.06×10^{-5}	2.00×10^{-5} (d_3) = 1.97×10^{-6})	$(2 \times 10^{-5})_{PD}$
^{60}Fe	^{56}Fe	0.241	0.28	1.06×10^{-7}	8.43×10^{-8}	1.37×10^{-8}	$(4 \times 10^{-9})_{CK}$
^{41}Ca	^{40}Ca	1.26×10^{-2}	0.38	3.36×10^{-10}	1.2×10^{-11}	~ 0	$(1.5 \times 10^{-8})_{CAI}$
^{182}Hf	^{180}Hf	0.022	58	2.41×10^{-6}	2.31×10^{-6}	1.66×10^{-6}	$(3.4 \times 10^{-4})_{PD}$
^{36}Cl	^{35}Cl	0.100	0.50	3.07×10^{-8}	9.63×10^{-9}	~ 0	—
^{26}Al(*)	^{27}Al	0.012	1.5	1.09×10^{-5}	6.71×10^{-6}	1.33×10^{-7}	$(5 \times 10^{-5})_{CAI}$

aFor abbreviations, see footnotes to Table 1.
b(*) Production in the envelope (E) from Lattanzio & Forestini 1998. M_E/M_{He} = 500 from the FRANEC code case with Reimers mass loss η = 10.
cd_0 is the dilution factor derived from ^{107}Pd.

of late addition to the ESS comes from an AGB star, it must have been dominated by neutrons from the ^{22}Ne source in the convective He shell alone and have only a very small to negligible contribution from any ^{13}C pocket. The question of a fully quantitative agreement cannot yet be answered.

As for more detailed characteristics of a possible AGB source, we thus consider that any model source that grossly (within an order of magnitude) underproduces any of the nuclei ^{41}Ca, ^{26}Al, and ^{60}Fe must be excluded. The same is true if the model requires a time scale (as determined by ^{41}Ca) shorter than a reasonable free-fall time ($\sim$3 $\times$ 10^5 years, see Wasserburg et al 1995). The overproduction of a nuclide such as ^{60}Fe can, in principle, be accounted for by changing $\Delta_1 + \Delta_2$. The most serious question then is regarding the production of ^{26}Al. As discussed earlier, our assessment of the ^{26}Al yields has uncertainties caused both by the production/destruction and the details of transport. Insofar as the TDU is properly described in the present models, we are thus left with the issue of the reaction rates, especially for the ^{26}Al(p,γ) reaction.

An IMS with hot bottom burning appears to be excluded because of low ^{41}Ca. The remaining choices are LMS between 1.5 and 3 $M_\odot$. The 1.5-$M_\odot$ case has a value for ^{26}Al that is too low by a factor of $\sim$3, and the 3-$M_\odot$ case is too low by a factor of 8. This matter can be resolved only if there is a significant decrease in the ^{26}Al destruction rate by proton capture as must be determined experimentally, or if there is an additional mechanism that may produce ^{26}Al without major destruction (e.g. CBP). The production of ^{26}Al has been found to be problematical for stars of 3 $M_\odot$. However, if the production were found to be more efficient than currently calculated values or if CBP were effective in producing ^{26}Al, then another polluting source may also be an AGB star of perhaps 3 $M_\odot$ within the cloud. This would enable a self-contaminating mechanism within the cloud without requiring an encounter.

The ^{60}Fe remains an issue owing to the matter of early solar system time scales ($\Delta_1 + \Delta_2$). A reliable determination must be made of ^{60}Fe/^{56}Fe on samples with a known ^{26}Al/^{27}Al (Wasserburg et al 1998). This would eliminate the floating parameter (Δ_2) and come close to establishing a quantitative scale for assessing the legitimacy of an AGB source. The experimental works on ^{60}Fe by Shukolyukov & Lugmair (1993a,b) are of great importance. However, we note that the available data on ^{60}Fe/^{56}Fe is very limited, and the experimental difficulties of measuring and correlating ^{60}Ni excesses with Fe are formidable. Further, there are at present no data on ^{60}Fe in CAIs in which ^{26}Al was present or in the rare chondrules with some small levels of ^{26}Al. In addition, although ^{53}Mn/^{55}Mn in planetary differentiates is now very well established, there has been no serious effort to determine this on samples showing the presence of ^{26}Al although the original work by Birck & Allègre (1985, 1988) on CAIs suggests a connection. A quantitative relationship between ^{53}Mn and ^{26}Al for CAIs would provide a clear estimate of the ^{53}Mn inventory of the ESS that has been postulated to be from the ISM as a product of longer-term UP (Wasserburg et al 1996). These data on ^{53}Mn and ^{26}Al would also be used to eliminate the uncertainty in $\Delta_1 + \Delta_2$, using the connection through

^{53}Mn in both CAIs and planetary differentiates. There is always the problem of multistage metamorphism (e.g. see Figure 18d,e) so that a simple assignment of Δ_2 to an object is not a priori evident.

Can Contamination Happen? We now try to address the plausibility that an AGB star with suitable characteristics could encounter a protosolar cloud at the right distance to pollute it with short-lived nuclei. It was argued by Kastner & Myers (1994) that today the probability of a close encounter between a mass-losing AGB star and a molecular cloud in the solar neighborhood is one encounter in 10^6 years. In the scenario emphasized in the present section, the AGB star having suitable nucleosynthesis yields is of low mass ($\sim 1.5\ M_\odot$) and is therefore certainly older than the molecular cloud out of which the sun may have formed. A chance encounter is therefore necessary. We underline, however, a series of issues that have to be considered. As noted by Kastner & Myers (1994), the number of AGB stars at the time of the solar system formation was probably much larger than today (perhaps a factor e^2; see Wyse & Silk 1987). Among these AGBs, a low-mass star would certainly be favored in the initial mass function as compared with an IMS. Moreover, while cloud cores of large mass, forming clusters, are associated with high-density molecular regions (Myers 1998), low-mass cores like the one from which the sun plausibly formed are often found in relatively low-density molecular clouds [e.g. in Taurus (Codella et al 1997)]. This increases the probability of encounter consistently and may possibly not require the pollution of a large cloud mass, but only of the cloud core itself.

Basic dynamical questions to be answered are (*a*) can the shock wave from the blown-off envelope provide the required momentum to trigger the collapse of a dense cloud core on a short time scale without disrupting it; and (*b*) can there be entrapment ($\sim 1\%$) of the ejected material in the collapsing SC? These problems also pertain to the suggestion of a supernova trigger (Cameron & Truran 1977) and have received recent attention in a number of works (Boss 1995, Foster & Boss 1998, Boss & Foster 1998, Vanhala & Cameron 1998). It was found that, for shock velocities of 10–25 km/s, the collapse of the dense core can be obtained, with an $\sim 1\%$ entrapment of fresh debris. It appears that this may be a plausible mechanism for both triggering collapse and injecting freshly synthesized matter. However, the dynamical problem is complex and will undoubtedly be the object of intense study in the near future.

8. CONCLUSIONS

In recent years there has been significant progress in studies of the late stages of evolution of low and intermediate mass stars. In some stellar evolutionary codes, using computational schemes at high resolution and improved opacities and requiring only the Schwarzschild criterion for convection, third dredge-up is found to self-consistently occur in AGB stars of low mass and nearly solar metallicity.

It remains to be seen whether this approach will provide the generally accepted solution to the third dredge-up problem. These new models allow the formation of low-mass carbon stars at the appropriate luminosities in accordance with observations. It was further found that the ^{13}C$(\alpha,n)^{16}$O reaction takes place under radiative conditions (not convective) during the interpulse periods in a layer below the H-burning zone. It is in this region that the dominant s processing takes place in AGB stars at a relatively low neutron density ($\sim 10^7$ n cm^{-3}). Subsequent entrainment of this s-processed material into the thermal pulses exposes the mixed matter to a weak neutron source from the ^{22}Ne$(\alpha,n)^{25}$Mg reaction at somewhat higher neutron densities ($\lesssim 5 \times 10^{10}$ n cm^{-3}) than were previously obtained. With these recent stellar models, full s-process calculations have been done for both LMS and IMS. Based on the burning of ^{13}C under radiative conditions, the distribution of neutron exposures during the s process cannot be approximated by an exponential form, as is instead assumed in the phenomenological approach (and was achieved in stellar models in which the ^{13}C neutron source was also assumed to operate within the convective pulse). It appears that the solar system s-process abundances are not the result of a unique s process, but rather of galactic chemical evolution, which mixes the products of s processing in stars of different metallicity and with a range of ^{13}C pockets. The mechanism of formation of the ^{13}C pocket still remains a fundamental problem. As such, it is selected to fit the observational data. Current models show that the third dredge-up establishes a sharp discontinuity between the H-rich and ^{12}C-rich layers. This suggests that the ^{13}C pocket is the result of transport mechanisms that mix protons from the convective envelope into the radiative ^{12}C-rich zone, where they are subsequently burned to ^{13}C. Dedicated hydrodynamical studies of this boundary region are required to establish the actual mechanism of formation of the ^{13}C pocket. Until this is accomplished and shown to be a natural consequence of stellar evolution models, the ^{13}C pocket remains a matter of parameterization in s-process nucleosynthesis.

There is also the problem of additional mixing of the envelope down to regions close to the H-burning shell. This is required by the observational data on LMS for C and O isotopes, and on the precise measurements of O on circumstellar dust grains extracted from meteorites. While slow convective mixing down to just above the H-burning shell [cool bottom processing (CBP)] is required by the observations, it does not appear as a natural consequence of the existing models of stellar evolution and may again require a very detailed hydrodynamical treatment. We also note that all of the stellar calculations discussed here are based on one-dimensional models and that three-dimensional dynamics may define transport mechanisms that are obscured by the one-dimensional treatment. This might be the cause of some level of transport near the H shell that is required by CBP. Three-dimensional hydrodynamics may involve the role of convective plumes, which could affect both the above mixing problems and also the transport of "delicate" nuclei such as Li to the stellar envelope. We further note that the problem of mass loss has so far also been treated parametrically; dynamical models are still in their infancy, particularly as related to high-mass-loss regimes and to sporadic-mass-loss enhancements. We recognize that there are great difficulties in treating the external boundary of the star, where there are complex interactions between neutral gas, dust, charged particles, radiation, and the associated dynamics of loss.

Insofar as the stellar evolution is now properly described by the major nuclear-reaction rates, the opacity function, and the mixing-length parameter α_p, s-process enhancements in stellar envelopes in the AGB models show that the envelope compositions are functions of the initial mass, mass loss rate, metallicity, and ^{13}C pocket. For a given initial stellar mass and metallicity, the composition is then dependent on two arbitrary parameters, the mass of the ^{13}C pocket and the mass loss rate. Current models appear to provide a good quantitative description of several observations and also a basis for estimating the reasons for the abundance scatter that is evident at any metallicity. There are many remaining problems, particularly concerning very low-metallicity stars ([Fe/H] < -2) and the Li-rich stars.

Whereas many radioactive isotopes in the ESS with intermediate lifetimes (^{244}Pu, ^{247}Cm, ^{182}Hf, ^{129}I, and ^{53}Mn) appear to derive from the standing inventory in the ISM provided by diverse supernova sources, the abundances of some short-lived nuclei (^{26}Al, ^{41}Ca, ^{60}Fe, and ^{107}Pd) are compatible with a contamination of the placental nebular cloud from which the sun formed with the ejecta of a close-by AGB source. The new AGB models show that the source must be a low-mass star (~ 1.5 $M_\odot$). Higher-mass stars appear to be excluded because of inadequate ^{26}Al and/or ^{41}Ca production. There appears to be some problem with the extent to which AGB stars can provide sufficient ^{26}Al. There is a marginal shortfall of ^{26}Al production in a 1.5-$M_\odot$ star and, for higher-mass stars, a substantial shortfall. The results on ^{26}Al are critically dependent on some nuclear-reaction rates that need to be established and on the details of transport at the third dredge-up, because the destruction of ^{26}Al in the thermal pulses controls the amount in the envelope. It is possible that CBP contributes to ^{26}Al production in the envelope and should be investigated. The question of a supernova or AGB star as a source of short-lived contaminants and as a possible trigger remains to be tested. A possible test may be the ^{26}Al/^{60}Fe ratio in the early solar system (Wasserburg et al 1998).

ACKNOWLEDGMENTS

We acknowledge valuable comments by A McWilliam concerning the manuscript and Allan Sandage's critical and constructive input. We thank O Straniero and A Chieffi for useful suggestions. We are deeply indebted to ME Johnson for her continuous help in the preparation of 6.023×10^{23} successive versions of this work. MB and RG are grateful to the Lunatic Asylum and to Caltech for their hospitality. This work was supported by NASA grants NAG5-4076 and NAG5-4083, and by a MURST Cofin98 Italian grant. Caltech Division Contribution 8594(1012).

Visit the Annual Reviews home page at http://www.AnnualReviews.org

LITERATURE CITED

Alexander DR, Ferguson JW. 1994. *Ap. J.* 437:879

Alexander EC Jr, Lewis RS, Reynolds JH, Michel MC. 1971. *Science* 172:837

Alongi M, Bertelli G, Bressan A, Chiosi C. 1991. *Astron. Astrophys.* 244:95

Anders E, Grevesse N. 1989. *Geochim. Cosmochim. Acta* 53:197

Anders E, Zinner E. 1993. *Meteoritics* 28:490

Anders E, Zinner E. 1994. *Icarus* 112:303

Aoki W, Tsuji T, Ohnaka K. 1998a. *Astron. Astrophys.* 333:L19

Aoki W, Tsuji T, Ohnaka K. 1998b. *Astron. Astrophys.* 340:222

Arndt TU, Fleischer AJ, Sedlmayr E. 1997. *Astron. Astrophys.* 327:614

Arnould M, Mowlavi N, Champagne AE. 1995. In *Stellar Evolution. What Should Be Done. XXXII Liège Astrophys. Colloq.* ed. A Noels, D Fraipont-Caro, M Gabriel, N Grevesse, B Demarque, p. 17. Liège, Belgium: Inst. d'Astrophys.

Bahcall J, Pinsonneault MH, Wasserburg GJ. 1995. *Rev. Mod. Phys.* 67:781

Bao ZY, Käppeler F. 1987. *At. Data Nucl. Data Tables* 36:411

Barnes CA. 1982. In *Essays in Nuclear Astrophysics*, ed. CA Barnes, DD Clayton, DN Schramm, p. 193. Cambridge, U.K.: Cambridge Univ. Press

Bazan G, Lattanzio JC. 1993. *Ap. J.* 409:726

Becker SA. 1981. In *Physical Processes in Red Giants*, ed. I Iben Jr, A Renzini, p. 121. Dordrecht, The Netherlands: Reidel

Becker SA, Iben I Jr. 1980. *Ap. J.* 232:831

Beer H, Voss F, Winters RR. 1992. *Ap. J. Suppl.* 80:403

Bernatowicz TJ, Amari S, Zinner EK, Lewis RS. 1991. *Ap. J.* 373:L73

Bernatowicz TJ, Cowsik R, Gibbons PC, Lodders K, Fegley B, et al. 1996. *Ap. J.* 472:760

Bertelli G, Bressan A, Chiosi C. 1985. *Ap. J.* 392:522

Birck J-L, Allègre CA. 1985. *Geophys. Res. Lett.* 12:745

Birck J-L, Allègre CA. 1988. *Nature* 331:579

Blanco VM, McCarthy M, Blanco BM. 1980. *Ap. J.* 242:938

Blöcker T. 1995a. *Astron. Astrophys.* 299:755

Blöcker T. 1995b. *Astron. Astrophys.* 297:727

Blöcker T, Schönberner D. 1991. *Astron. Astrophys.* 244:L43

Blöcker T, Schönberner D. 1996. *Mem. Soc. Sci. Liège* 32:455

Blommaert JADJ, Trams N, Okumura K, Groenewegen MAT, Cioni MR, et al. 1998. *Astrophys. Space Sci.* 255:399

Boothroyd AI, Sackmann I-J. 1988a. *Ap. J.* 328:632

Boothroyd AI, Sackmann I-J. 1988b. *Ap. J.* 328:641

Boothroyd AI, Sackmann I-J. 1988c. *Ap. J.* 328:653

Boothroyd AI, Sackmann I-J. 1988d. *Ap. J.* 328:672

Boothroyd AI, Sackmann I-J, Wasserburg GJ. 1994. *Ap. J.* 430:L77

Boothroyd AI, Sackmann I-J, Wasserburg GJ. 1995. *Ap. J.* 442:L21

Boss AP. 1995. *Ap. J.* 439:224

Boss AP, Foster PN. 1997. In *Proc. AIP Conf. 402, Astrophysical Implications Laboratory Study Presolar Materials, St. Louis*, p. 649. Woodbury, NY: Am. Inst. Phys.

Boss AP, Foster PN. 1998. *Ap. J.* 494:L103

Brazzle RH, Pravdivtseva OV, Meshik AP, Hohenberg CM. 1999. *Geochim. Cosmochim. Acta.* 63. In press

Bujarrabal V, Fuente A, Omont A. 1994a. *Astron. Astrophys.* 285:247

Bujarrabal V, Fuente A, Omont A. 1994b. *Ap. J.* 421:L47

Burbidge EM, Burbidge GR, Fowler WA, Hoyle F. 1957. *Rev. Mod. Phys.* 29:547

Busso M, Gallino R. 1997. *Nucl. Phys. A* 621: C431
Busso M, Gallino R, Raiteri CM, eds. 1995a. *Proc. AIP Conf. Nuclei in the Cosmos, III, Assergi, Italy. 327*, p. 625. Woodbury, NY: Am. Inst. Phys.
Busso M, Lambert DL, Beglio L, Gallino R. 1995b. *Ap. J.* 446:775
Busso M, Picchio G, Gallino R, Chieffi A. 1988. *Ap. J.* 326:196
Buzzoni A, Fusi Pecci F, Buonanno R, Corsi CE. 1983. *Astron. Astrophys.* 128:94
Cameron AGW. 1954. *Phys. Rep.* 93:932
Cameron AGW. 1957. *Chalk River Rep.* CRL-41
Cameron AGW. 1960. *A. J.* 65:485
Cameron AGW. 1993. In *Protostars and Planets III*, ed. EH Levy, JL Lunine, p. 47. Tucson: Univ. Arizona Press
Cameron AGW, Thielemann FK, Cowan JJ. 1993. *Phys. Rep.* 227:283
Cameron AGW, Truran JW. 1977. *Icarus* 30: 447
Cameron AGW, Vanhala H, Höflich P. 1997. In *Proc. AIP Conf. 402, Astrophysical Implications Laboratory Study Presolar Materials St. Louis*, ed. TJ Bernatowicz, E Zinner, p. 667. Woodbury, NY: Am. Inst. Phys.
Castellani V, Giannone P, Renzini A. 1969. *Astrophys. Space Sci.* 4:103
Caughlan GR, Fowler WA, Harris MJ, Zimmerman BA. 1985. *At. Data Nucl. Data Tables* 32:197
Caughlan GR, Fowler WA. 1988. *At. Data. Nucl. Data Tables* 40:283
Champagne AE, Wiescher M. 1992. *Annu. Rev. Nucl. Part. Sci.* 42:39
Chan SJ, Kwok S. 1988. *Ap. J.* 334:362
Charbonnel C. 1994. *Astron. Astrophys.* 282: 811
Charbonnel C. 1995. *Ap. J.* 453:L41
Chen JH, Wasserburg GJ. 1996. In *Earth Processes: Reading the Isotopic Code*, ed. A. Basu, S Hart. *Geophys. Monogr.* 95:1. Washington, DC: Am. Geophys. Union
Chen JH, Wasserburg GJ. 1981a. *Earth Planet. Sci. Lett.* 52:1

Chen JH, Wasserburg GJ. 1981b. *Anal. Chem.* 53:2060
Chiosi C, Bertelli G, Bressan A. 1992. *Annu. Rev. Astron. Astrophys.* 30:235
Choi B-G, Huss GR, Wasserburg GJ, Gallino R. 1998. *Science* 282:1284
Clayton DD. 1968. *Principles of Stellar Evolution and Nucleosynthesis*. Chicago: Univ. Chicago Press. 612 pp.
Clayton DD. 1988. *MNRAS* 234:1
Clayton DD, Fowler WA, Hull TE, Zimmerman BA. 1961. *Ann. Phys.* 12:331
Clayton DD, Jin LP. 1995. *Ap. J.* 451:681
Clayton DD, Ward RA. 1974. *Ap. J.* 193:397
Clegg RES, Lambert DL, Bell RA. 1979. *Ap. J.* 234:188
Codella C, Welser R, Henkel C, Benson PJ, Myers PC. 1997. *Astron. Astrophys.* 324:203
Cohen M. 1984. *MNRAS* 206:137
Cosner K, Iben I Jr, Truran JW. 1980. *Ap. J.* 238:L91
Cowan JJ, Pfeiffer B, Kratz K-L, Thielemann F-K, Sneden C, et al. 1999. *Ap. J.* In press
Cowan JJ, Sneden C, Truran JW, Burris DL. 1996. *Ap. J.* 460:L115
Crawford JL, Sneden C, King JR, Boesgaard AM, Deliyannis CP. 1998. *A. J.* 116:2489
D'Antona F, Mazzittelli I. 1996. *Ap. J.* 470: 1093
Dearborn DSP. 1992. *Phys. Rep.* 210:367
Decin L, Van Winckel H, Waelkens C, Bakker EJ. 1998. *Astron. Astrophys.* 332:929
Denissenkov PA, Weiss A. 1996. *Astron. Astrophys.* 308:773
Denker A, Drotleff HW, Grosse M, Knee H, Kunz R, et al. 1995. In *Proc. AIP Conf. 327, Nuclei Cosmos III Assergi Italy*, ed. M Busso, CM Raiteri, R Gallino, p. 255. New York: Am. Inst. Phys.
Despain KH. 1980. *Ap. J.* 236:648
Diehl R. Timmes FX. 1998. *PASP* 110:637
Dominy JF, Wallerstein G. 1986. *Ap. J.* 310: 371
Edvardsson B, Andersen J, Gustafsson B, Lambert DL, Nissen PE, Tomkin J. 1993. *Astron. Astrophys.* 275:101
El Eid MF. 1994. *Astron. Astrophys.* 285:915

Ferrini F, Matteucci F, Pardi C, Penco U. 1992. *Ap. J.* 387:138

Forestini M, Charbonnel M. 1997. *Astron. Astrophys. Suppl.* 123:241

Forestini M, Paulus G, Arnould M. 1991. *Astron. Astrophys.* 252:597

Foster PN, Boss AP. 1998. *Ap. J.* 494:L103

Fowler WA, Caughlan GR, Zimmermann BA. 1967. *Annu. Rev. Astron. Astrophys.* 5:525

Fowler WA, Caughlan GR, Zimmermann BA. 1975. *Annu. Rev. Astron. Astrophys.* 13:69

Fowler WA, Hoyle F. 1960. *Ann. Phys.* 10:280

Frogel JA, Mould J, Blanco VM. 1990. *Ap. J.* 352:96

Frost CA, Lattanzio JC. 1996. *Ap. J.* 473:383

Frost CA, Lattanzio JC, Wood PR. 1998. *Ap. J.* 500:355

Fuller GM, Fowler WA, Newman MJ. 1985. *Ap. J.* 293:1

Gallino R. 1989. In *Evolution of Peculiar Red Giant Stars*, ed. HR Johnson, B Zuckerman, *International Astronomical Union (IAU) Colloq.* 106:176. Cambridge, U.K.: Cambridge Univ. Press

Gallino R, Arlandini C, Busso M, Lugaro M, Travaglio C, et al. 1998. *Ap. J.* 497:388

Gallino R, Busso M, Lugaro M. 1997. In *Proc. AIP Conf. 402, Astrophysical Implications Laboratory Study Presolar Materials*, ed. TJ Bernatowicz, E Zinner, St. Louis Mo, p. 115. Woodbury, NY: Am. Inst. Phys.

Gallino R, Busso M, Lugaro M, Travaglio C, Vaglio P. 1999. In *Nuclei in the Cosmos V*, ed. N Prantzos, S Harissopulos, p. 216. Paris: Ed. Frontières.

Gallino R, Busso M, Picchio G, Raiteri CM, Renzini A. 1988. *Ap. J.* 334:L45

Gallino R, Raiteri CM, Busso M. 1993. *Ap. J.* 410:400

García-Berro E, Ritossa C, Iben I Jr. 1997. *Ap. J.* 485:765

Gilroy KK. 1989. *Ap. J.* 347:835

Gilroy KK, Brown JA. 1991. *Ap. J.* 371:578

Gilroy KK, Sneden C, Pilachowski CA, Cowan JJ. 1988. *Ap. J.* 327:298

Glassgold AE. 1996. *Annu. Rev. Astron. Astrophys.* 34:241

Glassgold AE, Feigelson ED, Montmerle T. 1998. In *Protostars & Planets*. Tucson: Univ. Arizona Press. In press

González G, Wallerstein G. 1992. *MNRAS* 254:343

Görres J, Mathews G, Wiescher M, Shore S, eds. 1997. *Nuclei in the Cosmos IV, Nucl. Phys. A* 621:1c–643c

Gratton RG. 1996. In *AGB Evolution and Nucleosynthesis*, ed. M Busso, R Gallino, C Arlandini, *Mem. Soc. Astron. It.* 67:777

Gratton RG, Ortolani S. 1986. *Astron. Astrophys.* 169:201

Gratton RG, Sneden C. 1994. *Astron. Astrophys.* 287:927

Greenstein JL. 1954. In *Modern Physics for Engineers*, ed. L Ridenour. New York: McGraw-Hill

Grevesse N, Noels A, Sauval AJ. 1996. In *Cosmic Abundances*, ed. SS Holt, G Sonneborn. ASP Conf. Ser. Vol. 99, p. 117. San Francisco: Astronomical Society of the Pacific

Groenewegen MAT, de Jong T. 1999. *Astron. Astrophys.* In press

Gustafsson B. 1989. *Annu. Rev. Astron. Astrophys.* 27:701

Harper CL, Jacobsen SB. 1994. *Lunar Planet. Sci.* 25:509

Harper CL, Jacobsen SB. 1996. *Geochim. Cosmochim. Acta* 60:1131

Hartwick FDA, Härm R, Schwarzschild M. 1968. *Ap. J.* 153:389

Herwig F, Blöcker T, Schönberner D, El Eid M. 1997. *Astron. Astrophys.* 324:L81

Herwig F, Schönberner D, Blöcker T. 1998. *Astron. Astrophys.* 340:L43

Hollowell D, Iben I Jr. 1988. *Ap. J.* 333:L25

Hollowell D, Iben I Jr. 1989. *Ap. J.* 340:966

Hollowell D, Iben I Jr. 1990. *Ap. J.* 349:208

Holmes JA, Woosley WA, Fowler WA, Zimmerman BA. 1976. *At. Data Nucl. Data Tables* 18:305

Hoppe P, Ott U. 1997. In *Proc. AIP Conf. 402, Astrophysical Implications Laboratory Study Presolar Materials*, St. Louis Mo, ed. TJ

Bernatowicz, E Zinner, p. 27. Woodbury, NY: Am. Inst. Phys.
Hoyle F, Schwarzschild M. 1955. *Ap. J. Suppl.* 2:1
Howard WM, Mathews GJ, Takahashi K, Ward RA. 1986. *Ap. J.* 309:633
Hron J, Loidl R, Hofner S, Jorgensen UG, Aringer B, Kerschbaum F. 1997. *Astron. Astrophys.* 335:L69
Hudson GB, Kennedy BM, Podosek FA, Hohenberg CM. 1989. *Proc. Lunar Planet. Sci. Conf.* 19:547
Huss GR, Fahey AJ, Gallino R, Wasserburg GJ. 1994. *Ap. J.* 430:L81
Huss GR, Lewis RS. 1995. *Geochim. Cosmochim. Acta* 59:115
Hutcheon ID, Krot AN, Keil K, Phinney DL, Scott ERD. 1998. *Science* 282:1865
Iben I Jr. 1973. *Ap. J.* 185:209
Iben I Jr. 1975. *Ap. J.* 196:525
Iben I Jr. 1976. *Ap. J.* 208:165
Iben I Jr. 1977. In *Advanced Stages of Stellar Evolution*, ed. P Bouvier, A Maeder, p. 1. Sauverny, Switzerland: Geneva Obs.
Iben I Jr. 1982. *Ap. J.* 260:821
Iben I Jr. 1983. *Ap. J.* 275:L65
Iben I Jr. 1984. *Ap. J.* 277:333
Iben I Jr. 1991. In *Evolution of Stars: the Photospheric Abundance Connection*, ed. G Michaud, A Tutukov. *IAU Symp.* 145:257. Dordrecht, The Netherlands: Kluwer
Iben I Jr, Renzini A. 1982a. *Ap. J.* 249:L79
Iben I Jr, Renzini A. 1982b. *Ap. J.* 263:L23
Iben I Jr, Renzini A. 1983. *Annu. Rev. Astron. Astrophys.* 21:271
Iben I Jr, Renzini A. 1984. *Phys. Rep.* 105:329
Iben I Jr, Ritossa C, García-Berro E. 1997. *Ap. J.* 489:772
Iben I Jr, Truran JW. 1978. *Ap. J.* 220:980
Israelian G, García López RJ, Rebolo R. 1998. *Ap. J.* 507:805
Izumiura H, Hashimoto O, Kawara K, Yamamura I, Waters LBFM. 1996. *Astron. Astrophys.* 315:L221
Jacobsen SB, Harper CL. 1996. In *Earth Processes: Reading the Isotopic Code*, ed. A Basu, S Hart. *Geophys. Monogr.* 95:47. Washington, DC: Am. Geophys. Union
Jacobsen SB, Wasserburg GJ. 1984. *Earth Planet. Sci. Lett.* 67:137
Jeffrey PM, Reynolds JH. 1961. *J. Geophys. Res.* 66:3582
Jorissen A, Mayor M. 1992. *Astron. Astrophys.* 260:115
Jura M. 1997. In *Proc. AIP Conf. 402, Astrophysical Implications Laboratory Study Presolar Materials*, ed. TJ Bernatowicz, E Zinner, p. 379. Woodbury, NY: Am. Inst. Phys.
Jura M, Kahane C, Omont A. 1988. *Astron. Astrophys.* 201:80
Jura M, Kleinmann SG. 1989. *Ap. J.* 341:359
Kahane C, Cernicharo J, Gómez-González J, Guélin M. 1992. *Astron. Astrophys.* 256:235
Käppeler F. 1997. In *Nuclei in the Cosmos IV*, ed. J Görres, G Mathews, M Wiescher, S Shore. *Nucl. Phys. A* 621:221c
Käppeler F. 1999. In *Nuclei in the Cosmos V*, ed. N Prantzos, S Harissopulos, p. 174. Paris: Ed. Frontières.
Käppeler F, Beer H, Wisshak K. 1989. *Rep. Prog. Phys.* 52:945
Käppeler F, Beer H, Wisshak K, Clayton DD, Macklin RL, Ward RA. 1982. *Ap. J.* 257:821
Käppeler F, Gallino R, Busso M, Picchio G, Raiteri CM. 1990. *Ap. J.* 354:630
Käppeler F, Wiescher M, Giesen U, Görres J, Baraffe I, et al. 1994. *Ap. J.* 437:396
Käppeler F, Wisshak K, eds. 1993. *Nuclei in the Cosmos II*, Bristol, U.K.: Inst. Phys.
Kastner JH, Myers PC. 1994. *Ap. J.* 421:605
Kavanagh RW. 1982. In *Essays in Nuclear Astrophysics*, ed. CA Barnes, DD Clayton, DN Schramm, p. 159. Cambridge, U.K.: Cambridge Univ. Press
Kelly WR, Wasserburg GJ. 1978. *Geophys. Res. Lett.* 5:1079
Kilston S. 1985. *PASP* 97:189
Kipper T. 1991. In *Evolution of Stars: the Photospheric Abundance Connection*, ed. G Michaud, A Tutukov, *IAU Symp.* 145:317. Dordrecht, The Netherlands: Kluwer

Knapp GR, Chang KM. 1985. *Ap. J.* 293:281
Kraft RP. 1994. *PASP* 106:553
Kraft RP, Sneden C, Langer GE, Shetrone MD. 1993. *Astron. J.* 106:1490
Kurucz RL. 1991. In *Stellar Atmospheres: Beyond Classical Models*, ed. L Crivellari, I Hubeny, DG Hummer, p. 141. Dordrecht, The Netherlands: Kluwer
Lambert DL. 1985. In *Cool Stars with Excesses of Heavy Elements*, ed. M Jaschek, PC Keenan, p. 191. Dordrecht, The Netherlands: Reidel
Lambert DL, Ries LM. 1981. *Ap. J.* 248:228
Lambert DL, Smith VV, Busso M, Gallino R, Straniero O. 1995. *Ap. J.* 450:302
Langer N. 1986. *Astron. Astrophys.* 164:45
Langer N, Heger A, Woosley SE, Herwig F. 1999. In *Nuclei in the Cosmos V*, ed. N Prantzos. S Harissopulos, p. 129. Paris: Ed. Frontières
LaTourrette T, Wasserburg GJ. 1998. *Earth Planet. Sci. Lett.* 158:91
Lattanzio JC. 1986. *Ap. J.* 311:708
Lattanzio JC. 1989. *Ap. J.* 344:L25
Lattanzio JC. 1995. In *Nuclei in the Cosmos III*, ed. M Busso, R Gallino, CM Raiteri, *AIP Conf. Proc.* 327:591. Woodbury, New York: AIP
Lattanzio JC, Boothroyd A. 1997. In *Proc. AIP Conf. 402, Astrophysical Implications Laboratory Study Presolar Materials*, ed. TJ Bernatowicz, E Zinner, St. Louis Mo, p. 85. Woodbury, NY: Am. Inst. Phys.
Lattanzio JC, Frost C, Cannon R, Wood P. 1996. In *AGB Evolution and Nucleosynthesis*, ed. M Busso, R Gallino, C Arlandini, *Mem. Soc. Astron. It.* 67:729
Le Bertre T. 1998. *Astron. Astrophys.* 203:85
Lee D-C, Halliday AN. 1995. *Nature* 378:771
Lee D-C, Halliday AN. 1996. *Science* 274:1876
Lee D-C, Halliday AN. 1997. *Nature* 388:854
Lee D-C, Halliday AN. 1998. *Lunar Planet. Sci.* 29:1416
Lee D-C, Halliday AN. 1999. *Chem. Geol.* In press
Lee T, Papanastassiou DA, Wasserburg GJ. 1977. *Ap. J.* 211:L107
Lee T, Papanastassiou DA, Wasserburg GJ. *Geophys. Res. Lett.* 3:109
Lee T, Shu FH, Shang H, Glassgold AE, Rehm KE. 1998. *Ap. J.* 506:898
Little SJ, Little-Marenin IR, Hagen Bauer W. 1987. *Ap. J.* 94:981
Little-Marenin IR. 1986. *Ap. J.* 307:L15
Little-Marenin IR. 1989. In *Evolution of Peculiar Red Giant Stars*, ed. HR Johnson, B Zuckerman, *IAU Coll.* 106:131. Cambridge, U.K.: Cambridge Univ. Press
Lodders K, Fegley B Jr. 1995. *Meteoritics* 30:661
Lodders K, Fegley B Jr. 1997. In *Proc. AIP Conf. 402, Astrophysical Implications Laboratory Study Presolar Materials*, ed. TJ Bernatowicz, E Zinner, p. 391. Woodbury, NY: Am. Inst. Phys.
Loup C, Forveille T, Omont A, Paul JF. 1993. *Astron. Astrophys. Suppl.* 99:291
Luck RE, Bond HE. 1981. *PASP* 93:211
Luck RE, Bond HE. 1984. *Ap. J.* 279:729
Luck RE, Bond HE. 1985. *Ap. J.* 292:559
Luck RE, Bond HE. 1991. *Ap. J. Suppl.* 77:515
Luck RE, Bond HE, Lambert DL. 1990. *Ap. J.* 357:188
Luck RE, Lambert DL. 1982. *Ap. J.* 256:189
Lugmair GW, Marti K. 1977. *Earth Planet. Sci. Lett.* 35:273
Lugmair GW, Shukolyukov A. 1998. *Geochim. Cosmochim. Acta* 62:2863
MacPherson GJ, Davis AM, Zinner EK. 1995. *Meteoritics* 30:365
Maeder A. 1975. *Astron. Astrophys.* 40:303
Malaney RA. 1986. *MNRAS* 223:683
Malaney RA. 1987. *Astrophys. Space Sci.* 137:251
Marigo P, Bressan A, Chiosi C. 1996. *Astron. Astrophys.* 313:545
Martin PG, Rogers C. 1987. *Ap. J.* 322:374
Mathews GJ, Bazan G, Cowan JJ. 1992. *Ap. J.* 391:719
Mathews GJ, Takahashi K, Ward RA. 1986. *Ap. J.* 302:410
Mathews GJ, Ward RA. 1985. *Rep. Prog. Phys.* 48:1371
McClure RD. 1984. *PASP* 96:117

McClure RD, Fletcher JM, Nemec JM. 1980. *Ap. J.* 238:L35
McCulloch MT, Wasserburg GJ. 1978. *Ap. J.* 220:L15
McWilliam A. 1995. *Annu. Rev. Astron. Astrophys.* 35:503
McWilliam A. 1998. *Ap. J.* 115:1640
McWilliam A, Lambert DL. 1988. *MNRAS* 230: 573
McWilliam A, Preston GW, Sneden C, Searle L. 1995. *Astron. J.* 109:2757
McWilliam A, Searle L. 1999. In *Galaxy Evolution: Connecting the Distant Universe with the Local Fossil Record*, Paris-Meudon Observatory. pp. 21–25, Sept. 1998. Dordrecht, The Netherlands: Kluwer. In press
Merrill PW. 1952. *Science* 115:484
Meyer BS. 1994. *Annu. Rev. Astron. Astrophys.* 32:153
Mould J, Reid N. 1987. *Ap. J.* 321:156
Mowlavi N, Jorissen A, Arnould M. 1996. *Astron. Astrophys.* 311:803
Mowlavi N, Jorissen A, Arnould M. 1998. *Astron. Astrophys.* 334:153
Murty SVS, Goswami JN, Shukolyukov A. 1997. *Ap. J.* 475:L65
Myers PC. 1998. *Ap. J.* 497:850
Nittler LR, Alexander CMO'D, Gao X, Walker RM, Zinner E. 1994. *Nature* 370:443
Nordlund A, Stein RS. 1995. In *Stellar Evolution. What Should Be Done. XXXII Liège Astrophys. Colloq.* ed. A Noels, D Fraipont-Caro, M Gabriel, N Grevesse, B Demarque, p. 75. Liège, Belgium: Inst. d'Astrophys.
Norris JE, Ryan SG, Beers T. 1997. *Ap. J.* 488: 350
North P, Berthet S, Lanz T. 1994. *Astron. Astrophys.* 281:775
Nyman LA, Olofsson H, Johansson LEB. 1993. *Astron. Astrophys.* 269:377
Olofsson H, Eriksson K, Gustafsson B, et al. 1993a. *Ap. J. Suppl.* 87:267
Olofsson H, Eriksson K, Gustafsson B, et al. 1993b. *Ap. J. Suppl.* 87:305
Ott U, Begemann F. 1990. *Ap. J.* 353:L57
Paczyński B. 1970. *Acta Astron.* 20:47
Paczyński B. 1971. *Acta Astron.* 21:417
Paczyński B. 1975. *Ap. J.* 202:558
Pilachowski CA. 1998. In *Unsolved Problems in Stellar Evolution*. Cambridge, U.K.: Cambridge Univ. Press. In press
Pilachowski CA, Sneden C, Hinkle K, Joyce R. 1997. *Ap. J.* 114:545
Plez B, Brett JM, Nordlund A. 1992. *Astron. Astrophys.* 256:551
Plez B, Smith VV, Lambert DL. 1993. *Ap. J.* 418:812
Podosek FA, Cassen P. 1994. *Meteoritics* 29:6
Podosek FA, Nichols RH Jr. 1997. In *Proc. AIP Conf. 402, Astrophysical Implications Laboratory Study Presolar Materials*, ed. TJ Bernatowicz, E Zinner, p. 617. Woodbury, NY: Am. Inst. Phys.
Prantzos N, Harissopulos S. eds. 1999. *Nuclei in the Cosmos V*. Paris: Ed. Frontières. 592 pp.
Prantzos N, Diehl R. 1996. *Phys. Rep.* 267:1
Prinzhofer A, Papanastassiou DA, Wasserburg GJ. 1989. *Ap. J.* 344:L81
Qian Y-Z, Vogel P, Wasserburg GJ. 1998. *Ap. J.* 506:868
Raiteri CM, Villata M, Gallino R, Busso M, Cravanzola A. 1999. *Ap. J.* 518:L91
Ramaty R, Kozlovsky B, Lingenfelter RE. 1996. *Ap. J.* 456:525
Reimers D. 1975. In *Problems in Stellar Atmospheres and Envelopes*, ed. B Baschek, H Kegel, G Traving, p. 229. Berlin: Springer-Verlag
Renzini A. 1977. In *Advanced Stages in Stellar Evolution*, ed. P Bouvier, A Maeder, p. 151. Sauverny, Switzerland: Geneva Obs.
Renzini A. 1984. In *Observational Tests of the Stellar Evolution Theory*, ed. A Maeder, I Renzini, p. 21. Dordrecht, The Netherlands: Reidel
Renzini A. 1987. *Astron. Astrophys.* 188:49
Renzini A, Fusi Pecci F. 1988. *Annu. Rev. Astron. Astrophys.* 26:199
Renzini A, Voli M. 1981. *Astron. Astrophys.* 94:175
Reynolds JH. 1960. *Phys. Rev. Lett.* 4:8
Ritossa C, García-Berro E, Iben I Jr. 1996. *Ap. J.* 460:489

Rogers GJ, Iglesias CA. 1992. *Ap. J.* 412:752
Rolfs CE, Rodney WS. 1988. *Cauldrons in the Cosmos.* Chicago: Univ. Chicago Press 561 pp.
Rotaru M, Birck J-L, Allègre CJ. 1992. *Nature* 358:465
Rowe MW, Kuroda PK. 1965. *J. Geophys. Res.* 70:709
Russell SS, Srinivasan G, Huss GR, Wasserburg GJ, MacPherson GJ. 1996. *Science* 273:757
Ryan SG, Norris JE, Bessell MS. 1991. *Astron. J.* 102:303
Sackmann IJ, Boothroyd AI. 1991. *Ap. J.* 366:529
Sackmann IJ, Boothroyd AI. 1992. *Ap. J.* 393:L21
Sahai R, Trauger JT, Watson AM, Stapelfeldt KR, Hester JJ, et al. 1998. *Ap. J.* 493:301
Sahijpal S, Goswami JN, Davis AM, Grossman L, Lewis RS. 1998. *Nature* 391:559
Sandage A, Schwarzschild M. 1952. *Ap. J.* 116:463
Sanders RH. 1967. *Ap. J.* 150:971
Scalo JM, Ulrich RK. 1973. *Ap. J.* 183:151
Schramm DN, Wasserburg GJ. 1970. *Ap. J.* 162:57
Schönberner D. 1979. *Astron. Astrophys.* 79:108
Schönberner D. 1983. *Ap. J.* 272:708
Schröder K-P, Winters JM, Arndt TU, Sedlmayr E. 1998. *Astron. Astrophys.* 335:L9
Schwarzschild M, Härm R. 1965. *Ap. J.* 142:885
Schwarzschild, M, Härm R. 1967. *Ap. J.* 150:961
Secchi A. 1868. *C. R. Acad. Sci. Paris* 66:124
Sedlmayr E, Dominik C. 1995. *Space Sci. Rev.* 73:211
Seeger PA, Fowler WA, Clayton DD. 1965. *Ap. J. Suppl.* 11:121
Sharp CM. 1992. *Astron. Astrophys. Suppl.* 94:1
Sharp CM, Wasserburg GJ. 1995. *Geochim. Cosmochim. Acta* 59:1632
Shaviv G, Salpeter EE. 1973. *Ap. J.* 184:191
Shukolyukov A, Lugmair GW. 1993a. *Science* 259:1138
Shukolyukov A, Lugmair GW. 1993b. *Earth Planet. Sci. Lett.* 119:159
Singh HP, Roxburgh IW, Chan KL. 1998. *Astron. Astrophys.* 340:178
Smith VV. 1984. *Astron. Astrophys.* 132:326
Smith VV, Lambert DL. 1985. *Ap. J.* 294:326
Smith VV, Lambert DL. 1986. *Ap. J.* 311:843
Smith VV, Lambert DL. 1989. *Ap. J.* 345:L75
Smith VV, Lambert DL. 1990. *Ap. J. Suppl.* 72:387
Smith VV, Plez B, Lambert DL, Lubowich DA. 1995. *Ap. J.* 441:735
Smith VV, Wallerstein G. 1983. *Ap. J.* 273:742
Sneden C, Cowan JJ, Burris DL, Truran JW. 1998. *Ap. J.* 496:235
Sneden C, Kraft R, Prosser CF, Langer GE. 1991. *Astron. J.* 102:2001
Sneden C, McWilliam A, Preston GW, Cowan JJ, Burris DL, Armosky BJ. 1996. *Ap. J.* 467:819
Sneden C, Parthasarathy M. 1983. *Ap. J.* 267:757
Srinivasan G, Sahijpal S, Ulyanov AA, Goswami GN. 1996. *Geochim. Cosmochim. Acta* 60:1823
Srinivasan G, Ulyanov AA, Goswami JN. 1994. *Ap. J.* 431:L67
Spite M, Spite F. 1991. *Astron. Astrophys.* 252:689
Stewart B, Papanastassiou DA, Wasserburg GJ. 1996. *Earth Planet. Sci. Lett.* 143:1
Straniero O, Chieffi A, Limongi M, Busso M, Gallino R, Arlandini C. 1997. *Ap. J.* 478:332
Straniero O, Gallino R, Busso M, Chieffi A, Raiteri CM, et al. 1995. *Ap. J.* 440:L85
Sugimoto D. 1971. *Prog. Theor. Phys.* 45:761
Sweigart AV, Gross PG. 1978. *Ap. J. Suppl.* 36:405
Sweigart AV, Mengel JG. 1979. *Ap. J.* 229:624
Takahashi K, Yokoi K. 1987. In *At. Data Nucl. Data Tables* 36:375
Travaglio C, Galli D, Gallino R, Busso M, Ferrini F, Straniero O. 1999. *Ap. J.* 520:
Truran JW. 1981. *Astron. Astrophys.* 97:371
Truran JW, Cameron AGW. 1978. *Ap. J.* 219:226

Truran JW, Iben I Jr. 1977. *Ap. J.* 216:797

Ulrich RK. 1973. In *Explosive Nucleosynthesis*, ed. DN Schramm, WD Arnett, p. 139. Austin: Univ. Texas Press

Utsumi K. 1985. In *Cool Stars with Excesses of Heavy Elements*, ed. M Jashek, PC Keenan, p. 243. Dordrecht, The Netherlands: Reidel

Vanhala HAT, Cameron AGW. 1998. *Ap. J.* 508:291

van der Veen WECJ, Habing HJ. 1988. *Astron. Astrophys.* 194:125

van Loon JTH, Zijlstra AA, Whitelock PA, Te Linkel Hekkert P, Chapman JM, Loup C. 1998. *Astron. Astrophys.* 329:169

Vanture AD. 1992. *Astron. J.* 104:1986

Vassiliadis E, Wood PR. 1993. *Ap. J.* 413:641

Voors RHM, Waters LBFM, Morris PW, Trams NR, DeKoter A, Bouwman J. 1999. *Astron. Astrophys.* 341:L67

Waelkens C, Van Winkel H, Boegaert E, Trams NR. 1991. *Astron. Astrophys.* 251:495

Wagenhuber J, Groenewegen MAT. 1998. *Astron. Astrophys.* 340:183

Wallerstein G, Dominy JF. 1988. *Ap. J.* 330:937

Wallerstein G, Iben I Jr, Parker P, Boesgaard AM, Hale GM, et al. 1997. *Rev. Mod. Phys.* 69:995

Wallerstein G, Knapp GR. 1998. *Annu. Rev. Astron. Astrophys.* 36:369

Ward RA. 1977. *Ap. J.* 216:540

Ward RA, Newman MJ. 1978. *Ap. J.* 219:195

Ward RA, Newman MJ, Clayton DD. 1976. *Ap. J. Suppl.* 31:33

Wasserburg GJ. 1987. *Earth Planet. Sci. Lett.* 86:129

Wasserburg GJ, Boothroyd AI, Sackmann I-J. 1995. *Ap. J.* 447:L37

Wasserburg GJ, Busso M, Gallino R. 1996. *Ap. J.* 466:L109

Wasserburg GJ, Busso M, Gallino R, Raiteri CM. 1994. *Ap. J.* 424:412

Wasserburg GJ, Fowler WH, Hoyle F. 1960. *Phys. Rev. Lett.* 4:112

Wasserburg GJ, Gallino R, Busso M. 1998. *Ap. J.* 500:L189

Wasserburg GJ, Gallino R, Busso M, Goswami JN, Raiteri CM. 1995. *Ap. J.* 440:L101

Waters LBFM, Loup C, Kester DJM, Bontekoe TJR, de Jong T. 1994. *Astron. Astrophys.* 281:L1

Waters LBFM, Beintema DA, Zijlstra AA, De Kroter A, Molster FJ, et al. 1998. *Astron. Astrophys.* 331:L61

Weigert A. 1996. *Z. Astrophys.* 64:395

Wheeler CJ, Sneden C, Truran JW. 1989. *Annu. Rev. Astron. Astrophys.* 27:391

Willems FJ, de Jong T. 1988. *Astron. Astrophys.* 196:173

Winters RR, Macklin RL. 1982. *Phys. Rev. C* 25:208

Wood PR. 1981. In *Physical Processes in Red Giants*, ed. I Iben Jr., A Renzini, p. 135. Dordrecht, The Netherlands: Reidel

Woosley SE, Weaver TA. 1995. *Ap. J. Suppl.* 101:181

Wyse RFG, Silk J. 1987. *Ap. J.* 313:L11

Yamamura I, De Jong T, Onaka T, Cami J, Waters LBFM. 1998. *Astron. Astrophys.* 341:L9

Zahn JP. 1992. *Astron. Astrophys.* 265:115

Zinner E. 1997. In *Proc. AIP Conf. 402, Astrophysical Implications Laboratory Study Presolar Materials*, ed. TJ Bernatowicz, E Zinner, p. 3. St. Louis Mo. Woodbury, NY: Am. Inst. Phys.

PHYSICAL CONDITIONS IN REGIONS OF STAR FORMATION

Neal J. Evans II
Department of Astronomy, The University of Texas at Austin, Austin, Texas

Key Words star formation, interstellar molecules, molecular clouds

■ **Abstract** The physical conditions in molecular clouds control the nature and rate of star formation, with consequences for planet formation and galaxy evolution. The focus of this review is on the conditions that characterize regions of star formation in our Galaxy. A review of the tools and tracers for probing physical conditions includes summaries of generally applicable results. Further discussion distinguishes between the formation of low-mass stars in relative isolation and formation in a clustered environment. Evolutionary scenarios and theoretical predictions are more developed for isolated star formation, and observational tests are beginning to interact strongly with the theory. Observers have identified dense cores collapsing to form individual stars or binaries, and analysis of some of these cores support theoretical models of collapse. Stars of both low and high mass form in clustered environments, but massive stars form almost exclusively in clusters. The theoretical understanding of such regions is considerably less developed, but observations are providing the ground rules within which theory must operate. The richest and most massive star clusters form in massive, dense, turbulent cores, which provide models for star formation in other galaxies.

1. INTRODUCTION

Long after their parent spiral galaxies have formed, stars continue to form by repeated condensation from the interstellar medium. In the process, parts of the interstellar medium pass through a cool, relatively dense phase with a great deal of complexity—molecular clouds. While both the diffuse interstellar medium and stars can be supported by thermal pressure, most molecular clouds cannot be thermally supported (Goldreich & Kwan 1974). Simple consideration would suggest that molecular clouds would be a very transient phase in the conversion of diffuse gas to stars, but in fact they persist much longer than expected. During this extended life, they produce an intricate physical and chemical system that provides the substrate for the formation of planets and life, as well as stars. Comparison of cloud masses to the total mass of stars that they produce indicates that most of the matter in a molecular cloud is sterile; stars form only in a small fraction of the mass of the cloud (Leisawitz et al 1989).

The physical conditions in the bulk of a molecular cloud provide the key to understanding why molecular clouds form an essentially metastable state along the path from diffuse gas to stars. Most of the mass of most molecular clouds in our Galaxy is contained in regions of modest extinction, allowing photons from the interstellar radiation field to maintain sufficient ionization for magnetic fields to resist collapse (McKee 1989); most of the molecular gas is, in fact, in a photon-dominated region (PDR) (Hollenbach & Tielens 1997). In addition, most molecular gas has supersonic turbulence (Zuckerman & Evans 1974). The persistence of such turbulence over the inferred lifetimes of clouds in the face of rapid damping mechanisms (Goldreich & Kwan 1974) suggests constant replenishment, most likely in a process of self-regulated star formation (Norman & Silk 1980, Bertoldi & McKee 1996), because star formation is accompanied by energetic outflows, jets, and winds (Bachiller 1996).

For this review, the focus is on the physical conditions in regions that are forming stars and likely precursors of such regions. Although gravitational collapse explains the formation of stars, the details of how it happens depend critically on the physical conditions in the star-forming region. The details determine the mass of the resulting star and the amount of mass that winds up in a disk around the star, in turn controlling the possibilities for planet formation. The physical conditions also control the chemical conditions. With the recognition that much interstellar chemistry is preserved in comets (Crovisier 1999, van Dishoeck & Blake 1998) and that interstellar chemistry may also affect planet formation and the possibilities for life (e.g. Pendleton 1997, Pendleton & Tielens 1997, Chyba & Sagan 1992), the knowledge of physical conditions in star-forming regions has taken on additional significance.

Thinking more globally, different physical conditions in different regions determine whether a few lightly clustered stars form (the isolated mode) or a tight grouping of stars forms (the clustered mode) (Lada 1992, Lada et al 1993). The star formation rates per unit mass of molecular gas vary by a factor of $>10^2$ in clouds within our own Galaxy (Evans 1991, Mead et al 1990), and starburst galaxies achieve even higher rates than are seen anywhere in our Galaxy (e.g. Sanders et al 1991). Ultimately, a description of galaxy formation must incorporate an understanding of how star formation depends on physical conditions, gleaned from careful study of our Galaxy and nearby galaxies.

Within the space limitations of this review, it is not possible to address all of the issues raised by the preceding overview. I generally avoid topics that have been recently reviewed, such as circumstellar disks (Sargent 1996, Papaloizou & Lin 1995, Lin & Papaloizou 1996, Bodenheimer 1995), as well as bipolar outflows (Bachiller 1996) and dense PDRs (Hollenbach & Tielens 1997). I also discuss the sterile parts of molecular clouds only as relevant to the process that leads some parts of the cloud to be suitable for star formation. While the chemistry and physics of star-forming regions are coupled, chemistry has been recently reviewed (van Dishoeck & Blake 1998). Astronomical masers are being concurrently reviewed

(Menten, in preparation); as with HII regions, they are discussed only as signposts for regions of star formation.

I will focus on star formation in our Galaxy. Nearby regions of isolated, low-mass star formation will receive considerable attention (Section 4) because we have made the most progress in studying them. Their conditions will be compared with those in regions forming clusters of stars, including massive stars (Section 5). These regions of clustered star formation are poorly understood, but they probably form the majority of stars in our Galaxy (Elmegreen 1985), and they are the regions relevant for comparisons to other galaxies.

Even with such a restricted topic, the literature is vast. I make no attempt at completeness in referencing. On relatively noncontroversial topics, I tend to give an early reference and a recent review; for more unsettled topics, more references, with different points of view, are given. Recent or upcoming publications with significant overlap include Hartmann (1998), Lada & Kylafis (1999), and Mannings et al (2000).

2. PHYSICAL CONDITIONS

The motivation for studying physical conditions can be found in a few simple theoretical considerations. Our goal is to know when and how molecular gas collapses to form stars. In the simplest situation—a cloud with only thermal support—collapse should occur if the mass exceeds the Jeans (1928) mass,

$$M_J = \left(\frac{\pi k T_K}{\mu m_H G}\right)^{1.5} \rho^{-0.5} = 18\ M_\odot T_K^{1.5} n^{-0.5}, \qquad (1)$$

where T_K is the kinetic temperature (kelvins), ρ is the mass density (g cm^{-3}), and n is the total particle density (cm^{-3}). In a molecular cloud, H nuclei are almost exclusively in H$_2$ molecules, and $n \simeq n(H_2) + n(He)$. Then $\rho = \mu_n m_H n$, where m_H is the mass of a hydrogen atom and μ_n is the mean mass per particle (2.29 in a fully molecular cloud with 25% by mass helium). Discrepancies between coefficients in the equations presented here and those in other references usually are traceable to a different definition of n. In the absence of pressure support, collapse will occur in a free-fall time (Spitzer 1978),

$$t_{ff} = \left(\frac{3\pi}{32 G \rho}\right)^{0.5} = 3.4 \times 10^7 n^{-0.5}\ \text{years}. \qquad (2)$$

If $T_K = 10$ K and $n \geq 50$ cm^{-3}, typical conditions in the sterile regions (e.g. Blitz 1993), $M_J \leq 80\ M_\odot$, and $t_{ff} \leq 5 \times 10^6$ years. Our Galaxy contains about 1–$3 \times 10^9\ M_\odot$ of molecular gas (Bronfman et al 1988, Clemens et al 1988, Combes 1991). The majority of this gas is probably contained in clouds with $M > 10^4\ M_\odot$

(Elmegreen 1985). It would be highly unstable on these grounds, and free-fall collapse would lead to a star formation rate, $M_* \geq 200$ $M_\odot$ year^{-1}, far in excess of the recent galactic average of 3 $M_\odot$ year^{-1} (Scalo 1986). This argument, first made by Zuckerman & Palmer (1974), shows that most clouds cannot be collapsing at free fall (see also Zuckerman & Evans 1974). Together with evidence of cloud lifetimes of about 4×10^7 years (Bash et al 1977, Leisawitz et al 1989), this discrepancy motivates an examination of other support mechanisms.

Two possibilities have been considered, magnetic fields and turbulence. Calculations of the stability of magnetized clouds (Mestel & Spitzer 1956, Mestel 1965) led to the concept of a magnetic critical mass (M_B). For highly flattened clouds (Li & Shu 1996),

$$M_B - (2\pi)^{-1} G^{-0.5} \Phi, \qquad (3)$$

where Φ is the magnetic flux threading the cloud,

$$\Phi \equiv \int B da. \qquad (4)$$

Numerical calculations (Mouschovias & Spitzer 1976) indicate a similar coefficient (0.13). If turbulence can be thought of as causing pressure, it may be able to stabilize clouds on large scales (e.g. Bonazzola et al 1987). It is not at all clear that turbulence can be treated so simply. In both cases, the cloud can only be metastable. Gas can move along field lines, and ambipolar diffusion will allow neutral gas to move across field lines with a timescale of (McKee et al 1993)

$$t_{AD} = \frac{3}{4\pi G \rho \tau_{ni}} \simeq 7.3 \times 10^{13} x_e \text{ years}, \qquad (5)$$

where τ_{ni} is the ion-neutral collision time. The ionization fraction (x_e) depends on ionization by photons and cosmic rays, balanced by recombination. It thus depends on the abundances of other species $[X(x) \equiv n(x)/n]$.

These two suggested mechanisms of cloud support (magnetic and turbulent) are not entirely compatible because turbulence should tangle the magnetic field (compare the reviews by Mouschovias 1999 and McKee 1999). A happy marriage between magnetic fields and turbulence was long hoped for; Arons & Max (1975) suggested that magnetic fields would slow the decay of turbulence if the turbulence was sub-Alfvénic. Simulations of MHD turbulence in systems with high degrees of symmetry supported this suggestion and indicated that the pressure from magnetic waves could stabilize clouds (Gammie & Ostriker 1996). However, more recent three-dimensional simulations indicate that MHD turbulence decays rapidly and that replenishment is still needed (Mac Low et al 1998, Stone et al 1999). The usual suggestion is that outflows generate turbulence, but Zweibel (1998) has suggested that an instability induced by ambipolar diffusion may convert magnetic energy into turbulence. Finally, the issue of supporting clouds assumes a certain stability and cloud integrity that may be misleading in a dynamic interstellar medium (e.g.

Ballesteros-Paredes et al 1999). For a current review of this field, see Vázquez-Semadini et al (2000).

With this brief and simplistic review of the issues of cloud stability and evolution, we have motivated the study of the basic physical conditions:

$$T_K, n, \vec{v}, \vec{B}, X. \qquad (6)$$

All of these are local variables; in principle, they can have different values at each point ($\vec{r}$) in space and can vary with time (t). In practice, we usually can measure only one component of vector quantities, integrated through the cloud. For example, we measure only line-of-sight velocities, usually characterized by the line width (Δv) or higher moments, and the line-of-sight magnetic field (B_z) through the Zeeman effect, or the projected direction (but not the strength) of the field in the plane of the sky ($B_\perp$) by polarization studies. In addition, our observations always average over finite regions, so we attempt to simplify the dependence on $\vec{r}$ by assumptions or models of cloud structure. In Section 3, I describe the methods used to probe these quantities and some overall results. Abundances have been reviewed recently (van Dishoeck & Blake 1998), so only relevant results will be mentioned, most notably the ionization fraction x_e.

In addition to the local variables, quantities that explicitly integrate over one or more dimensions are often measured. Foremost is the column density,

$$N \equiv \int n \, dl. \qquad (7)$$

The extinction or optical depth of dust at some wavelength is a common surrogate measure of N. If the column density is integrated over an area, one measure of the cloud mass within that area is obtained:

$$M_N \equiv \int N \, da. \qquad (8)$$

Another commonly used measure of mass is obtained by simplification of the virial theorem. If external pressure and magnetic fields are ignored,

$$M_V = C_v G^{-1} R \Delta v^2 = 210 \, M_\odot C_v R(pc) [\Delta v (\text{km s}^{-1})]^2, \qquad (9)$$

where R is the radius of the region, Δv is the FWHM line width, and the constant (C_v) depends on geometry and cloud structure, but is of order unity (e.g. McKee & Zweibel 1992, Bertoldi & McKee 1992). A third mass estimate can be obtained by integrating the density over the volume,

$$M_n \equiv \int n \, dv. \qquad (10)$$

M_n is commonly used to estimate f_v, the volume-filling factor of gas at density n, by dividing M_n by another mass estimate, typically M_V. Of the three methods of mass determination, the virial mass is the least sensitive to uncertainties in distance and size, but care must be taken to exclude unbound motions, such as outflows.

Parallel to the physical conditions in the gas, the dust can be characterized by a set of conditions:

$$T_D, n_D, \kappa(\nu), \tag{11}$$

where T_D is the dust temperature, n_D is the density of dust grains, and $\kappa(\nu)$ is the opacity at a given frequency, ν. If we look in more detail, grains have a range of sizes (Mathis et al 1977) and compositions. For smaller grains, T_D is a function of grain size. Thus, we would have to characterize the temperature distribution as a function of size, the composition of grains, both core and mantle, and many more optical constants to capture the full range of grain properties. For our purposes, T_D and $\kappa(\nu)$ are the most important properties, because T_D affects gas energetics and T_D and $\kappa(\nu)$ control the observed continuum emission of molecular clouds. The detailed nature of the dust grains may come into play in several situations, but the primary observational manifestation of the dust is its ability to absorb and emit radiation. For this review, a host of details can be ignored. The optical depth is set by

$$\tau_D(\nu) = \kappa(\nu) N, \tag{12}$$

where it is convenient to define $\kappa(\nu)$ so that N is the gas column density rather than the dust column density. Away from resonances, the opacity is usually approximated by $\kappa(\nu) \propto \nu^\beta$.

3. PROBES OF PHYSICAL CONDITIONS

The most fundamental fact about molecular clouds is that most of their contents are invisible. Neither the H_2 nor the He in the bulk of the clouds is excited sufficiently to emit. Although fluorescent emission from H_2 can be mapped over the face of clouds (Luhman & Jaffe 1996), the UV radiation needed to excite this emission does not penetrate the bulk of the cloud. In shocked regions, H_2 emits rovibrational lines that are useful probes of T_K and $\vec{v}(\vec{r})$ (e.g. Draine & McKee 1993). Absorption by H_2 of background stars is also difficult: the dust in molecular clouds obnubilates the UV that would reveal electronic transitions; the rotational transitions are so weak that only huge N would produce absorption, and the dust again obnubilates background sources; only the vibrational transitions in the near-infrared have been seen in absorption in only a few molecular clouds (Lacy et al 1994). Gamma rays resulting from cosmic-ray interactions with atomic nuclei do probe all the material in molecular clouds (Bloemen 1989, Strong et al 1994). So far, gamma-ray studies have suffered from low spatial resolution and uncertainties in the cosmic ray flux; they have been used mostly to check consistency with other tracers on large scales.

In the following subsections, I discuss probes of different physical quantities, including some general results, concluding with a discussion of the observational and analytical tools. Genzel (1992) has presented a detailed discussion of probes of physical conditions.

3.1 Tracers of Column Density, Size, and Mass

Given the reticence of the bulk of the H_2, essentially all probes of physical conditions rely on trace constituents, such as dust particles and molecules other than H_2. Dust particles (Mathis 1990, Pendleton & Tielens 1997) attenuate light at short wavelengths (UV to near infrared) and emit at longer wavelengths (far infrared to millimeter). Assuming that the ratio of dust extinction at a fixed wavelength to gas column density is constant, one can use extinction to map N in molecular clouds, and early work at visible wavelengths revealed locations and sizes of many molecular clouds before they were known to contain molecules (Barnard 1927, Bok & Reilly 1947, Lynds 1962). There have been more recent surveys for small clouds (Clemens & Barvainis 1988) and for clouds at high latitude (Blitz et al 1984). More recently, near-infrared surveys have been used to probe much more deeply; in particular, the $H - K$ color excess can trace N to an equivalent visual extinction, $A_V \sim 30$ mag (Lada et al 1994, Alves et al 1998). This method provides many pencil beam measurements through a cloud toward background stars. The very high-resolution but very undersampled data require careful analysis but can reveal information on mass, large-scale structure in N, and unresolved structure (Alves et al 1998). Padoan et al (1997) interpret the data of Lada et al (1994) in terms of a lognormal distribution of density, but Lada et al (1999) show that a cylinder with a density gradient $n(r) \propto r^{-2}$ also matches the observations.

Continuum emission from dust at long wavelengths is complementary to absorption studies (e.g. Chandler & Sargent 1997). Because the dust opacity decreases with increasing wavelength [$\kappa(\nu) \propto \nu^\beta$, with $\beta \sim 1$–2], emission at long wavelengths can trace large column densities and provide independent mass estimates (Hildebrand 1983). The data can be fully sampled and have reasonably high resolution. The dust emission depends on the dust temperature (T_D), linearly if the observations are firmly in the Rayleigh-Jeans limit, but exponentially on the Wien side of the blackbody curve. Observations on both sides of the emission peak can constrain T_D. Opacities have been calculated for a variety of scenarios including grain mantle formation and collisional concrescence (e.g. Ossenkopf & Henning 1994). For grain sizes much less than the wavelength, $\beta \sim 2$ is expected from simple grain models, but observations of dense regions often indicate lower values. By observing at a sufficiently long wavelength, one can trace N to very high values. Recent results indicate that $\tau_D(1.2 \text{ mm}) = 1$ only for $A_V \sim 4 \times 10^4$ mag (Kramer et al 1998a) in the less dense regions of molecular clouds. In dense regions, there is considerable evidence for increased grain opacity at long wavelengths (Zhou et al 1990, van der Tak et al 1999), suggesting grain growth through collisional concrescence in addition to the formation of icy mantles. Further growth of grains in disks is also likely (Chandler & Sargent 1997).

The other choice is to use a trace constituent of the gas, typically molecules that emit in their rotational transitions at millimeter or submillimeter wavelengths. By using the appropriate transitions of the appropriate molecule, one can tune the probe to study the physical quantity of interest and the target region along the line of sight. This technique was first used with OH (Barrett et al 1964), and it has been pursued intensively in the 30 years since the discovery of polyatomic interstellar molecules (Cheung et al 1968, van Dishoeck & Blake 1998).

The most abundant molecule after H_2 is carbon monoxide; the main isotopomer ($^{12}C^{16}O$) is usually written simply as CO. It is the most common tracer of molecular gas. On the largest scales, CO correlates well with the gamma-ray data (Strong et al 1994), suggesting that the overall mass of a cloud can be measured even when the line is quite opaque. This stroke of good fortune can be understood if the clouds are clumpy and macroturbulent, with little radiative coupling between clumps (Wolfire et al 1993); in this case, the CO luminosity is proportional to the number of clumps, hence total mass. Most of the mass estimates for the larger clouds and for the total molecular mass in the Galaxy and in other galaxies are in fact based on CO. On smaller scales and in regions of high column density, CO fails to trace column density, and progressively rarer isotopomers are used to trace progressively higher values of N. Dickman (1978) established a strong correlation of visual extinction A_V with ^{13}CO emission for $1.5 \leq A_V \leq 5$. Subsequent studies have used $C^{18}O$ and $C^{17}O$ to trace still higher N (Frerking et al 1982). These rarer isotopomers will not trace the outer parts of the cloud, where photodissociation affects them more strongly than the common isotopomers, but we are concerned with the more opaque regions in this review.

When comparing N measured by dust emission with N traced by CO isotopomers, it is important to correct for the fact that emission from low-J transitions of optically thin isotopomers of CO decreases with T_K, whereas dust emission increases with T_D (Jaffe et al 1984). Observations of many transitions can avoid this problem but are rarely done. To convert $N(CO)$ to N requires knowledge of the abundance, $X(CO)$. The only direct measure gave $X(CO) = 2.7 \times 10^{-4}$ (Lacy et al 1994), three times greater than inferred from indirect means (e.g. Frerking et al 1982). Clearly, this area needs increased attention, but at least a factor of 3 uncertainty must be admitted. Studies of some particularly opaque regions in molecular clouds indicate severe depletion (Kuiper et al 1996, Bergin & Langer 1997), raising the concern that even the rare CO isotopomers may fail to trace N. Indeed, Alves et al (1999) find that $C^{18}O$ fails to trace column density above $A_V = 10$ in some regions, and Kramer et al (1999) argue that this failure is best explained by depletion of $C^{18}O$.

Sizes of clouds, characterized by either a radius (R) or diameter (l), are measured by mapping the cloud in a particular tracer; for nonspherical clouds, these are often the geometric mean of two dimensions, and the aspect ratio (a/b) characterizes the ratio of long and short axes. The size along the line of sight (depth) can be constrained only by making geometrical assumptions (usually of a spherical or cylindrical cloud). One possible probe of the depth is H_3^+, which is unusual in having a calculable, constant density in molecular clouds. Thus, a measurement of $N(H_3^+)$ can yield a measure of cloud depth (Geballe & Oka 1996).

With a measure of size and a measure of column density, the mass (M_N) may be estimated (Equation 8); with a size and a line width (Δv), the virial mass (M_V) can be estimated (Equation 9). On the largest scales, the mass is often estimated from integrating the CO emission over the cloud and using an empirical relation between mass and the CO luminosity, $L(CO)$. The mass distribution has been estimated for both clouds and clumps within clouds, primarily from CO, ^{13}CO, or C^{18}O, using a variety of techniques to define clumps and estimate masses (Blitz 1993, Kramer et al 1998b, Heyer & Terebey 1998, Heithausen et al 1998). These studies have covered a wide range of masses, with Kramer et al extending the range down to $M = 10^{-4}$ M$_\odot$. The result is fairly well agreed on: $dN(M) \propto M^{-\alpha} dM$, with $1.5 \leq \alpha \leq 1.9$. Elmegreen & Falgarone (1996) have argued that the mass spectrum is a result of the fractal nature of the interstellar gas, with a fractal dimension $D = 2.3 \pm 0.3$. There is disagreement over whether clouds are truly fractal or have a preferred scale (Blitz & Williams 1997). On one hand, the latter authors suggest a scale of 0.25–0.5 pc in Taurus based on ^{13}CO. On the other hand, Falgarone et al (1998), analyzing an extensive data set, find evidence for continued structure down to 200 AU in gas that is not forming stars. The initial mass function (IMF) of stars is steeper than the cloud mass distribution for $M_* > 1$ M$_\odot$ but is flatter than the cloud mass function for $M_* < 1$ M$_\odot$ (e.g. Scalo 1998). Understanding the origin of the differences is a major issue (see Williams et al 2000, Meyer et al 2000).

3.2 Probes of Temperature and Density

The abundances of other molecules are so poorly constrained that CO isotopomers and dust are used almost exclusively to constrain N and M_N. Of what use are the >100 other molecules? Although many are of interest only for astrochemistry, some are very useful probes of physical conditions like T_K, n, v, B_z, and x_e.

Density (n) and gas temperature (T_K) are both measured by determining the populations in molecular energy levels and comparing the results to calculations of molecular excitation. A useful concept is the excitation temperature (T_{ex}) of a pair of levels, defined to be the temperature that gives the actual ratio of populations in those levels, when substituted into the Boltzmann equation. In general, collisions and radiative processes compete to establish level populations; when lines are optically thick, trapping of line photons enhances the effects of collisions. For some levels in some molecules, radiative rates are unusually low, collisions dominate, $T_{ex} = T_K$, and observational determination of these "thermalized" level populations yields T_K. Unthermalized level populations depend on both n and T_K; with a knowledge of T_K, observational determination of these populations yields n, although trapping usually needs to be accounted for. While molecular excitation probes the local n and T_K in principle, the observations themselves always involve some average over the finite beam and along the line of sight. Consequently, a model of the cloud is needed to interpret the observations. The simplest model is of course a homogeneous cloud, and most early work adopted this model, either explicitly or implicitly.

Tracers of temperature include CO, with its unusually low dipole moment, and molecules in which transitions between certain levels are forbidden by selection rules. The latter include different K ladders of symmetric tops like NH_3, CH_3CN, etc. (Ho & Townes 1983, Loren & Mundy 1984). Different K_{-1} ladders in H_2CO also probe T_K in dense, warm regions (Mangum & Wootten 1993). A useful feature of CO is that its low-J transitions are both opaque and thermalized in most parts of molecular clouds. In this case, observations of a single line provide the temperature, after correction for the cosmic background radiation and departures from the Rayleigh-Jeans approximation (Penzias et al 1972, Evans 1980). Early work on CO (Dickman 1975) and NH_3 (Martin & Barrett 1978) established that $T_K \simeq 10$ K far from regions of star formation and that sites of massive star formation are marked by elevated T_K, revealed by peaks in maps of CO (e.g. Blair et al 1975).

The value of T_K far from local heating sources can be understood by balancing cosmic-ray heating and molecular cooling (Goldsmith & Langer 1978), while elevated values of T_K in star-forming regions have a more intricate explanation. Stellar photons, even when degraded to the infrared, do not couple well to molecular gas, so the heating goes via the dust. The dust is heated by photons, and the gas is heated by collisions with the dust (Goldreich & Kwan 1974); above a density of about 10^4 cm^{-3}, T_K becomes well coupled to T_D (Takahashi et al 1983). Observational comparison of T_K to T_D, determined from far-infrared observations, supports this picture (e.g. Evans et al 1977, Wu & Evans 1989).

In regions where photons in the range of 6 to 13.6 eV impinge directly on molecular material, photoelectrons ejected from dust grains can heat the gas very effectively, and T_K may exceed T_D. These PDRs (Hollenbach & Tielens 1997) form the surfaces of all clouds, but the regions affected by these photons are limited by dust extinction to about $A_V \sim 8$ mag (McKee 1989). However, the CO lines often do form in the PDR regions, raising the question of why they indicate that $T_K \sim 10$ K. Wolfire et al (1993) explain that the optical depth in the lower-J levels usually observed reaches unity at a place where the T_K and n combine to produce an excitation temperature (T_{ex}) of about 10 K. Thus, the agreement of T_K derived from CO with the predictions of energetics calculations for cosmic-ray heating may be fortuitous. The T_K derived from NH_3 refer to more opaque regions and are more relevant to cosmic-ray heating. Finally, in localized regions, shocks can heat the gas to very high T_K; values of 2000 K are observed in H_2 rovibrational emission lines (Beckwith et al 1978). It is clear that characterizing clouds by a single T_K, which is often done for simplicity, obscures a great deal of complexity.

Density determination requires observations of several transitions that are not in local thermodynamic equilibrium (LTE). Then the ratio of populations, or equivalently T_{ex}, can be used to constrain density. A useful concept is the critical density for a transition from level j to level k,

$$n_c(jk) = A_{jk}/\gamma_{jk}, \qquad (13)$$

where A_{jk} is the Einstein A coefficient and $n\gamma_{jk}$ is the collisional deexcitation rate per molecule in level j. In general, both H_2 and He are effective collision partners,

TABLE 1 Properties of density probes

Molecule	Transition	ν (GHz)	E_{up} (K)	n_c(10 K) (cm^{-3})	n_{eff}(10 K) (cm^{-3})	n_c(100 K) (cm^{-3})	n_{eff}(100 K) (cm^{-3})
CS	$J = 1 \to 0$	49.0	2.4	4.6×10^4	7.0×10^3	6.2×10^4	2.2×10^3
CS	$J = 2 \to 1$	98.0	7.1	3.0×10^5	1.8×10^4	3.9×10^5	4.1×10^3
CS	$J = 3 \to 2$	147.0	14	1.3×10^6	7.0×10^4	1.4×10^6	1.0×10^4
CS	$J = 5 \to 4$	244.9	35	8.8×10^6	2.2×10^6	6.9×10^6	6.0×10^4
CS	$J = 7 \to 6$	342.9	66	2.8×10^7	...	2.0×10^7	2.6×10^5
CS	$J = 10 \to 9$	489.8	129	1.2×10^8	...	6.2×10^7	1.7×10^6
HCO$^+$	$J = 1 \to 0$	89.2	4.3	1.7×10^5	2.4×10^3	1.9×10^5	5.6×10^2
HCO$^+$	$J = 3 \to 2$	267.6	26	4.2×10^6	6.3×10^4	3.3×10^6	3.6×10^3
HCO$^+$	$J = 4 \to 3$	356.7	43	9.7×10^6	5.0×10^5	7.8×10^6	1.0×10^4
HCN	$J = 1 \to 0$	88.6	4.3	2.6×10^6	2.9×10^4	4.5×10^6	5.1×10^3
HCN	$J = 3 \to 2$	265.9	26	7.8×10^7	7.0×10^5	6.8×10^7	3.6×10^4
HCN	$J = 4 \to 3$	354.5	43	1.5×10^8	6.0×10^6	1.6×10^8	1.0×10^5
H$_2$CO	$2_{12} \to 1_{11}$	140.8	6.8	1.1×10^6	6.0×10^4	1.6×10^6	1.5×10^4
H$_2$CO	$3_{13} \to 2_{12}$	211.2	17	5.6×10^6	3.2×10^5	6.0×10^6	4.0×10^4
H$_2$CO	$4_{14} \to 3_{13}$	281.5	30	9.7×10^6	2.2×10^6	1.2×10^7	1.0×10^5
H$_2$CO	$5_{15} \to 4_{14}$	351.8	47	2.6×10^7	...	2.5×10^7	2.0×10^5
NH$_3$	(1,1)inv	23.7	1.1	1.8×10^3	1.2×10^3	2.1×10^3	7.0×10^2
NH$_3$	(2,2)inv	23.7	42	2.1×10^3	3.6×10^4	2.1×10^3	4.3×10^2

... means no value; inv means inversion transition.

with comparable collision rates, so that excitation techniques measure the total density of collision partners, $n \simeq n(H_2) + n(He)$. In some regions of high x_e, collisions with electrons may also be significant. Detection of a particular transition is often taken to imply that $n \geq n_c(jk)$, but this statement is too simplistic. Lines can be seen over a wide range of n, depending on observational sensitivity, the frequency of the line, and the optical depth (e.g. Evans 1989). Observing high frequency transitions, multilevel excitation effects, and trapping all tend to lower the effective density needed to detect a line. Table 1 contains information for some commonly observed lines, including the frequency, energy in K above the effective ground state [$E_{up}(K)$], and the critical densities at $T_K = 10$ K and 100 K. For comparison, the Table also has n_{eff}, the density needed to produce a line of 1 K, easily observable in most cases. The values of n_{eff} were calculated with a large velocity gradient (LVG) code (Section 3.5) to account for trapping, assuming $\log(N/\Delta v) = 13.5$ for all species but NH$_3$, for which $\log(N/\Delta v) = 15$ was used. $N/\Delta v$ has units of cm^{-2} (km s^{-1})$^{-1}$. These column densities are typical and produce modest optical depths. Note that n_{eff} can be as much as a factor of

1000 less than the critical density, especially for high excitation lines and high T_K. Clearly, the critical density should be used as a guideline only; more sophisticated analysis is necessary to infer densities.

Assuming that we have knowledge of T_K, at least two transitions with different $n_c(jk)$ are needed to determine both n and the line optical depth, $\tau_{jk} \propto N_k/\Delta v$, which determines the amount of trapping, and more transitions are desirable. Because $A_{J,J-1} \propto J^3$, where J is the quantum number for total angular momentum, observing many transitions up a rotational energy ladder provides a wide range of $n_c(jk)$. Linear molecules, like HCN and HCO$^+$, have been used in this way, but higher levels often occur at wavelengths with poor atmospheric transmission. Relatively heavy species, like CS, have many accessible transitions, and up to five transitions ranging up to $J = 10$ have been used to constrain density (e.g. Carr et al 1995, van der Tak et al 1999). More complex species provide more accessible energy levels; transitions within a single K_{-1} ladder of H$_2$CO provide a valuable density probe (Mangum & Wootten 1993). Transitions of H$_2$CO with $\Delta J = 0$ are accessible to large arrays operating at centimeter wavelengths (e.g. Evans et al 1987). The lowest few of these H$_2$CO transitions have the interesting property of absorbing the cosmic background radiation (Palmer et al 1969), T_{ex} being cooled by collisional pumping (Townes & Cheung 1969).

Application of these techniques to the homogeneous cloud model generally produces estimates of density exceeding 10^4 cm^{-3} in regions forming stars, while the sterile regions of the cloud are thought to have typical $n \sim 10^2 - 10^3$ cm^{-3}, although these are less well constrained. Theoretical simulations of turbulence have predicted lognormal (Vázquez-Semadini 1994) or power-law (Scalo et al 1998) probability density functions. Studies of multiple transitions of different molecules with a wide range of critical densities often reveal evidence for density inhomogeneities; in particular, pairs of transitions with higher critical densities tend to indicate higher densities (e.g. Evans 1980, Plume et al 1997). Both density gradients and clumpy structure have been invoked to explain these results (see Sections 4 and 5 for detailed discussion). Because lines with high $n_c(jk)$ are excited primarily at higher n, one can avoid to some extent the averaging over the line of sight by tuning the probe.

3.3 Kinematics

In principle, information on $\vec{v}(\vec{r})$ is contained in maps of the line profile over the cloud. In practice, this message has been difficult to decode. Only motions along the line of sight produce Doppler shifts, and the line profiles average over the beam and along the line of sight. Maps of the line center velocity generally indicate that the typical cloud is experiencing neither overall collapse (Zuckerman & Evans 1974) nor rapid rotation (Arquilla & Goldsmith 1986, Goodman et al 1993). Instead, most clouds appear to have velocity fields dominated by turbulence, because the line widths are usually much greater than expected from thermal broadening. Although such turbulence can explain the breadth of the lines, the line profile is not easily matched. Even a homogeneous cloud will tend to develop an excitation gradient in unthermalized lines because of trapping, and gradients in T_K or n toward embedded sources should exacerbate this tendency. Simple microturbulent models with decreasing $T_{\text{ex}}(r)$ predict that self-reversed line profiles should be seen more commonly than they are. Models with many small clumps and macroturbulence have had some success in avoiding self-reversed line profiles (Martin et al 1984, Wolfire et al 1993, Falgarone et al 1994, Park & Hong 1995).

The average line widths of clouds are larger for larger clouds, the line width-size relation: $\Delta v \propto R^\gamma$ (Larson 1981). For clouds with the same N, the virial theorem would predict $\gamma = 0.5$, consistent with the results of many studies of clouds as a whole (e.g. Solomon et al 1987). Myers (1985) summarized the different relations and distinguished between those comparing clouds as a whole and those studying trends within a single cloud. The status of line width-size relations within clouds, particularly in star-forming regions, will be discussed in later sections.

3.4 Magnetic Field and Ionization

The magnetic field strength and direction are important but difficult to measure. Heiles et al (1993) review the observations, and McKee et al (1993) review theoretical issues. The only useful measure of the strength is the Zeeman effect, which probes the line-of-sight field B_z. Observations of HI can provide some useful probes of B_z in PDRs (e.g. Brogan et al 1999), but cannot probe the bulk of molecular gas. Molecules suitable for Zeeman effect measurements have unpaired electrons, and their resulting reactivity tends to decrease their abundance in the denser regions (e.g. Sternberg et al 1997). Almost all work has been done with OH, along with some work with CN (Crutcher et al 1996), but future prospects include CCS and excited states of OH and CH (Crutcher 1998). Measurements of B_z have been made with thermal emission or absorption by OH (e.g. Crutcher et al 1993), mostly probing regions with $n \sim 10^3$ cm^{-3}, where $B_z \simeq 20$ μG or with OH maser emission, probing much denser gas, but with less certain conditions. As reviewed by Crutcher (1999a), the results for 14 clouds of widely varying mass with good Zeeman detections indicate that M_B is usually within a factor of 2 of the cloud mass. Given uncertainties, this result suggests that clouds with measured B_z lie close to the critical-subcritical boundary (Shu et al 1999). The observations can be fit with $B \propto n^{0.47}$, remarkably consistent with predictions of ambipolar-diffusion calculations (e.g. Fiedler & Mouschovias 1993). However, if turbulent motions in clouds are constrained to be comparable to the Alfvén velocity, $v_a \propto Bn^{-0.5}$, this result is also expected (Myers & Goodman 1988, Bertoldi & McKee 1992).

The magnetic-field direction, projected on the plane of the sky, can be measured because spinning, aspherical grains tend to align their spin axes with the magnetic-field direction (see Lazarian et al 1997 for a list of mechanisms). Then the dust grains absorb and emit preferentially in the plane perpendicular to the field. Consequently, background starlight will be preferentially polarized along

$B_\perp$, and thermal emission from the grains will be polarized perpendicular to $B_\perp$ (Hildebrand 1988). Goodman (1996) has shown that the grains that polarize background starlight do not trace the field very deeply into the cloud, but maps of polarized emission at far-infrared, submillimeter (Schleuning 1998) and millimeter (Akeson & Carlstrom 1997, Rao et al 1998) wavelengths are beginning to provide maps of field direction deep into clouds. Line emission may also be weakly polarized under some conditions (Goldreich & Kylafis 1981), providing a potential probe of $B_\perp$ with velocity information. After many attempts, this effect has been detected recently (Greaves et al 1999).

The ionization fraction (x_e) is determined by chemical analysis and has been discussed by van Dishoeck and Blake (1998). Theoretically, x_e should drop from about 10^{-4} near the outer edge of the cloud to about 10^{-8} in interiors shielded from UV radiation. Observational estimates of x_e are converging on values around 10^{-8} to 10^{-7} in cores (de Boisanger et al 1996, Caselli et al 1998, Williams et al 1998, Bergin et al 1999).

3.5 Observational and Analytical Tools

Having discussed how different physical conditions are probed, I will end this section with a brief summary of the observational and analytical tools that are used. Clearly, most information on physical conditions comes from observations of molecular lines. Most of these lie at millimeter or submillimeter wavelengths, and progress in this field has been driven by the development of large single-dish telescopes operating at submillimeter wavelengths and by arrays of antennas operating interferometrically at millimeter wavelengths (Sargent & Welch 1993). The submillimeter capability has allowed the study of high-J levels for excitation analysis and increased sensitivity to dust continuum emission, which rises with frequency ($S_\nu \propto \nu^2$ or faster). Studies of millimeter and submillimeter emission from dust have been greatly enhanced recently with the development of cameras on single dishes, both at millimeter wavelengths (Kreysa 1992) and at submillimeter wavelengths, with SHARC (Hunter et al 1996) and SCUBA (Cunningham et al 1994). Examples of the maps that these cameras are producing are the color plates showing the 1.3-mm emission from the ρ Ophiuchi region (Motte et al 1998), shown in Figure 1 (color), and the 850-μm and 450-μm emission from the ridge in Orion (L1641) (Johnstone & Bally 1999), shown in Figure 2 (color).

Interferometric arrays, operating at millimeter wavelengths, have provided unprecedented angular resolution (now better than $1''$) maps of both molecular-line and continuum emission. They are particularly critical for separating the continuum emission from a disk and the envelope and for studying deeply embedded binaries (e.g. Looney et al 1997, Figure 3). Complementary information has been provided in the infrared, with near-infrared star counting (Lada et al 1994, 1999), near-infrared and mid-infrared spectroscopy of rovibrational transitions (e.g. Mitchell et al 1990, Evans et al 1991, Carr et al 1995, van Dishoeck et al 1998), and far-infrared continuum and spectral-line studies (e.g. Haas et al 1995). Early results from the Infrared Space Observatory can be found in Yun & Liseau (1998).

The analytical tools for molecular-cloud studies have grown gradually in sophistication. Early studies assumed LTE excitation, an approximation that is still used in some studies of CO isotopomers, but it is clearly invalid for other species. Studies of excitation require solution of the statistical equilibrium equations (Goldsmith 1972). Goldreich & Kwan (1974) pointed out that photon trapping will increase the average T_{ex} and provided a way of including its effects that was manageable with the limited computer resources of that time: the LVG approximation. Tied originally to their picture of collapsing clouds, this approximation allowed one to treat the radiative transport locally. Long after the overall collapse scenario had been discarded, the LVG method has remained in use, providing a quick way to include trapping, at least approximately. In parallel, more computationally intensive codes were developed for microturbulent clouds, in which photons emitted anywhere in the cloud could affect excitation anywhere else (e.g. Lucas 1974).

The microturbulent and LVG assumptions are the two extremes, and real clouds probably lie between. For modest optical depths, the conclusions of the two methods differ by factors of ~ 3, comparable to uncertainties caused by uncertain geometry (White 1977, Snell 1981). These methods are still useful in some situations, but they are gradually being supplanted by more flexible radiative transport codes, using either the Monte Carlo technique (Bernes 1979, Choi et al 1995, Park & Hong 1998, Juvela 1997) or Λ-iteration (Dickel & Auer 1994, Yates et al 1997, Wiesemeyer 1999). Some of these codes allow variations in the velocity, density, and temperature fields, nonspherical geometries, clumps, etc. Of course, increased flexibility means more free parameters and the need for more extensive observations to constrain them.

Similar developments have occurred in the area of dust continuum emission. Since stellar photons are primarily at wavelengths at which dust is quite opaque, a radiative transport code is needed to compute dust temperatures as a function of distance from a stellar heat source (Egan et al 1988, Manske & Henning 1998). For clouds without embedded stars or protostars, only the interstellar radiation field heats the dust; T_D can get very low (5–10 K) in centers of opaque clouds (Leung 1975). Embedded sources heat clouds internally; in clouds opaque to the stellar radiation, it is absorbed close to the source and reradiated at longer wavelengths. Once the energy is carried primarily by photons at wavelengths where the dust is less opaque, the temperature distribution relaxes to the optically thin limit (Doty & Leung 1994):

$$T_D(r) \propto L^{q/2} r^{-q}, \qquad (14)$$

where L is the luminosity of the source and $q = 2/(\beta + 4)$, assuming $\kappa(\nu) \propto \nu^\beta$.

4. FORMATION OF ISOLATED LOW-MASS STARS

Many low-mass stars actually form in regions of high-mass star formation, where clustered formation is the rule (Elmegreen 1985, Lada 1992, McCaughrean & Stauffer 1994). The focus here is on regions where we can isolate the individual star-forming events, and these are almost inevitably forming low-mass stars.

4.1 Theoretical Issues

The theory of isolated star formation has been developed in some detail. It relies on the existence of relatively isolated regions of enhanced density that can collapse toward a single center, although processes at smaller scales may cause binaries or multiples to form. One issue then is whether isolated regions suitable for forming individual, low-mass stars are clearly identifiable. The ability to separate these from the rest of the cloud underlies the distinction between sterile and fertile parts of clouds (Section 1).

Because of the enormous compression needed, gravitational collapse plays a key role in all star formation theories. In most cases, only a part of the cloud collapses, and theories differ on how this part is distinguished from the larger cloud. Is it brought to the verge of collapse by an impulsive event, like a shock wave (Elmegreen & Lada 1977) or a collision between clouds or clumps (Loren 1976), or is the process gradual? Among gradual processes, the decay of turbulence and ambipolar diffusion are leading contenders. If the decay of turbulence leaves the cloud in a subcritical state ($M < M_B$), then a relatively long period of ambipolar diffusion is needed before dynamical collapse can proceed (Section 2). If the cloud is supercritical ($M > M_B$), then the magnetic field alone cannot stop the collapse (e.g. Mestel 1985). If turbulence does not prevent it, a rapid collapse ensues, and fragmentation is likely. Shu et al (1987a,b) suggested that the subcritical case describes isolated low-mass star formation, while the supercritical case describes high-mass and clustered star formation. Recently, Nakano (1998) has argued that star formation in subcritical cores via ambipolar diffusion is implausible; instead he favors dissipation of turbulence as the controlling factor. For the present section, the questions are whether there is evidence that isolated, low-mass stars form in subcritical regions and what the status of turbulence in these cores is.

Rotation could in principle support clouds against collapse, except along the rotation axis (Field 1978). Even if rotation does not prevent the collapse, it is likely to be amplified during collapse, leading at some point to rotation speeds able to affect the collapse. In particular, rotation is usually invoked to produce binaries or multiple systems on small scales. What do we know about rotation rates on large scales and how the rotation is amplified during collapse? Is there any correlation between rotation and the formation of binaries observationally? If rotation controls whether binaries form, can we understand why collapse leads to binary formation roughly half the time? It is clear that both magnetic flux and angular momentum must be redistributed during collapse to produce stars with reasonable fields and rotation rates, and these processes will affect the formation of binaries and protoplanetary disks. Some of these questions will be addressed in the next sections on globules and cores; while others will be discussed in the context of testing specific theories.

4.2 Globules and Cores: Overall Properties

Nearby small dark clouds or globules are natural places to look for isolated star formation (Bok & Reilly 1947). A catalog of 248 globules (Clemens & Barvainis 1988) has provided the basis for many studies. Yun & Clemens (1990) found that 23% of the CB globules appear to contain embedded infrared sources, with spectral-energy distributions typical of star-forming regions (Yun 1993). About one-third of the globules with embedded sources have evidence of outflows (Yun & Clemens 1992, Henning & Launhardt 1998). Clearly, star formation does occur in isolated globules.

Within the larger dark clouds, one can identify numerous regions of high opacity (e.g. Myers et al 1983), commonly called cores (Myers 1985). Surveys of such regions in low-excitation lines of NH_3 (e.g. Benson & Myers 1989) led to the picture of an isolated core within a larger cloud, which then might pursue its course toward star formation in relative isolation from the rest of the cloud. Most intriguing was the fact that the NH_3 line widths in many of these cores indicated that the turbulence was subsonic (Myers 1983); in some cores, thermal broadening of NH_3 lines even dominated over turbulent broadening (Myers & Benson 1983, Fuller & Myers 1993). Although later studies in other lines indicated a more complex dynamical situation (Zhou et al 1989, Butner et al 1995), the NH_3 data provided observational support for theories describing the collapse of isothermal spheres (Shu 1977). The discovery of *IRAS* sources in half of these cores (Beichman et al 1986) indicated that they were indeed sites of star formation. The observational and theoretical developments were synthesized into an influential paradigm for low-mass star formation (Shu et al 1987a; Section 4.5 below).

Globules would appear to be an ideal sample for measuring sizes because the effects of the environment are minimized by their isolation, but distances are uncertain. Based on the angular size of the optical images and an assumed average distance of 600 pc (Clemens & Barvainis 1988), the mean size $\langle l \rangle = 0.7$ pc. A subsample of these with distance estimates were mapped in molecular lines, yielding much smaller average sizes: $\langle l \rangle = 0.33 \pm 0.15$ pc for a sample of 6 "typical" globules mapped in CS (Launhardt et al 1998); maps of the same globules in $C^{18}O$ $J = 2 \rightarrow 1$ (Wang et al 1995) give sizes smaller by a factor of 2.9 ± 1.6. A sample of 11 globules in the southern sky mapped in NH_3 (Bourke et al 1995) have $\langle l \rangle = 0.21 \pm 0.08$ pc.

Cores in nearby dark clouds have the advantage of having well-determined distances; the main issue is how clearly they stand out from the bulk of the molecular cloud. Gregersen (1998) found that some known cores are barely visible above the general cloud emission in the $C^{18}O$ $J = 1 \rightarrow 0$ line. The mean size of a sample of 16 cores mapped in NH_3 is 0.15 pc, whereas CS $J = 2 \rightarrow 1$ gives 0.27 pc, and $C^{18}O$ $J = 1 \rightarrow 0$ gives 0.36 pc (Myers et al 1991). These differences may reflect the effects of opacity, chemistry, and density structure.

Globules are generally not spherical. By fitting the opaque cores with ellipses, Clemens & Barvainis (1988) found a mean aspect ratio (a/b) of 2.0. Measurements of aspect ratio in tracers of reasonably dense gas toward globules give $\langle a/b \rangle \sim$ 1.5 − 2 (Wang et al 1995, Bourke et al 1995), and cores in larger clouds have $\langle a/b \rangle \sim 2$ (Myers et al 1991). Myers et al (1991) and Ryden (1996) have argued that the underlying three-dimensional shapes were more likely to be prolate, with axial ratios of ~ 2, than oblate, for which axial ratios of 3–10 were needed. However, toroids may also be able to match the data because of their central density minima (Li & Shu 1996).

The uncertainties in size are reflected directly into uncertainties in mass. For the sample of globules mapped by Bourke et al (1995) in NH_3, $\langle M_N \rangle = 4 \pm 1\ M_\odot$, compared with $10 \pm 2\ M_\odot$ for cores. Larger masses are obtained from other tracers. For the sample of globules studied by Launhardt et al (1998), $\langle M_V \rangle = 26 \pm 12$ based on CS $J = 2 \rightarrow 1$ and 10 ± 6, based on $C^{18}O\ J = 2 \rightarrow 1$. Studies of the cores in larger clouds found similar ranges and differences among tracers (e.g. Fuller 1989, Zhou et al 1994a). A series of studies of the Taurus cloud complex has provided an unbiased survey of cores with known distance identified by $C^{18}O$ $J = 1 \rightarrow 0$ maps. Starting from a large-scale map of $^{13}CO\ J = 1 \rightarrow 0$ (Mizuno et al 1995), Onishi et al (1996) covered 90% of the area with $N > 3.5 \times 10^{21}$ with a map of $C^{18}O\ J = 1 \rightarrow 0$ with 0.1 pc resolution. They identified 40 cores with $\langle l \rangle = 0.46$ pc, $\langle a/b \rangle = 1.8$, and $\langle M_N \rangle = 23\ M_\odot$. The sizes extended over a range of a factor of 6 and masses over a factor of 80. Comparing these cores to the distribution of T Tauri stars, infrared sources, and $H^{13}CO^+$ emission, Onishi et al (1998) found that all cores with $N > 8 \times 10^{21}$ cm^{-2} are associated with $H^{13}CO^+$ emission and/or cold *IRAS* sources. In addition, the larger cores always contained multiple objects. They concluded that the core mass per star-forming event is relatively constant at 11 $M_\odot$.

It is clear that characterizing globules and cores by a typical size and mass is an oversimplification. First, they come in a range of sizes that is probably just the low end of the general distribution of cloud sizes. Second, the size and mass depend strongly on the tracer and method used to measure them. If we ignore these caveats, it is probably fair to say that most of these regions have sizes measured in tracers of reasonably dense gas in the range of a few tenths of a pc and masses of $<100\ M_\odot$, with more small, low-mass cores then massive ones. The larger cores tend to be fragmented, so that the mass of gas with $n \geq 10^4$ cm^{-3} tends toward 10 $M_\odot$, within a factor of 2. Star formation may occur when the column density exceeds 8×10^{21} cm^{-2}, corresponding in this case to $n \sim 10^4$ cm^{-3}.

4.3 Globules and Cores: Internal Conditions

If the sizes and masses of globules and cores are poorly defined, at least the temperatures seem well understood. Early CO observations (Dickman 1975) showed that the darker globules are cold ($T_K \sim 10$ K), as expected for regions with only cosmic-ray heating. Similar results were found from NH_3 (Martin & Barrett 1978, Myers & Benson 1983, Bourke et al 1995). Clemens et al (1991) showed the distribution of CO temperatures for a large sample of globules; the main peak corresponded to $T_K = 8.5$ K, with a small tail to higher T_K. Determination of the dust temperature was more difficult, but Keene (1981) measured $T_D = 13$–16 K in B133, a starless core. More recently, Ward-Thompson et al (1998) used Infrared Space Observatory data to measure $T_D = 13$ K in another starless core. These results are similar to predictions for cores heated by the interstellar radiation field (Leung 1975), although T_D is expected to be lower in the deep interiors.

A sample of starless globules with $A_V \sim 1$–2 mag produced no detections of NH_3 (Kane et al 1994), but surveys of H_2CO (Wang et al 1995) and CS (Launhardt et al 1998, Henning & Launhardt 1998) toward more opaque globules indicate that dense gas is present in some. The detection rate of CS $J = 2 \to 1$ emission was much higher in globules with infrared sources. Butner et al (1995) analyzed multiple transitions of DCO^+ in 18 low-mass cores, finding $\langle \log n(cm^{-3}) \rangle \simeq 5$, with a tendency to slightly higher values in the cores with infrared sources. Thus, gas still denser than the $\langle \log n(cm^{-3}) \rangle \simeq 4$ gas traced by NH_3 exists in these cores.

Discussion of the kinematics of globules and cores has often focused on the relationship between line width (Δv) and size (l or R). This relation is much less clearly established for cores within a single cloud than is the relation for clouds as a whole (Section 3.3), and it may have a different origin (Myers 1985). Goodman et al (1998) have distinguished four types of line width-size relationships. Most studies have used Goodman Type-1 (multitracer, multicore) relations, but it is difficult to distinguish different causes in such relations. Goodman Type-2 (single-tracer, multicore) relations within a single cloud would reveal whether the virial masses of cores are reliable. Interestingly, the most systematic study of cores in a single cloud found no correlation in 24 cores in Taurus mapped in $C^{18}O$ $J = 1 \to 0$ ($\gamma = 0.0 \pm 0.2$), but the range of sizes (0.13 to 0.4 pc) may have been insufficient (Onishi et al 1996).

To study the kinematics of individual cores, the most useful relations are Goodman Types-3 (multitracer, single-core) and -4 (single-tracer, single-core) relations. In these, a central position is defined, either by an infrared source or a line peak. A Type-3 relationship using NH_3, $C^{18}O$ $J = 1 \to 0$, and CS $J = 2 \to 1$ lines was explored by Fuller & Myers (1992), who found $\Delta v \propto R^\gamma$, with R the radius of the half-power contour. Caselli & Myers (1995) added ^{13}CO $J = 1 \to 0$ data and constructed a Type-1 relation for eight starless cores, after removing the thermal broadening. The mean γ was 0.53 ± 0.07 with a correlation coefficient of 0.81. However, both these relationships depend strongly on the fact that the NH_3 lines have small Δv and R. Some other species [e.g. HC_3N (Fuller & Myers 1993)] also have narrow lines and small sizes, but DCO^+ emission has much wider lines over a similar map size (Butner et al 1995), raising the possibility that chemical effects cause different molecules to trace different kinematic regimes within the same overall core. Goodman et al (1998) suggest that the DCO^+ is excited in a region outside the NH_3 region and that the size of the DCO^+ region is underestimated. On the other hand, some chemical simulations indicate that NH_3 will deplete in dense cores while ions like DCO^+ will not (Rawlings et al 1992), suggesting the opposite solution.

A way to avoid such effects is to use a Type-4 relation, searching for a correlation between Δv_{NT} as spectra are averaged in rings of larger size, although line of sight confusion cannot be avoided as one gives up the ability to tune the density sensitivity. This method has almost never been applied to tracers of dense gas, but Goodman et al (1998) use an indirect method to obtain a Type-4 relation for three clouds in OH, $C^{18}O$, and NH_3. The relations are very flat ($\gamma = 0.1$–0.3), and $\gamma \neq 0$ has statistical significance only for OH, which traces the least dense gas. In particular, NH_3 shows no significant Type-4 relation, having narrow lines on every scale (see also Barranco & Goodman 1998). Goodman et al (1998) interpret these results in terms of a "transition to coherence" at the scale of 0.1–0.2 pc from the center of a dense core. Inside that radius, the turbulence becomes subsonic and no longer decreases with size (Barranco & Goodman 1998). Although this picture accords nicely with the idea that cores can be distinguished from the surroundings and treated as "units" in low-mass star formation, the discrepancy between values of Δv measured in different tracers of the dense core (cf Butner et al 1995) indicates that caution is required in interpreting the NH_3 data.

Rotation can be detected in some low-mass cores, but the ratio of rotational to gravitational energy has a typical value of 0.02 on scales of 0.1 pc (Goodman et al 1993). The inferred rotation axes are not correlated with the orientation of cloud elongation, again suggesting that rotation is not dynamically important on the scale of 0.1 pc. Ohashi et al (1997a) find evidence in several star-forming cores for a transition at $r \sim 0.03$ pc, inside of which the specific angular momentum appears to be constant at $\sim 10^{-3}$ km s^{-1} pc down to scales of ~ 200 AU.

Knowledge of the magnetic-field strength in cores and globules would be extremely valuable in assessing whether they are subcritical or supercritical. Unfortunately, Zeeman measurements of these regions are extremely difficult because they must be done with emission, unless there is a chance alignment with a background radio source. Crutcher et al (1993) detected Zeeman splitting in OH in only 1 of 12 positions in nearby clouds. Statistical analysis of the detection and the upper limits, including the effects of random orientation of the field, led to the conclusion that the data could not falsify the hypothesis that the clouds were subcritical. Another problem is that the OH emission probes relatively low densities ($n \sim 10^3$ cm^{-3}). Attempts to use CN to probe denser gas produced upper limits that were less than expected for subcritical clouds, but the small sample size and other uncertainties again prevented a definitive conclusion (Crutcher et al 1996). Improved sensitivity and larger samples are crucial to progress in this field. At present, no clear examples of subcritical cores have been found (Crutcher 1999a), but uncertainties are sufficient to allow this possibility.

Onishi et al (1996) found that the major axis of the cores they identified in Taurus tended to be perpendicular to the optical polarization vectors and hence $B_\perp$. Counterexamples are known in other regions (e.g. Vrba et al 1976) and the fact that optical polarization does not trace the dense portions of clouds (Goodman 1996) suggests that this result be treated cautiously. Further studies of $B_\perp$ using dust emission at long wavelengths are clearly needed.

The median ionization fraction of 23 low-mass cores is 9×10^{-8}, with a range of $\log x_e$ of -7.5 to -6.5, with typical uncertainties of a factor of 0.5 in the log (Williams et al 1998). Cores with stars do not differ significantly in x_e from cores without stars, consistent with cosmic-ray ionization. For a cloud with $n = 10^4$ cm^{-3}, the ambipolar diffusion timescale is $t_{AD} \sim 7 \times 10^6$ years $\sim 20 t_{ff}$. If the cores are subcritical, they will evolve much more slowly than a free-fall time. Recent comparisons of the line profiles of ionized and neutral species have been interpreted as setting an upper limit on the ion-neutral drift velocity of 0.03 km s^{-1}, consistent with that expected from ambipolar diffusion (Benson et al 1998).

To summarize the last two subsections, there is considerable evidence that distinct cores can be identified, both as isolated globules and as cores within larger clouds. Although there is a substantial range of properties, scales of 0.1 pc in size and 10 M$_\odot$ in mass seem common. The cores are cold ($T_K \sim T_D \sim 10$ K) and contain gas with $n \sim 10^4$ cm^{-3}, extending in some cases up to $n \sim 10^5$ cm^{-3}. Although different molecules differ in the magnitude of the effect, these cores seem to be regions of decreased turbulence compared with the surroundings. Although no clear cases of subcritical cores have been found, the hypothesis that low-mass stars form in subcritical cores cannot be ruled out observationally. How these cores form is beyond the scope of this review, but again one can imagine two distinct scenarios: ambipolar diffusion brings an initially subcritical core to a supercritical state, or dissipation of turbulence plays a similar role in a core originally supported by turbulence (Myers & Lazarian 1998). In the latter picture, cores may build up from accretion of smaller diffuse elements (Kuiper et al 1996), perhaps the structures inferred by Falgarone et al (1998).

4.4 Classification of Sources and Evolutionary Scenarios

The *IRAS* survey provided spectral energy distributions over a wide wavelength range for many cores (e.g., Beichman et al 1986), leading to a classification scheme for infrared sources. In the original scheme (Lada & Wilking 1984, Lada 1987), the spectral index between 2 μm and the longest observed wavelength was used to divide sources into three Classes, designated by roman numerals, with Class I indicating the most emission at long wavelengths. These classes rapidly became identified with stages in the emerging theoretical paradigm (Shu et al 1987a): Class-I sources are believed to be undergoing infall with simultaneous bipolar outflow, Class-II sources are typically visible T Tauri stars with disks and winds, and Class-III sources have accreted or dissipated most of the material, leaving a pre–main-sequence star, possibly with planets (Adams et al 1987, Lada 1991).

More recently, submillimeter continuum observations have revealed a large number of sources with emission peaking at still longer wavelengths. Some of these new sources also have infrared sources and powerful bipolar outflows, indicating that a central object has formed; these have been designated Class 0 (André et al 1993). André & Montmerle (1994) argued that Class-0 sources represent the primary infall stage, in which there is still more circumstellar than stellar matter.

Outflows appear to be most intense in the earliest stages, declining later (Bontemps et al 1996). Other cores with submillimeter emission have no *IRAS* sources and probably precede the formation of a central object. These were found among the "starless cores" of Benson & Myers (1989), and Ward-Thompson et al (1994) referred to them as "pre-protostellar cores." The predestination implicit in this name has made it controversial, and I use the less descriptive (and somewhat tongue-in-cheek) term, Class -1. There has also been some controversy over whether Class-0 sources are really distinct from Class-I sources or just more extreme versions. The case for Class-0 sources as a distinct stage can be found in André et al (2000).

While classification has an honored history in astronomy, serious tests of theory are facilitated by continuous variables. Myers & Ladd (1993) suggested that we characterize the spectral energy distribution by the flux-weighted mean frequency, or, more suggestively, by the temperature of a black body with the same mean frequency. The latter (T_{bol}) was calculated by Chen et al (1995) for many sources, and the following boundary lines in T_{bol} were found to coincide with the traditional classes: $T_{bol} < 70$ K for Class 0, $70 \leq T_{bol} \leq 650$ K for Class I, and $650 < T_{bol} \leq 2800$ K for Class II. In a crude sense, T_{bol} captures the "coolness" of the spectral energy distribution, which is related to how opaque the dust is, but it can be affected strongly by how much mid-infrared and near-infrared emission escapes and thus by geometry. Other measures, such as the ratio of emission at a submillimeter wavelength to the bolometric luminosity (L_{smm}/L_{bol}), may also be useful. One of the problems with all of these measures is that the bulk of the energy for classes earlier than II emerges at far-infrared wavelengths, where resolution has been poor. Maps of submillimeter emission are showing that many near-infrared sources are displaced from the submillimeter peaks and may have been falsely identified as Class-I sources (Motte et al 1998). Ultimately, higher spatial resolution in the far-infrared will be needed to sort out this confusion.

4.5 Detailed Theories

The recent focus on the formation of isolated, low-mass stars is at least partly due to the fast that it is more tractable theoretically than the formation of massive stars in clusters. Shu (1977) argued that collapse begins in a centrally condensed configuration and propagates outward at the sound speed (a); matter inside $r_{inf} = at$ is infalling after a time t. Because the collapse in this model is self-similar, the structure can be specified at any time from a single solution. This situation is called inside-out collapse. In Shu's picture, the precollapse configuration is an isothermal sphere, with $n(r) \propto r^{-p}$ and $p = 2$. Calculations of core formation via ambipolar diffusion gradually approach a configuration with an envelope that is close to a power law, but with a core of ever-shrinking size and mass in which $p \sim 0$ (e.g. Mouschovias 1991). It is natural to identify the ambipolar diffusion stage with the Class -1 stage and the inside-out collapse with Class 0 to Class I. As collapse proceeds, the material inside r_{inf} becomes less dense, with a power law approaching $p = 1.5$ in the inner regions, after a transition region where the density is not a power law. In addition, $v(r) \propto r^{-0.5}$ at small r.

There are many other solutions to the collapse problem (Hunter 1977, Foster & Chevalier 1993), ranging from the inside-out collapse to overall collapse (Larson 1969, Penston 1969). Henriksen et al (1997) have argued that collapse begins before the isothermal sphere is established; the inner ($p = 0$) core undergoes a rapid collapse that they identify with Class-0 sources. They suggest that Class-I sources represent the inside-out collapse phase, which appears only when the wave of infall has reached the $p = 2$ envelope.

Either rotation or magnetic fields will break spherical symmetry. Terebey et al (1984) added slow rotation of the original cloud at an angular velocity of Ω to the inside-out collapse picture, resulting in another characteristic radius, the point where the rotation speed equals the infall speed. A rotationally supported disk should form somewhere inside this centrifugal radius (Shu et al 1987a)

$$r_c = \frac{G^3 M^3 \Omega^2}{16 a^8}, \tag{15}$$

where M is the mass already in the star and disk. Because disks are implicated in all models of the ultimate source of the outflows, the formation of the disk may also signal the start of the outflow. Once material close to the rotation axis has accreted (or been blown out), all further accretion onto the star should occur through the disk.

Core formation in a magnetic field should produce a flattened structure (e.g. Fiedler & Mouschovias 1993), and Li & Shu (1996) argue that the equilibrium structure equivalent to the isothermal sphere is the isothermal toroid. Useful insights into the collapse of a magnetized cloud have resulted from calculations in spherical geometry (Safier et al 1997, Li 1998) or for thin disks (Ciolek & Königl 1998). Some calculations of collapse in two dimensions have been done (e.g. Fiedler & Mouschovias 1993). A magnetically channeled, flattened structure may appear; this has been called a pseudodisk (Galli & Shu 1993a,b) to distinguish it from the rotationally supported disk. In this picture, material would flow into a pseudodisk at a scale of ~ 1000 AU before becoming rotationally supported on the scale of r_c. The breaking of spherical symmetry on scales of 1000 AU may explain some of the larger structures seen in some regions (see Mundy et al 2000 for a review).

Ultimately, theory should be able to predict the conditions that lead to binary formation, but this is not currently possible. Steps toward this goal can be seen in numerical calculations of collapse with rotation (e.g. Bonnell & Bastien 1993, Truelove et al 1998, Boss 1998). Rotation and magnetic fields have been combined in a series of calculations by Basu & Mouschovias (1995 and references therein); Basu (1998) has considered the effects of magnetic fields on the formation of rotating disks.

Theoretical models make predictions that are testable, at least in principle. In the simplest picture of the inside-out collapse of the rotationless, nonmagnetic isothermal sphere, the theory predicts all of the density and velocity structure with only the sound speed as a free parameter. Rotation adds Ω, and magnetic fields add a reference field or equivalent as additional parameters. Departure from spherical symmetry adds the additional observational parameter of viewing angle. Consequently, observational tests have focused primarily on testing the simplest models.

4.6 Tests of Evolutionary Hypotheses and Theory

Both the empirical evolutionary sequence based on the class system and the detailed theoretical predictions can be tested by detailed observations. One can compute the expected changes in the continuum emission as a function of time for a particular model. Examples include plots of L versus A_V (Adams 1990), L versus T_{bol} (Myers et al 1998), and L_{mm} versus L (Saraceno et al 1996). For example, as time goes on, A_V decreases, T_{bol} increases, and L reaches a peak and declines. At present the models are somewhat idealized and simplified, and different dust opacities need to be considered. Comparison of the number of objects observed in various parts of these diagrams with those expected from lifetime considerations can provide an overall check on evolutionary scenarios and provide age estimates for objects in different classes (e.g. André et al 2000). One can also test the models against observations of particular objects. However, models of source structure constrained only by the spectral-energy distribution are not unique (Butner et al 1991, Men'shchikov & Henning 1997). Observational determination of $n(r)$ and $v(r)$ can apply more stringent tests.

Maps of continuum emission from dust can trace the column density as a function of radius quite effectively, if the temperature distribution is known. With an assumption about geometry, this information can be related to $n(r)$. For the Class -1 sources, with no central object, the core should be isothermal ($T_D \sim 5$–10 K) or warmer on the outside if exposed to the interstellar radiation field (Leung 1975, Spencer & Leung 1978). New results from the Infrared Space Observatory will put tighter limits on possible internal energy sources (e.g. Ward-Thompson et al 1998, André et al 2000). Based on small maps of submillimeter emission, Ward-Thompson et al (1994) found that Class -1 sources were not characterized by single power laws in column density; they fit the distributions with broken power laws, indicating a shallower distribution closer to the center. This distribution appears consistent with the interpretation that these are cores still forming by ambipolar diffusion. Maps at 1.3 mm (André et al 1996, Ward-Thompson et al 1999) confirm the early results: the column density is quite constant in an inner region ($r < 3000$–4000 AU), with $M \sim 0.7$ $M_\odot$. SCUBA maps of these sources are just becoming available, but they suggest similar conclusions (D Ward-Thompson, personal communication; Shirley et al 1998). In addition, some of the cores are filamentary and fragmented (Ward-Thompson et al 1999). Citing these observed properties, together with a statistical argument regarding lifetimes, Ward-Thompson et al (1999) now argue that ambipolar diffusion in subcritical cores does not match the data (see also André et al 2000).

For cores with central sources (Class > -1), $T_D(r)$ will decline with radius from the source. If the emission is in the Rayleigh-Jeans limit, $I_\nu \propto T_D \kappa(\nu) N$. If $\kappa(\nu)$ is not a function of r, $I_\nu(\theta) \propto \int T_D(r) n(r) dl$, where the integration is performed along the line of sight (dl). If one avoids the central beam and any outer cut-off, assumes the optically thin expression for $T_D(r) \propto r^{-q}$, and fits a power law to $I_\nu(\theta) \propto \theta^{-m}$, then $p = m + 1 - q$ (Adams 1991, Ladd et al 1991). With a proper calculation of $T_D(r)$ and convolution with the beam, one can include all of the information (Adams 1991). This technique has been applied in the far-infrared (Butner et al 1991), but it is most useful at longer wavelengths. Ladd et al (1991) found $p = 1.7 \pm 0.3$ in two cores, assuming $q = 0.4$, as expected for $\beta = 1$ (Equation 14). Results so far are roughly consistent with theoretical models, but results in this area from the new submillimeter cameras will "explode" about the time this review goes to press.

Another issue arises for cores with central objects. If a circumstellar disk contributes significantly to the emission in the central beam, it will increase the fitted value of m. Disks contribute more importantly at longer wavelengths, so the disk contribution is more important at millimeter wavelengths (e.g. Chandler et al 1995). Luckily, interferometers are available at those wavelengths, and observations with a wide range of antenna spacings can separate the contributions of a disk and envelope. Application of this technique by Looney et al (1997) to L1551 IRS5, a Class-I source, reveals very complex structures (Figure 3): binary circumstellar disks (cf Rodríguez et al 1998), a circumbinary structure (perhaps a pseudodisk), and an envelope with a density distribution consistent with a power law ($p = 1.5$–2). This technique promises to be very fruitful in tracing the flow of matter from envelope to disk. Early results indicated that disks are more prominent in more evolved (Class-II) systems (Ohashi et al 1996), but compact structures are detectable in some younger systems (Terebey et al 1993). Higher-resolution observations and careful analysis will be needed to distinguish envelopes, pseudodisks, and Keplerian disks (see Mundy et al 2000 for a review). At the moment, one can say only that disks in the Class-0 stage are not significantly more massive than disks in later classes (Mundy et al 2000). Meanwhile, the interferometric data confirm the tendency of envelope mass to decrease with class number inferred from single-dish data (Mundy et al 2000).

Similar techniques have been used for maps of molecular line emission. For example, ^{13}CO emission has been used to trace column density in the outer regions of dark clouds. With an assumption of spherical symmetry, the results favor $p \sim 2$ in most clouds (Snell 1981, Arquilla & Goldsmith 1985). The ^{13}CO lines become optically thick in the inner regions; studies with higher spatial resolution in rarer isotopomers, like $C^{18}O$ or $C^{17}O$, tend to show somewhat more shallow density distributions than expected with the standard model (Zhou et al 1994b, Wang et al 1995). Depletion in the dense, cold cores may still confuse matters (e.g. Kuiper et al 1996, Section 3.1). Addressing the question of evolution, Ladd et al (1998) used two transitions of $C^{18}O$ and $C^{17}O$ to show that N toward the central source declines with T_{bol}, with a power between 0.4 and 1.0. To reproduce the inferred

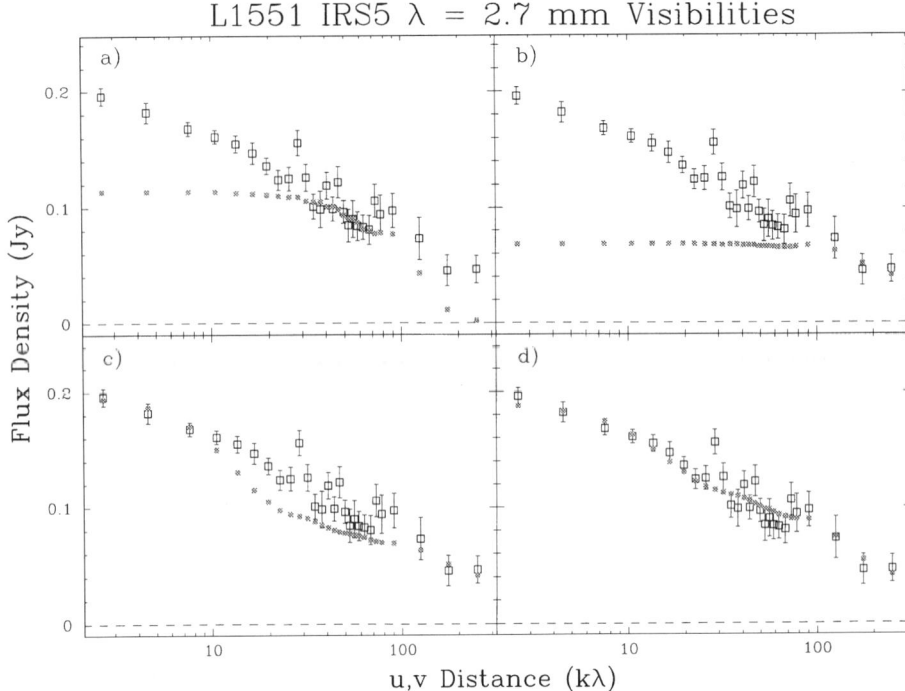

Figure 3 Observed visibilities of L1551 IRS 5 at 2.7 mm, binned in annuli (*open squares* with *error bars*), are plotted versus the projected baseline in units of 10^3 times the wavelength. Different models are shown in each panel by the small boxes. Panel *a* has a model with only a Gaussian source of radius 80 AU; panel *b* has a model with two point sources constrained to match the map; panel *c* adds a truncated power law ($p = 1.5$, $q = 0.5$) envelope to the two point sources; panel *d* adds to the previous components a circumbinary structure, represented by a Gaussian. In the final optimization (panel *d*), the envelope has a mass of 0.28 $M_\odot$ and an outer radius of 1100 AU, the circumbinary structure has a mass of 0.04 $M_\odot$, and the circumstellar disk masses are 0.024 and 0.009 $M_\odot$ (Looney et al 1997).

rapid decrease in mass with time, they suggest higher early mass loss than predicted by the standard model.

By observing a series of lines of different critical density, modeling those lines with a particular cloud model and appropriate radiative transport, and predicting the emission into the beams used for the observations, one can constrain the run of density more directly. Studies using two transitions of H_2CO have again supported $p = 2 \pm 0.5$ on relatively large scales (Loren et al 1983, Fulkerson & Clark 1984). When interferometry of H_2CO was used to improve the resolution on one core, p appeared to decrease at small r (Zhou et al 1990), in agreement with the model of Shu (1977). Much of the recent work on this topic has involved testing of detailed collapse models, including velocity fields and the complete density law, rather than a single power law, as described in the next section.

4.7 Collapse

The calculation of line profiles as a function of time (Zhou 1992) for the collapse models of Shu (1977), or Larson (1969) and Penston (1969), along with claims of collapse in a low-mass–star-forming region (Walker et al 1986), reinvigorated the study of protostellar collapse. Collapsing clouds will depart from the line width–size relation (Section 4.3), having systematically larger line widths for a given size (Zhou 1992). Other simulations of line profiles range from a simple two-layer model (Myers et al 1996) to detailed calculations of radiative transport (Choi et al 1995; Walker et al 1994; Wiesemeyer 1997, 1999).

Zhou et al (1993) showed that several lines of CS and H_2CO observed towards B335, a globule with a Class-0 source, could be fitted very well by using the exact $n(r)$ and $v(r)$ of the inside-out collapse model. Using a more self-consistent radiative transport code, Choi et al (1995) found slightly different best-fit parameters. With a sound speed determined from lines away from the collapse region, the only free parameters were the time since collapse began and the abundance of each molecule. With several lines of each molecule, the problem is quite constrained (Figure 4). This work was important in gaining acceptance for the idea that collapse had finally been seen.

Examination of the line profiles in Figure 4 reveals that most are strongly self-absorbed. Recall that the overall collapse idea of Goldreich & Kwan (1974) was designed to avoid self-absorbed profiles. The difference is that Goldreich & Kwan assumed that $v(r) \propto r$, so that every velocity corresponded to a single point along the line of sight. In contrast, the inside-out collapse model predicts $v(r) \propto r^{-0.5}$ inside a static envelope. If the line has substantial opacity in the static envelope, it will produce a narrow self-absorption at the velocity centroid of the core (Figure 5). The other striking feature of the spectra in Figure 4 is that the blue-shifted peak is stronger than the red-shifted peak. This blue profile occurs because the $v(r) \propto r^{-0.5}$ velocity field has two points along any line of sight with the same Doppler shift (Figure 6). For a centrally peaked temperature and density distribution, lines with high critical densities will have higher T_{ex} at the point closer to the center. If the line has sufficient opacity at the relevant point in the cloud, the high T_{ex} point in the red peak will be obscured by the lower T_{ex} one, making the red peak weaker than the blue peak (Figure 6). Thus a collapsing cloud with a velocity and density gradient similar to those in the inside-out collapse model will produce blue profiles in lines with suitable excitation and opacity properties. A double-peaked profile with a stronger blue peak or a blue-skewed profile relative to an optically thin line then becomes a signature of collapse. These features were discussed by Zhou & Evans (1994) and, in a more limited context, by Snell & Loren (1977) and Leung & Brown (1977).

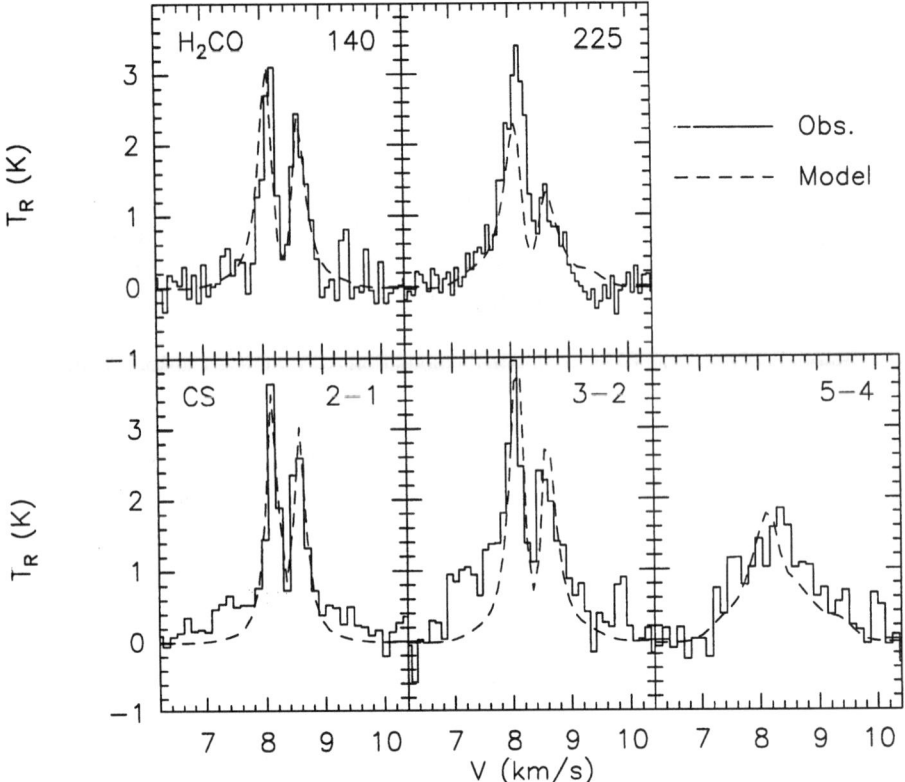

Figure 4 Line profiles of H_2CO and CS emission (*solid histograms*) toward B335 and the best-fitting model (*dashed lines*). The model line profiles were calculated with a Monte Carlo code, including non-LTE excitation and trapping, with an input density and velocity field taken directly from the collapse model of Shu (1977) and a temperature field calculated with a separate dust radiation transport code. The best-fitting model has an infall radius of 0.03 pc (Choi et al 1995).

Of course, the collapse interpretation of a blue profile is not unique. Such profiles can be produced in a variety of ways. To be a plausible candidate for collapse, a core must also show these features: an optically thin line must peak between the two peaks of the opaque line; the strength and skewness should peak on the central source; and the two peaks should not be caused by clumps in an outflow. The optically thin line is particularly crucial, because two cloud components, for example, colliding fragments, could produce the double-peaked blue profile, but they would also produce a double-peaked profile in the optically thin line.

Rotation, combined with self-absorption, can create a line profile like that of collapse (Menten et al 1987, Adelson & Leung 1988), but toward the center of rotation, the line would be symmetric (Zhou 1995). Rotating collapse can cause

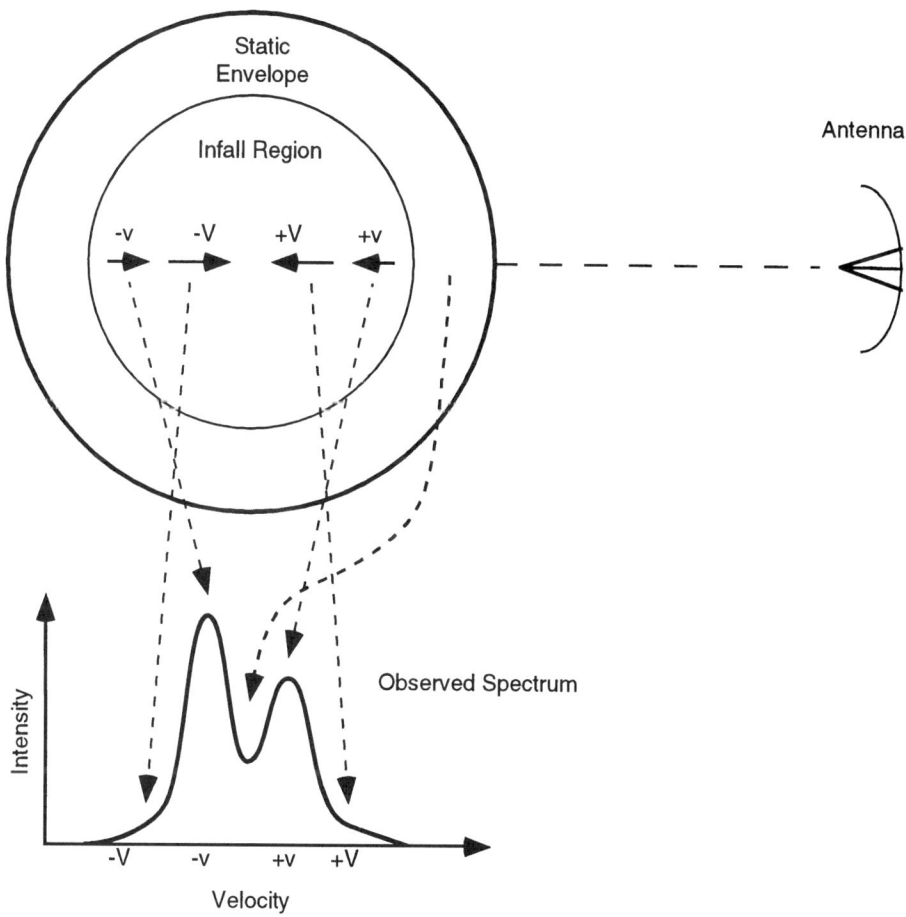

Figure 5 The origin of various parts of the line profile for a cloud undergoing inside-out collapse. The static envelope outside r_{inf} produces the central self-absorption dip, the blue peak comes from the back of the cloud, and the red peak from the front of the cloud. The faster collapse near the center produces line wings, but these are usually confused by outflow wings.

the line profiles to shift from blue to red-skewed on either side of the rotation axis, with the sign of the effect depending on how the rotation varies with radius (Zhou 1995). Maps of the line centroid can be used to separate rotation from collapse (Adelson & Leung 1988, Walker et al 1994).

To turn a collapse candidate into a believable case of collapse, one has to map the line profiles, account for the effects of outflows, model rotation if present, and show that a collapse model fits the line profiles. To date this has been done only for a few sources: B335 (Zhou et al 1993, Choi et al 1995), L1527 (Myers et al 1995,

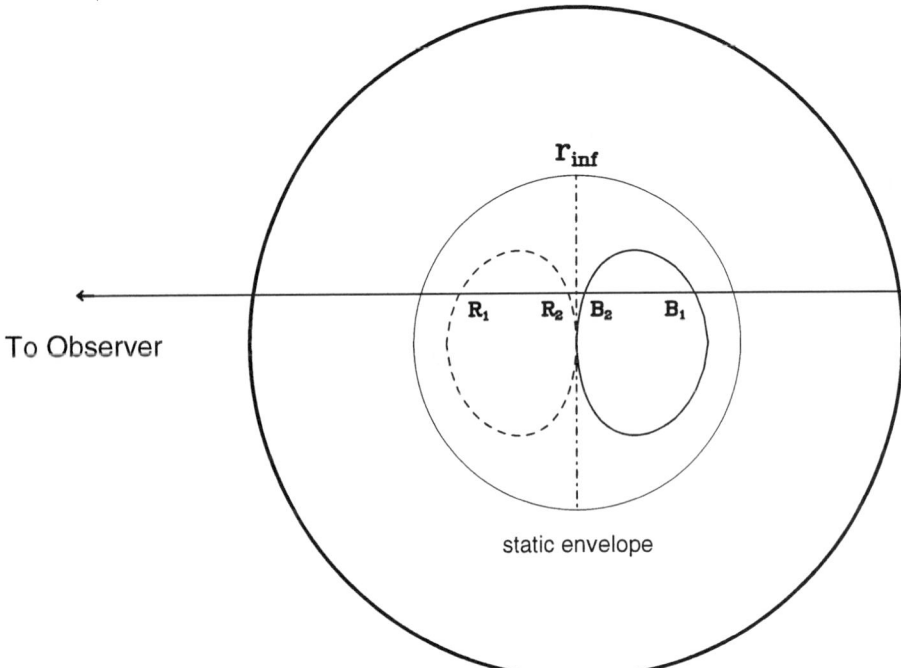

Figure 6 A schematic explanation of why line profiles of optically thick, high-excitation lines are skewed to the blue in a collapsing cloud. The *ovals* are loci of constant line-of-sight velocity, for $v(r) \propto r^{-0.5}$. Each line of sight intersects these loci at two points. The point closer to the center will have a higher T_{ex}, especially in lines that are hard to excite, so that $T_{ex}(R_2) > T_{ex}(R_1)$ and $T_{ex}(B_2) > T_{ex}(B_1)$. If the line is sufficiently opaque, the point R_1 will obscure the brighter R_2, but B_2 lies in front of B_1. The result is a profile with the blue peak stronger than the red peak (Zhou & Evans 1994).

Zhou et al 1996, Gregersen et al 1997), and *IRAS* 16293 (Zhou 1995, Narayanan et al 1998). Of this group, only *IRAS* 16293, rotating about 20 times faster than B335, is known to be a binary (Wootten 1989, Mundy et al 1992), supporting the idea that faster rotation is more likely to produce a binary. Mathieu (1994) reviews binarity in the pre–main-sequence stage, and Mundy et al (2000) discuss recent evidence on the earlier stages.

Interferometric observations have also revealed infall motions and rotational motions on scales of ∼1000 AU in several sources (e.g. Ohashi et al 1997a, Momose et al 1998). Such studies can reveal how matter makes the transition from infall to a rotating disk. Inevitably, irregularities in the density and velocity fields will confuse matters in real sources, and these may be more noticeable with interferometers. Outflows are particularly troublesome (Hogerheijde et al 1998). Extreme blue/red ratios are seen in interferometric observations of HCO^+ and HCN $J = 1 \to 0$ lines, which are difficult to reproduce with standard models

(Choi et al 1999). Even in B335, the best case for collapse, Velusamy et al (1995) found evidence for clumpy structure within the overall gradients. In addition, very high-resolution observations of CS $J = 5 \rightarrow 4$ emission toward B335 are not consistent with predicted line profiles very close to the forming star (DJ Wilner, PC Myers, D Mardones, M Tafalla, manuscript in preparation); either CS is highly depleted in the infalling gas, or the velocity or density fields depart from the model.

If, for the sake of argument, we accept a blue profile as a collapsing core, can we see any evolutionary trends? A number of surveys for blue profiles have been undertaken recently. Gregersen et al (1997) found 9 blue profiles in 23 Class-0 sources, using the $J = 3 \rightarrow 2$ and $J = 4 \rightarrow 3$ lines of HCO^+ and $H^{13}CO^+$. After consideration of possible confusion by outflows etc, they identified six sources as good candidates for collapse. Mardones et al (1997) extended the search to Class-I sources with $T_{bol} \lesssim 200$ K, using CS and H_2CO lines. They introduced the line asymmetry as a collapse indicator:

$$\delta V = (V_{thick} - V_{thin})/\Delta V_{thin}, \qquad (16)$$

where V_{thick} is the velocity of the peak of the opaque line, V_{thin} is the velocity of the peak of the optically thin line, and ΔV_{thin} is the line width of the thin line. They confirmed many of the collapse candidates found by Gregersen et al (1997) and identified six more, but they found very few collapse candidates among the Class-I sources. The difference could be caused by using different tracers, because the CS and H_2CO lines are less opaque than the HCO^+ lines. To remove this uncertainty, Gregersen (1998) surveyed the Class-I sources of Mardones et al (1997) in HCO^+. Using δV as the measure, the fraction of blue profiles did not decrease substantially between Class 0 and Class I. Most of these line profiles need further observations before they become bona fide candidates.

When does collapse begin? Surveys of the Class -1 cores might reveal very early stages of collapse. In the inside-out collapse picture, blue profiles should appear, if at all, only toward the center. In fact, blue profiles have been seen in a substantial number of these cores (Gregersen 1998, Lee et al 1999), and maps of one, L1544, show that the blue profiles are very extended spatially (Tafalla et al 1998). Clearly, extended blue profiles in Class -1 cores do not fit expectations for early collapse, and Tafalla et al argue that the velocities are higher than expected for ambipolar diffusion. If the regions producing the blue and red peaks are indeed approaching one another, they are forming a core more rapidly than in standard models, suggesting that some new ideas may be necessary (e.g. Nakano 1998, Myers & Lazarian 1998). For a current review of this field, see Myers et al (2000).

4.8 Summary of Isolated Star Formation

Distinct cores can be identified in tracers of dense ($n \geq 10^4$ cm^{-3}) gas; these cores are frequently associated with star formation. There is no clear evidence that they are magnetically subcritical, and some kinematic evidence suggests that the decay of turbulence, rather than ambipolar diffusion, is the critical feature.

An empirical evolutionary sequence, based on the spectral appearance of dust emission, and detailed theoretical models are now being tested by observations. The spatial distribution of dust emission is providing important tests by probing $n(r)$. Predictions of the evolution of spectral lines during collapse are available for the simplest theory, and observations of some sources match predictions of theory quite well. Evidence of collapse is now strong in a few cases, and surveys for distinctive line profiles have revealed many more possible cases.

5. CLUSTERED STAR FORMATION AND MASSIVE STARS

In this section, we address the issues of clustered formation, regardless of mass, and high-mass star formation, which seems to occur exclusively in clusters. In this review, the term cluster refers to a group of forming stars, regardless of whether the eventual outcome is a bound cluster. Since high-mass stars are rare, the nearest examples are more distant than is the case for low-mass-star formation. Together with the fact that they form in clusters, the greater distance makes it difficult to isolate individual events of star formation. On the other side of the balance, massive stars are more easily detectable at large distances because luminosity is such a strong function of mass. Heating of the surroundings makes them strong emitters in both dust and many spectral lines; the spectra of regions forming massive stars are often very rich, reflecting both a high excitation state and, in many cases, enhanced abundances, as complex chemistry is driven by elevated temperatures (van Dishoeck & Blake 1998). These features led early studies, when sensitivity was poor, to concentrate on regions forming massive stars. However, most of these advantages arise because the star is strongly influencing its surroundings; if we wish to know the preconditions, we will be misled. This problem is aggravated by the fast evolution of massive stars to the main sequence. Reviews of the topic of clustered star formation can be found in Elmegreen (1985), Lada & Lada (1991), and Lada (1999). Reviews focusing on the formation of massive stars include Churchwell (1993, 1999), Walmsley (1995), Stahler et al (2000), and Kurtz et al (2000).

5.1 Theoretical Issues

Some of the primary theoretical issues regarding the formation of massive stars have been reviewed by Stahler et al (2000). First, what is the relevant dividing line between low-mass and high-mass stars? For the star formation problem, the question is how far in mass the scenario for low-mass stars can be extended. Theoretically, the limit is probably about 10 $M_\odot$, where stars reach the main sequence before the surrounding envelope is dissipated. Observations of the physical conditions in regions forming intermediate mass stars (Herbig Ae/Be stars and their more embedded precursors) can reveal whether modifications are needed at even lower masses. Because accretion through disks plays a crucial role in the standard model, it is important to know the frequency and properties of disks around more massive stars.

Stars as massive as 100 $M_\odot$ seem to exist (Kudritzki et al 1992), but radiation pressure from the rapidly evolving stellar core should stop accretion before such masses can be built (e.g. Wolfire & Cassinelli 1987). In addition, massive stars produce very strong outflows (Shepherd & Churchwell 1996, Bachiller 1996). Because standard accretion theory cannot produce rates of mass accretion high enough to overwhelm these dispersive effects, some new effects must become important.

Related questions concern the formation of clusters. To what extent can the ideas of isolated core collapse be applied if there are competing centers of collapse nearby? If the collapse to form massive stars is supercritical, a whole region may collapse and fragment to form many stars.

The fact that the most massive stars are found near the centers of forming clusters has led to the suggestion that massive stars are built by collisional coalescence of stars or protostars (Bonnell et al 1997, 1998). This scheme requires high densities $(n_* \geq 10^4$ stars pc$^{-3})$; for a mean stellar mass of 1 $M_\odot$, this corresponds to $n \geq 2 \times 10^5$ cm^{-3}. Even higher stellar densities are seen in the core of the Orion Nebula Cluster (Hillenbrand & Hartmann 1998).

The special problems of making the most massive stars are a subset of the larger question of explaining the mass distribution of all stars. There may be variations in the IMF between clusters (Scalo 1998), which can test theories. Hipparcos observations of nearby OB associations have extended the membership to lower-mass stars (de Zeeuw et al 1999), suggesting total masses of a few 10^3 $M_\odot$.

The main questions these issues raise for observations are whether the mass and density of cores are sufficient for forming clusters and massive stars and whether the mass distribution of clumps in a star-forming region can be related to the mass distribution of stars ultimately formed.

5.2 Overall Cloud and Core Properties

What do we know about the general properties of the galactic clouds? The broadest picture is provided by surveys of CO and ^{13}CO. Surveys of significant areas of the galaxy indicate that the power-law distribution in mass seen for small clouds $[dN(M) \propto M^{-\alpha}dM$; Section 3.1] continues up to a cutoff at $M \sim 6 \times 10^6$ $M_\odot$ (Williams & McKee 1997). Studies by Casoli et al (1984), Solomon & Rivolo (1989), and Brand & Wouterloot (1995) find $1.4 \leq \alpha \leq 1.8$ over both inner and outer areas of the galaxy. Extinction surveys find flatter slopes (Scalo 1985), but Scalo & Lazarian (1996) suggest that cloud overlap affects the extinction surveys. The fact that $\alpha < 2$ implies that most of the mass is in the largest structures, although there are issues of how to separate clouds at the largest scales. Because the star formation rate per unit mass, measured by CO, appears not to depend on cloud mass (Mead et al 1990, Evans 1991), the mass distribution supports the idea that most stars form in massive clouds (Elmegreen 1985). However, the enormous spread ($>10^2$) in star formation rate per unit mass at any given mass, together with the fact that most of the molecular gas is sterile, suggests that comparisons to overall cloud masses, measured by CO, are not particularly relevant.

Surveys in molecular lines indicative of denser gas have generally been biased towards signposts of star formation. Exceptions are the CS $J = 2 \to 1$ survey of L1630 (Lada et al 1991) and the CS $J = 1 \to 0$ and $J = 2 \to 1$ surveys of L1641 (Tatematsu et al 1993, 1998). These two clouds, also called Orion B and Orion A, are adjacent. In both cases, the CS $J = 2 \to 1$ maps showed more contrast than maps of ^{13}CO (Bally et al 1987), with CS $J = 1 \to 0$ being somewhat intermediate. Maps of higher-J transitions have been less complete, but show still less area covered by emission. The CS surveys detected <20% of the total mass in both clouds (Lada et al 1991; Tatematsu, personal communication). The $J = 2 \to 1$ emission in L1641 is somewhat smoother than the $J = 2 \to 1$ emission from L1630 (Tatematsu et al 1998), and this difference may be reflected in the distribution of star formation. Star formation in L1641 appears to include a distributed component (Strom et al 1993); in contrast, star formation in L1630 is tightly concentrated in clusters associated with massive cores of dense gas (Lada 1992, Li et al 1997).

Excitation analysis of CS lines with higher critical density in L1630 shows that the star-forming regions all contain gas with $n \geq 10^5$ cm^{-3} (Lada et al 1997). These results suggest that surveys in lines of high $n_c(jk)$ are relevant for characterizing star formation regions. Of the possible tracers, CS (e.g. Tatematsu et al 1993) and NH$_3$ (e.g. Harju et al 1993) have been most widely surveyed. In comparison to cores in Taurus, where only low-mass stars are forming, cores in the Orion clouds tend to be more massive and to have larger line widths when observed with the same tracer. The differences are factors of 2–4 for the majority of the cores, but larger for the cores near the Orion Nebula and those forming clusters in L1630 (Lada et al 1991).

CS transitions have been surveyed toward ultracompact (UC) HII regions, H$_2$O masers, or luminous *IRAS* sources (see Kurtz et al 2000). Since the *IRAS* survey became available, most samples are drawn from the *IRAS* catalog with various color selection criteria applied. The most complete survey (Bronfman et al 1996) was toward *IRAS* sources with colors characteristic of UC HII regions (Wood & Churchwell 1989) over the entire galactic plane. Bronfman et al found CS $J = 2 \to 1$ emission (see Table 1 for density sensitivity) in 59% of 1427 *IRAS* sources, and the undetected sources were either weak in the far-infrared or had peculiar colors. Searches toward H$_2$O masers have used the catalogs of Braz & Epchtein (1983) and Cesaroni et al (1988). Surveys of CS $J = 2 \to 1$ (Zinchenko et al 1995, Juvela 1996) toward southern H$_2$O masers found detection rates of ∼100%, suggesting that dense, thermally excited gas surrounds the compact, ultradense regions needed to produce H$_2$O masers. The detection rate drops in higher-J transitions of CS (Plume et al 1992, 1997) but is still 58% in the CS $J = 7 \to 6$ line, which probes higher densities (Table 1). An LVG, multitransition study of CS lines found

$\langle \log n(\text{cm}^{-3}) \rangle = 5.9$ for 71 sources and a similar result from a smaller sample using $C^{34}S$ data (Plume et al 1997). Densities derived assuming LVG fall between the average and maximum densities in clumpy models with a range of densities (Juvela 1997, 1998).

Maps of the cores provide size and mass information. Based on the sizes and masses of 28 cores mapped in the CS $J = 2 \rightarrow 1$ line (Juvela 1996), one can compute a mean size, $\langle l \rangle = 1.2 \pm 0.5$ pc; mean virial mass, $\langle M_V \rangle \sim 5500$ M$_\odot$; and $\langle M_N \rangle \sim 4900$ M$_\odot$. Although the two mass estimates agree on average, there can be large differences in individual cases. Remarkably, cloud structure does not introduce a big uncertainty into the cloud masses; using a clumpy cloud model, (see Section 5.4), Juvela (1998) found that M_N increased by a factor of 2 on average compared with homogeneous models and agreed with M_V to within a factor of 2. Plume et al (1997) obtained similar results from strip maps of CS $J = 5 \rightarrow 4$: $\langle l \rangle = 1.0 \pm 0.7$ pc (average over 25 cores); $\langle M_V \rangle = 3800$ M$_\odot$ (16 cores). As usual mean values must be regarded with caution; there is a size distribution. As cores with weaker emission are mapped, the mean size decreases; an average of 30 cores with full maps of $J = 5 \rightarrow 4$ emission gives $\langle l \rangle = 0.74 \pm 0.56$ pc, with a range of 0.2–2.8 pc (Y Shirley, unpublished results).

Churchwell et al (1992) surveyed 11 UC HII regions for CS $J = 2 \rightarrow 1$ and $J = 5 \rightarrow 4$ emission, leading to estimates of $n \geq 10^5$ cm^{-3}. Cesaroni et al (1991) surveyed 8 UC HII in three transitions of CS and $C^{34}S$ and estimated typical sizes of 0.4 pc, masses of 2000 M$_\odot$, and densities of 10^6 cm^{-3}. More extensive surveys have been made in NH$_3$; Churchwell et al (1990) found NH$_3$ $(J, K) = (1, 1)$ and (2,2) emission from 70% of a sample of 84 UC HII regions and *IRAS* sources with similar colors. They derived T_K, finding a peak in the distribution around 20 K, but a significant tail to higher values. Further studies in the (4,4) and (5,5) lines toward 16 UC HII regions with strong (2,2) emission (Cesaroni et al 1992) detected a high fraction. Estimates for T_K ranged from 64 to 136 K and sizes of 0.5 pc. Two sources indicated much higher densities and NH$_3$ abundances. Follow-up studies with the VLA (Cesaroni et al 1994a, 1998) revealed small (~ 0.1 pc), hot ($T_K \sim 50$–200 K), dense ($n = 10^7$ cm^{-3}) regions with enhanced NH$_3$ abundances. These hot cores (discussed below) are slightly displaced from the UC HII, but coincide with H$_2$O masers.

Magnetic-field strengths have been measured with the Zeeman effect toward about 10 regions of massive-star formation. The fields are substantially stronger than the fields seen in isolated, low-mass cores, but the masses are also much higher. In most cases the mass-to-flux ratio is comparable to the critical ratio, once geometrical effects are considered (Crutcher 1999b). Given the uncertainties and sample size, it is too early to decide whether regions forming massive stars are more likely to be supercritical than regions forming only low-mass stars.

The ionization fraction in the somewhat more massive Orion cores appears to be very similar to that in low-mass cores: $-6.9 < \log x_e < -7.3$ (Bergin et al 1999). The most massive cores in their sample have $x_e \leq 10^{-8}$, as do some of the massive cores studied by de Boisanger et al (1996). Expressed in terms of column density, the decline in x_e appears around $N \sim 3 \times 10^{22}$ cm^{-2}. At $x_e = 10^{-8}$, $t_{AD} = 7 \times 10^5$ years, $\sim 0.1 t_{AD}$ in isolated, low-mass cores. Even if the massive cores are subcritical, their evolution should be faster than that of low-mass cores.

To summarize, the existing surveys show ample evidence that massive star formation usually takes place in massive ($M > 10^3$ M$_\odot$), dense ($n \sim 10^6$ cm^{-3}) cores, consistent with the requirements inferred from the study of young clusters and associations, and with conditions needed to form the most massive stars by mergers. Cores with measured B_z seem to be near the boundary between subcritical and supercritical cores.

5.3 Evolutionary Scenarios and Detailed Theories

To what extent can an evolutionary scenario analogous to the class system be constructed for massive-star formation? Explicit attempts to fit massive cores into the class system have relied on surveys of *IRAS* sources. Candidates for massive Class-0 objects, with $L > 10^3$ L$_\odot$, have been found (Wilner et al 1995, Molinari et al 1998). One difficulty with using the shape of the spectral energy distribution for massive-star formation is that dense, obscuring material usually surrounds objects even after they have formed, and a single star may be quite evolved but still have enough dust in the vicinity to have the same spectral energy distribution as a much younger object. The basic problem is the difficulty in isolating single objects. Also, the role of disks in massive regions is less clear, and they are unlikely to dominate the spectrum, as they do in low-mass Class-II sources. Other markers, such as the detection of radio continuum emission, must be used as age indicators. Hot cores provide obvious candidates to be precursors of UC HII regions, but some have embedded UC HII regions and may be transitional (Kurtz et al 2000). The chemical state of massive cores may also provide an evolutionary sequence; Helmich et al (1994) suggested an evolutionary ordering of three sources in the W3 region based on their molecular spectra and chemical models (see van Dishoeck & Blake 1998).

Although theories for clustered- and massive-star formation are much less developed, some steps have been taken (e.g. Bonnell et al 1998, Myers 1998). The larger Δv in regions forming massive stars imply that turbulence must be incorporated into the models. Myers & Fuller (1992) suggested a "thermal-non-thermal" (TNT) model, in which $n(r)$ is represented by the sum of two power laws and the term with $p = 1$ dominates outside the radius (r_{TNT}) at which turbulent motions dominate thermal motions. McLaughlin & Pudritz (1997) develop the theory of a logatropic sphere, which has a density distribution approximated by $p = 1$. Collapse in such a configuration leads to power laws in the collapsing region with similar form to those in the collapsing isothermal sphere, but with higher densities and lower velocities. These ideas lead to accretion rates that increase with time and timescales for forming massive stars that are much weaker functions of the final stellar mass than is the case for the isothermal sphere (Section 4.5). Recent simulations of unmagnetized fragmentation that follow the interaction of clumps find that the mass spectrum of fragments steepens from $\alpha = 1.5$ to a lognormal distribution of the objects likely to form stars (e.g. Klessen et al 1998). To avoid an excessive global star formation rate (Section 2) and distortion of the clump mass spectrum in the bulk of the cloud, this process must be confined to star-forming regions in clouds.

5.4 Filaments, Clumps, Gradients, and Disks

An important issue is whether the dense cores have overall density gradients or internal structures (clumps) that are likely to form individual stars and whether the mass distribution is like that of stars. Unfortunately, the terms "clumps" and "cores" have no standard usage; I generally use "cores" to refer to regions that appear in maps of high-excitation lines and "clumps" to mean structures inside cores, except where a different usage is well established (e.g. "hot cores"). Myers (1998) has suggested the term "kernels" to describe clumps within cores. Cores themselves are usually embedded in structures traced by lines with lower critical density, and these are also called clumps by those who map in these lines. Many of these low-density structures are quite filamentary in appearance; examples include the ^{13}CO maps of L1641 (Orion B) of Bally et al (1987). In some cases, this filamentary structure is seen on smaller scales in tracers of high density or column density (Figure 2; e.g. Johnstone & Bally 1999).

The clumpy structure of molecular clouds measured in low-excitation lines suggests that dense cores will be clumpy as well. Suggestions of clumpy structure came from early work comparing densities derived from excitation analysis in different tracers, but smooth density gradients provided an alternative (e.g. Evans 1980). Multitransition studies of three cores in CS (Snell et al 1984), $C^{34}S$ (Mundy et al 1986), and H_2CO (Mundy et al 1987) found no evidence for overall density gradients; the same high densities were derived over the face of the core, while the strength of the emission varied substantially. This was explained in a clumpy model with clump-filling factors of the dense gas $f_v \sim$ 0.03–0.3, based on a comparison of M_V with M_n (Snell et al 1984). This comparison forms the basis for most claims of unresolved clumps.

Observations at higher resolution support the idea of clumps postulated by Snell et al (1984). For example, Stutzki & Güsten (1990) deconvolved 179 clumps from a map of $C^{18}O$ $J = 2 \rightarrow 1$ emission near M17. Because of overlap, far fewer clumps are apparent to the eye; assumptions about the clump shape and structure may affect the deconvolution. Maps of the same source in several CS and $C^{34}S$ lines (Wang et al 1993) could be reproduced with the clump catalog of Stutzki & Güsten, but only with densities about 5 times higher than they found. Thus the clumps themselves must have structure, either a continuation of clumpiness or smooth gradients. Because the inferred clumps are now similar in size to the cores forming low-mass stars, a natural question is whether massive cores are fragmented into many clumps which can be modeled as if they were isolated cores. In favor of this view, Stutzki & Güsten (1990) noted that the Jeans length was similar to the size of their clumps.

A significant constraint on this picture is provided by the close confinement of the clumps; unlike the picture of isolated-core formation, the sphere of influence of each clump will be limited by its neighbors. A striking example is provided by the dust continuum maps of the ρ Ophiuchi cloud (Figure 1), our nearest example of cluster formation, albeit with no very massive stars. Within about six cores of size 0.2 pc, Motte et al (1998) find about 100 structures with sizes of 1000–4000 AU. They deduce a fragmentation scale of 6000 AU, five times smaller than isolated cores in Taurus. Thus the "feeding zone" of an individual clump is considerably less, and the evolution must be more dynamic, with frequent clump-clump interactions, than is the case for isolated-star formation. This picture probably applies even more strongly to the more massive cores. In ρ Ophiuchi, the clump mass spectrum above 0.5 $M_\odot$ steepens to $\alpha = 2.5$, close to the value for stars (Motte et al 1998), in agreement with predictions of Klessen et al (1998). A similar result ($\alpha = 2.1$) is found in Serpens, using millimeter interferometry (Testi & Sargent 1998). Although more such studies are needed, these results suggest that dust continuum maps do trace structures that are likely precursors of stars, opening the study of the origin of the IMF to direct observational study.

Some of the less massive, relatively isolated cores, such as NGC2071, S140, and GL2591, have been modeled with smooth density and temperature gradients (Zhou et al 1991, 1994b; Carr et al 1995; van der Tak et al 1999). Models with gradients can match the relative strengths of a range of transitions with different excitation requirements, improving on homogeneous models. Zhou et al (1994) summarized attempts to deduce gradients and found preliminary evidence that, as core mass increases, the tendency is first toward smaller values of p. The most massive cores showed little evidence for any overall gradient and more tendency toward clumpy substructure. This trend needs further testing, but it is sensible if more massive cores form clusters. Lada et al (1997) found that the L1630 cores forming rich embedded clusters with high efficiency tended to have larger masses of dense ($n > 10^5$ cm^{-3}) gas, but a lower volume filling factor of such gas, indicating more fragmentation.

However, the line profiles of optically thick lines predicted by models with gradients are usually self-absorbed, while the observations rarely show this feature in massive cores. Clumps within the overall gradients are a likely solution. The current state of the art in modeling line profiles in massive cores is the work of Juvela (1997, 1998), who has constructed clumpy clouds from both structure tree and fractal models, performed three-dimensional radiative transport and excitation, and compared the model line profiles to observations of multiple CS and $C^{34}S$ transitions in massive cores. He finds that the clumpy models match the line profiles much better than nonclumpy models, especially if macroturbulence dominates microturbulence. Structure trees (Houlahan & Scalo 1992) match the data better than fractal models, but overall density and/or temperature gradients with $p + q \approx 2$ are needed in addition to clumps.

The study of gradients versus clumps in regions forming intermediate-mass stars could help to determine whether conditions change qualitatively for star formation above some particular mass and how this change is related to the outcome. Using near-infrared observations of regions with Herbig Ae/Be stars, Testi et al (1997) found that the cluster mode of star formation becomes dominant when the most massive star has a spectral type earlier than B7. Studies of the far-infrared emission from dust remaining in envelopes around Herbig Ae/Be stars found values of p ranging from 0.5 to 2 (Natta et al 1993). Maps of dust continuum emission illustrate the difficulties—the emission may not peak on the visible star, but on nearby, more embedded objects (Henning et al 1998, Di Francesco et al 1998). Detailed models of several suitable sources yield $p = 0.75$ to 1.5 (Henning et al 1998, Colomé et al 1996). Further work is needed to determine whether a change in physical conditions can be tied to the change to cluster mode.

Many Herbig Ae stars have direct evidence for disks from interferometric studies of dust emission (Mannings & Sargent 1997), although fewer Herbig Be stars have such direct evidence (Di Francesco et al 1997). For a review, see Natta et al (2000). Disks may be more common during more embedded phases of B star formation; Shepherd & Kurtz (1999) have found a large (1000-AU) disk around an embedded B2 star. The statistics of UC HII regions may provide indirect evidence for disks around more massive stars. Because such regions should expand rapidly unless confined, the large number of such regions posed a puzzle (Wood & Churchwell 1989). Photoevaporating disks have been suggested as a solution (Hollenbach et al 1994). Such disks have also been used to explain very broad recombination lines (Jaffe & Martín-Pintado 1999). Kinematic evidence for disks will be discussed in Section 5.5.

A particular group of clumps (or cores) deserves special mention, hot cores (e.g. Ohishi 1997). First identified in the Orion cloud (Genzel & Stutzki 1989), ~20 are now known (see Kurtz et al 2000). They are small regions ($l \sim 0.1$ pc), characterized by $T_K > 100$ K, $n > 10^7$ cm^{-3}, and rich spectra, probably reflecting enhanced abundances, as well as high excitation (van Dishoeck & Blake 1998). Theoretical issues have been reviewed by Millar (1997) and Kaufman et al (1998), who argue that hot cores are likely to be heated internally, but they often lack radio continuum emission. Because dynamical timescales for gas at such densities are short, these may plausibly be precursors to the UC HII regions.

The evidence for flatter density distribution in regions of intermediate mass supports the relevance of models like the TNT or logatropic sphere models in massive regions (Section 5.3), but it will be important to study this trend with the same methods now being applied to regions forming low-mass stars, with due regard for the greater distance to most regions forming massive stars. The increasingly fragmented structure in more massive cores and the increased frequency of clusters above a certain mass are consistent with a switch to a qualitatively different mode of star formation, for which different theories are needed. Finally, the common appearance of filaments may support a continuing role for turbulence in dense regions, because simulations of turbulence often produce filamentary structures (e.g. Scalo et al 1998).

5.5 Kinematics

Lada et al (1991) found only a weak correlation between line width and size, disappearing entirely for a different clump definition, in the L1630 cores (Goodman Type-2 relation, see Section 4.3). Caselli & Myers (1995) also found that the Type-1 line width-size relation (with only nonthermal motions included, $\Delta v_{NT} \propto R^\gamma$) is flatter in massive cloud cores ($\gamma = 0.21 \pm 0.03$) than in low-mass cores ($\gamma = 0.53 \pm 0.07$); in addition, the correlation is poor (correlation coefficient of 0.56) although the correlation is better for individual cores (Type-3 relations). Caselli & Myers (1995) also noted that nonthermal (turbulent) motions are much more dominant in more massive cores and find good agreement with predictions of the TNT model. The "massive" cores in the Caselli & Myers study are mostly the cores in Orion with masses between 10 and 100 $M_\odot$. The much more massive ($\langle M_V \rangle = 3800$ $M_\odot$) cores studied by Plume et al (1997) exhibit no statistically significant line width-size relation at all (correlation coefficient is 0.26) and the line widths are systematically higher (by factors of 4–5) for a given size than would be predicted by the relationships derived for low- and intermediate-mass cores (Caselli & Myers 1995). In addition, the densities of these cores exceed by factors of 100 the predictions of density-size relations found for less massive cores (Myers 1985). The regions forming truly massive stars are much more dynamic, as well as much denser, than would be expected from scaling relations found in less massive regions. The typical line width in massive cores is 6–8 km s^{-1}, corresponding to a one-dimensional velocity dispersion of 2.5–3.4 km s^{-1}, similar to that of the stars in the Orion Nebula Cluster (Hillenbrand & Hartmann 1998). Larson (1981) noted that regions of massive-star formation, like Orion and M17, did not follow his original line width-size relation, suggesting that gravitational contraction would decrease size while keeping Δv roughly constant or increasing it.

Searching for collapse in massive cores is complicated by the turbulent, clumpy structure, with many possible centers of collapse, outflow, etc. A collapse signature may indicate an overall collapse of the core, with accompanying fragmentation. In fact, self-absorbed line profiles from regions forming massive stars are rather rare (Plume et al 1997). A possible collapse signature has been seen in CS emission toward NGC 2264 IRS, a Class-I source with $L \sim 2000$ $L_\odot$ (Wolf-Chase & Gregersen 1997). If an HII region lies at the center of a collapsing core, absorption lines should trace only the gas in front and should be redshifted relative to the emission lines. Failure to see this effect, comparing H_2CO absorption to CO emission, supported an early argument against the idea that all clouds were collapsing (Zuckerman & Evans 1974). A more recent application of this technique to dense cores, using high-excitation lines of NH_3, showed no preference for red-shifted absorption (Olmi et al 1993) overall, but a few dense cores do show this kind of effect. These sources include W49 (Welch et al 1988, Dickel & Auer 1994), G10.6–0.4 (Keto et al 1988), and W51 (Zhang & Ho 1997, Zhang et al 1998a).

Dickel & Auer (1994) tested different collapse scenarios against observations of HCO^+ and favored free-fall collapse, with $n \propto r^{-1.5}$ and $v \propto r^{-0.5}$ throughout

W49A North; they noted that more complex motions are present on small scales. Keto et al (1988) used NH_3 observations with 0.3″ resolution to separate infall from rotational motions toward the UC HII region, G10.6–0.4. Zhang & Ho (1997) used NH_3 absorption and Zhang et al (1998a) added CS $J = 3 \rightarrow 2$ and CH_3CN observations to identify collapse onto two UC HII regions, W51e2 and W51e8, inferring infall velocities of about 3.5 km s^{-1} on scales of 0.06 pc. Young et al (1998) have tested various collapse models against the data on W51e2 and favor a nearly constant collapse velocity ($v \sim 5$ km s^{-1}) and $n(r) \propto r^{-2}$. Mass infall rates of about 6×10^{-3} (Zhang et al 1998a) to 5×10^{-2} (Young et al 1998) $M_\odot$ year^{-1} were inferred for W51e2. Similar results were found in G10.6–0.4 (Keto et al 1988), and even more extreme mass infall rates (10^{-2} to 1 $M_\odot$ year^{-1}) have been suggested for W49A, a distant source with enormous mass ($\sim 10^6 M_\odot$) and luminosity ($L \sim 10^7 L_\odot$) (Welch et al 1988). These high infall rates may facilitate formation of very massive stars (Section 5.1) and help confine UC HII regions (Walmsley 1995).

Large transverse velocity gradients have been seen in some hot cores, including G10.6–0.4 (Keto et al 1988), W51 (Zhang & Ho 1997), and *IRAS*20126 + 4104 (Cesaroni et al 1997, Zhang et al 1998b). For example, gradients reach 80 km s^{-1} pc^{-1} in the NH_3 $(J, K) = (4, 4)$ line and 400 km s^{-1} pc^{-1} in the CH_3CN $J = 6 \rightarrow 5$ line toward G29.96−0.02 and G31.41+0.31 (Cesaroni et al 1994b, 1998). Cesaroni et al (1998) interpret the gradients in terms of rotating disks extending to about 10^4 AU. By using the dust continuum emission at 3 mm, they deduce disk masses up to 4200 $M_\odot$ for G31.41+0.31.

5.6 Implications for Larger Scales

Since most stars form in the massive cores discussed in this section (Elmegreen 1985), they are most relevant to issues of star formation on a galactic scale. If the luminosity is used to trace star formation rate (e.g. Rowan-Robinson et al 1997, Kennicutt 1998), the star formation rate per unit mass is proportional to L/M. Considering clouds as a whole, $\langle L/M \rangle \sim 4$ in solar units (e.g. Mooney & Solomon 1988), with a spread exceeding a factor of 10^2 (Evans 1991). Using CS $J = 5 \rightarrow 4$ emission to measure M in the dense cores, Plume et al (1997) found $\langle L/M \rangle = 190$ with a spread of a factor of 15. The star formation rate per unit mass is much higher and less variable if one avoids confusion by the sterile gas. The average L/M seen in dense cores in our Galaxy is similar to the highest value seen in luminous infrared galaxies, where M is measured by CO (Sanders et al 1991, Sanders & Mirabel 1998). The most dramatic starburst galaxies behave as if their entire interstellar medium has conditions like those in the most active, massive dense cores in our Galaxy. Some studies (e.g. Mauersberger & Henkel 1989) indeed found strongly enhanced CS $J = 5 \rightarrow 4$ emission from starburst galaxies. It will be interesting to observe high-J CS lines in the most luminous galaxies to compare to the conditions in massive cores in our Galaxy. Perhaps the large scatter in L/M seen in galaxies will be reduced if CS, rather than CO, is used as a measure of the gas, ultimately leading to a better understanding of what controls galactic star formation rates (see Kennicutt 1998).

5.7 Summary of Clustered Star Formation

The cloud mass distribution found for lower-mass objects continues to massive clouds, but less is known about the distribution for dense cores. Very massive cores clearly exist, with sufficient mass ($M > 10^3$ $M_\odot$) to make the most massive clusters and associations. These cores are denser and much more dynamic than cores involved in isolated-star formation, with typical $n \sim 10^6$ cm^{-3} and line widths about 4–5 times larger than predicted from the line width-size relation. Pressures in massive cores (both thermal and turbulent) are substantially higher than in lower-mass cores. The densities match those needed to form the densest clusters and the most massive stars by coalescence. There are some cores with evidence of overall collapse, but most do not show a clear pattern. There is some evidence that more massive regions have flatter density profiles and that fragmentation increases with mass, but more studies are needed. High-resolution studies of nearby regions of cluster formation are finding many clumps, limiting the feeding zone of a particular star-forming event to $l \sim 6000$ AU, much smaller than the reservoirs available in the isolated mode. In some cases, the clump mass distribution approaches the slope of the IMF, suggesting that the units of star formation have been identified. Studies of intermediate-mass stars indicate that a transition to clustered mode occurs at least by a spectral type of B7.

6. CONCLUSIONS AND FUTURE PROSPECTS

The probes of physical conditions have been developed and are now fairly well understood. Some physical conditions have been hard to measure, such as the magnetic-field strength, or hard to understand, notably the kinematics. Star-forming structures or cores within primarily sterile molecular clouds can be identified by thresholds in column density or density. Stars form in distinct modes, isolated and clustered, with massive stars forming almost exclusively in a clustered mode. The limited number of measurements of magnetic field leave open the question of whether cores are subcritical or supercritical and whether this differs between the isolated and clustered mode. Cores involved in isolated-star formation may be distinguished from their surroundings by a decrease in turbulence to subsonic levels, but clustered-star formation occurs in regions of enhanced turbulence and higher density, compared with isolated star formation.

Evolutionary scenarios and detailed theories exist for the isolated mode. The theories assume cores with extended, power-law density gradients, leading to the mass accretion rate as the fundamental parameter. Detailed tests of these ideas are providing overall support for the picture, but also raising questions about the detailed models. Notably, kinematic evidence of gravitational collapse has finally been identified in a few cases. The roles of turbulence, the magnetic field, and rotation must be understood, and the factors that bifurcate the process into single- or multiple-star formation must be identified. Prospects for the future include a less biased census for cores in early stages and improved information on density gradients, both facilitated by the appearance of cameras at millimeter and submillimeter wavelengths on large telescopes. Antenna arrays operating at these wavelengths will provide more detailed information on the transition region between envelope and disk and will be used to study early disk evolution and binary fraction. Future, larger antenna arrays will probe disk structure to scales of a few AU. Finally, a closer coupling of physical and chemical studies with theoretical models will provide more pointed tests of theory.

Theories and evolutionary scenarios are less developed for the clustered mode, and our understanding of the transition between isolated and clustered mode is still primitive. Current knowledge suggests that more massive cores have flatter density distributions and a greater tendency to show substructure. At some point, the substructure dominates, and multiple centers of collapse develop. With restricted feeding zones, the mass accretion rate gives way to the mass of clump as the controlling parameter, and some studies of clump mass spectra suggest that the stellar IMF is emerging in the clump mass distribution. The most massive stars form in very turbulent regions of very high density. The masses and densities are sufficient to form the most massive clusters and to explain the high stellar density at the centers of young clusters. They are also high enough to match the needs of coalescence theories for the formation of the most massive stars. As with isolated regions, more and better measurements of magnetic fields are needed, along with a less biased census, particularly for cool cores that might represent earlier stages. Larger antenna arrays will be able to separate clumps in distant cores and determine mass distributions for comparison with the IMF. Larger airborne telescopes will provide complementary information on luminosity sources in crowded regions. Observations with high spatial resolution and sensitivity in the mid-infrared will provide clearer pictures of the deeply embedded populations, and mid-infrared spectroscopy with high spectral resolution could trace kinematics close to the forming star. Deeper understanding of clustered star formation in our Galaxy will provide a foundation for understanding the origin and evolution of galaxies.

ACKNOWLEDGMENTS

I am grateful to P André, J Bally, M Choi, F Motte, D Johnstone, and L Looney for supplying figures. Many colleagues sent papers in advance of publication and/or allowed me to discuss results in press or in progress. A partial list includes P André, R Cesaroni, R Crutcher, D Jaffe, L Mundy, E Ostriker, Y Shirley, J Stone, D Ward-Thompson, and D Wilner. I thank R Cesaroni, Z Li, P Myers, F Shu, F van der Tak, and M Walmsley for detailed, helpful comments on an earlier version. This work has been supported by the State of Texas and NASA, through grants NAG5-7203 and NAG5-3348.

Visit the Annual Reviews home page at http://www.AnnualReviews.org

LITERATURE CITED

Adams FC. 1990. *Ap. J.* 363:578–88
Adams FC. 1991. *Ap. J.* 382:544–54
Adams FC, Lada CJ, Shu FH. 1987. *Ap. J.* 312:788–806
Adelson L, Leung CM. 1988. *MNRAS* 235:349–64
Akeson RL, Carlstrom JE. 1997. *Ap. J.* 491:254–66
Alves J, Lada CJ, Lada EA. 1999. *Ap. J.* 515:265–74
Alves J, Lada CJ, Lada EA, Kenyon S, Phelps R. 1998. *Ap. J.* 506:292–305
André P, Barsony M, Ward-Thompson D. 2000. In *Protostars and Planets IV*, ed. V Mannings, A Boss, S Russell. Tucson: Univ. Arizona. In press
André P, Montmerle T. 1994. *Ap. J.* 420:837–62
André P, Ward-Thompson D, Barsony M. 1993. *Ap. J.* 406:122–41
André P, Ward-Thompson D, Motte F. 1996. *Astron. Astrophys.* 314:625–35
Arons J, Max CE. 1975. *Ap. J. Lett.* 196:L77–81
Arquilla R, Goldsmith PF. 1985. *Ap. J.* 297:436–54
Arquilla R, Goldsmith PF. 1986. *Ap. J.* 303:356–74
Bachiller R. 1996. *Annu. Rev. Astron. Astrophys.* 34:111–54
Ballesteros-Paredes J, Vázquez-Semadini E, Scalo J. 1999. *Ap. J.* 515:286–303
Bally J, Stark A, Wilson RW, Langer WD. 1987. *Ap. J. Lett.* 312:L45–49
Barnard EE. 1927. In *A Photographic Atlas of Selected Regions of the Milky Way*, ed. EB Frost, MR Calvert. Washington D.C.: Carnegie Inst. 134 pp.
Barranco JA, Goodman AA. 1998. *Ap. J.* 504:207–22
Barrett AH, Meeks ML, Weinreb S. 1964. *Nature* 202:475–76
Bash FN, Green E, Peters WL III. 1977. *Ap. J.* 217:464–72
Basu S. 1998. *Ap. J.* 509:229–37
Basu S, Mouschovias TC. 1995. *Ap. J.* 453:271–83
Beckwith S, Persson SE, Neugebauer G, Becklin EE. 1978. *Ap. J.* 223:464–70
Beichman CA, Myers PC, Emerson JP, Harris S, Mathieu R, et al. 1986. *Ap. J.* 307:337–49
Benson PJ, Caselli P, Myers PC. 1998. *Ap. J.* 506:743–57
Benson PJ, Myers PC. 1989. *Ap. J. Suppl.* 71:89–108
Bergin EA, Langer WD. 1997. *Ap J.* 486:316–28
Bergin EA, Plume R, Williams JP, Myers PC. 1999. *Ap. J.* 512:724–39
Bernes C. 1979. *Astron. Astrophys.* 73:67–73
Bertoldi F, McKee C. 1992. *Ap. J.* 395:140–57
Bertoldi F, McKee C. 1996. In *Amazing Light*, ed. RY Chiao, pp. 41–53. New York: Springer-Verlag
Blair GN, Peters WL, Vanden Bout PA. 1975. *Ap. J. Lett.* 200:L161–64
Blitz L. 1993. In *Protostars and Planets III*, ed. EH Levy, JI Lunine, pp. 125–61. Tucson: Univ. Arizona Press
Blitz L, Magnani L, Mundy L. 1984. *Ap. J. Lett.* 282:L9–12
Blitz L, Williams JP. 1997. *Ap. J. Lett.* 488:L145–48
Bloemen H. 1989. *Annu. Rev. Astron. Astrophys.* 27:469–516
Bodenheimer P. 1995. *Annu. Rev. Astron. Astrophys.* 33:199–238
Bok BJ, Reilly EF. 1947. *Ap. J.* 105:255–57
Bonazzola S, Falgarone E, Heyvaerts J, Pérault M, Puget JL. 1987. *Astron. Astrophys.* 172:293–98
Bonnell I, Bastien P. 1993. *Ap. J.* 406:614–28
Bonnell IA, Bate MR, Clarke CJ, Pringle JE. 1997. *MNRAS* 285:201–8
Bonnell IA, Bate MR, Zinnecker H. 1998. *MNRAS* 298:93–102
Bontemps S, André P, Terebey S, Cabrit S.

1996. *Astron. Astrophys.* 311:858–72
Boss A. 1998. *Ap. J. Lett.* 501:L77–81
Bourke TL, Hyland AR, Robinson G, James SD, Wright CM. 1995. *MNRAS* 276:1067–84
Brand J, Wouterloot JGA. 1995. *Astron. Astrophys.* 303:851–71
Braz MA, Epchtein N. 1983. *Astron. Astrophys. Suppl.* 54:167–85
Brogan CL, Troland TH, Roberts DA, Crutcher RM. 1999. *Ap. J.* 515:304–22
Bronfman L, Cohen RS, Alvarez H, May J, Thaddeus P. 1988. *Ap. J.* 324:248–66
Bronfman L, Nyman LA, May J. 1996. *Astron. Astrophys. Suppl.* 115:81–95
Butner HM, Evans NJ II, Lester DF, Levreault RM, Strom SE. 1991. *Ap. J.* 376:636–53
Butner HM, Lada EA, Loren RB. 1995. *Ap. J.* 448:207–25
Carr JS, Evans NJ II, Lacy JH, Zhou S. 1995. *Ap. J.* 450:667–90
Caselli P, Myers PC. 1995. *Ap. J.* 446:665–86
Caselli P, Walmsley CM, Terzieva R, Herbst E. 1998. *Ap. J.* 499:234–49
Casoli F, Combes F, Gerin M. 1984. *Astron. Astrophys.* 133:99–109
Cesaroni R, Churchwell E, Hofner P, Walmsley CM, Kurtz S. 1994a. *Astron. Astrophys.* 288:903–20
Cesaroni R, Felli M, Testi L, Walmsley CM, Olmi L. 1997. *Astron. Astrophys.* 325:725–44
Cesaroni R, Hofner P, Walmsley CM, Churchwell E. 1998. *Astron. Astrophys.* 331:709–25
Cesaroni R, Olmi L, Walmsley CM, Churchwell E, Hofner P. 1994b. *Ap. J. Lett.* 435:L137–40
Cesaroni R, Palagi R, Felli M, Catarzi M, Comoretto G, et al. 1988. *Astron. Astrophys. Suppl.* 76:445–58
Cesaroni R, Walmsley CM, Churchwell E. 1992. *Astron. Astrophys.* 256:618–30
Cesaroni R, Walmsley CM, Kömpe C, Churchwell E. 1991. *Astron. Astrophys.* 252:278–90
Chandler CJ, Koerner DW, Sargent AI, Wood DOS. 1995. *Ap. J. Lett.* 449:L139–42
Chandler CJ, Sargent AI. 1997. In *From Stardust to Planetesimals*, ed. YJ Pendleton, AGGM Tielens, 122:25–36. San Francisco: Astron. Soc. Pacific
Chen H, Myers PC, Ladd EF, Wood DOS. 1995. *Ap. J.* 445:377–92
Cheung AC, Rank DM, Townes CH, Thornton DD, Welch WJ. 1968. *Phys. Rev. Lett.* 21:1701–5
Choi M, Evans NJ II, Gregersen E, Wang Y. 1995. *Ap. J.* 448:742–47
Choi M, Panis J-F, Evans NJ II. 1999. *Ap. J. Suppl.* 122:519–56
Churchwell E, Walmsley CM, Cesaroni R. 1990. *Astron. Astrophys. Suppl.* 83:119–44
Churchwell E, Walmsley CM, Wood DOS. 1992. *Astron. Astrophys.* 253:541–56
Churchwell EB. 1993. In *Massive Stars: Their Lives in the Interstellar Medium*, ed. JP Cassinelli, EB Churchwell, 35:35–44. San Francisco: Astron. Soc. Pacific
Churchwell EB. 1999. In *The Origin of Stars and Planetary Systems*, ed. CJ Lada, ND Kylafis. Dordrecht: Kluwer. In press
Chyba CF, Sagan C. 1992. *Nature* 335:125–32
Ciolek GE, Königl A. 1998. *Ap. J.* 504:257–79
Clemens DP, Barvainis R. 1988. *Ap. J. Suppl.* 68:257–86
Clemens DP, Sanders DB, Scoville NZ. 1988. *Ap. J.* 327:139–55
Clemens DP, Yun JL, Heyer MH. 1991. *Ap. J. Suppl.* 75:877–904
Colomé C, Di Francesco J, Harvey PM. 1996. *Ap. J.* 461:909–19
Combes F. 1991. *Annu. Rev. Astron. Astrophys.* 29:195–237
Crovisier J. 1999. In *Formation and Evolution of Solids in Space*, ed. JM Greenberg and A Li, p. 389. Dordrecht: Kluwer
Crutcher RM. 1998. *Astro. Lett. Comm.* 37:113–24
Crutcher RM. 1999a. *Ap. J.* 520:706–13
Crutcher RM. 1999b. In *Interstellar Turbulence, Proceedings of the 2nd Guillermo Haro Conference*, ed. J Franco, A Carraminana. Cambridge, U.K.: Cambridge Univ. Press. In press

Crutcher RM, Troland TH, Goodman AA, Heiles C, Kazès I, Myers PC. 1993. *Ap. J.* 407:175–84

Crutcher RM, Troland TH, Lazareff B, Kazès I. 1996. *Ap. J.* 456:217–24

Cunningham CR, Gear WK, Duncan WD, Hastings PR, Holland WS. 1994. *Proc. SPIE* 2198:638–49

de Boisanger C, Helmich FP, van Dishoeck EF. 1996. *Astron. Astrophys.* 310:315–27

de Zeeuw PT, Hoogerwerf R, de Bruijne JHJ, Brown AGA, Blaauw A. 1999. *Astron. J.* 117:354–99

Dickel HR, Auer LH. 1994. *Ap. J.* 437.222–38

Dickman RL. 1975. *Ap. J.* 202:50–57

Dickman RL. 1978. *Ap. J. Suppl.* 37:407–27

Di Francesco J, Evans NJ II, Harvey PM, Mundy LG, Butner HM. 1998. *Ap. J.* 509:324–49

Di Francesco J, Evans NJ II, Harvey PM, Mundy LG, Guilloteau S, Chandler CJ. 1997. *Ap. J.* 482:433–41

Doty SD, Leung CM. 1994. *Ap. J.* 424:729–47

Draine BT, McKee CF. 1993. *Annu. Rev. Astron. Astrophys.* 31:373–432

Egan MP, Leung CM, Spagna GR. 1988. *Comput. Phys. Comm.* 48:271–92

Elmegreen BG. 1985. In *Protostars and Planets II*, ed. DC Black, MS Matthews, pp. 33–58. Tucson: Univ. Arizona Press

Elmegreen BG, Falgarone E. 1996. *Ap. J.* 471:816–21

Elmegreen BG, Lada CJ. 1977. *Ap. J.* 214:725–41

Evans NJ II. 1980. In *Interstellar Molecules*, ed. BH Andrew, pp. 1–19. Dordrecht: Reidel

Evans NJ II. 1989. *Rev. Mex. Astr. Astrof.* 18:21–28

Evans NJ II. 1991. In *Frontiers of Stellar Evolution*, ed. DL Lambert, 20:45–95. San Francisco: Astron. Soc. Pacific

Evans NJ II, Blair GN, Beckwith S. 1977. *Ap. J.* 217:448–63

Evans NJ II, Kutner ML, Mundy LG. 1987. *Ap. J.* 323:145–53

Evans NJ II, Lacy JH, Carr JS. 1991. *Ap. J.* 383:674–92

Falgarone E, Lis DC, Phillips TG, Pouquet A, Porter DH, Woodward PR. 1994. *Ap. J.* 436:728–40

Falgarone E, Panis J-F, Heithausen A, Pérault M, Stutzki J, et al. 1998. *Astron. Astrophys.* 331:669–96

Fiedler RA, Mouschovias TC. 1993. *Ap. J.* 415:680–700

Field GB. 1978. In *Protostars and Planets*, ed. T Gehrels, pp. 243–64. Tucson: Univ. Ariz. Press

Foster PN, Chevalier RA. 1993. *Ap. J.* 416:303–11

Fierking MA, Langer WD, Wilson RW. 1982. *Ap. J.* 262:590–605

Fulkerson SA, Clark FO. 1984. *Ap. J.* 287:723–27

Fuller GA. 1989. *Molecular Studies of Dense Cores.* PhD thesis. Berkeley, Calif.: Univ. Calif. 280 pp.

Fuller GA, Myers PC. 1992. *Ap. J.* 384:523–27

Fuller GA, Myers PC. 1993. *Ap. J.* 418:273–86

Galli D, Shu FH. 1993a. *Ap. J.* 417:220–42

Galli D, Shu FH. 1993b. *Ap. J.* 417:243–58

Gammie CF, Ostriker EC. 1996. *Ap. J.* 466:814–30

Geballe T, Oka T. 1996. *Nature* 384:334–35

Genzel R, Stutzki J. 1989. *Annu. Rev. Astron. Astrophys.* 27:41–85

Genzel R. 1992. In *The Galactic Interstellar Medium*, ed. D Pfenniger, P Bartholdi, pp. 275–391. Berlin: Springer-Verlag

Goldreich P, Kwan J. 1974. *Ap. J.* 189:441–53

Goldreich P, Kylafis ND. 1981. *Ap. J. Lett.* 243:L75–78

Goldsmith PF, Langer WD. 1978. *Ap. J.* 222:881–95

Goldsmith PF. 1972. *Ap. J.* 176:597–610

Goodman AA. 1996. In *ASP Conf. Ser. 97, Polarimetry of the Interstellar Medium* ed. WG Roberge, DCB Whittet, pp. 325–44. San Francisco: Astron. Soc. Pacific

Goodman AA, Barranco JA, Wilner DJ, Heyer MH. 1998. *Ap. J.* 504:223–46

Goodman AA, Benson PJ, Fuller GA, Myers PC. 1993. *Ap. J.* 406:528–47

Greaves JS, Holland WS, Friberg P, Dent WRF. 1999. *Ap. J. Lett.* 512:L139–42
Gregersen EM, Evans NJ II, Zhou S, Choi M. 1997. *Ap. J.* 484:256–76
Gregersen EM. 1998. *Collapse and Beyond: An Investigation of the Star Formation Process.* PhD thesis. Univ. Texas, Austin. 187 pp.
Haas MR, Davidson JA, Erickson EF, eds. 1995. *Airborne Astronomy Symposium on the Galactic Ecosystem: From Gas to Stars to Dust.* San Francisco: Astron. Soc. Pacific. Vol. 73, 737 pp.
Harju J, Walmsley CM, Wouterloot JGA. 1993. *Astron. Astrophys. Suppl.* 98:51–76
Hartmann L. 1998. *Accretion Processes in Star Formation.* Cambridge U.K.: Cambridge Univ. Press. 237 pp.
Heiles C, Goodman AA, McKee CF, Zweibel EG. 1993. In *Protostars and Planets III*, ed. EH Levy, JI Lunine, pp. 279–326. Tucson: Univ. Arizona Press.
Heithausen A, Bensch F, Stutzki J, Falgarone E, Panis J-F. 1998. *Astron. Astrophys.* 331:L65–68
Helmich FP, Jansen DJ, de Graauw TH, Groesbeck TD, van Dishoeck EF. 1994. *Astron.* 283:626–34
Henning T, Burkert A, Launhardt R, Leinert Ch, Stecklum B. 1998. *Astron. Astrophys.* 336:565–86
Henning T, Launhardt R. 1998. *Astron. Astrophys.* 338:223–42
Henriksen RN, André P, Bontemps S. 1997. *Astron. Astrophys.* 323:549–65
Heyer M, Terebey S. 1998. *Ap. J.* 502:265–77
Hildebrand RH. 1983. *Q. J. R. Astron. Soc.* 24:267–82
Hildebrand RH. 1988. *Q. J. R. Astron. Soc.* 29:327–51
Hillenbrand LA, Hartmann LW. 1998. *Ap. J.* 492:540–53
Ho PTP, Townes CH. 1983. *Annu. Rev. Astron. Astrophys.* 21:239–70
Hogerheijde MR, van Dishoeck EF, Blake GA, van Langevelde HJ. 1998. *Ap. J.* 502:315–36
Hollenbach DJ, Tielens AGGM. 1997. *Annu. Rev. Astron. Astrophys.* 35:179–215
Hollenbach DJ, Johnstone D, Lizano S, Shu F. 1994. *Ap. J.* 428:654–69
Houlahan P, Scalo J. 1992. *Ap. J.* 393:172–87
Hunter C. 1977. *Ap. J.* 214:488–97
Hunter TR, Benford DJ, Serabyn E. 1996. *PASP* 108:1042–50
Jaffe DT, Hildebrand RH, Keene J, Harper DA, Loewenstein RF, Moran JM. 1984. *Ap. J.* 281:225–36
Jaffe DT, Martín-Pintado J. 1999. *Ap. J.* 520:162–72
Jeans JH. 1928. *Astronomy and Cosmogony,* p. 340. Cambridge, U.K.: Cambridge Univ. Press
Johnstone D, Bally J. 1999. *Ap. J. Lett.* 510:L49–53
Juvela M. 1996. *Astron. Astrophys. Suppl.* 118:191–226
Juvela M. 1997. *Astron. Astrophys.* 322:943–61
Juvela M. 1998. *Astron. Astrophys.* 329:659–82
Kane BD, Clemens DP, Myers PC. 1994. *Ap. J. Lett.* 433:L49–52
Kaufman MJ, Hollenbach DJ, Tielens AGGM. 1998. *Ap. J.* 497:276–87
Keene J. 1981. *Ap. J.* 245:115–23
Kennicutt RC Jr. 1998. *Annu. Rev. Astron. Astrophys.* 36:189–231
Keto ER, Ho PTP, Haschick AD. 1988. *Ap. J.* 324:920–30
Klessen RS, Burkert A, Bate M. 1998. *Ap. J. Lett.* 501:L205–8
Kramer C, Alves J, Lada C, Lada E, Sievers A, et al. 1998a. *Astron. Astrophys.* 329:L33–36
Kramer C, Alves J, Lada CJ, Lada EA, Sievers A, et al. 1999. *Astron. Astrophys.* 342:257–70
Kramer C, Stutzki J, Rohrig R, Corneliussen U. 1998b. *Astron. Astrophys.* 329:249–64
Kreysa E. 1992. In *ESA Symp. Photon Detectors Space Instrumentation, Noodwijk, ESA SP-356,* p. 356
Kudritzki RP, Hummer DG, Pauldrich AWA, Puls J, Najarro F, Imhoff C. 1992. *Astron. Astrophys.* 257:655–62
Kuiper TBH, Langer WD, Velusamy T. 1996. *Ap. J.* 468:761–73
Kurtz S, Cesaroni R, Churchwell EB, Hofner P,

Walmsley M. 2000. In *Protostars and Planets IV*, ed. V Mannings, A Boss, S Russell. Tucson: Univ. Ariz. In press

Lacy JH, Knacke R, Geballe TR, Tokunaga AT. 1994. *Ap. J. Lett.* 428:L69–72

Lada CJ. 1987. In *Star Forming Regions, IAU Symp. 115*, ed. M Peimbert J Jugaku, pp. 1–18. Dordrecht: Reidel

Lada CJ. 1991. In *The Physics of Star Formation and Early Stellar Evolution*, ed. CJ Lada, ND Kylafis, pp. 329–63. Dordrecht: Kluwer

Lada CJ. 1999. In *The Origin of Stars and Planetary Systems*, ed. CJ Lada, ND Kylafis. Dordrecht. Kluwer. In press

Lada CJ, Alves J, Lada EA. 1999. *Ap. J.* 512:250–59

Lada CJ, Kylafis ND, eds. 1999. *The Origin of Stars and Planetary Systems*. Dordrecht: Kluwer. In press

Lada CJ, Lada EA. 1991. In *The Formation and Evolution of Star Clusters*, ed. K Janes, 13: 3–22. San Francisco: Astron. Soc. Pacific

Lada CJ, Lada EA, Clemens DP, Bally J. 1994. *Ap. J.* 429:694–709

Lada CJ, Wilking BA. 1984. *Ap. J.* 287:610–21

Lada EA. 1992. *Ap. J. Lett.* 393:L25–28

Lada EA, Bally J, Stark AA. 1991. *Ap. J.* 368:432–44

Lada EA, DePoy DL, Evans NJ II, Gatley I. 1991. *Ap. J.* 371:171–82

Lada EA, Evans NJ II, Falgarone E. 1997. *Ap. J.* 488:286–306

Lada EA, Strom KM, Myers PC. 1993. In *Protostars and Planets III*, ed. EH Levy, JI Lunine, pp. 245–77. Tucson: Univ. Ariz.

Ladd EF, Adams FC, Casey S, Davidson JA, Fuller GA, et al. 1991. *Ap. J.* 382:555–69

Ladd EF, Fuller GA, Deane JR. 1998. *Ap. J.* 495:871–90

Larson RB. 1969. *MNRAS* 145:271–95

Larson RB. 1981. *MNRAS* 194:809–26

Launhardt R, Evans, NJ II, Wang YS, Clemens DP, Henning T, Yun J. 1998. *Ap. J. Suppl.* 119:59–74

Lazarian A, Goodman AA, Myers PC. 1997. *Ap. J.* 490:273–80

Lee CW, Myers PC, Tafalla M. 1999. In press

Leisawitz D, Bash FN, Thaddeus P. 1989. *Ap. J. Suppl.* 70:737–812

Leung CM. 1975. *Ap. J.* 199:340–60

Leung CM, Brown RL. 1977. *Ap. J. Lett.* 214:L73–78

Li W, Evans NJ II, Lada EA. 1997. *Ap. J.* 488:277–85

Li ZY. 1998. *Ap. J.* 497:850–58

Li ZY, Shu FH. 1996. *Ap. J.* 472:211–24

Lin DNC, Papaloizou JCB. 1996. *Annu. Rev. Astron. Astrophys.* 34:703–47

Looney LW, Mundy LG, Welch WJ. 1997. *Ap. J. Lett.* 484:L157–60

Loren RB. 1976. *Ap. J.* 209:466–88

Loren RB, Mundy LG. 1984. *Ap. J.* 286:232–51

Loren RB, Sandqvist A, Wootten A. 1983. *Ap. J.* 270:620–40

Lucas R. 1974. *Astron. Astrophys.* 36:465–67

Luhman ML, Jaffe DT. 1996. *Ap. J.* 463:191–204

Lynds BT. 1962. *Ap. J. Suppl.* 7:1–52

Mac Low M-M, Klessen RS, Burkert A, Smith MD. 1998. *Phys. Rev. Lett.* 80:2754–57

Mangum JG, Wootten A. 1993. *Ap. J. Suppl.* 89:123–53

Mannings V, Boss A, Russell S, eds. 2000. *Protostars and Planets IV*. Tucson: Univ. Ariz. In press

Mannings V, Sargent AI. 1997. *Ap. J.* 490:792–802

Manske V, Henning T. 1998. *Astron. Astrophys.* 337:85–95

Mardones D, Myers PC, Tafalla M, Wilner DJ, Bachiller R, Garay G. 1997. *Ap. J.* 489:719–33

Martin HM, Hills RE, Sanders DB. 1984. *MNRAS* 208:35–55

Martin RN, Barrett AH. 1978. *Ap. J. Suppl.* 36:1–51

Mathieu RD. 1994. *Annu. Rev. Astron. Astrophys.* 32:465–530

Mathis JS. 1990. *Annu. Rev. Astron. Astrophys.* 28:37–70

Mathis JS, Rumpl W, Nordsieck KH. 1977. *Ap. J.* 217:425–33

Mauersberger R, Henkel C. 1989. *Astron. Astrophys.* 223:79–88

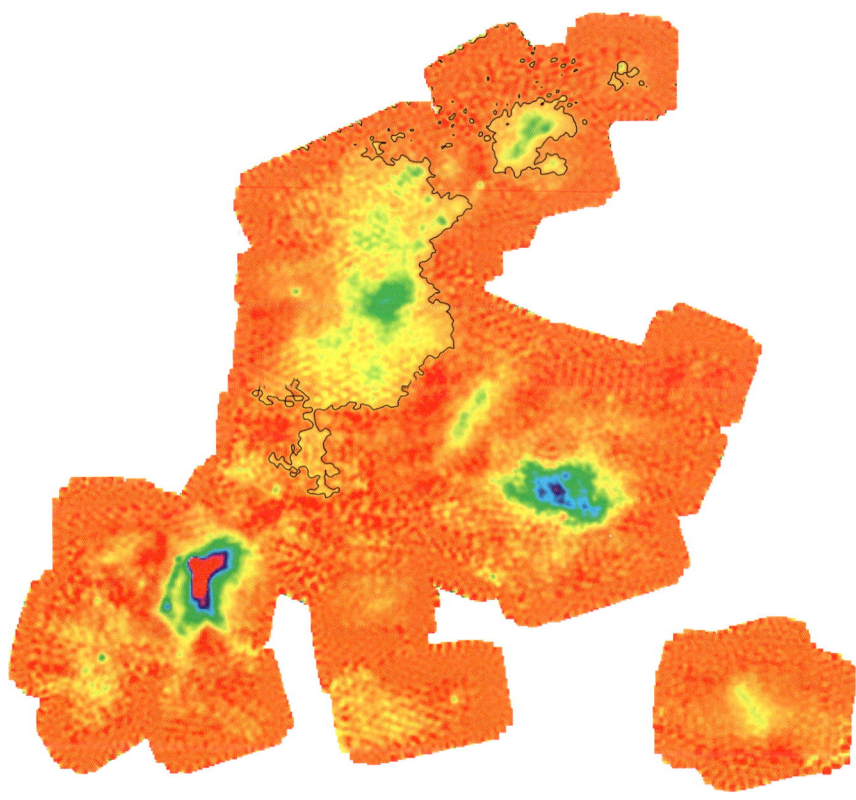

Figure 1 Map of 1.3-mm dust continuum emission covering about 2 pc in the ρ Ophiuchi cloud, with a resolution of 15" (Motte et al 1998). The emission is proportional to column density, and the lowest contour corresponds to $N \sim 1.4 \times 10^{22}$ cm^{-2} or $A_V \sim 14$ mag. Contour levels increase in spacing at higher levels (see Motte et al 1998 for details). The highest contour implies $A_V \sim 780$ mag.

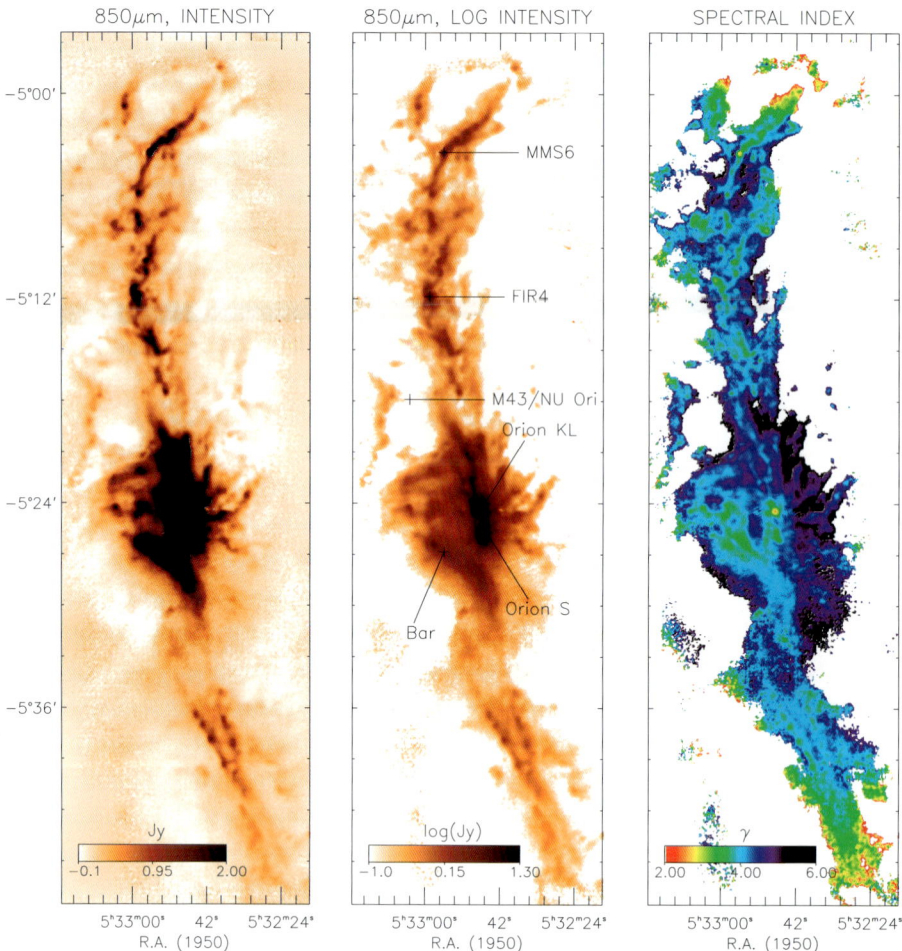

Figure 2 The 850-μm emission and 450- to 850-μm spectral index distribution, covering about 7 pc with a resolution of 14" at the northern end of the Orion A (L1641) molecular cloud (Johnstone & Bally 1999). The Orion Nebula is located directly in front of the strong emission at the center of the image. (*Left*) The 850-μm image showing the observed flux from −0.1 to 2 Jy/beam with a linear transfer function. The highest flux level corresponds roughly to $A_V \sim 320$ mag if $T_K = 20$ K. (*Center*) The 850-μm image showing the observed flux from 100 mJy/beam to 20 Jy/beam with a logarithmic transfer function. (*Right*) The 450- to 850-μm spectral index in the range 2–6 with a linear transfer function.

McCaughrean MJ, Stauffer JR. 1994. *Astron. J.* 108:1382–97

McKee C. 1999. In *The Origin of Stars and Planetary Systems*, ed. CJ Lada, ND Kylafis. Dordrecht: Kluwer. In press

McKee CF. 1989. *Ap. J.* 345:782–801

McKee CF, Zweibel EG. 1992. *Ap. J.* 399:551–62

McKee CF, Zweibel EG, Goodman AA, Heiles C. 1993. In *Protostars and Planets III*, ed. EH Levy, JI Lunine, pp. 327–66. Tucson: Univ. Ariz.

McLaughlin DE, Pudritz RE. 1997. *Ap. J.* 476:750–65

Mead KN, Kutner ML, Evans, NJ II. 1990. *Ap. J.* 354:492–503

Men'shchikov AB, Henning T. 1997. *Astron. Astrophys.* 318:879–907

Menten KM, Serabyn E, Güsten R, Wilson TL. 1987. *Astron. Astrophys.* 177:L57–60

Mestel L. 1965. *Q. J. R. Astron. Soc.* 6:265–98

Mestel L. 1985. In *Protostars and Planets II*, ed. DC Black, MS Matthews, pp. 320–39. Tucson: Univ. Ariz. Press

Mestel L, Spitzer L. 1956. *MNRAS* 116:503–14

Meyer MR, Adams FC, Hillenbrand LA, Carpenter JM, Larson RB. 2000. In *Protostars and Planets IV*, ed. V Mannings, A Boss, S Russell. Tucson: Univ. Ariz. In press

Millar T. 1997. In *Molecules in Astrophysics: Probes and Processes, IAU Symp. 178*, ed. EF van Dishoeck, pp. 75–88. Dordrecht: Reidel

Mitchell GF, Maillard J-P, Allen M, Beer R, Belcourt K. 1990. *Ap. J.* 363:554–73

Mizuno A, Onishi T, Yonekura Y, Nagahama T, Ogawa H, Fukui Y. 1995. *Ap. J.* 445:L161–65

Molinari S, Testi L, Brand J, Cesaroni R, Palla F. 1998. *Ap. J. Lett.* 505:L39–42

Momose M, Ohashi N, Kawabe R, Nakano T, Hayashi M. 1998. *Ap. J.* 504:314–33

Mooney TJ, Solomon PM. 1988. *Ap. J.* 334:L51–54

Motte F, André P, Neri R. 1998. *Astron. Astrophys.* 336:150–72

Mouschovias TCh. 1991. *Ap. J.* 373:169–86

Mouschovias TCh. 1999. In *The Origin of Stars and Planetary Systems*, ed. CJ Lada, ND Kylafis. Dordrecht: Kluwer. In press

Mouschovias TCh, Spitzer L. 1976. *Ap. J.* 210:326–27

Mundy LG, Evans NJ II, Snell RL, Goldsmith PF. 1987. *Ap. J.* 318:392–409

Mundy LG, Looney LW, Welch WJ. 2000. In *Protostars and Planets IV*, ed. V Mannings, A Boss, S Russell. Tucson: Univ. Ariz. In press

Mundy LG, Snell RL, Evans NJ II, Goldsmith PF, Bally J. 1986. *Ap. J.* 306:670–81

Mundy LG, Wootten A, Wilking BA, Blake GA, Sargent AI. 1992. *Ap. J.* 385:306–13

Myers PC. 1983. *Ap. J.* 270:105–18

Myers PC. 1985. In *Protostars and Planets II*, ed. DC Black, MS Matthews, pp. 81–103. Tucson: Univ. Ariz.

Myers PC. 1998. *Ap. J. Lett.* 496:L109–12

Myers PC, Adams FC, Chen H, Schaff E. 1998. *Ap. J.* 492:703–26

Myers PC, Bachiller R, Caselli P, Fuller GA, Mardones D, et al. 1995. *Ap. J. Lett.* 449:L65–68

Myers PC, Benson PJ. 1983. *Ap. J.* 266:309–20

Myers PC, Evans NJ II, Ohashi N. 2000. In *Protostars and Planets IV*, ed. V Mannings, A Boss, S Russell. Tucson: Univ. Ariz. In press

Myers PC, Fuller GA. 1992. *Ap. J.* 396:631–42

Myers PC, Fuller GA, Goodman AA, Benson PJ. 1991. *Ap. J.* 376:561–72

Myers PC, Goodman A. 1988. *Ap. J. Lett.* 326:L27–30

Myers PC, Ladd EF. 1993. *Ap. J. Lett.* 413:L47–50

Myers PC, Lazarian A. 1998. *Ap. J. Lett.* 507:L157–60

Myers PC, Linke RA, Benson PJ. 1983. *Ap. J.* 264:517–37

Myers PC, Mardones D, Tafalla M, Williams JP, Wilner DJ. 1996. *Ap. J. Lett.* 465:L133–36

Nakano T. 1998. *Ap. J.* 494:587–604

Narayanan G, Walker CK, Buckley HD. 1998. 496:292–310
Natta A, Grinin VP, Mannings V. 2000. In *Protostars and Planets IV*, ed. V Mannings, A Boss, S Russell. Tucson: Univ. Ariz. In press
Natta A, Palla F, Butner HM, Evans NJ II, Harvey PM. 1993. *Ap. J.* 406:674–91
Norman CA, Silk J. 1980. *Ap. J.* 238:158–74
Ohashi N, Hayashi M, Ho PTP, Momose M. 1997a. *Ap. J.* 475:211–23
Ohashi N, Hayashi M, Ho PTP, Momose M, Tamura M, et al. 1997b. *Ap. J.* 488:317–29
Ohashi N, Hayashi M, Kawabe R, Ishiguro M. 1996. *Ap. J.* 466:317–37
Ohishi M. 1997. In *Molecules in Astrophysics: Probes and Processes, IAU Symp. 178*, ed. EF van Dishoeck, pp. 61–74. Dordrecht: Reidel
Olmi L, Cesaroni R, Walmsley CM. 1993. *Astron. Astrophys.* 276:489–506
Onishi T, Mizuno A, Kawamura A, Ogawa H, Fukui Y. 1996. *Ap. J.* 465:815–24
Onishi T, Mizuno A, Kawamura A, Ogawa H, Fukui Y. 1998. *Ap. J.* 502:296–314
Ossenkopf V, Henning Th. 1994. *Astron. Astrophys.* 291:943–59
Padoan P, Jones BJT, Nordlund AP. 1997. *Ap. J.* 474:730–34
Palmer P, Zuckerman B, Buhl D, Snyder LE. 1969. *Ap. J. Lett.* 156:L147–50
Papaloizou JCB, Lin DNC. 1995. *Annu. Rev. Astron. Astrophys.* 33:505–40
Park YS, Hong SS. 1995. *Astron. Astrophys.* 300:890–902
Park YS, Hong SS. 1998. *Ap. J.* 494:605–13
Pendleton YJ. 1997. *Ori. Life Evol. Biosph.* 27:53–78
Pendleton YJ, Tielens AGGM, eds. 1997. *From Stardust to Planetesimals*. San Francisco: Astron. Soc. Pacific. Vol. 132, 480 pp.
Penston MV. 1969. *MNRAS* 144:425–48
Penzias AA, Solomon PM, Jefferts KB, Wilson RW. 1972. *Ap. J. Lett.* 174:L43–48
Plume R, Jaffe DT, Evans NJ II. 1992. *Ap. J. Suppl.* 78:505–15
Plume R, Jaffe DT, Evans NJ II, Martín-Pintado J, Gómez-González J. 1997. *Ap. J.* 476:730–49

Rao R, Crutcher RM, Plambeck RL, Wright MCH. 1998. *Ap. J. Lett.* 502:L75–78
Rawlings JMC, Hartquist TW, Menten KM, Williams DA. 1992. *MNRAS* 255:471–85
Rodríguez LF, D'Alessio P, Wilner DJ, Ho PTP, Torrelles JM, et al. 1998. *Nature* 395:355–57
Rowan-Robinson M, Mann RG, Oliver SJ, Efstathiou A, Eaton N, et al. 1997. *MNRAS* 289:490–96
Ryden BS. 1996. *Ap. J.* 471:822–31
Safier PN, McKee CF, Stahler SW. 1997. *Ap. J.* 485:660–79
Sanders DB, Scoville NZ, Soifer BT. 1991. *Ap. J.* 370:158–71
Saraceno P, André P, Ceccarelli C, Griffin M, Molinari S. 1996. *Astron. Astrophys.* 309:827–39
Sargent AI. 1996. In *Disks and Outflows from Young Stars*, ed. SVW Beckwith, J Staude, A Quetz, A Natta, pp. 1–23. Berlin: Springer
Sargent AI, Welch WJ. 1993. *Annu. Rev. Astron. Astrophys.* 31:297–343
Scalo J. 1985. In *Protostars and Planets II*, ed. DC Black, MS Matthews, pp. 201–96. Tucson: Univ. Ariz.
Scalo J. 1986. *Fund. Cosmic Phys.* 11:1–278
Scalo J. 1998. In *The Stellar Initial Mass Function*, ed. G Gilmore, I Parry, S Ryan, 142:201–36. San Francisco: Astron. Soc. Pacific
Scalo J, Lazarian A. 1996. *Ap. J.* 469:189–93
Scalo J, Vázquez-Semadini E, Chappell D, Passot T. 1998. *Ap. J.* 504:835–53
Schleuning DA. 1998. *Ap. J.* 493:811–25
Shepherd DS, Churchwell E. 1996. *Ap. J.* 472:225–39
Shepherd DS, Kurtz SE. 1999. *Ap. J.* In press
Shirley Y, Evans NJ, Rawlings JMC, Gregersen E. 1998. *Bull. Am. Astron. Soc.* 30:1358
Shu FH. 1977. *Ap. J.* 214:488–97
Shu FH, Adams FC, Lizano S. 1987a. *Annu. Rev. Astron. Astrophys.* 25:23–81
Shu FH, Allen A, Shang H, Ostriker EC, Li ZY. 1999. In *The Origin of Stars and Planetary Systems*, ed. CJ Lada, ND Kylafis. Dordrecht: Kluwer. In press
Shu FH, Lizano S, Adams FC. 1987b. In *Star*

Forming Regions, IAU Symp. 115, ed. M Peimbert, J Jugaku, pp. 417–34. Dordrecht: Reidel

Snell RL. 1981. *Ap. J. Suppl.* 45:121–75

Snell RL, Loren RB. 1977. *Ap. J.* 211:122–27

Snell RL, Mundy LG, Goldsmith PF, Evans NJ II, Erickson NR. 1984. *Ap. J.* 276:625–45

Solomon PM, Rivolo AR. 1989. *Ap. J.* 339:919–25

Solomon PM, Rivolo AR, Barrett J, Yahil A. 1987. *Ap. J.* 319:730–41

Spencer RG, Leung CM. 1978. *Ap. J.* 222:140–52

Spitzer L Jr. 1978. *Physical Processes in the Interstellar Medium*, p. 282. New York: Wiley

Stahler SW, Palla F, Ho PTP. 2000. In *Protostars and Planets IV*, ed. V Mannings, A Boss, S Russell. Tucson: Univ. Ariz. In press

Sternberg A, Yan M, Dalgarno A. 1997. In *Molecules in Astrophysics: Probes and Processes, IAU Symposium 178*, ed. EF van Dishoeck. Dordrecht: Reidel. pp. 141–53

Stone JM, Ostriker EC, Gammie CF. 1999. In *Interstellar Turbulence, Proc. Guillermo Haro Conf., 2nd*, ed. J Franco, A Carraminana. Cambridge, U.K.: Cambridge Univ. Press. In press

Strom KM, Strom SE, Merrill KM. 1993. *Ap. J.* 412:233–53

Strong AW, Bennett K, Bloemen H, Diehl R, Hermsen W, et al. 1994. *Astron. Astrophys.* 292:82–91

Stutzki J, Güsten R. 1990. *Ap. J.* 356:513–33

Tafalla M, Mardones D, Myers PC, Caselli P, Bachiller R, Benson PJ. 1998. *Ap. J.* 504:900–14

Takahashi T, Silk J, Hollenbach DJ. 1983. *Ap. J.* 275:145–62

Tatematsu K, Umemoto T, Heyer MH, Hirano N, Kameya O, Jaffe DT. 1998. *Ap. J. Suppl.* 118:517–39

Tatematsu K, Umemoto T, Kameya O, Hirano N, Hasegawa T, et al. 1993. *Ap. J.* 404:643–62

Terebey S, Chandler CJ, André P. 1993. *Ap. J.* 414:759–72

Terebey S, Shu FH, Cassen P. 1984. *Ap. J.* 286:529–51

Testi L, Palla F, Prusti T, Natta A, Maltagliati S. 1997. *Astron. Astrophys.* 320:159–66

Testi L, Sargent AI. 1998. *Ap. J. Lett.* 508:L91–94

Townes CH, Cheung AC. 1969. *Ap. J. Lett.* 157:L103–8

Truelove JK, Klein RI, McKee CF, Holliman JH II, Howell LH, et al. 1998. *Ap. J.* 495:821–52

van der Tak FFS, van Dishoeck EF, Evans NJ II, Bakker EJ, Blake GA. 1999. *Ap. J.* In press

van Dishoeck EF, Blake GA. 1998. *Annu. Rev. Astron. Astrophys.* 36:317–68

van Dishoeck EF, Helmich FP, Schutte WA, Ehrenfreund P, Lahuis F, et al. 1998. In *Star Formation with the Infrared Space Observatory (ISO)*, ed. JL Yun, R Liseau, 132:54–65. San Francisco: Astron. Soc. Pacific

Vázquez-Semadini E. 1994. *Ap. J.* 423:681–92

Vázquez-Semadini E, Ostriker EC, Passot T, Gammie CF, Stone JM. 2000. In *Protostars and Planets IV*, ed. V Mannings, A Boss, S Russell. Tucson: Univ. Ariz. In press

Velusamy T, Kuiper TBH, Langer WD. 1995. *Ap. J. Lett.* 451:L75–78

Vrba FJ, Strom SE, Strom KM. 1976. *Astron. J.* 81:958–69

Walker CK, Lada CJ, Young ET, Maloney PR, Wilking BA. 1986. *Ap. J. Lett.* 309:L47–61

Walker CK, Narayanan G, Boss AP. 1994. *Ap. J.* 431:767–82

Walmsley M. 1995. Rev. Mex. Astron Astrof. Conf. Ser. 1, pp. 137–48

Wang Y, Evans NJ II, Zhou S, Clemens DP. 1995. *Ap. J.* 454:217–32

Wang Y, Jaffe DT, Evans NJ II, Hayashi M, Tatematsu K, Zhou S. 1993. *Ap. J.* 419:707–24

Ward-Thompson D, André P, Motte F. 1998. In *Star Formation with the Infrared Space Observatory (ISO)*, ed. JL Yun, R. Liseau, 132:195–202. San Francisco: Astron. Soc. Pacific

Ward-Thompson D, Motte F, André P. 1999. *MNRAS*. 305:143–150

Ward-Thompson D, Scott PF, Hills RE, André P. 1994. *MNRAS* 268:276–90

Welch WJ, Dreher JW, Jackson JM, Terebey S, Vogel SN. 1988. *Science* 238:1550–55

White RE. 1977. *Ap. J.* 211:744–54

Wiesemeyer H. 1997. *The Spectral Signature of Accretion in Low-Mass Protostars*. PhD dissertation. Univ. Bonn., Bonn, Germany. 153 pp.

Wiesemeyer H. 1999. *Astron. Astrophys.* In press

Williams JP, Bergin EA, Caselli P, Myers PC, Plume R. 1998. *Ap. J.* 503:689–99

Williams JP, Blitz L, McKee CF. 2000. In *Protostars and Planets IV*, ed. V Mannings, A Boss, S Russell. Tucson: Univ. Ariz. In press

Williams JP, McKee CF. 1997. *Ap. J.* 476:166–83

Wilner DJ, Welch WJ, Forster JR. 1995. *Ap. J. Lett.* 449:L73–76

Wolf-Chase GA, Gregersen E. 1997. *Ap. J. Lett.* 479:L67–70

Wolfire MG, Cassinelli JP. 1987. *Ap. J.* 319:850–67

Wolfire MG, Hollenbach D, Tielens AGGM. 1993. *Ap. J.* 402:195–215

Wood DOS, Churchwell EB. 1989. *Ap. J.* 340:265–72

Wootten A. 1989. *Ap. J.* 337:858–64

Wu Y, Evans NJ II. 1989. *Ap. J.* 340:307–13

Yates JA, Field D, Gray MD. 1997. *MNRAS* 285:303–16

Young LM, Keto E, Ho PTP. 1998. *Ap. J.* 507:270–80

Yun JL. 1993. *A Characterization of Young Stellar Objects in Bok Globules: Infrared Imaging, Spectral Energy Distributions, and Molecular Outflows*. PhD dissertation. Boston Univ. 230 pp.

Yun JL, Clemens DP. 1990. *Ap. J. Lett.* 365:L73–76

Yun JL, Clemens DP. 1992. *Ap. J. Lett.* 385:L21–25

Yun JL, Liseau R, eds. 1998. *Star Formation with the Infrared Space Observatory (ISO)*. San Francisco: Astron. Soc. Pacific. Vol. 132, 448 pp.

Zhang Q, Ho PTP. 1997. *Ap. J.* 488:241–57

Zhang Q, Ho PTP, Ohashi N. 1998a. *Ap. J.* 494:636–56

Zhang Q, Hunter TR, Sridharan TK. 1998b. *Ap. J. Lett.* 505:L151–54

Zhou S. 1992. *Ap. J.* 394:204–16

Zhou S. 1995. *Ap. J.* 442:685–93

Zhou S, Butner HM, Evans NJ II, Güsten R, Kutner ML, Mundy LG. 1994b. *Ap. J.* 428:219–32

Zhou S, Evans NJ II. 1994. In *Clouds, Cores, and Low Mass Stars*, ed. DP Clemens, R. Barvainis, 65:183–91. San Francisco: Astron. Soc. Pacific

Zhou S, Evans NJ II, Butner HM, Kutner ML, Leung CM, Mundy LG. 1990. *Ap. J.* 363:168–79

Zhou S, Evans NJ II, Güsten R, Mundy LG, Kutner ML. 1991. *Ap. J.* 372:518–613

Zhou S, Evans NJ II, Kömpe C, Walmsley CM. 1993. *Ap. J.* 404:232–46

Zhou S, Evans NJ II, Wang Y. 1996. *Ap. J.* 466:296–308

Zhou S, Evans NJ II, Wang Y, Peng R, Lo KY. 1994a. *Ap. J.* 433:131–48

Zhou S, Wu Y, Evans NJ II, Fuller GA, Myers PC. 1989. *Ap. J.* 346:168–79

Zinchenko I, Mattila K, Toriseva M. 1995. *Astron. Astrophys. Suppl.* 111:95–114

Zuckerman B, Evans, NJ II. 1974. *Ap. J. Lett.* 192:L149–52

Zuckerman B, Palmer P. 1974. *Annu. Rev. Astron. Astrophys.* 12:279–313

Zweibel E. 1998. *Ap. J.* 499:746–53

HIGH-ENERGY PROCESSES IN YOUNG STELLAR OBJECTS

Eric D. Feigelson
Department of Astronomy & Astrophysics, Pennsylvania State University, University Park, Pennsylvania 16802; e-mail: edf@astro.psu.edu

Thierry Montmerle
Service d'Astrophysique, CEA/DSM/DAPNIA/SAp, Centre d'Études de Saclay, 91191, Gif-sur-Yvette Cedex, France; e-mail: montmerle@cea.fr

Key Words pre-main sequence stars, magnetic flares, circumstellar disks, stellar X rays, meteorites

■ **Abstract** Observational studies of low-mass stars during their early stages of evolution, from protostars through the zero-age main sequence, show highly elevated levels of magnetic activity. This activity includes strong fields covering much of the stellar surface and powerful magnetic reconnection flares seen in the X-ray and radio bands. The flaring may occur in the stellar magnetosphere, at the star-disk interface, or above the circumstellar disk. Ionization from the resulting high-energy radiation may have important effects on the astrophysics of the disk, such as promotion of accretion and coupling to outflows, and on the surrounding interstellar medium. The bombardment of solids in the solar nebula by flare shocks and energetic particles may account for various properties of meteorites, such as chondrule melting and spallogenic isotopes. X-ray surveys also improve our samples of young stars, particularly in the weak-lined T Tauri phase after disks have dissipated, with implications for our understanding of star formation in the solar neighborhood.

1. INTRODUCTION

The formation and early evolution of low-mass[1] stars is generally discussed in terms of gravitational and hydrodynamical processes. A molecular cloud core collapses, a protostar emerges at the center while high-angular-momentum material forms a circumstellar accretion disk and bipolar outflows are produced. The star becomes visible, contracting quasistatically along the Hayashi and radiative tracks

---

[1] This review is restricted to solarlike stars with masses around $M_* \sim 0.2 - 2\ M_\odot$ and omits discussion of OB stars and intermediate-mass Herbig Ae/Be stars (Waelkens & Waters 1998). We further apply the expression "young stellar objects" (YSOs) to all phases of low-mass evolution before arrival at the main sequence.

in the T Tauri phase to the zero-age main sequence (ZAMS). Magnetic fields are believed to play a central role in regulating collapse through ambipolar diffusion and in transferring disk orbital motion to collimated outflows. Coupling between the neutral material and magnetic fields is thought to be provided by an approximate 10^{-8} fractional ionization produced by low-energy, galactic cosmic ray interaction with the molecular gas. Ancient meteorites indicate that additional high-energy processes affected the solar nebula, but the origins of these processes cannot be ascertained and are often omitted from astrophysical discussions of star formation and YSOs. (A notable exception is the "Protostars & Planets" series of meetings—Levy & Lunine 1993, Mannings et al 1999.)

However, empirical studies of YSOs in the 1980s and 1990s provide direct and ample evidence for kilo electron volt-energy radiation and MeV particles produced within protostellar and T Tauri systems. The X-ray emission is understood to be thermal emission from gas rapidly heated to temperatures 10^7 K by violent magnetohydrodynamical (MHD) reconnection events, analogous to solar magnetic flaring but elevated by factors of 10^1 to 10^6 above levels seen on the contemporary Sun. Other indicators of strong magnetic activity include large star spots revealed by optical photometry, an enhanced chromosphere and Zeeman effects seen in optical and ultraviolet spectroscopy, and powerful nonthermal radio-continuum flares.

These results reveal a deep relationship between high-energy phenomena observed in YSOs, the Sun, and other magnetically active late-type stars such as dMe flare stars and RS CVn binary systems. For 50 years, the Sun has been known to be an X-ray emitter, and the X-ray emission of the solar surface is now monitored in great spatial and spectral detail. Images from solar X-ray satellites show hot plasma, largely confined within magnetic loops, which can be suddenly heated by magnetic reconnection events (flares), often accompanied by ejection of substantial amounts of matter (coronal mass ejections and solar energetic particles). Circularly polarized gyrosynchrotron emission at radio wavelengths indicates the acceleration of electrons to mildly relativistic energies, and nuclear gamma-ray lines are produced by spallation reactions of energetic protons in the photosphere. All of these phenomena are observed or inferred to be present in YSOs at elevated levels.

These observations raise many astrophysical issues. The magnetic activity in YSOs may arise, as in the Sun, from a magnetic dynamo generated in the deep convection zones of the stellar interior. But the magnetic-field topologies cannot easily be predicted or deduced from activity tracers. The circumstellar material of YSOs may allow unusual magnetic configurations—such as star-disk, star-envelope, or disk-disk fields—that are not seen in older cool stars. Furthermore, magnetic activity may provide a crucial astrophysical link between the central star, circumstellar envelope, disk, and bipolar ejecta in YSOs.

X-ray ionization of circumstellar material will typically dominate galactic cosmic ray ionization on scales to approximately 10^4 AU and thus should be a major player in the dynamics and evolution of protoplanetary disks. The integrated effects of particle irradiation from YSO flaring may account for a variety of characteristics in the meteoritic record of the solar nebula. Finally, YSO magnetic activity also leads to insights regarding the population of young stars. X-ray surveys of large areas in the sky considerably enlarge the census of YSOs, both in star-forming regions and away from them, addressing such diverse problems as the star formation histories of molecular clouds, the longevity of circumstellar disks, and the kinematic dispersal of young stars into the galaxy.

After covering some background material on YSOs (Section 2), this review describes the evidence for YSO magnetic activity revealed in multiwavelength studies (Section 3). Astrophysical models of YSO magnetic fields are outlined in Section 4, followed by discussion of high-energy irradiation of interstellar and circumstellar material (Section 5). Section 6 reviews magnetic activity as a tracer of YSO populations, and Section 7 gives concluding remarks.

2. BACKGROUND

The first concepts of star formation via gravitational collapse were established by Pierre-Simon Laplace, whose *Exposition du Système du Monde* (1796) gives a vivid scenario of the formation of the Sun and the solar system from a rotating nebula. In the first half of this century, Edward Barnard and others speculated that dark clouds were the sites of stellar birth (Trimble 1996). Alfred Joy (1945) reported a class of unusual emission line and variable stars near dark clouds, representing newly formed stars, now called classical T Tauri (CTT) stars, which were found to be frequently grouped in "T associations" (Ambartsumian & Mirzoyan 1982).

During the 1980s and 1990s, a revolution took place in infrared and millimeter astronomy, which led to direct observational searches for the earlier infall and protostellar phases embedded deep in the clouds. Although the simple Laplacian picture lies at the foundation of contemporary models, the detailed astrophysics of YSOs proved to be much more complex than a simple self-gravitating nebula, and our understanding of YSOs has not proceeded in a simple fashion. Much of the thinking can be viewed as the interweaving of five themes.

2.1 Pre-Main Sequence Stellar Evolution

The basic picture established by early analytical treatments of YSO stellar interiors (Hayashi 1966) is still in force. T Tauri stars have fully convective stellar interiors with monotonically decreasing luminosities but nearly constant surface temperatures, powered principally by gravitational contraction rather than nuclear reactions. Models based on different assumptions regarding atomic opacities and convective structure give somewhat different isochrones on the Hertzsprung-Russell diagram. The rotational evolution of stars along the Hayashi tracks may be complex, and the resulting magnetic field generation is not well established. A YSO with a differentially rotating interior could generate a core field as high as approximately 10^6 G in 10^3 years, but turbulence without differential rotation would dissipate the field (Levy et al 1991).

2.2 Outflows

The broad emission lines of CTT stars often exhibit P Cygni-type profiles and were originally interpreted as dense hot winds (Herbig 1962, Kuhi 1964). Collimated outflow bow shocks seen as Herbig-Haro objects, small-scale optical jets, and molecular bipolar outflows were later found to be common among the younger YSOs (see reviews by Bachiller 1996, Reipurth & Bertout 1997). It was readily perceived that the YSO outflows are not accelerated by radiation or coronal gas pressure but that they required the intervention of magnetic fields and the circumstellar disk (e.g. DeCampli 1981, Pudritz & Norman 1983, Uchida & Shibata 1984). Although there is a consensus that magnetic fields confine and accelerate outflows, the detailed acceleration mechanism close to the star is still the subject of lively discussion (Pudritz et al 1999).

2.3 Disks

Although the optical obscuration of most YSOs was originally attributed to a spherical cocoon of unaccreted dusty material, the discovery of intense emission from the micrometer to the millimeter bands from CTT stars over the past 15 years requires that large amounts of dust be present in a flattened disk. These protoplanetary disks can now be directly imaged in emission with millimeter interferometers (e.g. Dutrey et al 1994), and the Hubble Space Telescope in visible light silhouette (McCaughrean & O'Dell 1996) or in near-infrared emission (Stapelfeldt et al 1998).

2.4 Accretion

Whereas star formation theory predicts stellar growth by accretion from a large circumstellar envelope (Shu et al 1987), direct evidence for infall proved elusive for many years. Doppler signatures of gas infall are now seen in the earliest protostellar phases (e.g. Zhou et al 1993, Mardones et al 1997). Ballistic infall models, including rotation, indicate that the envelope feeds a central accretion disk several hundred astronomical units in size (Terebey et al 1984), consistent with recent disk imaging. Some molecular line evidence of disk accretion has also been found (Ohashi et al 1996). After the envelope is cleared by outflows, this disk remains, and the material accretes onto the central star on time scales of 10^5–10^6 years. This process can proceed in a relatively steady fashion—seen as CTT stars—or in short-lived episodes of high accretion with associated ejection of material—seen in YY Ori stars (Bertout et al 1996) and at very high levels as FU Orionis stars (Hartmann et al 1993). Detailed models of T Tauri-permitted line profiles and continuum veiling—historically attributed to outflows or star-disk boundary layers—are now explained as emission in magnetically confined accretion columns

(Hartmann et al 1994, Hartmann 1998, Calvet & Gullbring 1998). Accretion also affects pre-main sequence evolutionary tracks (Siess et al 1997).

2.5 Magnetic Activity

Several independent lines of evidence indicated that YSOs exhibit unusually high levels of magnetic activity. The original perception by Joy (1945) that YSO spectra share characteristics with the solar chromosphere led to the development of T Tauri stellar atmospheres with large plage regions and deep chromospheres (Herbig & Soderblom 1980, Calvet et al 1984, Finkenzeller & Basri 1987). Hundreds of flash variable stars in star-forming regions were found, which suggests an analogy with dMe flare stars (Haro & Chavira 1966, Ambartsumian & Mirzoyan 1982). Powerful X-ray flares from T Tauri stars were found with the first satellite-borne imaging X-ray telescope, and these flares were generally interpreted as enhanced solar type emission (Feigelson & DeCampli 1981, Walter & Kuhi 1981, Montmerle et al 1983, Feigelson et al 1991). Gyrosynchrotron radio-continuum flares, similar in character but again orders-of-magnitude stronger than solar levels, were found in some YSOs (Feigelson & Montmerle 1985, White et al 1992b).

2.6 Stages of YSD Evolution

Although this review concentrates on the theme of magnetic activity, we trace recent efforts to synthesize all previous approaches into a coherent understanding of YSO astrophysics. For example, we discuss the possible origin of flares in magnetic field lines connecting the star to the disk and the possible roles of energetic radiation for ionizing the disk and promoting accretion and outflow. We summarize here some central concepts that underlie our discussion of magnetic and high-energy processes, recognizing that our treatment does not adequately present the rich phenomenology and astrophysics of YSOs (cf. Levy & Lunine 1993, Mannings et al 1999).

Figure 1 illustrates the principal phases of YSO evolution: protostars, CTT stars, and weak-lined T Tauri (WTT) stars. To reduce the effects of foreground extinction (T Tauri stars are often optically visible, whereas protostars are deeply embedded in their parent cloud), the evolutionary phase of YSOs is generally classified by their infrared-millimeter spectral energy distributions (Lada 1987, André & Montmerle 1994).

Class 0 infrared-millimeter sources are young protostars with massive, cold (~ 30 K) envelopes that collapse toward the central regions. A collimated outflow and a disk rapidly form within the envelope, which is 10^3 to 10^4 AU in size (Figure 2 *left*). The age of Class 0 sources is approximately 10^4 years.

Class I sources have ages of approximately 10^5 years. Most of the material in the envelope has accreted onto the disk or star, and the disk is a few hundred astronomical units in extent (Figure 2 *middle*). Outflow activity is still present but with a larger opening angle and a lower mass-loss rate than at the Class 0 stage (Bontemps et al 1996).

Properties	Infalling Protostar	Evolved Protostar	Classical T Tauri Star	Weak-lined T Tauri Star	Main Sequence Star
Sketch	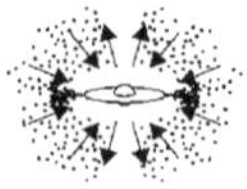	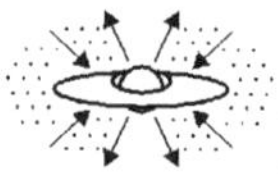	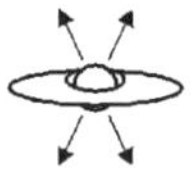	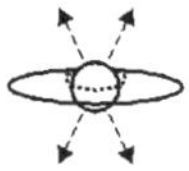	
Age (years)	10^4	10^5	$10^6 - 10^7$	$10^6 - 10^7$	$> 10^7$
mm/Infrared Class	Class 0	Class I	Class II	Class III	(Class III)
Disk	Yes	Thick	Thick	Thin or Non-existent	Possible Planetary System
X-ray	?	Yes	Strong	Strong	Weak
Thermal Radio	Yes	Yes	Yes	No	No
Non-Thermal Radio	No	Yes	No ?	Yes	Yes

Figure 1 The stages of low-mass young stellar evolution. This review chiefly addresses the bottom three rows of the chart. (Adapted from Carkner 1998.)

Class II is the infrared designation of CTT stars. Most of their complex phenomenology can be modeled as a star interacting with a circumstellar accretion disk (Figure 2 *right*). The youngest members of the class drive outflows, and all drive strong winds with $\dot{M} \approx 10^{-7} M_\odot$ year^{-1} and $v_w \sim 200$ km s^{-1}. Contemporary models are based on magnetically confined accretion from a magnetosphere extending out to the corotation radius (Figure 3). When Class II sources are unobscured, they can be placed on the Hertzsprung-Russell diagram and compared with theoretical evolutionary tracks. The derived ages are mostly between 0.5 and 3 million years (Myr), although some stars retain CTT characteristics as long as approximately 20 Myr.

Class III infrared sources, or WTT stars, have simple blackbody spectral-energy distributions, which implies little or no accretion disk (Wolk & Walter 1996). Many WTT stars occupy the same region on the Hertzsprung-Russell diagram as do CTT stars, although some are approaching the ZAMS. It is tantalizing to surmise that the loss of disks from the Class II–III phases is accompanied by planet formation, because roughly one-third of CTT stars have disks sufficiently massive to produce the primitive solar nebula (e.g. Beckwith et al 1990, André & Montmerle 1994). For stellar ages more than 20 to 30 Myr, all indications of circumstellar disks

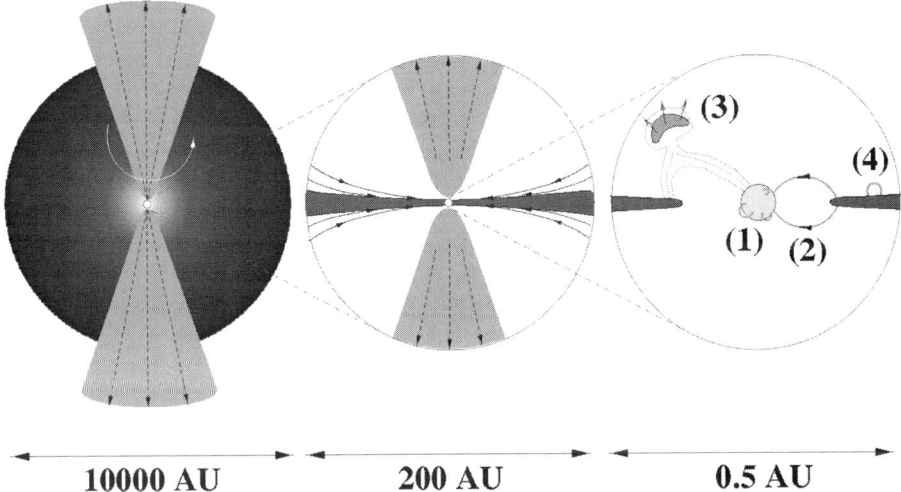

Figure 2 Four magnetic-field configurations that may be responsible for the magnetic activity of Class I protostars. The X rays come from the inner region of a complex structure comprising a collapsing extended envelope (*left*), an inner disk and outflow (*center*), and a star-disk magnetic-interaction region (*right*). (Courtesy of N. Grosso.)

disappear, and we enter the regime of "post-T Tauri" stars. These stars, long missing from YSO samples (Herbig 1978), are now emerging from wide-field X-ray surveys (see Section 6). They are distinguished by their location above the ZAMS (although absence of accurate distances frequently impedes accurate placement on the Hertzsprung-Russell diagram) and by photospheric lithium abundances above those seen in ZAMS stars, because the initial lithium is easily destroyed on the way to the main sequence as a result of convective mixing (Martín 1997).

The interested reader can consult a number of related reviews. Broad treatments of YSOs can be found in *Annual Review of Astronomy and Astrophysics* articles by Shu et al (1987) and Bertout (1989); a monograph by Hartmann (1998); and in conference volumes edited by Lada & Kylafis (1991), Levy & Lunine (1993), and Mannings et al (1999). Various aspects of magnetic activity and flaring in YSOs have been reviewed by Feigelson et al (1991), Montmerle (1991), Montmerle et al (1993), and Glassgold et al (1999). Radio emission is reviewed by André (1996), and recent X-ray results are summarized by Neuhäuser (1997). Some meteoritic implications are discussed by Woolum & Hohenberg (1993).

3. EVIDENCE FOR MAGNETIC ACTIVITY IN YSOs

3.1 Tracers of Magnetic Fields

It is difficult to study YSO magnetic fields directly, and in most cases, indirect tracers of magnetic activity such as cool star spots or high-energy radiation

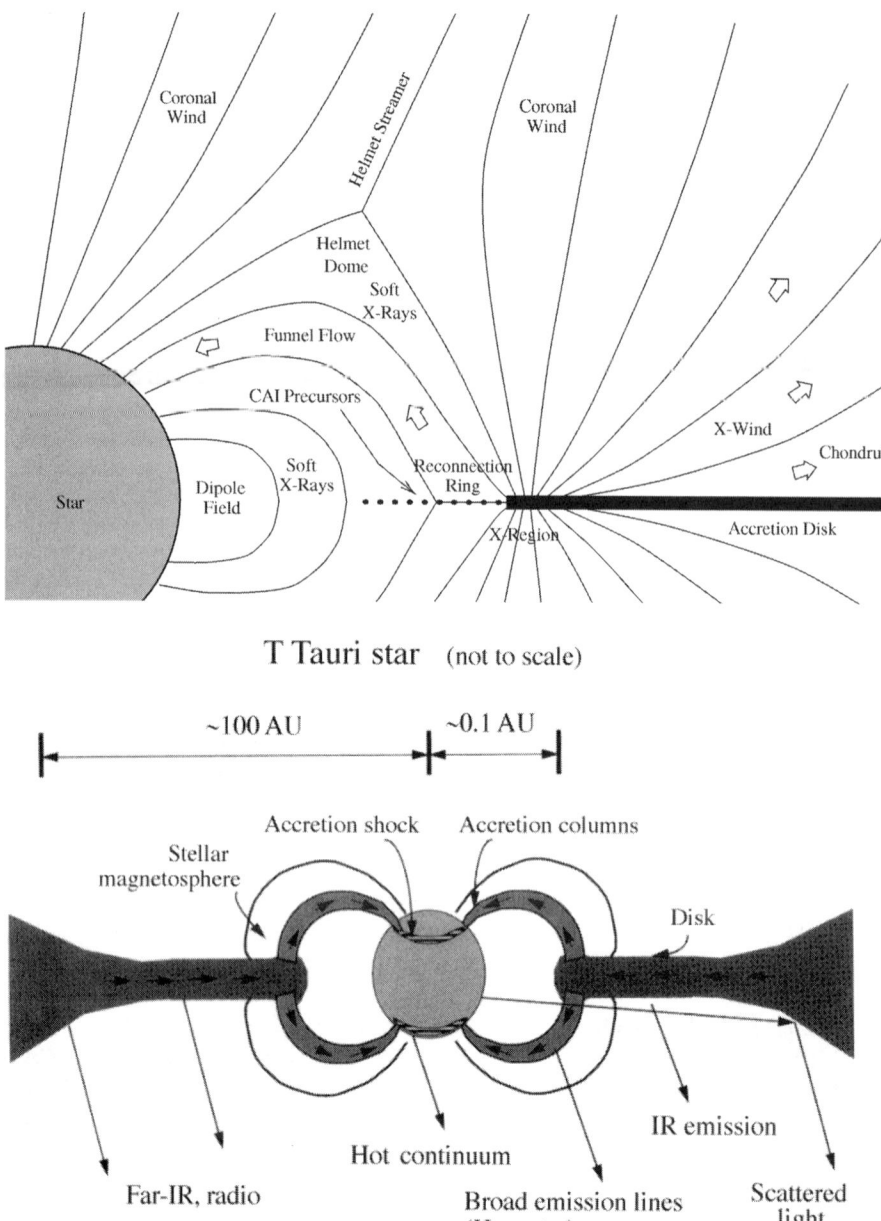

Figure 3 Two contemporary models for Class I–II YSOs, in which magnetic fields play crucial roles: (*top*) the x-wind model of YSOs showing magnetically collimated accretion and outflows with irradiated meteoritic solids (Shu et al 1997); (*bottom*) magnetically funneled accretion streams producing broadened emission lines (Hartmann 1998).

produced by violent field reconnection must be used. Table 1 provides a bibliographic guide to the observational literature on magnetic-activity tracers and is organized by waveband and star-formation regions. We have attempted to be complete in the X-ray and radio listings through 1998, but only a representative selection of the large optical literature is included. The latter includes studies of chromospheres, flares, Zeeman effects, photospheric spots, distances, kinematics, binarity, masses, ages, and other properties of magnetically active YSOs. We briefly examine here what can be learned about the magnetic fields and their reconnection from these observations.

The traditional method for measuring magnetic fields on the surfaces of late-type stars is the detection of Zeeman splitting in magnetically susceptible absorption lines. Applying this method to YSOs has been difficult because of their faintness and (except for WTT/Class III stars) profusion of emission features. Success has recently been achieved in a few cases, indicating fields around 1 to 3 kG covering a large fraction of the photosphere. Photometric and Doppler imaging shows a patchy distribution of large cool star spots, which suggests that the surface fields are complex and multipolar, as in the Sun. These results, however, tell us little about large-scale fields important for star-disk interactions (Figure 2 *right*).

Satellite spectral measurements show that YSO X rays are optically thin thermal bremsstrahlung with associated ionized metal emission lines from multitemperature plasmas with $1 < T_x < 100$ MK (Montmerle et al 1991). The spectral parameters of the emitting plasma—temperature distribution, foreground column density N_H, and metallicity—are similar to those of other magnetically active late-type stars. The importance of flaring is revealed by the X-ray variability. Virtually all YSOs examined at different epochs are variable, and, at any given moment, several percent exhibit luminous flare events with a fast rising curve followed by a slower decay over several hours.

For simple flare models, quantitative properties of the magnetic structures can be inferred from these X-ray flares (e.g. Montmerle et al 1983, Walter & Kuhi 1984). Assuming that radiative cooling dominates conductive cooling in the emitting plasma (and assuming a uniform temperature and density), one can calculate that the plasma density $n_e \simeq 10^{10}$–10^{12} cm^{-3}, which is similar to solar values. The equipartition magnetic field strength $B_{eq}^2/8\pi = 2n_e kT$ (which is the minimum possible value for the field strength dynamic flare loops) is approximately 10^2 G.

The morphology of the magnetic structures is a key question that will be raised several times in this review. X-ray flares give information on the emitting volume, not their geometry. If one considers solar-type cylindrical magnetic tubes with length ℓ and aspect ratio approximately 1:10, luminous YSO X-ray flares require $\ell \sim 10^{11}$–10^{12} cm $> R_*$ (note that typically R_* is approximately 2–3 $R_\odot$), if one assumes radiative cooling. If, however, reheating occurs during the flare decline (as is sometimes seen in solar and stellar flares), smaller values may be invoked (Reale et al 1997). However, the magnetic-field morphology may be far more complicated. For example, Skinner et al (1997) consider cooling loops, two-ribbon

TABLE 1 Observational studies of YSO magnetic activity

Star formation region	Non-imaging satellites	X-ray Einstein Observatory	X-ray ROSAT	X-ray ASCA	Radio	Optical
Orion	1 2 3 4 5 7 16 27 33	6 8 9 11 15 23 26 30	40 41 46 47 48 66 73 75	58 87	8 9 2) 21	4 24 25 47 53
Taurus-Auriga		8 9 10 12 14 17 18 20 38 39 58	29 32 44 50 57 61 66 73 82 86	50 69 76 78 82 86	1 2 4 5 11 13 15 18 19 22 23 24 25 27 29 31	2 3 5 7 8 9 10 11 12 15 17 18 19 22 24 27 30 31 37 38 46 48 51 55 57
Ophiuchus	19 24	13 14 17	36 60 66 73	31 62	3 6 12 14 15 16 17 18 19	2 14 59
Chamaeleon		22	26 35 59 66 73 77	85	26	6 21 23 29 36 39
Lupus		8 42	64 71 72 73			6 40 54
Corona Australis		12 17 70	65 73	52	7 28 32	2
Perseus		18	28 55 67 73 79			50
Sco-Cen		33	80		26	20 34 52 56 58
Other		12 21 37	53 54 63 64 66 68 73 74 81 83 88	63	10 17 19 33	2 32 41
Dispersed			35 43 45 48 51 89 90		17 24 27 30	1 13 16 25 28
YSOs			55 59 72 84			31 35 37 42 43 44 45 49

X-ray references: 1 Giacconi et al 1972; 2 White & Ricketts 1977; 3 Cooke et al 1978; 4 den Boggende et al 1978; 5 Bradt & Kelley 1979; 6 Ku & Chanan 1979; 7 Markert et al 1979; 8 Gahm 1980; 9 Feigelson & DeCampli 1981; 10 Feigelson & Kriss 1981; 11 Pravdo & Marshall 1981; 12 Walter & Kuhi 1981; 13 Montmerle et al 1983; 14 Walter & Kuhi 1984; 15 Stone & Taam 1985; 16 Agrawal et al 1986; 17 Walter 1986; 18 Feigelson et al 1987; 19 Koyama 1987; 20 Walter et al 1987; 21 Tagliaferri et al 1988; 22 Feigelson & Kriss 1989; 23 Strom et al 1990; 24 Koyama et al 1992; 25 Feigelson et al 1993; 26 Pravdo & Angelini 1993; 27 Yamauchi & Koyama 1993; 28 Preibisch et al 1993; 29 Feigelson et al 1994; 30 Gagné & Caillault 1994; 31 Koyama et al 1994; 32 Strom & Strom 1994; 33 Walter et al 1994; 34 Yamauchi et al 1994; 35 Alcalá et al 1995; 36 Casanova et al 1995; 37 Caillault et al 1995; 38 Damiani et al 1995a; 39 Damiani et al 1995b; 40 Gagné et al 1995; 41 Geier et al 1995; 42 Giovannelli et al 1995; 43 Neuhäuser et al 1995a; 44 Neuhäuser et al 1995b; 45 Neuhäuser et al. 1995c; 46 Pravdo & Angelini 1995; 47 Preibisch et al 1995; 48 Sterzik et al 1995; 49 Alcalá et al 1996; 50 Carkner et al 1996; 51 Jeffries et al 1996; 52 Koyama et al 1996; 53 Magnani et al 1996; 54 Park & Finley 1996; 55 Preibisch et al 1996; 56 van den Ancker et al. 1996; 57 Wichmann et al. 1996; 58 Yamauchi et al 1996; 59 Alcalá et al 1997; 60 Grosso et al 1997; 61 Gullbring et al 1997; 62 Kamata et al 1997; 63 Kastner et al 1997; 64 Krautter et al 1997; 65 Neuhäuser & Preibisch 1998; 66 Preibisch 1997a; 67 Preibisch 1997b; 68 Schulz et al 1997; 69 Skinner et al 1997; 70 Walter et al 1997; 71 Wichmann et al 1997a; 72 Wichmann et al 1997b; 73 Carkner et al 1998; 74 Gregorio-Hetem et al 1998; 75 Neuhäuser et al 1998; 76 Skinner & Waltner 1998; 77 Neuhäuser & Comerón 1998; 78 Tsuboi et al 1998; 79 Preibisch et al 1998a; 80 Sciortino et al 1998; 81 Hoff et al 1998; 82 Favata et al 1998; 83 Preibisch 1998; 84 Jensen et al 1998; 85 Yamauchi et al 1998; 86 Favata et al 1999; 87 Nakano et al 1999; 88 Naylor & Fabian 1999; 89 Webb et al 1999; 90 Mamajek et al 1999.

Radio references: 1 Bieging et al 1984; 2 Becker & White 1985; 3 Feigelson & Montmerle 1985; 4 Kutner et al 1986; 5 Cohen & Bieging 1986; 6 André et al 1987; 7 Brown 1987; 8 Churchwell et al 1987; 9 Garay et al 1987; 10 Becker & White 1988; 11 Bieging & Cohen 1989; 12 Stine et al 1988; 13 O'Neal et al 1990; 14 Leous et al 1991; 15 Phillips et al 1991; 16 André et al 1992; 17 Rucinski 1992; 18 White et al 1992a; 19 White et al 1992b; Felli et al 1993a; 21 Felli et al 1993b; 22 Phillips et al 1993; 23 Skinner 1993; 24 Feigelson et al 1994; 25 Skinner & Brown 1994; 26 Brown et al 1996; 27 Phillips et al 1996; 28 Suter et al 1996; 29 Chiang et al 1996; 30 Carkner et al 1997; 31 Ray et al 1997; 32 Feigelson et al 1998; 33 Stine & O'Neal 1998.

Optical references: 1 Herbig 1973; 2 Herbig & Soderblom 1980; 3 Mundt et al 1983; 4 Smith et al 1983; 5 Rydgren & Vrba 1983; 6 Finkenzeller & Basri 1987; 7 Hartmann et al 1987; 8 Vrba et al 1988; 9 Walter et al 1988; 10 Herbst 1989; 11 Strom et al 1989; 12 Mathieu et al 1989; 13 Pasquini et al 1991; 14 Bouvier & Appenzeller 1992; 15 Feigelson et al 1994; 16 Henry & Hall 1994; 17 Joncour et al 1994; 18 Strassmeier et al. 1994; 19 Strom & Strom 1994; 20 Walter et al 1994; 21 Alcalá et al 1995; 22 Bouvier et al 1995; 23 Gahm et al 1995; 24 Hatzes 1995; 25 Sciortino et al 1995; 26 Sterzik et al 1995; 27 Gullbring et al 1996; 28 Jeffries et al 1996; 29 Lawson et al 1996; 30 Rice & Strassmeier 1996; 31 Wichmann et al. 1996; 32 Bouvier, Forestini & Allain 1997; 33 Covino et al 1997; 34 Feigelson & Lawson 1997; 35 Magazzu et al 1997; 36 Covino et al 1997; 37 Donati et al 1997; 38 Frink et al 1997; 39 Guenther & Emerson 1997; 40 Johns-Krull & Hatzes 1997; 41 Martín 1997; 42 Mcceia et al 1997; 43 Motch et al 1997; 44 Neuhäuser et al 1997; 45 Sterzik et al. 1997; 46 Strassmeier et al 1997; 47 Alcalá et al 1998; 48 Guenther et al 1999; 49 Guillout et al 1998; 50 Herbig 1998; 51 Johns-Krull et al 1998; 52 Martín 1998; 53 Neuhäuser et al 1998; 54 Wichmann et al 1998; 55 Köhler & Leinert 1998; 56 Preibisch et al 1998b; 57 Luhman et al 1998; 58 Brandner & Köhler 1998; 59 Martín et al 1998.

flares, interbinary flares, and star-disk magnetic reconnection in a discussion of a giant X-ray flare on the Class II–III star V773 Tau. Solar two-ribbon flares (Schmitt 1994, Güdel et al 1999) and rising magnetic arches associated with eruptive solar flares (Svestka et al 1995), which involve complex evolving magnetic geometries and continuous injection of energy, may be valuable analogies for YSO flares.

Radio-continuum flares also provide clues to the magnetic fields in YSOs. The emission, seen in many Class III stars and one Class I object, is highly variable and sufficiently bright to have detectable circular polarization in a few cases. The emission mechanism is quite clearly gyrosynchrotron radiation (as seen in the Sun and in other late-type magnetically active stars), produced by mildly relativistic electrons with energies around 1 MeV spiraling in $\simeq 1$ G fields (Dulk 1985). In one dramatic case, linear polarization is also present, which implies electron acceleration up to several MeVs, which has not been seen in other stellar flares (Phillips et al 1996). Simple single-loop models of a rapid radio flare on the Class III star DoAr 21 might indicate a radio loop size significantly larger than X-ray loops (Montmerle et al 1993, André et al 1987).

We proceed with a more detailed presentation of magnetic tracers in YSOs, first by treating the X-ray emission—for which CTT and WTT stars have many features in common—and then by considering separately CTT and WTT properties at other wavelengths.

3.2 X-ray Properties of T Tauri Stars

A typical X-ray image of a nearby ($d < 500$ pc) star-forming region shows dozens (or, for the Orion Trapezium cluster, hundreds) of faint X-ray sources (e.g. Montmerle et al 1983, Strom & Strom 1994, Gagné et al 1995, Preibisch et al 1996; Figure 4). Most of these X-ray sources are associated with Class III WTT stars, but we will refer here to the entire T Tauri population (CTT and WTT) because X-ray properties demonstrate little or no dependence on disk interactions. Repeated imaging shows that most X-ray T Tauri stars vary on time scales of days or longer and, at any given moment, five to ten percent of the stars are caught in a high-amplitude flare with time scales of hours. The X-ray spectra show multitemperature plasma and are often modeled as a soft component with $T_x \simeq 2$–5 MK and a hard component with $T_x \simeq 15$–30 MK, although weak emission at higher temperatures may be present (Preibisch 1997a).

Without the flares, the YSO soft X-ray luminosity function is traced from approximately $10^{28.5}$ to approximately 10^{31} erg s^{-1}. It is difficult to attribute any single average X-ray luminosity to an X-ray population. For example, a particular ROSAT observation of the Chamaeleon I cloud gives a median value of $L_x \simeq 10^{29.4}$ erg s^{-1} for previously identified cloud members (Feigelson et al 1993), $L_x \simeq 10^{29.0}$ erg s^{-1} when X-ray–discovered stars are included (Lawson et al 1996), and a lower value when new low-mass ISO-discovered stars are included (Persi et al 1998). Median luminosities also appear higher when more distant regions (e.g. $d \sim 1$ kpc; Schulz et al 1997, Gregorio-Hetem et al 1998)

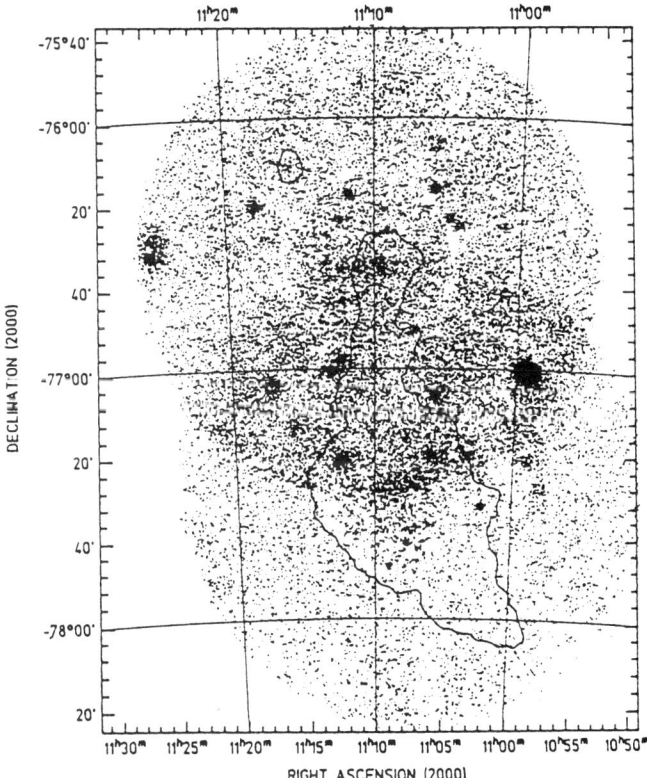

Figure 4 Two ROSAT soft X-ray images of the Chamaeleon I star-forming cloud showing dozens of X-ray–emitting YSOs (Feigelson et al 1993). The contour traces the outline of the cloud from the 100-μm IRAS map.

or less sensitive X-ray data (such as the ROSAT All-Sky Survey) are considered, which is clearly a selection effect.

A number of relationships between X-ray and other stellar properties are found. Most evidently, L_x scales roughly linearly with stellar bolometric luminosity L_{bol}. This relationship is frequently summarized as a constant ratio $L_x/L_{bol} \approx 10^{-4}$ (the exact value depends on the sensitivity and sample definition), which lies well below the rotational saturation level of $L_x/L_{bol} \approx 10^{-3}$ seen in late-type stellar populations (Fleming et al 1989). L_x is also correlated with stellar mass, photospheric temperature, stellar radius, and rotation (e.g. Feigelson 1993). The average T Tauri X-ray luminosity appears to be constant with age or, equivalently, L_x/L_{bol} increases with age (Kastner et al 1997), which implies that the active regions of the stellar surface occupy a roughly constant total area as stars contract. Thus, the X-ray surface flux $F_x = L_x/4\pi R_*^2$ increases with age as stars descend the Hayashi tracks.

The astrophysical origins of these relationships are poorly understood, and it is unclear which relationships represent causal links rather than ancillary statistical dependencies. The only relation expected from standard magnetic-activity theory is between L_x and some rotation indicator—reflecting a more powerful magnetic dynamo in rapidly rotating convective stars—and a link between X-ray surface flux and X-ray temperature ($\sim F_x \propto T_x^2$; Preibisch 1997a), which is expected from standard solar-type magnetic-loop flare models. It is possible, for example, that the L_x mass correlation represents a fundamental connection between YSO surface activity and the incorporation of fossil magnetic fields from the star formation process. The dissipation of such fields by ohmic and turbulent decay can be sufficiently slow that they may be present throughout the T Tauri phase (Tayler 1987).

Large numbers of previously unknown WTT stars appear in X-ray images of nearby star-forming regions (e.g. Feigelson & Kriss 1981, Lawson et al 1996, Neuhäuser 1997), which often increases the catalogued YSO population by factors of two or more (Section 6). However, many well-studied CTT stars are also X-ray luminous. Roughly half of the catalogued CTT stars in Taurus-Auriga and Chamaeleon I clouds are detected in X-ray surveys sensitive to $L_x \simeq 10^{29.0}$ erg s^{-1} (Feigelson et al 1993, Damiani et al 1995a). CTT X-ray properties are similar, if not identical, to those of WTTs. Their X-ray luminosities range from less than $10^{28.5}$ to $10^{30.5}$ erg s^{-1} in the soft X-ray bands, and they are uncorrelated with either Hα luminosity or infrared excess. Their luminosity function sometimes appears slightly diminished compared with that of WTT stars, but that difference can be attributed to low-L_x WTT stars that are not yet identified.

Most CTT X-ray emissions vary by factors of 2 to 10 on timescales of months (Montmerle et al 1983, Walter & Kuhi 1984), and they can occasionally exhibit rapid flares. Figure 5 shows the extraordinary flare in LHα92, reaching 5×10^{32} erg s^{-1} on timescales of an hour. The X-ray plasma temperature rose from about 15 MK during quiescence to 40 MK at the peak during this event (Preibisch et al 1993). A flare with a similar light curve but a more modest peak $L_x \simeq 2 \times 10^{30}$ erg s^{-1} was seen in DD Tau (Strom & Strom 1994). The deeply embedded source SVS 16 in the NGC 1333 star-forming region, which is probably a CTT star, is anomalous: its X-ray emission is constant at a level around 2×10^{32} erg s^{-1}, far higher than the quiescent level of other T Tauri stars (Preibisch 1998).

3.3 Weak-Lined T Tauri Stars

Unresolved radio-continuum emission is seen in several dozen WTT stars at levels of 10^{15} to 10^{18} erg s^{-1} Hz^{-1} or three to six orders of magnitude brighter than powerful contemporary solar flares (e.g. Garay et al 1987, Stine et al 1988, White et al 1992a, Chiang et al 1996). Roughly three-fourths of observed stars are undetected because of instrumental-sensitivity limits; the NRAO Very Large Array has been the most effective telescope. All sources exhibit high-amplitude variability on long time scales, and a few flares on time scales of hours have been caught in two nearby WTT stars, HD 283447 and DoAr 21 (Feigelson & Montmerle 1985,

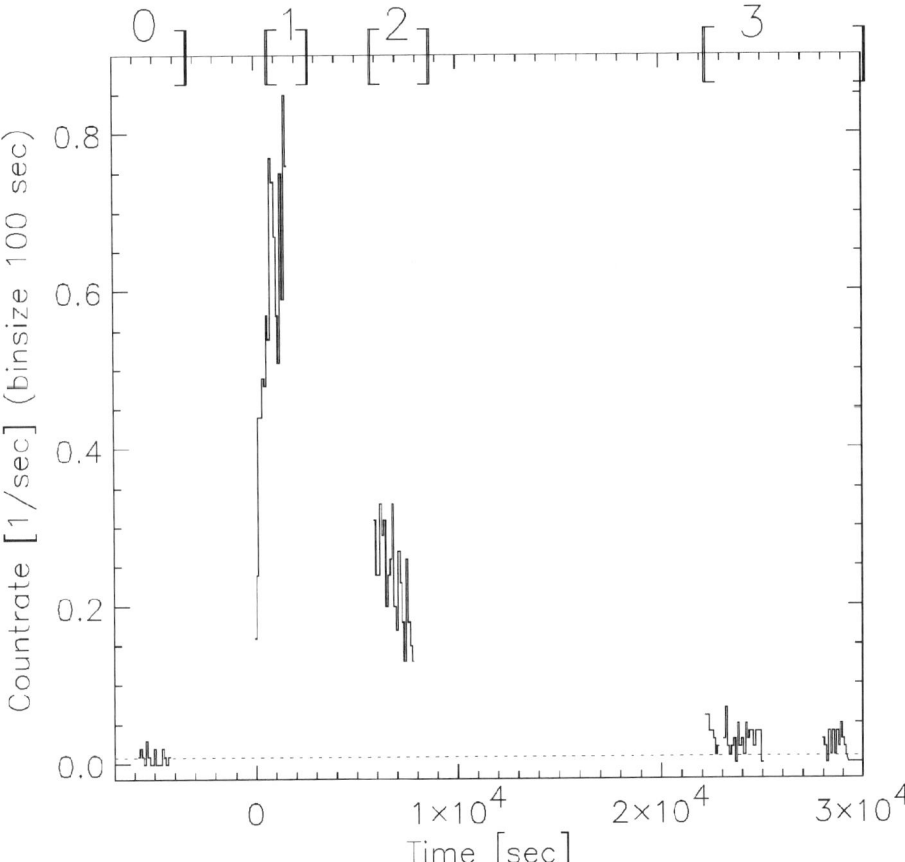

Figure 5 ROSAT light curve of a powerful flare in the Class II CTT star LkHα92 (Preibisch et al 1993).

Phillips et al 1996). Although radio emission from Class I–II objects can be either thermal or nonthermal, the absence of circumstellar disks and ejecta from Class III objects points to a nonthermal mechanism (Montmerle 1991). This possibility is confirmed in several of the stronger sources, in which circular polarization at a level of a few percent is seen (White et al 1992b, Skinner 1993) and VLBI measurements show that the emitting region is too small for thermal processes (Phillips et al 1991). In nearly all respects, WTT radio emission closely resembles that seen in the class of RS CVn magnetically active late-type binary stars.

Optical and UV studies reveal magnetic effects on the stellar surface in several ways. First, a large body of literature from photometric and Doppler imaging studies shows rotational modulations of star spots that cover less than five percent to approximately 50 percent of the surface with temperatures of approximately 500 to 1000 K below the photospheric temperature (e.g. Rydgren & Vrba 1983, Bouvier et al 1995, Joncour et al 1994, Strassmeier et al 1994). The WTT star V410 Tau, for example, has both large high-latitude and low-latitude spots; one large active region may have persisted for more than 1000 rotations (Vrba et al 1988, Rice & Strassmeier 1996). Second, magnetic flares can be seen photometrically or spectroscopically, despite the overwhelming photospheric emission. The X-ray–discovered star V826 Tau exhibited an excursion of $\Delta U = 0.5$ in 40 min (Rydgren & Vrba 1983) and later, a 1-h X-ray flare (Carkner et al 1996). Sudden increases in Balmer continuum and line emission have also been seen with total power around 10^{33}–10^{34} ergs (Gahm et al 1995, Guenther & Emerson 1997). Third, there has been recent success in measurements of Zeeman effects on photospheric absorption lines. Magnetic enhancement of the equivalent width of Fe I lines in the WTT star LkCa 16 is clearly seen and can be interpreted as $B = 2.4$ kG fields covering $f = 0.6$ of the stellar surface (Guenther et al 1999). Spectropolarimetric observations have detected fields in V410 Tau, HD 283472, and HD 155555 (Donati et al 1997).

To illustrate the interwoven phenomenology of magnetic activity at different wavelengths, we briefly describe the most active T Tauri star in the Taurus-Auriga cloud complex, HD 283447 = V773 Tau (Feigelson et al 1994, Phillips et al 1996, Skinner et al 1997, Tsuboi et al 1998). It is a hierarchical triple system comprising a K2–K3 close binary and a distant K3 companion. The close binary has $L_{bol} \simeq 5\, L_\odot$, age 1 Myr, and rapid rotation with $v \sin i = 44$ km s^{-1} and $P_{rot} = 3.4$ days. The system has a truncated circumstellar disk and very faint broad Hα emission and thus might be considered an intermediate WTT/CTT object. About 17 percent of the surface is covered with a rotationally modulated cool spot, and the MgII chromospheric emission is strongly elevated. The continuous quiescent X-ray emission from the system is unusually strong with $L_x \simeq 1 \times 10^{31}$ erg s^{-1}, and the star repeatedly exhibits powerful day-long flares with peak $L_x \simeq 2 - 10 \times 10^{32}$ erg s^{-1} with peak plasma temperatures of more than 100 MK. Both inner companions are radio loud. The centimeter radio emission is highly variable at around several times 10^{17} erg s^{-1} Hz^{-1}. Unique among late-type stars, the emission exhibits both circular and linear polarization, which indicates electron acceleration to energies significantly more than one MeV. The unusually high levels of magnetic activity in V773 Tau might be attributed to the early loss of its interacting disk, which left the star with a rapid rotation rate and consequently strong magnetic dynamo. Similar systems include V410 Tau and HDE 283572 in Taurus-Auriga (Rice & Strassmeier 1996, Walter et al 1987), DoAr 21 in Ophiuchus (Feigelson & Montmerle 1985), and Par 1724 in Orion (Neuhäuser et al 1998).

The more typical solar-mass WTT star, however, is 2–20 Myr old with slow rotation ($P_{rot} > 10$ days), $L_x \simeq 10^{28.5}$–$10^{29.5}$ erg s^{-1}, no known X-ray flares, and undetectable radio emission or photometric spots. By studying the initial mass function, we know that the most common Class III stars must be pre–main-sequence M stars with $M_* \sim 0.2 - 0.6\, M_\odot$ and that significant numbers of brown dwarfs exist. Because of the (unexplained) L_x-mass correlation, these brown dwarfs are underrepresented in existing X-ray studies. At least two brown dwarfs have been detected in X rays to date at low levels (Neuhäuser et al 1999).

WTT stars extend empirical relations between magnetic activity tracers seen in older solarlike stars. A sample of 1 $M_\odot$ main-sequence stars with a wide range of ages shows several consistent patterns: As one moves from the oldest disk population to ZAMS stars, radio luminosity rises by 10^4, X-ray luminosity rises by 10^3, and the fraction of plasma at $T_x > 10$ MK rises by 10^2 in emission measure (Gudel et al 1997). These trends are explained by a scaling of magnetic activity with rotational velocity, in which X-ray (but not radio) luminosity is limited by saturation at the stellar surface and average plasma temperatures rise with increased fraction of plasma in microflares rather than in the quiescent corona. Although the data are still fragmentary, T Tauri stars appear to follow this pattern: X-ray luminosities are generally below saturation levels and their L_r/L_x ratios and average plasma temperatures are several times higher than in ZAMS stars (Skinner & Walter 1998).

3.4 Classical T Tauri Stars

In contrast with WTT stars, evidence for CTT magnetic activity at non–X-ray wavebands is sparse. A few CTT stars are detected in continuum at centimeter wavelengths (e.g. LkHα101, DG Tau, T Tau North), but this emission is caused by thermal bremsstrahlung in partially ionized winds (Cohen & Bieging 1986). The apparent absence of nonthermal radio emission in CTTs may not mean that they are magnetically inactive. André (1987) convincingly argues that, under reasonable assumptions of ionization and geometry, YSO winds are more than sufficiently dense to free-free absorb nonthermal radio-continuum emissions produced close to the star.

In the optical band, the manifestations of magnetic activity seen in WTT stars are intermixed with the powerful effects of star-disk interactions such as emission lines, hotspots, continuum veiling, and high-amplitude aperiodic photometric variations (Smith et al 1999). For example, BP Tau shows complex emission line variations that are more readily interpreted in terms of magnetically channeled accretion hotspots than of magnetic reconnection flares (Gullbring et al 1996). However, valuable results have emerged from studies of CTT stars that have relatively weak winds and veiling but still show indications of star-disk interactions. Doppler imaging of the CTT/WTT star Sz 68 reveals near-polar star spots several hundred degrees cooler than the photosphere (Johns-Krull & Hatzes 1997). The inferred spotted surface is similar to that seen in WTT stars but is accompanied by variable red-shifted Hα and Na I D absorption likely to be caused by a magnetospheric accretion flow. Magnetic activity in Sz 68 is also evident from its X-ray emission at $L_x \simeq 6 \times 10^{29}$ erg s^{-1} (Krautter et al 1997).

Efforts to detect direct manifestations of magnetic fields in CTT optical-infrared spectra have recently borne fruit. The spectrum of BP Tau shows broadening in the Zeeman-sensitive Ti I line at 22.2 μm, corresponding to a total magnetic flux of $Bf \simeq 3.3 \pm 0.3$ kG (Johns-Krull et al 1999). Enhancements in equivalent widths of photospheric lines sensitive to the Zeeman effect have been found in the CTT star T Tau North and CTT/WTT star LkCa 15, indicating fields of $Bf \simeq 2.5$ and 1.4 kG, respectively (Guenther et al 1999). These field strengths are comparable with those of solar active regions but cover large fractions of the stellar surface.

The simplest conclusion we can draw is that CTT stars are magnetically very similar to WTT stars. The X-ray emission and flaring, the X-ray loop sizes, and the photospheric magnetic fields inferred from Zeeman effects seem largely unaffected by the complex star-disk interactions that produce a complex combination of accretion and outflows. The absence of radio gyrosynchrotron emission can be explained as an absorption effect, combined with smaller magnetic radio loops. It is thus tempting to conclude, by analogy with WTT stars where the multiwavelength data are more complete, that CTT stars are basically characterized by enhanced solar-type magnetic activity and do not exhibit any radical differences in magnetic geometries or behaviors.

3.5 Protostars

Detection of X rays from magnetic activity in Class I or 0 stars is inherently very difficult; these YSOs are almost always deeply embedded in their nascent molecular clouds, are surrounded by infalling envelopes and outflowing jets, and may be additionally obscured by their circumstellar disks. Typical extinctions from these components are $10 \leq A_V \leq 50$ for Class I sources and may reach $A_V \simeq 1000$ for Class 0 sources. Nonthermal radio emission is not absorbed by intervening cold material but is likely to suffer free-free absorption from ionized gas associated with the base of bipolar outflows, as in CTT stars (André 1996). Such gas is shown to be frequently present by elongated radio-continuum structures associated with the YSOs producing Herbig-Haro objects (Rodríguez 1997).

In light of these observational difficulties, evidence for protostellar magnetic activity did not emerge until the mid-1990s (Table 2). The first indication was the tentative report of several Class I stars among the dozens of faint X-ray sources found in a deep ROSAT exposure of the rich YSO cluster in the ρ Ophiuchi cloud, which was later confirmed with ASCA satellite observations (Casanova et al 1995, Kamata et al 1997). Unequivocal X-ray detection of a cluster of Class I sources emerged from an ASCA study of the Corona Australis cloud core (Koyama et al 1996). Remarkably, these sources were seen in the 4- to 10-keV band, whereas the T Tauri stars in the same field were found only in the 0.5- to 2-keV band (Figure 6). Note, however, that these detections probably represent the most magnetically active protostars, as most of the protostars falling in ROSAT PSPC fields are not detected at these levels (Carkner et al 1998).

Several protostars exhibit quite extraordinary X-ray properties that distinguish them from T Tauri and other late-type stars. The strongest X-ray source in the Corona Australis cloud core is likely (although crowding raises some doubt) associated with the extremely young Class 0–I YSO CrA IRS 7, which exhibited a flare similar to those seen in T Tauri stars but with a spectrum extended to 10 keV and with a strong and curiously broadened iron emission line complex around 6 to 7 keV (Figure 6; Koyama et al 1996). YLW 15 in the ρ Ophiuchi cloud core,

TABLE 2 X-ray detected protostars

Cluster	Source	A_v^a	Detector	L_x^b	Variability	Ref.[c]
R CrA	IRS1 (TS2.6)	19†	SIS	3	—	1
			PSPC	3	—	2
	IRS2 (TS13.1)	19†	SIS	2	—	1
			PSPC	9	—	2
			HRI	4	—	2
	IRS5 (TS2.4)	19†	SIS	3	—	1
			PSPC	5	—	2
	IRS7 (R1)	19†	SIS	$\to 12$	Flare	1
	IRS9 (R2) ?	19†	SIS	2	—	1
ρ Oph	YLW15 (IRS43)	56*	PSPC	160	—	3
		30*	HRI	$\to 10^{3-5}$	Flare	4
		15†	SIS	150	Periodic?	5
	EL29	17†	SIS	$3 \to 54$	Flare	6
	IRS44	52*	PSPC	<80	—	3
		13†	SIS	$3 \to 10$	Flare?	6
	IRS46	37*	PSPC	<20	—	3
		13†	SIS	$3 \to 11$	Flare?	6
	WL6	22†	SIS	$3 - 14$	Periodic?	6
Cha I	Ced110 IRS6	30*	PSPC	2	—	7
Orion	SSV63	50*	SIS	$80 \to 720$	Flare	8

[a]Visual absorption in magnitudes estimated from infrared colors (∗) or from the X-ray spectrum (†).

[b]L_x is given in units of 10^{30} erg s^{-1} cm^{-2} for the energy range of the detector: 0.1–2.4 keV for ROSAT PSPC and HRI; 0.5–10 keV (ref. 2) and 2–10 keV (ref. 5) for ASCA SIS. Arrows point to the peak luminosity of flares.

[c]References:
1. Koyama et al 1996
2. Neuhaüser & Preibisch 1997
3. Casanova et al 1995
4. Grosso et al 1997
5. Tsuboi et al 1999 (submitted for publication)
6. Kamata et al 1997
7. Carkner et al 1998
8. Ozawa et al 1999 (submitted for publication)

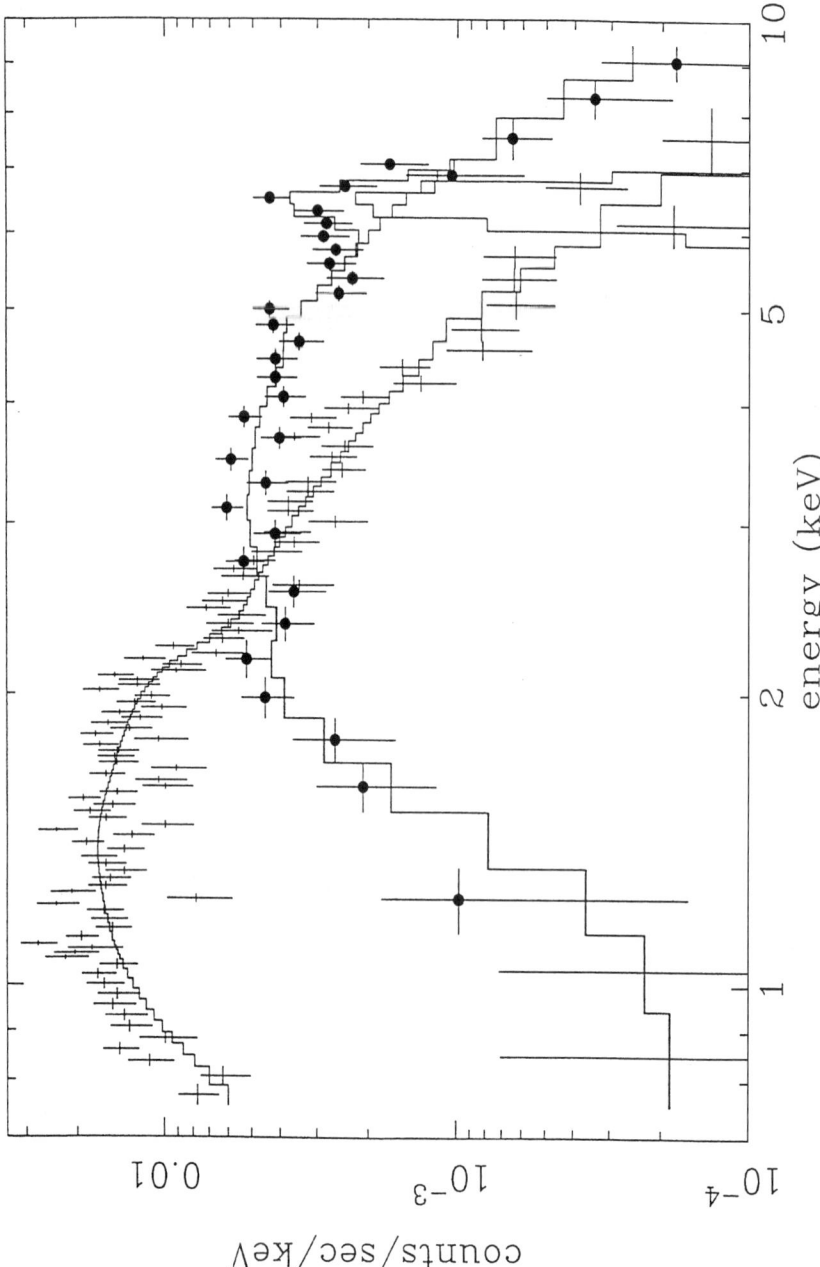

Figure 6 ASCA spectrum of a hotter absorbed Class 0–I protostars (*filled circles*) and a cooler unabsorbed Class III WTT star in the Corona Australis cloud core (Koyama et al 1996).

a luminous and heavily obscured Class I protostar, exhibited a powerful X-ray flare with a 5-h decay in a ROSAT HRI observation. Depending on the unknown plasma temperature and uncertain absorption ($A_V \sim 30$), this event reached a peak $L_x \sim 10^{33}$–10^{35} erg s^{-1}, perhaps the most powerful X-ray flare ever seen in a late-type star (Grosso et al 1997). The associated sizes for the magnetically confined plasma are of order 0.1 AU for this event. It is possible that such superflares were responsible for enigmatic variations seen years earlier with nonimaging Tenma and Ginga satellite observations of the ρ Ophiuchi cloud, with a total unresolved emission of $L_x \sim 10^{32}$ erg s^{-1} (Koyama 1987, Koyama et al 1992). Last, one of a dense group of heavily embedded ($A_V \simeq 30$) infrared sources in the Serpens cloud core, which are likely to be protostars, was found to have $L_x \simeq 10^{32.5}$–$10^{33.5}$ erg s^{-1} in each of two ROSAT observations separated by 2.5 years (Preibisch 1998). This finding may be the most luminous level of quiescent X ray emission yet seen in a YSO.

Although only one tentative Class 0 source in X rays has been detected to date (CrA IRS 7; Koyama et al 1996), it is unclear whether X-ray emission or other magnetic activity indicators are present at this early stage. Class 0 protostars produce unusually powerful outflows (Bontemps et al 1996), but they could be collimated by nonmagnetic mechanisms (Henriksen et al 1997). It is also conceivable that high-energy processes occur far from the central protostar, caused by the compression and reconnection of magnetic fields in the collapsing envelope (Norman & Heyvaerts 1985).

Very Large Array studies demonstrate that protostars typically exhibit extended thermal radio-continuum emission at the same luminosity levels as seen in WTT stars, 10^{15}–10^{17} erg s^{-1} Hz^{-1} (Anglada 1996). This emission pattern is attributed to shock-induced ionization at the base of the outflow. However, two unusual cases of nonthermal radio-continuum emission from protostars have been found. First, T Tau South, the infrared companion to the optically bright CTT prototype T Tau North, has circularly polarized centimeter wavelength emission. T Tau South has been spatially resolved into two lobes of opposite helicity with the MERLIN interferometer on a scale of 10 AU (Ray et al 1997). This result may arise from magnetic shocks and particle acceleration in the bipolar outflow. (A linearly polarized triple-radio source associated with a protostar in Serpens may have a similar origin; Henriksen et al 1991.) The second case, the X-ray–emitting IRS 5 in the Corona Australis cloud, more closely resembles a magnetically active WTT. Its centimeter emission is 10^{16}–10^{17} erg s^{-1} Hz^{-1}, varying by factors of 2–10, and its circular polarization fraction jumped from 10 to 37 percent in a day (Feigelson et al 1998). CrA IRS 5 does not appear to be powering an outflow, so its environment may be relatively free of absorbing ionized material.

With optical/UV band studies precluded by obscuration, and only a handful of cases with detected X-ray and nonthermal radio emission, our knowledge of protostellar magnetic activity is still fragmentary. There are nonetheless tantalizing suggestions that protostellar X-ray flares can be more powerful and can produce hotter plasma temperature components than seen in T Tauri flares.

4. ORIGIN OF THE MAGNETIC ACTIVITY

From the observations outlined here, we can distinguish three lines of evidence for magnetically generated high-energy processes in YSOs: X-ray emission with high levels of continuous emission and powerful 10^4-s flares of $T \simeq 10^6$ to 10^8 K plasma; gyrosynchrotron radio-continuum emission and flares from MeV electrons spiraling along large-scale and well-ordered magnetic-field lines; and various optical studies (photometric and Doppler reconstructed star spots, spectroscopic chromospheric indicators, Zeeman line broadening, and photometric and spectroscopic flares) demonstrating that photospheric surfaces have elevated magnetic fields and associated active regions and flares. The X-ray and radio observations provide explicit evidence for high-energy photons and particles, which probably can be produced only in explosive magnetic reconnection events and confined in closed large-scale magnetic structures.

Together, these lines of evidence suggest that magnetic fields rooted at the stellar surface are responsible for the observed flaring. This reasoning led to the model of YSOs as enhanced solar-type magnetic activity (e.g. Montmerle et al 1983, Feigelson et al 1991). The magnetic activity of CTT stars appears similar in most respects, although most characteristics are masked by the signatures of accretion and wind ejection. Alternative models, such as CTT X rays arising from the release of gravitational energy in magnetic accretion columns (Lamzin et al 1996) or from colliding CTT winds (Zhekov et al 1994), cannot account for the hard X-ray spectra or the lack of correlation between X-ray and optical variability and between X-ray and emission lines.

Enhanced solar-type magnetic activity is not surprising in cool stars with deep convection zones and relatively rapid rotation, because a magnetic dynamo should be active unless the star rotates like a solid body. For the quiet Sun, the large-scale magnetic structure can be explained—and analytically computed without a dynamo—combining an azimuthal equatorial-current sheet and an axisymmetric multipole field representing the internal magnetic field (Banaskiewicz et al 1998). The most popular theory, the α-ω dynamo, combines convection and differential rotation to amplify fields, which then erupt through the surface and produce the observed magnetic tracers described above (Gilman 1983). This model is an important element in successfully explaining various solar phenomena such as the butterfly diagram and convective amplification of photospheric fields. Although the dynamo mechanism probably also operates on late-type stars, no detailed calculations or models exist for YSOs. Strictly construed, the standard α-ω dynamo breaks down for fully convective T Tauri stars. It is also disturbing that the available T Tauri data show rather noisy correlations between L_x-rotation diagrams. The current evidence for enhanced solar-type magnetic activity in YSOs is thus based more on empirical analogies than astrophysical insights.

The solar-type activity interpretation of high-energy emission in the earlier YSO stages is under challenge, mainly by growing arguments for field structures far larger than those seen on the Sun. Recent models of Class I–II systems require that stellar magnetic fields have a strong bipolar component that extends out several stellar radii, where it couples to the disk at the star-disk corotation radius, which is typically around 3 to 10 R_* from the stellar surface (e.g. Königl 1991, Shu et al 1994, Ferreira & Pelletier 1995, Paatz & Camenzind 1996, Spruit et al 1997). These models have considerable explanatory power, accounting for the generation of winds and collimated outflows (Figure 3 *top*), the distribution of YSO surface rotational velocities (Cameron & Campbell 1993, Armitage & Clarke 1996, Li 1996, Bouvier et al 1997), and the unusual shape of the broad optical emission lines as magnetically funneled accretion streams (Hartmann et al 1994, Figure 3 *bottom*). The real situation, however, is likely to be far more complex than YSO models based on dipole magnetic geometry and steady-state assumptions (Safier 1998). For example, time-dependent MHD numerical calculations illustrate rapid changes of any initial magnetic field configuration within a few disk orbital periods (Miller & Stone 1998).

From a broad perspective, we can consider four possible magnetic geometries in YSOs (Figure 2, *right panel*):

1. Solar-type multipolar fields with both footprints rooted in the stellar photosphere. As in the Sun, reconnection would arise from differential rotation and convection under the photosphere. This model is well adapted to the WTT and CTT stars. The fields might be generated by an α-ω-type dynamo, or might be "fossil fields" inherited from the parent molecular cloud.

2. Field lines connecting the star to the circumstellar disk at its corotation radius. This configuration is supported by models of rotational spindown (e.g. Königl 1991, Edwards et al 1993), magnetically funneled accretion, and collimated outflows in CTT stars. Here, reconnection occurs in field lines near but not directly at the corotation radius, perhaps because of the passage of accretion inhomogeneities through the disk (Shu et al 1997). There is also the possibility of reconnection inside the corotation radius if the star is still rotating rapidly with respect to the inner disk. This may be the case for young protostars before magnetic braking is complete. The quasi-periodic X-ray flares in the Class I protostar YLW 15 may arise from such a situation (Tsuboi et al 1999).

3. Field lines above the corotation radius. Two-dimensional, time-dependent MHD calculations of a stellar field threading a Keplerian circumstellar disk predict that plasmoids filled with X-ray–emitting gas will be ejected away from the disk by reconnection events caused by star-disk differential rotation (Figure 7; Hayashi et al 1996).

4. Magnetic loops with both feet in the disk. The combination of differential rotation and convective motions within the circumstellar disk may produce a self-amplifying magnetic dynamo in the disk. Magnetized disk models of

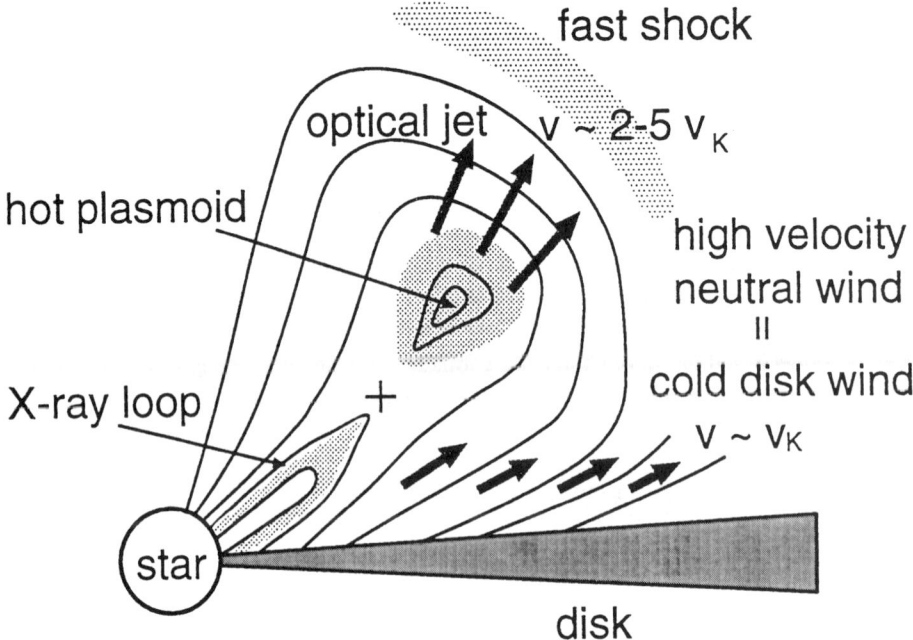

Figure 7 The result of an MHD calculation of magnetic reconnection and plasmon ejection above the star-disk corotation radius in a YSO (Hayashi et al 1996).

this type have been discussed in the contexts of both astrophysical disks (e.g. Field & Rogers 1993, Romanova et al 1998) and the solar nebula, in the context of meteoritics (Levy & Araki 1989; also see Section 5.3).

A fifth possible configuration can be imagined involving magnetic connections between companions in short-period binary YSOs. However, only a small fraction of YSOs reside in binaries such as RS CVn systems, in which the orbits are sufficiently small for this possibility to be considered (Mathieu 1994). RS CVn magnetic activity is usually explained by individually enhanced fields produced by internal dynamos caused by tidally locked rapid rotation, whereas, in YSOs, the rotation is inherited from the star formation process (either directly or through a disk). Therefore, binarity is not expected to play an important role in explaining the magnetic activity of YSOs.

Option 1 has been extensively studied in the solar context, and is the basic explanation for magnetic activity in older late-type stars such as dMe and RS CVn binary stars. For YSOs, the possibility of disk-based fields (options 2–4) has been the subject of considerable recent, detailed study. The geometry of star-disk fields depends critically on the magnetic resistivity of the disk: a fully ionized disk will tend to exclude external fields, whereas a neutral disk will be fully threaded but uncoupled from such fields (Miller & Stone 1998). Outflows may be accelerated along open stellar poloidal field lines by complex interactions at the star-disk field boundary and/or by disk poloidal fields (e.g. Blandford & Payne 1982, Uchida & Shibata 1984, Shu et al 1994, Hirose et al 1997, Ouyed & Pudritz 1997, Kudoh & Shibata 1997, Shang et al 1998). Time-dependent MHD calculations of the star-disk field interaction (Figure 7) show a sequence of twisting of closed stellar field lines, current sheet formation, reconnection, and ejection of a magnetic island with plasma heated to $\leq 10^8$ K, consistent with X-ray flare observations (Hayashi et al 1996; Goodson et al 1997). The plasma physics of one specific magnetic reconnection process—the resistive tearing instability—has been investigated in this context, and this process shows growth rates on time scales of hours (Birk 1998). If field lines thread the disk at different disk radii, differential rotation will cause twisting and reconnection, perhaps leading to a perpetually X-ray emitting, flaring disk corona (Tout & Pringle 1996, Romanova et al 1998). Given that resulting X rays are likely to promote ionization and turbulence in the disk (see Section 5.1), it seems plausible that this situation may commonly occur.

We conclude that magnetic reconnection processes occur very frequently in YSOs. The fundamental conditions for reconnection, continual displacement of magnetic footprints, is even more likely to occur in YSOs than in other stellar situations due to the many possible magnetic configurations offered by star-star, star-disk, and disk-disk interactions.

5. EFFECTS ON THE CIRCUMSTELLAR AND INTERSTELLAR ENVIRONMENT

Because YSOs are surrounded by circumstellar gas and dust and may be deeply embedded in molecular cores (at least in their early stages of evolution), X rays will induce a variety of effects, including ionization, heating, and modification of gas chemistry and dust grain composition. The issues outlined here are reviewed in more detail by Glassgold et al (1999).

5.1 X-ray Ionization

X-ray ionization of a fraction of the primarily molecular gas within and around YSOs is particularly important because of its role in coupling gas and magnetic fields. Unlike high-mass YSOs emitting copious UV photons, which create a fully ionized HII region terminated by a thin transition region, a low-mass YSO emitting X rays produces an extended region of low ionization and X-ray heating. The effects of the X rays gradually become negligible because of absorption by intervening material and geometric dilution. Research on these X-ray dissociation regions began many years ago (e.g. Dalgarno & McCray 1972, Halpern & Grindlay 1980, Lepp & McCray 1983, Krolik & Kallman 1983) and has recently undergone a vigorous renewal.

The most important X-ray ionization process of cold material is photoionization of the inner K and L shells of heavy elements: a $\sim$1-keV photoelectron is generated, which produces a cascade of secondary electrons that are responsible for most subsequent ionizations in the medium. The mean energy to create an ion pair in a cosmic abundance gas is approximately 35 eV, so that approximately 30 secondary electrons are created by an initial 1-keV photoelectron. The atomic structure of the heavy atom rearranges itself with the ejection of a few Auger electrons and fluorescent photons (Kaastra & Mewe 1993). In addition, X rays with $E_x > 2$ keV can ionize when they scatter through a large angle (Halpern & Grindlay 1980), an effect known as Compton ionization. The energy transfer is much smaller than in the photoelectric effect, especially for the heavy atoms responsible for most of the absorption cross-section, and Compton energy losses do not become competitive until $E_x \sim 20$ keV.

The total photoionization cross-section decreases rapidly with photon energy, roughly as $\sigma \propto E_x^{-2.5}$. It also depends on the atomic weight of the material as $\sigma \propto Z^3$. In a gas with cosmic abundance, most of the ionizations occur in light atoms (H and He); but metals are responsible for the absorption of higher-energy X rays. For 1-keV photons, the optical depth is unity for a hydrogen column density of $N_H \simeq 4.4 \times 10^{21}$ cm^{-2}, which, for a normal interstellar dust-to-gas ratio, is equivalent to $A_V \simeq 2$ or $A_J \simeq 1$ (Ryter 1996). Therefore, 5-keV photons, which are detected in YSOs with the ASCA satellite, can penetrate to $A_V \simeq 100$ and thus have a potential effect even in deeply embedded environments.

The principal competing source of ionizing radiation in the vicinity of YSOs is ambient UV starlight, which dominates the outer $A_V \simeq 3$ regions of star-forming clouds and galactic cosmic rays, which produce $\approx 10^{-17}$ ionizations s^{-1} (McKee 1989). The cosmic-ray ionization rate, which corresponds to ionization fractions of $\approx 10^{-9}$ throughout cloud interiors, is particularly uncertain; for instance, low-energy cosmic rays may be excluded from dense cloud cores by magnetic scattering (Lepp 1992). For a typical YSO with $L_x = 10^{29}$ erg s^{-1} and photon energies of approximately 1 keV, the total X-ray ionization may (depending on intervening absorption) dominate cosmic-ray ionization out to distances $r_{max} \approx 0.02\, L_x/(10^{29}$ erg s$^{-1})$ pc $\sim$ 4000 AU (Krolik & Kallman 1983, Glassgold et al 1999).

The X-ray ionization effects in the molecular cloud environment may be considerably greater than this estimate because the recombination time scale is approximately 10^1 years, considerably longer than the $<10^1$-year timescale of flare recurrence in YSOs. For cloud ionization, the episodic YSO flares appear essentially as a continuous high-intensity process. Thus, a single YSO with flare luminosities around 10^{30}–10^{31} erg s^{-1} may be the dominant source of ionization for a 0.1-pc cloud core (Glassgold et al 1999). Note that the low ionization fractions involved are sufficient to couple the neutral matter with any magnetic fields and require ambipolar diffusion for gas passage through the field (e.g. Ciolek & Mouchovias 1995).

On smaller spatial scales, the implications of YSO X-ray ionization for circumstellar disks have recently been calculated by Glassgold et al (1997) and Igea &

Glassgold (1999). They consider a model disk similar to the solar nebula illuminated by an X-ray source elevated a few stellar radii above the corotation radius (Figure 3 *top*). Figure 8 shows the ionization of X rays at various energies in terms of the column density of the disk measured perpendicularly from its surface, N_{perp}. These particular curves were calculated at a radial distance of 1 AU from the star. YSO X rays penetrate to some distance towards the midplane in the inner disk, whereas they reach all disk material in the outer disk. The inner disk thus has a midplane neutral dead zone surrounded by an ionized zone (Gammie 1996).

Although the resulting disk ionization level is always low, the YSO X rays impact a far larger volume in the disk than is ionized by cosmic rays. The X-ray ionization will couple the disk material to magnetic fields and MHD processes. In particular, this weakly ionized differentially rotating disk is thought to stimulate the Balbus-Hawley magnetorotational instability (Balbus & Hawley 1991), which will induce an MHD turbulent viscosity and promote flow towards the inner boundary of the disk (Gammie 1996). By this means, YSO X rays may regulate the supply of material accreting onto the protostar and for ejection in high-velocity winds, Herbig-Haro jets, or FU Orionis outbursts.

In an analogous fashion, X-ray ionization may be important for the coupling between disk and outflow required for outflow acceleration. Models of highly collimated Herbig-Haro jets and weakly collimated molecular bipolar flows often require that the rotational energy of the disk be converted into a radial acceleration by some kind of magneto-centrifugal process (Königl & Ruden 1993, Pudritz et al 1999). No calculations of the X-ray effects on outflow physics have been made to date.

5.2 X-ray Effects on Ambient Chemistry and Dust

X-irradiation of molecular gas will produce a complicated sequence of chemical reactions of molecular material at various distances within and around YSOs (e.g. Krolik & Kallman 1983, Maloney et al 1991, Lepp & Dalgarno 1996, Yan & Dalgarno 1997, Aikawa et al 1998). The detailed consequences are quite difficult to calculate in a fully self-consistent model, and they depend on the assumed chemical reaction network. In the warmer regions, a variety of neutral reactions are promoted by X-ray heating of the gas, whereas in cooler regions the X rays stimulate molecular synthesis by ion-molecule reactions.

Kastner et al (1997) report possible evidence for X-ray–induced chemistry in the disk of the nearby ($d \simeq 50\,\text{pc}$) CTT star TW Hya. They attribute the high CN/HCN ratio and HCO^+ abundance to X-ray illumination. The isotopes of HCO^+ have long served as tracers of electron fractions in molecular clouds; HCO^+ itself may be destroyed near the X-ray source by dissociative recombination with electrons, as observed in the molecular cloud close to the microquasar 1E1740.7−2942 (Yan & Dalgarno 1997). In YSO environs, X-ray ionization can thus be probed with high-resolution millimeter interferometers. Other potentially observable tracers of X-ray–induced chemistry are located in the far-IR and include the 149-μm

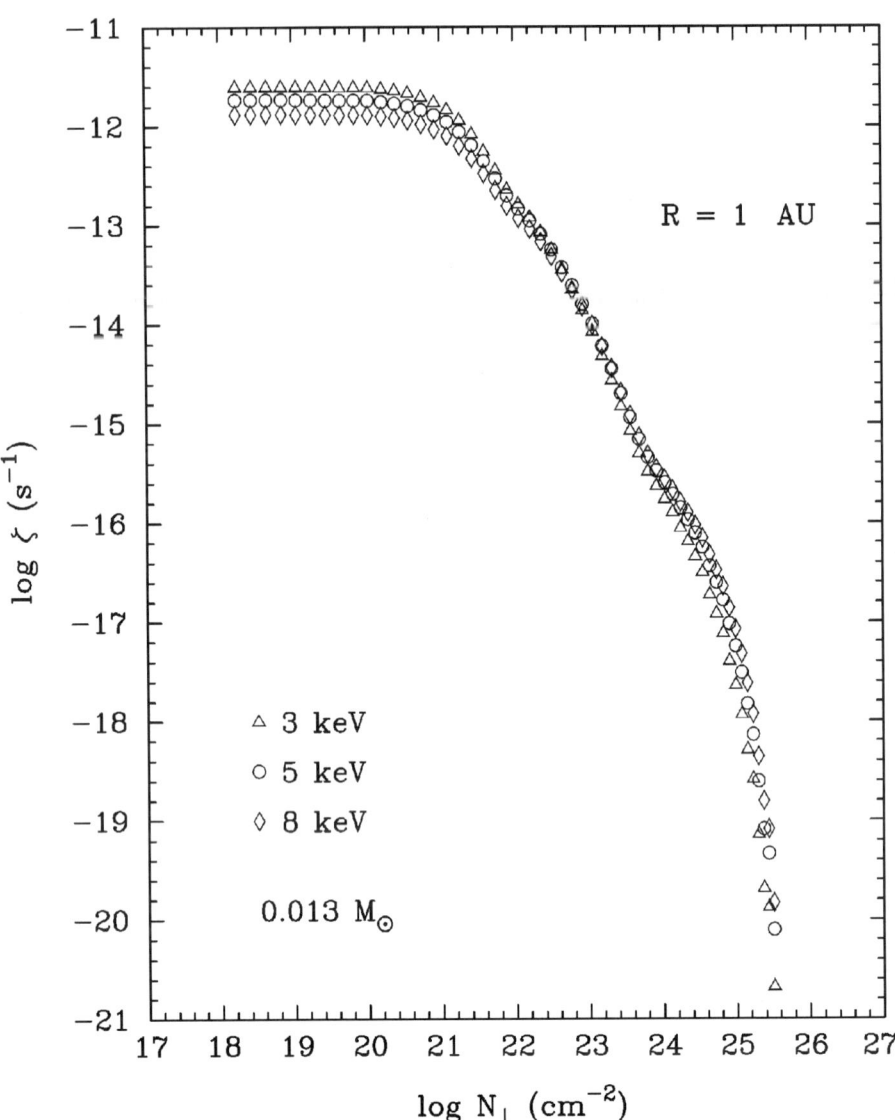

Figure 8 X-ray ionization of a circumstellar disk for X rays of several energies, calculated at r = 1 AU for a solar nebular model (Igea & Glassgold 1999). For comparison, the ionization from galactic cosmic rays is estimated to be log $\xi \sim -17$ s^{-1}.

rotational transition of HeH$^+$, and the [OI] 63-μm, [SII] 35-μm, [FeII] 26-μm, and [CII] 158-μm fine-structure lines (Maloney et al 1996). Such lines are being sought with the ISO satellite.

A dust particle subject to an X-ray will absorb most of the secondary electrons, resulting in increased temperature. For example, a small grain with $a = 5$ nm, absorbs all photons with energy $E_x < 0.1$ keV, whereas a large grain with $a = 200$ nm absorbs all photons with $E_x < 2$ keV (Dwek & Smith 1996). Very small grains may evaporate completely, which suggests that their unidentified infrared emission features in the 3- to 13-μm band may disappear near X-ray luminous YSOs (Voit 1992). This effect may have been seen in active galactic nuclei, in which ISO spectra show the disappearance of the near-infrared PAH features (Genzel et al 1998).

Laboratory accelerator X-ray exposures of various materials relevant to interstellar grains (carbonaceous compounds, silicates, etc) have recently been conducted to study the structural damage produced by X-irradiation (Gougeon 1998). Radiation doses equivalent to those obtained at 0.1 pc from a TTS having an L_X of approximately 10^{30} erg s^{-1} for a period of 10^8 years cause dehydrogenation and breakage of aromatic rings in hydrocarbons but have little effect on silicates. X-irradiation will also cause photodissociation and chemical changes in the ices of dust grain mantles (Cornelison et al 1998). These possible effects of YSO X rays on dust mainly affect line profiles, and their detection must await high-resolution spectroscopic study with future instruments, such as those planned for the SIRTF satellite.

5.3 Meteorite Exposure to Energetic Particles and Shocks

In addition to the readily observed X rays from kilo electron volt plasmas and radio-band gyrosynchrotron radiation from MeV electrons produced by magnetic reconnection events in YSOs, it is reasonable to suppose that, as in solar flares, these events are accompanied by the release of MeV nuclei and strong shocks that propagate into the circumstellar matter. Although no direct astronomical evidence for the bombardment of disk material by protons and shock can be inferred from studies of YSOs, considerable evidence for such processes in the early solar nebula has been found in the meteoritic record.

Meteorites are typically small bodies from the asteroid belt whose orbits are perturbed to collide with Earth; they survive passage through the Earth's atmosphere and are retrieved for study. They give a unique—but incomplete and complex—view of solid materials throughout the early phases of solar evolution (Kerridge & Matthews 1989). Whereas many meteorites show evidence of extensive processing (elemental differentiation, impact shocks, aqueous chemistry, etc), carbonaceous chondrites and portions of other types appear to be relatively pristine remnants of the pre–main-sequence solar nebula. Carbonaceous chondrites are relatively loose conglomerates of early materials, including spheroidal lumps of melted rock (chondrules), rare melted Ca-Al–rich inclusions (CAIs), coal-like carbonaceous matrix, and trace fractions of organic molecules (such as amino acids) and presolar (interstellar) carbonaceous grains. Radioactive dating indicates they were formed

4.55 to 4.56 Gyr ago and are thus contemporaneous within approximately 10 Myr (Russell & Wadhwa 1999).

Four avenues of meteoritic research relate to the magnetic activity and associated high-energy processes discussed in the astronomical contexts above. In some cases, the meteoritic evidence and astronomical evidence clearly support each other, while in others the relationship is only one of several possible interpretations of the meteoritic evidence.

Chondrule Formation   The majority of meteorites are composed largely of chondrules, round millimeter-size igneous rocks. Laboratory investigations have established that they were flash-melted to temperatures of approximately 2000 K, followed by slow cooling over hours. The chondrule precursors are probably loose balls of interstellar dust that coagulated in the disk. The exact cause of the melting has been an enigma for more than a century, and many models have been proposed (see reviews by Wasson 1993 and Boss 1996). Such models include melting in accretion shocks caused by inhomogeneous infall onto the circumstellar disk, impact shocks from planetesimal collisions, bow shocks when planetesimals acquire supersonic velocities caused by resonances with Jupiter, ablation when planetesimals enter planetary atmospheres, FU Orionis outbursts, lightning caused by charge separation in the disk, and magnetic-reconnection flares. Melting by nebular shock waves successfully explains chondrule petrologic properties, but it is unclear whether these shocks are produced by solid-body kinetics within the solar nebula or from external agents such as magnetic flaring (Connolly & Love 1998).

The astronomical findings discussed here clearly support a model of chondrule melting by flare shocks. Several variants of this idea have been discussed. Levy & Araki (1989) propose that the YSO disk continually generates magnetic fields, which erupt into the disk corona, are pulled and twisted by disk Keplerian motions, reconnect, and produce shocks. Magnetic-field amplification by an α-ω dynamo or Balbus-Hawley instability in the differentially rotating disk supports the concept of an MHD disk that continuously creates and reconnects fields (Section 4). Cameron (1995) proposes that ordinary chondrules were melted during the WTT/Class III phase by magnetic reconnection flares produced in a bow shock between a stellar wind and a truncated disk. Shu et al (1997) propose that dust balls are lifted from the disk by the outflowing wind and are melted by reconnection flares around the complex star-disk magnetic interface, and the resulting chondrules are cast outward to orbit further out in the disk.

Natural Remanent Magnetism   Polarization analysis of a wide variety of chondrules indicates that they were formed in the presence of a strong $\sim$1-G magnetic field (Levy 1978; review by Saguira & Strangway 1988). Field orientations differ between chondrules in a meteorite, indicating that the field was embedded during chondrule melting and cooling—not during later consolidation or processing. Because the location of chondrule melting in the YSO system is not certain, it is not clear whether this natural-remanent magnetism was caused by stellar fields around the disk corotation radius, interstellar cloud fields that had not been fully extruded from the disk, or fields amplified within the disk. The ubiquity of magnetism in many types of meteorites and the strength of the implied field seem to support the generation of fields throughout the disk.

^{21}Ne and Particle-Track Excesses in Meteoritic Grains   For some grains in carbonaceous chondrites, the exposure to energetic particles has two manifestations: excesses of spallogenic ^{21}Ne and high densities of particle tracks from heavy nuclei (reviews by Woolum & Hohenberg 1993, Rao et al 1997). Surrounding grains do not show these effects, indicating they arose when the grains were freely floating in the solar nebula before incorporation into the meteorite. Detailed arguments convincingly demonstrate that irradiation is caused by an active early Sun, with energetic particle fluxes elevated at least a factor of ten above contemporary levels, and cannot be caused by long-duration exposure to galactic cosmic rays. This case is perhaps the clearest in which meteorites show direct effects of the flaring activity seen in YSOs by X-ray and radio telescopes.

Short-Lived Isotopic Anomalies in Ca-Al–Rich Inclusions   CAIs exhibit very puzzling isotopic anomalies, including excess short-lived nuclides such as ^{41}Ca and ^{26}Al with radioactive half-lifetimes of 0.15 Myr and 1.1 Myr, respectively (review Goswami & Vanhala 1999). This finding indicates that either the young solar system produced these nuclei in situ by nuclear spallation, or that the solar nebula was contaminated with freshly synthesized nuclear products from stellar sources immediately before the formation of certain inclusions and chondrules. Virtually all simple models for isotopic anomalies encounter difficulties. For example, external seeding by a Type-II supernova could account for the ^{41}Ca, but it produces too much ^{60}Fe, and its distance must be tuned to avoid disrupting the molecular cloud core that collapsed to form the solar system. An asymptotic giant branch red giant star during dredge-up phase passing through the molecular cloud could account for the ^{41}Ca and ^{26}Al but not the observed ^{53}Mn.

The high levels of magnetic activity found in YSOs support models in which at least some meteoritic isotopic anomalies are produced within the YSO system by MeV particle spallation of meteoritic material. This idea has a long heritage (e.g. Fowler et al 1962, Clayton et al 1977, Lee 1978, Feigelson 1982, Levy & Araki 1989, Lee et al 1998) but has generally not gained wide acceptance. In the most thorough analysis to date, Lee et al (1998) calculate that the ^{41}Ca anomaly can arise from reactions such as ^{42}Ca (p, pn) ^{41}Ca and ^{40}Ca$(\alpha, ^{3}$He$)$ ^{41}Ca, in which the protons and alpha particles are accelerated in YSO flares. The high abundance of ^{26}Al is attributed to irradiation by ^{3}He-rich events as seen in solar impulsive flares, giving reactions such as ^{24}Mg$(^{3}$He$, p)$ ^{26}Al. Overproduction of ^{41}Ca is avoided by particle absorption in thick rims. These ideas have the advantage of residing within the comprehensive x-wind model, which also explains YSO magnetic activity, accretion, winds and Herbig-Haro jets, chondrule melting, and dispersal (Shu et al 1997). However, spallation models fail to produce sufficient ^{60}Fe and may overproduce other isotopes.

5.4 Shock Effects in the Interstellar Medium

Other high-energy processes linked with YSOs may also affect the dense interstellar medium beyond the immediate YSO environment, owing to shocks from energetic outflows and winds. The nonthermal radio emission of the outflows of T Tauri South and the Serpens triple source indicates that electron acceleration can be efficient in YSO jets (Section 3.5). Diffusive shock acceleration can also accelerate protons and heavier particles. Such effects from many mass-losing YSOs might produce excesses of low-energy cosmic rays in a molecular cloud complex, producing spallation reactions near the energy threshold and resulting in diffuse nuclear gamma-ray and X-ray emission lines with identifiable spectral properties (Ip 1995, Tatischeff et al 1998).

6. MAGNETIC ACTIVITY AS A TRACER OF YSO POPULATIONS

In addition to opening a new window into YSO astrophysics, astronomical surveys of magnetically active YSOs can serve as a powerful tool for finding previously missed YSOs and studying their collective properties. An unbiased census of YSO populations from different star-forming clouds is essential in addressing a variety of important issues concerning star formation and pre–main-sequence evolution. Does a molecular cloud form stars continuously over 10^7 years, in several episodes, or in a brief terminal starburst? Is the star formation efficiency of a cloud, M_{stars}/M_{gas}, low (perhaps five percent) or high (50 percent) over its lifetime? What is the distribution of longevities of circumstellar disks, and what does this distribution imply for the time scales available for planet formation? What can be learned about the dynamics of star formation from the kinematics of YSOs?

X-ray surveys are particularly effective for locating Class III WTT stars in crowded stellar fields around nearby star-forming regions. Such stars are largely missed by infrared surveys, because their cool blackbody spectral energy distributions cannot be distinguished from those of main-sequence stars. Before X-ray surveys, YSO samples were dominated by Class 0–II stars with ages less than or equal to two Myr, owing to their strong infrared and Hα excesses. Yet, because they last longer than earlier evolutionary phases, Class III stars must outnumber all other YSOs when averaged over all current and past star-forming regions. Thus, a large population of older WTT stars and post-T Tauri stars must exist (Herbig 1978, Feigelson 1996). The characteristics that most readily distinguish older YSOs from ordinary main-sequence stars are high L_x/L_{bol} (Section 3) and high-surface lithium abundance revealed through the Li 670.7-nm absorption line

(Martín 1997). Because the YSO X-ray luminosity function is relatively unchanged along the Hayashi tracks from less than 1 to $\simeq 50$ Myr, X-ray surveys locate considerable numbers of both older WTT stars and previously unrecognized young WTT stars, which are coeval with the more prominent CTT stars. One major problem is that, as older YSOs stars are considered, one must examine ever larger regions around star-forming clouds because of their kinematic dispersion. Efforts to address the astrophysical questions above should thus be viewed with caution, as the YSO samples—even after enhancement by X-ray surveys—are still spatially incomplete and flux limited.

6.1 Census of YSOs Within Star-Forming Regions

The most reliable census of X-ray–emitting T Tauri stars comes from a few deep ROSAT images of nearby active star-forming clouds, in which the newly discovered X-ray–emitting stars have been characterized spectroscopically and photometrically and placed on the Hertzsprung-Russell diagram. These stars include L1495W in Taurus-Auriga (Strom & Strom 1994), Chamaeleon I (Figure 9; Feigelson et al 1993, Lawson et al 1996), and IC348 in Perseus (Preibisch et al 1996, Herbig 1998). These X-ray surveys, combined with previous optical-infrared YSO surveys, show that Class III stars outnumber Class II stars within active star-forming regions at least by factors of 2–3:1. It is surprising that, in the Chamaeleon I and Taurus-Auriga clouds, the age distribution of the Class III stars is substantially identical to the Class II age distribution (Figure 9).

The large population of X-ray–discovered Class III stars has two immediate consequences. First, the star formation rate of a cloud, measured in stars/Myr, is higher than that inferred from early samples dominated by emission line Class II stars. The inferred star formation efficiency is probably more than doubled, but this quantity is more difficult to measure, as older Class III stars may have drifted far from the cloud and could be missing from the census. Also, molecular material is likely to decrease over time because of star formation and cloud disruption or evaporation. Nonetheless, it is likely that star-forming clouds that were once thought to have star formation efficiencies of approximately ten percent (Cohen & Kuhi 1979) may actually have efficiencies of approximately 20 to 30 percent.

Second, the presence of many Class III stars near the stellar birth line implies that, at least in Chamaeleon I, about half of low-mass stars cease interacting with their circumstellar disks (and usually lose any detectable disk) within one Myr. Millimeter studies of CTT and WTT populations also indicate that early disk dissipation is common and can occur over timescales of 10^5 years or less (André & Montmerle 1994). Earlier estimates that disks dissipate on timescales of two to three Myr based on less complete Einstein Observatory data (Strom et al 1989) are probably not reliable. Indeed, it is not clear that there is any preferred disk lifetime, because Class II and III stars coexist along most of the Hayashi track. Rather, the data suggest the distribution of disk longevities ranges smoothly from 10^5 to $\simeq 2 \times 10^7$ years (Lawson et al 1996). This result may be an important

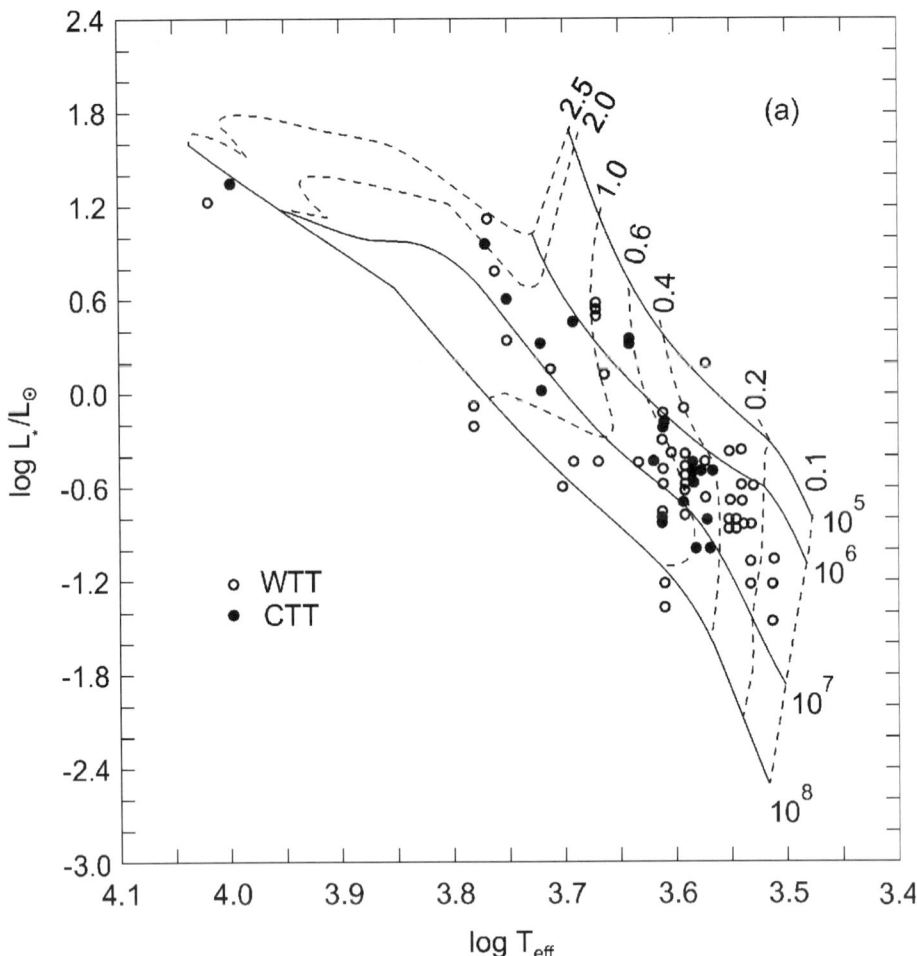

Figure 9 Hertzsprung-Russell diagram of 80 well-characterized YSOs in the Chamaeleon I star-forming cloud. *Filled circles* are CTTs, *open circles* are WTTs, and *dotted lines* are theoretical evolutionary tracks labeled by mass (in $M_\odot$) and by age (in yrs). (Adapted from Lawson et al 1996.)

input into models of planet formation in the solar nebula and other circumstellar disks. This empirical evidence for a wide range of disk lifetimes agrees nicely with theoretical analyses indicating that star-disk coupling can efficiently transfer momentum from the star so that a broad distribution of disk lifetimes explains the wide dispersion of rotational velocities among T Tauri stars and open cluster stars (Bouvier et al 1997).

An early unbiased survey of lithium-rich stars in a young stellar cluster suggests that X-ray–selected samples are reasonably complete for intermediate-mass stars ten Myr old or younger (Preibisch et al 1998b). If the census of all YSOs from a given cloud is complete, then the cloud's star formation history can be established by counting stars vertically along the Hayashi tracks. One such analysis in the Ophiuchus cloud complex suggests that star formation starts slowly and accelerates towards a terminal starburst (Martín et al 1999), as may be expected if ambipolar diffusion delays gravitational collapse (Palla & Galli 1997). However, the distribution of stellar ages in the Chamaeleon I cloud suggests a more continuous star formation history (Lawson et al 1996).

6.2 Census of Dispersed YSOs

These efforts to develop complete YSO samples are compromised by the likelihood that many of the older WTT and post-T Tauri stars have kinematically dispersed from the cloud vicinity (Feigelson 1996). Unless corrections for dispersion are made, all existing T Tauri samples, even with X-ray–discovered Class III stars, show a steep decline in star formation rates for stars older than 2 Myr. Stars inheriting velocities of ~ 1 km s^{-1}, which is the characteristic velocity dispersion within molecular clouds on small scales, produce populations that disperse more than 10 to 20 degrees if star formation continues for ten Myr in nearby star formation regions ($d \simeq 150$ pc).

This hypothesis is supported by the discoveries of what appear to be many Class III stars at large distances from active star-forming clouds (Figure 10). The ROSAT All-Sky Survey (RASS) has revealed thousands of X-ray sources associated with 9- to 15-magnitude stars at low galactic latitudes, many of which are late-type stars with strong Li 670.7 nm lines (0.01 to 0.05 nm equivalent widths) characteristic of WTT stars (e.g. Neuhäuser et al 1995a, Alcalá et al 1995, Sterzik et al 1995, Wichmann et al 1996). A vigorous debate has grown over these findings. Some argued that the stars are 50- to 100-Myr-old ZAMS and not 1- to 30-Myr-old stars on the Hayashi tracks (Briceño 1997, Favata et al 1997). Many recent studies are extending the survey to the entire celestial sphere and measuring various properties of these stars.

X-ray surveys have also uncovered a previously unrecognized nearby pre-main sequence star cluster, the eta Cha cluster (Mamajek et al 1999), and have located new members of the TW Hya association (Webb et al 1999). These are very nearby (d = 50 − 100 pc) open clusters about 10 Myr old, lying far from significant molecular material and exhibiting high proper motions. They are distinctly different in one respect: TW Hya stars are spread over a 30-degree (30 pc) region, while the known eta Cha members lie within a 0.5 degree (1 pc) region. Both the eta Cha cluster and RASS stars spread throughout the Chamaeleon region have proper motion similar to that of the Sco-Cen association. It thus seems possible that the Sco-Cen star-forming region extends tens of parsecs further into the southern sky than previously thought.

Although our understanding of these widely dispersed magnetically active Li-rich young stars is still incomplete, some results are emerging. Observationally,

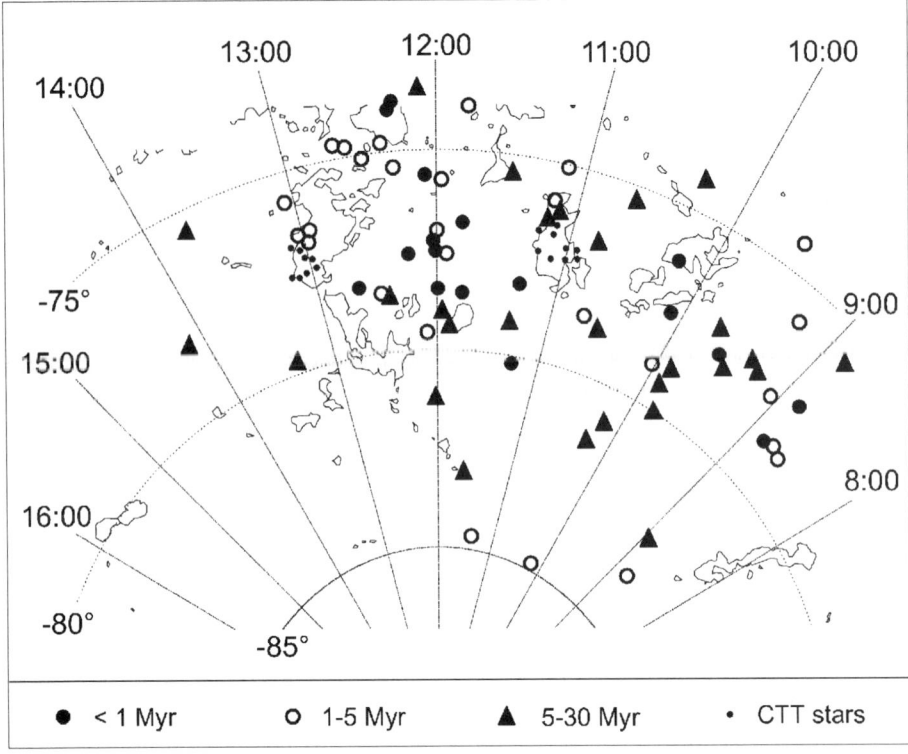

Figure 10 Spatial distribution of Li-rich magnetically active stars that appear to be Class III YSOs, found in the ROSAT All-Sky Survey around the Chamaeleon star-forming clouds. Contours indicate small molecular clouds, and small dots represent previously known Class II stars. (Adapted from Neuhäuser 1997.)

several recent studies indicate that the RASS stars are a mixture of older ZAMS stars and younger but widely dispersed WTT stars, in which the ratio between the types varies across the sky (Martín 1997, Covino et al 1997, Motch et al 1997, Alcalá et al 1998). These studies, however, depend on the subtle discrimination of WTT, post-T Tauri, and ZAMS stars from their location on the T_{eff}-lithium equivalent-width diagram. A cross-correlation between the Tycho parallax and RASS X-ray catalogs produced a sample of approximately 9000 stars, showing that a considerable fraction of the RASS stars appear to be distributed in a disk associated with Gould's Belt (Guillout et al 1998), a well-known ring of star-forming clouds, OB associations, and open clusters lying between 150 and 500 pc from the Sun. As the RASS is severely flux limited, detecting only 20 percent of the youthful stellar population at $d \simeq 150$ pc, tens of thousands of low-mass stars may be dispersed throughout Gould's Belt.

Astrophysically, several possible origins of isolated WTT stars lying far from a star-forming cloud are viable. First, they may be dynamically ejected at high velocity from star-forming regions caused by gravitational scattering from close binaries (Sterzik & Durisen 1995). Although energetically feasible, the efficiency of ejection would have to be implausibly high to explain the large observed population of dispersed WTT stars. Second, stars may inherit high velocities from cloudlets within giant molecular clouds (Feigelson 1996). Molecular spectroscopy shows that cloud complexes typically exhibit supersonic turbulence with large-scale velocity dispersions on the order of ten km s^{-1}, which is sufficient to project stars tens of parsecs while still on the Hayashi tracks. However, the highest-velocity cloudlets are typically not associated with the densest star-forming cloud cores. Third, a starburst may occur near the end of molecular cloud lifetimes (perhaps star formation is delayed by ambipolar diffusion), so that the molecular material dissipates soon after star formation resulting in groups of isolated WTT stars (Palla & Galli 1997).

However incomplete the current samples are, X-ray surveys are leading to a resolution of the long-standing mystery of the missing post-T Tauri stars (Herbig 1978) and are simultaneously uncovering a widely distributed population of low-mass young stars associated with Gould's Belt and/or the Local Association (Eggen 1999). As the properties of the dispersed magnetically active young stars are elucidated, significant insights into the history and dynamics of star formation in the solar neighborhood over the past 10^8 years may emerge.

7. CONCLUDING REMARKS

A confluence of astronomical techniques—imaging X-ray astronomy, polarization-sensitive radio interferometry, highly accurate optical spectroscopy and photometry, and laboratory analysis of meteoritic materials—persuasively show that high-energy processes are prevalent in low-mass YSOs. The rapid heating and cooling of plasma to 10^7 to 10^8 K and acceleration of particles to MeV energies are almost certainly not the product of hydrodynamical phenomena such as gravitational collapse and accretion. These violent phenomena must arise from efficient MHD processes such as solar-type magnetic reconnection flares. The X-ray manifestation of this enhanced magnetic activity is a ubiquitous characteristic of low-mass YSOs, present from the Class I protostellar phase with ages of approximately 10^5 years to stars approaching the ZAMS with ages of 10^7 years. X-ray tracers of YSOs have led to substantial increases in YSO samples, which in turn provide new insights into the history of star formation in the solar neighborhood.

These findings show that the traditional hydrodynamic paradigm for understanding the earliest stages of stellar evolution is not complete. The most important astrophysical effect may be the ionization produced by flare X rays, which, unlike the HII regions of early-type stars, produces extended regions of partially ionized molecular gas. If X-ray emission begins in the earliest Class 0 phase, then YSO ionization may crucially affect the gravitational collapse of star formation. It is not clear today whether this occurs, because there is only one tentative X-ray detection of a Class 0–I YSO (CrA IRS7). Because X-ray emission is definitely prevalent in the Class I–II phases, X-ray ionization is quite likely to play a central role in the astrophysics and evolution of the circumstellar disk. One important effect is the induction of MHD turbulence and viscosity, thereby regulating accretion onto the star. X-ray ionization may also be necessary for the magnetic coupling of the central star to the inner disk, and coupling of the disk to magnetically accelerated and collimated outflows.

The energetic photons and particles produced in YSO flares may furthermore account for long-standing findings in meteorites that are quite difficult to explain in the quiet hydrodynamic model of the solar nebula. MeV protons, presumed to accompany the MeV electrons detected in YSO gyrosynchrotron radio flares, may account for excess particle tracks and spallogenic isotopes in meteorites. Flare shocks may melt chondrules.

In the near future, empirical investigations of hot energetic processes in cold YSOs are likely to thrive. The Chandra X-ray Observatory and the X-ray Multimirror Mission, to be launched soon, represent a considerable improvement in X-ray observational capability, particularly in the harder bands that are much less attenuated by foreground material. A number of problems raised in this review may be solved with these new observatories. For example,ced Chandra, with its subarcsecond resolution, will be able to detect thousands of T Tauri stars in the Orion giant molecular cloud. Combined with optical studies placing the stars on the Hertzsprung-Russell diagram (Hillenbrand 1997), one can disentangle L_x-mass and L_x-age correlations to study the causes of YSO magnetic activity (see Section 3.2). The excellent high throughput of XMM will allow the detection of very faint and low-mass YSOs deeply embedded in molecular cores, including young brown dwarfs, and will provide direct measurements of extinction needed to obtain accurate X-ray luminosities. The search for X rays from Class 0 protostars has particularly important astrophysical implications, such as deciding whether infall is braked by the onset of YSO-ionizing radiation or whether Class 0 outflows have a gravitational or magnetic origin.

Proposed radio astronomical facilities like the Atacama Large Millimeter Array, Very Large Array upgrade, and Square Kilometer Array would significantly boost the study of MeV particles in YSOs, because YSO fluxes are close to the sensitivity limits of current telescopes. Insight should emerge on the magnetic structures, disk gas chemistry, and dust properties in magnetically active YSOs. The proliferation of 8-m–class optical/infrared telescopes, with techniques such as Doppler imaging and high-resolution IR spectroscopy and interferometry, as well as planned infrared space missions such as SIRTF and NGST, will permit more accurate characterization of the magnetic properties of the central stars and the accretion process in YSO systems.

There are urgent needs for more theoretical and laboratory development of the issues discussed here. The concept of a magnetically active but thermodynamically cold molecular disk linked to other structures within YSO systems must be investigated in greater detail. This is a challenging problem requiring time-dependent, three-dimensional MHD calculations with reconnection, ionization and heating.

The effects of X rays on circumstellar gas and disks, which so far have been studied primarily for disks, should be extended to protostellar envelopes. These efforts are closely related to the study of high-energy effects on cold material around active galactic nuclei. There has also been inadequate study of the effects of flare shocks and energetic particles on solids. Further accelerator experiments with X rays and particles should be pursued actively. The final evaluation of the astrophysical importance of energetic processes in YSOs awaits these future investigations.

ACKNOWLEDGMENTS

The authors thank Philippe André, Claude Bertout, George Herbig, Eugene Levy, Roberto Pallavicini, Jean-Claude Pecker, and Frank Shu for their valuable comments on the manuscript. Lee Carkner and Nicholas Grosso kindly provided figures. We also benefited from discussions with many other colleagues, particularly at the 1998 Protostars and Planets IV conference. EDF was supported by NASA grants NAS8-38252 and NAG5-8422.

Visit the Annual Reviews home page at http://www.AnnualReviews.org

LITERATURE CITED

Agrawal PC, Koyama K, Matsuoka M, Tanaka Y. 1986. *PASJ* 38:723–79

Aikawa Y, Umebayashi T, Nakano T, Miyama S. 1998. *Faraday Discuss.* 109:281–301

Alcalá JM, Chavarria-KC, Terranegra L. 1998. *Astron. Astrophys.* 330:1017–28

Alcalá JM, Krautter J, Schmitt JHMM, Covino E, Wichmann R, Mundt R. 1995. *Astron. Astrophys. Suppl.* 114:109–34

Alcalá JM, Terranegra L, Wichmann R, Chavarriá-KC, Krautter J, et al. 1996. *Astron. Astrophys. Suppl.* 119:7–24

Alcalá JM, Krautter J, Covino E, Neuhäuser R, Schmitt JHMM, Wichmann R. 1997. *Astron. Astrophys.* 319:184–200

Alcalá JM, Chavarria–KC, Terranegra L. 1998. *Astron. Astrophys.* 330:1017–28

Ambartsumian VA, Mirzoyan LY. 1982. *Astrophys. Sp. Sci.* 84:317–30

André P. 1987. In *Protostars and Molecular Clouds*, eds. T. Montmerle, C. Bertout, 143–87, Saclay, France: CEA

André P, Montmerle T, Feigelson ED. 1987. *Astron. J.* 93:1182–98

André P, Deeney BD, Phillips RB, Lestrade J-F. 1992. *Astrophys. J.* 401:667–77

André P, Montmerle T. 1994. *Astrophys. J.* 420:837–62

André P. 1996. In *Radio Emission from the Stars and the Sun*, ed. AR Taylor, JM Paredes, Vol. 93, pp. 273–84. San Francisco: ASP

Anglada G. 1996. In *Radio Emission from the Stars and the Sun*, ed. AR Taylor and JM Paredes, Vol. 93, pp. 3–14. San Francisco: ASP

Armitage PJ, Clarke CJ. 1996. *MNRAS* 280:458–68

Bachiller R. 1996. *Annu. Rev. Astron. Astrophys.* 34:111–54

Balbus SA, Hawley, JF. 1991. *Astrophys. J.* 376:223–33

Banaskiewicz M, Axford WI, McKenzie JF. 1998. *Astron. Astrophys.* 337:940–94

Becker RH, White RL. 1985. *Astrophys. J.* 297:649–51

Becker RH, White RL. 1988. *Astrophys. J.* 324:893–98

Beckwith SVW, Sargent AI, Chini RS, Guesten R. 1990. *Astron. J.* 99:924–45

Bertout C. 1989. *Annu. Rev. Astron. Astrophys.* 27:351–95

Bertout C, Harder S, Malbet F, Mennessier C, Regev O. 1996. *Astron. J.* 112:2159

Bieging JH, Cohen M. 1989. *Astron. J.* 98:1686–92

Bieging JH, Cohen M, Schwartz PR. 1984. *Astrophys. J.* 282:699–708

Birk GT. 1998. *Astron. Astrophys.* 330:1070–76

Blandford RD, Payne DR. 1982. *MNRAS* 199:883–903

Bontemps S, André P, Terebey S, Cabrit S. 1996. *Astron. Astrophys.* 311:858–72

Boss AP. 1996. In *Chondrules and the Protoplanetary Disk*, eds. RH Hewins, RH Jones, ERD Scott, 257–63, New York: Cambridge Univ. Press

Bouvier J, Appenzeller, I. 1992. *Astron. Astrophys. Suppl.* 92:481–516

Bouvier J, Covino E, Kovo O, Martín EL, Matthews JM, et al. 1995. *Astron. Astrophys.* 299:89–107

Bouvier J, Forestini M, Allain S. 1997. *Astron. Astrophys.* 326:1023–43

Bradt H, Kelley RL. 1979. *Astrophys. J. Lett.* 228:L33

Brandner W, Köhler R. 1998. *Astrophys. J.* 499:L79–82

Briceño C, Hartmann LW, Stauffer JR, Gagné M, Stern RA. 1997. *Astron. J.* 113:740–51

Brown A. 1987. *Astrophys. J.* 322:L31–34

Brown A, Walter FM, Ambruster C, Stewart RT, Jeffries R. 1996. In *Radio Emission from the Stars and the Sun*, ed. AR Taylor, JM Paredes, Vol. 93, pp. 294–296, San Francisco: Astron. Soc. Pacific

Busso M, Gallino R, Wasserburg GJ. 1999. *Annu. Rev. Astron. Astrophys.* 37:239–309

Caillault J-P, Magnani L, Fryer CA. 1995. *Astrophys. J.* 441:261–69

Calvet N, Basri G, Kuhi LV. 1984. *Astrophys. J.* 277:725–37

Calvet N, Gullbring E. 1998. *Astrophys. J.* 509:802–18

Cameron AGW. 1995. *Meteoritics* 30:133–61

Cameron AC, Campbell CG. 1993. *Astron. Astrophys.* 274:309–18

Carkner L, Feigelson ED, Koyama K, Montmerle T, Reid IN. 1996. *Astrophys. J.* 464:286–305

Carkner L, Mamajek E, Feigelson ED, Neuhäuser R, Wichmann R, et al. 1997. *Astrophys. J.* 490:735–43

Carkner L, Kozak JA, Feigelson ED. 1998. *Astron. J.* 116:1933–39

Casanova S, Montmerle T, Feigelson ED, André P. 1995. *Astrophys. J.* 439, 752–70

Chiang E, Phillips RB, Lonsdale CJ. 1996. *Astron. J.* 111:355–64

Churchwell E, Wood DOS, Felli M, Massi M. 1987. *Astrophys. J.* 321:516–29

Ciolek GE, Mouchovias TC. 1995. *Ap. J.* 454:194–216

Clayton DD, Dwek E, Woosley SE. 1977. *Astrophys. J.* 214:300–15

Cohen M, Bieging JH. 1986. *Astron. J.* 92:1396–1402

Cohen M, Kuhi LV. 1979. *Astrophys. J. Suppl.* 41:743–843

Connolly HC Jr., Love SG. 1998. *Science* 280:62–67

Cooke BA, Ricketts MJ, Maccacaro T, Pye JP, Elvis M, et al. 1978. *MNRAS* 182:489

Cornelison DM, Dillingham TR, Tegler SC, Galle K, Miller GA, Lutz BL. 1998. *Astrophys. J.* 505:443–51

Covino E, Alcalá JM, Allain S, Bouvier J, Terranegra L, et al. 1997. *Astron. Astrophys.* 328:187–202

Dalgarno A, McCray RA. 1972. *Annu. Rev. Astron. Astrophys.* 10:375–426

Damiani F, Micela G, Sciortino S, Harnden FR Jr. 1995a. *Astrophys. J.* 446:331–40

Damiani F, Micela G. 1995b. *Astrophys. J.* 446:341–49

DeCampli WM. 1981. *Astrophys. J.* 244:124–46

den Boggende AJF, Mewe R, Gronenschild EHBW, Heise J, Grindlay JE. 1978. *Astron. Astrophys* 62:1–7

Donati J-F, Semel M, Carter BD, Rees DE, Cameron AC. 1997. *MNRAS* 291:658–82

Dulk GA. 1985. *Annu. Rev. Astron. Astrophys.* 23:169–224

Dutrey A, Guilloteau S, Simon M. 1994. *Astron. Astrophys.* 286:149–59
Dwek E, Smith RK. 1996. *Astrophys. J.* 459:686–700
Edwards S, Strom SS, Hartigan P, Strom KM, Hillenbrand LA, et al. 1993. *Astron. J.* 106:372–82
Eggen O. 1999. *Astron. J.* In press
Favata F, Micela G, Sciortino S. 1997. *Astron. Astrophys.* 326:647–54
Favata F, Micela G, Sciortino S. 1998. *Astron. Astrophys.* 337:413–20
Favata F, Micela G, Sciortino S. 1999. *Astron. Astrophys.* In press
Feigelson ED. 1982. *Icarus* 51:155–63
Feigelson ED. 1996. *Astrophys. J.* 468:306–22
Feigelson ED, Carkner L, Wilking BA. 1998. *Astrophys. J. Lett.* 494:L215–18
Feigelson ED, Casanova S, Montmerle T, Guibert JR. 1993. *Astrophys. J.* 416:623–46
Feigelson ED, DeCampli WM. 1981. *Astrophys. J. Lett.* 243:L89–93
Feigelson ED, Giampapa MS, Vrba FJ. 1991. In *The Sun in Time*, ed. CP Sonett, MS Giampapa, MS Matthews, p. 658–81. Tucson, AZ: Univ. Ariz. Press
Feigelson ED, Jackson JM, Mathieu RD, Myers PC, Walter FM. 1987. *Astron. J.* 94:1251–59
Feigelson ED, Kriss, GA. 1981. *Astrophys. J. Lett.* 248:L35–38
Feigelson ED, Kriss GA. 1989. *Astrophys. J.* 338:262–76
Feigelson ED, Lawson WA. 1997. *Astron. J.* 113:2130–33
Feigelson ED, Montmerle T. 1985. *Astrophys. J.* 289:L19–23
Feigelson ED, Welty AD, Imhoff C, Hall JC, Etzel PB, Phillips RB, Lonsdale CJ. 1994. *Astrophys. J.* 432:373–85
Felli M, Churchwell E, Wilson TL, Taylor GB. 1993a. *Astron. Astrophys. Suppl.* 98:137–64
Felli M, Taylor GB, Catarzi M, Churchwell E, Kurtz S. 1993b. *Astron. Astrophys. Suppl.* 101:127–51
Ferreira J, Pelletier G. 1995. *Astron. Astrophys.* 295:807

Field GB, Rogers RD. 1993. *Astrophys. J.* 403:94–109
Finkenzeller U, Basri G. 1987. *Astrophys. J.* 318:832–43
Fleming TA, Gioia IM, Maccacaro T. 1989. *Astrophys. J.* 340:1011–23
Fowler W., Greenstein J, Hoyle F. 1962. *Geophys. J. R. Astron. Soc.* 6:148–200
Frink S, Röser S, Neuhäuser R, Sterzik MF. 1997. *Astron. Astrophys.* 325:613–22
Gagné M, Caillault J-P. 1994. *Astrophys. J.* 437:361–83
Gagné M, Caillault J-P, Stauffer JR. 1995. *Astrophys. J.* 445:280–313
Gahm GF. 1980. *Astrophys. J. Lett.* 242:L163–66
Gahm GF, Loden KL, Loden K, Gullbring E, Hartstein D. 1995. *Astron. Astrophys.* 301:89–104
Gammie CF. 1996. *Astrophys. J.* 457:355–62
Garay G, Moran JM, Reid MJ. 1987. *Astrophys. J.* 314:535–50
Geier S, Wendker HJ, Wisotzki L. 1995. *Astron. Astrophys.* 299:39–52
Genzel R, Lutz D, Sturm E, Egami E, Kunze D, et al. 1998. *Ap. J.* 498:579–605
Giacconi R, Murray S, Gursky H, Kellogg E, Schreier E, Tananbaum H. 1972. *Astrophys. J.* 178:281–308
Gilman PA. 1983. In *Solar and Stellar Magnetic Fields*, IAU Symp. 102, ed. JO Stenflo, p. 247. Boston: Reidel
Giovannelli F, Vittone AA, Rossi C, Errico L, Bisnovatyi-Kogan GG. 1995. *Astron. Astrophys. Suppl.* 114:341–61
Glassgold AE, Feigelson ED, Montmerle T. 1999. See Mannings et al 1999.
Glassgold AE, Najita J, Igea J. 1997. *Astrophys. J.* 480:344–50 (Erratum *Astrophys. J.* 485:920)
Goodson AP, Winglee RM, Böhm K-H. 1997. *Astrophys. J.* 489:199–209
Goswami JN, Vanhala H. 1999. See Mannings, et al. 1999
Gougeon S. 1998. Thèse Doct., Univ. Paris VII
Gregorio-Hetem J, Montmerle T, Casanova

S, Feigelson ED. 1998. *Astron. Astrophys.* 331:193–210

Grosso N, Montmerle T, Feigelson ED, André P, Casanova S, Gregorio-Hetem J. 1997. *Nature* 387:56–58

Güdel M, Guinan EF, Skinner SL. 1997. *Astrophys. J.* 483:947–60

Güdel M, Linsky JL, Brown A, Nagase F. 1999. *Astrophys. J.* 511:405–21

Guenther EW, Emerson JP. 1997. *Astron. Astrophys.* 321:803–10

Guenther EW, Lehmann H, Emerson JP, Staude J. 1999. *Astron. Astrophys.* 341:768–83

Guillout P, Sterzik MF, Schmitt JHMM, Motch C, Neuhäuser R. 1998. *Astron. Astrophys.* 337:113–24

Gullbring E, Barwig H, Chen PS, Gahm GF, Bao MX. 1996. *Astron. Astrophys.* 307:791–802

Gullbring E, Barwig H, Schmitt JHMM. 1997. *Astron. Astrophys.* 324:155

Halpern JP, Grindlay JE. 1980. *Astrophys. J.* 242:1041–55

Haro G, Chavira E. 1966. *Vistas in Astronomy,* 8:89–107

Hartmann L. 1998. *Accretion Processes in Star Formation*, Cambridge, UK: Cambridge Univ. Press

Hartmann L, Hewett R, Calvet N. 1994. *Astrophys. J.* 426:669

Hartmann L, Kenyon S, Hartigan P. 1993. In Levy & Lunine 1993, pp. 497–518

Hartmann LW, Soderblom DR, Stauffer JR. 1987. *Astron. J.* 93:907–12

Hatzes AP. 1995. *Astrophys. J.* 451:784

Hayashi C. 1966. *Annu. Rev. Astron. Astrophys.* 4:171–92.

Hayashi M, Shibata K, Matsumoto R. 1996. *Astrophys. J. Lett.* 468:L37–40

Henriksen RN, Mirabel IF, Ptuskin VS. 1991. *Astron. Astrophys.* 248:221–26

Henriksen R, André P, Bontemps S. 1997. *Astron. Astrophys.* 323:549–65

Henry GW, Hall DS. 1994. *Astrophys. J. Lett.* 425:L25

Herbig G. 1962. *Adv. Astron. Astrophys.* 1:47

Herbig GH. 1973. *Astrophys. J.* 182:129

Herbig GH. 1978. In *Problems of Physics and Evolution of the Universe*, ed. LV Mirozyan, pp. 171–79, Yerevan: Acad. Sci. Armenian SSR

Herbig GH. 1998. *Astrophys. J.* 497:736–58

Herbig GH, Soderblom DR. 1980. *Astrophys. J.* 242:628–37

Herbst W. 1989. *Astron. J.* 89:2268–74

Hillenbrand LA. 1997. *Astron. J.* 113:1733–68

Hirose S, Uchida Y, Shibata K, Matsumoto R. 1997. *Pub. Astron. Soc. Japan* 49:193–205

Hoff W, Henning T, Pfau W. 1998. *Astron. Astrophys.* 336:242–50

Igea J, Glassgold AE. 1999. *Astrophys. J.* In press

Ip W-H. 1995. *Astron. Astrophys.* 300:283–88

Jeffries RD, Buckley DAH, James DJ, Stauffer JR. 1996. *MNRAS* 281:100–15

Jensen ELN, Cohen DH, Neuhäuser R. 1998. *Astron. J.* 116:414–23

Johns-Krull CM, Hatzes AP. 1997. *Astrophys. J.* 487:896–915

Johns-Krull CM, Valenti JA, Hatzes AP, Kanaan A. 1999. *Astrophys. J.* 510:L41–44

Joncour I, Bertout C, Ménard F. 1994. *Astron. Astrophys.* 285:L25–28

Joy AH. 1945. *Astrophys. J.* 102:168–95

Kaastra JS, Mewe R. 1993. *Astron. Astrophys. Suppl.* 97:443–82

Kamata Y, Koyama K, Tsuboi Y, Yamauchi S. 1997. *PASJ* 49:461–70

Kastner JH, Zuckerman B, Weintraub DA, Forveille T. 1997. *Science* 277:67–71

Kerridge J, Matthews M., eds. 1989. *Meteorites and the Early Solar System*, Tucson: Univ. Ariz. Press

Köhler R, Leinert C. 1998. *Astron. Astrophys.* 331:977–98

Königl A. 1991. *Astrophys. J.* 370:L39–43

Königl A, Ruden SP. 1993. In Levy & Lunine 1993, 641–87

Koyama K. 1987. *PASJ* 39:245–52

Koyama K, Asaoka I, Kuriyama T, Tawara Y. 1992. *PASJ* 44:L255–57

Koyama K, Hamaguchi K, Ueno S, Kobayashi N, Feigelson ED. 1996. *PASJ* 48:L87–92

Koyama K, Maeda Y, Ozaki M, Ueno S,

Kamata Y, Tawara Y, Skinner S, Yamauchi S. 1994. *PASJ* 46:L125–29
Krautter J, Wichmann R, Schmitt JHMM, Alcalá JM, Neuhäuser R, Terranegra L. 1997. *Astron. Astrophys. Suppl.* 123:329–52
Krolik JH, Kallman TR. 1983. *Astrophys. J.* 267:610–24
Ku WH-M, Chanan GA. 1979. *Astrophys. J. Lett.* 234:L59–63
Kudoh T, Shibata K. 1997. *Astrophys. J.* 476:632–48
Kuhi LV. 1964. *Astrophys. J.* 140:1409–33
Kutner ML, Rydgren AE, Vrba FJ. 1986. *Astron. J.* 92:895–97
Lada CJ. 1987. In *Star Forming Regions*, IAU Symp. 115, eds. M Peimbert, J. Jugaku. Dordrecht: Kluwer
Lada CJ, Kylafis ND. 1991. In *The Physics of Star Formation and Early Stellar Evolution*. Dordrecht: Kluwer
Lamzin SA, Bisnovatyi-Kogan GS, Errico L, Giovannelli F, Katysheva NA, Rossi C, Vittone AA. 1996. *Astron. Astrophys.* 306:877–91
Lawson WA, Feigelson ED, Huenemoerder DP. 1996. *MNRAS* 280:1071–88
Lee T. 1978. *Astrophys. J.* 224:217–26
Lee T, Shu FH, Shang H, Glassgold AE, Rehm KE. 1998. *Astrophys. J.* 506:898–912
Leous JA, Feigelson ED, André P, Montmerle T. 1991. *Astrophys. J.* 379:683–88
Lepp S. 1992. In *Astrochemistry of Cosmic Phenomena*, ed. P. Singh, p. 471–75. Dordrecht: Rcidel
Lepp S, Dalgarno A. 1996. *Astron. Astrophys.* 306:L21–24
Lepp S, McCray R. 1983. *Astrophys. J.* 269:560–67
Levy EH. 1978. *Nature* 276:481
Levy EH, Ruzmaikin AA, Ruzmaikina TV. 1991. In *The Sun in Time*, eds. CP Sonnett, MS Giampapa, MS Matthews, pp. 589–632. Tucson: Univ. Arizona Press
Levy EH, Lunine JI eds. 1993. *Protostars and Planets III*, Tucson: Univ. Arizona Press
Levy E, Araki S. 1989. *Icarus* 81:74–91 (and discussion in *Icarus* 87:241–6)

Li J. 1996. *Astrophys. J.* 456:696–707
Luhman KL, Briceño C, Rieke GH, Hartmann L. 1998. *Astrophys. J.* 493:909–13
Magazzù A, Martín EL, Sterzik MF, Neuhäuser R, Covino E, et al. 1997. *Astron. Astrophys. Suppl.* 124:449–67
Magnani L, Caillault J-P, Hearty T, Stauffer J, Schmitt JHMM. 1996. *Astrophys. J.* 465:825–39
Maloney PR, Hollenbach DR, Tielens AG. 1991. *Astrophys. J.* 466:561–84
Mamajek EE, Lawson WA, Feigelson ED. 1999. *Astrophys. J.* 516:L77–80
Mannings V, Boss A, Russell S. eds. 1999. *Protostars and Planets IV*, Tucson, AZ: Univ. Ariz. Press. In press
Mardones D, Myers PC, Tafalla M, Wilner DJ, Bachiller R, Garay G. 1997. *Astrophys. J.* 489:719–33
Markert TH, Winkler PF, Laird FN, Clark GW, Hearn DR, et al. 1979. *Astrophys. J. Suppl.* 39:573
Martín EL. 1997. *Astron. Astrophys.* 321:492–96
Martín EL. 1998. *Astron. J.* 115:351–57
Martín EL, Montmerle T, Gregorio–Hetem J, Casanova S. 1998. *MNRAS* 300:733–46
Mathieu RD, Walter FM, Myers PC. 1989. *Astron. J.* 98:987–1001
Mathieu RD. 1994. *Annu. Rev. Astron. Astrophys.* 32:465–530
McCaughrean MJ, O'Dell CR. 1996. *Astron. J.* 111:1977–86
McKee CF. 1989. *Astrophys. J.* 345:782–801
Micela G, Favata F, Sciortino S. 1997. *Astron. Astrophys.* 326:221
Miller KA, Stone JM. 1998. *Astrophys. J.* 489:890–902
Montmerle T, Koch-Miramond L, Falgarone E, Grindlay JE. 1983. *Astrophys. J.* 269:182–201
Montmerle T, Feigelson ED, Bouvier J, André P. 1993. In Levy & Lunine 1993, 689–717
Montmerle T. 1991. In *The Physics of Star Formation and Early Stellar Evolution*, ed. CJ Lada, ND Kylafis, p. 675. Dordrecht: Kluwer
Motch C, Guillout P, Haberl F, Pakuli M,

Piesch W, Reinsch K. 1997. *Astron. Astrophys.* 318:111–133
Mouschovias TC. 1976. *Astrophys. J.* 207:141–58
Mundt R, Walter FM, Feigelson ED, Finkenzeller U, Herbig GH, et al. 1983. *Astrophys. J.* 269:229–38
Nakano M, Yamauchi S, Sugitani K, Ogura K, Kogure T. 1999. *Publ. Astron. Soc. Japan* 51:1–12
Naylor T, Fabian AC. 1999. *MNRAS.* 302:714–22
Neuhäuser R. 1997. *Science* 276:1363–70
Neuhäuser R, Briceno C, Comeron F, Hearty T, Martín E, et al. 1999. *Astron. Astrophys.* 343:883–93
Neuhäuser R, Preibisch T. 1997. *Astron. Astrophys.* 322:L37–40
Neuhäuser R, Sterzik MF, Schmitt JHMM, Wichmann R, Krautter J. 1995a. *Astron. Astrophys.* 295:L5–8
Neuhäuser R, Sterzik MF, Schmitt JHMM, Wichmann R, Krautter J. 1995b. *Astron. Astrophys.* 297:391–417
Neuhäuser R, Sterzik MF, Torres G, Martín E. 1995c. *Astron. Astrophys.* 299:L13–16
Neuhäuser R, Torres G, Sterzik MF, Randich S. 1997. *Astron. Astrophys.* 325:647–63
Neuhäuser R, Wolk SJ, Torres G, Preibisch T, Stout-Batalha NM, et al. 1998. *Astron. Astrophys* 334:873–94
Norman C, Heyvaerts J. 1985. *Astron. Astrophys.* 147:247–56
Ohashi N, Hayashi M, Ho PTP, Momose M, Hirano N. 1996. *Astrophys. J.* 466:957–63
O'Neal D, Feigelson ED, Mathieu RD, Myers PC. 1990. *Astron. J.* 100:1610–20
Ouyed R, Pudritz RE. 1997. *Astrophys. J.* 482:712–32
Paatz G, Camenzind M. 1996. *Astron. Astrophys.* 308:77–90
Palla F, Galli D. 1997. *Astrophys. J.* 476:L35–38
Park S, Finley JP. 1996. *Astron. J.* 112:693–99
Pasquini L, Cutispoto G, Gratton R, Major M. 1991. *Astron. Astrophys.* 248:72–80

Persi P, Olofsson G, Kaas AA, Nordh L, Bontemps S, et al. 1998. In *Star Formation with the Infrared Space Observatory*, eds. JL Yun, R. Liseau, Vol. 132, pp. 158–162. San Francisco: ASP
Phillips RB, Lonsdale CJ, Feigelson ED. 1991. *Astrophys. J.* 382:261–69
Phillips RB, Lonsdale CJ, Feigelson ED. 1993. *Astrophys. J. Lett.* 403:L43–45
Phillips RB, Lonsdale CJ, Feigelson ED, Deeney BD. 1996. *Astron. J.* 111:918–29
Pravdo SH, Angelini L. 1993. *Astrophys. J.* 407:232–36
Pravdo SH, Angelini L. 1995. *Astrophys. J.* 447:342–52
Pravdo SH, Marshall FE. 1981. *Astrophys. J.* 248:591–95
Preibisch T. 1997a. *Astron. Astrophys.* 320:525–39
Preibisch T. 1997b. *Astron. Astrophys.* 324:690–98
Preibisch T. 1998. *Astron. Astrophys.* 338:L25–28
Preibisch T, Zinnecker H, Herbig GH. 1996. *Astron. Astrophys.* 310:456–73
Preibisch T, Zinnecker H, Schmitt JHMM. 1993. *Astron. Astrophys.* 279:L33–36
Preibisch T, Neuhäuser R, Alcalá JM. 1995. *Astron. Astrophys.* 304:L13–16
Preibisch T, Neuhäuser R, Stanke T. 1998a. *Astron. Astrophys.* 338:923
Preibisch T, Guenther E, Zinnecker H, Sterzik M, Frink S, et al. 1998b. *Astron. Astrophys.* 333:619–28
Pudritz RE, Norman CA. 1983. *Astrophys. J.* 274:677–97
Pudritz RE, et al. 1999. In Mannings, et al. 1999. In press
Rao MN, Garrison DH, Palma RL, Bogard DD. 1997. *Meteorit. Plan. Sci.* 32:531–43
Ray TP, Muxlow TWB, Axon DJ, Brown A, Corcoran D, et al. 1997. *Nature* 385:415–17
Reale F, Betta R, Peres G, Serio S, McTiernan J. 1997. *Astron. Astrophys.* 325:782–90
Reipurth B, Bertout C, eds. 1997. *Herbig-Haro Flows and the Birth of Stars*, IAU Symp. 182, Dordrecht: Kluwer

Rice JB, Strassmeier KG. 1996. *Astron. Astrophys.* 316:164–72

Rodríguez LF. 1997. In *Herbig-Haro Flows and the Birth of Low Mass Stars*, eds. B Reipurth, C. Bertout, p. 83. Dordrecht: Kluwer

Romanova MM, Ustyugova GV, Koldoba AV, Chechetkin VM, Lovelace RVE. 1998. *Astrophys. J.* 500:703–13

Rucinski SM. 1992. *PASP* 104:311–13

Russell S, Wadhwa M. 1999. See Mannings et al. 1999.

Rydgren AE, Vrba FJ. 1983. *Astrophys. J.* 267:191–98

Ryter CE. 1996. *Astrophys. Sp. Sci.* 236:285–91

Safier PN. 1998. *Astrophys. J.* 494:336–41

Saguira N, Strangway D. 1988. In *Meteorites and the Early Solar System*, eds. JF Kerridge, MS Matthews, p. 595. Tucson, AZ: Univ. Ariz. Press

Schmitt JHMM. 1994. *Astrophys. J. Suppl.* 90:735–42

Schulz N, Berghöfer TW, Zinnecker H. 1997. *Astron. Astrophys.* 325:1001–12

Sciortino S, Favata F, Micela G. 1995. *Astron. Astrophys.* 296:370–79

Sciortino S, Damiani F, Favata F, Micela G. 1998. *Astron. Astrophys.* 332:825–41

Shang H, Shu FH, Glassgold AE. 1998. *Astrophys. J.* 493:L91–94

Shu FH, Adams FC, Lizano S. 1987. *Annu. Rev. Astron. Astrophys.* 25:23–81

Shu F, Najita J, Ruden SP, Lizano S. 1994. *Astrophys. J.* 429:781–96

Shu FH, Shang H, Glassgold AE, Lee T. 1997. *Science* 277:1475–79

Siess L, Forestini M, Dougados C. 1997. *Astron. Astrophys.* 324:556–65

Skinner SL. 1993. *Astrophys. J.* 408:660–67

Skinner SL, Brown A. 1994. *Astron. J.* 107:1461–68

Skinner SL, Güdel M, Koyama K, Yamauchi S. 1997. *Astrophys. J.* 486:886–902

Skinner SL, Walter FM. 1998. *Astrophys. J.* 509:761–67

Smith MA, Pravdo SH, Ku WH-M. 1983. *Astrophys. J.* 272:163–67

Smith KW, Lewis GF, Bonnell IA, Bunclark PS, Emerson JP. 1999. *MNRAS* 304:367–88

Spruit HC, Foglizzo T, Stehle R. 1997. *MNRAS* 288:333–42

Stapelfeldt KR, Krist JE, Ménard F, Bouvier J, Padgett DL, Burrows CJ. 1998. *Astrophys. J.* 502:L65–69

Sterzik MF, Alcalá JM, Neuhäuser R, Schmitt JHMM. 1995. *Astron. Astrophys.* 297:418–26

Sterzik MF, Durisen RH. 1995. *Astron. Astrophys.* 304:L9–12

Sterzik MF, Durisen RH, Brandner W, Jurcevic J, Honeycutt RK. 1997. *Astron. J.* 114:1555–66

Stine PC, Feigelson ED, André P, Montmerle T. 1988. *Astron. J.* 96:1394–1406

Stine PC, O'Neal D. 1998. *Astron. J.* 116:890–94

Stone RC, Taam RE. 1985. *Astrophys. J.* 291:183–87

Strassmeier KG, Welty AD, Rice JB. 1994. *Astron. Astrophys.* 285:L17–20

Strassmeier KG, Bartus J, Cutispoto G, Rodono M. 1997. *Astron. Astrophys. Suppl.* 125:11–63

Strom KM, Strom SE. 1994. *Astrophys. J.* 424:237–56

Strom KM, Strom SE, Edwards S, Cabrit S, Skrutskie MF. 1989. *Astron. J.* 97:1451–70

Strom K, Strom SE, Wilkin FP, Carrasco L, Cruz-González I, et al. 1990. *Astrophys. J.* 362:168–90

Suter M, Stewart RT, Brown A, Zealey W. 1996. *Astron. J.* 111:320–26

Svestka Z, Fárník F, Hudson HS, Uchida Y, Hick P, et al. 1995. *Sol. Phys.* 161:331–63

Tagliaferri G, Giommi P, Angelini L, Osborne JP, Pallavicini R. 1988. *Astrophys. J.* 331:L113–16

Tatischeff V, Ramaty R, Kozlovsky B. 1998. *Astrophys. J.* 504:874–88

Tayler RJ. 1987. *MNRAS* 227:553–61

Terebey S, Shu FH, Cassen P. 1984. *Astrophys. J.* 286:529–51

Tout CA, Pringle JE. 1996. *MNRAS* 281:219–25
Trimble V. 1996. In *Star Formation Near and Far*, ed. SS Holt, LG Mundy, p. 15–37. Woodbury, NY: AIP
Tsuboi Y, Koyama K, Murakami H, Hayashi M, Skinner S, et al. 1998. *Astrophys. J.* 503:894–901
Uchida T, Shibata K. 1984. *Pub. Astron. Soc. Japan* 37:515–35
van den Ancker ME, de Winter D, Thé PS. 1996. *Astron. Astrophys.* 313:517–22
Voit GM. 1992. *MNRAS* 258:841–48
Vrba FJ, Herbst W, Booth JF. 1988. *Astron. J.* 96:1032–39
Waelkens C, Waters LBFM. 1998. *Annu. Rev. Astron. Astrophys.* 36:233–66
Walter FM. 1986. *Astrophys. J.* 306:573–86
Walter FM, Brown A, Linsky JL, Rydgren AE, Vrba F, et al. 1987. *Astrophys. J.* 314:297–307
Walter FM, Brown A, Mathieu RD, Myers PC, Vrba FJ. 1988. *Astron. J.* 96:297–325
Walter FM, Kuhi LV. 1981. *Astrophys. J.* 250:254–61
Walter FM, Kuhi LV. 1984. *Astrophys. J.* 284:194–201
Walter FM, Vrba FJ, Mathieu RD, Brown A, Myers PC. 1994. *Astron. J.* 107:692–719
Walter FM, Vrba FJ, Wolk SJ, Mathieu RD, Neuhäuser R. 1997. *Astron. J.* 114:1544
Wasson J. 1993. *Meteoritics* 28:14
Webb RA, Zuckerman B, Platais I, Patience J, White RJ, et al. 1999. *Astrophys. J.* 512:L63–67

White GJ, Ricketts MJ. 1977. *Astrophys. Lett.* 18:79
White SM, Pallavicini R, Kundu MR. 1992b. *Astron. Astrophys.* 257:557–66
White SM, Pallavicini R, Kundu MR. 1992a. *Astron. Astrophys.* 259:149–54
Wichmann R, Krautter J, Schmitt JHMM, Neuhäuser R, Alcalá J, et al. 1996. *Astron. Astrophys.* 312:439–54
Wichmann R, Bouvier J, Allain S, Krautter J. 1998. *Astron. Astrophys.* 330:521–32
Wichmann R, Krautter J, Covino E, Alcalá JM, Neuhäuser R, Schmitt JHMM. 1997a. *Astron. Astrophys.* 320:185–95
Wichmann R, Sterzik M, Krautter J, Metanomski A, Voges W. 1997b. *Astron. Astrophys.* 326:211–17
Wolk SJ, Walter FM. 1996. *Astron. J.* 111:2066–76
Woolum DS, Hohenberg C. 1993. See Levy & Lunine 1993, pp. 903–19
Yamauchi S, Koyama K. 1993. *Astrophys. J.* 405:268–72
Yamauchi S, Koyama K, Inda-Koide M. 1994. *PASJ* 46:473–78
Yamauchi S, Koyama K, Sakano M, Okada K. 1996. *PASJ* 48:719–37
Yamauchi S, Hamaguchi K, Koyama K, Murakami H. 1998. *PASJ* 50:465–74
Yan M, Dalgarno A. 1997. *Astrophys. J.* 481:296–301
Zhekov SA, Palla F, Myasnikov AV. 1994. *MNRAS* 271:667–75
Zhou S, Evans NJ, Koempe C, Walmsley CM. 1993. *Astrophys. J.* 404:232–76

SOURCES OF RELATIVISTIC JETS IN THE GALAXY

I. F. Mirabel
Centre d'Etudes de Saclay, CEA/DSM/DAPNIA/Sap, F-91191 Gif-sur-Yvette, France, and Instituto de Astronomía y Física del Espacio C.C. 67, Suc. 28. 1428, Buenos Aires, Argentina; mirabel@discovery.saclay.cea.fr

L. F. Rodríguez
Instituto de Astronomía, UNAM, Apdo. Postal 70-264, 04510 México, D.F., México; luisfr@astrosmo.unam.mx

Key Words   radio continuum stars, superluminal motion, X-rays binaries

■ **Abstract**   Black holes of stellar mass and neutron stars in binary systems are first detected as hard X-ray sources using high-energy space telescopes. Relativistic jets in some of these compact sources are found by means of multiwavelength observations with ground-based telescopes. The X-ray emission probes the inner accretion disk and immediate surroundings of the compact object, whereas the synchrotron emission from the jets is observed in the radio and infrared bands, and in the future could be detected at even shorter wavelengths. Black-hole X-ray binaries with relativistic jets mimic, on a much smaller scale, many of the phenomena seen in quasars and are thus called microquasars. Because of their proximity, their study opens the way for a better understanding of the relativistic jets seen elsewhere in the Universe. From the observation of two-sided moving jets it is inferred that the ejecta in microquasars move with relativistic speeds similar to those believed to be present in quasars. The simultaneous multiwavelength approach to microquasars reveals in short timescales the close connection between instabilities in the accretion disk seen in the X-rays, and the ejection of relativistic clouds of plasma observed as synchrotron emission at longer wavelengths. Besides contributing to a deeper understanding of accretion disks and jets, microquasars may serve in the future to determine the distances of jet sources using constraints from special relativity, and the spin of black holes using general relativity.

1. JETS IN ASTROPHYSICS

While the first evidence of jet-like features emanating from the nuclei of galaxies goes back to the discovery by Curtis (1918) of the optical jet from the elliptical galaxy M87 in the Virgo cluster, the finding that jets can also be produced in smaller scale by binary stellar systems is much more recent. The detection by Margon et al (1979) of large, periodic Doppler drifts in the optical lines of SS 433 resulted in the proposition of a kinematic model (Fabian & Rees 1979; Milgrom 1979) consisting of two precessing jets of collimated matter with velocity of 0.26c. High angular radio imaging as a function of time showed the presence of outflowing radio jets and fully confirmed the kinematic model (Spencer 1979; Gilmore & Seaquist 1980; Gilmore et al 1981; Hjellming & Johnston 1981). The early history of SS 433 has been reviewed by Margon (1984).

Since the detection of Sco X-1 at radio wavelengths (Ables 1969), some X-ray binaries had been known to be strong, time-variable non-thermal emitters. Ejection of synchrotron-emitting clouds was suspected from those days, but the actual confirmation of radio jets came only with the observations of SS 433. At present, there are about 200 known galactic X-ray binaries (van Paradijs 1995), of which about 10 percent are radio-loud (Hjellming & Han 1995). Of these radio-emitting X ray binaries, 10 have shown evidence of relativistic jets of synchrotron emission, and this review focuses on this set of objects. After the definition of Bridle & Perley (1984) for extragalactic jets, we use the term "jets" to designate collimated ejecta that have opening angles $\leq 15°$.

In the last years it has become clear that collimated ejecta can be produced in several stellar environments when an accretion disk is present. Jets with terminal velocities in the order of a few hundred to a few thousand km s^{-1} are now known to emanate from objects as diverse as very young stars (Reipurth & Bertout 1997), nuclei of planetary nebulae (López 1997), and accreting white dwarfs that appear as supersoft X-ray sources (Motch 1998, Cowley et al 1998). These types of stellar jets have, however, non-relativistic velocities (~ 100–10000 km s^{-1}) and their associated emission is dominantly thermal (i.e. free-free continuum emission in the radio as well as characteristic near-IR, optical and UV lines). Interestingly, in all known types of jet sources a disk is believed to be present. This review concentrates on synchrotron jets with velocities that can be considered relativistic ($v \geq 0.1c$), which are observed in X-ray binaries that contain a compact object, that is, a neutron star or a black hole. Our emphasis is on the radio characteristics of these sources. For detailed reviews of the X-ray properties of these sources we refer the reader to the reviews by Tanaka & Shibazaki (1996) and Zhang et al (1997).

2. MICROQUASARS

At first glance it may seem paradoxical that relativistic jets were first discovered in the nuclei of galaxies and distant quasars, and that for more than a decade SS 433 was the only known object of its class in our Galaxy (Margon 1984). The reason for this is that disks around supermassive black holes emit strongly at optical and UV wavelengths. Indeed, the more massive the black hole, the cooler the surrounding accretion disk is. For a black hole accreting at the Eddington limit, the characteristic black body temperature at the last stable orbit in the surrounding accretion disk will be given approximately by $T \sim 2 \times 10^7 M^{-1/4}$ (Rees 1984), with T in K and the mass of the black hole, M, in solar masses. Then, while accretion disks in AGNs have strong emission in the optical and ultraviolet with distinct broad emission lines, black hole and neutron star binaries usually are identified for the first time by their X-ray emission. Among these sources, SS 433 is unusual given its broad optical emission lines and its brightness in the visible. Therefore, it is understandable that there was an impasse in the discovery of new stellar sources of relativistic jets until the recent developments in X-ray astronomy.

Observations in the two extremes of the electromagnetic spectrum, in the domain of the hard X-rays on one hand (Sunyaev et al 1991; Paul et al 1991), and in the domain of radio wavelengths on the other hand, revealed the existence of new stellar sources of relativistic jets known as *microquasars* (Mirabel et al 1992; Mirabel & Rodríguez 1998). These are stellar-mass black holes in our Galaxy that mimic, on a smaller scale, many of the phenomena seen in quasars. The microquasars combine two relevant aspects of relativistic astrophysics: accreting black holes (of stellar origin) which are a prediction of general relativity and are identified by the production of hard X-rays and gamma-rays from surrounding accretion disks, and relativistic jets of particles that are understood in terms of special relativity and are observed by means of their synchrotron emission.

Multi-wavelength studies of the X-ray and gamma-ray sources in the galactic center region led in the year 1992 to the discovery of two microquasars: 1E1740.7-2942 and GRS 1758-258 (Mirabel et al 1992, Rodríguez et al 1992). The X-ray luminosity, the photon spectrum, and the time variability of these two sources are comparable to those of the black hole binary Cygnus X-1 (Churazov et al 1994; Kuznetsov et al 1997), and it is unlikely that they are extragalactic since no such persistent hard X-ray ultraluminous AGNs are observed (Mirabel et al 1993). In Figure 1 we show the radio counterpart of 1E1740.7-2942. As in Cygnus X-1, the centimeter radio counterpart of 1E1740.7-2942 is a weak core source that exhibits flux variations of the order of $\sim$50% which at epochs appear anticorrelated with the X-ray flux (Mirabel et al 1992). At radio wavelengths these two X-ray persistent sources located near the galactic center have a striking morphological resemblance with distant radio galaxies; they consist of compact components at the center of two-sided jets that end in weak, extended lobes with no significant radio flux variations observed in the last 6 years (Rodríguez & Mirabel 1999b). 1E1740.7-2942 and GRS 1758-258 seem to be persistent sources of both X-rays and relativistic jets. Mirabel et al (1993) have argued why it would be unlikely that the radio sources are radio galaxies accidentally superposed on the X-rays sources. For 1E1740.7-2942 no counterpart in the optical or near infrared wavelengths has been found so far, although there is a report of a marginal detection at $\lambda 3.8$ μm by Djorgovski et al (1992). GRS 1758-258 has two possible faint candidate counterparts (Martí et al 1998).

In these binaries of stellar-mass are found the three basic ingredients of quasars; a black hole, an accretion disk heated by viscous dissipation, and collimated jets of high energy particles. But in microquasars the black hole is only a few solar masses instead of several millon solar masses; the accretion disk has mean thermal temperatures of several millon degrees instead of several thousand degrees; and the particles ejected at relativistic speeds can travel up to distances of a few light years only, instead of several millon light years as in giant radio galaxies (Mirabel &

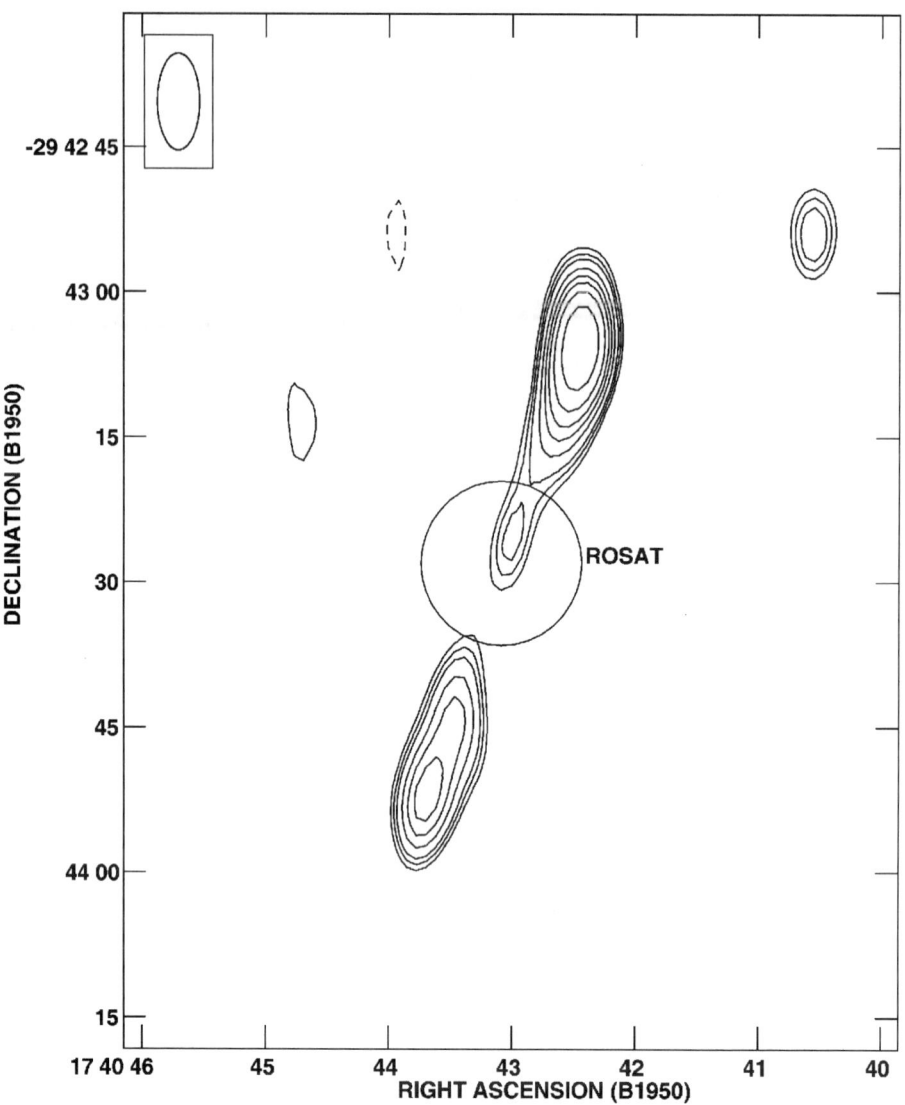

Figure 1 Contour map of the 6-cm emission from the radio counterpart of 1E1740.7-2942, as observed with the Very Large Array (Mirabel et al 1992; Rodríguez & Mirabel 1999c). The error circle of the ROSAT position (Heindl et al 1995), that includes the core source, is also shown. At a distance of 8 kpc the length of the jet structure would be $\sim$5 pc. The half power contour of the beam is shown in the top left corner. Contours are $-4, 4, 5, 6, 8, 10, 12, 15,$ and 20 times 28 μJy beam^{-1}.

Rodríguez 1998). Indeed, simple scaling laws govern the physics of flows around black holes, with length and time scales being proportional to the mass of the black holes (Sams et al 1996; Rees 1998). The word *microquasar* was chosen to suggest that the analogy with quasars is more than morphological, and that there is an underlying unity in the physics of accreting black holes over an enormous range of scales, from stellar-mass black holes in binary systems, to supermassive black holes at the center of distant galaxies. Strictly speaking and if it had not been for the historical circumstances described above, the acronym *quasar* ("quasi-stellar-radio-source") would have suited better the stellar mass versions rather than their super-massive analogs at the centers of galaxies.

3. SUPERLUMINAL SOURCES

Expansions at up to ten or more times the speed of light have been observed in quasars for more than 20 years (Pearson & Zensus 1987; Zensus 1997). At first these superluminal motions provoked concern because they appeared to violate relativity, but they were soon interpreted as illusions due to relativistic aberration (Rees 1966). However, the ultimate physical interpretation had remained uncertain. In the extragalactic case the moving jets are observed as one-sided (because strong Doppler favoritism renders the approaching ejecta detectable) and it is not possible to know if superluminal motions represent the propagation of waves through a slowly moving jet, or if they reflect the actual bulk motion of the sources of radiation.

In the context of the microquasar analogy, one may ask if superluminal motions could be observed from sources known to be in our own Galaxy. Among the handful of black holes of stellar mass known so far, three transient X-ray sources have indeed been identified at radio waves as sporadic sources of superluminal jets. The first superluminal source to be discovered (Mirabel & Rodríguez 1994) was GRS 1915+105, a recurrent transient source of hard X-rays first found and studied with the satellite GRANAT (Castro-Tirado et al 1994; Finoguenov et al 1994). The discovery of superluminal motions in GRS 1915+105 stimulated a search for similar relativistic ejecta in other transient hard X-ray sources. Soon after, the same phenomenon was observed by two different groups (Tingay et al 1995; Hjellming & Rupen 1995) in GRO J1655-40, a hard X-ray nova found with the Compton Gamma Ray Observatory (Zhang et al 1994). A third superluminal source may be XTE J1748-288 (Hjellming et al 1998), a transient source with a hard X-ray spectrum recently found with XTE (Smith et al 1998).

GRS 1915+105 is at ∼12 kpc from the Sun (Rodríguez et al 1995; Chaty et al 1996) on the opposite side of the galactic plane and cannot be studied in the optical. Given the large extinction by dust along the line of sight (Mirabel et al 1994; Chaty et al 1996), the precise nature of the binary has been elusive. Castro-Tirado et al (1996) proposed that GRS 1915+105 is a low mass binary, while Mirabel et al (1997) proposed that it is a long period binary with a companion star of transitional spectral type. From the nature of the line variability in the infrared, Eikenberry et al (1998b) propose that the emission lines in GRS 1915+105 arise in an accretion disk rather than in the circumstellar disk of an Oe/Be companion (Mirabel et al 1997). GRS 1915+105 has similarities in the X-rays and gamma-rays with GRO J1655-40 and other black hole binaries, and it is also likely to harbor a black hole (Greiner et al 1996). The X-ray luminosity of GRS 1915+105 (reaching 2×10^6 solar luminosities) far exceeds the Eddington limit (above which the radiation pressure will catastrophically blow out the external layers of the source) for a 3 solar mass object, which is 10^5 solar luminosities. Furthermore, it shows the typical hard X-ray tail beyond 100 keV seen in black hole binaries (Cordier et al 1993; Finoguenov et al 1994; Grove et al 1998). Finally, it is known that the absolute hard X-ray luminosities in black hole systems are systematically higher than in neutron star systems (Ballet et al 1994, Barret et al 1996), another result that points to a black hole in GRS 1915+105.

GRO J1655-40 is at a distance of 3.2 kpc and the apparent transverse motions of its ejecta in the sky are the largest yet observed (Tingay et al 1995; Hjellming & Rupen 1995) until now from an object beyond the solar system. It has a bright optical counterpart and consists of a star of 1.7–3.3 solar masses orbiting around a collapsed object of 4–7 solar masses (Orosz & Bailyn 1997, Phillips et al 1999). The compact object is certainly a black hole, since its mass is beyond the theoretical maximum mass limit of $\sim$3 solar masses for neutron stars (Kalogera & Baym 1996).

King (1998) proposes that the superluminal sources are black hole binaries with the secondary in the Hertzsprung-Russell gap, which provides super-Eddington accretion into the black hole. In the Galaxy there would have been $\geq 10^3$ systems of this class with a lifetime for the jet phase of $\leq 10^7$ years, which is the duration of the spin-down phase of the black hole.

3.1 Superluminal Motions in GRS 1915+105

Figure 2 shows a pair of bright radio condensations emerging in opposite directions from the compact, variable core of GRS 1915+105. Before and after the remarkable ejection event shown in Figure 2, the source ejected other pairs of condensations

Figure 2 Pair of radio condensations moving away from the hard X-ray source GRS 1915+105 (Mirabel & Rodríguez 1994). These uniform-weight VLA maps were made at λ3.5-cm for the 1994 epochs on the right side of each map. The position of the stationary core is indicated with a small cross. The maps have been rotated 60° clockwise for easier display. The cloud to the left appears to move away from the stationary core at 125% the speed of light. Contours are 1, 2, 4, 8, 16, 32, 64, 128, 256 and 512 times 0.2 mJy/beam^{-1} for all epochs except for March 27 where the contour levels are in units of 0.6 mJy/beam^{-1}. The half power beam width of the observations, 0.2 arc sec, is shown in the top right corner.

RELATIVISTIC JETS IN THE GALAXY 415

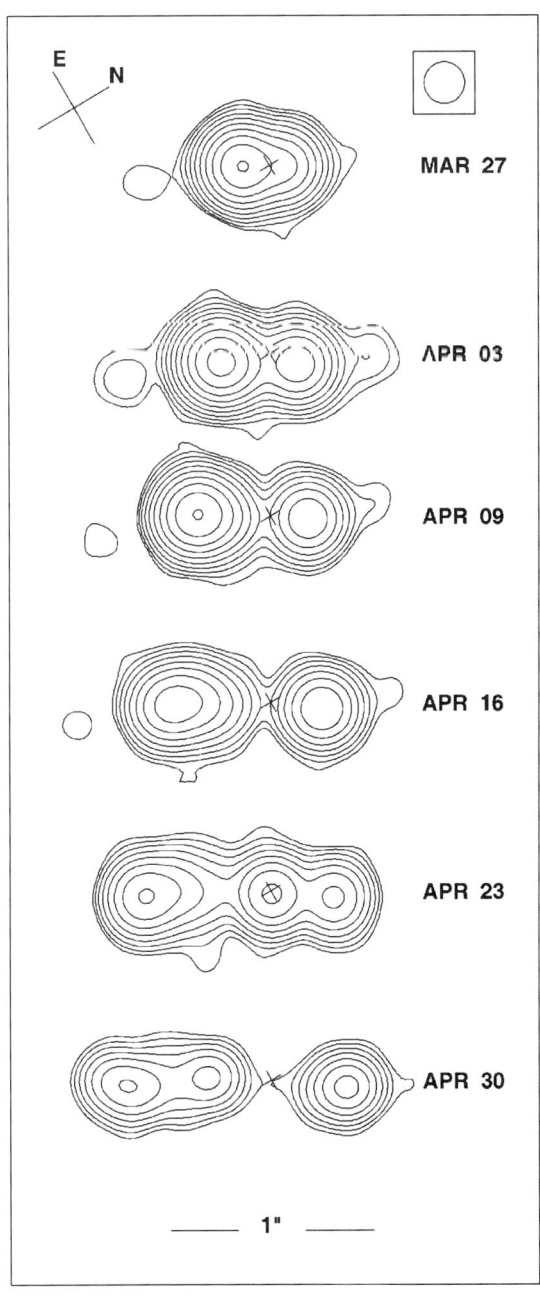

but with flux densities one to two orders of magnitude weaker. One of these weaker pairs can be seen in the first four maps of Figure 2, as a fainter pair of condensations moving ahead of the bright ones at about the same proper motion and direction.

In Figure 3 we show the proper motions of the condensations detected from four ejection events in 1994. The angular displacements from the stationary core are consistent with unaccelerated motions. The time separation between ejections suggests a quasi-periodicity at intervals in the range of 20–30 days. Although the clouds in each event appear to move ballistically, always in the same general region of the sky, their position angles suggest changes by $\sim 10°$ in the direction of ejection in one month.

Figures 2 and 3 show two asymmetries: one in apparent transverse motions, another in brightness. The cloud that appears to move faster also appears brighter. It has been shown that both asymmetries, in proper motions and in brightness, are consistent with the hypothesis of an anti-parallel ejection of twin clouds moving at relativistic velocities (Mirabel & Rodríguez 1994), as discussed in Section 4. At a distance of 12 kpc the proper motions measured with the VLA in 1994 of the approaching (17.6 ± 0.4 mas d^{-1}) and receding (9.0 ± 0.1 mas d^{-1}) condensations shown in Figure 2 imply apparent velocities on the plane of the sky of 1.25c and 0.65c, respectively. From the analysis of relativistic distorsion effects using the equations in the next section and the VLA data, it is inferred that the ejecta move with a speed of 0.92c at an angle $\theta = 70°$ to the line of sight.

Within the errors of the measurements and a precession of $\leq 10°$, relativistic ejections with a stable jet axis at scales of 500–5000 AU and larger were later observed from GRS 1915+105 over a time span of four years (Mirabel et al 1996a, Fender et al 1999, Dhawan et al 1999). The VLBA images of GRS 1915+105 show that the jets are already collimated at milliarcsec scales (Dhawan et al 1999), namely, at about 10 AU from the compact source (Figure 4). The core appears as a synchrotron jet of length ~ 100 AU before and during optically thin flares, and at those scales it already exhibits Doppler boosting. Discrete ejecta have appeared

Figure 3 Angular displacements as a function of time for four ejection events observed in 1994 in GRS 1915+105 (Rodríguez & Mirabel 1999a). Top: Angular displacements as a function of time for four approaching condensations corresponding to ejections that took place on (from left to right) 1994 January 29 (triangles), February 19 (squares), March 19 (circles), and April 21 (crosses). Bottom: Angular displacements as a function of time for three receding condensations corresponding to ejections that took place on (from left to right) 1994 February 19 (squares), March 19 (circles), and April 21 (crosses). The clouds of the 1994 January 29 ejection were relatively weak and the receding component could not be detected unambiguously. The dashed lines are the least squares fit to the angular displacements of the 1994 March 19 event, the brighter and better studied. Note that the motions appear to be ballistic (that is, unaccelerated).

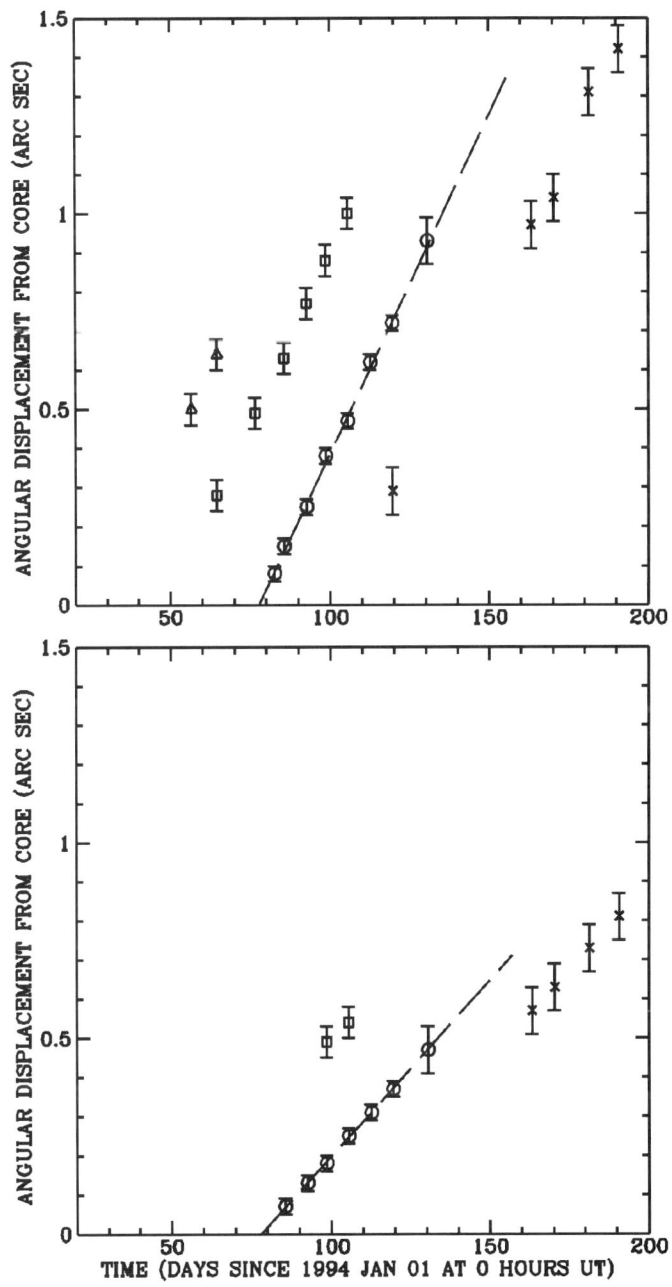

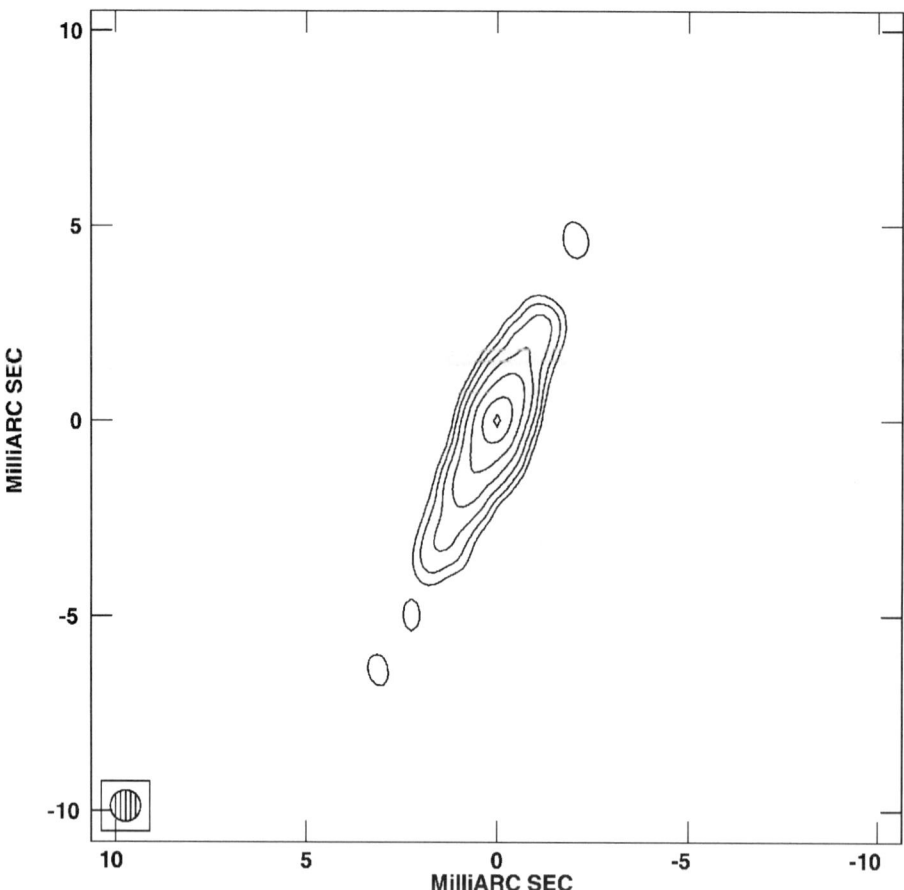

Figure 4 Contour map of the 2-cm emission from the core of GRS 1915+105, as observed on April 11, 1997 with the Very Long Baseline Array at milliarcsecond angular resolution (Dhawan et al 1999). The angular resolution corresponds to about 10 AU at GRS 1915+105. The half power contour of the beam is shown in the bottom left corner. Contours are -1, 1, 2, 4, 8, 16, 32, 64, and 96 times 0.26 mJy beam^{-1}. The position angle of this ejection at milliarcsec scale is the same as that seen at the arcsec scales three years before.

at about 500 AU. Both, the observations with MERLIN (Fender et al 1999) and with the VLBA (Dhawan et al 1999) in the years 1997 and 1998 have shown faster apparent superluminal motions at 1.3c–1.7c at scales of hundreds of AU, and intrinsic expansions of the expelled clouds mostly in the direction of their bulk motions. At present it is not clear if the faster motions measured with the higher resolution observations of MERLIN and VLBA in 1997 relative to the VLA observations in 1994 are due to intrinsic faster ejections, changes in the angle to the line of sight, or to resolution effects between the arrays as suggested by Fender et al (1999).

A secular proper motion of 5.8 ± 1.5 mas yr^{-1} in the galactic plane, in rough agreement with the HI distance of 12 kpc (Rodríguez et al 1995), has been measured with the VLBA (Dhawan et al 1999).

3.2 Superluminal Motions in GRO J1655-40

The relativistic ejections observed in the radio in GRO J1655-40 have striking similarities as well as differences with those in GRS 1915+105. Bright components moving apart with proper motions in the range of 40 to 65 mas d^{-1} were independently observed with the Southern Hemisphere VLBI Experiment array (Tingay et al 1995), and the VLA and VLBA (Hjellming & Rupen 1995). In Figure 5 is shown a sequence of seven VLBA radio images of GRS J1655-40 from Hjellming & Rupen (1995). At a distance of 3.2 kpc the motions of the ejecta have been fit—using a kinematic model—with a velocity of 0.92c, and a jet axis inclined 85° to the line of sight at a position angle of 47°, about which the jets rotate every three days at an angle of 2°.

In contrast to what has been observed in the repeated ejections of GRS 1915+105, the flux ratios of the blobs on either side of GRO J1655-40 cannot be ascribed to relativistic Doppler boosting. In GRO J1655-40 the asymmetry in brightness appears to flip from side to side (Hjellming & Rupen 1995). Not only do the jets appear to be intrinsically asymmetric, but also the sense of that asymmetry changes from event to event. Therefore, although similar intrinsic velocities greater than 0.9c are found in both superluminal sources, due to the asymmetries in GRO J1655-40, the ultimate physical interpretation of the superluminal expansions in this source remains uncertain.

We point out that in SS 433 flux asymmetries between knots ejected simultaneously on both sides have also been observed (Fejes 1986). This asymmetry could be due to intrinsic variations, so perhaps GRO J1655-40 is not unusual in this respect. However, VLBA multiwavelength monitoring of SS 433 (Paragi et al 1998) shows that it is always the receding part of the core-complex which is fainter compared to the approaching one, and that this effect cannot be explained simply by Doppler beaming. It is possible that free-free absorption and the different pathlengths through an absorbing medium could explain some of these asymmetries in SS 433 and other jet sources. Furthermore, in SS 433 more than 90% of the radio emission is in knots rather than in continuous jets, and the core complex disappears after large outbursts, as observed in GRS 1915+105 during the years 1992 and 1993 (Mirabel & Rodríguez 1994).

3.3 Superluminal Motions in XTE J1748-288

Two major relativistic ejection sequences moving at least 20 mas day^{-1} were observed in June 1998 (Hjellming et al 1998) from the hard X-ray transient XTE

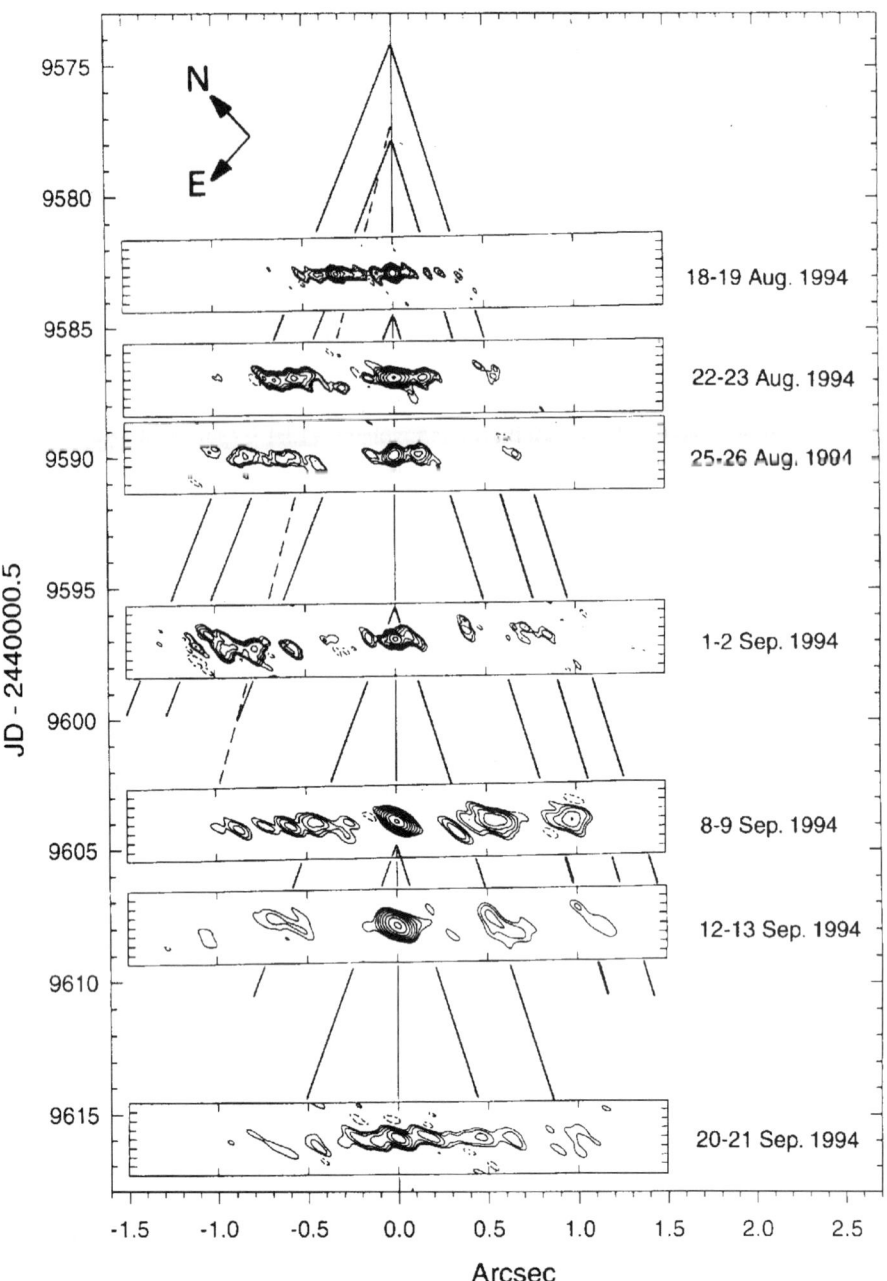

Figure 5 A sequence of seven VLBA images of GRO J1655-40 at 1.6 GHz, each rotated anticlockwise by 43°, and each having an angular resolution of 3.0 × 0.4 arcsec (Hjellming & Rupen 1995). Each image is labeled with the date of the observations. The solid lines between images identify motions of 54 mas day^{-1} (left) and 45.5 mas day^{-1} (right). The vertical line marks the position of the central source, assumed to be the brightest point on each image.

J1748-288 (Smith et al 1998). Each sequence appeared to begin with a one-sided relativistic ejection. The ejecta are highly linearly polarized, and at a distance of 8 kpc, derived from the HI λ21cm absorption line, their motions would imply apparent speeds of 0.9c and 1.5c, and intrinsic velocities of more than 0.9c (Hjellming et al 1998). This is the first galactic source of relativistic jets where it has been observed in real time that the jets collide with environmental material, being decelerated while brightening at the leading edge of the jet.

4. SPECIAL RELATIVITY EFFECTS

4.1 Parameters of the Ejection

The main characteristics of the superluminal ejections can be understood in terms of the simultaneous ejection of a pair of twin condensations moving at velocity β ($\beta = v/c$), with v being the velocity of the condensations and c the speed of light), with the axis of the flow making an angle θ ($0° \leq \theta \leq 90°$) with respect to the line of sight of a distant observer (Rees 1966; see Figure 6). The apparent proper motions in the sky of the approaching and receding condensations, μ_a and μ_r, are given by:

$$\mu_a = \frac{\beta \sin \theta}{(1 - \beta \cos \theta)} \frac{c}{D}, \qquad (1)$$

$$\mu_r = \frac{\beta \sin \theta}{(1 + \beta \cos \theta)} \frac{c}{D}, \qquad (2)$$

where D is the distance from the observer to the source. These two equations can be transformed to the equivalent pair of equations:

$$\beta \cos \theta = \frac{\mu_a - \mu_r}{\mu_a + \mu_r}, \qquad (3)$$

$$D = \frac{c \tan \theta}{2} \frac{(\mu_a - \mu_r)}{\mu_a \mu_r}. \qquad (4)$$

If only the proper motions are known, an interesting upper limit for the distance can be obtained from Equations (3) and (4):

$$D \leq \frac{c}{\sqrt{\mu_a \mu_r}}. \qquad (5)$$

In all equations we use cgs units and the proper motions are in radians s^{-1}. In the case of the bright ejection event of 1994 March 19 for GRS 1915+105, the proper motions measured were $\mu_a = 17.6 \pm 0.4$ mas day^{-1} and $\mu_r = 9.0 \pm 0.1$ mas day^{-1}. Using Equation (5), we derive an upper limit for the distance, $D \leq 13.7$ kpc, confirming the galactic nature of the source.

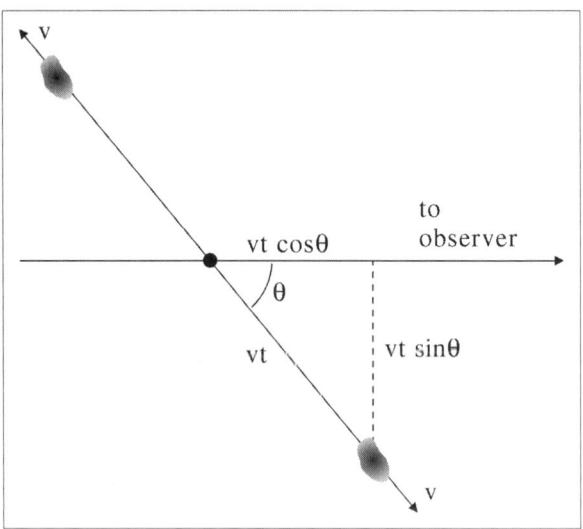

Figure 6 Geometry of the two-sided ejection. The emission is symmetric, but when the emitting clouds move at relativistic speeds the approaching component of the pair appears to move faster and to be brighter than the receding component.

The distance to GRS 1915+105 is found to be, from HI absorption studies, 12.5 ± 1.5 kpc (Rodríguez et al 1995; Chaty et al 1996). Then, the proper motions of the approaching and receding condensations measured with the VLA in 1994 and 1995 imply apparent velocities on the plane of the sky of $v_a = 1.25c$ and $v_r = 0.65c$ for the approaching and receding components respectively. The ejecta move with a true speed of $v = 0.92c$ at an angle $\theta = 70°$ with respect to the line of sight (Mirabel & Rodríguez 1994). The faster proper motions of 24 mas day^{-1} measured with MERLIN (Fender et al 1999) and the VLBA (Dhawan et al 1999) in 1997 would imply a true speed of $0.98c$ at an angle of $66°$ to the line of sight.

4.2 A Relativistic Distance Determination

We note that the detection of a known line from either of the condensations would allow a precise determination of the distance. The Doppler factors, namely, the ratios of observed to emitted frequency (v_o) for the approaching and receding condensations are given by

$$\delta_a = \frac{v_a}{v_o} = \gamma^{-1}(1 - \beta \cos\theta)^{-1}, \quad (6)$$

$$\delta_r = \frac{v_r}{v_o} = \gamma^{-1}(1 + \beta \cos\theta)^{-1}. \quad (7)$$

In these last two equations $\gamma = (1 - \beta^2)^{-1/2}$ is the Lorentz factor. Since we know $\beta \cos\theta$, a determination of either v_a/v_o or v_r/v_o will allow the determination of β and thus the determination of θ and of the distance from Equation (4). In the case of cosmologically distant objects, the Equations 1, 2, 4, and 5 are valid replacing the distance D by the angular size distance D_a (Peebles 1993), and the rest frequency v_o by $v_o/(1+z)$, with z being the observed redshift of the central source. The angular size distance is given by $D_a = (cz/H_0)[1-(1+q_0)z/2+\cdots]$, where H_0 is Hubble's constant and q_0 is the dimensionless acceleration (or deceleration) parameter. Then, the observations of proper motions and frequency shifts in extragalactic relativistic ejecta pairs could potentially be used to test between different cosmological models.

4.3 Doppler Boosting

The ratios of observed to emitted flux density S_o, from a twin pair of optically-thin, isotropically emitting jets are:

$$\frac{S_a}{S_o} = \delta_a^{k-\alpha}, \tag{8}$$

$$\frac{S_r}{S_o} = \delta_r^{k-\alpha}, \tag{9}$$

where α is the spectral index of the emission ($S_\nu \propto \nu^\alpha$), and k is a parameter that accounts for the geometry of the ejecta, with $k = 2$ for a continuous jet and $k = 3$ for discrete condensations. Then, the ratio of observed flux densities (measured at equal separations from the core) will be given by

$$\frac{S_a}{S_r} = \left(\frac{1+\beta\cos\theta}{1-\beta\cos\theta}\right)^{k-\alpha}, \tag{10}$$

Since for the 1994 March 19 event $\beta\cos\theta = 0.323$ and $\alpha = -0.8$ the flux ratio in the case of discrete condensations would be 12, whereas for a continuous jet it would be 6. For a given angular separation it was found that the observed flux ratio between the approaching and receding condensations is 8 ± 1. Similar results were found using the MERLIN observations by Fender et al (1999). Therefore, irrespective of the distance to the source, the flux ratios for equal angular separations from the core are consistent with the assumption of a twin ejection at relativistic velocities. Atoyan & Aharonian (1997) have considered the observable effects in the flux density ratio of asymmetries between the jet and counterjet. Bodo & Ghisellini (1995) have proposed that there could be a contribution of wave propagation in the pattern motions, but that most of the observed displacements are true bulk plasma velocities.

5. ACCRETION DISK INSTABILITIES AND JET FORMATION

Collimated jets seem to be systematically associated with the presence of an accretion disk around a star or a collapsed object. In the case of black holes, the characteristic dynamical times in the flow of matter are proportional to the black hole's mass, and the events with intervals of minutes in a microquasar could correspond to analogous phenomena with duration of thousands of years in a quasar of 10^9 $M_\odot$ (Sams et al 1996; Rees 1998). Therefore, the variations with minutes of duration observed in a microquasar in the radio, IR, optical, and X rays could sample phenomena that we have not been able to observe in quasars.

X-rays probe the inner accretion disk region, radio waves the synchrotron emission from the relativistic jets. The long term multiwavelength light curves of the superluminal sources show that the hard X-ray emission is a necessary but not sufficient condition for the formation of collimated jets of synchrotron radio emission. In GRS 1915+105 the relativistic ejection of pairs of plasma clouds have always been preceded by unusual activity in the hard X-rays (Harmon et al 1997), more specifically, the onset of major ejection events seems to be simultaneous to the sudden drop from a luminous state in the hard X-rays (Foster et al 1996; Mirabel et al 1996a). However, not all unusual activity and sudden drops in the hard X-ray flux appear to be associated with radio emission from relativistic jets. In fact, in GRO J1655-40 there have been several hard X-ray outbursts without following radio flare/ejection events. A more detailed summary of the long term multifrequency studies of black hole binaries can be found in Zhang et al (1997).

The episodes of large amplitude X-ray flux variations in time-scales of seconds and minutes, and in particular, the abrupt dips observed (Greiner et al 1996, Belloni et al 1997, Chen et al 1997) in GRS 1915+105 are believed to be evidence for the presence of a black hole, as discussed below. These variations could be explained if the inner (≤ 200 km) part of the accretion disk goes temporarily into an advection-dominated mode (Abramowicz et al 1995; Narayan et al 1997). In this mode, the time for the energy transfer from ions (that get most of the energy from viscosity) to electrons (that are responsible for the radiation) is larger than the time of infall to the compact object. Then, the bulk of the energy produced by viscous dissipation in the disk is not radiated (as happens in standard disk models), but instead is stored in the gas as thermal energy. This gas, with large amounts of locked energy, is advected (transported) to the compact object. If the compact object is a black hole, the energy quietly disappears through the horizon. In constrast, if the compact object is a neutron star, the thermal energy in the superheated gas is released as radiation when it collides with the surface of the neutron star and heats it up. The cooling time of the neutron star photosphere is relatively long, and in this case a slow decay in the X-ray flux is observed. Thus, one would expect the luminosity of black hole binaries to vary over a much wider range than that of neutron star binaries (Barret et al 1996). The idea of advection-dominated flow has also been proposed (Hameury et al 1997) to explain the X-ray delay in an optical outburst (Orosz et al 1997) of GRO J1655-40.

During large-amplitude variations in the X-ray flux of GRS 1915+105, remarkable flux variations on time-scales of minutes have also been reported at radio (Pooley & Fender 1997, Rodríguez & Mirabel 1997, Mirabel et al 1998) and near-infrared wavelengths (Fender et al 1997, Fender & Pooley 1998, Eikenberry et al 1998a, Mirabel et al 1998). The rapid flares at radio and infrared waves are thought to come from expanding magnetized clouds of relativistic particles. This idea is supported by the observed time shift of the emission at radio waves as a function of wavelength and the finding of infrared synchrotron precursors to the follow-up radio flares (Mirabel et al 1998). Sometimes the oscillations at radio waves appear as isolated events composed of twin flares with characteristic time shifts of 70 ± 20 minutes (e.g. Pooley & Fender 1997, Dhawan et al 1999). The time shift between the twin peaks seems to be independent of wavelength (Mirabel et al 1998), and no Doppler boosting is observed. This suggests that these quasiperiodic flares may come from expanding clouds moving in opposite directions with non-relativistic bulk motions.

In Figure 7 are shown simultaneous light curves in the X-rays, infrared, and radio wavelengths, together with the X-ray photon index during a large amplitude oscillation. These light curves can be consistently interpreted to imply that the relativistic clouds of plasma emerge at the time of the dips and follow-up recovery of the X-ray flux. In adiabatically expanding clouds the maximum flux density at short wavelengths (i.e. the near infrared) should be observed very shortly after the ejection (10^{-3} sec), and it is only in the radio wavelengths that significant time delays occur (Mirabel et al 1998). Figure 7 shows that the onset of the infrared flare occured ≥ 200 sec after the drop of the X-ray flux, during its recovery from the dip, probably at the time of the appearance of an X-ray spike ($t = 13$ min) which is associated to a sudden softening of the (13–60 keV)/(2–13 keV) photon index due to the drop in the hard X-ray flux. Similar phenomena have been observed in this source by Eikenberry et al (1998a). In the context of the unstable accretion disk model of Belloni et al (1997), these observations suggest that the ejection of plasma clouds takes place during the subsequent replenishment of the inner accretion disk, well after the disappearance of the soft component at the sudden drop. The ejection of the clouds seems to be coincident with the soft X-ray peak at the dip. Furthermore, the slow rise of the infrared flux to maximum seen in Figure 7 indicates that the injection of relativistic particles is not instantaneous and that it could last up to tens of minutes.

Mirabel et al (1998) have estimated that the minimum mass of the clouds that are ejected every few tens of minutes is $\sim 10^{19}$ g. On the other hand, the estimated total mass that is removed from the inner accretion disk in one cycle of a few tens of minutes is of the order of $\sim 10^{21}$ g (Belloni et al 1997). Given the uncertainties in the estimation of these masses, it is still unclear what is the fraction of mass of the inner accretion disk that disappears through the horizon of the black hole. Anyway, it seems plausible that during accretion disk instabilities consisting of

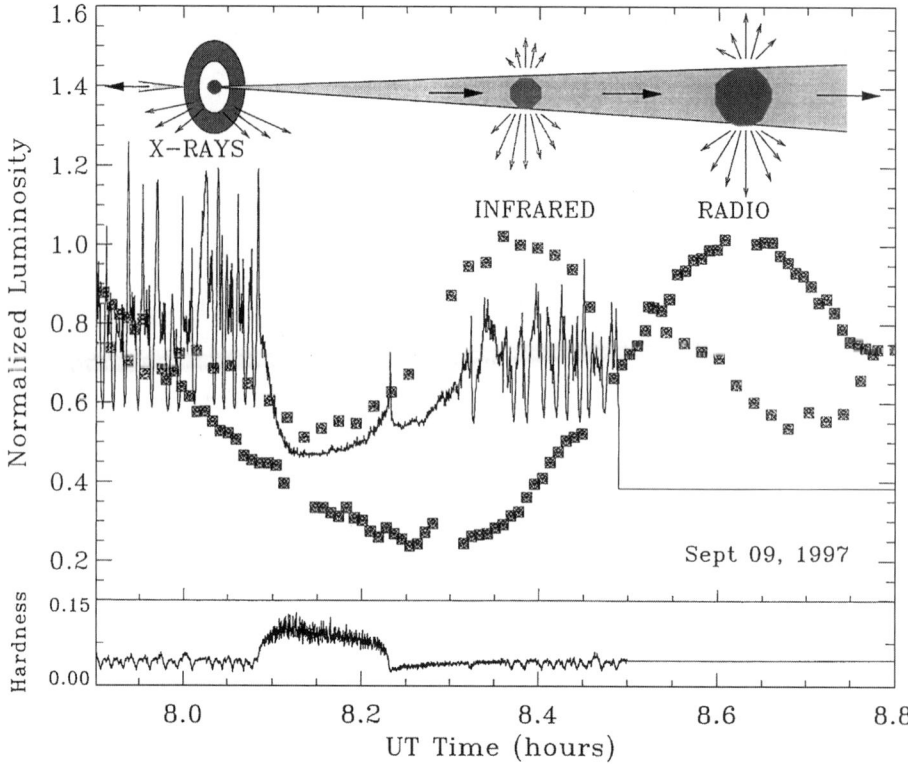

Figure 7 Radio, infrared, and X-ray light curves for GRS 1915+105 at the time of quasi-periodic oscillations on 1997 September 9 (Mirabel et al 1998). The infrared flare starts during the recovery from the X-ray dip, when a sharp, isolated X-ray spike is observed. These observations show the connection between the rapid disappearance and follow-up replenishment of the inner accretion disk seen in the X-rays (Belloni et al 1997), and the ejection of relativistic plasma clouds observed as synchrotron emission at infrared wavelengths first and later at radio wavelengths. A scheme of the relative positions where the different emissions originate is shown in the top part of the figure. The hardness ratio (13–60 keV)/(2–13 keV) is shown at the bottom of the figure.

the sudden disappearance of its inner part, most of it is advected into the black hole, and only some fraction is propelled into synchrotron-emitting clouds of plasma.

Energy outbursts in the flat synchrotron spectrum over at least four decades of frequency have also been observed in Cygnus X-3 (Fender et al 1996). The optical polarization observed in GRO J1655-40 (Scaltriti et al 1997) could also be related to the presence of synchrotron emission at optical wavelengths. The study of GRS 1915+105 led to the realization that besides the energy invested in the acceleration of the plasma clouds to their bulk motions, the oscillations of the type shown in Figure 7 require synchrotron luminosities of at least 10^{36} erg s^{-1}. This synchrotron luminosity is not negligible with respect to the thermal luminosity radiated in the X-rays. These results give support to the observation of synchrotron infrared jets reaching distances of a few thousand AU from GRS 1915+105 (Sams et al 1996).

6. JET FORMATION

The processes by which the jets are accelerated and collimated are still not clearly understood, but it is believed that several of the concepts proposed for extragalactic jets can be extended to galactic jets.

Blandford & Znajek (1977) take advantage of the fact that, in principle, it is possible to extract energy and angular momentum from a rotating black hole (Penrose 1969), to produce electric and magnetic fields and possibly fast outflowing jets. A magnetized accretion disk around the Kerr black hole brakes it electromagnetically. However, Ghosh & Abramowicz (1997) and Livio et al (1997) have called into question that the Blandford-Znajek process can provide the primary power in the jets.

A seminal idea that has been followed by many researchers in the field is that of the magnetohydrodynamical model of Blandford & Payne (1982). These authors proposed that the angular momentum of a magnetized accretion disk around the collapsed object is the responsible for the acceleration of the plasma. The magnetic field lines are taken to be frozen into the disk and the plasma is assumed to follow them like a "bead on a wire", at least close to the disk. If the field line forms an angle with the plane of the disk smaller than $60°$, the displacements of the plasma from its equilibrium position become unstable. This happens because along these field lines the component of the centrifugal force will be larger than the component of the gravitational force and the plasma will be accelerated outwards. Then, in its origin, the outflow motion has an important "equatorial" component, while on larger scales the jets are observed to have a motion that is dominantly "poloidal". In other words, after the acceleration a collimating mechanism is required to change the wide-angle centrifugal outflow into a collimated jet.

This collimation is proposed to be achieved as follows. Inside an inner region, the magnetic field energy density is larger than the kinetic energy density of the flow but at some distance from the disk (the Alfvén surface), this situation reverses and the flow stops corotating with the disk. This causes a loop of toroidal (azimuthal) field to be added to the flow for each rotation of the footpoint of the field line. The tension of this wound-up toroidal field that is formed external to the Alfvén surface produces a force directed toward the axis (the "hoop stress") that eventually collimates the flow into a jet. Most models for the production of jets in the astrophysical context use elements of MHD acceleration and collimation.

Recently, several groups (Spruit et al 1997, Lucek & Bell 1997, Begelman 1998) have pointed out that the toroidal field traditionally held responsible for collimating jets in the MHD mechanism is unstable and cannot collimate the jets effectively. It has been proposed alternatively that the collimating agent is the poloidal component of the magnetic field.

Koide et al (1998) have performed for the first time full general relativistic MHD numerical simulations of the formation of jets near a black hole. Their results suggest that the ejected jet has a two-layer structure with an inner, fast gas-pressure driven component and an outer, slow magnetically driven component. The presence of the inner, fast gas-pressure driven component is a result of the strong pressure increase produced by shocks in the disk through fast advection flows inside the last stable orbit around a black hole. This feature is not seen in non-relativistic calculations.

Within the uncertainties of the small sample, the velocity of the jets seems to show a bimodal distribution, with some sources having $v_{jet} \simeq 0.3c$ and others having $v_{jet} \geq 0.9c$. Two explanations have been offered in the literature. On one hand, Kudoh & Shibata (1995) suggest that the terminal velocity of the jet is of order of the Keplerian velocity at the footpoint of the jets, that is, that the fastest jets probably come from the deepest gravitational wells (Livio 1997). However, recent observations suggest that Sco X-1 which is a neutron star binary has $v_{jet} \sim 0.5c$ (Fomalont 1999), departing from the bimodal distribution. On the other hand, Meier et al (1997) propose that the velocity of the jets is regulated by a magnetic "switch", with highly relativistic velocities achieved only above a critical value of the ratio of the Alfvén velocity to the escape velocity. The determination of the mass of the collapsed object in a larger number of jet sources would discriminate between these two models.

While it seems that a steady state MHD model can account for the formation of continuous relativistic jets, the events discussed by Mirabel et al (1998), Belloni et al (1998), and Fender & Pooley (1998) that seem to involve a connection between the disappearance of the inner accretion disk and the sudden ejection of condensations may require a different mechanism. Clearly, the time seems to be ripe for new theoretical advances on the models of formation of relativistic jets that take into account the observational features found in stellar jets.

Another characteristic that the jet models must account for is the production of relativistic particles that will produce the synchrotron emission that is observed in several sources. As in other astrophysical contexts, it is believed that the acceleration of electrons to relativistic speeds takes place in shocks (Blandford & Ostriker 1978). On the other hand, most of the X-ray binaries are "radio-quiet", implying that relativistic electrons and/or magnetic fields are not always present in sufficient amounts.

7. SYNCHROTRON EMISSION

The high brigtness temperature, rapid variability, and linear polarization observed in the radio emission from X-ray binaries indicates a synchrotron origin. The time evolution of the radio emission has been modeled in terms of conical jets or expanding clouds of magnetized plasma (Hjellming & Johnston 1988, Martí et al 1992, Seaquist 1993).

In the simplest case of an adiabatically expanding spherical cloud in the optically thin regime, the van der Laan (1966) model is used, where the flux density is given by $S_\nu \propto \nu^{(1-p)/2} r^{-2p}$, and the relativistic electrons have an energy distribution given by $N(E) = KE^{-p}$, with K being a constant that is related to the density of the relativistic electrons. In this equation r is the radius of the cloud. Assuming that the cloud expands linearly with time, the flux density is given by $S_\nu \propto \nu^{(1-p)/2} t^{-2p}$. Assuming a typical value of $p = 2.4$, one obtains $S_\nu \propto \nu^{-0.7} t^{-4.8}$. This simple model fits the flux decrease reasonably well for several of the radio-emitting X-ray binaries (Ball 1996). However, in some of the best studied jet sources (SS 443, Hjellming & Johnston 1988, Vermeulen et al 1993, GRS 1915+105, Rodríguez & Mirabel 1999a), much less steep decreases are observed. This situation can be accounted for by making modifications to the simple expanding model. One possibility is to attribute this shallower drop of flux density with time to constrained expansion (the source cannot expand in 3 dimensions but only in 1 or 2 dimensions). In fact, the GRS 1915+105 maps with milliarcsec resolution by Dhawan et al (1999) show that the expansion of the clouds at hundreds of AU from the compact source is mostly in one direction. The flux density can be then approximately described as $S_\nu \propto \nu^{-0.7} t^{-(2/3)pn}$, where n is the number of dimensions where expansion is allowed. Both in SS 433 (Hjellming & Johnson 1988) and in GRS 1915+105 (Rodríguez & Mirabel 1999a), a break in the power law that describes the decrease in flux as a function of time is observed. Remarkably, in both sources the decrease close to the source can be described with $S_\nu \propto t^{-1.3}$, while after a distance of $\sim 2 \times 10^{17}$ cm, $S_\nu \propto t^{-2.6}$ is observed. Hjellming & Johnston (1988) have proposed that these power laws can be explained as a result of an initial slowed expansion followed by free expansion in two dimensions. This steepening of the decrease in flux density with angular separation could be related to the similar tendency observed in the jets of some radio galaxies, where the intensity I declines with angular distance ϕ as $I_\nu \propto \phi^{-x}$, with $x = 1.2$–1.6 in the inner regions and $x \sim 4$ in the outer regions of the jet (Bridle & Perley 1984).

It is also possible that continued injection of relativistic particles and/or magnetic field into the emitting plasma can produce shallower decreases with time of the flux density (Mirabel et al 1998). The particle injection could result from in situ acceleration as the moving gas shocks and entrains ambient gas or could result from beams or winds from the central energy source. The optically thick rise occurs very rapidly and has yet to be observed in detail for a proper comparison with the theoretical expectations.

It is possible to estimate the parameters of the ejected condensations using the formulation of Pacholczyk (1970) for minimum energy, correcting for relativistic effects and integrating the radio luminosity over the observed range of frequencies. Rodríguez & Mirabel (1999a) estimate for the bright 1994 March 19 event in GRS 1915+105 a magnetic field of about 50 mGauss and an energy of about 4×10^{43} ergs in the relativistic electrons. Assuming that there is one (non-relativistic) proton per (relativistic) electron, one gets a proton mass estimate in the order of 10^{23} g. To estimate the peak mechanical power during the ejection we need a value for the time over which the acceleration and ejection took place. Mirabel & Rodríguez (1994) conservatively estimate that the ejection event must have lasted ≤ 3 days, requiring a minimum power of $\sim 5 \times 10^{38}$ erg s^{-1}, a value comparable with the maximum observed steady photon luminosity of GRS 1915+105, which is $\sim 3 \times 10^{38}$ erg s^{-1} (Harmon et al 1994).

The ejection events that preceded and followed the 1994 March 19 outburst are estimated to have masses in the order of 10^{21-22} g (Rodríguez & Mirabel 1999a, Gliozzi et al 1999). Finally, if the repetitive events observed with periods of tens of minutes in GRS 1915+105 (Rodríguez & Mirabel 1997, Pooley & Fender 1997, Mirabel et al 1998, Eikenberry et al 1998a) are interpreted as mini-ejection episodes, the mass associated with them is of order 10^{19} g. We crudely estimate that, on the average, GRS 1915+105 injects energy in the order of 10^{23} g year^{-1} in the form of relativistic (0.92c–0.98c), collimated outflows. This corresponds to an average mechanical energy of $L_{mech} \sim 10^3 \, L_{\odot}$. In contrast, SS 433 as a result of its more continuous jet flow, has $L_{mech} \sim 10^5 \, L_{\odot}$ (Margon 1984) despite having a lower flow velocity than GRS 1915+105. The GRS 1915+105 bursts are thus very energetic but more sporadic.

Recently, there has been evidence that during some events the synchrotron emission in GRS 1915+105 extends from the radio into at least the near-infrared (Mirabel et al 1998; Fender & Pooley 1997). Then the synchrotron luminosity becomes significant, reaching values of 10^{36} erg s^{-1}.

As emphasized by Hjellming & Han (1995), relativistic plasmas are difficult to confine and synchrotron radiation sources in stellar environments will tend to be variable in time. Then, one of the behaviors most difficult to account for is the relative constancy of the radio flux in some sources, of which Cyg X-1 is the extreme example. The presence of a steady outflow that is too faint to be followed up in time as synchrotron-emitting ejecta could be consistent with the lack of large variability in this type of source.

8. POSSIBLE LABORATORIES FOR GENERAL RELATIVITY

The X-ray power of the superluminal sources exhibits a large variety of quasi-periodic oscillations (QPOs) of high frequency. Of particular interest is the class of fast oscillations with a maximum stable frequency of 67 Hz observed many times in GRS 1915+105, irrespective of the X-ray luminosity of the source (Morgan et al 1997). A QPO with maximum fix frequency of 300 Hz has been observed in GRO J1655-40 (Remillard et al 1999). These stable maximum frequencies are not seen at times of strong radio flares or jet injection. They are believed to be a function of the fundamental properties of the black holes, namely, their mass and spin.

One possible interpretation is that these frequencies correspond to the last stable circular orbit around the black hole. This frequency depends on the black hole's mass and spin, as well as on the rotation direction of the accretion disk, and offers the prospect of inferring the spin of black holes with masses independently determined. Since from optical observations the mass of the hole in GRO J1655-40 is known to be in the range of 4-7 solar masses, one can conclude that GRO J1655-40 contains a Kerr black hole rotating at $\geq 70\%$ of the maximum spin possible (Zhang et al 1997).

Alternatively, the maximum QPO stable frequency could be related to general relativity disk seismology, more specifically, to the maximum radial epicyclic frequency (Nowak et al 1997), which also depends on the spin of the black hole.

A third interpretation has been proposed in terms of the relativistic dragging of the inertial frame around the spinning black hole (Cui et al 1998). By comparing the computed disk precession frequency with that of the QPO, the spin can be derived if the mass is known. The two sources of sporadic superluminal jets are found to be the black holes that spin at rates close to maximum limit. Obviously, theoretical work to distinguish between these three alternative interpretations will be important to estimate the spin of the black holes with known masses.

X-ray spectroscopy of the two superluminal sources obtained with the satellite ASCA (Ebisawa 1996, Ueda et al 1998) has shown K_α H and He like iron absorption lines, whereas the observations with SAX have only shown emission features from the relativistic accretion disk around 7 keV, which have been interpreted as iron lines (Matt et al 1998). One expects that with greater sensitivity these lines will show a profile reminiscent of that of the asymmetric iron lines observed in Seyfert galaxies (Tanaka et al 1995). The accretion disks of GRS 1915+105 and GRO J1655-40 are viewed obliquely, and the blueshifted side of the lines should look much stronger due to the Doppler beaming effect. In addition, the center of the line should be redshifted as expected from general relativity effects on radiation escaping from the surroundings of a strongly gravitating object. In the future, perhaps these lines could be used as probes of general relativity effects in the innermost parts of the accretion flows into black holes.

General relativity theory in weak gravitational fields has been successfully tested by observing in the radio the expected decay in the orbit of a binary pulsar, an effect produced by gravitational radiation damping (Taylor & Weisberg 1982). Observations of binary pulsars have also been used to constrain the nature of gravity in the strong-field regime (Taylor et al 1992). Although the interpretation of the maximum stable frequency of the X-ray power spectrum in the superluminal sources is still uncertain, these frequencies are known to originate close to the horizon of the black hole, and perhaps they could be used in the future to test the physics of accretion disks and black holes in the strong field limit.

9. OTHER SOURCES OF RELATIVISTIC JETS IN THE GALAXY

In X-ray binaries there is a general correlation between the X-ray properties and the jet properties. The time interval and flux amplitude of the variations in radio waves seems to correspond to the time and amplitude variations in the X-ray flux. More specifically, persistent X-ray sources are also persistent radio sources, and the transient X-ray sources produce at radio waves sporadic outburst/ejection events. Persistent sources of hard X-rays (e.g. 1E1740.7-2942, GRS 1758-258) are usually associated to faint, double-sided radio structures that have sizes of several arcmin (parsec scales). The radio core of these two persistent sources are weak (≤ 1 mJy) and do not exhibit high amplitude variability. On the contrary, rapidly variable hard X-ray transients (e.g. GRS 1915+105, GRO J1655-40, XTE J1748-288) may exhibit variations in the X-rays and radio fluxes of several orders of magnitude in short intervals of time. Because these black-hole X-ray transients produce sporadic ejections of discrete, bright plasma clouds, the proper motions of the ejecta can be measured.

Probably all hard X-ray sources that accrete at super-Eddington rates produce relativistic jets. However, the observational study of these jets presents in practice several difficulties. Persistent hard X-ray sources like Cygnus X-1 are surrounded by faint non-thermal radio features extending several arcmin (Martí et al 1996), and even in the cases where they are well aligned with the variable compact radio counterpart it is very difficult to prove conclusively that the faint and extended radio features are actually associated with the X-ray source. This was the case of Sco X-1, where possible large-scale radio "lobes" were found to be extragalactic sources symmetrically located in the plane of the sky with respect to Sco X-1 (Fomalont & Geldzahler 1991). On the other hand, in transient black hole binaries one may observe transient sub-arcsec jets, but unless the interferometric observations are conveniently scheduled, the evolution is too rapid and it may not be possible to follow up the proper motions of discrete clouds. This may have been the case in the radio observations of the X-ray sources Nova Oph 93 (Dela Valle et al 1994) and Nova Muscae (Ball et al 1995), among others.

We list in Table 1 the sources of relativistic jets in the Galaxy known so far. The first six are transients, whereas the next four are persistent X-ray sources. Proper motions of the relativistic ejecta have been determined with accuracy in GRS 1915+105, GRO J1655-40, XTE J1748-288, and SS 433. Besides these four sources, proper motions were also measured—but with less accuracy—for moving features in Cygnus X-3 (Schalinski et al 1995, Martí et al 1999), Scorpius X-1 (Fomalont 1999), Circinus X-1 (Fender et al 1998), and CI Cam (XTE J0421+560; Hjellming & Mioduszewski 1998; Mioduszewski et al 1998). Jet structures have been reported to be associated to Cygnus X-1, but these results are still uncertain.

It is interesting that the ejecta from the black hole binaries GRS 1915+105, GRO J1655-40, and probably also XTE J1748-288 have velocities greater than $0.9c$, while the ejecta from the four sources believed to be neutron star binaries have velocities $\leq 0.5c$. From their models of magnetically driven jets, Kudoh & Shibata (1995) have proposed that jet velocities such as those listed in Table 1 are comparable to the Keplerian rotational velocities expected at the base of the jets, close to neutron stars and black holes, respectively. Livio (1997) has also stressed the similarity between the velocity of jets and the escape velocity of the gravitational well from where they were ejected. If this notion is confirmed, jet velocities could then be used to discriminate between neutron stars and black

TABLE 1 Sources of Relativistic Jets in the Galaxy[1]

Source	Compact object	V_{app}[2]	V_{int}[3]	Θ[4]	References
GRS 1915+105	black hole	1.2c–1.7c	0.92c–0.98c	66°–70°	Mirabel & Rodríguez (1994); Fender et al (1999); Dhawan, Mirabel, Rodríguez (1999)
GRO J1655-40	black hole	1.1c	0.92c	72°–85°	Tingay et al (1995); Hjellming & Rupen (1995); Orosz & Bailyn (1997)
XTE J1748-288	black hole	1.3c	>0.9c		Hjellming et al (1998)
SS 433	neutron star?	0.26c	0.26c	79°	Margon (1984); Spencer (1984)
Cygnus X-3	neutron star?	~0.3c	~0.3c	>70°	Schalinski et al (1993); Martí et al (1999)
CI Cam	neutron star?	~0.15c	~0.15c	>70°	Mioduszewski et al (1998); García et al (1998)
Scorpio X-1	neutron star	~0.5c			Fomalont (1999)
Circinus X-1	neutron star	≥0.1c	≥0.1c	>70°	Stewart et al (1993); Fender et al (1998)
1E1740.7-2942	black hole				Mirabel et al (1992); Rodríguez & Mirabel (1999c)
GRS 1758-258	black hole				Rodríguez et al (1994)
Sgr A*	black hole				Lo et al (1998)

[1]Sources reported as of December 1998.
[2]V_{app} is the apparent speed of the highest velocity component of the ejecta.
[3]V_{int} is the intrinsic velocity of the ejecta.
[4]Θ is the angle between the direction of motion of the ejecta with the line of sight.

holes, with jet velocities close to the speed of light been produced only in black hole binaries.

Another possible source of relativistic jets in the Galaxy is, of course, Sgr A*, the presumed black hole of 2.5 million solar masses at the galactic center (Eckart & Genzel 1997). The radio source is always present at about the 1 Jy level and exhibits a flat spectrum with relative small variations, a behavior similar to that of the faint compact mJy radio sources associated with Cygnus X-1 (Martí et al 1996) and GRS 1915+105 in its plateau state at times when no strong outburst/ejection events take place, a state that in the latter source can last from days to weeks (Pooley & Fender 1997). This type of radio emission could arise from a jet in a coupled jet-disk system (Falcke et al 1993), from electrons in an advection dominated flow (Narayan et al 1998, Mahadevan 1998), or from shocks in massive winds (Blandford & Begelman 1999). Despite heavy interstellar scattering at radio wavelengths, recent VLBA observations at 7-mm may have resolved Sgr A* in an elongated radio source of 72 Schwarzschild radii suggesting the presence of a jet (Lo et al 1998).

10. INTERACTION OF RELATIVISTIC JETS WITH THE ENVIRONMENT

If a compact source (black hole or neutron star) injects collimated relativistic jets into its cold environment, it is expected that some fraction of the injected power will be dissipated by shocks in the circumstellar gas and dust. The collision of relativistic ejecta with environmental material has been observed in real time in XTE J1748-288 (Hjellming et al 1998), where the leading edge of the jet decelerates while strongly brightening. The interaction of the mildly relativistic jets from CI Cam (Hjellming & Mioduszewski 1998) with an HII and dust shell nebula has been reported by García et al (1998). Other signatures of the interaction of relativistic jets with the environment are the radio lobes of 1E 1740.7-2942 (Mirabel et al 1992) and GRS 1758-258 (Rodríguez et al 1992), the twisted arcmin jets of Circinus X-1 (Steward et al 1993; Fender et al 1998), and the two lateral extensions of tens of pc in the radio shell W50 that hosts at its center SS 433. The interaction of SS 433 with the nebula W50 has been studied in the X-rays (Brinkmann et al 1996), infrared (Mirabel et al 1996b), and radio wavelengths (Dubner et al 1998 and references therein).

SS 433 is a high mass X-ray binary at a distance of $\sim$3 kpc near the centre of the radio shell W50 (Margon 1984). The latter may be either the supernova remnant from the formation of the compact object (Velusamy & Kundu 1974), or a bubble evacuated by the energy outflow of SS 433 (Begelman et al 1980). Besides the well-known relativistic jets seen at sub-arcsec scales in the radio, large-scale jets become visible in the X-rays at distances $\sim$30 arcmin ($\sim$25 pc) from the compact source (Brinkmann et al 1996). In the radio, the lobes reach distances of up to 1° ($\sim$50 pc). These large-scale X-ray jets and radio lobes are the result of the interaction of the mass outflow with the interstellar medium. From optical and

X-ray emission lines it is found that the sub-arcsec relativistic jets have a kinetic energy of $\sim 10^{39}$ erg s^{-1} (Margon 1984, Spencer 1984), which is several orders of magnitude larger than the energy radiated in the X-rays and in the radio.

In Figure 8 is shown the $\lambda 20$ cm map with 55 arcsec resolution by Dubner et al (1998). It shows the connection between the subarcsec relativistic jets and the extended nebula over $\sim 10^5$ orders of magnitude in distance scales. Dubner et al (1998) estimate that the kinetic energy transferred into the ambient medium is $\sim 2 \ 10^{51}$ ergs, thus confirming that the relativistic jets from SS 433 represent an important contribution to the overall energy budget of the surrounding nebula W50. Begelman et al (1980) characterized W50 as a "beambag," interpreting the elongated shape and filled-in radio structure of W50 as evidence for continuing injection of magnetic field and high-energy particles from SS 433.

Evidences for the interaction of jets with the environmental medium have also been searched in the two well-established superluminal sources. In GRS 1915+105 Chaty, Mirabel & Rodríguez (1999) searched at millimeter, infrared, and X-rays for evidences of the physical association between the relativistic jets and two IRAS sources projected symmetrically on each side at ~ 15 arcmin of angular distance from the compact source that at first glance could be lobes caused by the impact of the jets in interstellar molecular clouds (Rodríguez & Mirabel 1999b). Besides the good alignment of the IRAS sources with the subarcsec jets and the presence of an intriguing non-thermal jet-like source in the SE IRAS source (Rodríguez & Mirabel 1999b), no conclusive physical evidence for association with GRS 1915+105 has been found, with the IRAS sources most probably being normal HII regions. On the other hand, Hunstead et al (1999) find regions of extended low-surface-brightness emission aligned with the radio jets of GRO J1655-40, but their real association with the high energy source has not been confirmed. The jets in GRS 1915+105 and GRO J1655-40 are faster than those in SS 433, but much more sporadic, and this probably accounts for the lack of obvious lobes associated with them.

It has been proposed that the interaction of relativistic jets with the environment may induce high energy radiation. Positrons released impulsively from the compact source could annihilate locally in the hot plasma producing a broad 511 keV spectral feature (Sunyaev et al 1991, Ramaty et al 1992). Alternatively, a fraction of the positrons could stream up to the interstellar gaseous environment, slowing down and annihilating in such cold medium, thus emitting 511 keV narrow line emission, and inducing radio lobe synchrotron emission and bremsstrahlung gamma-ray continuum emission (Laurent & Paul 1994).

11. MICROBLAZARS AND GAMMA-RAY BURSTS

It is interesting that in all three sources where θ (the angle between the line of sight and the axis of ejection) has been determined, a large value is found (that is, the axis of ejection is close to the plane of the sky). These values are $\theta \simeq 79°$ (SS 433, Margon 1984), $\theta \simeq 66° - 70°$ (GRS 1915+105, Mirabel & Rodríguez

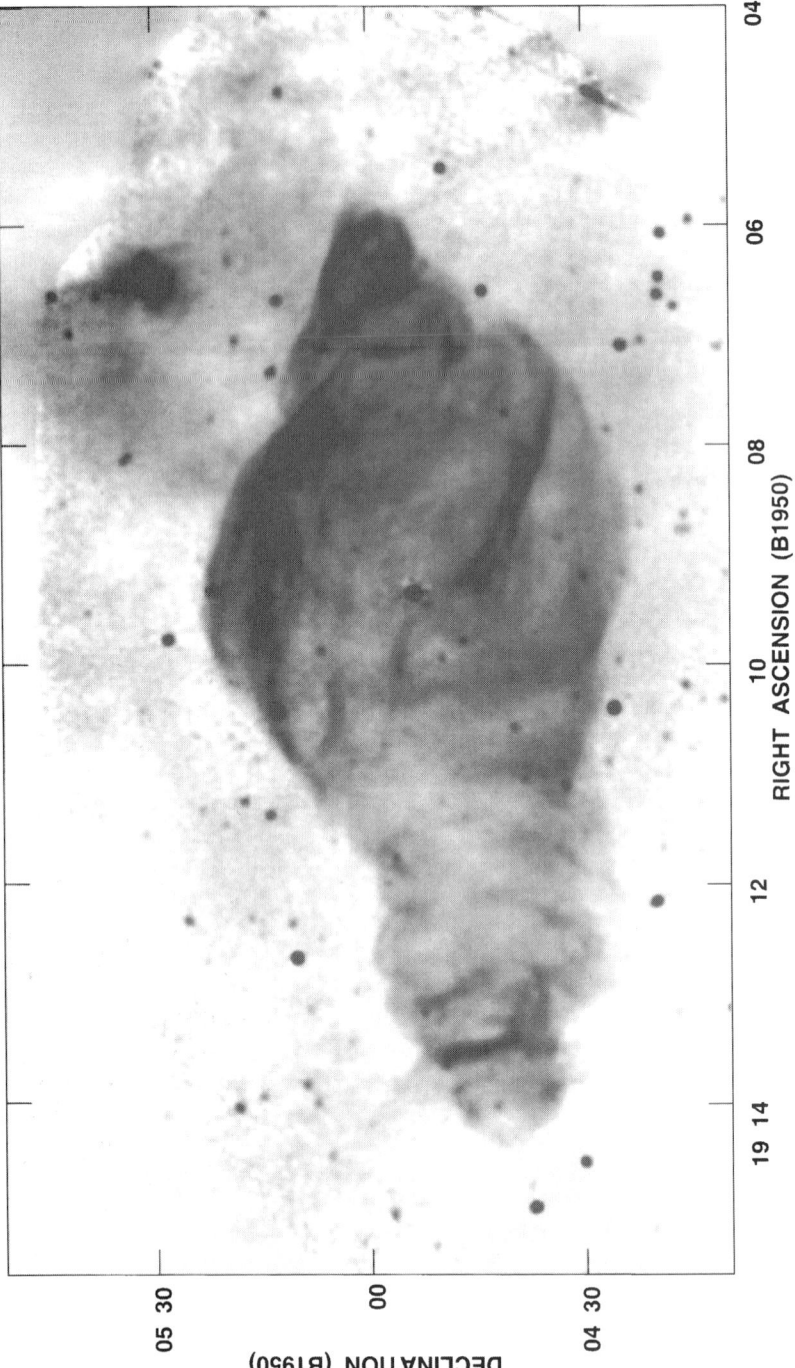

Figure 8 Very Large Array continuum mosaic of W50 at 1.5 GHz (Dubner et al 1998). The radio counterpart of SS 433 is the bright unresolved source at the center of the image. The lateral E-W extension of the nebula over ~1° (~50 pc) is caused by the injection of the relativistic jets from SS 433. The greyscale varies from 1 to 25 mJy beam^{-1}. The angular resolution is 56 × 54 arcsec.

1994, Fender et al 1999), $\theta \simeq 85°$ (GRO J1655-40, Hjellming & Rupen 1995), and $\theta \geq 70°$ for the remaining sources. This result is not inconsistent with the statistical expectation since the probability of finding a source with a given θ is proportional to $\sin \theta$. We then expect to find as many objects in the $60° \leq \theta \leq 90°$ range as in the $0° \leq \theta \leq 60°$ range. However, this argument suggests that we should eventually detect objects with a small θ. For objects with $\theta \leq 10°$ we expect the timescales to be shortened by 2γ and the flux densities to be boosted by $8\gamma^3$ with respect to the values in the rest frame of the condensation. For instance, for motions with $v = 0.98c$ ($\gamma = 5$), the timescale will shorten by a factor of ~ 10 and the flux densities will be boosted by a factor of $\sim 10^3$. Then, for a galactic source with relativistic jets and small θ we expect fast and intense variations in the observed flux. These microblazars may be quite hard to detect in practice, both because of the low probability of small θ values and because of the fast decline in the flux.

Gamma-ray bursts are at cosmological distances and ultra-relativistic bulk motion and beaming appear as essential ingredients to solve the enormous energy requirements (e.g. Kulkarni et al 1999, Castro-Tirado et al 1999). Beaming reduces the energy release by the beaming factor $f = \Delta\Omega/4\pi$, where $\Delta\Omega$ is the solid angle of the beamed emission. Additionally, the photon energies can be boosted to higher values. Extreme flows from collapsars with bulk Lorentz factors >100 have been proposed as sources of γ-ray bursts (Mészáros & Rees 1997). High collimation (Dar 1998, Pugliese et al 1999) can be tested observationally (Rhoads 1997), since the statistical properties of the bursts will depend on the viewing angle relative to the jet axis.

Recent studies of gamma-ray afterglows suggest that they are highly collimated jets. The brightness of the optical transient associated to GRB 990123 showed a break (Kulkarni et al 1999), and a steepening from a power law in time t proportional to $t^{-1.2}$, ultimately approaching a slope $t^{-2.5}$ (Castro-Tirado et al 1999). The achromatic steepening of the optical light curve and early radio flux decay of GRB 990510 are inconsistent with simple spherical expansion, and well fit by jet evolution (Harrison et al 1999). It is interesting that the power laws that describe the light curves of the ejecta in microquasars show similar breaks and steepening of the radio flux density (Section 7, Rodríguez & Mirabel 1999a). In microquasars, these breaks and steepenings have been interpreted (Hjellming & Johnston 1988) as a transition from slow intrinsic expansion followed by free expansion in two dimensions. Besides, linear polarizations of about 2% were recently measured in the optical afterglow of GRB 990510 (Covino et al 1999), providing strong evidence that the afterglow radiation from gamma-ray bursters is, at least in part, produced by synchrotron processes. Linear polarizations in the range of 2–10% have been measured in microquasars at radio (Rodríguez et al 1995, Hannikainen et al 1999), and optical (Scaltriti et al 1997) wavelengths.

In this context, microquasars in our own Galaxy seem to be less extreme local analogs of the super-relativistic jets associated to the more distant γ-ray bursters. However, γ-ray bursters are different to the microquasars found so far in our own Galaxy. The former do not repeat and seem to be related to catastrophic events, and have much larger super-Eddington luminosities. Therefore, the scaling laws in terms of the black hole mass that are valid in the analogy between microquasars and quasars do not seem to apply in the case of γ-ray bursters.

12. CONCLUSIONS AND PERSPECTIVES

The study of relativistic jets from X-ray binaries in our own galaxy sets on a firmer basis the relativistic ejections seen elsewhere in the Universe. The analogy between quasars and microquasars led to the discovery of superluminal sources in our own galaxy, where it is possible to follow the motions of the two-sided ejecta. This permits astronomers to overcome the ambiguities that had dominated the physical interpretation of one-sided moving jets in quasars, and conclude that the ejecta consist mainly of matter moving with relativistic bulk motions, rather than waves propagating through a slowly moving jet. The Lorentz factors of the bulk motions in the jets from microquasars seem to be similar to those believed to be common in quasars. From the study of the two-sided moving jets in one microquasar, an upper limit for the distance to the source was derived, using constraints from special relativity.

Because of the relative short timescales of the phenomena associated with the flows of matter around stellar mass black holes, one can sample phenomena that we have not been able to observe in quasars. Of particular importance is to understand the connection between accretion flow instabilities observed in the X-rays, with the ejection of relativistic clouds of plasma observed in the radio, infrared, and possibly in the optical. The detection of synchrotron infrared flares implies that the ejecta in microquasars contain very energetic particles with particle Lorentz factors of at least 10^3.

The discovery of microquasars opens several new perspectives that could prove to be particularly productive:

1. They provide a new method to determine distances using special relativity constraints. If the proper motions of the two-sided ejecta and the Doppler factor of a spectral line from one ejecta are measured, the distance to the source can be derived. With the rapid advance of technological capabilities in astronomy, this relativistic method to determine distances may be applied first to black hole jet sources in galactic binaries, and in the decades to come, to quasars.

2. Microquasars are nearby laboratories that can be used to gain a general understanding of the mechanism of ejection of relativistic jets. The multiwavelength observations of GRS 1915+105 during large-amplitude oscillations suggest that the clouds are ejected during the replenishment of the inner accretion disk that follows its sudden disappearance beyond the last stable orbit around the black hole. In the context of these new data, the time seems to be ripe for new theoretical advances on the models of formation of relativistic jets.

3. High sensitivity X-ray spectroscopy of jet sources with future X-ray space observatories may clarify the phenomena in accretion disks that are associated with the formation of jets.

4. More microquasars will be discovered in the future. Among them, microblazars should appear as sources with fast and large amplitude variations in the observed flux. Depending on the beaming angle and bulk Lorentz factor they will be observed up to very high photon energies. Microquasars in our own Galaxy may be less extreme local analogs of the super-relativistic jets that seem to be associated with distant gamma-ray bursters.

5. The spin of stellar mass black holes could be derived from the observed maximum stable frequency of the QPOs observed in the X-rays, provided the mass has been independently determined. However, theoretical work is needed to distinguish between the alternative interpretations that in the context of general relativity have been proposed for the maximum stable frequency of QPOs.

6. Finally, microquasars could be test grounds for general relativity theory in the strong field limit. General relativity theory in weak gravitational fields has been successfully tested by observing in the radio wavelengths the expected decay in the orbit of a binary pulsar, an effect produced by gravitational radiation damping. We expect that phenomena observed in microquasars could be used in the future to investigate the physics of strong field relativistic gravity near the horizon of black holes.

ACKNOWLEDGMENTS

We are most grateful to Vivek Dhawan for permission to include in this review unpublished results from our VLBA observations, and to Sylvain Chaty and Josep Martí for help in producing Figure 7. We thank Ralph Spencer, Jacques Paul, Josep Martí, and Alan Harmon for comments on the original manuscript. We are also grateful to Philippe Durouchoux for the organization of the workshop on *Relativistic Jet Sources in the Galaxy* held in Paris on December 12–13, 1998, from which we have benefitted. During this work, LFR received partial support from CONACyT, México and DGAPA, UNAM.

Visit the Annual Reviews home page at http://www.AnnualReviews.org

LITERATURE CITED

Ables JG. 1969. *Ap. J.* 155:L27–30

Abramowicz MA, Chen X, Kato S, Lasota JP, Reguev O. 1995. *Ap. J.* 438:L37–40

Atoyan AM, Aharonian FA. 1997. *Ap. J.* 490: L149–52

Ball L, Kesteven MJ, Campbell-Wilson D, Turtle AJ, Hjellming RM. 1995. *MNRAS* 273: 722–30

Ball L. 1996. In *Radio Emission from the Stars and the Sun, ASP Conf. Ser.* 93:219–27

Ballet J, et al. 1994. *AIP Conference Proceedings* 308:131–34
Barret D, McClintock JE, Grindlay JE. 1996. *Ap. J.* 473:963–73
Begelman MC. 1998. *Ap. J.* 493:291–300
Begelman MC, Hatchett SP, McKee CF, Sarazin CL, Arons J. 1980. *Ap. J.* 238:722–30
Belloni T, Méndez M, King AR, van der Klis M, van Paradijs J. 1997. *Ap. J.* 479:L145–48
Blandford RD, Begelman MC. 1999. *MNRAS* 303:L1–5
Blandford RD, Ostriker JP. 1978. *Ap. J.* 221: L29–32
Blandford RD, Payne DG. 1982. *MNRAS* 199: 883–94
Blandford RD, Znajek RL. 1977. *MNRAS* 179: 433–40
Bodo G, Ghisellini G. 1995. *Ap. J.* 441:L69–71
Bridle AH, Perley RA. 1984. *Annu. Rev. Astr. Astrophys.* 22:319–58
Brinkmann W, Aschenbach B, Kawai N. 1996. *Astron. Astrophys.* 312:306–16
Castro-Tirado AJ, Brandt S, Lund N, Lapshov I, Sunyaev RA, et al. 1994. *Ap. J. Suppl.* 92: 469–72
Castro-Tirado AJ, Geballe TR, Lund N. 1996. *Ap. J.* 461:L99–102
Castro-Tirado AJ. et al. 1999. *Science* 283: 2069–73
Chaty S, Mirabel IF, Duc PA, Wink JE, Rodríguez LF. 1996. *Astron. Astrophys.* 310: 825–30
Chaty S, Mirabel IF, Rodríguez LF. 1999. In preparation
Chen X, Swank JH, Taam RE. 1997. *Ap. J.* 477: L41–44
Churazov E, Gilfanov M, Sunyaev R, Khavenson N, Novikov B, et al. 1994. *Ap. J. Suppl.* 92:381–85
Cordier B, Paul J, Ballet J, Goldwurm A, Bouchet L, et al. 1993. *Astron. Astrophys.* 275, L1–4
Covino S, et al. 1999. *IAU Circular 7172*
Cowley AP, Schmidtke PC, Crampton D, Hutchings JB. 1998. *Ap. J.* 504:854–65
Cui W, Zhang SN, Chen W. 1998. *Ap. J.* 492: L53–56
Curtis HD. 1918. *Publ. Lick Obs.* 13:9–42
Dar A. 1998. *Ap. J.* 500:L93–96
Dela Valle M, Mirabel IF, Rodríguez LF. 1994. *Astron. Astrophys.* 290:803–6
Dhawan V, Mirabel IF, Rodríguez LF. 1999. In preparation
Djorgovski S, Thompson D, Mazzarella J, Klemola A, Neugebauer G. 1992. *IAU Circular 5596*
Dubner GM, Holdaway M, Goss WM, Mirabel IF. 1998. *Astron. J.* 116:1842–55
Ebisawa K. 1996. In *X-ray Imaging and Spectroscopy of Cosmic Hot Plasmas*, ed. F Makino & K Mitsuda, Tokyo: University Academy Press, pp. 427–31
Eckart A, Genzel R. 1997. *MNRAS* 284:576–98
Eikenberry SS, Matthews K, Morgan EH, Remillard RA, Nelson RW. 1998a. *Ap. J.* 494: L61–64
Eikenberry SS, Matthews K, Murphy TW, Nelson RW, Morgan EH, Remillard RA, Muno M. 1998b. *Ap. J.* 506:L31–34
Fabian AC, Rees MJ. 1979. *MNRAS* 187:13p–16p
Falcke H, Mannheim K, Biermann PL. 1993. *Astron. Astrophys.* 278:L1–4
Fejes I. 1986. *Astron. Astrophys.* 168:69–71
Fender RP, Bell Burnell SJ, Williams PM, Webster AS. 1996. *MNRAS* 283:798–804
Fender RP, Garrington ST, McKay DJ, Muxlow TWB, Pooley GG, Spencer RE, Stirling AM, Waltman EB. 1999. *MNRAS* 304:865–76
Fender RP, Pooley GG. 1998. *MNRAS* 300:573–76
Fender RP, Pooley GG, Brocksopp C, Newell SJ. 1997. *MNRAS* 290:L65–69
Fender RP, Spencer R, Tzioumis T, Wu K, et al. 1998. *Ap. J.* 506:L121–25
Finoguenov A, et al. 1994. *Ap. J.* 424:940–42
Fomalont EB, Geldzahler BJ. 1991. *Ap. J.* 383: 289–94
Fomalont EB. 1999. *Private communication*
Foster RS, Waltman EB, Tavani M, Harmon BA, Zhang SN, et al. 1996. *Ap. J.* 467:L81–84
García MR, et al. 1998. In *Workshop on*

Relativistic Jet Sources in the Galaxy. Paris, December 12–13, 1998
Ghosh P, Abramowicz MA. 1997. *Bull. Am. Astron. Soc.* 191:66.05
Gilmore WS, Seaquist ER. 1980. *Astron. J.* 85:1486–95
Gilmore WS, Seaquist ER, Stocke JT, Crane PC. 1981. *Astron. J.* 86:864–70
Gliozzi M, Bodo G, Ghisellini G. 1999. *MNRAS* 303:L37–40
Greiner J, Morgan EH, Remillard RA. 1996. *Ap. J.* 473:L107–10
Grove JE, Johnson WN, Kroeger RA, McNaron-Brown K, Skibo JG, et al. 1998. *Ap. J.* 500:899–908
Hameury JM, Lasota JP, McClintock JE, Narayan R. 1997. *Ap. J.* 489:234–43
Hannikainen DC, Hunstead RW, Sault RJ, McKay DJ. 1999. 19th Texas Symposium on Relativistic Astrophysics and Cosmology. Paris, France, Dec 14–18, 1998. Eds.: J Paul, T Montmerle, and E Auburg
Harmon BA, Deal KJ, Paciesas WS, Zhang SN, Gerard E, Rodríguez LF, Mirabel IF. 1997. *Ap. J.* 477:L85–90
Harmon BA, Zhang SN, Wilson CA, Rubin BC, Fishman GJ, et al. 1994. In *AIP Conference Proceedings No. 304*, eds. Fichtel CE, Gehrels N, Norris JP. pp. 210–19. New York: AIP
Harrison FA, et al. 1999. *Ap. J. Lett.* In press
Heindl WA, Prince TA, Grunsfeld JM. 1995. *Ap. J.* 430:829–33
Hjellming RM, Han X. 1995. In *X-Ray Binaries*, ed. WHG Lewin, J van Paradijs, EPJ van den Heuvel, p. 308. Cambridge, UK Cambridge University Press
Hjellming RM, Johnston KJ. 1981. *Ap. J.* 246:L141–45
Hjellming RM, Johnston KJ. 1988. *Ap. J.* 328:600–9
Hjellming RM, Mioduszewski AM. 1998. *IAU Circular 6872*
Hjellming RM, Rupen MP. 1995. *Nature* 375:464–67
Hjellming RM, Rupen MP, Mioduszewski AM, et al. 1998. In *Workshop on Relativistic Jet Sources in the Galaxy*. Paris, December 12–13, 1998
Hunstead RW, Wu K, Campbell-Wilson D. 1999. In preparation
Kalogera V, Baym G. 1996. *Ap. J.* 470:L61–64
King A. 1998. In *Workshop on Relativistic Jet Sources in the Galaxy*. Paris, December 12–13, 1998
Koide S, Shibata K, Kudoh T. 1998. *Ap. J.* 495:L63–66
Kudoh T, Shibata K. 1995. *Ap. J.* 452:L41–44
Kulkarni SR, et al. 1999. *Nature* 398:389–94
Kuznetsov S, Gilfanov M, Churazov E, Sunyaev R, Korel I, et al. 1997. *MNRAS* 292:651–56
Laurent P, Paul J. 1994. *Ap. J. Suppl.* 92:375–79
Livio M. 1997. In *Accretion Flows and Related Phenomena*, ASP Conf. Ser. IAU Colloquium 163, eds. Wickramasinghe D, Ferrario L, Bicknell G. 121:845–60
Livio M, Ogilvie GI, Pringle JE. 1998. *Ap. J.* 512:100–4
Lo KY, Shen ZQ, Zhao JH, Ho PTP. 1998. *Ap. J.* 508:L61–64
López JA. 1997. *Planetary Nebulae*, IAU Symposium No. 180, pp. 197–203, eds. HJ Habing & HJGLM Lamers. Dordrecht: Kluwer
Lucek SG, Bell AR. 1997. *MNRAS* 290:327–33
Mahadevan R. 1998. *Nature* 394:651–53
Margon BA, Stone RPS, Klemola A, Ford HC, Katz JI, et al. 1979. *Ap. J.* 230:L41–45
Margon BA. 1984. *Annu. Rev. Astr. Astrophys.* 22:507–36
Martí J, et al. 1999. In preparation
Martí J, Mereghetti S, Chaty S, Mirabel IF, Goldoni P, et al. 1998. *Astron. Astrophys.* 338:L95–99
Martí J, Paredes JM, Estalella R. 1992. *Astron. Astrophys.* 258:309–15
Martí J, Rodríguez LF, Mirabel IF, Paredes JM. 1996. *Astron. Astrophys.* 306:449–54
Matt G, et al. 1998. In *Workshop on Relativistic Jet Sources in the Galaxy*. Paris, December 12–13, 1998
Meier DL, Edgington S, Godon P, Payne DG, Lind KR. 1997. *Nature* 388:350–52

Mészáros P, Rees MJ. 1997. *Ap. J.* 482:L29–32
Milgrom M. 1979. *Astron. Astrophys.* 79:L3–6
Mioduszewski AM, et al. 1998. In *Workshop on Relativistic Jet Sources in the Galaxy.* Paris, December 12–13, 1998
Mirabel IF, Bandyopadhyay R, Charles PA, Shahbaz T, Rodríguez LF. 1997. *Ap. J.* 477:L45–48
Mirabel IF, Claret A, Cesarsky CJ, Cesarsky DA, Boulade O. 1996b. *Astron. Astrophys.* 315:L113–16
Mirabel IF, Dhawan V, Chaty S, Rodríguez LF, Robinson C, Swank J, Geballe T. 1998. *Astron. Astrophys.* 330:L9–12
Mirabel IF, Duc P-A, Rodríguez LF, et al. 1994. *Astron. Astrophys.* 282:L17–20
Mirabel IF, Rodríguez LF. 1994. *Nature* 371:46–48
Mirabel IF, Rodríguez LF. 1998. *Nature* 392:673–76
Mirabel IF, Rodríguez LF, Chaty S, Sauvage M, Gerard E, et al. 1996a. *Ap. J.* 472:L111–14
Mirabel IF, Rodríguez LF, Cordier B, Paul J, Lebrun F. 1992. *Nature* 358:215–17
Mirabel IF, Rodríguez LF, Cordier B, Paul J, Lebrun F. 1993. In *Sub-arcsecond Radio Astronomy,* eds. RJ Davis & RS Booth, Cambridge University Press, 47–49
Morgan EH, Remillard RA, Greiner J. 1997. *Ap. J.* 482:993–1010
Motch C. 1998. *Astron. Astrophys.* 338:L13–16
Narayan R, García MR, McClintock JE. 1997. *Ap. J.* 478:L79–82
Narayan R, Mahadevan R, Grindlay JE, Popham RG, Gammie C. 1998. *Ap. J.* 492:554–68
Nowak MA, Wagoner RV, Begelman MC, Lehr DE. 1997. *Ap. J.* 477:L91–94
Orosz JA, Bailyn CD. 1997. *Ap. J.* 477:876–96
Orosz JA, Remillard RA, Bailyn CD, McClintock JE. 1997. *Ap. J.* 478:L83–86
Pacholczyk AG. 1970. *Radio Astrophysics.* Freeman, San Francisco
Paragi Z, Vermeulen RC, Fejes I, Schilizzi RT, Spencer RE, Stirling AM. 1998. In *Workshop on Relativistic Jet Sources in the Galaxy.* Paris, December 12–13, 1998

Paul J. et al. 1991. In *Advances in Space Research,* 11:289–302
Pearson TJ, Zensus JA. 1987. In *Superluminal Radio Sources,* Cambridge University Press, eds. Zensus JA & Pearson TJ, p. 1
Peebles PJE. 1993. *Principles of Physical Cosmology,* Princeton University Press, Princeton
Penrose R. 1969. *Nuovo Cimento* 1, 252–76
Phillips SN, Shahbaz T, Podsiadlowski Ph. 1999. *MNRAS* 304:839–44
Pooley GG, Fender RP. 1997. *MNRAS* 292:925–33
Pugliese G, Falcke H, Biermann PL. 1999. *Astron. Astrophys.* 344:L37–40
Ramaty R, Leventhal M, Chan KW, Lingenfelter RE. 1992. *Ap. J.* 392:L63–67
Rees MJ. 1966. *Nature* 211:468–70
Rees MJ. 1984. *Annu. Rev. Astr. Astrophys.* 22:471–506
Rees MJ. 1998. In *Black Holes and Relativistic Stars,* ed. Wald RM, Chicago: University of Chicago Press, pp. 79–101
Reipurth B, Bertout C. 1997. *Herbig-Haro Flows and the Birth of Stars,* IAU Symposium No. 182, Dordrecht: Kluwer
Remillard RA, Morgan EH, Levine A, et al. 1999. *Bull. Am. Astron. Soc.* HEAD meeting 31:28.08
Rhoads JE. 1997. *Ap. J.* 487:L1–4
Rodríguez LF, Gerard E, Mirabel IF, Gómez Y, Velázquez A. 1995. *Ap. J. Suppl.* 101:173–79
Rodríguez LF, Mirabel IF. 1997. *Ap. J.* 474:L123–25
Rodríguez LF, Mirabel IF. 1999a. *Ap. J.* 511:398–404
Rodríguez LF, Mirabel IF. 1999b. *Astron. Astrophys.* 340:L47–50
Rodríguez LF, Mirabel IF. 1999c. In preparation
Rodríguez LF, Mirabel IF, Martí J. 1992. *Ap. J.* 401:L15–18
Sams BJ, Eckart A, Sunyaev R. 1996. *Nature* 382:47–49
Scaltriti F, Bodo G, Ghisellini G, Gliozzi M, Trussoni E. 1997. *Astron. Astrophys.* 325:L29–31

Schalinski CJ, Johnston KJ, Witzel A, Spencer RE, Fiedler R, et al. 1995. *Ap. J.* 447:752–59

Seaquist ER. 1993. *Reports on Progress in Physics* 56:1145–1208

Smith DA, Levine A, Wood A. 1998. *IAU Circular 6932*

Spencer RE. 1979. *Nature* 282:483–84

Spencer RE, 1984. *MNRAS* 209:869–79

Spruit HC, Foglizzo T, Stehle R. 1997. *MNRAS* 288:333–42

Stewart RT, Caswell JL, Haynes RF, Nelson GJ. 1993. *MNRAS* 261:593–98

Sunyaev R, Churazov E, Giltanov M, et al. 1991. *Ap. J.* 383:L49–53

Tanaka Y, Nandra K, Fabian AC, Inoue H, Otani C. et al. 1995. *Nature* 375:659–61

Tanaka Y, Shibazaki N. 1996. *Annu. Rev. Astr. Astrophys.* 34:607–44

Taylor JH, Weisberg JM. 1982. *Ap. J.* 253:908–20

Taylor JH, Wolszczan A, Damour T, Weisberg JM. 1992. *Nature* 355:132–36

Tingay SJ, Jauncey DL, Preston RA, Reynolds JE, Meier DL, et al. 1995. *Nature* 374:141–43

van der Laan H. 1966. *Nature* 211:1131–33

van Paradijs J. 1995. In *X-Ray Binaries*, ed. WHG Lewin, J. van Paradijs, EPJ van den Heuvel, Cambridge University Press: Cambridge, p. 536

Velusamy T, Kundu MR. 1974. *Astron. Astrophys.* 32:375–90

Vermeulen RC, Schilizzi RT, Spencer RE, Romney JD, Fejes I. 1993. *Astron. Astrophys.* 270:177–88

Ueda Y, et al. 1998. In *Workshop on Relativistic Jet Sources in the Galaxy.* Paris, December 12–13, 1998

Zensus JA. 1997. *Annu. Rev. Astr. Astrophys.* 35:607–36

Zhang SN, Mirabel IF, Harmon BA, Kroeger RA, Rodríguez LF, et al. 1997. *Proceedings of the Fourth Compton Symposium*, ed. CD Dermer, MS Strickman, JD Kurfess, AIP: New York, pp. 141–62

Zhang SN, Wilson CA, Harmon BA, Fishman GJ, Wilson RB. 1994. *IAU Circular 6046*

THE FIRST 50 YEARS AT PALOMAR: 1949–1999 The Early Years of Stellar Evolution, Cosmology, and High-Energy Astrophysics

Allan Sandage
The Observatories of the Carnegie Institution of Washington, 813 Santa Barbara Street, Pasadena, CA 91101

Key Words stellar evolution, observational cosmology, radio astronomy, high energy astrophysics

PROLOGUE

In 1999 we celebrate the 50th anniversary of the initial bringing into operation of the Palomar 200-inch Hale telescope. When this telescope was dedicated, it opened up a much larger and clearer window on the universe than any telescope that had gone before.

Because the Hale telescope has played such an important role in twentieth century astrophysics, we decided to invite one or two of the astronomers most familiar with what has been achieved at Palomar to give a scientific commentary on the work that has been done there in the first fifty years.

The first article of this kind which follows is by Allan Sandage, who has been an active member of the staff of what was originally the Mount Wilson and Palomar Observatories, and later the Carnegie Observatories for the whole of these fifty years.

The article is devoted to the topics which covered the original goals for the Palomar telescope, namely observational cosmology and the study of galaxies, together with discoveries that were not anticipated, but were first made at Palomar and which played a leading role in the development of high energy astrophysics.

The Palomar work first showed how optical astronomy would be the key to our understanding of observations made in other parts of the electromagnetic spectrum, particularly at radio wavelengths and at X-ray energies.

—*Geoffrey Burbidge, Editor*

■ **Abstract** An account is given of the history of two observational programs set for the Palomar 200-inch telescope, one by Walter Baade and the other by Edwin Hubble near the start of the scheduled operation of the telescope 50 years ago. The review is partly an assessment of whether, and how well, these programs have been carried to completion, and partly an account of the response of Palomar to new discoveries and developments not foreseen in 1950. Stellar evolution, the discovery of variations in the metallicity of stars of different populations, the chemical evolution of the Galaxy, the Cepheid P-L relation, the redshift-distance relation of the expanding universe, and the extragalactic distance scale are discussed as they relate to the predictions for progress on the programs set out by Baade and Hubble. Not foreseen was the invention and development of radio astronomy and high energy astrophysics, leading to the discovery of radio galaxies, quasars, and the gradual realization of violent events, both in stars and in galaxies. The review is highly restricted to these subjects, covering only three areas among the totality of the work in observational astrophysics studied during the first 50 years at Palomar.

1. THE BEGINNING

The 200-inch Palomar reflector, shown in the famous drawing by Russell W. Porter in Figure 1, was commissioned for regular scheduled observations on November 12, 1949, fifty years to the month of the distribution of this volume of Annual Reviews. The Hale telescope, planned since 1929, had enormous publicity surrounding both its construction and the hopes for astronomy as to what it might accomplish. The purpose of this review is to discuss the extent to which those hopes have been realized. Palomar, together with Mount Wilson in the joint operation known at first as the Mount Wilson and Palomar Observatories, often lead the way in the explosive developments that have characterized astronomy in the period.

The formal dedication of the Palomar Observatory and of the Hale telescope took place on June 3, 1948, led by Vannevar Bush, president of the Carnegie Institution of Washington, and Lee Du Bridge, president of the California Institute of Technology.

A scientific dedication took place a month later during the joint meeting of the Astronomical Society of the Pacific and the American Astronomical Society (Richardson 1948). The principal scientific address was by Walter Baade (1948) titled "A Program of Extragalactic Research for the 200-inch Telescope." This was a prescient document, outlining a research program that was to take 30 years. Much of Baade's lecture will be discussed later.

The telescope was not released to the astronomers for another 16 months. Ira Bowen (Figure 2), hero of that period and for the following two decades, keeping his head when all others were losing theirs, knew that if he released the telescope to the astronomers, even for a few months, he would never get it back. As director and as one of the world's foremost experts on optics, he was responsible for bringing the telescope to as high a state of perfection as the design of the engineers permitted.

By 1948, two problems had surfaced (Bowen 1948, 1949). (*1*) The lever system that grabbed the honeycomb ribbings at the back of the five-meter mirror had too much friction by a factor of 30 to keep the mirror at its proper figure for all gravity loadings. (*2*) As it left the optical shop, the mirror had a slight turned-up edge, purposely, so that its sag in the cell beyond the radius of the back squeeze levers

Figure 1 Drawing by Russell W. Porter of the 200-inch telescope and its dome made in 1936. Note the roller bearings at the north pier, replaced in the real telescope by the oil-pad flotation invention. Note also that the girders of the dome structure are not covered here as they are in the dome. The dome was finally completed with a double sheathing for thermal control.

was expected to compensate for the raised outer 20 inches. But the mirror did not compensate as was predicted. It was too stiff.

Thus, Bowen made the very difficult decision to take the mirror out of the telescope and polish down the turned-up edge on the dome floor. The final figuring of the outer raised 20 inches was done by Don Hendricks, the chief optician of the Mount Wilson observatory who had a magical touch, together with Mel Johnson, one of the opticians who had originally worked on the mirror from 1936 to 1946 in the Caltech optical shop (with four years out for World War II).

Figure 2 The leading players in the early days of stellar evolution and cosmology at Palomar. *Left to right, top to bottom*: Ira S. Bowen, Walter Baade, Edwin Hubble, Milton Humason, Rudolph Minkowski.

A long series of optical tests were made after this fix and were analyzed by Bowen from measurements on the Hartmann test plates. They revealed that the mirror was nearly perfect. The matter-of-fact description of these trying tests and his initial decision to remove the mirror for the fix is described by Bowen (1950) in a classic paper entitled "Final Adjustments of the 200-inch Telescope." He had waited to publish this account until Hubble (1949) had written an account called "First Photographs With the 200-Inch Hale Telescope" in which Bowen is not mentioned, yet the needed mirror correction is discussed as if the photographs had first revealed the problem. However, the first photographs had in fact been taken by Bowen with the telescope, months before, as part of the testing regime. As great an astronomer as Hubble was, he could never overcome his disappointment that he had not been chosen as the director of Palomar.

Nevertheless, Hubble's fame, importance, and close association with the 200-inch project through cosmology, which in some sense justified its construction following his seminal discoveries at Mount Wilson from 1922 to 1936, were the reasons he was given the opportunity to use the telescope in four observing runs between January 1949 and April 1949, interspersed with Bowen's Hartmann-testing regime.

The first scheduled run in which the telescope was officially assigned to an astronomer was on November 12, 1949. That observer was not Hubble. He had suffered a heart attack in June 1949 before the completion of Bowen's extensive work.

As important as the 200-inch reflector has been for astrophysics and cosmology, the Palomar wide field 48-inch Schmidt survey telescope was in some ways as important in the early years for mapping the northern hemisphere sky. That era is so long ago that it is difficult to remember the state of our ignorance of the deep sky before the Palomar Schmidt Survey. As it turned out, without that mapping, the 200-inch would have been vision impaired.

The primary 72-inch mirror for the big Schmidt had been completed by 1941 in the Mount Wilson optical shop. The difficult 48-inch correcting plate, ground and polished to its non-spherical figure by Hendricks, was completed in the summer of 1948 and the telescope went into routine operation in January 1949 (Bowen 1948). Walter Baade had been instrumental in bringing the principle of the Schmidt optical train back from Hamburg in the early 1930s, having been a colleague of Bernhard Schmidt, a taciturn Estonian, in the late 1920s when Baade was still on the staff of the Hamburg Observatory. They had been members of the Hamburg eclipse expeditions to Lapland in 1927 and the Philippines in 1929, and were close friends.

The first official Schmidt plate (recorded in the record book on September 29, 1948) was taken by Hendricks, who was in charge of the completion of the optics and collimations of the telescope. The 14-by-14-inch plate was of M31, seen for the first time on a single plate with superior definition and faint limiting magnitude in its full 4° extent. A reproduction of the Hendricks' plate is in Panel 18 of the Hubble Atlas (Carnegie Publication No. 618). The first published photographs from the 48-inch Schmidt were by Minkowski (1949) who showed them in his description and analysis of the nebula surrounding the young galactic cluster NGC 2244.

Four wondrous books on Hale and the making of the Hale telescope, of Palomar itself, and of its predecessor observatories of Yerkes and Mount Wilson are *The Glass Giant of Palomar* (Woodbury 1940), *Explorer of the Universe: A Biography of George Ellery Hale* (Wright 1966, 1994), *Palomar: The World's Largest Telescope* (Wright 1952), and *The Perfect Machine* (Florence 1994).

2. THE EARLY YEARS

Today astronomy and astrophysics is vastly different than it was in 1950. Although at that time much of the astrophysics of the stars was known through the spectacular advances in the basic physics (reviews by Stromgren 1951, Chandrasekhar 1951), little was yet understood about origins or evolution or the grand synthesis involving stars and galaxies that dominates the current climate. In that period the emphasis was on physical processes (see Aller 1956, Osterbrock 1974 for reviews of the physics of gaseous nebulae following Menzel, Aller, Goldberg, and Baker in the 1940s), classification systems, absolute magnitude calibrations (cf. Adams et al 1935 for the HR diagram), and surveys for the extant types of astronomical objects. Furthermore, observational cosmology had only just begun with galaxy classification, proof that galaxies and their distribution form the last great hierarchy in the organization of matter, and that the expansion exists (Hubble 1925, 1926a, 1926b, 1929a, 1929b, 1934, 1936a, 1936b, 1936c; cf. Sandage 1998b).

All this was the prelude to the 50 years that has now just ended. The tremendous change that we have witnessed can only be compared with the paradigm shift in geology and biology in the Darwinian era of the 1850s and the developments in relativity and quantum physics from 1900 to 1940. Astronomy's turn has been the last half century.

Of course, Mount Wilson and Palomar were not the only places where the first great advances into the new astronomy would take place, but the Observatories did have an enormous advantage in the first twenty years of Palomar. Radio astronomy hardly existed in 1950 and X-ray astronomy was not to be for 15 years. The first orbiting Astronomical Observatories (OAOs) were not operational until 1969. Computers had just been invented using von Neumann's concepts of stored programs.

In addition, most of today's large telescopes did not exist before approximately 1970. Those in existence before 1970 were not highly productive until after the 1960s. Hence, in 1950, Mount Wilson, Palomar, Lick, and Yerkes and McDonald were still the major centers of observational astronomy and astrophysics in America, with Radcliffe (Pretoria) and Stromlo just beginning in the southern hemisphere.

In this review I cannot cover even a fair fraction of the research done in the past half century, either at Palomar or worldwide. This account is necessarily highly restricted to stellar evolution, observational cosmology, and the beginnings of high energy astrophysics. It is primarily a retrospective on the programs set for the Palomar telescopes by Baade and Hubble 50 years ago.

2.1 The First Palomar Discoveries

Among the first of the interesting discoveries that often made the newspapers, even in the 1950s, occurred at Palomar during the initial trials with the 48-inch Schmidt prior to the beginning of the National Geographic-Palomar Sky Survey mapping.

Baade, having estimated how many asteroids were brighter than 19th magnitude,[1] discovered a fast moving asteroid on a Schmidt plate on June 26, 1949. It turned out to have a highly eccentric orbit whose perihelion distance was smaller than that of Mercury. It passed within 17 million miles of the sun, and within four million miles of the earth near the time of discovery (Richardson 1949, 1965). Because its perihelion distance was closer to the sun than any of the planets, Baade agreed to the name of Icarus.

It was the type of discovery that the public could understand, even if its cosmic importance did not rival the more central discoveries made soon thereafter. Nevertheless its significance for solar system astronomy and tests of general relativity soon became evident. The variations in its orbital parameters can be used to determine the mass of Mercury and the advance of the line of apsides due to space-time curvature caused by the solar mass (Herrick 1953, Gilvarry 1953, Dicke 1965, Francis 1965).

2.1.1 Stellar Evolution and Observational Cosmology

When the astronomical history of this century is written, the two central advances that will be cited 400 years from now are (*1*) the final understanding of stellar evolution as summarized by the HR diagram, and (*2*) the understanding of the universe as a whole through the development of observational and theoretical cosmology. Work with the 200-inch dominated both of these fields.

Our understanding of stellar evolution has progressed through the interaction of theoretical advances in studies of stellar interiors with the observational results concerning stellar populations, as defined by Baade (1944a, 1944b) in his resolution of the centers of M31, NGC 205, NGC 147, and NGC 185.

The population concept had crystallized in Baade's mind between 1939 and 1944, following the crucial clue provided by the discovery of globular clusters in the Fornax dwarf dE galaxy. In his many teaching sessions with H.C. Arp and this writer between 1949 and 1953, Baade was explicit in making the connection between the resolution into stars (Baade & Hubble 1939) of the main body of the Fornax dwarf and in its globular clusters. The Fornax globular clusters resolved into stars at the same apparent magnitude as the resolved stars in the main body of the Fornax dwarf itself. Thus it was natural to connect the HR diagram of globular clusters (Baade's population II) with the previously unresolved central region of M31 and also in the early type dE galaxies NGC 147 and NGC 185 whose stars were postulated to be of the same type as in the body of the Fornax dwarf.

A review of this approach by Baade to his population concept through the Fornax dwarf and then to contrasting the HR diagrams of globular clusters and of open clusters has been given elsewhere (Sandage 1986).

---

[1] He did this following a rebuff by the IAU on the naming of an asteroid for his wife. The excuse of the name givers, following an early IAU convention, was that only Greek names could be used for asteroids, but Baade sometime in the 1930s set out to show that there were not enough Greek names to accommodate his estimate of appx. 10^5 asteroids brighter than $m = 19$.

2.2 Baade's Program Proposed

Hubble's distances to M31, M33, and NGC 6822 depended both on the absolute magnitude calibration of the P-L relation of classical Cepheids and the accuracy of their apparent magnitudes. Both parts of this dependence were uncertain in 1948.

Baade discussed both of these problems in his paper at the July 1, 1948 scientific dedication of the 200-inch. Among other points he said:

> "Our first concern will be the extension of the magnitude scales down to magnitude 22.5, the limit attainable with the 200-inch. ... With the photoelectric cell at the 200-inch, we ... shall be able to reach magnitude 18.5, leaving to photographic photometry the faint end from 18 mag, to 22.5 mag. which can be comfortably bridged in a single step."

The most fascinating part of his lecture is:

> "Unfortunately ... there remains considerable uncertainty regarding the adopted zero point of the period-luminosity relation of the Cepheids. ... We intend to derive the zero point both of the Cepheids and the cluster-type variables in an entirely [new] manner. There is every indication that beginning with spectral type F the dwarf branches in the populations I and II coincide. To mention just one piece of the evidence: the dwarfs nearer than 10 parsecs with well determined trigonometric parallaxes are scattered along the same line in the Hertzsprung-Russell diagram whether they are slow moving stars or are of the high-velocity type. Hence by extending the H-R diagram in a globular cluster like Messier 3 to dwarfs of solar brightness we shall be able to connect directly the cluster-type variables with G dwarfs, the mean absolute magnitudes of which are ultimately based on well-determined trigonometric parallaxes. We thus obtain the mean absolute magnitude of the cluster-type variables. Now with the 200-inch we shall be able to reach in the Andromeda nebula the cluster-type variables (and the long-period variables associated with the population II). Since we know their absolute magnitudes, a comparison with the classical Cepheids in the outer spiral regions will furnish the absolute magnitude of the latter."

> "... Now we do not expect that this program, simple and straightforward as it may seem, can be carried out within the next three to four years, if necessary by resorting to so-called hard work."

In this last sentence, Baade was pointing out the importance of rare excellent seeing, rare enough that the work will be slowed because of it. But what actually happened is that the precept that the population I and II main sequences coincide (which was assumed by Baade) is not correct because of chemical composition differences—a difference not dreamed of in 1948.

When variations in the chemical compositions of stars did become apparent, a new program was required to solve the many problems caused by such variations. Those solutions were to occupy us for the first 20 years, and yet today, certain of the calibrations are still in doubt (many would say controversial). They include (*a*) the precise position of the main sequence as a function of metallicity, (*b*) the putative effect of chemical composition variations on the absolute magnitudes of the classical Cepheids (Sandage, Bell, Tripicco 1999), and (*c*) even yet, the absolute magnitude calibration of the RR Lyrae variables (Sandage 1993a, 1993b; Reid 1999).

The story I tell next is how Baade (1952) concluded even from his first observations of M31 with the 200-inch that the Cepheid P-L relation that was first calibrated by Hertzsprung (1913), copied by Shapley (1918), used by Hubble (1925, 1926a, 1929a), and apparently confirmed by Wilson (1939), was undoubtedly too faint by about 1.5 magnitudes.

2.2.2 Baade's Program Realized

The four parts of the program outlined by Baade (1948) were (*a*) redetermine the magnitude scales to m = 22.5, (*b*) recalibrate the Cepheid P-L relation by the new method that was, (*c*) to find the main sequence of globular clusters and calibrate therewith the absolute magnitudes of RR-Lyrae variables, and (*d*) discover RR-Lyrae variables in the disk of M31 to compare with the apparent magnitudes of the M31 classical Cepheids, thus calibrating them.

All parts of this program were eventually carried out using both the Palomar and Mount Wilson telescopes. Much was learned in the process about stellar evolution, the chemical evolution of the Galaxy, methods to age-date the stars, and the relation between kinematics and chemical composition among the different stellar populations.

First Revisions of the Magnitude Scales Made at Mount Wilson   In the 1930s Baade had been engaged in redetermining the faint magnitude scales in particular Selected Areas of the original Kapteyn program (cf. Blaauw & Elvius 1964). He had found significant magnitude scale corrections to the Mount Wilson Catalog of Photographic Magnitudes in Selected Areas (Seares et al 1930). However, although Baade's photographic photometry had shown the urgent need for corrections to the Mount Wilson Catalog fainter than $m_{pg} = 16$, more accurate results could be obtained using the new photomultiplier cells developed during the war. At the request of Baade and Hubble, Stebbins, Whitford, and Johnson (1950) began a photoelectric campaign to test the faint magnitude scales of the Mount Wilson Catalog and the North Polar Sequence (Seares 1922).

Their important 1950 paper had a profound effect in showing that the Mount Wilson scales needed corrections at the faint end beyond m = 16 that sometimes reached 1.5 mag at m_{pg} (Seares) = 19. The result was a sober realization that a vast work in precision photometry for cosmology and stellar population studies lay ahead. That work, once begun, was to occupy the next 30 years.

Later Revisions of the Magnitude Scales at Mount Wilson and Palomar   A program of photoelectric photometry to determine faint magnitude sequences was begun soon after the 200-inch was operational. W.A. Baum, an excellent experimental physicist, was appointed to the Carnegie staff to begin the work. Baum (1946) had worked in the rocket group led by Richard Tousey at the U.S. Naval Research Laboratory. With this experience and the photometric expertise he had gained in his Caltech PhD thesis in measuring the extinction of the atmosphere near the 3000 A cutoff, Baum came to the Observatories as the resident photometrist.

He supervised the construction of a prime-focus focal plane, pulse-counting photometer (Baum 1953) and started a program of three-color photometry of particular Selected Areas that were of importance for the galaxy programs of Hubble and Baade. Eventually, photometric sequences were determined to B = 22, well beyond the limit of the Mount Wilson Catalog.

Although never published, Baum's magnitude sequences in SA 61, SA 68, and SA 57 were used internally for many years at the Observatories for various programs where photographic transfers were still used to calibrate particular fields.

However, after about 1970 the observing programs that required photometry were planned around the modern practice of establishing photoelectric sequences directly in the fields of interest. Nevertheless, it is still of interest to determine the general level of corrections needed to the Mount Wilson Catalog even as bright as m = 18. To this end, the Selected Area program begun by Baum was continued into the 1980s using both the Mount Wilson and Palomar telescopes. The results for the 11 Selected Areas of SA 28, 29, 45, 55, 57, 71, 82, 94, 106, 107, and 118 are now in the literature (Sandage 1999).

The two main reasons why Hubble's extragalactic distance scale needed the large revision that is now nearly completed were (*a*) all his magnitude scales fainter than $m_{pg} = 18$ were too bright by as much as 2 mag at his limit, and (*b*) his adopted Cepheid P-L calibration was too faint by 1.5 mag.

2.2.3 Baade's Factor of Two in the Distance Scale

Failure to Resolve the RR Lyraes in the Disk of M31 Hubble's (1929a) distance modulus of M31 was m − M = 22.0. Baade (1944a) revised Hubble's modulus to m − M = 22.4, based on his preliminary revision of the Mount Wilson Catalog magnitudes in Selected Area 68, but still using Hubble's adopted zero-point of the Classical Cepheids.

The assumed absolute magnitude of RR Lyrae variables was $M_{pg}(RR) = 0.0$ at that time, based on the most recent calibration by Wilson (1939). This apparently confirmed an earlier statistical parallax determination by Bok and Boyd (1933). Wilson's analysis combined the more recent and more accurate proper motions with radial velocities of 55 RR Lyraes measured by Joy at Mount Wilson. Therefore, the calibration by Wilson of both the classical Cepheids and the RR Lyrae absolute magnitudes seemed secure.

Consequently, if m − M = 22.4 for M31, and with M(RR) = 0.0 for the RR Lyraes, Baade should have resolved RR Lyrae variables in the disk of M31 near the 200-inch plate limit of $m_{pg} = 22.5$. Even more telling, the giant branch of the M31 globular clusters should resolve easily beginning at $m_{pg} = 21$. Their absolute magnitudes were believed to be $M_{pg} = -1.5$ ($M_{vis} = -2.8$) according to the HR diagrams known at the time (Baade 1944a, Figure 1).

What Baade actually observed is described in the report of IAU Commission 28 for the Rome meeting (Hoyle 1952):

> "Baade then went on to describe several results of great cosmological significance. He pointed out that, in the course of his work on the two stellar populations in M31, it had become more and more clear that either the zero-point of the classical cepheids or the zero-point of the cluster variables must be in error. Data obtained recently—Sandage's colour-magnitude diagram of M3—supported the view that the error lay with the zero-point of the classical cepheids, not with the cluster variables."

Following a request by Shapley for Baade to expand on the evidence, Baade replied, again at the Commission 28 IAU session:

> "According to the present zero-points we should expect to find the cluster variables of the Andromeda nebula at $m_{pg} = 22.4$ since the distance modulus of this system, derived from the classical cepheids, is $m - M = 22.4$. The very first exposures on M31, taken with the 200-inch, showed at once that something was wrong. Tests had shown that we reach with this instrument, using the f/3.67 correcting lens, stars of $m_{pg} = 22.4$ in an exposure of 30 min. Hence, we should just reach in such an exposure the cluster-type variables in M31, at least at their maximum phases. Actually we reach only the brightest stars of population II in M31 with such an exposure. Since, according to the latest colour-magnitude diagrams of globular clusters, the brightest stars of population II are photographically about 1.5 mag. brighter than the cluster-type variables, we must conclude that the latter are to be found in M31 at $m_{pg} = 23.9$, and not at $m_{pg} = 22.4$ as predicted on the basis of our present zero-points.
>
> "We have also convincing proof that the brightest stars of population II in M31 are properly identified because when they emerge above the plate limit the globular clusters in M31 begin to be resolved into stars. It should be emphasized that these are rough first data indicating the order of corrections which the present constants require."

To make these points secure was to occupy the next 20 years at Palomar.

Although Baade's public announcement was not made until the Rome meeting of the IAU in 1952, Hubble (1951) had already discussed Baade's discovery in his Penrose Lecture before the American Philosophical Society in April 1951. He wrote:

> "But now we find that something is wrong somewhere. The luminosities assigned to the Cepheids led us to expect that the globular clusters in M31 would be readily resolved with the 200-inch, and that their brightest stars, as

well as comparable stars in the main body of the nebula, could be studied individually with ease. It was found, however, that the stars in question were fainter than expected by a considerable fraction of a magnitude, and that they could be recorded and studied only with difficulty. The discrepancy is important because M31 has been explored on the basis of Cepheids while our system has been explored, in a sense, on the basis of globular clusters."

The Route Through Globular Clusters: Discovery of the Main Sequence
Baade organized the work in two parts. First he began a long campaign on the Cepheids in M31 itself, mounting a discovery program for variable stars in four fields at varying distances from the center. He invited Henrietta Swope, a Harvard expert in photographic photometry, to join him in the analysis of his 200-inch plates. A first summary of the Cepheids found in the inner three fields is given by Baade & Swope (1955).

By 1963 Swope had completed the analysis of the crucial outlying Field IV (Baade & Swope 1963) where the absorption and crowding is minimal. Baade and Swope used a local photoelectric sequence that had been measured by Arp with the 200-inch and published as part of the Baade-Swope paper. A well-defined P–L relation was found in B and V. The mean magnitudes of the Cepheids ranged from B = 20.5 to 22.8. These data, together with the new Cepheid P–L calibrations by Arp (1960a) and by Kraft (1961), which we discuss in a later section, gave distance moduli of 24.84 in B and 24.68 in V. These are close to the modern value of $(m - M)_o = 24.44$ (Madore & Freedman 1991) as corrected for E (B-V) = 0.10 mag reddening.

The second part of Baade's program was to tie the classical Cepheids to the RR Lyraes via observations in the disk of M31. As was discussed earlier, this required discovering the main sequences of globular clusters and fitting the resulting HR diagrams with the main sequence stars with known trigonometric parallaxes. The vertical fit of the fitted HR diagrams calibrated the luminosities of all the cluster stars, including the RR Lyrae variables.

The latter was the problem Baade posed to Arp and this writer, his two graduate students at the time. Baade proceeded to train us well in the then extant large-telescope techniques of observing on the Newtonian platforms of such telescopes as the Mount Wilson 60-inch at first, and later at the 100-inch Hooker reflector. Using plates of M92 calibrated by photographic transfers to Selected Areas 61 and 57, close to the two target globular clusters of M92 and M3, we just reached the main sequence in M92 near V = 19.

Near the end of the two-year observing program we were joined by Baum who supplied a photoelectric calibration of our photographic sequences, and therefore of all the cluster stars in M92.

At the same time, Baade had taken a series of plates of the cluster M3 in B and V to carry the work to the limit of the Palomar telescope. The M3 main sequence was reached easily, and the photometry could be extended for several magnitudes beyond the turn-off point. With the main sequences in two globular clusters now identified, we were confident enough to announce the result at the Cleveland meeting of the American Astronomical Society at a symposium on the HR diagram (Arp, Baum, & Sandage 1952; ABS). The details for both clusters were published a year later (ABS, 1953; Sandage 1953).

We made the fit of the M92 and M3 main sequences to the main sequence of the nearby trigonometric stars in the manner suggested by Baade in his 1948 lecture. Our result that $M_V(RR) = 0.0$ seemed to confirm Baade's conclusion, stated in Rome, that the error was not in Wilson's absolute magnitude of the RR Lyraes but rather that the distance to M31 must be larger than Hubble's (1929a) value. As a consequence the error must be in the adopted P–L relation of the Cepheids by the reasoning that Baade had set out at Rome. Baade's result is generally considered to be the first major discovery with the Palomar telescope.

But convincing as our main sequence fits appeared to be, the problem of calibrating the RR Lyrae variables was eventually found to be considerably more complicated. The positions of the population I and II main sequences in the HR diagram were soon shown not to be the same, contrary to Baade's assumption. Involved were (*a*) the central discovery of the chemical variations among the stars, (*b*) its meaning for stellar evolution and Galactic structure, (*c*) the relation of the discovery to the work by Burbidge et al (1957) on the formation of the chemical elements, and (*d*) the theoretical expectation and the observational proof that the luminosity of stars at a given temperature depends on the metallicity, i.e. that main sequence positions are functions of [Fe/H].

This was all heady stuff at the time. The daily excitement, fervor, and transcendental character of the work, together with its promise for a cosmological synthesis of origins and astronomical evolution in the broadest sense, is impossible to describe. All who lived through the development from 1950 well into the 1980s will recognize these years as being as close to ineffable perfection as a scientific life can get.

3. STELLAR EVOLUTION

3.1 Stars Move Off the Main Sequence Rather Than Up or Down

With the discovery of the main sequence turn-off of the globular clusters and the heretofore unseen connection of the main sequence to the subgiant and the giant branches, the path of stellar evolution had suddenly become clear. Stars move off the main sequence rather than either down or up it as they age. Although trivial now, it was a revelation in 1952.

By 1940 it was well known that the main sequence could be explained as the locus of stars with different fractions of hydrogen throughout a chemically homogeneous stellar interior. Hence, a fully convective star, mixing the products of hydrogen burning uniformly throughout its interior, will progressively increase the mean molecular weight of its gas owing to the progressive increase in the helium content, and will move up the main sequence (Gamow 1940, Figure 36; Oke & Schwarzschild 1952, Figure 3; Bok 1946; Stromgren 1952). Likewise, stars on the bright part of the main sequence will move down the sequence if they lose significant mass as they age.

Even as late as 1946, these two processes were believed to be the keys to an understanding of the HR diagram and the time-dependent evolution of stars (see Section 8 of Bok 1946). Stromgren (1952) also discussed the process. Soviet astronomers continued to discuss the process even into the 1960s (Ambartsumian 1952; Massevich 1954, 1959; Fessenkov 1952; Fessenkov & Idlis 1959). Cowling (1958) provides an excellent review.

However, with the discovery of the globular cluster main sequence, its turn off near $M_V = +4$, and the identification of the turn-off with the Schonberg-Chandrasekhar 10 percent mass limit for a hydrogen exhausted isothermal core (Schonberg & Chandrasekhar 1942, Sandage & Schwarzschild 1952), it had become clear that a mapping off the MS was the principal feature of the evolution. This mapping also gave a satisfactory explanation of Trumpler's (1925) system of classification for the variety of HR diagrams of open clusters, and his related discovery of the turn-up of the main sequences of open clusters near their MS termination points (Oke 1955; Sandage 1958a,b, 1988).

3.2 The Chemical Composition Connection

A most significant development in the decade of the 1950s was the emergence of what is now perhaps the most important idea in stellar evolution and whose ramifications cover parts of cosmology as well. The mantra, summarized from its various versions, would read:

> "The chemical elements heavier than H and He are made in stellar interiors and are spread throughout the interstellar medium by supernovae explosions. Because this is a continuous process, the metallicity of all stars subsequently made from the ISM will increase with time. Hence, on average, the oldest stars will be metal deficient compared with newly forming stars. The age–metallicity relation is expected to vary from place to place. Study of metallicity as a function of age, kinematics, and position can give clues to the problem of the formation of the Galaxy in particular, and the chemical evolution of galaxies in general, at all redshifts."

The opening campaign to reach this advanced form of the mantra was first set out by Hoyle (1946, 1954 in his Figure 1). The Hoyle program was, of course, brought to its classic maturity in the monumental paper by Burbidge et al (1957, B^2FH), from which the concept took flight. A proper history of how the ideas of varying metallicity and its implications, as developed by the early 1960s, requires a separate article. Here, we simply give the barest of the historical highlights of how, within less than two decades, the subject of the chemical evolution of cosmic objects grew into the present maelstrom. For the purposes here, the theme is to show only why our early main-sequence fits were incorrect. The history in greater detail has been given elsewhere (Sandage 1986, Sections 4 and 6).

The early study that began the study of variations in the chemical abundances of stars was by Chamberlain and Aller (1951). They analyzed Mount Wilson 100-inch high dispersion Coude spectra of the extreme subdwarfs HD 140283 and HD 19445, finding metallicity deficiencies of factors between 6 and 10 relative to the sun.

The famous underground story,[2] often told, is that the referee of the paper was so astounded that he disbelieved the factor of approximately 100 originally derived by Chamberlain and Aller.

It is said that the referee suggested that their stellar temperatures should be adjusted upward so as to decrease the metallicity deficiencies. Chamberlain and Aller did this partially, and finally published metallicities of [Fe/H] $= -0.8$ and -1.0 for HD19445 and HD 140283 respectively (their Table 8). But even they were astounded at this result. They wrote:

> "The one possible undesirable factor in our interpretation is the prediction of abnormally small amounts of Ca and Fe. As Greenstein suggests, the observed deficiency of some elements could possibly be caused by an *excess* second ionization—i.e., much higher ionization than that predicted by the Saha formula."

Hence, even as late as 1951 there was no clear idea of the extent (or even the existence) of gross deviations of the chemical abundances relative to the sun. However, the Chamberlain & Aller paper was the beginning of the current remarkable era of abundance analyses (see McWilliam 1997 for a review). Their paper was followed by Burbidge & Burbidge (1956) where the abundance variations, taken by them to be real, began to be put in the context of chemical evolution of the galaxy. The later detailed analysis of HD 140283 by Baschek (1959) using high dispersion plates taken in 1957 with the Mount Wilson 100-inch Coude spectrograph by Unsold, was definitive and gave [Fe/H] $= -2.32$. The value was closely confirmed by Aller & Greenstein (1960) using new, highly widened plates taken both with the Mount Wilson and the Palomar Coude spectrographs.

In the meantime, Roman (1954) had made the central discovery of the general ultraviolet excess in the spectral energy distribution of particular high-velocity stars. In a prescient paper she had isolated a homogeneous group of 17 "ultra" high-velocity stars that had ultraviolet excesses (relative to the UBV two-color locus defined by stars of low velocity with the same B-V colors) reaching 0.2 mag in delta U-B. The excess UV radiation was suggested by Stromgren (unpublished) to be due to lower Fraunhofer line absorption shortward of 4000 A in the spectrum,

---

[2]The nearly universal opinion of the time was that there was a fixed chemical composition of all stars throughout the universe, a kind of cosmic palimpsest containing some hidden initial, but universal, mystery relating to origins and evolution.

presumably due to lower metallicity. This was shown to be the case for less extreme metal deficiencies (Schwarzschild et al 1955). A model with the introduction of "blanketing lines" in the UBV two-color diagram (Sandage & Eggen 1959) generalized the case. Roman (1955) greatly added to the study of the UV excess for high velocity stars for which she had made a catalog that contained both photometric and kinematic data.

The first calibration of the UV excess in terms of the [Fe/H] metallicity deficiency was made by Wallerstein & Carlson (1960). They used the small number of abundance analyses that were just coming from the stellar abundance program of Greenstein and his team of postdoctoral fellows, using both the Mount Wilson and the Palomar Coude spectrographs. Wallerstein (1962) made a later calibration of the UV excess/metallicity relation that has been widely used because of the much larger number of calibrating stars it contains. Examples of the early literature are the papers by Helfer et al (1959), Wallerstein (1962), Wallerstein et al (1963), Wallerstein & Greenstein (1964), and Wallerstein & Helfer (1966).

The slope of the blanketing lines in the UBV two-color diagram was calibrated by Wildey et al (1962) based on high dispersion spectrograms obtained by the Burbidges with the Mount Wilson 100-inch Coude. A table of blanketing-line slopes and the resulting corrections to colors was made using the Wallerstein (1962) calibration of the excess itself. Later, account was taken (Sandage 1969) of the effect of the guillotine on the observed excess due to the convergence of the slope values of the blanketing and intrinsic lines in the U-B, B-V diagram for subdwarfs redder than B-V = 0.7. A normalized UV excess was defined, as reduced to B-V = 0.6, to account for the guillotine. Subsequent calibrations of the normalized UV excess include those of Carney (1979) and Cameron (1985).

After Roman's discovery of the UV excess in the 17 field high-velocity subdwarfs, the same type of excess was sought in globular cluster stars. It was soon found in the giant stars of the cluster NGC 4147 (Sandage & Walker 1955), and soon thereafter in M3 (Johnson & Sandage 1956), M13 (Baum et al 1959), M5 (Arp 1962), and M15 and M92 (Sandage 1970a). This closed the last test for the identification of the field high-velocity F and G stars with stars in globular clusters that had been suggested by Baade (1944a), based on Oort's (1926) PhD thesis on high-velocity stars.

As a consequence, the work also established the great age of the field subdwarfs because, from the age-dating of the globular clusters (Sandage & Schwarzschild 1952) it was known that these clusters, as a whole, were as old as the first stars formed in the galaxy. By extension, this identification led naturally to the model of the formation of the galaxy through collapse from a larger volume (Eggen et al 1962, Sandage 1990a).

With the knowledge gained that there is a large difference in the metallicity between globular cluster stars and the common field F and G stars of the population I, we became aware, but only gradually, that our initial fit of the globular cluster main sequence to the nearby trigonometric field stars discussed earlier was not correct. The suspicion was that the MSs of metal-poor subdwarfs were fainter than the population I MS by a progressive amount that depends on [Fe/H].

In hindsight, it turns out that the first evidence of that fact was in fact the position of HD 19445 and HD 140283 in the HR diagram of Adams et al (1935) where these two stars were among the six "intermediate white dwarfs" found in the Mount Wilson spectroscopic catalog of 4367 spectroscopic parallaxes (look carefully at their diagram: the six outriders at spectral type near A5 and absolute magnitude near +5 are easily missed).

The second reason to suspect an offset between the population I and II MSs was theoretical. Stromgren (1952) had set out homology relations from stellar interior theory that showed that MS stars of low metal abundance are expected to have fainter MS positions than those of higher metallicity. Although now a well-known proven proposition, the premise had not been proved observationally or indeed by calculated stellar models (rather than homology transformations) even in the mid 1950s. The first calculated model with zero metals was by Reiz (1954), where his model star was indeed a star lying fainter than the high metallicity MS (i.e. it was a subdwarf).

Recall that all we had to work with in the 1950s were stellar parallaxes that were sufficiently inaccurate for the few very high velocity field stars that the observational test for the existence of a faint subdwarf sequence was uncertain. However, analyzing the available data on less extreme stars that had only mild subdwarf characteristics (intermediate velocities and small but definite UV excesses), we could provide an observational proof of sufficient weight that was almost convincing that a relation does exist between UV excess and depression of the MS from that of the population I high metal stars (Sandage & Eggen 1959, Figs. 1 and 2).

However, the extreme case for the very low metallicity stars, similar to those in globular cluster stars, was only given by the moving group parallax of the Groombridge 1830 group (Eggen & Sandage 1959), a result not widely accepted at the time.

The problem of the MS position was discussed again three years later using the totality of known trigonometric parallaxes and the rapidly growing data on UBV photometry of the trigonometric parallax stars. By combining these data we could show that the departure from the main sequence of F and G stars was a progressive function of [Fe/H] (Eggen & Sandage 1962), reaching the order of 1 mag departure faintward for subdwarfs with [Fe/H] $= -2$. From this calibration, a new MS-fitting procedure could be used, as illustrated for the fit to M13 in Figure 10 of the cited paper (Eggen & Sandage 1962). Later fits were made (Sandage 1970a), the most complete being of 19 cluster diagrams to the continuum of MS positions as a function of [Fe/H] (Sandage & Cacciari 1990).

With this development, the RR Lyrae stars could now be calibrated by the MS fits, bringing to a close, in principle, that phase of Baade's Palomar program originally proposed in 1948. The most recent use of the method is reviewed by Reid (1999), based on post-Hipparcos analyses of the relevant new trigonometric parallaxes.

4. CALIBRATION OF THE CEPHEID P–L RELATION

Baade's plan to recalibrate the zero-point of the Cepheid P-L relation using the difference in apparent magnitude between the Cepheids and RR Lyrae stars in M31 (Baade 1956a) was frustrated by the strong dependence of M(RR) on [Fe/H] just discussed. The calibration of the effect is still in contention at the 0.3 mag level, and until the calibration of M(RR) is beyond doubt, Baade's route to the Cepheid P-L calibration will remain uncertain.

Remarkably, an independent method of calibration of the P–L relation of classical Cepheids came into play in 1958. Peter Doig (1925) first pointed out that the Cepheid U Sgr was probably a member of the galactic cluster M 25. Irwin (1955) rediscovered this fact, which, together with Eggen's knowledge that NGC 7790 contains three Cepheids, opened the way to use photometric parallaxes of the parent open clusters to calibrate the Cepheids.

A program of photometry and spectroscopy was begun by Arp, Kraft, and the present writer in 1958, both at Mount Wilson and at Palomar, to obtain the necessary data on distances and reddenings to carry a new calibration of the P–L relation to a preliminary conclusion.

By 1961, six calibrators had been completed—CF Cas (Sandage 1958c), EV Sct (Arp 1958), DL Cas (Arp et al 1959), CV Mon (Arp 1960b), U Sgr (Sandage 1960, Johnson 1960, Wampler et al 1961), and S Nor (Irwin 1958, Fernie 1961). Kraft (1961) discussed all of the photometry of the parent open clusters that existed to 1961, redetermining the cluster reddenings and distance moduli on a uniform basis.

Kraft's new reductions, plus four additional long-period Cepheids in the h and chi Per complex adopted as members on the basis of Eggen's (1965) discussion, were used in a calibration of the P–L relation whose shape was redefined by combining relative relations from M31, NGC 6822, LMC, and SMC (Sandage & Tammann 1968). The double Cepheid CE Cas a and b in NGC 7790 was added later (Sandage & Tammann 1969) following the difficult photometry of the individual components that are separated by only 2.3 arcsec.

The resulting 1968 zero-point, based on only nine fundamental calibrators is, remarkably, within 0.02 mag of the recent Hipparcos zero-point as defined by Feast & Catchpole (1998). It is also brighter by between 0.07 and 0.10 mag than the Feast & Walker (1987) calibration, which itself is an average of 0.11 mag fainter than the Hipparcos calibration (Sandage & Tammann 1998, Seggewiss 1998), suggesting a small systematic error in that calibration of order 0.1 mag.

Hence, the 1961 and 1969 Cepheid calibrations fully supported and refined Baade's 1952 discovery of the error in the previous calibrations from Hertzsprung (1913), to Shapley (1918), to Wilson (1939). Hence, by 1970 Baade's 1948 program was complete.

The dream of the astronomers at the Palomar dedication in 1948, expressed with conviction by Vannevar Bush in his address "Two Observatories Operate as One," had in fact come true.

5. COSMOLOGY

Hubble had outlined his vision of cosmological research with the Palomar telescopes in his 1951 Penrose lecture. His principal plans concerned four programs:

(1) the large scale distribution of galaxies
(2) the law of redshifts
(3) the cosmic distance scale
   (a) Cepheid P-L relation
   (b) distance to M31
   (c) novae as distance indicators
   (d) distances to M81 group and M101 group galaxies
   (e) the "average" galaxy as a statistical distance indicator
   (f) brightest stars
   (g) brightest galaxies in clusters
(4) Cosmological theory

Many of these programs have eventually been carried to completion, yet the final solution of the distance scale remains at this writing in contention, even if only at the 25 percent level.

Other programs not mentioned by Hubble but which have also been central to the Palomar work concern Galaxy morphology as it relates to galaxy formation and evolution. Here Baade's population concept has become intertwined with Hubble's purely geometrical cosmology.[3]

In particular, the problem of the earliest stages of galaxy formation, not even conceived in the 1930s, is now centered on studies of the Lyman alpha forest that was seen in high dispersion quasar spectra, detected at sufficient spectral resolution by Weymann et al (1981). Many Palomar studies followed, where the telescope was host to the efficient instrumentation developed by the UK group led by Boksenberg and used in many collaborations with Caltech astronomers. Important reviews include those by Weymann (1993), Lanzetta (1993), Bajtlik (1993), Boksenberg (1995), and Rauch (1998).

How much of Hubble's program has in fact been completed?

5.1 Large-Scale Distribution

Hubble had dreamed of beginning at Palomar a galaxy count program of the type he had pioneered in the 1930s (Hubble 1934). The data had formed the basis of

---

[3]Baade's dictum, often stated with vigor when, in his opinion, too large a fraction of the scheduled 200-inch time was being assigned to the redshift programs of Hubble and Humason, was "You will never understand the spatial geometry (world model) until you understand the galaxies that you are using as markers of the space."

his work with Tolman on world models (Hubble & Tolman 1935) and his famous analysis papers (Hubble 1936a,b) on the log N(m) relation (see Sandage 1998b for a review). To this end he began to accumulate plates in the spring of 1949 for galaxy counts with the 48-inch Schmidt camera. The very large 14-by-14-inch plates had exquisite definition over a $7° \times 7°$ field compared with the Mount Wilson large reflector plates that covered only 0.25 square degrees which Hubble had used for the 1934 study.

My first association with Hubble and the research part of Mount Wilson and Palomar was as a summer assistant in 1949 in measuring the plates and beginning the galaxy counts. The program was not continued for several reasons, one of which was Hubble's heart attack in July 1949, and another was the press of the more urgent programs that Hubble had set out.

It is fortunate that the count program did not become a central project at Palomar because it would not have been carried to the necessary conclusions to have made a major contribution. Hubble acknowledges this in his Penrose lecture, stating:

> "The distribution of nebulae over the sky is not included in the Palomar program. Such an investigation is being carried on at the Lick Observatory by counting the nebulae on survey plates with the fine 20-inch camera which in time will cover the entire sky observable from that latitude. [The program by Shane and Wirtanen] will present the distribution of more than a million nebulae to about 18th magnitude. The data will test the current assumption of large-scale uniformity over the sky ('isotropy' or 'no favored direction') and will describe the small-scale distribution, or tendency toward clustering."

Indeed, the Lick survey did just that (Shane 1975).

The program on the large-scale distribution, although not done at Palomar, is central to modern cosmology and includes the early papers by Seldner et al (1977), Gregory & Thompson (1978, 1982), Chincarini & Rood (1979), and Tarenghi et al (1979), with reviews by Oort (1983) and Rood (1988).

A most important survey of galaxy distribution that was done at Palomar and that has had a lasting influence on observational astronomy is the catalog of galaxy clusters by Abell (1958, 1975). Abell's survey was made using the 48-inch Palomar Observatory–National Geographic Sky Survey plates (POSS) as his Ph.D. thesis from the California Institute of Technology, under the direction of Rudolph Minkowski, director of the Sky Survey. More will be said about this important Abell catalog in the next sections, but before leaving the large-scale distribution problem, it should be pointed out that Abell, already in 1959, anticipated the discovery of the filaments, sheets, and voids, which are now known to be present by his first identification of what he called superclusters, or clusters of clusters.

5.2 The Velocity-Distance Relation

5.2.1 The Mount Wilson Years

Following Hubble's announcement of the primitive form of a velocity-distance relation (Hubble 1929b), anticipated by Robertson (1928) and Lemaitre (1927, 1931), Humason began an extended program of redshift measurements with the Mount Wilson 100-inch Hooker reflector. By 1931 these data were combined (Hubble & Humason 1931) with apparent magnitude estimates to define a redshift-apparent magnitude relation for field galaxies and for the few galaxy clusters known at the time.

By 1936 Humason (1936) had carried the redshift data to 40,000 km s^{-1} using a total of 10 great clusters known at the time (found by serendipitous discoveries in other survey programs). These data permitted Hubble (1936b) to define a velocity–"distance" relation for clusters to what was the enormous redshift of $z = 0.13$ for the Bootis cluster. It was here that the analysis and further observations from Mount Wilson stopped in 1936, awaiting the completion of the 200-inch.

Hence, at the time of the 1948 Palomar dedication, the cluster redshift-magnitude relation was defined by only ten clusters, and the field galaxy data numbered fewer than 200 objects. Essentially no data existed on (*1*) the isotropy of the expansion (the same in all directions?), (*2*) systematic velocity deviations, if any, from a linear velocity–distance relation that is required in all expanding models where large-scale homogeneity is maintained throughout cosmic time, and (*3*) the coldness of the flow itself as measured by the mean random motion about the systematic linear expansion.

These and other problems are what Hubble either meant, anticipated, or implied in 1951 in his program to study "the law of redshifts." That study was to take 30 years at Palomar.

5.2.2 The Humason-Mayall Redshift Catalog and its Analysis

By 1935 Mayall, having designed and put into operation a fast spectrograph at the Lick Observatory for the 36-inch Crossley reflector, began a redshift program on the brighter Shapley-Ames (1932) galaxies, mostly of spiral type. At the same time, Humason continued his 1931 redshift determinations using spectrographs at the Cassegrain focus of the Mount Wilson 100-inch Hooker telescope. Humason and Mayall continued observations through 1941, and began again in 1945 at the close of World War II (see the dates in Table V of Mayall in HMS 1956).

Humason began his extended redshift program at Palomar immediately upon the commissioning of the 200-inch, using a highly efficient new solid-block Schmidt camera that had been designed by Hendricks (1939) and Hendricks & Christie (1939), and whose difficult optics were eventually made by Hendricks in the Mount Wilson optical shop (Bowen 1960).

By 1956, Humason had obtained redshifts of 472 field galaxies at Mount Wilson and at Palomar, and 152 galaxies at Palomar in 26 clusters whose redshifts had reached the remarkable value of $z = 0.2$ for the Hydra cluster.

Most of the distant clusters were new. As mentioned earlier, many had been found in the systematic visual scanning of the Sky Survey plates as they were brought down to Pasadena as the survey progressed. This special search for clusters by Humason and the writer is described by Bowen (1954). At the time of HMS, the Abell Catalog was yet several years in the future.

Once candidate clusters had been found on the deep Schmidt plates, 200-inch prime focus photographs of the candidates were taken to prepare for Humason's later 200-inch spectroscopic observations. Because even the first-ranked galaxies in most of these clusters were not bright enough to be seen on the spectrograph slit, the plates, taken several months prior to the spectrographic observations, were measured in Pasadena for coordinate offsets relative to bright stars that Humason could see. Using these "offset" coordinates, Humason could move the prime focus spectrograph with precision screws to the invisible galaxy whose redshift was sought. The offset plates also were made with star trails superposed on them, made by stopping the telescope drive so that the brighter stars would show trails, from which the precise cardinal directions at the position could be determined. All of this was necessary during the 20 years before television viewing came to a data room that was remote from the telescope itself.

Mayall had also determined redshifts for 300 field galaxies at Lick. There were 114 overlaps between the lists of Humason and Mayall, which, together with many duplicate measures of a given galaxy at each observatory, permitted a thorough evaluation of the accuracy of the redshifts.

In the meantime, Pettit (1954) had measured photoelectric magnitudes at the Mount Wilson 60-inch of most of the field galaxies in the two lists, with a few also measured by Stebbins and Whitford (1952), also at Mount Wilson. For the extensive list of Virgo cluster members, individual photoelectric magnitudes of large-diameter galaxies had been measured by Whitford (1936). Others had been measured by Bigay (1951).

The entire photometry was combined with the Humason and Mayall redshifts to discuss the velocity–"distance" relation (Humason et al 1956, HMS), but now with a much larger database than was available to Hubble (1936) and Humason (1935) in the decade of the 1930s.

The Hubble diagram (redshift versus apparent magnitude, corrected for aperture effects and for the effects of redshift), now for 18 clusters reaching 60,000 km s^{-1}, showed that the velocity–"distance" relation was indeed linear, as had been suspected from Hubble and Humason's earlier results. Linearity is, of course, the most fundamental requirement of the theory and all models if the expansion is real (Heckmann 1942). Every place in the manifold appears to be the center of the expansion, but only if the velocity–"distance" relation is linear. The importance of this linearity requirement is the reason why such a large effort was put into the observational proof at Palomar of linearity from 1950 to well into the 1970s.

Besides testing for linearity, the new feature of the work was the early attempt to obtain the second-order term (the deceleration) from the cluster data, but the result was clearly premature, requiring a much larger data sample.

Following Humason et al (1956), the work was then expanded to eventually include more than 100 clusters over many directions to test for isotropy and for deviations from a pure Hubble flow (Sandage 1970b, 1975).

5.2.3 The Expanded Palomar Program on the Hubble Diagram

It was clear that a large campaign on clusters was needed to search for any anisotropy in the expansion rate and to attempt to measure the second-order (deceleration) term in the Hubble diagram (Robertson 1955, Hoyle & Sandage 1956). The program was begun before Mattig (1958) had found the exact solution of the Friedmann scale factor valid for arbitrarily large redshifts (see Sandage 1995, 1998b).

The observational campaign to extend the Hubble diagram using a large sample of remote clusters lasted from 1955 into the 1980s. The beginning of the program and the results to 1969 were outlined in a progress report (Sandage 1970b). The principal limitation was again the discovery and subsequent photometry and spectroscopy of suitable remote galaxy clusters beyond the limit of the Abell Catalog and of the 1956 HMS program.

Candidate clusters were found by two methods. The first was to exploit the discoveries of clusters that contained radio sources. The second, described in Section 5.2.4, was again a search with the 48-inch Schmidt. We discuss first the radio sources.

By the mid-1960s a number of identifications of the radio sources in the 3CR Cambridge radio source catalog had been made with relatively bright galaxies. These were generally the first or second ranked E galaxies in clusters. Following the earlier work of Minkowski (1960), Schmidt (1965) had begun a 200-inch observing program in the early 1960s for the redshifts of many of the identified radio galaxies.

Photoelectric photometry was also begun on many of the clusters and radio sources then known. The observations were made with the 200-inch prime focus photometer for the faint sources. The brighter radio galaxies were measured with the two Mount Wilson reflectors.

The results were published from 1972 to 1975 in a series of eight papers. The theme of the program was to continue to test the linearity of the velocity-distance relation and its isotropy (the same in all directions?), and to attempt again a measurement of the second-order deceleration term. The contents of the papers are too detailed to describe adequately here. Table 1 gives an outline of the problems discussed in each paper.

5.2.4 The Extension of the Hubble Diagram to Higher Redshifts

By 1975 the available cluster candidates from the Abell catalog and from southern clusters and groups had been nearly exhausted for the high redshifts necessary to determine q_0. The redshift range was limited to less than $z = 0.5$, the largest being Minkowski's (1960) redshift for 3C 295. The conclusion from Paper VII was that $q_0 = 1 \pm 1$, hardly a useful result.

TABLE 1 Summary of the eight papers of the Redshift Distance series

Paper	Subject	Reference
I	Distinction is made between metric and isophotal diameters. Method to correct aperture photometry to a standard metric diameter is derived as used in the remaining papers of the series.	Sandage (1972a)
II	Hubble diagram for 84 first-ranked galaxies in clusters is derived, including data from Peterson (1970a) and Westerland & Wall (1969). $q_0 = 0.94 \pm 0.4$ derived. Evolutionary correction is discussed (see Sandage & Tammann 1983).	Sandage (1972b)
III	Hubble diagram of 128 radio galaxies shows the equality of $\langle M \rangle$ for 3CR radio galaxies with that of first-ranked cluster E galaxies. A form of the Spaenhauer (1978) diagram is first used to illustrate selection bias.	Sandage (1972c)
IV	Position in the Hubble diagram of QSO relative to first-ranked cluster galaxies is discussed with a two-component luminosity model with QSO as the central component combined with the distributed luminosity of the host E galaxy. Color decomposition is given of the two components. Model is tested by Kristian (1973). Veron-Cetty & Veron (1985) and Hewitt & Burbidge (1987) show the continuum of QSO luminosities relating QSO, N galaxies (Matthews et al 1964) with Seyferts and LINERS.	Sandage (1973a)
V	V-r colors introduced as a photometric system (Sandage & Smith (1963) that proved to be identical (Sandage 1997) to $(V-R)_J$ by Johnson (1964, 1965). Hubble diagram of first-ranked galaxies first given in r magnitudes.	Sandage (1973b)
VI	Deviations from the mean line of the Hubble diagram studied as functions of cluster richness, and Bautz-Morgan (1970) luminosity contrast. Debate on the meaning of the constancy of the luminosity function for the first few ranked cluster members as a function of cluster richness (Peebles 1969, Peterson 1970a, 1970b).	Sandage (1973c)
VII	Luminosity functions of first three ranked cluster E galaxies. Comparison of known luminosity functions (Abell 1975 from 1968, Rood 1969, Oemler 1974, Krupp 1974, Sandage 1976). Argument by Geller & Peebles (1976), Schechter & Peebles (1976), Oemler (1976), and Dressler (1978) based on cD galaxies (Matthews et al 1964, Morgan & Lesh 1965). q_0 derived again.	Sandage & Hardy (1973)
VIII	Redshifts (Sandage 1978) given for part of the remaining galaxies observed from Stromlo necessary to complete the redshift coverage of the Revised Shapley-Ames Catalog (Sandage & Tammann 1981, 1987). Isotropy of the local velocity expansion field studied with the result that the Rubin-Ford effect does not exist (see Rubin et al 1976).	Sandage (1975a)

Two parallel programs to discover clusters more remote than those in the Abell catalog were then begun at Palomar. One was carried out by the writer and described in Westphal et al (1975), and the other was by Gunn and Oke (1975).

The first search was made with the 48-inch Palomar Schmidt using photographic plates taken on the fine grain IIIaJ and 127-04 red plates that had recently been developed at Eastman Kodak by Millikan. These plates reached 0.5 mag fainter than the Sky Survey plates. Thirty fields were surveyed with a total area of 1500 square degrees. About 200 new clusters were found which, together with the clusters found between 1952 and 1957 by Humason & Sandage (1957) and Bowen (1954, 1957) but not observed by Humason, constituted the sample that we began to observe with a new sky-subtracting digital spectrograph invented by Westphal et al (1975). No catalog was published.

The survey by Gunn & Oke (1975) was begun using the 48-inch Schmidt, but they also went to fainter magnitudes by making a blind photographic search with the Hale telescope, eventually covering 11.3 square degrees and finding thereby 76 faint clusters (Gunn, Hoessel, & Oke 1986). The catalog from this program, augmented by the search made by Hoessel with the Kitt Peak four-meter Mayall reflector, was eventually published by Gunn, Hoessel, & Oke (1986). It still contains today the faintest sample of clusters found by optical searches.

The photometry and redshifts from the first survey were published in three papers (Westphal et al 1975, WKS, Sandage et al 1976, SKW, Kristian et al 1978, KSW). In the final paper we could extend the Hubble diagram to $z = 0.75$. Many of the new clusters had measured redshifts between 0.25 and 0.50. From the measurements of the colors as a function of redshift, no evolutionary effects in B-V or V-R were detected (KSW 1978) at the 0.05 mag level over the entire redshift range. This result heavily constrains theoretical models of E galaxy evolution in the relevant look-back times to those with only passive evolution (Sandage 1961, 1963; Oke 1971; Wilkinson & Oke 1978; Sandage & Tammann 1983).

After 1975 the program was carried entirely by Gunn, Oke, and their students. A series of papers were published on all aspects of the Hubble diagram, confirming the correlations in Papers VI and VII (Table 1) of luminosity of first-ranked cluster galaxies with cluster richness and Bautz-Morgan types, and extending the work to larger redshifts. Not only did they find new distant clusters, but Gunn and Oke individually developed powerful new instruments for the 200-inch that benefited all observers.

Oke (1969) designed and oversaw the construction of a 32-channel spectrum scanner used at the Cassegrain focus of the 200-inch. Gunn designed and oversaw construction of two instruments that also saw major use. They were (*a*) a combined photometric and spectroscopic instrument named PFUEI for Prime Focus Universal Extragalactic Instrument (Gunn & Westphal 1981), and (*b*) the prototype instrument for the Hubble Space Telescope WFPC camera, called the "four-shooter," built for Palomar at Jet Propulsion Laboratory under the leadership of Gunn and Westphal (Gunn et al 1984).

In their first paper on the Hubble diagram, Gunn & Oke (1975) initiated a new technique for the reduction of the aperture magnitudes, restricting the final observed raw magnitude to a fixed metric aperture of 16 Kpc (for $H_o = 60$). They then corrected for aperture effect using a correction that depended on the slope of the growth curve at their standard fixed metric diameter. This method differs from the procedure invented in HMS and adopted in the Redshift-Distance series (Table 1) where a standard growth curve was used to correct aperture magnitudes to essentially "total" magnitudes (Sandage 1972a, Paper I). The procedure of Gunn & Oke has also been used in more modern times by Postman & Lauer (1995) in their study of first-ranked cluster galaxies. The small differences in the conclusions of KSW 78 and Gunn & Oke (1975) concerning q_o can probably be traced to these different reduction procedures, showing the extreme sensitivity of the conclusions to the minuteness of the q_o effect.

New photometry on the intermediate band photometric system of Thuan & Gunn (1976) of first-ranked cluster galaxies in many Abell clusters was set out by Hoessel, Gunn, & Thuan (1980, HGT). They solved for the deceleration parameter giving $q_o = -0.55 \pm 0.45$. This, of course, is an accelerating universe, indicating a finite value of the cosmological constant, but the errors are too large to secure the result. As part of the same program Hoessel (1980) presented the surface photometry determined from two-dimensional detectors then available, for the galaxies studied by HGT.

In a major study, Schneider, Gunn, & Hoessel (1983a, SGH) analyzed the Hubble diagram for redshifts between 0.04 and 0.3, using the CCD photometry from the previous data papers. They obtained results similar to those of Sandage & Hardy (1973, SH) which showed that there are systematic variations of the absolute magnitude of first several ranked cluster galaxies with cluster richness and Bautz-Morgan contrast class. The work was extended to the first three ranked cluster members by SGH (1983b) where they also confirmed the result of SH concerning cannibalism.

Hoessel & Schneider (1985) published their surface photometry of the first-ranked cluster galaxies done at Palomar, again confirming the previous correlations with cluster richness (their Figure 4) and Bautz-Morgan contrast class (their Figure 3).

An important advance was made by Gunn, Hoessel, & Oke (1986) with the publication of their extensive catalog of remote clusters, discussed earlier. The catalog lists 418 clusters whose redshifts range from 0.15 to 0.92. It has provided the candidate lists for many of the current programs on distant clusters (Postman et al 1996, Oke et al 1998, Postman et al 1998, Lubin et al 1998).

This long narrative of the Palomar program on the "law of the redshifts" shows the central importance of the Palomar 48-inch and the Hale 200-inch telescopes in the development of practical (observational) cosmology during the last 50 years. The fleshing out of the Hubble diagram by the many varied programs shows how Hubble's proposed program was in fact carried to a level that could not have been foreseen in 1951.

Hubble's penultimate program, (see Section 5), was the recalibration of the extragalactic distance scale, which we discuss next. We have no space to discuss his last proposed program of "Cosmological Theory," which in fact now fills many current textbooks such as Narlikar's (1983) exemplar.

6. THE EXTRAGALACTIC DISTANCE SCALE

6.1 The First Step from the Local Group to NGC 2403

By the end of Baade's 200-inch campaign on M31, Baade & Swope (1963) had determined its moduli to be $(m - M)_{AB} = 24.84$ and $(m - M)_{AV} = 24.68$, fully 2.7 mag larger than Hubble's (1929a) value.

In a program parallel to Baade's, Hubble started an observing program to discover Cepheids in galaxies just beyond the Local Group. The targets were galaxies in the M81 and M101 groups (Holmberg 1950), eventually narrowing to M81, NGC 2403, and M101. The principal observers were Hubble, Humason, and the writer, with occasional plates taken by Arp, Baum, and Minkowski.

By 1963, 59 variables had been found in NGC 2403, of which 17 were later confirmed to be Cepheids. Some 10 Cepheid candidates were found in M81 together with 24 normal novae and a number of luminous blue variables (LBVs) of the kind discussed by Hubble & Sandage (1953) in M31 and M33. No Cepheids were found in M101, although nine LBVs were discovered (Sandage & Tammann 1974c, Sandage 1983b) using the total plate material that extended from 1909 (with the Mount Wilson 60-inch) to 1963. The variables in M81 were considerably less conspicuous than those in NGC 2403. Consequently the data for NGC 2403 were the first to be analyzed.

The long collaboration between G.A. Tammann and the writer began in 1963, starting with the analysis of the 200-inch plates of NGC 2403. By 1968 Tammann had obtained light curves for the 17 Cepheids in NGC 2403, with periods between 87 and 20 days, and light curves based on a photoelectric sequence that had been set up in the field of the galaxy. The distance modulus was $(m - M)_0 = 27.56$ (Tammann & Sandage 1968), based on the Cepheid calibration of the P–L relation discussed in Section 4.

The result was a major shock at the time because of its implied consequences for the revision of Hubble's "remote" distance scale, and therefore for the value of the Hubble constant. As late as 1950, Hubble's distance modulus for NGC 2403 was $(m - M) = 24.0$ ($D = 0.6$ Mpc). If we were right that the modulus was $(m - M)_0 = 27.56$ ($D = 3.2$ Mpc), then even at the very local distance of NGC 2403, Hubble's distance scale would be too small by a factor of five. This was much larger than Baade's original factor of two, as well as the final factor of three from Baade & Swope (1963). Therefore, by 1965 it was clear that a much larger attack on the distance scale was necessary than had been anticipated in 1948. The program was expanded into what ultimately became the series of ten papers called "Steps Toward the Hubble Constant."

6.2 The Steps Series from NGC 2403 to the Global Expansion Field

The Steps series has been reviewed elsewhere (Sandage 1998a). Only a few aspects need be summarized here in Table 2, but with a few added comments.

The five distance indicators listed by Hubble were Cepheids, brightest stars, novae, average galaxy luminosities, and brightest cluster galaxies. These were all eventually used in the programs to determine the distances to M31, the M81 and M101 groups, and ultimately to galaxies farther away in the remote expansion field.

Cepheids and the brightest cluster galaxies, together with the Palomar and Mount Wilson programs to calibrate them, were described earlier. The Steps program added the new distance indicators of (*a*) the angular size of HII regions and the linear calibration of the first three largest, and (*b*) supernovae of type Ia. In addition, novae, brightest stars, and galaxy luminosity functions that calibrated the van den Bergh luminosity classes in late type spirals were used, thereby completing Hubble's (1951) list. A few details of the Steps series are given in Table 2.

The definitive solution to the distance scale problem via the traditional ladder approach has had to await the results of the Hubble Space Telescope (HST) for Cepheid distances of galaxies that have produced supernovae of type Ia. The reviews by Teerikorpi (1997), and Branch (1998), and the calibrations of SNe Ia at maximum light via Cepheids (Sandage, Tammann, & Saha 1998; Saha et al 1999) continue to favor the long distance scale with $H_o = 55 \pm 5$, although other solutions still occur in the literature (Madore et al 1998).

7. EARLY DEVELOPMENTS IN HIGH ENERGY ASTROPHYSICS

7.1 Initial Identification of Radio Sources

What would develop into the present major discipline of relativistic astrophysics began with the "rediscovery" of cosmic radio waves. The initial discovery of Jansky (1932, 1933, 1937, 1939), and the genius and tenacity of Reber (1940a,b, 1942, 1944) in advancing the discovery are too well known to again recount here.

However, the major advances came after World War II when radio telescopes were first brought into operation. For the history of this early period we can refer the reader to articles and books by Hey (1946, 1973), Hey, Parsons, & Phillips (1946), Moffett (1975), and Sullivan (1982, 1984).

The connection with Walter Baade and Rudolph Minkowski at Palomar came after several groups of radio astronomers began to detect discrete sources of radio emission that needed to be identified with optical objects in the sky. Interferometric measurements in Australia and at Cambridge, England led to the discovery of several strong sources. The first definitive optical identifications were made by Baade and Minkowski (1954a, 1954b) with the 200-inch telescope, based on radio data provided by Bolton and Stanley (1948) who found the diameter of the Cygnus

TABLE 2 Summary of the content of the ten papers of the "Steps Toward the Hubble Constant" series

Paper	Subject	Reference
I	Data on angular sizes of the first three largest HII regions are given in galaxies in the Local Group. Distance-degenerate effects are discussed.	Sandage & Tammann (1974a)
II	Identification and photometry of the brightest resolved stars are given in 11 nearby galaxies with known Cepheid distances. Later data of the same kind for M81 and M101 are in Sandage (1983a, 1984).	Sandage & Tammann (1974b)
III	The stellar content of M101 is given based on a long series of 200-inch photometrically calibrated photographs. The distance modulus of m − M = 29.3 is 11 times the distance of m − M = 24.0 given by Hubble (Holmberg 1950). Kelson et al (1996), with m − M = 29.34, confirms the Steps value. $H_o = 56 \pm 9$ km s^{-1} Mpc^{-1}.	Sandage & Tammann (1974c)
IV	Calibration of the absolute magnitude of the van den Bergh luminosity classes. Modulus of the Virgo cluster derived as m − M = 31.45 ± 0.09. $H_o = 57 \pm 6$.	Sandage & Tammann (1974d)
V, VI	The quietness of the local velocity field is derived from the photometric distances of nearby galaxies compared with kinematic distances. Redshift data in Steps VI for Sc I galaxies to v = 20,000 km s^{-1} are combined with Zwicky et al (1961–1968) magnitudes to give a Hubble diagram requiring $H_o = 57 \pm 3$. Observational selection bias, using a form of a Spaenhauer diagram, was discussed for the first time.	Sandage & Tammann (1975a, 1975b)
VII	Virgo cluster distance modulus of m − M = 31.70 ± 0.08 derived using Tully-Fisher method. $H_o = 50 \pm 4$.	Sandage & Tammann (1976)
VIII	Type Ia supernovae calibrated via brightest stars in IC 4182 gives $H_o = 50 \pm 7$.	Sandage & Tammann (1982)
IX	A new method is introduced to tie the Virgo Cluster redshift to the remote expansion field devoid of all local velocity anomalies. The method was later made more precise by Jerjen & Tammann (1993) and Federspiel et al. (1998). Virgo modulus derived as m − M = 31.70 ± 0.09. $H_o = 52 \pm 2$.	Sandage & Tammann (1990)
X	Globular clusters are used to give a Virgo Cluster modulus revising an earlier value by Harris et al. (1991) who used an incorrect calibration of M_V (RR). Using the Oosterhoff–Preston-Arp period-metallicity relation for RR Lyrae stars (Arp 1955; Kinmann 1959) as calibrated from pulsation equations (Sandage 1990b,c, 1993b,c) gives m − M (Virgo) = 31.64 ± 0.25. $H_o = 57 \pm 5$.	Sandage & Tammann (1995)

source to be less than eight arcmin; Bolton (1948) who added to the list of discrete sources; Ryle & Smith (1948) with a good interferometric position of Cas A; Bolton, Stanley, & Slee (1949) who had an early good radio position of Virgo A; and Smith (1951) who provided a very accurate radio position that coincided with the Crab nebula.

The three principal competing radio astronomy groups were strong rivals, each attempting to play a dominant role in the highly charged identification game, but each trusted Baade and Minkowski to play fairly. These optical astronomers kept all the radio-source-position information, communicated to them privately, strictly compartmentalized. It is clear from the content of the 1954 papers by Baade & Minkowski that each radio group had sent their new radio positions to Pasadena before publication, permitting the optical identifications to be made using the Palomar 48-inch and 200-inch telescopes.

In the early 1950s Caltech began a radio astronomy program, and John Bolton, an English physicist who was leading a radio astronomy group at CSIRO in Australia, was invited to develop a Caltech radio observatory. Starting from nothing in 1954, he built the Owens Valley Radio Observatory.

But before choosing that radio-quiet site, Bolton mounted his first antenna on the Palomar grounds, some 500 yards west of the 200-inch dome. He had recruited from Australia Gordon Stanley and J.A. Roberts, and also attracted graduate students and postdoctoral fellows to begin a dominant research program at Palomar before the completion of the permanent site in Owens Valley.

At the beginning, Bolton and his associates were still radio physicists, not astronomers. This led to wonderful accounts of their early encounters with the astronomy of the celestial sphere. The same problems were encountered by Ryle and his colleagues in Cambridge. Baade told of times when he and Minkowski began receiving radio positions from England and Australia from the several radio physics groups. Minkowski would write back asking for the equinox used in reporting the positions, to which questions were asked back as to "what do you mean by the equinox?" Not only were the astronomers rapidly being educated in the new world of radio physics, but the radio scientists soon learned about the celestial sphere.[4]

What was the connection of this new radio astronomy with relativistic astrophysics and, as it turned out, again with Palomar? The connection came through

---

[4]A story as remarkable as it is true, verified by those who worked regularly on Palomar mountain, is the day when Bolton aligned the polar axis of his equatorially mounted, temporary, 25-foot Palomar dish. The antenna could be adjusted in azimuth once the direction of true north was established. The alignment was accomplished one sunny day when Bolton pounded a stake in the ground near the antenna and watched the direction of the shadow as the sun rose toward noon. When his watch, set to Pacific standard time, read 12 noon, Bolton marked the line of the shadow and made his polar axis parallel to it, clamping the antenna azimuth adjustments home. It was soon thereafter that Bolton decided he should have a few astronomy postdoctoral astronomical fellows join the project. One of these was T.A. Matthews, a recent Harvard Ph.D., who later played such an important role in the discovery of quasars.

the fact that the bulk of the radio emission is incoherent synchrotron radiation emitted by high energy electrons (typical energies of approximately 1 Gev), spiraling in weak magnetic fields (approximately 10^{-3} to 10^{-5} gauss). The developments that led to that conclusion are well summarized by Ginzburg & Syrovatskii (1964).

If synchrotron emission is the predominant process, it requires that (1) relativistic electrons and magnetic fields be present in the objects that are parents to the radio sources, and (2) that parts of the radiation must be polarized. Concerning item (*1*), Ginzburg (1953) had shown that relativistic electrons must be formed continuously in the interstellar gas of the galaxy by collisions of relativistic protons, already known in the cosmic rays reaching the earth, on interstellar atoms. Furthermore, in a very important paper, Pikelner (1953) had emphasized that magnetic fields exist in the interstellar medium throughout the galactic system. Concerning item (*2*), polarization of the galactic radio noise was believed to have been measured by the Soviet radio astronomer Razin (1956, 1957).

The Russians had also predicted that if this theory was correct, it might be possible to detect linear polarization of the optical light which in a few cases might be emitted by the synchrotron process. They first applied the argument to the radiation from the Crab. Optical polarization was first discovered in the Crab Nebula (Vashakidze 1954, Dombrovsky 1954). Oort & Walraven (1956) then published what has become a classic paper on observations and theory of the Crab radiation, settling the question of the physical process. Subsequently, Baade (1956c) published exquisite photographs made with the 200-inch that dramatically showed the polarization. These photographs were analyzed by Woltjer (1957). The result showed that the Crab Nebula contains a significant magnetic field and a reservoir of relativistic electrons.

A second prediction was made by Shklovsky (1955) that the jet in the center of M87 (the radio source Virgo A) would show optical polarization demonstrating that the synchrotron mechanism was ubiquitous in galaxies as well as in galactic objects such as the Crab. Acting on Shklovsky's suggestion, Baade (1956b), using the 200-inch, discovered polarization in the jet of that galaxy using a polarizing filter and photographic plates.

The importance of the discovery and an analysis not only of the energetics but also of the origin of the relativistic electrons were set forth in an important paper by Burbidge (1956b). This was the second of a number of papers on the energetics of radio sources by Burbidge (1956a, 1958, 1959) who was by then a Carnegie Fellow of the Mount Wilson and Palomar Observatories. Burbidge was the first theoretician appointee to the Observatories fellowship program, beginning a long and distinguished career in activities connected in many ways with the Observatories and with Caltech.

7.2 The Discovery of Quasars and the AGN Phenomena

The discovery of quasars is too tangled a story to yet be described fully here. A summary of the principal events leading to the discovery of the redshift of 3C 273 is given in the important account by Schmidt (1975, from an original manuscript of 1969). Fuller accounts are given in the monograph by Burbidge & Burbidge (1967) and the reviews by Sullivan (1982, 1984), and Hartwick & Schade (1990).

Soon after the 3C and 3CR Cambridge catalogs of the brightest radio sources were published, efforts were made using interferometry to determine the angular sizes of the various radio sources. Particularly important for the quasar history was the long base-line interferometry made by the Jodrell Bank radio astronomers. The group, led by Henry Palmer, moved three portable cylindrical parabolic antennas with effective diameters of 28 feet over a number of baselines in England. They performed interferometry with the Jodrell Bank 250-foot antenna, measuring the fringe visibilities of more than 350 3CR sources. With baselines ranging from 2200 to 61,100 wavelengths (Allen et al 1962, Rowson 1962), they compiled a list of unresolved 3C sources whose angular diameters were smaller than one arcsec, and therefore had "brightness temperatures" larger than 10^7 K, a telling indication of nonthermal radiation (Palmer 1961). Also from these data, Allen, Hanbury Brown, & Palmer (1962) surmised that many resolved 3CR sources were double, following the previous demonstration by Hanbury Brown & Das Gupta that Cygnus A is double.

T.A. Matthews, a recent addition to Bolton's Caltech radio astronomy group, was fascinated with the possibilities of optical identifications of the 3C sources using the Owens Valley radio interferometer, composed of two 90-foot movable radio dishes. Matthews also had good communications with the Manchester group of Palmer, and he also knew of the long baseline work they were doing.

Palmer had given an account of the Jodrell Bank work at one of the first Texas symposia on high energy astrophysics, which Matthews also attended (circa 1959; my records are incomplete). At that meeting (or shortly thereafter) Palmer gave more details privately to Matthews, including a rather complete list of 3C sources that were unresolved at 61,100 wavelength separation. Palmer's list was a subset of the complete list published later by Allen et al (1962).

In late 1959 Matthews began a program of position determinations in RA, and supervised a Ph.D. thesis program by Read (1963) for declination positions. He also made a working list of positions determined by others, including those by the Palmer group. In 1960 Matthews suggested a collaboration with optical astronomers with access to the 200-inch to begin the final phases of an optical identification program. It was through the invitation of Matthews that I became involved in the optical identification program that was to last far beyond the quasar discovery era, until most of the 3CR radio sources had in fact been identified.

At first we concentrated on Palmer's list of unresolved sources at 61,000-wavelength baseline. There were at least 20 such sources on our initial observing list (which has been lost), all of which later turned out to be radio-loud quasars. I began to take plates with the 200-inch in 1959, centered on Matthews' radio positions. The plates were measured by Matthews in Pasadena, from which the first three star-like identifications of 3C 48, 3C 196, and 3C 286 (Matthews & Sandage 1963) had been made by early 1960. All three were abnormally blue in the ultraviolet as determined from photoelectric photometry begun with the 200-inch in 1960, showing that 3C 48 varied in intensity by 0.4 mag over a one year period.

The object also had a totally abnormal spectrum that I could not decipher. The rest (almost) is history, as set out by Schmidt (1975).

"Almost" refers to the crucial radio position determined from a series of lunar occultations by Hazard, Mackey, & Shimmins (1963) for 3C 273. This position permitted Schmidt to identify the optical image on one of the 200-inch plates that was in Matthews' possession, which Matthews had evidently turned over to Schmidt. The Hazard et al position of the two radio components (one centered on the optical star-like object and the other at the end of the obvious jet) were accurate to 1 arcsec, making the optical identification secure.

The heroes of the quasar discoveries were clearly Palmer for his brilliant measurements with his Jodrell Bank colleagues of the radio angular diameters, Matthews for all aspects of the identification work and for the organization of the joint Owens Valley–Palomar collaboration, Hazard for the position of 3C 273, and Oke (1963) and Schmidt (1963) for their joint discovery of the redshift of 3C 273.[5,6]

Beginning in 1963 the quasar program became frenzied with the 3C 273 redshift discovery, not only at Palomar, but also at Kitt Peak, Lick, and Hawaii, with rivalry between all groups and within each group often leading to severe tension.

Quasars come in all radio and optical luminosities. The most luminous in both optical and radio absolute magnitudes turned out, in fact, to be those identified from the 3CR catalog, which is not surprising when understood based on the ideas of bias in flux-limited catalogs (e.g. Teerikorpi 1997).

After 1963, radio weak (or quiet) quasars were discovered (Sandage 1965). The least luminous of these in both optical and radio luminosity blend with the N galaxies discussed in an earlier section. The continuum continues to the Seyfert galaxies, and finally to the LINERS mentioned earlier.

After the first few identifications of objects in the 3CR radio catalog had been made, the program developed into an attempt to identify all objects in the catalog. It was known by 1970 that perhaps 25 percent of the 3CR sources were not visible to the plate limits of the Palomar 48-inch Schmidt, and a number of these were also not seen on routine exposures at the known radio position with the 200-inch. A review of the remaining problems up to 1974 was given by Kristian & Minkowski (1975).

To complete the work, a cooperative program was begun with Campbell Wade using the new three-element interferometer composed of three movable 85 foot

---

[5] The breaking of the redshift code in the spectrum of 3C 273 was done by Oke and Schmidt together when they combined their wavelength data and realized that they were seeing the Balmer series.

[6] The joys of doing science are boundless. Sometime during the long subsequent campaign with Kristian to optically identify all the remaining 3C sources, I was in communication with D.O. Edge, lead author of the 3CR Cambridge Catalog. Although we had never met, we had written in earlier letters about something concerning 3C 273 (the correspondence is lost). Edge, who is a devout churchgoer, confided that indeed 3C 273 was the radio source he liked best. The reason was that 273 is the number of his favorite anthem in the Methodist Hymnal.

dishes at the National Radio Astronomy Observatory (NRAO) in Greenbank, West Virginia, to obtain highly precise radio positions, from which Kristian and the writer would measure deep 200-inch plates to attempt to complete the 3CR identifications. The program was highly successful (Wade 1970; Wade, Sandage, & Kristian 1970; Brosche, Wade, & Hjellming 1973; Kristian & Sandage 1970; Kristian, Sandage, & Katem 1974, 1978), providing exceptionally faint galaxy candidates for the spectrographic observers of the 1980s, mostly at Lick and Kitt Peak. The redshifts and subsequent photometry extended the Hubble diagram to the new limits that were achieved in the 1980s.

7.3 M82 and NGC 1275 as Prototypes of Starburst Galaxies and Other Violent Events

We conclude this review with a glimpse at a few of the most bizarre objects found over the years with the 200-inch and other large telescopes. The types of galaxies in this category are now commonplace as prototypes of violent processes, but were not so at their discovery.

The subject, which is also a part of the discipline of high energy astrophysics, was reviewed early by Burbidge, Burbidge, & Sandage (1963), where photographs of some of the unusual galaxies are shown and analyses of the physics given.

Most of the types now recognized (Seyfert galaxies such as NGC 1068 and NGC 4151), the new amorphous galaxy type characterized by M82, possible colliding or even merging galaxies as in NGC 1275, jets as in M87, and starburst galaxies, probably with M82 again as a prototype) were shown.

One of the more spectacular examples is the enormous outflow of material from the "starburst" galaxy M82, shown in a composite photograph in Figure 3. The photographs making up this composite were made with the 200-inch using special filters. The energetics in the M82 system have proved to be important in understanding the general class of starburst galaxies, which have been found to be ubiquitous.

We finally comment on either a new type of violent event or a radical type of starburst galaxy where "violent star formation" is taking place. The type examples are NGC 625, NGC 1569, NGC 1705, and M82 in which copious star formation is clearly taking place and in which one or more very bright compact (essentially unresolved) embedded objects exist. A new morphological galaxy class for the type was invented (Sandage and Brucato 1979), called "amorphous." The class is illustrated in the Carnegie Atlas of Galaxies (panels 333–340) where descriptions of the many examples of the bright embedded objects are given.

The importance of these bright objects became apparent when high dispersion spectra were taken by Arp with the 200-inch of the two luminous knots in NGC 1569. The spectra show narrow absorption lines of hydrogen and a definite Balmer discontinuity, characteristic of supergiant stars of type A0. However, the luminosities of the knots are $M_B = -14$, showing that the objects are very young superluminous star clusters (Arp and Sandage 1985). They are probably protoglobular clusters, just formed and with their main sequences intact all the way

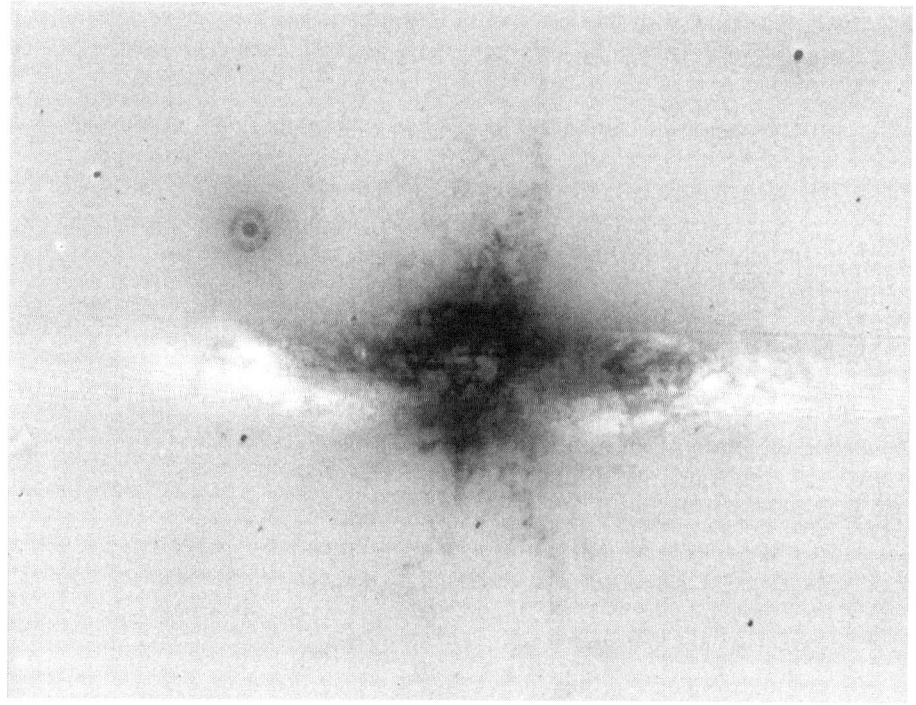

Figure 3 Composite photograph of M82 made by photographically subtracting a continuum image in broad-band yellow wavelengths, from a narrow-band H alpha photograph showing the H alpha emission strung along the minor axis. (Photograph from Lynds & Sandage 1963. Also in the Carnegie Atlas of Galaxies, Panel 333.)

to high luminosities. The phenomenon is ubiquitous in the amorphous types, as described in the Carnegie Atlas (Sandage & Bedke 1994).

The high-energy properties of the violent star formation in NCG 1705 and the formation of its supercluster at absolute magnitude near -14 in NGC 1705 is discussed by Mauer et al (1988, 1992). The two young globular clusters in NGC 1569 were known already in 1952 by Baade and by Hubble from their early 200-inch plates, but the ubiquitous character of the phenomenon has only recently become known, based on the Carnegie/Las Campanas morphological survey of Shapley-Ames galaxies that led to the Carnegie Atlas of Galaxies.

8. CONCLUSION

As a balanced account, this review is dreadfully incomplete concerning all that was accomplished at Palomar in its first 50 years. It has been organized in large measure by showing how the plans for research with the 200-inch telescope that were set out by Baade (1948) and Hubble (1951) were eventually completed—

slowed, to be sure, by the discovery of new subjects not even known when these giants walked the earth. Baade's prediction and final dictum in his prescient 1948 paper came to pass. He wrote:

> "We expect, therefore, that [the program outlined here] will spread over quite a number of years. At times it may even look as if it had been forgotten entirely because everyone is in hot pursuit of a new lead which opened up suddenly. But it will be carried out, because without a secure base we will go astray, and finally become lost."

Clearly, new leads were opened up, and the astronomers at Palomar and elsewhere were in hot pursuit. The result, seen in hindsight in these last months of the old century, is one of wonder, spectacular new understanding, and a knowledge of great accomplishment, made a reality by the insight of dreamers in the 1930s that made the great telescope possible.

ACKNOWLEDGMENTS

This review has been greatly improved by the scientific comments and detailed editing by G.R. Burbidge. However, I claim ownership of the errors that remain. I am also grateful to J.B. Oke for an important conversation concerning the discovery of the redshift in the spectrum of 3C 273.

Visit the Annual Reviews home page at http://www.AnnualReviews.org

LITERATURE CITED

Abell GO. 1958. *Ap. J.* 3:211 (Suppl.)
Abell GO. 1975. In *Galaxies and the Universe*, ed. A Sandage, M Sandage, J Kristian. Chicago: Univ. Chicago Press
Adams W, Joy AH, Humason ML, Brayton AM. 1935. *Ap. J.* 64:225
Allen LR, Anderson B, Conway RG, Palmer HP, Reddish VC, Rowson B. 1962. *MNRAS* 124:477
Allen LR, Hanbury Brown R, Palmer HP. 1962. *MNRAS* 125:57
Aller LH. 1956. *Gaseous Nebulae*, London: Chapman & Hall LTD
Aller LH, Greenstein JL. 1960. *Ap. J.* 5:139 (Suppl.)
Ambartsumian VA. 1952. *Trans. IAU VIII*, p. 673. Cambridge, UK: Univ. Cambridge Press
Arp HC. 1955. *Astron. J.* 60:317
Arp HC. 1958. *Ap. J.* 128:166 (EV Sct)
Arp HC. 1960a. *Astron. J.* 65:404
Arp HC. 1960b. *Ap. J.* 131:322 (CV Mon)
Arp HC. 1962. *Ap. J.* 135:311
Arp HC, Baum WA, Sandage A. 1952. *Astron. J.* 57:4
Arp HC, Baum WA, Sandage A. 1953. *Astron. J.* 58:4
Arp HC, Sandage A. 1985. *Astron. J.* 90:1163
Arp HC, Sandage A, Stephens C. 1959. *Ap J.* 130:80 (DL Cas)
Baade W. 1944a. *Ap. J.* 100:137
Baade W. 1944b. *Ap. J.* 100:147
Baade W. 1948. *PASP* 60:230
Baade W. 1952. *Trans. IAU VIII, Report to Commission 28 of IAU, Rome 1952*, p. 397. Cambridge, UK: Cambridge Univ. Press
Baade W. 1956a. *PASP* 68:5
Baade W. 1956b. *Ap. J.* 123:550
Baade W. 1956c. *B.A.N.* 12:312

Baade W, Hubble E. 1939. *PASP* 51:40
Baade W, Minkowski R. 1954a. *Ap. J.* 119:206
Baade W, Minkowski R. 1954b. *Ap. J.* 119:215
Baade W, Swope HH. 1955. *Astron. J.* 60:151
Baade W, Swope HH. 1963. *Astron. J.* 68:435
Bajtlik S. 1993. In *The Environment and Evolution of Galaxies*, ed. JM Shull, HA Thronson, p. 191. Dordrecht: Kluwer
Baschek B. 1959. *Zs. f. Ap.* 48:95
Baum WA. 1953. *Astron. J.* 58:211
Baum WA, Hiltner WA, Johnson HL, Sandage A. 1959. *Ap. J.* 130:749
Baum WA, Johnson JJ, Oberly CC, Rockwood CC, Strain CV, Tousey R. 1946. *Phys. Rev* 70:781
Bautz LP, Morgan WW. 1970. *Ap. J.* 162:L49
Bigay J. 1951. *Ann. d'Astrophys.* 14:319
Blaauw A, Elvius T. 1964. In *Galactic Structure*, ed. A Blaauw, M Schmidt, p. 589. Chicago: Univ. Chicago Press
Bok BJ. 1946. *MNRAS* 106:61 (Sec. 8)
Bok PF, Boyd CD. 1933. *Bull. Harv. Coll. Obs.* 893:1
Boksenberg A. 1995. In *QSO Absorption Lines*, ed. G Meylan, p. 253. Berlin: Springer Verlag
Bolton JG. 1948. *Nature* 162:141
Bolton JG, Stanley GJ. 1948. *Nature* 161:312
Bolton JG, Stanley GJ, Slee OB. 1949. *Nature* 164:101
Bowen IS. 1948. *Director's Report: Yearbook of the Carnegie Institution of Washington.* 48:3
Bowen IS. 1949. *Director's Report: Yearbook of the Carnegie Institution of Washington.* 49:3
Bowen IS. 1950. *PASP* 62:91
Bowen IS. 1954. *Director's Report: Yearbook of the Carnegie Institution.* 53:22
Bowen IS. 1957. *Director's Report: Yearbook of the Carnegie Institution.* 56:62
Bowen IS. 1960. *Telescopes*, ed. GP Kuiper, BM Middlehurst Chicago: Univ. Chicago Press
Branch D. 1998. *Annu. Rev. Astron. Astropys.* 36:17
Brosche P, Wade CM, Hjellming RM. 1973. *Ap. J.* 183:805
Burbidge GR. 1956a. *Ap. J.* 123:416
Burbidge GR. 1956b. *Ap. J.* 124:416
Burbidge GR. 1958. *Ap. J.* 128:1
Burbidge GR. 1959. *Ap. J.* 129:849
Burbidge EM, Burbidge GR. 1956. *Ap. J.* 124:116
Burbidge GR, Burbidge EM. 1967. *Quasi-stellar Objects.* San Francisco: W.H. Freeman & Co.
Burbidge EM, Burbidge GR, Fowler WA, Hoyle F. 1957. *Rev. Mod. Phys.* 23:547
Burbidge GR, Burbidge EM, Sandage A. 1963. *Rev. Mod. Phys.* 35:947
Cameron LM. 1985. *Astron. Astropys.* 146:59
Carney BW. 1979. *Ap. J.* 233:211
Chamberlain JW, Aller LH. 1951. *Ap. J.* 114:52
Chandrasekhar S. 1951. *Astrophysics*, ed. JA Hynek, p. 598. New York: McGraw Hill
Chincarini G, Rood HJ. 1979. *Ap. J.* 230:648
Cowling T. 1958. *Trans. IAU. X, Report to the Commission on Stellar Constitution*, p. 551. Cambridge, UK: Univ. Cambridge Press
Dicke RH. 1965. *Astron. J.* 70:395
Doig P. 1925. *BAAA* 35:201
Dombrovsky VA. 1954. On the nature of the radiation of the Crab Nebula. *Dokl. Akad. Nauk. SSSR* 94:1021 (In Russian)
Dressler A. 1978. *Ap. J.* 222:23
Eggen OJ. 1965. *Annu. Rev. Astron. Astrophys.* 3:235
Eggen OJ, Lynden-Bell D, Sandage A. 1962. *Ap. J.* 136:748
Eggen OJ, Sandage A. 1959. *MNRAS* 119:254
Eggen OJ, Sandage A. 1962. *Ap. J.* 136:735
Eggen OJ, Sandage A. 1969. *Ap. J.* 158:699
Feast MW, Catchpole RM. 1997. *MNRAS* 286:L1
Feast MW, Walker A. 1987. *Annu. Rev. Astron. Astrophys.* 25:345
Federspiel M, Tammann GA, Sandage A. 1998. *Ap. J.* 495:115
Fernie JD. 1961. *Ap. J.* 133:64
Fessenkov VG. 1952. *Trans. IAU. VIII*, p. 707. Cambridge, UK: Univ. Cambridge Press
Fessenkov VG, Idlis GM. 1959. *IAU Symp. 10* 8:115 (Suppl.)

Freedman WL, Madore BF. 1988. *Ap. J.* 332: L63
Florence R. 1994. *The Perfect Machine.* New York: HarperCollins
Francis MP. 1965. *Astron. J.* 70:449
Gamow G. 1940. In *The Birth and Death of the Sun*, p. 136. New York: Viking
Geller MJ, Peebles PJE. 1976. *Ap. J.* 206:939
Gilvarry JJ. 1953. *Phys. Rev.* 89:1046
Ginzburg VL. 1953. The origin of cosmic rays and radio astronomy. *Uspekhi Fiz. Nauk.* 51:343 (In Russian)
Ginzburg VL, Syrovatskii SI. 1964. *The Origin of Cosmic Rays.* Oxford: Pergamon Press (Trans. 1963 Russian ed.)
Gregory SA, Thompson LA. 1978. *Ap. J.* 222:784
Gregory SA, Thompson LA. 1982. *Sci. Amer.* 246(3):106
Gunn JE, Westphal JA. 1981. *Proc. SPIE.* 290:16
Gunn JE, et al. 1984. *BAAS* 16:477
Gunn JE, Hoessel JG, Oke JB. 1986. (Cluster Catalog) *Ap. J.* 306:30
Gunn JE, Oke JB. 1975. *Ap. J.* 195:255
Harris WE, Allwright JWB, Pritchet CJ, van den Bergh S. 1991. *Ap. J.* 76:115 (Suppl.)
Hartwick FDA, Schade D. 1990. *Annu. Rev. Astron. Astrophys.* 28:437
Heckmann O. 1942. *Theorien der Kosmologie.* Berlin: Springer
Hazard C, Mackey MB, Shimmins AJ. 1963. *Nature* 197:1037
Hertzsprung E. 1913. *Astron. Nach.* 196:201
Helfer HL, Wallerstein G, Greenstein JL. 1959. *Ap. J.* 129:700
Hendricks D. 1939. *PASP* 51:158
Hendricks D, Christie WH. 1939. *Sci. Am.* 161:118
Henyey LG, Keenan PC. 1940. *Ap. J.* 91:625
Herrick S. 1953. *Astron. J.* 58:156
Hey JS. 1946. *Nature* 157:47
Hey JS. 1973. *The Evolution of Radio Astronomy.* New York: Neale Watson Acad. Publ.
Hey JS, Parsons SJ, Phillips JW. 1946. *Nature* 158:234
Hewitt A, Burbidge G. 1987. *Ap. J.* 63:1 (Suppl.)
Hoessel JG. 1980. *Ap. J.* 241:493
Hoessel JG, Gunn JE, Thuan TX. 1980. *Ap. J.* 241:486
Hoessel JG, Schneider DP. 1985. *Astron. J.* 90:1648
Holmberg E. 1950. *Medd. Lunds Obs.* Ser. 2, No. 128
Hoyle F. 1946. *MNRAS* 106:343
Hoyle F. 1952. *Trans. IAU VIII*, Report to Commission 28:397
Hoyle F. 1954. *Ap. J.* Suppl. 1:121
Hoyle F, Sandage A. 1958. *PASP* 68:301
Hubble E. 1925. *Ap. J.* 62:409 (NGC 6822)
Hubble E. 1926a. *Ap. J.* 63:236 (M33)
Hubble E. 1926b. *Ap. J.* 64:321 (classification)
Hubble E. 1929a. *Ap. J.* 69:103 (M31)
Hubble E. 1929b. *Proc. Nat. Acad. Sci.* 15:168
Hubble E. 1934. *Ap. J.* 79:8
Hubble E. 1936a. *Ap. J.* 84:158
Hubble E. 1936b. *Ap. J.* 84:517
Hubble E. 1949. *PASP* 61:121
Hubble E. 1951. *Proc. Am. Phil. Soc.* 51:461 (The Penrose Lecture)
Hubble E, Humason ML. 1931. *Ap. J.* 74:43
Hubble E, Sandage A. 1953. *Ap. J.* 118:353
Hubble E, Tolman RC. 1935. *Ap. J.* 82:302
Humason ML. 1936. *Ap. J.* 83:10
Humason ML, Mayall NU, Sandage A. 1956. *Astron. J.* 61:97 (HMS)
Humason ML, Sandage A. 1957. In *Director's Report: Yearbook of the Carnegie Institution.* 56:62
Irwin JB. 1955. *M. N. Astr. Soc. So. Africa.* 14:38
Irwin JB. 1958. *Astron. J.* 63:197 (U Sgr)
Jansky KG. 1932; 1933; 1937; 1939. *Proc. Inst. Rad. Eng.* 20:1920; 21:1387; 25:1517; 27:763
Jerjen H, Tammann GA. 1993. *Astron. Astrophys.* 276:1
Johnson HL. 1960. *Ap. J.* 131:620
Johnson HL. 1964. *Bol. Tonantzintla Tacubaya* 3:305
Johnson HL. 1965. *Ap. J.* 141:170
Johnson HL, Sandage A. 1956. *Ap. J.* 124:379

Kelson DD, et al. 1996. *Ap. J.* 463:26
Kinman TD. 1959. *MNRAS* 119:538
Kraft RP. 1961. *Ap. J.* 134:616
Kristian J. 1973. *Ap. J.* 179:L61
Kristian J, Minkowski R. 1975. In *Galaxies and the Universe*, eds. A Sandage, M Sandage, J Kristian, p. 199. Chicago: Univ. Chicago Press
Kristian J, Sandage A. 1970. *Ap. J* 162:391
Kristian J, Sandage A, Katem B. 1974. *Ap. J.* 191:43
Kristian J, Sandage A, Katem B. 1978. *Ap. J.* 219:803
Kristian J, Sandage A, Westphal JA. 1978. *Ap. J.* 221:383
Krupp EC. 1974. *PASP* 86:385
Lanzetta KM. 1993. In *The Environment and Evolution of Galaxies*, ed. JM Shull, HA Thronson, p. 237. Dordrecht: Kluwer
Lemaitre G. 1927. *Ann. Soc. Sci. Bruxelles* 47:349
Lemaitre G. 1931. *MNRAS* 91:483 (Translation of the 1927 paper)
Lynds CR, Sandage A. 1963. *Ap. J.* 137:1005
Lubin LM, et al. 1998. *Astron. J.* 116:584
Madore BF, et al. 1998. *Nature* 395:47
Madore BF, Freedman WL. 1991. *PASP* 103:933
Matthews TA, Morgan WW, Schmidt M. 1964. *Ap. J.* 140:35
Matthews TA, Sandage A. 1963. *Ap. J.* 138:30
Mattig W. 1958. *Astron. Nach.* 285:1
Massevich AG. 1954. *Mem. Soc. Roy. Sci. Liege. Ser. IV* 14:267
Massevich AG. 1959. *IAU Symp. 10* (8):89 (Suppl.)
McWilliam A. 1997. *Annu. Rev. Astron. Astrophy.* 35:503
Meauer GH, Freeman KC, Dopita MA. 1988. *Astrophys. Space Sci.* 156:141
Meauer GH, Freeman KC, Dopita MA, Cacciari C. 1992. *Astron. J.* 103:60
Minkowski R. 1949. *PASP* 61:151
Minkowski R. 1960. *Ap. J.* 132:908
Moffet AT. 1975. *Galaxies and the Universe*, ed. A Sandage, M Sandage, J Kristian. Chicago: Univ. Chicago Press
Morgan WW, Lesh J.R. 1965. *Ap. J.* 142:1364
Narlikar JV. 1983. *Introduction to Cosmology*. Boston: Johns and Bartlett
Oemler A. 1974. *Ap. J.* 194:1
Oemler A. 1976. *Ap. J.* 209:693
Oke JB. 1955. *J. R. Astron. Soc. Can.* 49:8
Oke JB. 1963. *Nature* 197:1040
Oke JB. 1969. *PASP* 81:11
Oke JB. 1971. *Ap. J.* 170:193
Oke JB, Postman M, Lubin L. 1998. *Astron. J.* 116:549
Oke JB, Schwarzschild M. 1952. *Ap. J.* 116:317
Oort JH. 1926. *Groningen Publ. No. 40*
Oort JH. 1983. *Annu. Rev. Astron. Astrophys.* 21:373
Oort JH, Walraven Th. 1956. *B.A.N.* 12:285
Osterbrock DE. 1974. *Astrophysics of Gaseous Nebulae*. New York: WH. Freeman & Co.
Palmer HP. 1961. In *Problems of Extragalactic Reasearch, IAU Symp. 15*, ed. GC McVitte, p. 315. New York: Macmillan
Peach J. 1969. *Nature* 223:1140
Peebles PJE. 1969. *Nature* 224:1093
Peterson BA. 1970a. *Astron. J.* 75:695
Peterson BA. 1970b. *Ap. J.* 159:333
Peterson BA. 1970c. *Nature* 227:54
Pettit E. 1954. *Ap. J.* 120:413
Pikelner SB. 1953. The kinematic properties of the interstellar gas in connection with the isotropy of cosmic rays. *Dokl. Akad. Nauk. SSSR* 88:229 (In Russian)
Postman M, Lauer TR. 1995. *Ap. J.* 440:28
Postman M, Lubin L, Oke JB. 1998. *Astron. J.* 116:560
Postmann M, et al. 1996. *Astron. J.* 111:615
Rauch M. 1998. *Ann. Rev. Astron. Astrophys.* 36:267
Razin VA. 1956. Preliminary results of the measurement of the polarization of cosmic radio emission at a wavelength of 1.45 m. *Radiotek. Elek.* 1:846 (In Russian)
Razin VA. 1957. *Uch. Zap. Gork. Gos. Univ. Ser. Fiz.* 35:35
Read RB. 1963. *Ap. J.* 138:1
Reber G. 1940a; 1942. *Proc. Inst. Rad. Eng.* 28:68; 30; 367

Reber G. 1940b; 1944. *Ap. J.* 91:621:100; 297
Reid IN. 1999. *Annu. Rev. Astron. Astrophys.* 37:000
Reiz A. 1954. *Ap. J.* 120:342
Richardson RS. 1948. *PASP* 60:215
Richardson RS. 1949. *PASP* 61:162
Richardson RS. 1965. *Sci. Amer.* 212:106
Robertson HP. 1928. *Phil. Mag.* 5:835
Robertson HP. 1955. *PASP* 67:82
Roman NG. 1954. *Astron. J.* 59:307
Roman NG. 1955. *Ap. J.* 2:195 (Suppl.)
Rood HJ. 1969. *Ap. J.* 158:657
Rood HJ. 1988. *Annu. Rev. Astron. Astrophys.* 26:245
Rowson B. 1963. *MNRAS* 125:177
Rubin VC, Ford WK, Thonnard N, Roberts MS, Graham JA. 1976. *Astron. J.* 81:687
Ryle M, Smith FG. 1948. *Nature* 162:462
Saha A, Sandage A, Tammann GA, Labhardt L, Macchetto FD, Panagia N. 1999. *Ap. J.* (In press)
Sandage A. 1953. *Astron. J.* 58:61
Sandage A. 1957. *Ap. J.* 126:326
Sandage A. 1958a. *Stellar Populations: The Vatican Conference*, ed. DJK O'Connell, 5: 41. Specola Vatican: Ricerche Astronomiche
Sandage A. 1958b. *Ap. J.* 127:513
Sandage A. 1958c. *Ap. J.* 128:150
Sandage A. 1960. *Ap. J.* 131:610
Sandage A. 1961. *Ap. J.* 134:916
Sandage A. 1963. *Ap. J.* 138:863
Sandage A. 1965. *Ap. J.* 141:1560
Sandage A. 1969. *Ap. J.* 158:1115
Sandage A. 1970a. *Ap. J.* 162:841
Sandage A. 1970b. *Phys. Today* 23, Feb issue, p. 34
Sandage A. 1972a. *Ap. J.* 173:485
Sandage A. 1972b. *Ap. J.* 178:1
Sandage A. 1972c. *Ap. J.* 178:25
Sandage A. 1973a. *Ap. J.* 180:687
Sandage A. 1973b. *Ap. J.* 183:711
Sandage A. 1973c. *Ap. J.* 183:731
Sandage A. 1975a. *Ap. J.* 202:563
Sandage A. 1975b. *Galaxies and the Universe*, ed. A Sandage, M Sandage, J Kristian, Ch. 19. Chicago: Univ. Chicago Press
Sandage A. 1976. *Ap. J.* 205:6
Sandage A. 1978. *Astron. J.* 83:904
Sandage A. 1983a. In *Kinematics, Dynamics, & Structure of the Milky Way*, ed. WLH Shuter. p. 315. Dordrecht: Reidel
Sandage A. 1983b. *Astron. J.* 88:1569
Sandage A. 1984. *Astron. J.* 89:621
Sandage A. 1986. *Annu. Rev. Astron. Astrophys.* 24:421
Sandage A. 1988. *PASP* 100:281
Sandage A. 1990a. *J. R. Astron. Soc. Can.* 84:70
Sandage A. 1990b. *Ap. J.* 350:631
Sandage A. 1990c. *Ap. J.* 350:645
Sandage A. 1993a. *Astron. J.* 106:687
Sandage A. 1993b. *Astron. J.* 106:703
Sandage A. 1993c. *Astron. J.* 106:719
Sandage A. 1994. *Ap. J.* 430:1
Sandage A. 1995. *The Deep Universe (Sass Fee Lectures for 1993).* New York: Springer
Sandage A. 1997. *PASP* 109:1193
Sandage A. 1998a. In *Supernovae and Cosmology; Festschrift for G.A. Tammann in Celebration of His 65th Birthday*, ed. L Labhardt, B Binggeli, R Buser, p. 201. Basel: Astron. Inst. Basel
Sandage A. 1998b. In *The Hubble Deep Field* (Workshop report on the ST ScI Deep Field Conference), ed. M Livio, p. 1. Cambridge: Cambridge Univ. Press
Sandage A. 1999. *PASP* (In press)
Sandage A, Bedke J. 1994. Carnegie Atlas of Galaxies, Publ. 638. Washington: Carnegie Institution of Washington
Sandage A, Brucato R. 1979. *Astron. J.* 84:472
Sandage A, Bell RA, Tripicco MJ. 1999. *Ap. J.* (In press)
Sandage A, Cacciari C. 1990. *Ap. J.* 350:645
Sandage A, Eggen OJ. 1959. *MNRAS* 119:278
Sandage A, Hardy E. 1973. *Ap. J.* 183:743
Sandage A, Kristian J, Westphal JA. 1976. *Ap. J.* 205:688
Sandage A, Schwarzschild M. 1952. *Ap. J.* 116:463
Sandage A, Smith LL. 1963. *Ap. J.* 137:1057
Sandage A, Tammann GA. 1968. *Ap. J.* 151:531
Sandage A, Tammann GA. 1969. *Ap. J.* 157:683
Sandage A, Tammann GA. 1974a. *Ap. J.* 190: 295

Sandage A, Tammann GA. 1974b. *Ap. J.* 191: 603
Sandage A, Tammann GA. 1974c. *Ap. J.* 194: 223
Sandage A, Tammann GA. 1974d. *Ap. J.* 194: 559
Sandage A, Tammann GA. 1975a. *Ap. J.* 196: 313
Sandage A, Tammann GA. 1975b. *Ap. J.* 197: 265
Sandage A, Tammann GA. 1976. *Ap. J.* 210:7
Sandage A, Tammann GA. 1987. *A Revised Shapley-Ames Catalog of Bright Galaxies*, Publ. 635. Washington: Carnegie Inst. Wash.
Sandage A, Tammann GA. 1982. *Ap. J.* 256:339
Sandage A, Tammann GA. 1983. In *Large Scale Structure of the Universe, Cosmology, and Fundamental Physics*, ed. G Setti, L Van Hove, p. 127. Geneva: CERN
Sandage A, Tammann GA. 1990. *Ap. J.* 365:1
Sandage A, Tammann GA. 1995. *Ap. J.* 446:1
Sandage A, Tammann GA, Saha A. 1998. In *Third Dark Matter Conference, (UCLA Feb. 1998)*, ed. D. Cline. Physics Reports, 307:1
Sandage A, Walker M. 1955. *Astron. J.* 60:230
Schechter PL, Peebles PJE. 1976. *Ap. J.* 209: 670
Schmidt M. 1963. *Nature* 197:1040
Schmidt M. 1965. *Ap. J.* 141:1
Schmidt M. 1969. 1975. In *Galaxies and the Universe*, ed. A Sandage, M Sandage, J Kristian, Ch. 8. Chicago: Univ. Chicago Press
Schneider DP, Gunn JE, Hoessel JG. 1983a. *Ap. J.* 264:337
Schneider DP, Gunn JE, Hoessel JG. 1983b. *Ap. J.* 268:476
Schonberg M, Chandrasekhar S. 1942. *Ap. J.* 96:161
Schwarzschild M, Howard RF, Searle L. 1955. *Ap. J.* 122:353
Seares FH. 1922. *Trans. IAU, Report to Commission 25 on Photometry*, p 69. London: Imperial College Bookstall
Seares FH, Kapteyn JC, van Rhijn PJ. 1930, *Mount Wilson Catalog of Photographic Magnitudes in Selected Areas 1-139*, Publ. 402. Carnegie Inst.
Seggewiss W. 1998. In *Supernovae and Cosmology; Festschrift for GA. Tammann in celebration of his 65th Birthday*, ed. L Labhardt, B Binggeli, R Buser, p. 101. Basel: Astron. Inst. Univ. Basel
Seldner M, Siebers B, Groth EJ, Peebles PJE. 1977. *Astron. J.* 82:249
Shane CD. 1975. In *Galaxies and the Universe*, eds. A Sandage, M Sandage, J Kristian, Ch. 16, Chicago: Univ. Chicago Press
Shapley H. 1918. *Ap. J.* 48:89
Shapley H, Ames A. 1932. *Ann. Har. Coll. Obs.* 88:43
Shklovsky IS. 1955. *Astron. J. USSR* 32:215
Shklovsky IS. 1956. *Cosmic Radio Waves*. Cambridge: Harvard Univ. Press (Repr. 1960)
Smith FG. 1951. *Nature* 168:555
Spaenhauer A. 1978. *Astron. Astrophys.* 65:313
Stebbins J, Whitford AE. 1952. *Ap. J.* 115:284
Stebbins J, Whitford AE, Johnson HL. 1950. *Ap. J.* 112:469
Stromgren B. 1951. In *Astrophysics*, ed. JA Hynek, p. 172. New York: McGraw Hill
Stromgren B. 1952. *Astron. J.* 57:65
Sullivan WT. 1982. *Classics in Radio Astronomy*. Boston: Reidel-Kluwer
Sullivan WT. 1984. *The Early Years of Radio Astronomy*. New York: Cambridge Univ. Press
Tammann GA, Sandage A. 1968. *Ap. J.* 151:825
Tarengghi M, Tifft WG, Chincarini G, Rood HJ, Thompson LA. 1979. *Ap. J.* 234:793
Teerikorpi P. 1997. *Annu. Rev. Astron. Astrophys.* 35:101
Thuan TX, Gunn JE. 1976. *PASP* 88:543
Trumpler RJ. 1925. *PASP* 37:307
Vashakidze MA. 1954. On the degree of polarization of the radiation of nearby extragalactic nebulae and of the Crab Nebula. *Astr. Cir.* 147:11 (In Russian)
Veron-Cetty M-P, Veron P. 1985. *ESO Sci. Rep. No. 4* Garching: ESO
Wade CM, Sandage A, Kristian J. 1970. *Ap. J.* 162:399
Wade CM. 1970. *Ap. J.* 162:381
Wallerstein G. 1962. *Ap. J.* 6:407 (Suppl.)

Wallerstein G, Carlson M. 1960. *Ap. J.* 160:276

Wallerstein G, Greenstein JL. 1964. *Ap. J.* 139:1163

Wallerstein G, Helfer HL. 1966. *Astron. J.* 71:350

Wallerstein G, Helfer HL, Greenstein JL. 1963. *Ap. J.* 138:97

Wampler EJ, Peach P, Hiltner WA, Kraft RP. 1961. *Ap. J.* 133:895

Westerlund BE, Wall JV. 1969. *Astron. J.* 74:335

Westphal JA, Kristian J, Sandage A. 1975. *Ap. J.* 197:L95

Weymann RJ. 1993. In *The Environment and Evolution of Galaxies*, ed. JM Shull, HA Thronson, p. 213. Dordrecht: Kluwer

Whitford AE. 1936. *Ap. J.* 83:424

Wildey RL, Burbidge EM, Sandage A, Burbidge G. 1962. *Ap. J.* 135:94

Wilkinson A, Oke JB. *Ap. J.* 178:376

Wilson RE. 1939. *Ap. J.* 89:218

Woodbury DO. 1940. *The Glass Giant of Palomar*. New York: Dodd

Woltjer L. 1957. *B.A.N.* 13:301

Wright H. 1966. *Explorer of the Universe: A Biography of George Ellery Hale*. Dutton; Repr. 1994. Woodbury: Amer. Phys. Soc. New York

Wright H. 1952. *Palomar: The World's Largest Telescope*. New York: Macmillan

Zwicky F, Hertzog E, Wild P, Karpowicz M, Kowal CT. 1961–1968 (6 vol. ser.). *Catalog of Galaxies and Clusters of Galaxies*, Pasadena: Calif. Inst. Technol.

ELEMENTAL ABUNDANCES IN QUASISTELLAR OBJECTS: Star Formation and Galactic Nuclear Evolution at High Redshifts

Fred Hamann
Department of Astronomy, University of Florida, 211 Bryant Space Sciences Center, Gainesville, FL 32611-2055; e-mail: hamann@astro.ufl.edu and Center for Astrophysics and Space Sciences, University of California, San Diego, La Jolla, California 92093-0424

Gary Ferland
Department of Physics and Astronomy, University of Kentucky, Lexington, KY 40506-0055; e-mail: gary@pa.uky.edu and Canadian Institute for Theoretical Astrophysics, University of Toronto, Toronto, ON, Canada, M5S 3H8

Key Words quasars, metallicity, emission lines, absorption lines, cosmology

■ **Abstract** Quasar (QSO) elemental abundances provide unique probes of high-redshift star formation and galaxy evolution. There is growing evidence from both the emission and intrinsic absorption lines that QSO environments have roughly solar or higher metallicities out to redshifts >4. The range is not well known, but solar to a few times solar metallicity appears to be typical. There is also evidence for higher metallicities in more luminous objects and for generally enhanced N/C and Fe/α abundances compared with solar ratios.

These results identify QSOs with vigorous, high-redshift star formation—consistent with the early evolution of massive galactic nuclei or dense protogalactic clumps. However, the QSOs offer new constraints. For example, (*a*) most of the enrichment and star formation must occur before the QSOs "turn on" or become observable, on time scales of $\lesssim 1$ Gyr at least at the highest redshifts. (*b*) The tentative result for enhanced Fe/α suggests that the first local star formation began at least ~ 1 Gyr before the QSO epoch. (*c*) The star formation must ultimately be extensive to reach high metallicities; that is, a substantial fraction of the local gas must be converted into stars and stellar remnants. The exact fraction depends on the shape of the initial mass function (IMF). (*d*) The highest derived metallicities require IMFs that are weighted slightly more toward massive stars than in the solar neighborhood. (*e*) High metallicities also require deep gravitational potentials. By analogy with the well-known mass–metallicity relation among low-redshift galaxies, metal-rich QSOs should reside in galaxies (or protogalaxies) that are minimally as massive (or as tightly bound) as our own Milky Way.

1. INTRODUCTION

Quasistellar objects (QSOs or quasars) are valuable probes of the high-redshift universe (Schneider 1998). Their most distant representatives are now measurable out to redshifts of $z \sim 5$ (Schneider et al 1991, Fan et al 1999). In Big Bang cosmologies, these redshifts correspond to times when the universe itself was just ~ 1 gigayear (Gyr) old (see Figure 1).

Understanding the elemental abundances in these distant, early-epoch environments is a major goal of quasar research. Some of the first spectroscopic studies noted simply that quasars contain the usual array of "metals" (elements C, N, O and heavier) produced by stellar nucleosynthesis (Shklovskii 1965, Burbidge & Burbidge 1967). More quantitative estimates of the abundances came later from theoretical work on the broad emission lines, culminating in the important review by Davidson & Netzer (1979; also Baldwin & Netzer 1978, Shields 1976). Those studies inferred solar or slightly higher metal abundances, with large uncertainties. The past two decades have seen considerable progress. Today we have a better theoretical understanding of quasar environments, and greater abilities to both observe and model a range of abundance diagnostics.

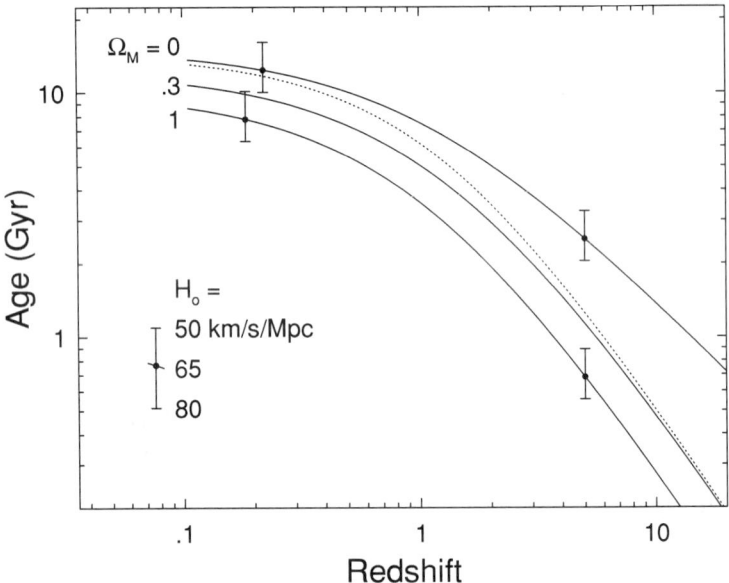

Figure 1 Redshift versus age of the universe in Big Bang cosmologies. The three *solid curves* correspond to $H_o = 65$ km s^{-1} Mpc^{-1}, $\Omega_\Lambda = 0$, and $\Omega_M = 0, 0.3$, or 1. The *dotted curve* uses the same H_o but with $\Omega_\Lambda = 0.7$ and $\Omega_M = 0.3$. The "error" bars show the range of ages possible for H_o between 50 and 80 km s^{-1} Mpc^{-1} (see Carroll et al 1992).

We also have renewed motivation from the growing evidence that links quasars to galaxies. See, for example, Kormendy et al (1998) and Magorrian et al (1998) Laor (1998) for black hole–host galaxy mass correlations; Chatzichristou et al (1999), Hines et al (1999), McLeod et al (1999), Boyce et al (1999), McLure et al (1999), Aretxaga et al (1998), Carballo et al (1998), Bahcall et al (1997), Miller et al (1996), and McLeod & Rieke (1995) for direct observations of QSO hosts; Cavaliere & Vittorini (1998), Shaver et al (1998), Terlevich & Boyle (1993), Boyle & Terlevich (1998), and Osmer (1998) for arguments based on QSO number-density evolution; McCarthy (1993), Saikia & Kulkarni (1998), Haas et al (1998), Brotherton et al (1998a) for radio galaxy-radio quasar unification schemes; and Turner (1991), Haehnelt & Rees (1993), Loeb & Rasio (1994), Katz et al (1994), Haehnelt et al (1998), Haiman & Loeb (1998), and Taniguchi et al (1999) for theoretical links between QSOs and galaxy evolution. If quasars reside, as expected, in galactic nuclei or dense protogalactic clumps, their abundances could yield unique constraints on the evolution of those environments. In particular, quasar abundances can indirectly probe the star formation that came before QSOs, possibly the first stars forming in massive collapsed structures after the Big Bang. Other studies, involving for example the "Lyman-break" objects (Steidel et al 1998, Connolly et al 1997) or the damped-Lyα or Lyα "forest" absorbers in QSO spectra (Pettini et al 1997, Lu et al 1998, Rauch 1998), probe galaxies and metal enrichment on much larger scales. The quasar results (on galactic nuclei) should therefore provide an important piece to the overall puzzle of high-redshift star formation and galaxy evolution.

Here we review the status and implications of quasar abundance work. We regret that many interesting related topics must be excluded; in particular, we will consider the quasars themselves to be simply light sources surrounded by emitting and absorbing gas. We discuss three abundance diagnostics that are readily observable in QSOs at all redshifts: the broad emission lines (BELs), the broad absorption lines (BALs), and the intrinsic narrow absorption lines (NALs). We include just these intrinsic spectral features to measure the abundances near QSO engines. We thereby exclude measures of more distant environments, such as the halos of the host galaxies, nearby cluster galaxies or cosmologically intervening material. If we could choose a maximum radius for the location of intrinsic gas, it would correspond to the size of the putative star clusters or galactic spheroids surrounding QSOs—perhaps a few hundred pc to a few kpc. Any material ejected from this region would also qualify as intrinsic. Ultimately, our interpretation of the abundance data depends critically on the location of the emitting/absorbing gas and its relationship to the quasar/host galaxy environment.

We begin with separate discussions of each abundance probe (Sections 2–3), followed by a summary of the overall results (Section 4). We then consider the plausible enrichment schemes, making a case for normal chemical evolution by stars in galactic nuclei (Section 5). Within that scheme, we use results from galactic studies (Section 6) to derive further implications of the QSO abundances (Section 7). We close with a brief outline for future work (Section 8).

In several sections below we present results of photoionization calculations performed with the numerical code Cloudy (version 90.05, Ferland et al 1998). This code is freely available on the World Wide Web (http://www.pa.uky.edu/~gary/cloudy/). Finally, we define solar abundances based on the meteoritic results in Grevesse & Anders (1989).

2. EMISSION LINE DIAGNOSTICS

2.1 Overview

Quasars are surprisingly alike in their emission line spectra (Osmer & Shields 1999 and references therein); for example, the range of intensity ratios is far less than in galactic nebulae. Figure 2 shows a composite UV spectrum that is fairly typical of QSOs without strong BALs. The object-to-object similarities span the full range of QSO redshifts, $0.1 \lesssim z \lesssim 5$, more than 4 orders of magnitude in luminosity, and billions of years in cosmological look-back time. The emission lines are either insensitive to the metal abundances, or QSOs have similar abundances across enormous ranges in other parameters. We argue that the truth involves a bit of both explanations.

We focus on the BELs in the rest-frame UV because they are present and relatively easy to measure in all QSOs at all redshifts. Furthermore, unlike the narrow emission lines, there is no ambiguity about their close physical connection to QSO engines (Davidson & Netzer 1979).

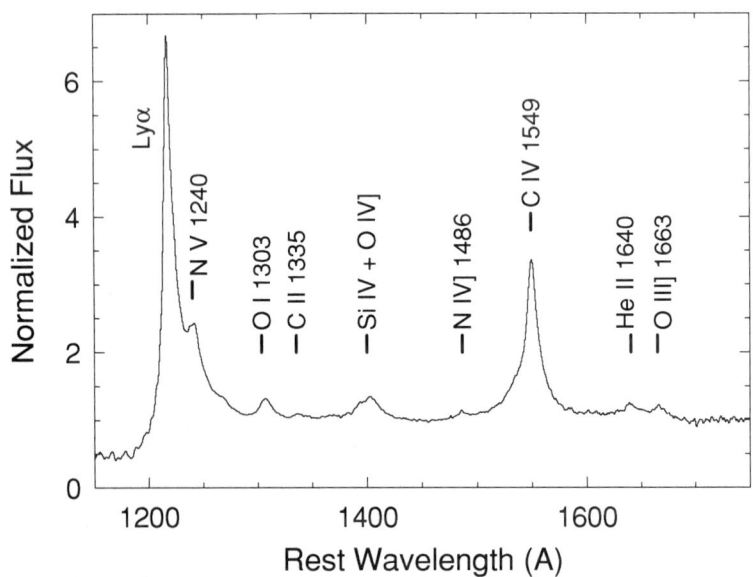

Figure 2 Normalized mean spectrum of 13 QSOs at $z > 4$ (from Shields et al 1997). Prominent BELs are labeled.

2.2 Origin of the Broad Emission Lines

Quasar emission-line research is an example of the inverse problem in astrophysics. We know the answer—the observed spectrum of a quasar, and we are trying to understand the question—the conditions that created it. Any model of the line-forming regions will have uncertainties related to uniqueness, but these can be minimized by considering the astrophysical context and by limiting the models to essential properties. The essential properties of the BEL region (BELR) are as follows:

1. The BELR is photoionized. The main evidence for photoionization is that the emission-line spectra change in response to changes in the continuum, with lag times corresponding to characteristic radii of the BELR (Peterson 1993). The shape of the ionizing continuum is a fundamental parameter and is in itself an area of active research (e.g. Zheng et al 1997, Korista et al 1997a, Brunner et al 1997, Laor 1998). Below we present calculations using simple power laws between 1 μm and 100 keV, and we describe results that do not depend strongly on the continuum shape.

2. The BELR spans a range of distances from the central object. The line variability or reverberation studies just mentioned find different lag times for different ions. Highly ionized species tend to lie closer to the continuum source. Overall, the radial distances scale with luminosity, such that $R \approx 0.1(L/10^{46} \text{ ergs s}^{-1})^{1/2}$ pc is a typical value (Peterson 1993).

3. The BELR has a wide range of densities and ionization states. The range in ionization follows simply from the lines detected, from OI λ1303 to at least NeVIII λ774 (Hamann et al 1998). The range in density comes mainly from the estimated radii and photoionization theory (e.g. Ferland et al 1992). Clouds with densities from 10^8 to $>10^{12}$ cm^{-3} may be present. [We use the term "cloud" loosely, referring to some localized part of the BELR but not favoring any particular model or geometry (see Arav et al 1998, Mathews & Capriotti 1985)]. Any given object could have a broad mixture of BELR properties (Baldwin et al 1995, 1996).

4. The BELR probably has large column densities. Large column densities, typically $N_H \gtrsim 10^{23}$ cm^{-2}, were originally used in BELR simulations to produce a wide range of ionizations in single clouds (Kwan & Krolick 1981, Ferland & Persson 1989). These large column densities might not apply globally because we now know that different lines form in different regions. In our calculations below, we truncate the clouds at the hydrogen recombination front, with the result that different clouds/calculations can have different total column densities. However, the truncation depths are in all cases large enough to include the full emission regions of the relevant lines.

5. Thermal velocities within clouds are believed to dominate the local line broadening and radiative transfer. The observed line widths are thus caused

entirely by bulk motions of the gas. This issue is important because (a) continuum photoexcitation (pumping) can overwhelm other excitation processes if the local line broadening (e.g. microturbulence) is large, and (b) the line optical depths and thus photon escape probabilities (see below) vary inversely with the amount of line broadening. The interplay between these factors makes it hard to predict the behavior of a given line without explicit calculations. Shields et al (1995) plot some line-strength behaviors for the particular case of low-column-density clouds. One argument against significant microturbulence involves the Lyα/Hβ intensity ratio. Simple recombination theory predicts a ratio of $\sim$34 (Osterbrock 1989), although the observed value is far smaller, closer to 10 (Baldwin 1977a). This discrepancy is worsened by microturbulence (Ferland 1999). The solution probably requires severely trapped Lyα photons resulting from large optical depths at thermal line widths (see also Netzer et al 1995).

2.3 Strategies for Abundance Work

There is much that is unknown about QSO line-forming regions. We do not, for example, have a clear picture of the overall geometry or the spatial variations of key parameters. But we do not need this information for abundance work. The emission lines from photoionized clouds are controlled fundamentally by the energy balance and microphysics. The strategy for abundance studies is to identify line ratios that have significant abundance sensitivities and minimal dependence on other unknown or uncertain parameters. For example, we can minimize the sensitivity to large-scale geometric effects by comparing lines that form as much as possible in the same gas. Detailed simulations are often needed to identify useful line ratios and quantify their parameter sensitivities. Simple analytic expressions can be used for some applications, and they can help, in any case, provide physical insight into the emission-line behaviors.

Below we review some of the basic principles of photoionization and emission line formation. See Osterbrock (1989) and Mihalas (1978) for further reviews, Davison & Netzer (1979), Kwan & Krolick (1981), Ferland & Shields (1985), and Netzer (1990) for applications to QSOs, and Ferland et al (1998) for more on the numerical simulations and input atomic data.

2.4 Basics of Abundance Analysis

2.4.1 Collisionally Excited Lines Collisionally excited lines form by the internal excitation of an ion after electron impact. Their emissivities, that is, the energy released per unit volume and time, follow from the statistical equilibrium of the energy levels. For example, the equilibrium (detailed balance) equation for a 2-level atom is

$$n_l n_e q_{lu} = n_u(\beta A_{ul} + n_e q_{ul}) \quad [\text{cm}^{-3}\ \text{s}^{-1}], \qquad (1)$$

where n_e is the electron density, β is the probability for line photons escaping the local region ($0 \leq \beta \leq 1$), A_{ul} is the spontaneous decay rate, n_u and n_l are the number densities in the upper and lower states, and q_{lu} and q_{ul} are the upward and downward collisional-rate coefficients, respectively. Note that $\beta \sim \tau^{-1}$ when $\tau \gg 1$, where τ is the line-center optical depth (Frisch 1984). For most applications, the ions are mainly in their ground state and n_l is approximately the ionic density. The line emissivity is

$$\epsilon_{coll} = n_u \beta A_{ul} h\nu_o = n_l \beta A_{ul} h\nu_o \left(\frac{n_e q_{lu}}{\beta A_{ul} + n_e q_{ul}} \right) \quad [\text{ergs cm}^{-3} \text{ s}^{-1}], \quad (2)$$

where ν_o is the line frequency. This emissivity has a strong temperature dependence because $q_{ul} \propto T^{-1/2}$ and $(q_{lu}/q_{ul}) = (g_u/g_l) \exp(-h\nu_o/kT)$, where g_u and g_l are the statistical weights. In the high-density limit we have

$$\epsilon_{coll} = n_l \beta A_{ul} h\nu_o \frac{g_l}{g_u} \exp\left(-\frac{h\nu_o}{kT}\right), \quad (3)$$

and the levels are said to be thermalized. Line thermalization, where ϵ_{coll} no longer depends on the transition strength, additionally requires $\tau \gg 1$. (A_{ul} and τ are both proportional to the oscillator strength, which therefore drops out of the factor $\beta A_{ul} \approx A_{ul}/\tau$ in Equation 3 if $\tau \gg 1$.) At low densities we have

$$\epsilon_{coll} = n_l n_e q_{lu} h\nu_o \propto n_l n_e T^{-1/2} \exp\left(-\frac{h\nu_o}{kT}\right). \quad (4)$$

Note that ϵ_{coll} scales here like the density squared, compared with the linear dependence in Equation 3. The critical density, n_{crit}, between these two limits is the density at which the two terms in the denominator of Equation 2 are equal,

$$n_{crit} = \frac{\beta A_{ul}}{q_{ul}} \approx \frac{A_{ul}}{\tau q_{ul}}, \quad (5)$$

where the approximate relation holds only if $\tau \gg 1$. Physically, n_{crit} is the density at which the upper level is as likely to be de-excited by collisions as by radiative decay. Note that significant optical depths have the effect of lowering n_{crit}. Also note that transitions with very different oscillator strengths (but similar collision strengths) will have similar n_{crit} in the limit $\tau \gg 1$ (because A_{ul}/τ is independent of oscillator strength).

2.4.2 Recombination Lines The most prominent recombination lines belong to HI, HeI, and HeII, with HI Lyα being typically strongest. These lines form by the capture of free electrons into excited states, followed by radiative decay to lower states. In the simplest case, in which every photon escapes freely and competing processes are unimportant, the emissivity is

$$\epsilon_{rec} = n_i n_e \alpha_{rad} h\nu_o \propto n_i n_e T^{-1} \quad [\text{ergs cm}^{-3} \text{ s}^{-1}], \quad (6)$$

where α_{rad} is the radiative recombination coefficient into the upper energy state and n_i is the number density of parent ions. The temperature dependence is approximate and derives from α_{rad} (see Osterbrock 1989).

2.4.3 Deriving Abundance Ratios
These two types of lines can be combined to form three types of ratios for abundance analysis. The general idea is that, for any element a in ion stage i, the observed line intensity $I(a_i)$ is proportional to the density in that ion, $n(a_i)$, times a function of the overall gas density and temperature $F(a_i, T, n)$, such that $I(a_i) = n(a_i) F(a_i, T, n)$. The ionic abundance ratios are then given by

$$\frac{n(a_i)}{n(b_j)} = \frac{I(a_i)}{I(b_j)} \frac{F(b_j, T, n)}{F(a_i, T, n)}. \tag{7}$$

Abundance studies require line pairs for which the ratio of the two functions F is nearly constant or has a limiting behavior that still allow for abundance constraints. The last step is to convert the ionic abundances into elemental abundances, which we express logarithmically relative to solar ratios as

$$\left[\frac{a}{b}\right] = \log\left(\frac{n(a_i)}{n(b_j)}\right) + \log\left(\frac{f(b_j)}{f(a_i)}\right) + \log\left(\frac{b}{a}\right)_\odot, \tag{8}$$

where $f(a_i)$ is the fraction of element a in ion stage i, etc. (Our notation here is based on the usual definition of logarithmic abundances normalized to solar ratios, $[a/b] \equiv \log(a/b) - \log(a/b)_\odot$.) The middle term on the right-hand side is the ionization correction (IC), which can be deduced from numerical simulations or set to zero (in the log) based on the similarity of the species (Peimbert 1967). Another strategy is to compare summed combinations of lines from different ion stages so that IC tends to zero on average (Davidson 1977).

Ratios of pure recombination lines are simplest because they are least sensitive to the temperature and density. In principle, we could derive the He/H abundance from these ratios. However, in practice, all of the strong HI and HeI recombination lines in QSOs, most notably Lyα, are affected by collisions and thermalization effects. Moreover, because H^0, He^+ and He^{+2} have different ionization energies, they need not be cospatial in the BELR, and their levels of ionization depend on the different numbers of photons available to produce each ion (Williams 1971). As a result, the H and He recombination spectra are most useful as indicators of the shape of the ionizing continuum (e.g. Korista et al 1997a). We do not expect substantial deviations from solar He/H abundances anyway, based on normal galactic chemical evolution, and the BEL data are grossly consistent with that expectation.

The second possibility involves the ratio of collisional to recombination lines. These ratios have strong temperature dependences (compare Equations 3 and 4 to Equation 6). Nonetheless, they can still be used for abundance work if the temperature sensitivities are quantified by explicit calculations. For example, there is an upper limit on the line ratio NV λ1240/HeII λ1640 related to the maximum temperature attained in photoionized BELRs. That upper limit sets a firm lower limit on the N/He abundance (Section 2.6.3 below).

The last ratio and the one most often used involves two collisionally excited lines. Roughly a dozen collisionally excited BELs are routinely measured in the UV spectra of quasars, so there is a variety of possibilities. The ideal collisionally excited line pair would have similar excitation energies, so their ratio has a small $h\Delta v_o/kT$ and thus a small temperature dependence (Equations 3 and 4). Similar values of n_{crit} and similar ionization energies further minimize the sensitivities to density and BELR structure. Well-chosen ratios that meet these criteria can sometimes provide abundance estimates without recourse to detailed simulations (e.g. Shields 1976; see Section 2.6.1 below).

2.5 Photoionization Simulations

A photoionized cloud is essentially a large-scale fluorescence problem. Energy comes into the cloud via continuum radiation, is converted into kinetic energy by the photoejection of electrons, and then leaves the cloud by various emission processes —mainly line radiation. The lines are thus the primary coolants; their total intensity depends on energy conservation and not at all on particular cloud properties.

In general situations, for example dense environments like BELRs, individual line strengths can be governed by a number of competing processes and by feedback related to the cloud structure and energy balance. Detailed calculations are needed to simultaneously consider a complex network of coupled processes. Here we describe some basic results for the line formation and ionization structure in realistic BELR clouds.

2.5.1 Parameters of Photoionization Equilibrium
The fundamental parameters in photoionization simulations are the shape and intensity of the ionizing continuum, and the space density, column density, and chemical composition of the gas. The flux of hydrogen-ionizing photons at the illuminated face of a cloud is

$$\Phi(H) \equiv \int_{v_{LL}}^{\infty} \frac{f_v}{hv} dv \quad [\text{photons cm}^{-2}\,\text{s}^{-1}], \tag{9}$$

where f_v is the energy flux density and v_{LL} is the frequency corresponding to 1 Rydberg. A dimensionless ionization parameter $U \equiv \Phi(H)/cn_H$ is often used instead, where c is the speed of light and n_H is the total hydrogen density (H^0+H^+). U is proportional to the level of ionization and has the advantage of stressing homology relations between clouds with the same U but different $\Phi(H)$ and n_H. This simplification is appropriate if we are interested in just the gross ionization structure or in emission lines that are not collisionally suppressed. More generally, we can use either $\Phi(H)$ or U as long as the density is also specified.

2.5.2 A Computed Structure
Figure 3 shows the ionization structure of a typical BELR cloud photoionized by a power-law spectrum with $\alpha = -1.5$, where $f_v \propto v^\alpha$. The hydrogen recombination front occurs at a depth of $\sim 10^{12}$ cm, whereas the He^{+2}–He^+ front is near 10^{11} cm. Note that there is significant ionization beyond the nominal H^0–H^+ front, owing to penetrating X-rays and Balmer continuum photoionizations out of the $n = 2$ level in H^0 (Kwan & Krolick 1981).

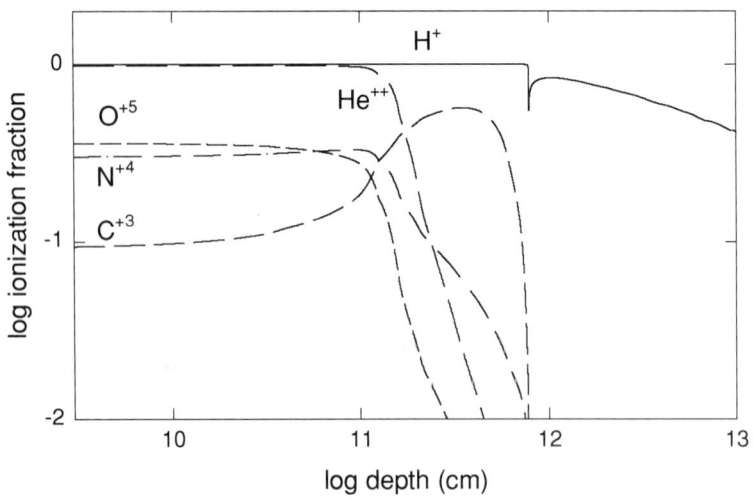

Figure 3 Ionization structure for a nominal BELR cloud with $n_H = 10^{10}$ cm^{-3}, $\log U = -1.5$, and solar abundances.

Some important low-ionization lines like FeII form in that region. The ionization fractions in plots like Figure 3 help us identify ions, such as O^{+5}, N^{+4} and He^{+2}, that are roughly co-spatial and thus good candidates for abundance comparisons.

2.5.3 An Example: the CIV λ1549 Equivalent Width   CIV λ1549 is one of the strongest collisionally excited lines in quasar spectra. The left panel of Figure 4

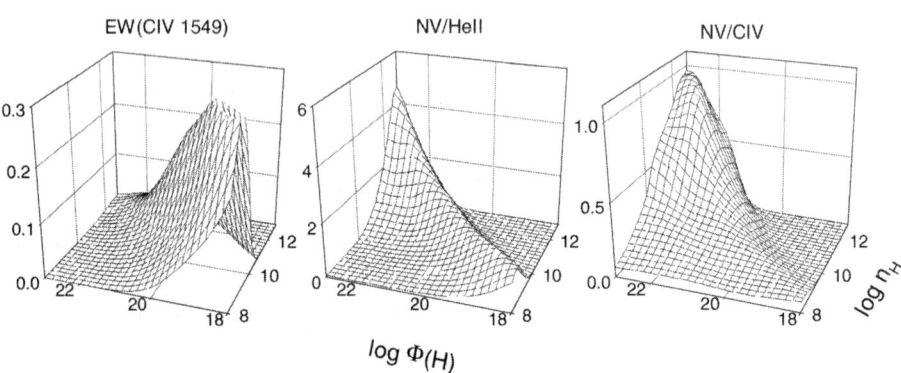

Figure 4 Predicted equivalent width (EW) of CIV λ1549 as a function of the cloud density, n_H, and incident ionizing flux $\Phi(H)$. The equivalent width here is dimensionless (line flux/$\nu_o f_{\nu_o}$ in the continuum) and applies for the hypothetical case of global covering factor unity. Flux ratios for NV λ1240/HeII λ1640 and NV/CIV are also shown. Other parameters are the same as those in Figure 3.

shows how its predicted equivalent width changes with the density (n_H) and ionizing flux [$\Phi(H)$; see Korista et al 1997b for many more similar plots]. Powerful selection effects are clearly at work; the line radiates efficiently over just a narrow range of parameters. Varying $\Phi(H)$ is equivalent to moving the cloud closer or farther from the continuum source. The line is weak at large values of $\Phi(H)$ because carbon is too highly ionized, and at low values of $\Phi(H)$ because carbon is too neutral. The line strength also changes with the gas density. When the density is above n_{crit}, the line is collisionally suppressed and other permitted lines take over the cooling. When the density is low, the line weakens as the many forbidden and semiforbidden lines become efficient coolants, and the gas temperature declines. The line is most prominent at $n_H \approx 10^{10}$ cm^{-3} and $\log U \approx -1.5$, which are the canonical BELR parameters deduced over 20 years ago from analysis of the CIV emission (Davisoun & Netzer 1979).

These selection effects exist whenever we observe an emission line. Baldwin et al (1995) showed that a typical quasar BEL spectrum might result simply from selection effects operating in BELRs that have simultaneously a wide range of cloud properties (e.g. density and distance from the QSO). Numerical simulations can identify pairs of lines with similar selection behaviors so that their ratios are insensitive to the ranges or specific values of the parameters.

2.5.4 Line Dependence on Continuum Shape Figure 5 shows a series of calculations with different incident spectral shapes. The actual shape of the ionizing

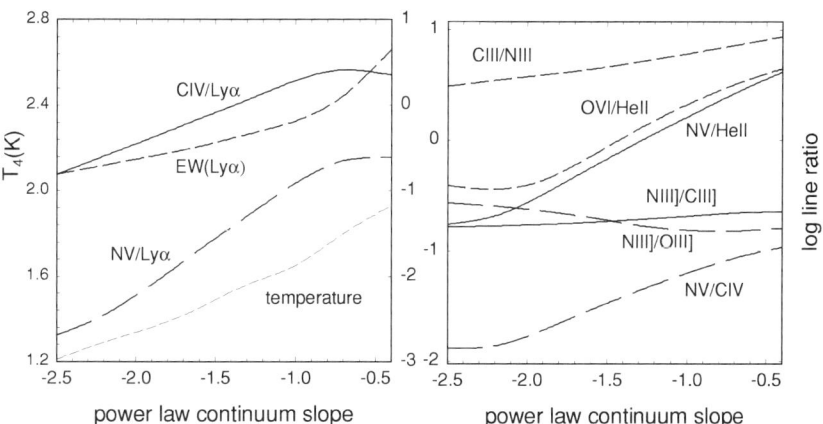

Figure 5 Predicted line flux ratios, gas temperatures ($T_4 = T/10^4$ K in the O^{+2} zone, i.e. weighted by the O^{+2} fraction), and dimensionless equivalent widths in Lyα (EW, as in Figure 4) are plotted for clouds photoionized by different power-law spectra. Other parameters are the same as those in Figure 3. The lines are CIII λ977, NIII λ991, OVI λ1034, NV λ1240, CIV λ1549, HeII λ1640, OIII] λ1664, NIII] λ1750, and CIII] λ1909.

continua in QSOs is a complicated issue, but the UV-to-X-ray slopes are roughly consistent with $\alpha \sim -1.5$, near the center of the range shown (see Laor 1999, Korista et al 1997a for recent discussions). The results in Figure 5 mainly reflect the conservation of energy in the cloud. Harder spectra (less negative α) provide more heating per photoionization, leading to higher temperatures. The increased heating requires more line cooling via collisionally excited lines like CIV. The ratio of a collisionally excited line to a recombination line, such as CIV/Lyα, is proportional to the cooling per recombination or equivalently the heating per photoionization (Davison & Netzer 1979). Such ratios therefore have a strong continuum-shape dependence. The strengths of collisionally excited lines relative to the adjacent continuum (i.e. their equivalent widths) also depend on the spectral slope because of the temperature sensitivity and because the continuum below the lines might be very different from that controlling the ionization. Ratios of collisionally excited lines, such as NV/CIV, can similarly depend on the spectral shape if their ionization or excitation energies are different. In dense BELRs, these simple behaviors can be moderated by other effects. For example, the Lyα equivalent width increases with spectral hardening at fixed U (Figure 5) because it has a significant collisional (temperature-sensitive) contribution.

2.5.5 Line Dependence on Abundances The left-hand panel of Figure 6 shows a series of calculations for clouds with different metallicities, Z (scaled from solar and preserving solar ratios among the metals). The strengths of the collisionally excited lines relative to Lyα change little with Z. In particular, CIV/Lyα varies negligibly for $0.1 \lesssim Z \lesssim 30\ Z_\odot$ (see also Hamann & Ferland 1993a). We have already noted that these ratios are more sensitive to the continuum shape (Section

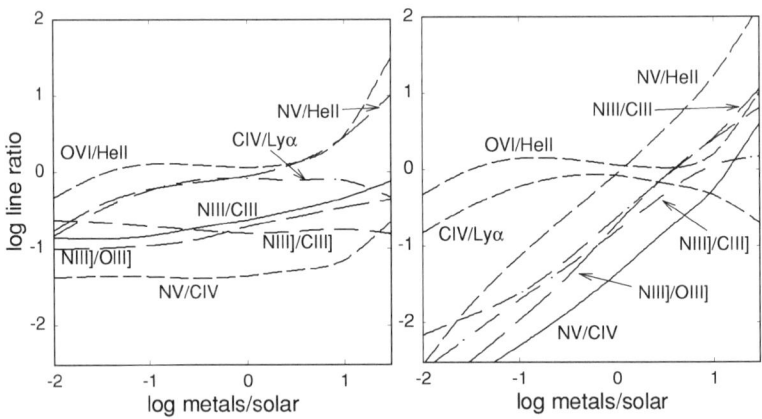

Figure 6 Predicted line flux ratios for photoionized clouds with different metallicities Z. All of the metals are scaled together (preserving solar ratios) in the left-hand panel, whereas nitrogen is scaled selectively like Z^2 in the right panel. Other parameters are the same as those in Figure 3. See Figure 5 for line notations.

2.5.4). Their lack of sensitivity to Z can be traced to feedback in the energy balance. As the metal abundances grow, the line cooling increases. The growing metallicities, which might otherwise increase the metal line strengths, are thus balanced in real clouds by lower temperatures—with the result that the total metal line flux stays constant. This feedback is especially important for strong lines, like CIV, that by themselves control a large fraction of the cooling. Weak lines respond better to abundance changes. At low metallicities ($Z \lesssim 0.02\,Z_\odot$), none of the metal lines are important coolants and their overall strengths do scale with Z.

Another factor in the line behaviors at high Z is the increasing bound-free continuum absorption by metal ions. The metals absorb a larger fraction of the far-UV flux at high Z, such that the H and He recombination lines become somewhat weaker. This effect dominates the high-Z rise in OVI/HeII and NV/HeII in Figure 6.

The right-hand panel in Figure 6 shows the same line ratios as before, but in this case nitrogen is scaled such that N/H $\propto Z^2$ (where N/H is solar at $Z = Z_\odot$). This selective scaling is based on the expected secondary nucleosynthesis of nitrogen (Section 6 below). Shields (1976) noted that this abundance behavior should occur in QSOs by analogy with its direct observation in galactic HII regions. Figure 6 shows that it leads to a strong metallicity dependence for line ratios involving nitrogen. This strong dependence is possible because the N lines do not control the cooling.

2.6 Abundance Diagnostics and Results

2.6.1 Intercombination Lines Shields (1976) proposed using various collisionally excited intercombination (semiforbidden) lines to derive metal-to-metal abundance ratios in QSOs. He emphasized the strengths of NIII] $\lambda1750$ and NIV] $\lambda1486$ compared with OIII] $\lambda1664$, CIII] $\lambda1909$ and CIV $\lambda1549$ as potential diagnostics of the overall metallicity. As noted above, the metallicity dependence stems from the expected Z^2 scaling of N via secondary nucleosynthesis (also Section 6 below). Shields selected lines with similar ionization and excitation energies, so that their ratios are insensitive to the uncertain temperature, ionization, and geometry. Comparisons with the measured line ratios in QSOs (see also Davidson 1977, Baldwin & Netzer 1978, Davidson & Netzer 1979, Osmer 1980, Uomoto 1984) suggested that N/C and N/O are often solar or higher, consistent with solar or higher metallicities. Gaskell et al (1981) extended this analysis to SiIII] $\lambda1892$ and other lines to show that the refractory elements cannot be substantially depleted by dust in BELRs.

One drawback of the intercombination lines is that most of them are weak and therefore difficult to measure. Nonetheless, the best recent measurements (Wills et al 1995, Laor et al 1995, Boyle 1990, Baldwin et al 1996) support the earlier results. It is now possible to gather even more data for these lines over a range of redshifts. A note of caution is that the strong feature generally attributed to CIII] $\lambda1909$ can have large contributions from other lines (Laor et al 1995, 1997;

Baldwin 1996), so that ratios like NIII]/CIII] might systematically underestimate N/C if line blending is not accounted for.

A more serious concern is that the early theoretical work did not consider the range of high densities now believed to be present in the BELR (Section 2.2). The intercombination lines probably form at or near their critical densities (typically 3×10^9 to 3×10^{11} cm^{-3} for $\beta = 1$ in Equation 5). Lines with different n_{crit} could have different degrees of collisional suppression. For example, the calculated results with $n_e \approx n_H = 10^{10}$ cm^{-3} in Figures 5 and 6 favor large NIII]/CIII] at a given N/C abundance because CIII] is collisionally suppressed above its $n_{crit} \approx 3 \times 10^9$ cm^{-3}. If there is a range of densities, lines with different n_{crit} might form in different regions (even if they have similar ionizations), leading to a geometry dependence. Nonetheless, line ratios involving similar n_{crit} and similar sensitivities to other parameters, such as NIII]/OIII], could still be robust abundance indicators when they are measurable. More theoretical work is needed to explore the parameter sensitivities and selection effects that can influence these lines in complex BELRs.

2.6.2 Permitted Lines   There are several possibilities for abundance diagnostics among the permitted UV lines. Figure 6 shows that NIII λ991/CIII λ977 and NV λ1240/CIV λ1549 should be good tracers of N/C. Another possibility is NV/HeII λ1640, or perhaps NV/(CIV+OVI λ1034). The NV, OVI and HeII lines form in overlapping regions (Figure 3), as do NIII and CIII, so their flux ratios should be insensitive to the global BELR structure. Also, as noted above, the N lines are not important coolants and are thus responsive to abundance changes. There are practical problems with most of these lines, however; NV is blended with Lyα; CIII, NIII, and HeII are weak; and CIII, NIII, and OVI lie in the "forest" of intervening Lyα absorption lines. Nonetheless, improvements in the quality of data (for example, high resolution and high signal-to-noise spectra in the Lyα forest) are permitting increasingly accurate measurements of these lines in large QSO samples.

2.6.3 NV/HeII and NV/CIV   Some of the first studies of QSO samples noted that NV λ1240 is often stronger than predicted by photoionization models using solar abundances (Osmer & Smith 1976, 1977). The NV/HeII and NV/CIV ratios have since received particular attention as abundance diagnostics (Hamann & Ferland 1992, Hamann & Ferland 1993a, Hamann & Ferland 1993b, Ferland et al 1996). Figure 7 shows the measured ratios in these lines for QSOs at different redshifts (from Hamann & Ferland 1993b and Hamann et al 1997a, with some new data and modifications based mainly on Wills et al 1995 and Baldwin et al 1996). NV/HeII is the ratio of a collisional to recombination line, with the expected strong temperature dependence (Section 2.4). Calculations similar to (but more exhaustive than) those shown in Figures 4–6, indicate that NV/HeII reaches a maximum value linked to the maximum temperature in photoionized clouds (Ferland et al 1996). The maximum NV/HeII ratio is ~2–3 for solar N/He abundances, depending on how "hard" a continuum shape one considers realistic

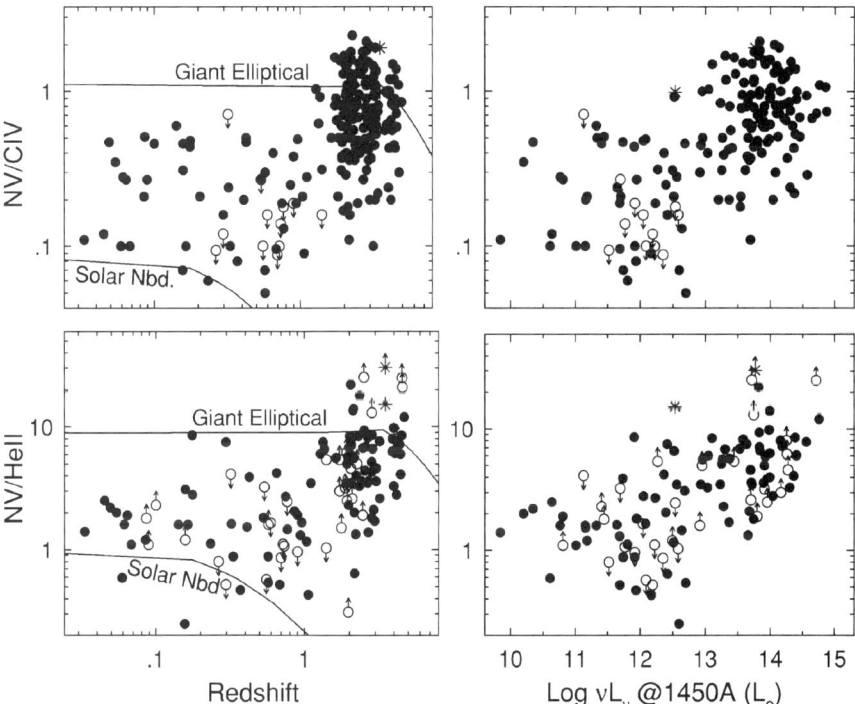

Figure 7 Measured NV/HeII and NV/CIV flux ratios versus redshift (*left panels*) and continuum luminosity (*right*). The upper and lower ranges might be undersampled (especially for NV/HeII at redshifts >1) because limits on weak lines (e.g. HeII) were often not available from the literature. The two *asterisks* in each panel represent mean values measured by Osmer et al (1994) for high- and low-luminosity QSOs at redshift >3. The *solid curves* are predictions based on chemical evolution models (discussed in Section 6).

for QSOs. (Beware that the highest ratios in Figure 4 occur for parameters where both lines are growing weak, cf the EW(CIV) plot or Korista et al 1997b.) Nominal BELR parameters predict NV/HeII near unity for solar N/He (Figures 4–6). These predictions fall well below most of the measured ratios (Figure 7), implying that QSOs typically have super-solar N/He. The ad hoc (high) temperatures that would be needed to explain the observed NV/HeII ratios with solar N/He are inconsistent with photoionization equilibrium and would lead to strong far-UV emission lines. The fact that these far-UV line strengths are not seen sets an upper limit on the temperature and supports the result for super-solar N/He (Ferland et al 1996).

The NV/CIV lines are collisionally excited with similar energies, so the temperature dependence in this ratio is smaller than NV/HeII. Nominal BELR parameters predict NV/CIV of ∼0.1 for solar N/C (Figures 4–6; see also Hamann & Ferland

1993a,b). Comparisons with the data in Figure 7 thus indicate super-solar N/C for most QSOs.

The two NV ratios together therefore imply that (*a*) quasar metallicities are often solar or higher, especially in high-redshift, high-luminosity objects, and (*b*) nitrogen (e.g. N/C) is typically enhanced compared with solar ratios (Ferland 1996, Hamann & Ferland 1993a,b). The conclusion for enhanced N/C is based largely on NV/CIV, but we note that the scaling of $N \propto Z^2$ leads to self-consistent estimates of Z based on NV/HeII and NV/CIV (Figure 7). The actual Z values are uncertain, but the main point is that many observed ratios require $Z \gtrsim Z_\odot$. Figures 6 and 7 combined suggest that the nominal metallicity range is $1 \lesssim Z \lesssim 10 \, Z_\odot$ for standard photoionization parameters and $N \propto Z^2$.

Hamann & Ferland 1993b noted that the observed NV ratios tend to be higher in more luminous sources (Figure 7). Most BELs exhibit the well-known Baldwin effect, that is, lower equivalent widths at higher luminosities (Baldwin 1977b). This effect is well-established in CIV and appears to be even stronger in OVI (Zheng et al 1995, Kinney et al 1990, Osmer & Shields 1999). Surprisingly, NV does not show this effect (Osmer et al 1994, Laor et al 1995, Francis & Koratkar 1995) even though its ionization is intermediate between CIV and OVI, and its electron structure is identical. We proposed that the peculiar NV behavior is caused by generally higher metallicities and more enhanced N abundances in more luminous QSOs. The recent theoretical study of the Baldwin effect by Korista et al (1998) gives quantitative support to that proposal. In Section 7, we argue that this proposed metallicity-luminosity trend in QSOs could naturally result from a mass-metallicity correlation among their host galaxies.

The abundance results based on NV have been questioned by Turnshek et al (1988, 1996) and Krolik & Voit (1998), who argue that the NV BEL forms largely by resonance scattering in an outflowing BAL region. NV emission might be selectively enhanced by this mechanism because it can scatter both the continuum and the underlying Lyα BEL. However, explicit calculations of the line scattering (Hamann et al 1993; Hamann & Korista 1996; F Hamann, KT Korista, & GJ Ferland, manuscript in preparation) do not support this scenario. For example, (*a*) the amount of NV scattering estimated by Krolik & Voit (1998) is too large by a factor of $\sim$3 on average, because BALRs do not generally have the right velocity/optical-depth structure to scatter all of the incident Lyα photons. In particular, NV BALs are not usually black across the Lyα emission-line wavelengths (see Figure 1 in Hamann et al 1999a). (*b*) It is difficult for NV ions in high-velocity BAL winds to scatter Lyα photons into simple BEL profiles with the observed half-widths of typically 2000–2500 km s^{-1}. For example, isotropic scattering of the Lyα flux would produce BEL half-widths of $\sim$6000 km s^{-1} (the velocity separation between the NV and Lyα lines). Anisotropic scattering (e.g. in BALRs with equatorial or bipolar geometries) would lead to strong orientation effects and systematically broader BEL profiles in BAL versus non-BAL QSOs. These differences are not observed (Weymann et al 1991). (*c*) It is not clear why, in individual spectra, the NV emission profiles should closely resemble those of other BELs if the former are produced by scattering in a high-velocity BAL wind whereas the latter are collisionally excited in a separate region [i.e. the usual BELR—whose velocity field is not mostly radial based on the reverberation studies (Türler & Courvoisier 1997, Korista et al 1995)]. Finally, (*d*) large scattering contributions to NV would minimally require much larger global BALR covering factors (the fraction of the sky covered by the BALR as seen from the central QSO) than expected from their observed detection frequency in (randomly oriented) QSO samples. Goodrich (1997) and Krolik & Voit (1998) argue that larger global covering factors could occur, but that issue is not settled.

Another concern is that complex BELR geometries might cause the NV/HeII and NV/CIV abundance indicators to fail—but they would fail in opposite directions. Specifically, clouds that are truncated at different physical depths (see Figure 3) could produce strong HeII with little or no NV and CIV emission, or strong HeII and NV with little or no CIV. For a given abundance set, this type of truncation could therefore either lower the observed NV/HeII ratio or increase NV/CIV. Comparing the data with simulations that do not take truncation into account (Figures 4–6) might then lead to underestimated N/He abundances or overestimated N/C. However, we have already shown that these two line ratios yield similar metallicities when compared with the nontruncated simulations, so we are not likely being misled by complex BELR geometries. Moreover, the NV/HeII ratio provides in any case a secure lower limit on N/He.

2.6.4 FeII/MgII The broad FeII emission lines pose unique problems because the atomic physics is complex and many blended lines contribute to the spectrum, particularly at the wavelengths $\sim$2000–3000 Å and $\sim$4500–5500 Å. Nonetheless, FeII is worth the effort because a delay of $\sim$1 Gyr in the Fe enrichment, relative to α elements such as O, Mg, or Si, might provide a clock for constraining the ages of QSOs and the epoch of their first star formation (see Sections 6–7 below; Hamann & Ferland 1993b).

A series of important papers on FeII emission (Osterbrock 1977; Phillips 1977, 1978; Grandi 1981) culminated with Wills & Netzer (1983) and Wills et al (1985). They performed sophisticated calculations showing that the large observed FeII fluxes, in particular FeII(UV)/MgII λ2799, require that either Fe is several-fold overabundant (compared with solar ratios) or some unknown process dominates the FeII excitation. One process that might selectively enhance FeII emission is photoexcitation by Lyα photons (Johansson & Jordan 1984). The absorption of Lyα radiation can pump electrons from the lower (metastable) energy levels of Fe^+ into specific high-energy states, leading to fluorescent cascades. Wills et al (1985) discounted this mechanism because it appeared insignificant in their simulations, but Penston (1987) noted that Lyα pumping is known to be important in some emission line stars, such as the symbiotic star RR Tel, and therefore might be important in QSOs (also Graham et al 1996, Laor et al 1997). More recent FeII simulations with better atomic data and exploring a wider range of physical conditions (Sigut & Pradhan 1998, Verner et al 1999) suggest that Lyα

can be important in some circumstances, but it is not yet clear whether those circumstances occur significantly in QSOs.

Recent observations have renewed interest in this question by showing that the FeII(UV)/MgII emission fluxes can be larger than the Wills et al (1985) predictions even at $z > 4$, with the tentative conclusion that Fe/Mg is at least solar (and thus the objects are at least ~ 1 Gyr old) (Taniguchi et al 1997, Yoshii et al 1998, Thompson et al 1999, and references therein). New theoretical efforts, such as Sigut & Pradhan (1998) and Verner et al (1999), are needed to test these conclusions and quantify the uncertainties. However, a better way to measure Fe/α might be with the intrinsic NALs (see below).

3. ABSORPTION LINE DIAGNOSTICS

3.1 Overview: Types of Absorption Lines

Quasar absorption lines can form in a variety of locations, from near the QSO engine (which we call "intrinsic," Section 1) to intervening gas at cosmologically significant distances. We exclude from our discussion the damped-Lyα absorbers and the "forest" of many narrow Lyα systems with weak or absent metal lines, because they form in cosmologically intervening gas (Rauch 1998). The remaining metal line systems can be divided into two classes by their broad or narrow profiles. This division is a gross simplification, but still useful because it distinguishes the clearly intrinsic broad lines from the many others of uncertain origin. Here we briefly characterize the two (broad and narrow) line types.

3.1.1 Broad Absorption Lines (BALs) Broad absorption lines are blueshifted relative to the emission lines and have velocity widths of at least a few thousand kilometers per second (for example, Figure 8). They appear in 10%–15% of optically selected QSOs and clearly identify high-velocity winds from the central engines. The precise location of the absorbing gas is unknown, but there is little doubt that it is intrinsic—originating within at least a few tens of parsecs from the QSOs. See recent work by Weymann et al (1991), Barlow et al (1992), Korista et al (1993), Hamann et al (1993), Voit et al (1993), Murray et al (1995), Arav (1996), Turnshek et al (1997), and Brotherton et al (1998b) and the reviews by Turnshek (1988, 1994), Weymann et al (1985), and Weymann (1994, 1997).

3.1.2 Narrow Absorption Lines (NALs) A practical definition of NALs would limit their full widths at half minimum (FWHMs) to less than the velocity separation of important doublets (e.g. <500 km s^{-1} for CIV, <1930 km s^{-1} for SiIV, or <960 km s^{-1} for NV), because it is our ability to resolve these doublets that makes their analysis fundamentally different from the BALs (Section 3.2.2 below).

NALs can form in a variety of locations, ranging from very near QSOs, as in ejecta like the BALs, to unrelated gas or galaxies at cosmological distances

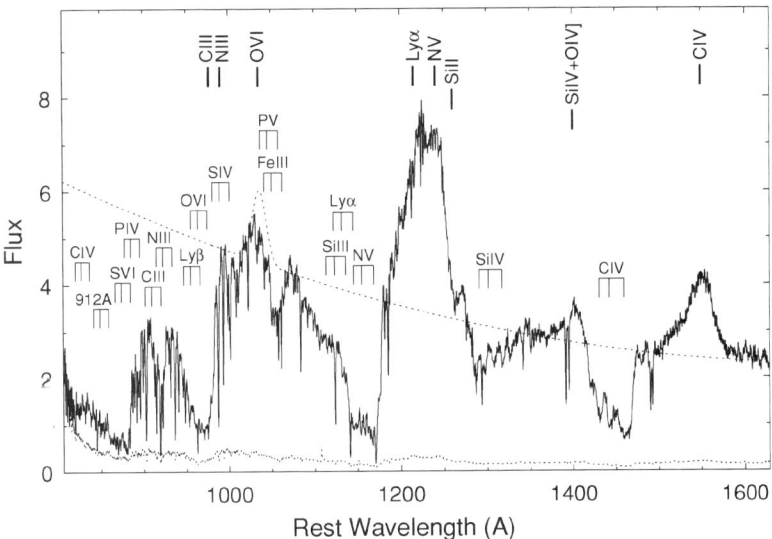

Figure 8 Spectrum of the BALQSO PG 1254+047 (emission redshift $z_e = 1.01$) with emission lines labeled across the top and possible BALs marked below at redshifts corresponding to the three deepest minima in the CIV trough. Not all of the labeled lines are detected. The *smooth dotted curve* is a power-law continuum fit extrapolated to short wavelengths (from Hamann 1998).

(Weymann et al 1979). It is not yet known what fraction of NALs at any velocity shift meet our definition of intrinsic (Section 1). Several studies have noted a statistical excess of NALs within a few thousand kilometers per second of the emission redshifts. These are the so-called "associated" or $z_a \approx z_e$ absorbers (with redshifts close to the emission redshift; Weymann et al 1979, 1981; Young et al 1982; Foltz et al 1986, 1988; Anderson et al 1987). Their strengths and frequency of occurrence appear to correlate with the QSO luminosities or radio properties, suggesting some physical relationship (see also Möller et al 1994, Aldcroft et al 1994, Wills et al 1995, Barthel et al 1997). These correlations may extend to NALs at blueshifts of $\geq 30,000$ km s^{-1} (Richards et al 1999). Nonetheless, we might expect a larger fraction of intrinsic NALs nearer the emission redshift and, if they are ejected from QSOs, they should appear at $z_a < z_e$ rather than $z_a > z_e$.

Several tests have been developed to help identify intrinsic NALs, including (*a*) time-variable line strengths, (*b*) multiplet ratios that imply partial line-of-sight coverage of the background light source(s), (*c*) high gas densities inferred from excited-state absorption lines, and (*d*) well-resolved line profiles that are smooth and broad compared with both thermal line widths and the velocity dispersions expected in intervening clouds (e.g. Bahcall et al 1967, Williams et al 1975, Young et al 1982, Barlow & Sargent 1997, Hamann et al 1997b, Hamann et al 1997c,

Petitjean & Srianand 1999, Ganguly et al 1999, and references therein). These criteria might not be definitive individually, but they sometimes appear in combination. Figure 9 shows a $z_a \approx z_e$ NAL system that is clearly intrinsic based on time-variable line strengths, partial line-of-sight coverage, and relatively broad profiles. High metallicities might be another indicator of intrinsic absorption (Section 3.4 below), but that criterion would bias abundance studies; we would like to determine the intrinsic versus intervening nature independently of the abundances. The other (nonabundance) tests indicate that bona fide intrinsic NALs can have velocity shifts out to $\gtrsim 24{,}000$ km s^{-1} and a wide range of FWHMs down to $\lesssim 30$ km s^{-1}. See the references above and the reviews of $z_a \approx z_e$ systems by Weymann et al (1981) and Foltz et al (1988).

3.2 General Abundance Analysis

Abundance estimates from absorption lines are, in principle, more straightforward than for emission lines because the absorption strengths are not sensitive to the temperatures or space densities in the gas. Moreover, absorption lines yield direct measures of the column densities in different ions. We need only apply appropriate ionization corrections to convert the column densities into relative abundances. For example, the abundance ratio of any two elements a and b can be written,

$$\left[\frac{a}{b}\right] = \log\left(\frac{N(a_i)}{N(b_j)}\right) + \log\left(\frac{f(b_j)}{f(a_i)}\right) + \log\left(\frac{b}{a}\right)_\odot, \quad (10)$$

which is identical to Equation 8 except that the Ns here are the column densities. Once again we define the ionization correction as $IC \equiv \log(f(b_j)/f(a_i))$. Abundance studies would ideally compare lines with similar ionizations to minimize IC and reduce the sensitivity to potentially complex geometries. Unfortunately, the lines available often require significant ionization corrections. In particular, we are often forced to compare highly ionized metals (such as CIV) to HI (Lyα) to derive the metallicity. We must therefore use ionization models.

The usual assumption is that the gas is photoionized by the QSO continuum flux. Collisional ionization would lead to lower derived metallicities because it creates less HI (and more HII) for a given level of ionization in the metals (cf Figures 4 and 5 in Hamann et al 1995; see also Turnshek et al 1996). However, collisional ionization has been generally dismissed for BAL regions (BALRs) because (a) it would be energetically hard to maintain, (b) it would produce excessive amounts of line emission (because of the much higher temperatures), and (c) it is hard to reconcile with the observed simultaneous variabilities in BAL troughs across a wide range of velocities (Weymann et al 1985, Junkkarinen et al 1987, Barlow 1993). In contrast, the strong radiative flux known to be present in QSOs provides a natural ionization source. We will assume that photoionization dominates in both BALRs and intrinsic NALRs.

Estimates of IC generally come from plots like Figure 10, which shows the ionization fractions of HI and various metal ions M_i in photoionized clouds (see Section 2.5 and Ferland et al 1998 for general descriptions of the calculations).

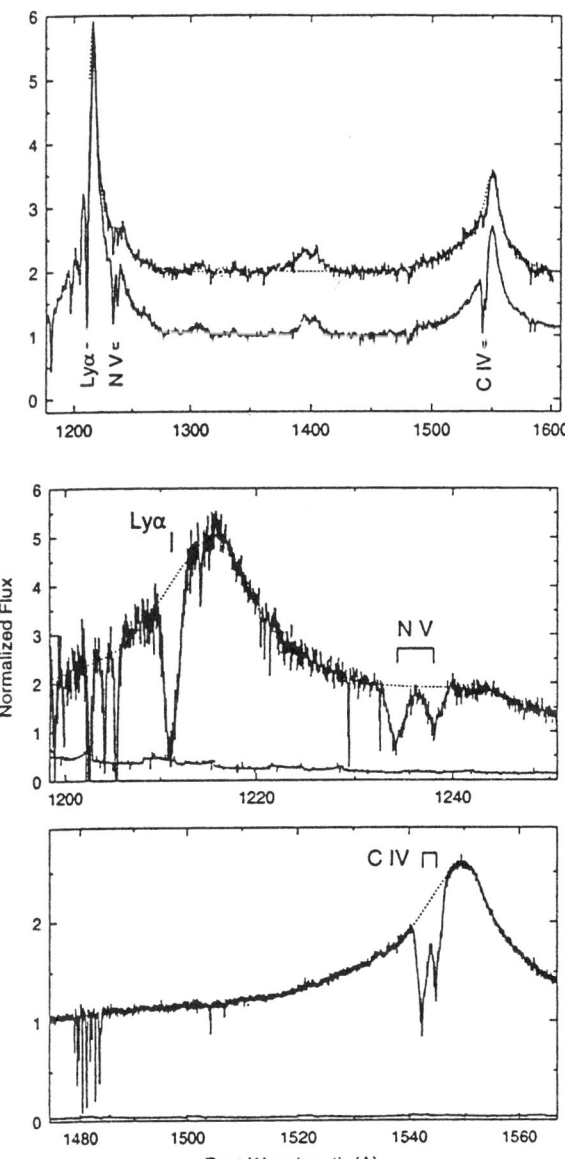

Figure 9 Spectra of the $z_a \approx z_e$ absorber in UM675 ($z_e = 2.15$) showing its time-variability in two epochs (*top panel*) and broad, smooth profiles at higher spectral resolution (9 km s^{-1}, *bottom panels*). From Hamann et al (1995, 1997b).

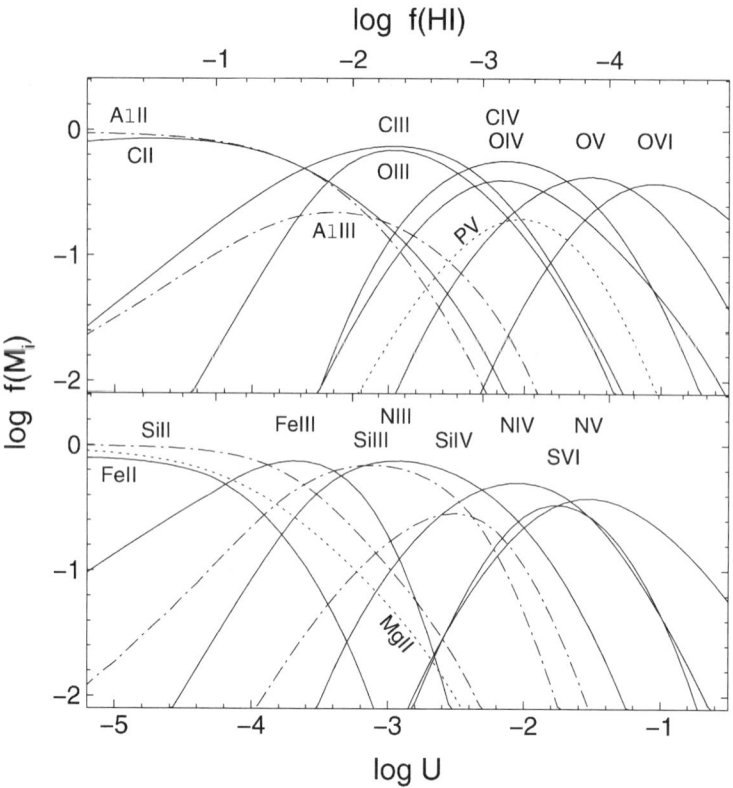

Figure 10 Ionization fractions in optically thin clouds photoionized by a power-law spectrum with $\alpha = -1.5$. Results are shown for a range of ionization parameters, U (see Section 2.5.1). The HI fraction appears across the top. The curves for the metal ions are labeled above or below their peaks whenever possible. The notation here is HI = H^0, CIV = C^{+3} etc.

Ideally, we would compare column densities in different ions of the same element to obtain abundance-independent constraints on the ionization and thus IC. Otherwise, column densities in different elements can also constrain IC with assumptions about their relative abundances.

Note that the results in Figure 10 are not sensitive to the particular abundances used in the calculations (in this case solar), so the figure is useful for general abundance/ionization estimates (see Hamann 1997). The model clouds are optically thin in the ionizing UV continuum, which means that gradients in the ionization are negligible across the cloud and the ionization fractions do not depend on the total column densities. This simplification appears to be appropriate for most intrinsic absorption line systems (based on their measured column densities), although shielding by many far-UV BALs might affect the ionization structure downstream in BALRs (Korista et al 1996, Turnshek 1997, Hamann 1997). Also, systems with low-ionization lines like FeII or MgII can be optically thick at the HI Lyman edge (Bergeron & Stasińska 1986, Voit et al 1993, Wampler et al 1995) and may require calculations with specific column densities that match the data.

3.2.1 Ionization Ambiguities The main theoretical uncertainties involve the shape of the ionizing spectrum, the frequent lack of ionization constraints (too few lines measured), and the possibility of inhomogeneous (multizone) absorbing regions. Hamann (1997) addressed these issues by calculating *IC* values for a wide range of conditions in photoionized clouds. He noted that, whenever there is or might be a multizone absorber with a range of ionization states, we can still make conservatively low estimates of the metal-to-hydrogen abundance ratios by assuming each metal line forms where that ion is most favored—that is, at the peak of its ionization fraction $f(M_i)$ in Figure 10. We can also place firm lower limits on the metal-to-hydrogen ratios by adopting the minimum values of *IC*, which correspond to minima in the $f(\text{HI})/f(M_i)$ ratios (see also Bergeron & Stasińska 1986). The lower limits are robust, even though they come from one-zone calculations, because the presence of different or additional zones can only mean that larger *IC* values are appropriate for the data. Figure 11 plots several minimum metal-to-hydrogen *IC*s for optically thin clouds photoionized by different power-law spectra. The results in this plot simply get added to the logarithmic column density ratios (Equation 10) to derive minimum metallicities. Note that some important metal-to-metal ratios also have minimum *IC* values, such as PV/CIV and FeII/MgII (Hamann 1998; F Hamann, R Chaffee, RJ Weymann, TA Barlow, VT Junkkarinen, manuscript in preparation).

3.2.2 Column Densities and Partial Coverage The final critical issue is deriving accurate column densities from the absorption troughs. In the simplest case, the optical depths in well-resolved lines are related to the observed intensities by

$$I_v = I_o \exp(-\tau_v), \tag{11}$$

where I_v and I_o are the observed and intrinsic (unabsorbed) intensities, respectively, and τ_v is the line optical depth, at each velocity shift v. The column densities follow from the optical depths by

$$N = \frac{m_e c}{\pi e^2 f \lambda_o} \int \tau_v \, dv, \tag{12}$$

where f is the oscillator strength and λ_o is the laboratory wavelength of the line. Column density derivations can involve line profile fitting or direct integration over the observed profiles (via Equations 11 and 12, Junkkarinen et al 1983, Grillmair & Turnshek 1987, Korista et al 1992, Savage & Sembach 1991, Jenkins 1996, Arav et al 1999). Very optically thick lines are not useful because the inferred values of

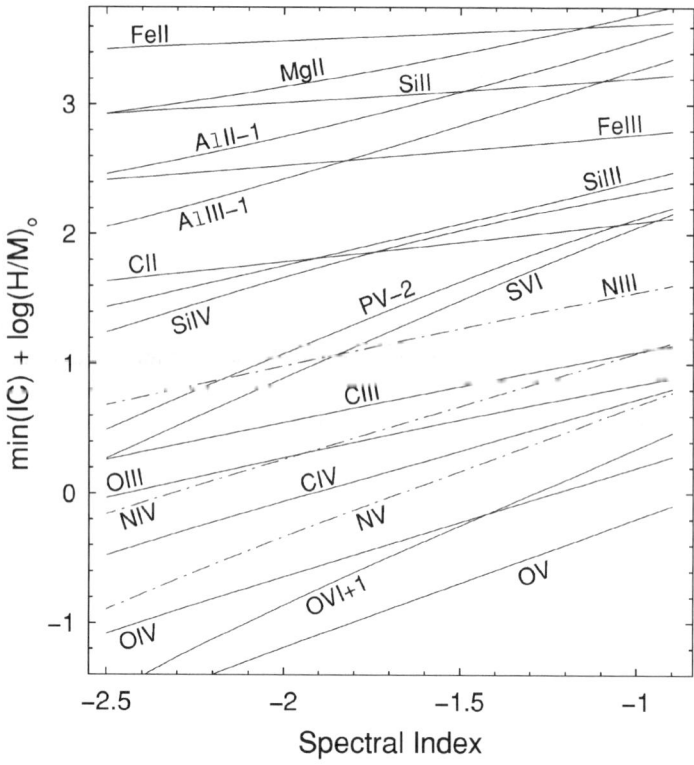

Figure 11 Minimum metal ion-to-HI ionization corrections (*IC*) normalized to solar abundances (the last two terms in Equation 10) are plotted for optically thin clouds photoionized by power-law spectra with different indices (α). The notation is the same as Figure 10. The curves have been shifted vertically by $+1$ for OVI, by -2 for PV, and by -1 for AlII and AlIII. The curves for nitrogen ions are *dash-dot*.

τ_v are far too sensitive to uncertainties in I_v. In other cases, the analysis might still be compromised by (*a*) unresolved absorption line components or (*b*) unabsorbed flux that fills in the bottoms of the observed troughs. If either of these possibilities occurs, the derived optical depths and column densities become lower limits, and the derived abundances become incorrect. Errors from the first possibility can always be reduced or avoided by higher-resolution spectroscopy.

The second possibility, of filled-in absorption troughs, is actually an asset for identifying intrinsic NALs (Section 3.1). We refer to this filling-in generally as partial coverage of the background emission source(s). Figure 12 shows several geometries that might produce partial coverage and filled-in troughs. (The situation can be potentially more complicated if the absorber itself is a source of emission. The analysis discussed below remains the same, however.) When partial coverage occurs, the observed intensities depend on both the optical depth and the

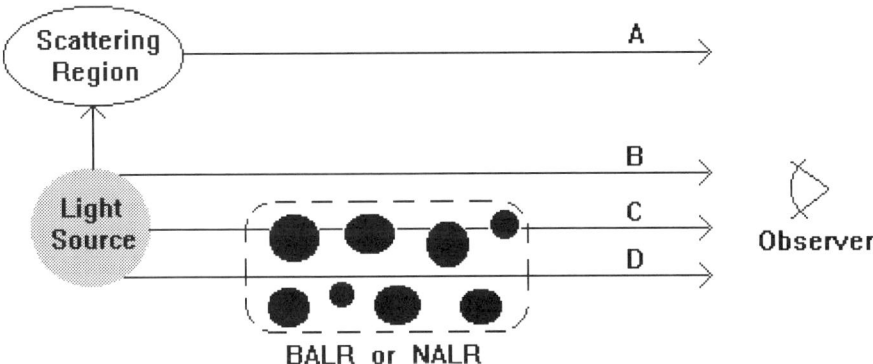

Figure 12 Possible "partial-coverage" geometries. Partial line-of-sight coverage occurs when light rays like C, which pass through absorption line clouds (indicated by *filled ellipsoids*), are combined with rays like A, B, or D, which do not. Ray A represents reflected light from a putative scattering region. Ray B simply misses the absorption line region. Ray D passes through the nominal absorbing zone but suffers no absorption because the region is porous.

line-of-sight coverage fraction, C_f, at each velocity,

$$I_v = (1 - C_f)I_o + C_f I_o \exp(-\tau_v), \tag{13}$$

where $0 \leq C_f \leq 1$, and the first term on the right side is the unabsorbed (or uncovered) contribution. Measured absorption lines can thus be shallow even when the true optical depths are large. In the limit $\tau_v \gg 1$, we have

$$C_f = 1 - \frac{I_v}{I_o}. \tag{14}$$

Outside of that limit, we can compare lines whose true optical-depth ratios are fixed by atomic physics, such as the HI Lyman lines or doublets like CIV $\lambda\lambda$1548,1550, SiIV $\lambda\lambda$1394,1403 etc, to determine uniquely both the coverage fractions and the true optical depths across the line profiles (Hamann 1997b, Barlow & Sargent 1997, Arav et al 1999, Srianand & Shankaranarayanan 1999, Ganguly et al 1999). For example, a little algebra shows that for doublets with true optical-depth ratios of ~2 (as in CIV, SiIV, etc), the coverage fraction at each absorption velocity is

$$C_f = \frac{I_1^2 - 2I_1 + 1}{I_2 - 2I_1 + 1}, \tag{15}$$

where I_1 and I_2 are the observed line intensities, normalized by I_o, at the same velocity in the weaker- and stronger-line troughs, respectively. The corresponding line optical depths are $\tau_2 = 2\tau_1$ and

$$\tau_1 = \ln\left(\frac{C_f}{I_1 + C_f - 1}\right). \tag{16}$$

It is a major strength of the NALs that we can resolve key multiplet lines and use this analysis to measure the coverage fractions, thus deriving reliable column densities and abundances. It is a great weakness of the BALs that this analysis is usually not possible because the lines are blended. We will argue below that BAL studies so far have been seriously compromised by unaccounted for partial-coverage effects.

The only drawback of partial coverage for the NALs is that there might be a range of coverage fractions in multizone-absorbing media. There is already evidence in some cases for coverage fractions that differ between ions or change with velocity across the line profiles (Barlow & Sargent 1997, Barlow et al 1997, Hamann et al 1997b). Variations in C_f with velocity can always be dealt with by analyzing limited velocity intervals in the line profiles (see also Arav 1997). But one can imagine complex geometries where ionization-dependent coverage fractions would jeopardize the simple abundance analysis described above, in particular for comparisons between high- and low-ionization species like CIV and HI. Abundance ratios based on disparate species like these might require specific models of the ionization-dependent coverage. On the other hand, this worst-case scenario is not known to occur, and there is no reason to believe it would lead to generally overestimated metallicities anyway.

3.3 Broad Absorption Line Results

One common characteristic of BAL spectra is that the metallic resonance lines like CIV λ1548,1951, SiIV λ1394,1403, NV λ1239,1243, and OVI λ1032,1038 are typically strong (deep) compared with Lyα (e.g. Figure 8). This result and the fact that low-ionization lines like MgII λ2796,2804 and FeII (UV) are usually absent indicates that the BALR ionization is generally high (Turnshek 1984; Weymann et al 1981, 1985). However, quantitative studies of the ionization have repeatedly failed to explain the measured line strengths with solar abundances. These difficulties were first noted by Junkkarinen (1980) and Turnshek (1981; see also Weymann & Foltz 1983), who showed that photoionization models with power-law ionizing spectra and solar abundances underpredict the metal ions, especially SiIV, by large factors relative to HI. A straightforward conclusion is that the metallicities are well above solar. Turnshek (1986, 1988) and Turnshek et al (1987) estimated metal abundances (C/H) of 10 to 100 times solar and provided tentative evidence for some extreme metal-to-metal abundance ratios such as P/C $\gtrsim$ 100 times solar.

Better data in the past 10 years have done nothing to change these startling results (e.g. Turnshek et al 1996). The early concerns about unresolved line components (Junkkarinen et al 1987, Kwan 1990) have gone away, thanks to spectroscopy with the Keck 10-m telescope at resolutions ($\sim$7 km s^{-1}) close to the thermal speeds (Barlow & Junkkarinen 1994, VT Junkkarinen, personal communication). The previously tentative detections of PV λ1118,1128 absorption, which led to the large P/C abundance estimates, have now been confirmed in two objects by excellent wavelength coincidences, by the predicted weakness of nearby lines like FeIII λ1122, and in one case by the probable presence of PIV λ951 absorption (Junkkarinen et al 1997, Hamann 1998; Figure 8). The commonality of PV absorption is not yet known (see also Korista et al 1992, Turnshek et al 1996), but its relative strength in just the two cases is surprising because the solar P/C ratio is only ~0.001.

More complex theoretical analyses, considering a range of ionizing spectral shapes or multiple ionization zones, also do not change the main result for metallicities and P/C ratios well above solar (Weymann et al 1985; Turnshek et al 1987, 1996, 1997; Korista et al 1996). Hamann (1997) used the analysis in Section 3.2 to determine how high the abundances must be, given the measured column densities and a photoionized BALR. He showed that average BALR column densities require [C/H] and [N/H] > 0 and [Si/H] > 1.0 for any range of ionizations and reasonable spectral shapes. The conservatively low values of IC [corresponding to the $f(M_i)$ at their peaks] indicate [C/H] and [N/H] $\gtrsim$ 1.0 and [Si/H] $\gtrsim$ 1.7. The results for individual BAL systems can be much higher. In PG1254+047 (Figure 8; Hamann 1998) the inferred minimum abundances are [C/H] and [N/H] $\gtrsim$ 1.0, [Si/H] $\gtrsim$ 1.8 and [P/C] $\gtrsim$ 2.2.

However, we now argue that all of these BAL abundance results are incorrect, because partial coverage effects have led to generally underestimated column densities.

3.3.1 Uncertainties and Conclusions There is now direct evidence for partial coverage in some BALQSOs based on widely separated lines of the same ion (Arav et al 1999) and resolved doublets in several narrow BALs and BAL components [Telfer et al 1999, Barlow & Junkkarinen 1994, Wampler et al 1995, Korista et al 1992—confirmed by VT Junkkarinen (personal communication)]. Although most of this evidence applies to narrow features, it is noteworthy that there are no counter-examples to our knowledge—in which narrow line components associated with BALs indicate complete coverage (also VT Junkkarinen, personal communication).

There is also circumstantial evidence for partial coverage in BAL systems; namely, (*a*) spectropolarimetry indicates that BAL troughs can be filled in by polarized flux (probably from an extended scattering region) that is not covered by the BALR (Figure 12; Goodrich & Miller 1995, Cohen et al 1995, Hines & Wills 1995, Schmidt & Hines 1999); (*b*) some BAL systems have a wide range of lines with suspiciously similar strengths or flat-bottom troughs that do not reach zero intensity (Arav 1997); (*c*) Voit et al (1993) made a strong case for low-ionization BALRs being optically thick at the Lyman limit, which implies large optical depths in Lyα; yet the Lyα troughs are not generally black in these systems; (*d*) the larger column densities that follow assuming partial coverage and saturated BALs [$N_H \gtrsim 10^{22}$ cm^{-2} (Hamann 1998)] are consistent with the large absorbing columns inferred from X-ray observations of BALQSOs (Green & Mathur 1996, Green et al 1997, Gallagher et al 1999).

More indirect evidence comes from the abundance results themselves. Voit (1997) noted that the derived overabundances tend to be greater for rare elements like P than for common elements like C. This is precisely what would occur if line saturation is not taken into account. The surprising detections of PV might actually be a signature of line saturation (and partial coverage) in strong lines like CIV, rather than extreme abundances (Hamann 1998). This assertion is supported by the one known NAL system with PV $\lambda\lambda 1118,1128$ absorption, where the doublet ratios in CIV, NV, and SiIV clearly indicate $\tau \gg 1$ (Barlow et al 1997; TA Barlow, personal communication).

We conclude that BAL column densities have been generally underestimated and that the true BALR abundances are not known. Observed differences between BAL profiles that resemble simple optical-depth effects are probably caused by a mixture of ionization, coverage fraction, and optical depth differences in complex, multizone BALRs. This conclusion paints a grim picture for BAL abundance work, but it might still be possible to derive accurate column densities and therefore abundances for some BALQSOs or some portions of BAL profiles (Wampler et al 1995, Turnshek 1997, Arav et al 1999). Most needed are spectra at shorter rest frame wavelengths to measure widely separated lines of the same ion and thereby diagnose the coverage fractions and true optical depths (Section 3.2.2, "Column Densities and Partial Coverage"; Arav 1997, Arav et al 1999).

3.4 Narrow Absorption Line Results

In contrast to the BALs, intrinsic NALs might be the best abundance probes we have for QSO environments. Resolved measurements of NAL multiplets allow us to measure both the coverage fractions and true column densities (Section 3.2.2). The NALs also allow separate measurements of important lines that are often blended in BAL systems, such as NV $\lambda 1239,1243$–Lyα, OVI $\lambda 1032,1038$–Lyβ, and many others. We therefore have potentially many more constraints on both the ionization and abundances.

Early NAL studies did not have the quality of data needed to derive column densities and abundances, but several groups noted a tendency for larger NV/CIV line strength ratios in $z_a \approx z_e$ systems compared with $z_a \ll z_e$ (Weymann et al 1981, Hartquist & Snijders 1982, Bergeron & Kunth 1983, Morris et al 1986, Bergeron & Boissé 1986). This trend is probably not caused simply by higher ionization in $z_a \approx z_e$ absorbers, because recent studies show that $z_a \ll z_e$ systems typically have strong OVI lines and therefore considerable high-ionization gas; NV appears to be weak relative to both CIV and OVI at $z_a \ll z_e$ (Lu & Savage 1993, Bergeron et al 1994, Burles & Tytler 1996, Kirkman & Tytler 1997, Savage et al 1998). The lower NV/OVI and NV/CIV line ratios at $z_a \ll z_e$ could be caused by an underabundance of nitrogen (compared with solar ratios) in metal-poor intervening gas (Bergeron et al 1994, Hamann et al 1997d, Kirkman & Tytler 1997). This would be the classic abundance pattern involving secondary nitrogen (Vila-Costas & Edmunds 1993). Relatively higher N abundances and thus stronger NV absorption lines should occur naturally in metal-rich environments near QSOs (see Section 2.6 above and Sections 6 and 7 below).

The first explicit estimates of $z_a \approx z_e$ metallicities were by Wampler et al (1993), Möller et al (1994), Petitjean et al (1994), and Savaglio et al (1994) for

QSOs at redshifts of ∼2–4. These studies found that $z_a \approx z_e$ systems often have $Z \gtrsim Z_\odot$, which is at least an order of magnitude larger than the $z_a \ll z_e$ systems measured in the same data. Several of the metal-rich $z_a \approx z_e$ systems have doublet ratios implying partial coverage and thus, very likely, an intrinsic origin (Wampler et al 1993, Petitjean et al 1994). The location of the other $z_a \approx z_e$ absorbers is not known, but Petitjean et al (1994) noted a marked change from [C/H] $\lesssim -1$ to [C/H] $\gtrsim 0$ at a blueshift of ∼15,000 km s^{-1} relative to the emission lines. If high abundances occur only in intrinsic systems, then these results suggest that most $z_a \approx z_e$ NALs are intrinsic (see also Möller et al 1994).

More recent studies support these findings. Petitjean & Srianand (1999) measured $Z \gtrsim Z_\odot$ and [N/C] > 0 in an intrinsic (partial-coverage) $z_a \approx z_e$ absorber. For $z_a \approx z_e$ systems of unknown origin, Savage et al (1998) estimated roughly solar metallicities and Tripp et al (1997) obtained [N/C] $\gtrsim 0.1$ and, very conservatively, [C/H] $\gtrsim -0.8$. (The lower limit on [C/H] for the latter system is −0.2 when more likely ionizing spectral shapes are used in the calculations.) Savaglio et al (1997) revised the metallicities downward slightly from those in their 1994 paper to −1 < [C/H] < 0, based on better data. Those systems are of special interest because of their high redshift ($z_a \approx 4.1$). Wampler et al (1996) estimated $Z \sim 2\,Z_\odot$ (based on a tentative detection of OI $\lambda 1303$) for the only other $z_a \approx z_e$ systems studied so far at $z > 4$.

Hamann (1997) and Hamann et al (1995, 1997b, 1997e, 1999b) used the analysis outlined in Section 3.2 to determine metallicities or establish lower limits for several $z_a \approx z_e$ systems, including some mentioned above and some that are clearly intrinsic by the indicators in Section 3.1. The results generally confirm the previous estimates and show further that, even when there are no constraints on the ionization (for example, when only Lyα and CIV lines are measured), the column densities can still require $Z \gtrsim Z_\odot$. A quick survey of those results suggests that bona fide intrinsic systems and most others with $Z \gtrsim Z_\odot$ have [N/C] $\gtrsim 0.0$.

3.4.1 Uncertainties and Conclusions Most of the NAL studies mentioned above would benefit from better data (higher signal-to-noise ratios and higher spectral resolutions) and more ionization constraints (wider wavelength coverage), but the frequent result for $Z \gtrsim Z_\odot$ is convincing. Unlike the BALs, there are no obvious systematic effects that might lead to higher abundance estimates for $z_a \approx z_e$ systems compared with $z_a \ll z_e$. The possibility of ionization-dependent coverage fractions presents an uncertainty for those systems with partial coverage, but we do not expect that to cause systematic overestimates of the metallicities (Section 3.2.2). We conclude that many $z_a \approx z_e$ NALs and, more importantly, all of the confirmed intrinsic systems have $Z \gtrsim 0.5\,Z_\odot$ and usually $Z \gtrsim Z_\odot$. The upper limits on Z are uncertain. The largest estimate for a well-measured system is $Z \sim 10\,Z_\odot$ (Petitjean et al 1994), but those data are also consistent with metallicities as low as solar because of ionization uncertainties (Hamann 1997). There are mixed and confusing reports in the literature regarding metal-to-metal abundance ratios, most notably N/C. In contrast to Franceschini & Gratton (1997), we find no tendency for subsolar N/C in $z_a \approx z_e$ systems. In fact, there is the general trend for stronger NV absorption at $z_a \approx z_e$ compared with $z_a \ll z_e$ systems, and the most reliable abundance data suggest solar or higher N/C ratios whenever $Z \gtrsim Z_\odot$.

The only serious problem is in interpreting the abundance results for absorbers of unknown origin. High metallicities might correlate strongly with absorption near QSOs, but the metallicities cannot define the absorber's location. For example, Tripp et al (1996) estimated $Z \gtrsim Z_\odot$ and [N/C] $\gtrsim 0$ for a $z_a \approx z_e$ system in which the lack of excited-state absorption in CII* $\lambda 1336$ (compared with the measured CII $\lambda 1335$) implies that the density is low, $\lesssim 7$ cm^{-3}, and thus the distance from the QSO is large, $\gtrsim 300$ kpc. [The relationship between density and distance follows from the flux requirements for photoionization (Section 2.5.1.)] Super-solar metallicities at these large distances are surprising. At $\gtrsim 300$ kpc from the QSO, we might have expected very low intergalactic or halo-like abundances. The solution might be that the absorbing gas was enriched much nearer the QSO and then ejected (Tripp et al 1996).

Unfortunately, the excited-state lines used for density and distance estimates are not generally available for $z_a \approx z_e$ systems (because they have low ionization energies, e.g. CII* and SiII*). Of the six $z_a \approx z_e$ absorbers known to be far ($\gtrsim 10$ kpc) from QSOs based on these indicators, three of them clearly have $z_a > z_e$ and are probably not intrinsic for that reason (Williams et al 1975, Williams & Weymann 1976, Sargent et al 1982, Morris et al 1986, Barlow et al 1997). Only one has a metallicity estimate—the system with $Z \gtrsim Z_\odot$ at $\gtrsim 300$ kpc distance studied by Tripp et al (1996).

4. GENERAL ABUNDANCE SUMMARY

The main abundance results are as follows.

1. There is a growing consensus from the BELs and NALs for $Z \gtrsim Z_\odot$ in QSOs out to $z > 4$. The upper limits on the metallicities are not well known, but none of the data require $Z > 10\, Z_\odot$. Solar to a few times solar metallicity appears to be typical. Based on very limited data, there is no evidence for a decline at the highest redshifts.

2. A trend in the NV/HeII and NV/CIV BEL ratios suggests that the metallicities are generally higher in more luminous QSOs.

3. The BELs and NALs both suggest that the relative nitrogen abundance (e.g. N/C and N/O) is typically solar or higher. We will argue below (Section 6) that this result corroborates the evidence for $Z \gtrsim Z_\odot$ (because of the likely secondary origin of nitrogen at these metallicities).

4. There is tentative evidence for super-solar Fe/Mg abundances out to $z > 4$ based on the FeII/MgII BEL strengths. Again, based on limited data, there is no evidence for a decline in this ratio at the highest redshifts.

5. The extremely high metallicities and large P/C ratios derived so far from the BALs are probably incorrect. In further support of that conclusion, we

note that BELR simulations with the nominally derived BAL abundances [including large enhancements in P and other odd-numbered elements like Al (Shields 1996)] are inconsistent with observed BEL spectra (based on unpublished work in collaboration with G Shields).

5. ENRICHMENT SCENARIOS

Several scenarios have been proposed for the production of heavy elements near QSOs, including (*a*) the normal evolution of stellar populations in galactic nuclei (Hamann & Ferland 1992, 1993b), (*b*) central star clusters with enhanced supernova (and perhaps nova) rates caused by mass accreted onto stars as they plunge through QSO accretion disks (Artymowicz et al 1993), (*c*) star formation inside QSO accretion disks (Silk & Rees 1998, Collin & Zahn 1999), and (*d*) nucleosynthesis without stars inside accretion disks (Jin et al 1989, Kundt 1996).

5.1 Occam's Razor: the Case for Normal Galactic Chemical Evolution

The first scenario listed above, for normal galactic chemical evolution, is most compelling because (*a*) it is the only one of these processes known to occur and (*b*) it is sufficient to explain the QSO data. In particular, the stars in the centers of massive galaxies today are (mostly) old and metal rich (Bica et al 1988, 1990; Gorgas et al 1990; Bruzual et al 1997; Vazdekis et al 1997; Jablonka et al 1992; Jablonka et al 1996; Feltzing & Gilmore 1998; Worthey et al 1992; Kuntschner & Davies 1997; Sansom & Proctor 1998; Ortolani et al 1996; Sil'chenko et al 1998; Idiart et al 1996; Fisher et al 1995; Bressan et al 1996). The exact ages are uncertain, but there is growing evidence for most of the star formation in massive spheroids (ellipticals and the bulges of large spiral galaxies) occurring at redshifts $z \gtrsim 2-3$, especially (but not only) for galaxies in clusters (see also Renzini 1997, 1998; Bernardi et al 1999; Bruzual & Magris 1997; Ellis et al 1997; Tantalo et al 1998; Ivison et al 1998; Kodama & Arimoto 1997; Ziegler & Bender 1997; Kauffmann 1996; Van Dokkum et al 1998; Mushotzky & Loewenstein 1997; Spinard et al 1997; Stanford et al 1998; Heap et al 1998; Barger et al 1998). The star-forming (Lyman break or Lyα emission) objects measured directly at $z \gtrsim 3$ might be galactic or protogalactic nuclei in the throes of rapid evolution (Friaca & Terlevich 1999; Baugh et al 1998; Steidel et al 1998, 1999; Connolly et al 1997; Lowenthal et al 1997; Trager et al 1997; Hu et al 1998; Franx et al 1997; Madau et al 1996; Giavalisco et al 1996). These objects are more numerous than QSOs and some have been measured at $z > 5$ (Dey et al 1998, Hu et al 1998, Weymann et al 1998), beyond the highest known QSO redshift of $z \approx 5.0$. On the theoretical side, recent cosmic-structure simulations show that protogalactic condensations can form stars and reach solar or higher metallicities at $z \gtrsim 6$ (Gnedin & Ostriker 1997).

These studies all suggest that there was considerable star formation at epochs preceding, or concurrent with, the QSOs. Quasars might form in the most massive and most dense of the early-epoch star-forming environments (Turner 1991, Loeb 1993, Haehnelt & Rees 1993, Miralda-Escude & Rees 1997, Haehnelt et al 1998, Spaans & Corollo 1997). They might also form preferentially in globally dense cluster environments, based on the higher detection rates of star-forming galaxies near high-z QSOs (Djorgovski 1998).

The gas in these environments might have been long ago ejected via galactic winds, consumed by central black holes, or diluted by subsequent gaseous infall, but its signature remains in the old stars today. The mean stellar metallicities in the cores of massive low-redshift galaxies are typically $\langle Z_{stars} \rangle \sim 1$–$3 \, Z_\odot$ (see references to metallicities listed above). [It is worth noting here that, because of a significant time-delay in the iron enrichment, O/H and Mg/H are better measures of the overall metallicity than Fe/H (see Section 6 and Wheeler et al 1989)]. Individual stars are distributed about the mean with metallicities reflecting the gas-phase abundance at the time of their formation. If the interstellar gas is well mixed and the abundances grow monotonically [as expected in simple enrichment schemes (Section 6)], the gas-phase metallicity, Z_{gas}, will always exceed $\langle Z_{stars} \rangle$. Only the most recently formed stars will have metallicities as high as the gas. Therefore, the most metal-rich stars today should reveal the gas-phase abundances near the end of the last major star-forming epoch.

In the bulge of our own Galaxy, the nominal value of $\langle Z_{stars} \rangle$ is $\sim 1 \, Z_\odot$ and the tail of the distribution reaches $Z_{stars} \gtrsim 3 \, Z_\odot$, with even higher values obtaining near the Galactic center (Rich 1988, 1990; Geisler & Friel 1992; McWilliam & Rich 1994; Minniti et al 1995; Tiede et al 1995; Terndrup et al 1995; Idiart et al 1996; Castro et al 1996; Bruzual et al 1997). The gas-phase metallicity should therefore have been $Z_{gas} \gtrsim 3 \, Z_\odot$ after most of the bulge star formation occurred. Simple chemical evolution models indicate more generally that Z_{gas} should be ~ 2 to three times $\langle Z_{stars} \rangle$ in spheroidal systems like galactic nuclei (Searle & Zinn 1978, Tinsley 1980, Rich 1990, Edmunds 1992, de Fretas Pacheco 1996). Thus the observations of $\langle Z_{stars} \rangle \sim 1$–$3 \, Z_\odot$ suggest that gas with $Z_{gas} \sim 2$–$9 \, Z_\odot$ once existed in these environments.

We might therefore expect to find $2 \lesssim Z \lesssim 9 \, Z_\odot$ in QSOs, as long as most of the local star formation occurred before the QSOs "turned on" or became observable. These expectations are consistent with the QSO abundance estimates reported above (Section 4). More exotic enrichment schemes are therefore not needed to explain the QSO data.

6. MORE INSIGHTS FROM GALACTIC CHEMICAL EVOLUTION

If we assume that QSO environments were indeed enriched by normal stellar populations, then we can use the results from galactic abundance and chemical-evolution studies to interpret the QSO data. Here we describe some relevant galactic results (see Wheeler et al 1989 for a general review).

6.1 The Galactic Mass-Metallicity Relation

One important result from galaxy studies is the well-known mass-metallicity relationship among ellipticals and spiral bulges (Faber 1973, Faber et al 1989, Bender et al 1993, Zaritsky et al 1994, Jablonka et al 1996, Cozial et al 1997). This relationship is attributed to the action of galactic winds; massive galaxies reach higher metallicities because they have deeper gravitational potentials and are better able to retain their gas against the building thermal pressures from supernovae (Larson 1974, Arimoto & Yoshii 1987, Franx & Illingworth 1990). Low-mass systems eject their gas before high Z's are attained. Quasar metallicities should be similarly tied to the gravitational binding energy of the local star-forming regions and, perhaps, to the total masses of their host galaxies (Section 7.1 below).

6.2 Specific Abundance Predictions

Another key result is the abundance behaviors of N and Fe relative to the α elements such as O, Mg, and Si. Hamann & Ferland (1993b) constructed one-zone infall models of galactic chemical evolution to illustrate these behaviors in different environments. Figure 13 plots the results for two scenarios at opposite extremes. Both use the same nucleosynthetic yields, but the "Giant Elliptical" model has much faster evolution rates and a flatter IMF (more favorable to high-mass stars)

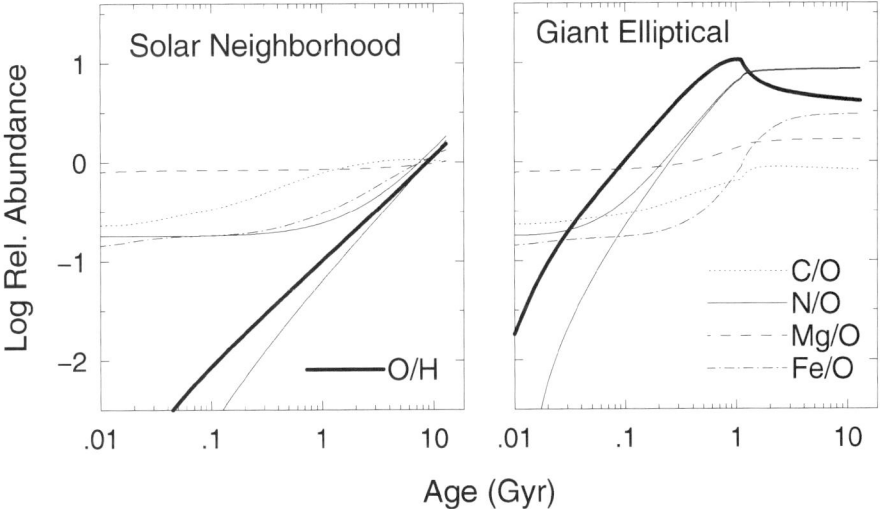

Figure 13 Logarithmic gas-phase abundance ratios normalized to solar for the two evolution models discussed in Section 6.2 of the text (adapted from Hamann & Ferland 1993b). Two scenarios for the N enrichment are shown (*thin solid lines*), one with secondary only and the other with secondary+primary (causing a plateau in N/O at low Z at early times). The thick solid curves represent O/H in both panels.

compared with the "Solar Neighborhood" (or spiral disk) case. The Giant Elliptical evolves passively (without further star formation) after ~ 1 Gyr, because the gas is essentially exhausted. The parameters used in these calculations were based on standard galactic infall models (e.g. Arimoto & Yoshii 1987, Matteucci & Tornambé 1987, Matteucci & Francois 1989, Matteucci & Brocato 1990, Köppen & Arimoto 1990). However, the results are only illustrative and not intended to match entire galaxies. For example, evolution like the Giant Elliptical model might occur in just the central cores of extreme high-mass galaxies (cf Friaca & Terlevich 1998).

6.3 Fe/α as a Clock

At early times the abundance evolution is controlled by short-lived massive stars, mainly via Type-II supernovae (SN IIs). The α elements, such as O and Mg, come almost exclusively from these objects, but Fe has a large delayed contribution from Type-Ia supernovae (SN Ias)—whose precursors are believed to be intermediate-mass stars in close binaries (Branch 1998). The predicted time delay is ~ 1 Gyr based on the IMF-weighted stellar lifetimes (Figure 13; Greggio & Renzini 1983, Matteucci & Greggio 1986). The actual delay is uncertain, but recent estimates are in the range of ~ 0.3 to 3 Gyr (Matteucci 1994, Yoshii et al 1996, Yoshii et al 1998). Because this delay does not depend on any of the global evolution time scales (e.g. the star formation rate, etc), Fe/α can serve as an absolute "clock" for constraining the ages of star-forming environments (Tinsley 1979, Thomas et al 1998).

Observations of metal-poor galactic stars suggest that the baseline value of [Fe/α] from SN IIs alone is nominally from -0.7 to -0.4 (Israelian et al 1998, Nissen et al 1994, King 1993, Gratton 1991, Magain 1989, Barbuy 1988; see also de Freitas Pacheco 1996), which is slightly larger than the prediction in Figure 13. The subsequent increase caused by SN Ias is a factor of a few or more. Note that the increase in Fe/α should be larger in rapidly evolving spheroidal systems because (*a*) by the time their SN Ia's turn on, there is relatively little gas left and each SN Ia has a greater effect; also (*b*) their rapid early star formation means that the SN Ia's occurring later are more nearly synchronized. The net result can be substantially super-solar Fe/α in the gas (even though Fe/α is sub-solar in most stars).

6.4 Nitrogen Abundances

Nitrogen also exhibits a delayed enhancement, although not on a fixed time scale like Fe/α. Nitrogen's selective behavior is caused by secondary CNO nucleosynthesis, in which N forms out of pre-existing C and O. Studies of galactic HII regions indicate that secondary processing dominates at metallicities above ~ 0.2 $Z_\odot$, resulting in N/O scaling like O/H (or $N \propto Z^2$) in that regime. At lower metallicities, primary N can be more important based on an observed plateau in [N/O] at roughly -0.7 (see Tinsley 1980, Vila-Costas & Edmunds 1993, Thurston et al 1996, Van Zee et al 1998, Kobulnicky & Skillman 1998, Thuan et al 1995, Izotov & Thuan 1999; but see also Garnett 1990, Lu et al 1998). The models in Figure 13 show two N/O behaviors, for secondary only and secondary plus primary; the latter has a low-Z plateau forced to match the HII region data. Notice that the secondary growth in N/O can be shifted down considerably from the simple theoretical relation [N/O] = [O/H], e.g. in the Giant Elliptical case, because of the delays related to stellar lifetimes. We therefore have a strong prediction, based on both observations and these simulations, that measured values of [N/O] $\gtrsim$ 0 imply $Z \gtrsim Z_\odot$—especially in quickly evolving spheroidal systems. This prediction was exploited above in the analysis of QSO BELs (Section 2.6; Shields 1976).

7. IMPLICATIONS OF QUASISTELLAR-OBJECT ABUNDANCES

7.1 High-Redshift Star Formation

We can conclude from the previous sections that QSOs are associated with vigorous star formation, consistent with the early-epoch evolution of massive galactic nuclei or dense protogalactic clumps (Section 5). However, QSO abundances provide new constraints. For example, the general result for $Z \gtrsim Z_\odot$ suggests that most of the enrichment and local star formation occur before QSOs turn on or become observable. The enrichment times can be so short in principle (Figure 13; Hamann & Ferland 1993b) that the star formation might also be coeval with QSO formation. In any event, the enrichment times cannot be much longer that $\sim$1 Gyr for at least the highest redshift objects [depending on the cosmology (Figure 1)].

If QSO metallicities are representative of a well-mixed interstellar medium, we can conclude further that the star formation was extensive. That is, a significant fraction of the initial gas must be converted into stars and stellar remnants to achieve $Z_{gas} \gtrsim Z_\odot$. The exact fraction depends on the IMF. A solar neighborhood IMF (Scalo 1990, as in the Solar Neighborhood model of Section 6) would lead to mass fractions in gas of only $\lesssim$15% at $Z \sim Z_\odot$ and would not be able to produce Z_{gas} above a few $Z_\odot$ at all. Flatter IMFs (favoring massive stars) could reach $Z_{gas} \gtrsim Z_\odot$ while consuming less of the gas. For example, the gas fraction when $Z_{gas} \sim Z_\odot$ in the Giant Elliptical model of Section 6.2 is nearly 70%.

Figure 7 in Section 2.6.3 illustrates the main star formation characteristics required by the QSO data. The solid curves on the right-hand side of that figure show theoretical BEL ratios from photoionization simulations that use nominal BELR parameters plus abundances from the two chemical-evolution models in Figure 13 (see Hamann & Ferland 1993b for more details). The evolution is assumed to begin with the Big Bang, and the conversion of time into redshift assumes a cosmology with $H_o = 65$ km s^{-1} Mpc^{-1}, $\Omega_M = 1$, and $\Omega_\Lambda = 0$. (Lower values of H_o or Ω_M would push the theoretical curves slightly toward the right in that figure, for example by $\sim$20% to $\sim$50% in z if $\Omega_M = 0.3$, see also Figure 1.) The main results are that the Solar Neighborhood evolution is too slow and, in any case, does not reach high enough metallicities or nitrogen enhancements to match most of the high-redshift QSOs. Much shorter time scales and usually higher metallicities, as in the Giant Elliptical simulation, are needed.

A trend in the NV BELs suggests further that the metallicities are typically higher in more luminous QSOs (Section 2.6.3). That result needs confirmation, but it could result naturally from a mass–metallicity relationship among QSO host galaxies that is similar (or identical) to the well-known relation in low-redshift galaxies (Section 6.1; Hamann & Ferland 1993b). By analogy with the galactic relation, the most luminous and metal-rich QSOs might reside in the most dense or massive host environments. This situation would be consistent with studies showing that QSO luminosities, QSO masses, and central black-hole masses in galactic nuclei all appear to correlate with the mass of the surrounding galaxies (McLeod et al 1999, McLeod & Rieke 1995, Bahcall et al 1997, Magorrian et al 1998, Laor 1998; see also Haehnelt & Rees 1993). Direct application of the galactic mass-metallicity relation suggests that metal-rich QSOs reside in galaxies (or protogalaxies) that are minimally as massive (or as tightly bound) as our own Milky Way.

7.2 Fe/α: Timescales and Cosmology

One of the most interesting predictions from galactic studies (Section 6.3) is that Fe/α ratios in QSOs might constrain the epoch of their first star formation and perhaps the cosmology. In particular, large Fe/α ratios (solar or higher) would suggest that the local stellar populations are at least ~ 1 Gyr old. At the highest QSO redshifts ($z \sim 5$), this age constraint would push the epoch of first star formation beyond the limits of current direct observation, to $z > 6$ (Figure 1). The $\gtrsim 1$ Gyr constraint would also be difficult to reconcile with $\Omega_M \approx 1$ in Big Bang cosmologies (because the age of the universe at $z \gtrsim 5$ in this cosmology is less than 1 Gyr). Conversely, measurements of low Fe/α would suggest that the local stellar populations are younger than ~ 1 Gyr (although we could not rule out the possibility that only SN IIs contributed to the enrichment for some other reason). Some BEL studies have already suggested that Fe/α is above solar in $z > 4$ QSOs (Section 2.6.4), implying (albeit tentatively) that these systems are already $\gtrsim 1$ Gyr old.

7.3 Comparisons to Other Results

Quasar abundances should be viewed in the context of other measures of the metallicity and star formation at high redshifts. Damped-Lyα absorbers in QSO spectra, which probe lines of sight through large intervening galaxies [probably spiral disks (Prochaska & Wolfe 1998)] have mean (gas-phase) metallicities of ~ 0.05 $Z_\odot$ at $z \gtrsim 2$ (Lu et al 1996, Pettini et al 1997, Lu et al 1998, Prochaska & Wolfe 1999). The Lyα forest absorbers, which presumably probe much more extended and tenuous intergalactic structures (Rauch 1998), typically have metallicities <0.01 $Z_\odot$ at high redshifts (Rauch et al 1997, Songalia & Cowie 1996, Tytler et al 1995). The much higher metal abundances near QSOs are consistent with the rapid and more extensive evolution expected in dense environments (Gnedin & Ostriker 1997). Perhaps this evolution is similar to that occurring in the many star-forming objects that are now measured directly at redshifts comparable to and greater than the QSOs (see references in Section 5.1).

The detections of strong dust and molecular gas emissions from QSOs support the evidence from their high abundances that considerable local star formation preceded the QSO epoch. The dust and molecules, presumably manufactured by stars, appear even in QSOs at $z \gtrsim 4$ (Isaac et al 1994, Omont et al 1996, Guilloteau et al 1997).

8. FUTURE PROSPECTS

We now have the observational and theoretical abilities to test and dramatically extend all of the QSO abundance work discussed above. The most pressing needs are to (a) develop more independent abundance diagnostics and (b) obtain more and better data to compare diagnostics in large QSO samples—spanning a range of redshifts, luminosities, radio properties, etc. Absorption line studies will benefit generally from higher spectral resolutions and wider wavelength coverage, providing more accurate column densities and more numerous constraints on the coverage fractions, ionizations, and abundances (Section 3). BEL studies should include more of the weaker lines, such as OVI $\lambda1034$, CIII $\lambda977$, NIII $\lambda991$, and the intercombination lines, whenever possible (Section 2). Theoretical analysis of the FeII/MgII emission ratios, in particular, is needed to test the tentative conclusion for high Fe/Mg abundances. This and other BEL results should be tested further by examining the same lines (or same elements) in intrinsic NAL systems. The steady improvement in our observational capabilities at all wavelengths will provide many more diagnostic opportunities.

Below are some specific issues that new studies might address.

1. More data at high redshifts will constrain better the epoch and extent of early star formation associated with QSOs.
2. Reliable measurements of Fe/α will further constrain the epoch of first star formation and, perhaps, the cosmology via the ~ 1 Gyr enrichment clock.
3. Better estimates of the metal-to-metal ratios generally will reveal more specifics of the star formation histories, via comparisons to well-studied galactic environments and theoretical nucleosynthetic yields.
4. Abundances for QSOs spanning a wide range of luminosities and redshifts will isolate any evolutionary (redshift) trends and test the tentative luminosity-Z relationship. This relationship might prove to be a useful indicator of the total masses or densities of the local stellar populations by analogy with the mass-Z trend in nearby galaxies.
5. The range of QSO metallicities at a given redshift and luminosity will help constrain the extent of star formation occurring before QSOs turn on or become observable. Are there any low-metallicity QSOs?
6. Combining the QSO abundances with direct-imaging studies of their host galaxies should test ideas about the chemical enrichment and help us

interpret data at the highest redshifts, where direct imaging is (so far) not possible. For example, are QSOs in large galaxies (e.g. giant ellipticals) more metal-rich than others?

7. Correlations between the abundances and other properties of QSOs, such as radio loudness or UV–X-ray continuum shape, might reveal new environmental factors in the enrichment or systematic uncertainties in our abundance derivations.

8. Observations with wide-wavelength coverage would allow us to compare abundances derived from the narrow emission lines (appearing in the rest frame optical) to BEL and NAL data in the same objects. These diverse diagnostics might provide crude abundance maps of QSO host galaxies.

9. How do QSO abundances compare with their low-redshift counterparts, the Seyfert galaxies and other low-luminosity active galactic nuclei? We might find that the metallicities at low redshifts are less than the QSOs owing to recent mergers or gaseous infall.

ACKNOWLEDGMENTS

We are grateful to G Burbidge for his help and encouragement. We also thank KT Korista, A Laor, A Sandage and JC Shields for comments on this manuscript, and TA Barlow, N Arav and VT Junkkarinen for helpful discussions. GF thanks the Canadian Institute for Theoretical Astrophysics for their hospitality during a sabbatical year, and acknowledges support from the Natural Science and Engineering Research Council of Canada through CITA. The work of FH was supported by NASA grant NAG 5-3234. Research in nebular astrophysics at the University of Kentucky is supported by the NSF through grant 96-17083 and by NASA through its ATP (award NAG 5-4235) and LTSA programs.

Visit the Annual Reviews home page at http://www.AnnualReviews.org

LITERATURE CITED

Aldcroft TL, Bechtold J, Elvis M. 1994. *Ap. J. Suppl.* 93:1

Anderson SF, Weymann RJ, Foltz CB, Chaffee FH. 1987. *Astron. J.* 94:278

Arav N. 1996. *Ap. J.* 465:617

Arav N. 1997. See Weymann et al. 1997. p. 208

Arav N, Barlow TA, Laor A, Sargent WLW, Blandford RD. 1998. *MNRAS* 297:990

Arav N, Korista KT, de Kool M, Junkkarinen VT, Begelman MC. 1999. *Ap. J.* 516:27

Aretxaga I, Terlevich RJ, Boyle BJ. 1998. *MNRAS* 296:643

Arimoto N, Yoshii Y. 1987. *Astron. Astrophys.* 173:23

Artymowicz P, Lin DNC, Wampler EJ. 1993. *Ap. J.* 409:592

Bahcall JN, Kirhakos S, Saxe DH, Schneider DP. 1997. *Ap. J.* 479:642

Bahcall JN, Sargent WLW, Schmidt M. 1967. *Ap. J.* 149:L11

Baldwin JA. 1977a. *MNRAS* 178:67P

Baldwin JA. 1977b. *Ap. J.* 214:679

Baldwin JA, Netzer H. 1978. *Ap. J.* 226:1

Baldwin JA, Ferland GJ, Korista KT, Carswell

RF, Hamann F, et al. 1996. *Ap. J.* 461:664

Baldwin JA, Ferland GJ, Korista KT, Verner D. 1995. *Ap. J.* 455:L119

Barbuy B. 1988. *Astron. Astrophys.* 191:121

Barger AA, Aragon-Salamanca A, Smail I, Ellis RS, Couch WJ, et al. 1998. *Ap. J.* 501:522

Barger AA, Cowie LL, Trentham N, Fulton E, Hu EM, et al. 1999. *Astron. J.* 117:102

Barlow TA. 1993. PhD dissertation, University of California, San Diego

Barlow TA, Hamann F, Sargent WLW. 1997. See Weymann et al, 1997, p. 13.

Barlow TA, Junkkarinen VT. 1994. *BAAS* 26:1339

Barlow TA, Junkkarinen VT, Burbidge EM, Weymann RJ, Morris SL, et al. 1992. *Ap. J.* 397:81

Barlow TA, Sargent WLW. 1997. *Astron. J.* 113:136

Barthel PD, Tytler DR, Vestergaard M. 1997. See Weymann et al, 1997, p. 48

Baugh CM, Cole S, Frank CS, Lacey CG. 1998. *Ap. J.* 498:504

Bender R, Burstein D, Faber SM. 1993. *Ap. J.* 411:153

Bergeron J, Boissé P. 1986. *Astron. Astrophys.* 168:6

Bergeron J, Kunth D. 1983. *MNRAS* 205:1053

Bergeron J, Petitjean P, Sargent WLW, Bahcall JN, Boksenberg A, et al. 1994. *Ap. J.* 436:33

Bergeron J, Stasińska G. 1986. *Astron. Astrophys.* 169:1

Bernardi M, Renzini A, da Costa LN, Wegner G, Victoria M, et al. 1999. *Ap. J.* 508:L413

Bica E, Alloin D, Schmidt AA. 1990. *Astron. Astophys.* 228:23

Bica E, Arimoto N, Alloin D. 1988. *Astron. Astrophys.* 202:8

Blades JC, Turnshek DA, Norman CA, eds. 1988. QSO Adsorption Lines: Probing the Universe. Cambridge, UK: Cambridge Univ. Press. p. 348

Boyce PJ, Disney MJ, Bleaken DG. 1999. *MNRAS* 302:39

Boyle B. 1990. *MNRAS* 243:231

Boyle BJ, Terlevich RJ. 1998. *MNRAS* 293:49

Branch D. 1998. *Annu. Rev. Astron. Astrophys.* 36:17

Bressan A, Chiosi C, Tantalo R. 1996. *Astron. Astrophys.* 311:425

Brotherton MS, Van Breugel W, Smith RJ, Boyle BJ, Shanks T. 1998b. *Ap. J.* 505:7

Brotherton MS, Wills BJ, Dey A, Van Breugal W, Antonucci R. 1998a. *Ap. J.* 501:110

Brunner H, Mueller C, Friedrich P, Doerrer T, Staubert R, et al. 1997. *Astron. Astrophys.* 326:885

Bruzual G, Barbuy B, Ortolani S, Bica E, Cuisinier F. 1997. *Astron. J.* 114:1531

Burbidge G, Burbidge M. 1967. In *Quasi-Stellar Objects*. New York: Freeman

Burles S, Tytler D. 1996. *Ap. J.* 460:584

Carballo R, Sanchez SF, Gonzalas-Serrano JI, Benn CR, Vigotti M. 1998. *Astron. J.* 115:1234

Carroll SM, Press WH, Turner EL. 1992. *Annu. Rev. Astron. Astrophys.* 30:499

Castro S, Rich RM, McWilliam A, Ho LC, Spinrad H, et al. 1996. *Astron. J.* 111:2439

Cavaliere A, Vittorini V. 1998. See D'Odorico et al, 1998, p. 28

Chatzichristou ET, Vanderriest C, Jaffe W. 1999. *Astron. Astrophys.* 343:407

Cohen MH, Ogle PM, Tran HD, Vermeulen RC, Miller JS, et al. 1995. *Ap. J.* 448:L77

Collin S, Zahn J-P. 1999. *Astron. Astrophys.* 344:433

Connolly AA, Szalay AS, Dickenson M, SubbaRao MU, Brunner RJ. 1997. *Ap. J.* 486:L11

Cozial R, Contini T, Davoust E, Considère S. 1997. *Ap. J.* 481:L67

Davidson K. 1977. *Ap. J.* 218:20

Davidson K, Netzer H. 1979. *Rev. Mod. Phys.* 51:715

de Freitas Pacheco JA. 1996. *MNRAS* 278:841

Dey A, Spinrad H, Stern D, Graham JR, Chaffee FH. 1998. *Ap. J.* 498:L93

Djorgovski SG. 1998. In *Fundamental Parameters of Cosmology*, ed. Y Giroud-Hèraud. Gif sur Yvette: Ed. Frontières. In press

D'Odorico S, Fontana A, Giallongo E, eds. 1998. *The Young Universe. Astron. Soc.*

Pacific Conf. Ser. vol. 146. San Francisco: PASP, p. 536

Edmunds MG. 1992. In *Elements and the Cosmos*, eds. MG Edmunds, R Terlevich. New York, Cambridge Univ. Press, p. 289

Ellis R, Smail I, Dressler A, Couch WJ, Oemler A, et al. 1997. *Ap. J.* 483:582

Faber SM. 1973. *Ap. J.* 179:423

Faber SM, Wegner G, Burstein D, Davies RL, Dressler A, et al. 1989. *Ap. J. Suppl.* 69:763

Fan X, Strauss MA, Schneider DP, Gunn JE, Lupton RH, et al. 1999. *Astron. J.* In press

Feltzing S, Gilmore G, 1998. In *Galaxy Evolution: Connecting the Distant Universe with the Local Fossil Record*, p. 71

Ferland GJ. 1999. See Ferland & Baldwin 1999, p. 147

Ferland GJ, Baldwin J, eds. 1999. *Quasars and Cosmology*, ASP Conf. Ser. Vol. 162

Ferland GJ, Baldwin JA, Korista KT, Hamann F, Carswell RF, et al. 1996. *Ap. J.* 461:683

Ferland GJ, Korista KT, Verner DA, Ferguson JW, Kingdon JB, et al. 1998. *PASP* 110:761

Ferland GJ, Persson SE. 1989. *Ap. J.* 347:656

Ferland GJ, Peterson BM, Horne K, Welsch WF, Nahar SN. 1992. *Ap. J.* 387:95

Ferland GJ, Shields GA. 1985. See Miller 1985, p. 157

Fisher D, Franx M, Illingworth G. 1995. *Ap. J.* 448:119

Foltz CB, Chaffee F, Weymann RJ, Anderson SF. 1988. See Blades et al, 1988, p. 53

Foltz CB, Weymann RJ, Peterson BM, Sun L, Malkan MA, et al. 1986. *Ap. J.* 307:504

Franceschini A, Gratton R. 1997. *MNRAS* 286:235

Francis PJ, Koratkar A. 1995. *MNRAS* 274:504

Franx M, Illingworth GD. 1990. *Ap. J.* 359:L41

Franx M, Illingworth GD, Kelson DD, Van Dokkum PG, Tran K. 1997. *Ap. J.* 486:75

Friaca ACS, Terlevich RJ. 1998. *MNRAS* 298:399

Friaca ACS, Terlevich RJ. 1999. *MNRAS* 298:399

Frisch H. 1984. In *Methods in Radiative Transfer*, ed. W Kalkofen, p. 65. Cambridge: Cambridge Univ. Press

Gallagher SC, Brandt WN, Sambruna RM, Mathur S, Yamasaki N. 1999. *Ap. J.* In press

Ganguly R, Eracleous M, Charlton JC, Churchill CW. 1999. *Astron. J.* 117:2594

Garnett DR. 1990. *Ap. J.* 363:142

Gaskell CM, Shields GA, Wampler EJ. 1981. *Ap. J.* 249:443

Geisler D, Friel DE. 1992. *Astron. J.* 104:128

Giavalisco M, Steidel CC, Macchetto FD. 1996. *Ap. J.* 470:189

Gnedin NY, Ostriker JP. 1997. *Ap. J.* 486:581

Goodrich RW. 1997. *Ap. J.* 474:606

Goodrich RW, Miller JS. 1995. *Ap. J.* 448:L73

Gorgas J, Efstathiou G, Aragón Salamanca AA. 1990. *MNRAS* 245:217

Graham MJ, Clowes RG, Campusano LE. 1996. *MNRAS* 279:1349

Grandi SA. 1981. *Ap. J.* 251:451

Gratton RG. 1991. In *Evolution of Stars: the Photospheric Abundance Connection, IAU Symp.* vol. 145, p. 27, eds. G Michaud, AV Tutukov. Montreal: Univ. Montreal

Green PJ, Aldcroft TL, Mathur S, Shartel N. 1997. *Ap. J.* 484:135

Green PJ, Mathur S. 1996. *Ap. J.* 462:637

Greggio L, Renzini A. 1983. *Astron. Astrophys.* 118:217

Grevesse N, Anders E. 1989. In *Proc. AIP Conf. Cosmic Abundances of Matter*, Minneapolis, Vol. 183, p. 1, ed. CI Waddington. New York: AIP

Grillmair CJ, Turnshek DA. 1987. In *QSO Absorption Lines: Probing the Universe, Poster Papers*, p. 1, eds. JC Blades, C Norman, DA Turnshek. Baltimore: Space Telescope Science Institute

Guilloteau S, Omont A, McMahon RG, Cox P, Petitjean P. 1997. *Astron. Astrophys.* 328:L1

Haas M, Chini R, Maisenheimer K, Stickel M, Lemke D, et al. 1998. *Ap. J.* 503:L109

Haehnelt MG, Natarajan P, Rees MJ. 1998. *MNRAS* 300:817

Haehnelt MG, Rees MJ. 1993. *MNRAS* 263:168

Haiman Z, Loeb A. 1998. *Ap. J.* 503:505

Hamann F. 1997. *Ap. J. Suppl.* 109:279

Hamann F. 1998. *Ap. J.* 500:798

Hamann F, Barlow TA, Beaver EA, Burbidge

EM, Cohen RD, et al. 1995. *Ap. J.* 443:606
Hamann F, Barlow TA, Cohen RD, Junkkarinen V, Burbidge EM. 1997b. *Ap. J.* 478:80
Hamann F, Barlow TA, Junkkarinen V. 1997c. *Ap. J.* 478:87
Hamann F, Beaver EA, Cohen RD, Junkkarinen V, Lyons RW, et al. 1997d. *Ap. J.* 488:155
Hamann F, Barlow TA, Cohen RD, Junkkarinen V, Burbidge EM. 1997e. See Weymann et al, 1997, p. 187
Hamann F, Cohen RD, Shields JC, Burbidge EM, Junkkarinen VT, et al. 1998. *Ap. J.* 496:761
Hamann F, Ferland GJ. 1992. *Ap. J.* 391:L53
Hamann F, Ferland GJ. 1993a. *Rev. Mex. Astron. Astrof.* 26:53
Hamann F, Ferland GJ. 1993b. *Ap. J.* 418:11
Hamann F, Korista KT. 1996. *Ap. J.* 464:158
Hamann F, Korista KT, Morris SL. 1993. *Ap. J.* 415:541
Hamann F, Shields JC, Cohen RD, Junkkarinen VT, Burbidge EM. 1997a. In *Emission Lines From Active Galaxies: New Methods and Techniques, IAU Col. 159, Astron. Soc. Pacific Conf. Ser. vol. 113*, p. 96, eds. BM Peterson, F-Z Cheng, AS Wilson
Hartquist TW, Snijders MAJ. 1982. *Nature* 299:783
Heap SR, Brown TM, Hubeny I, Landsman W, Yi S, et al. 1998. *Ap. J.* 492:L131
Hines DC, Low FJ, Thompson RI, Weymann RJ, Storrie-Lombardi LJ. 1999. *Ap. J.* 512:140
Hines DC, Wills BJ. 1995. *Ap. J.* 448:L69
Hu EM, Cowie LL, McMahon RG. 1998. *Ap. J.* 502:L99
Idiart TP, De Freitas Pacheco JA, Costa RDD. 1996. *Astron. J.* 112:2541
Isaac KG, McMahon RG, Hills RE, Withington S. 1994. *MNRAS* 269:28
Israelian G, López RJG, Rebolo R. 1998. *Ap. J.* 507:805
Ivison RJ, Dunlop JS, Hughes DH, Archibald EN, Stevens JA, et al. 1998, *Ap. J.* 494:211
Izotov YI, Thuan TX. 1999. *Ap. J.* 511:639
Jablonka P, Alloin D, Bica E. 1992. *Astron. Astrophys.* 260:97
Jablonka P, Martin P, Arimoto N. 1996. *Astron. J.* 112:1415
Johansson S, Jordan C. 1984. *MNRAS* 210:239
Jenkins EB. 1996. *Ap. J.* 471:292
Jin L, Arnett WD, Chakrabarti SK. 1989. *Ap. J.* 336:572
Junkkarinen VT. 1980. PhD dissertation, Univ. Calif. San Diego
Junkkarinen VT, Beaver EA. Burbidge EM, Cohen RC, Hamann F, et al. 1997. See Weymann et al 1997, p. 220
Junkkarinen VT. Burbidge EM, Smith HE. 1983. *Ap. J.* 265:51
Junkkarinen VT, Burbidge EM, Smith HE. 1987. *Ap. J.* 317:460
Katz N, Quinn T, Bertschinger E, Gelb JM. 1994. *MNRAS* 270:L71
Kauffmann G. 1996. *MNRAS* 281:487
King JR. 1993. *Astron. J.* 106:1206
Kinney AL, Rivolo AR, Koratkar AR. 1990. *Ap. J.* 357:338
Kirkman D, Tytler D. 1997. *Ap. J.* 489:L123
Kobulnicky HA, Skillman ED. 1998. *Ap. J.* 497:601
Kodama T, Arimoto N. 1997. *Astron. Astrophys.* 320:41
Köppen J, Arimoto N. 1990. *Astron. Astrophys.* 240:22
Korista KT, Alloin D, Barr P, Clavel J, Cohen RD, et al. 1995. *Ap. J. Suppl.* 97:285
Korista KT, Baldwin JA, Ferland GJ. 1998. *Ap. J.* 507:24
Korista KT, Baldwin JA, Ferland GJ, Verner D. 1997b. *Ap. J. Suppl.* 108:401
Korista KT, Ferland GJ, Baldwin BA. 1997a. *Ap. J.* 487:555
Korista KT, Hamann F, Ferguson J, Ferland GJ. 1996. *Ap. J.* 461:641
Korista KT, Voit GM, Morris SL, Weymann RJ. 1993. *Ap. J. Suppl.* 88:357
Korista KT, Weymann RJ, Morris SL, Kopko M, Turnshek DA, et al. 1992. *Ap. J.* 401:529
Kormendy J, Bender R, Evans AS, Richstone D. 1998. *Astron J.* 115:1823
Krolik J, Voit GM. 1998. *Ap. J.* 497:L5
Kundt W. 1996. *Astrophys. Sp. Sci.* 235:319

Kundtschner H, Davies RL. 1997. *MNRAS* 295: L29
Kwan J. 1990. *Ap. J.* 353:123
Kwan J, Krolik J. 1981. *Ap. J.* 250:478
Laor A. 1999. See Baldwin & Ferland 1999. In press
Laor A. 1998. *Ap. J.* 505:L83
Laor A, Bahcall JN, Jannuzi BT, Schneider DP, Green RF. 1995. *Ap. J. Suppl.* 99:1
Laor A, Jannuzi BT, Green RF, Boroson TA. 1997. *Ap. J.* 489:656
Larson RJ. 1974. *MNRAS* 169:229
Loeb A. 1993. *Ap. J.* 404:L37
Loeb A, Rasio FA. 1994. *Ap. J.* 432:52
Lowenthal JD, Koo DC, Guzman R, Gallego J, Phillips AC. 1997. *Ap. J.* 481:673
Lu L, Sargent WLW, Barlow TA. 1998. *Astron. J.* 115:55
Lu L, Sargent WLW, Barlow TA, Churchhill CW, Vogt SS. 1996. *Ap. J. Suppl.* 107:475
Lu L, Savage BD. 1993, *Ap. J.* 403:127
Madau P, Ferguson HC, Dickenson ME, Giavalisco M, Steidel CC, et al. 1996. *MNRAS* 283:1388
Magain P. 1989. *Astron. Astrophys.* 209:211
Magorrian J, Tremaine S, Richstone D, Bender R, Bower G, et al. 1998. *Astron. J.* 115:2285
Mathews WG, Capriotti ER. 1985. See Miller 1985, p. 183
Matteucci F. 1994. *Astron. Astrophys.* 288:57
Matteucci F, Brocato E. 1990. *Ap. J.* 365:539
Matteuci F, Francois P. 1989. *MNRAS* 239:885
Matteucci F, Greggio L. 1986. *Astron. Astrophys.* 154:279
Matteucci F, Tornambè A. 1987. *Astron. Astrophys.* 185:51
McCarthy PJ. 1993. *Annu. Rev. Astron. Astrophys.* 31:639
McLeod KK, Rieke GH. 1995. *Ap. J.* 454:L77
McLeod KK, Rieke GH, Storrie-Lombardi LJ. 1999. *Ap. J.* 511:L67
McLure RJ, Dunlop JS, Kukula MJ, Baum SA, O'Dea CP, et al. 1999. *Ap. J.* In press
McWilliam A, Rich RM. 1994. *Ap. J. Suppl.* 91:749
Mihalas D. 1978. *Stellar Atmospheres*, New York: Freeman

Miller J, Tran H, Sheinis A. 1996. *BAAS* 28:1031
Miller JS, ed. 1985. *Astrophysics of Active Galaxies and Quasi-Stellar Objects*. Mill Valley, CA: Univ. Sci. Books, 519 pp.
Minniti D, Olszewski EW, Liebert J, White SD, Hill JM, et al. 1995. *MNRAS* 277:1293
Miralda-Escudé J, Rees MJ. 1997. *Ap. J.* 478:L57
Möller P, Jakobsen P, Perryman MAC. 1994. *Astron. Astrophys.* 287:719
Morris SL, Weymann RJ, Foltz CB, Turnshek DA, Shectman S, et al. 1986. *Ap. J.* 310:40
Murray N, Chiang J, Grossman SA, Voit GM. 1995. *Ap. J.* 451:498
Mushotzky RF, Loewenstein M. 1997. *Ap. J.* 481:L63
Netzer H. 1990. In *Active Galactic Nuclei*, eds. RD Blandford, H Netzer, L Woltjer, p. 57. Berlin: Springer
Netzer H, Brotherton MS, Wills BJ, Han M, Baldwin JA, et al. 1995. *Ap. J.* 448:27
Netzer H, Wills BJ. 1983. *Ap. J.* 275:445
Nissen PE, Gustafsson B, Edvardsson B, Gilmore G. 1994. *Astron. Astrophys.* 285:440
Omont A, Petitjean P, Guilloteau S, McMahon RG, Solomon PM. 1996. *Nature* 382:428
Ortolani S, Renzini A, Gilmozzi R, Marconi G, Barbuy B, et al. 1996. In *Formation of the Galactic Halo... Inside and Out, Astron. Soc. Pacific Conf. Ser.* vol. 92, p. 96, eds. H Morrison, A Sarajedini
Osmer PS. 1980. *Ap. J.* 237:666
Osmer PS. 1998. See D'Odorico et al. 1998. p. 1
Osmer PS, Porter AC, Green RF. 1994. *Ap. J.* 436:678
Osmer PS, Shields JC. 1999. See Ferland & Baldwin 1999, p. 235
Osmer PS, Smith MG. 1976. *Ap. J.* 210:276
Osmer PS, Smith MG. 1977. *Ap. J.* 213:607
Osterbrock DE. 1977. *Ap. J.* 215:733
Osterbrock DE. 1989. *Astrophysics of Gaseous Nebulae and Active Galactic Nuclei*. Mill Valley, CA. Univ. Sci. Press, 408 pp.
Peimbert M. 1967. *Ap. J.* 150:825
Penston M. 1987. *MNRAS* 229:1P

Peterson BM. 1993. *PASP* 105:1084
Petitjean P, Rauch M, Carswell RF. 1994. *Astron. Astrophys.* 291:29
Petitjean P, Srianand R. 1999. *Astron. Astrophys.* 345:73
Pettini M, King DL, Smith LJ, Hunstead RW. 1997. *Ap. J.* 486:665
Phillips MM. 1977. *Ap. J.* 215:746
Phillips MM. 1978. *Ap. J.* 226:736
Prochaska JX, Wolfe AM. 1998. *Ap. J.* 507:113
Prochaska JX, Wolfe AM. 1999. *Ap. J.* In press
Rauch M. 1998. *Annu. Rev. Astron. Astrophys.* 36:267
Rauch M, Haehnelt MG, Steinmetz M. 1997. *Ap. J.* 481:601
Renzini A. 1997. *Ap. J.* 488:35
Renzini A. 1998. See D'Odorico et al, 1998, p. 298
Rich RM. 1988. *Astron. J.* 95:828
Rich RM. 1990. *Ap. J.* 362:604
Richards GT, York DG, Yanny B, Kollgaard RI, Laurent-Muehleisen SA, et al. 1999. *Ap. J.* 513:576
Saikia D, Kulkarni AR. 1998. *MNRAS* 298:L45
Sansom AE, Proctor RN. 1998. *MNRAS* 297:953
Sargent WLW, Boksenberg A, Young P. 1982. *Ap. J.* 252:54
Savage BD, Sembach KR. 1991. *Ap. J.* 379:245
Savage BD, Tripp TM, Lu L. 1998. *Astron. J.* 115:436
Savaglio S, Cristiani S, D'Odorico S, Fontana A, Giallong E, et al. 1997. *Astron. Astrophys.* 318:347
Savaglio S, D'Odorico S, Möller P. 1994. *Astron. Astrophys.* 281:331
Scalo JM. 1990. In *Windows on Galaxies*, eds. G Fabbiano, JS Gallagher, A Renzini, Dordrecht: kluwer, p. 125
Schmidt GD, Hines DC. 1999. *Ap. J.* 512:125
Schneider DP. 1998. In *Science with NGST, Astron. Soc. Pacific Conf. Ser. vol. 133. p. 106*, eds. EP Smith, A Koratkar, ASP
Schneider DP, Schmidt M, Gunn JE. 1991. *Astron. J.* 102:837
Searle L, Zinn R. 1978. *Ap. J.* 225:357
Shaver PA, Hook IM, Jackson CA, Wall JV,
Kellerman KI. 1998. In *Highly Redshifted Radio Lines*, eds. C Carilli, S Radford, K Menton, G Langston. PASP. In press
Shields GA. 1976. *Ap. J.* 204:330
Shields GA. 1996. *Ap. J.* 461:L9
Shields JC, Ferland GJ, Peterson BM. 1995. *Ap. J.* 441:507
Shields JC, Hamann F, Foltz CB, Chaffee FH. 1997. In *Emission Lines in Active Galaxies, Astron. Soc. Pacific Conf. Ser. vol. 113, p. 118*, eds. BM Peterson, F-Z Cheng
Shklovskii IS. 1965. *Sov. Astron.* 8:635
Sigut TAA, Pradhan AK. 1998 *Ap. J.* 499:L139
Sil'chenko OK, Burenkov AN, Vlasyuk VV. 1998. *Astron. Astrophys.* 337:349
Silk J, Rees MJ. 1998. *Astron. Astrophys.* 331:L1
Songalia A, Cowie LL. 1996. *Astron. J.* 112:335
Spaans M, Corollo CM. 1997. *Ap. J.* 482:L93
Spinard H, Dey A, Stern D, Dunlop J, Peacock J, et al. 1997. *Ap. J.* 484:581
Srianand R, Shankaranarayanan S. 1999. *Ap. J.* In press
Stanford SA, Eisenhardt PR, Dickenson M. 1998. *Ap. J.* 492:461
Steidel CC, Adelberger KL, Dickenson M, Giavalisco M, Pettini M, et al. 1998. *Ap. J.* 492:428
Steidel CC, Adelberger KL, Giavalisco M, Dickenson M, Pettini M. 1999. *Ap. J.* 519:1
Taniguchi Y, Arimoto N, Murayama T, Evans AS, Sanders DB, et al. 1997. In *Quasar Hosts, ESO-IAC Conf. Pro.* eds. DL Clements, I Perez-Fouron, p. 127
Taniguchi Y, Ikeuchi S, Shioya Y. 1999. *Ap. J.* 514:L12
Tantalo R, Chiosi C, Bressan A. 1998. *Astron. Astrophys.* 333:419
Telfer RC, Kriss GA, Zheng W, Davidsen AF, Green RF. 1998. *Ap. J.* 509:132
Terndrup DM, Sadler EM, Rich RR. 1995. *Astron. J.* 110:1774
Terlevich RJ, Boyle BJ. 1993. *MNRAS* 262:491
Thomas D, Greggio L, Bender R. 1998. *MNRAS* 302:537

Thompson KL, Hill GJ, Elston R. 1999. *Ap. J.* 515:487

Thuan TX, Izotov YI, Lipovetsky VA. 1995. *Ap. J.* 445:108

Thurston TR, Edmunds MG, Henry RB. 1996. *MNRAS* 283:990

Tiede GP, Frogel JA, Terndrup DM. 1995. *Astron. J.* 110:2788

Tinsley B. 1979. *Ap. J.* 229:1046

Tinsley B. 1980. *Fund. Cosmic Phys.* 5:287

Trager SC, Faber SM, Dressler A, Oemler A. 1997. *Ap. J.* 485:92

Tripp TM, Lu L, Savage BD. 1996. *Ap. J. Suppl.* 102:239

Tripp TM, Lu L, Savage BD. 1997. *Ap. J. Suppl.* 112:1

Türler M, Courvoisier TJ-L. 1997. *Astron. Astrophys.* 329:863

Turner EL. 1991. *Astron. J.* 101:5

Turnshek DA. 1981. PhD dissertation. Univ. Ariz.

Turnshek DA. 1984. 280:51

Turnshek DA. 1986. In *Quasars, IAU Symp. 119*, p. 317, ed. G Swarup, VK Kapahi. Dordrecht: Reidel

Turnshek DA. 1988. See Blades et al, 1988, p. 17

Turnshek DA. 1994. In *QSO Asborption Lines*, ed. G Meylan. Berlin: Springer-Verlag, p. 223

Turnshek DA. 1997. See Weymann et al 1997, p. 193

Turnshek DA, Briggs FH, Foltz CB, Grillmair CJ, Weymann RJ. 1987. In *QSO Absorption Lines: Probing the Universe, Poster Papers*, ed. JC Blades, C Norman, DA Turnshek, p. 1

Turnshek DA, Kopko M, Monier E, Noll D, Espey B, et al. 1996. *Ap. J.* 463:110

Turnshek DA, Monier EM, Christopher JS, Espey BR, 1997. *Ap. J.* 476:40

Tytler D, Fan XM, Burles S, Cottrell L, Davis C, et al. 1995. In *QSO Absorption Lines*, ed. G Meylan. Garching: ESO, p. 289

Uomoto A. 1984. *Ap. J.* 284:497

Van Dokkum PG, Franx M, Kelson DD, Illingworth GD. 1998. *Ap. J.* 504:L17

Van Zee L, Skillman ED, Salzer JJ. 1998. *Ap. J.* 497:L1

Vazdekis A, Peletier RF, Beckman JE, Casuso E. 1997. *Ap. J. Suppl.* 111:203

Verner EM, Verner DA, Korista KT, Ferguson JW, Hamann F, et al. 1999. *Ap. J. Suppl.* 120: 101

Vila-Costas MB, Edmunds MG. 1993. *MNRAS* 265:199

Voit GM. 1997. See Weymann et al, 1997, p. 200

Voit GM, Weymann RJ, Korista KT. 1993. *Ap. J.* 413:95

Wampler EJ, Bergeron J, Petitjean P. 1993. *Astron. Astrophys.* 273:15

Wampler EJ, Chugai NN, Petitjean P. 1995. *Ap. J.* 443:586

Wampler EJ, Williger GM, Baldwin JA, Carswell RF, Hazard C, et al. 1996. *Astron. Astrophys.* 316:33

Warren SJ, Hewett PC, Osmer PS. 1994. *Ap. J.* 421:412

Weymann R, Shlosman I, Arav N, eds. 1997. *Mass Ejection from AGN, Astron. Soc. Pacific Conf. Ser.* vol. 128

Weymann RJ. 1994. In *QSO Asborption Lines*, ed. G Meylan. Berlin: Springer-Verlag, p. 213

Weymann RJ. 1997. See Weymann et al, 1997, p. 3

Weymann RJ, Carswell RF, Smith MGA. 1981. *Annu. Rev. Astron. Astrophys.* 19:41

Weymann RJ, Foltz C. 1983. In *Quasars and Gravitational Lenses, 24th Liege Int. Astrophys. Coll.* Belgium: Univ. de Liege, p. 538

Weymann RJ, Morris SL, Foltz CB, Hewett PC. 1991. *Ap. J.* 373:23

Weymann RJ, Turnshek DA, Christiansen WA. 1985. See Miller 1985, p. 185

Weymann RJ, Stern D, Bunker A, Spinrad H, Chaffee FH, et al. 1998. *Ap. J.* 505: L95

Weymann RJ, Williams RE, Peterson BM, Turnshek DA. 1979. *Ap. J.* 234:33

Wheeler JC, Sneden C, Truran JW. 1989. *Annu. Rev. Astron. Astrophys.* 27:279

Williams RE. 1971. *Ap. J.* 167:L27

Williams RE, Strittmatter PA, Carswell RF, Craine ER. 1975. *Ap. J.* 202:296
Williams RE, Weymann RJ. 1976. *Ap. J.* 207:L143
Wills BJ, Netzer H. 1983. *Ap. J.* 275:445
Wills BJ, Netzer H, Wills D. 1985. *Ap. J.* 288:94
Wills BJ, Thompson KL, Han M, Netzer H, Wills D, et al. 1995. *Ap. J.* 447:139
Worthey G, Faber SM, Gonzalez J. 1992. *Ap. J.* 398:69
Yoshii Y, Tsujimoto T, Kawara K. 1998. *Ap. J.* 507:L113
Yoshii Y, Tsujimoto T, Nomoto K. 1996. *Ap. J.* 462:266
Young P, Sargent WLW, Boksenberg A. 1982. *Ap. J. Suppl.* 48:455
Zaritsky D, Kennicutt RC, Huchra JP. 1994. *Ap. J.* 420:87
Zheng W, Kriss GA, Davidsen AF. 1995. *Ap. J.* 440:606
Zheng W, Kriss GA, Telfer RC, Grimes JP, Davidsen AF. 1997. *Ap. J.* 475:469
Ziegler BL, Bender R. 1997. *MNRAS* 291:527

Origin and Evolution of the Natural Satellites

S. J. Peale
Department of Physics, University of California, Santa Barbara, California 93117; e-mail: peale@io.physics.ucsb.edu

Key Words tides, dissipation, dynamics, solar system, planets

1. INTRODUCTION

The natural satellites of planets in the solar system display a rich variety of orbital configurations and surface characteristics that have intrigued astronomers, physicists, and mathematicians for several centuries. As detailed information about satellite properties have become available from close spacecraft reconnaissance, geologists and geophysicists have also joined the study with considerable enthusiasm. This paper summarizes our ideas about the origin of the satellites in the context of the origin of the solar system itself, and it highlights the peculiarities of various satellites and the configurations in which we find them as a motivation for constraining the satellites' evolutionary histories. Section 2 describes the processes involved in forming our planetary system and points out how many of these same processes allow us to understand the formation of the regular satellites (those in nearly circular, equatorial orbits) as miniature examples of planetary systems. The irregular satellites, the Moon, and Pluto's satellite, Charon, require special circumstances as logical additions to the more universal method of origin. Section 3 gives a brief description of tidal theory and a discussion of various applications showing how dissipation of tidal energy effects secular changes in the orbital and spin configurations and how it deposits sufficient frictional heat into individual satellites to markedly change their interior and surface structures. These consequences of tidal dissipation are used in the discussions of the evolutions of satellite systems of each planet, starting with the Earth-Moon system in Section 4. Additional processes such as collisional breakup and reassembly of some of the smaller satellites associated with the ubiquitous equatorial rings of small particles are discussed, although ring properties and evolution are excluded (see Nicholson 1999). There is no attempt to include technical details of published explanations of the various phenomenologies. Rather, the explanations are described and the uncertainties in the assumptions and resulting conclusions are emphasized. The outstanding gaps in our understanding of each system are pointed out, where additional knowledge about a system or one of its parts always seems to generate more interesting problems than it resolves. This is the first time all of the satellite systems have been discussed in any detail in one place, and it is hoped that the reader will find in one or more of these systems problems to which he can apply his own expertise. The Galileo spacecraft has already uncovered some amazing properties of the Jupiter's satellites, which have shaken some of our long-held beliefs about the satellites; the Cassini spacecraft approaching Saturn promises to do the same. Let us build the context, which this and other new information will alter and refine.

2. ORIGINS

The origins of the natural satellites are of course closely linked to the origin of the solar system and the formation of planets therein. So it is appropriate to outline our current understanding and necessary speculations about solar system formation to understand the creation of the satellite systems. The origin of the satellite systems of the major planets can be understood in terms of similar processes and events, sometimes with extreme examples of some events for each system. The Moon and probably Pluto's satellite, Charon, required special circumstances within the broader set of processes that occurred during the early evolution of our planetary system.

It is generally accepted that our planetary system formed from a flat dissipative disk of gas and dust that surrounded the young Sun. The disk is the natural consequence of the collapse of a rotating cloud with angular momentum conservation, where dissipation leads to both the flat geometry and the circular orbits of the disk constituents. This general picture is observationally confirmed, since all young stars appear to have such disks of material for some part of their early existence (Strom 1995, Beckwith & Sargent 1998)—consistent with theoretical expectations. As the disk cools, nonvolatile elements and compounds will condense into small particles that settle to the midplane of the disk, where they collect into larger and larger sizes chiefly through collisional coagulation (Weidenschilling 1995). The nonvolatiles will consist of rock and iron-type materials in the terrestrial planet region close to the Sun but will include water and other ices beyond $\sim$4 AU from the Sun. The latter condensates increase the solids fraction of the nebular disk by a factor of $\sim$3 compared with the terrestrial zone. All of the details of the early coagulation process are not well understood, but the gravitational instabilities in a very thin disk of solid particles thought to dominate the process (Safronov 1969, Goldreich & Ward 1973, Ward 1976) apparently are completely frustrated by shear-induced turbulence and persistent velocity dispersions for the solid bodies (Weidenschilling 1995). Both the shear-induced turbulence and the persistent velocity dispersion result from the fact that partial support of the nebular gas by a radial pressure gradient causes it to orbit the Sun at an angular velocity that is slightly less than the Kepler velocity v_K (Whipple 1972). The deviation from the Kepler velocity is typically tens of meters per second (Weidenschilling 1977). As the solid particles are not supported by the pressure gradient, they tend to orbit at v_K. Hence, all particles large enough not to follow turbulent motions in the gas will face a headwind that causes them to spiral toward the Sun at rates inversely proportional to their linear size. It is therefore crucial that the planetesimals increase in size sufficiently rapidly to avoid being swept into the Sun while the gaseous part of the disk is still present. From observations of the disks around young stellar objects, the disk lifetimes—based on the theoretical ages of the central stars from their positions on the H-R diagram—are only 10 million–30 million years (e.g. Strom 1995).

Once the particles approach kilometer size and larger, the continued accumulation depends on a gravitationally enhanced cross section for two body encounters given by

$$\pi R_g^2 = \pi (R_1 + R_2)^2 \left(1 + \frac{v_e^2}{v^2}\right), \quad (1)$$

in which R_1, R_2 are the radii of planetesimals m_1, m_2; v is the magnitude of their relative velocity; $v_e = 2G(m_1 + m_2)/(R_1 + R_2)$ is an escape velocity from an equivalent planetesimal whose mass is $m_1 + m_2$ and whose radius is $R_1 + R_2$; and R_g is the maximum impact parameter that results in a collision. If we assume $m_1 = m_2 = m$ and $R_1 = R_2 = R$, we can write

$$\frac{dR}{dt} = \frac{\rho v}{\rho_p}\left(1 + \frac{8\pi G \rho_p R^2}{3v^2}\right), \quad (2)$$

where ρ is the mass density in the planetesimal disk, ρ_p is the density of the planetesimals, and v is now an average relative velocity.

What is interesting about Equation 2 is that when $v_e \gg v$, $dR/dt \propto R^2$, and the growth of the largest particle would run away to dominate all the nearby planetesimals. This runaway growth requires that the relative velocity of the accreting planetesimals be small compared with the escape velocity, and ordinarily one would believe that this condition would not prevail since gravitational scattering should lead to a velocity dispersion comparable with the escape velocity. However, there will always be a size distribution with larger numbers of smaller planetesimals, and the distribution of kinetic energies will tend toward equipartition. Stewart and Wetherill (1988) have shown that for reasonable size distributions, this equipartition is sufficient to keep the relative velocities of the larger particles small compared with their escape velocities and the runaway growth will continue until a planetary embryo has consumed most of the mass within its gravitational reach ($\Delta r/r = C(m/3M_\odot)^{1/3}$, with $C = 3$–4 and r being heliocentric distance). This notion has been verified by numerical three-body accretion calculations that include the effect of the Sun (Greenzweig & Lissauer 1990, 1992) and leads to planetary embryo masses of $m = 2 \times 10^{24}(Cr^2\sigma)^{3/2}$ (Lissauer 1987) with r in astronomical units and surface mass density of the nonvolatile fraction of the nebula σ in grams per square centimeter. For a minimum mass nebula (nonvolatile constituents of the planets distributed over heliocentric distance and augmented with sufficient hydrogen, helium, and other volatiles until solar composition is attained), planetary embryos in the Earth zone of a few lunar masses and of near an Earth mass in the Jupiter zone result. Inclusion of the effects of gas drag (Kary & Lissauer 1993) in resupplying depleted planetesimals can augment these embryo masses, as can increasing the surface mass density of the condensed part of the nebula above its minimum value. Recent N-body calculations by Kokubo and Ida (1998) arrive at somewhat larger embryos with wider separations after the runaway accretions have ceased. This latter result illustrates the continuing evolution of ideas for the accretion process. The planetesimals and embryos will consist of rock and iron in the terrestrial planet region, but, beyond 4 or 5 AU from the Sun, most of the embryo mass will consist of water and other ices while also including the refractory constituents.

In this scenario, the planetary embryos form by runaway accretion in a relatively short time ($\sim 10^5$ years), but continued growth into planetary bodies requires a much longer stochastic process [$\sim 10^8$ years (Wetherill 1990) in the terrestrial planet region], where neighboring embryos are perturbed into crossing orbits where they may collide and merge into larger bodies. Numerical Monte Carlo calculations of this process have been remarkably successful in accumulating sets of terrestrial planets that resemble the solar system distribution (Wetherill 1990, 1991, 1996), including the characteristics of the asteroid belt (Wetherill 1992). The qualititative results of these Monte Carlo calculations have been confirmed with numerical calculations using a modification of a symplectic integrator (Wetherill & Chambers 1997).

Because Jupiter and Saturn contain a higher fraction of heavy elements than does the Sun, their formation in a two-stage process by the accretion of a massive core of the solid material with subsequent accretion of the gas onto this core is argued. However, for a minimum-mass solar nebula it takes too long to accumulate the required 10–15 $M_\oplus$ core, and the nebula is gone before the gas can be attracted and held. This conclusion depends on the estimates of nebular lifetimes about young stars of 10 million–30 million years being correct and applying to the solar nebula. The necessary core can form quickly enough as an embryo by runaway accretion if the surface mass density of the nebula is increased by a sufficiently large factor over the minimum mass nebula as demonstrated by Lissauer (1987) and incorporated by Pollack et al (1996) in their scheme to form Jupiter. Although Pollack et al obtain the accretion of a complete Jupiter, including the gaseous part, several of the assumptions, such as maintenance of a very large gravitationally enhanced accretion cross section throughout the core formation, and neglected phenomena, such as the major scattering of the Jupiter core by nearby embryos of comparable size, mean that the two-stage formation of Jupiter and Saturn is still not convincingly demonstrated.

Boss (1997, 1998) has reopened the possibility that Jupiter and Saturn actually formed very quickly by a gravitational instability in the gaseous disk. In this case, a problem remains in supplying a sufficient amount of heavy elements through later accretion of solid planetesimals. Also, the perturbations by full-sized Jupiter and Saturn would stir the planetesimal disk sufficiently to hinder the runaway growth that created the large embryos needed for the successful modeling of terrestrial planet formation. Some also question whether the nebula could be simultaneously cool enough and dense enough for Boss's conditions of gravitational collapse to be met. (Toomre's 1964 Q factor must indicate instability.) We must conclude that there has been no robust method demonstrated for the formation of Jupiter or Saturn within observational constraints.

Uranus and Neptune apparently formed their cores more leisurely since the nebula was by then too thin to contribute much gas to their bulk. Still, core formation at this distance from the Sun requires $>10^9$ years in a minimum-mass solar nebula (Stewart & Levison 1998), so perhaps these planets should have received less gas than is apparent.

The remaining problems with planet formation aside, however the major planets formed—either by the two-stage process or by gravitational instability—the material going into these planets would have had significant angular momentum relative to the center of mass of the forming planet. For Jupiter and Saturn, the ultimate spin of the planet is determined largely by angular momentum contributed by the gaseous component. Neptune and Uranus accreted a relatively small fraction of their mass as hydrogen and helium and the terrestrial planets, essentially none. So the planetesimal accretion determined the angular momenta of these planet-satellite systems. Each accreting planetesimal contributes part of its angular momentum to the spin and part to the orbit of the combined planet-planetesimal mass, the relative contributions being determined by the details of the collision. Thus, both the magnitude and the direction of the spin vector undergo random walks as the accretion progresses, where the magnitude of the steps grows with the sizes of the planetesimals (Dones & Tremaine 1993). The wide scatter of planetary spin vectors testifies to the fact that the last stages of accretion involved very large planetesimals, although Mercury and Venus have had their spins altered by dissipative processes. This stage is especially dramatic for Uranus, whose spin axis is inclined by $97°$ relative to its orbit normal.

For Jupiter and Saturn, the cross-section for continued accretion of solid planetesimals will be enhanced by an extended gaseous atmosphere, where those planetesimals that do not plunge directly into the planet will be captured by gas drag or collisions into orbits whose eccentricities and inclinations relative to the equator plane are damped by the continued atmospheric gas drag and collisions with other orbiting debris, and the orbit semimajor axes will decrease at rates that depend on the size of the planetesimal. Even after the accretion is nearly complete and the atmosphere has continued its collapse toward the planet as it cools, any remaining debris in orbit will damp down to circular equatorial orbits. This damping can be understood by considering a hypothetical ring of orbiting planetesimals that is inclined to the equator plane of the planet. Because each planetesimal has a slightly different orbit, the orbits will precess at different rates owing to the oblate distribution of mass in the rotating planet. The ring will tend to spread into a cylindrical distribution where collisions in crossing orbits within the distribution will remove the components of velocity perpendicular to the equator plane. Eccentricities are similarly damped by dissipative collisions. The satellitesimals will then accumulate into satellites on much shorter time scales than for the analogous accretional accumulation of the planets in the solar nebula (Pollack 1985). This rapid accretion into large objects may be crucial to the survival of satellite material as the drag from the waning gaseous component of the disk causes it to spiral into the planet. The debris around Uranus and Neptune relaxed to the oblique equator planes before accumulating into their equatorial satellites. Consistent with this picture of satellite formation, the orbits of the closer satellites of the major planets are nearly circular and nearly equatorial as shown in Table 1.

The regular satellites of the major planets thus form naturally by the accretion of debris in a dissipative disk much like the process in the solar nebula leading to the planetary bodies. The irregular satellites were later captured from the remaining planetesimal swarm either by three-body interactions within the planetary sphere of influence, by collision with debris already orbiting the forming planet, or by gas drag in the extended primordial atmosphere of the forming planet. The last capture mechanism may be the least effective, since sufficient gas to capture the satellite is most probably also sufficient to cause the satellite to spiral into the planet before the atmosphere dissipates. The two families of distant irregular satellites orbiting Jupiter (one family in retrograde orbits) could be products of a disintegration in which single parent bodies were shattered either in collisions causing the captures or by later collisions with high-speed cometary bodies. Alternatively, gas drag from the extended atmosphere has been proposed for both the capture and the breakup of weakly bound parent bodies. Although the timing and the rate of atmosphere removal are critical in this scenario if the captured satellites are not to spiral into the planet, the occupancy of orbital resonances by two of the retrograde satellites implies that at least a thin atmosphere was necessarily in place at the time of their capture (Saha & Tremaine 1993). The capture of an intact satellite by tidal torques on the single first pass is possible for such a miniscule volume of the phase space of initial conditions that it most probably did not occur (see Boss & Peale 1986).

The origins of the Moon and of Pluto's satellite, Charon, require special consideration related to the large amount of specific angular momentum in each system. The origin and evolution of the Moon has understandably received more attention than has any other satellite. The analysis began with GH Darwin (1879, 1880) when he realized that the consequences of the tidal distortion of the Earth by the Moon would lead to an orbital evolution requiring the Moon to have once been very close to the Earth. Darwin suggested that the Moon formed from the outer layers of the Earth—a variation of the rotational fission type of origin elaborated during this century. Several other theories of origin have been proposed and opposed over the intervening years, including intact capture. Each variation of these theories is criticized in detail by Boss & Peale (1986), and all are rejected except formation from the debris resulting from the giant impact of a Mars-sized body (Hartmann

& Davis 1975). This idea has grown in popularity because it naturally provides the large angular momentum of the Earth-Moon system, it can easily account for the lack of volatiles and iron on the Moon if the impacting body is differentiated, and it is much more probable than any means of intact capture (Cameron & Ward 1976). The debris from a giant impact does not follow ballistic trajectories, which would automatically reimpact the Earth. Pressure gradients and the distribution of mass in the ejected debris cause accelerations after the impact that leave material in initially stable orbits. Recall that we concluded earlier that the final stages of accretion of the terrestrial planets involved planetary embryos from Moon-to-Mars size, so it is entirely reasonable that such an impact occurred.

Although the giant impact origin of the Moon has surfaced as the only viable scheme by a process of elimination, the consequences of the impact depend very much on the choices of many free parameters, such as the initial angular momentum and mass of the impactor. Large regions of the parameter space have been and are being explored numerically with various amounts of condensible silicate-like material being placed in orbit beyond the Roche radius, where it can collect into the Moon (Cameron 1997, Canup & Esposito 1996, Ida et al 1997, Cameron & Canup 1998a,b and references therein). (Inside the Roche radius, $(r \sim 2.5R_E(\rho_P/\rho_C)^{1/3} \sim 2.9R_E)$ tidal forces would break apart a fluid satellite and therefore inhibit accretion.) The orbiting debris from the impact would damp down to circular equatorial orbits through collisions, where it can accrete into the Moon. For many if not most of the sets of choices of the many parameters, insufficient material ends up in orbit beyond the Roche radius. Details along the route from impact to Moon are still uncertain, so improving the resolution and reliability of the numerical calculations remains an active area of research. These processes will be discussed further in Section 4.

A giant impact is also proposed as the origin of the Pluto-Charon binary system (McKinnon 1989, Stern 1991, Tancredi & Fernández 1991), and by the same process of elimation that was applied to the origin of the Moon, this method of origin again emerges as the only viable option (Dobrovolskis et al 1997).

The two satellites of Mars are in nearly circular equatorial orbits, which supports the argument of their accretion in situ from a debris disk in the equatorial plane left over from the planet's formation—like the regular satellites of the major planets. However, the Phobos mean density was estimated to be 2.0 ± 0.5 g/cm^3 from Viking orbiter data (Christensen et al 1977, Tolson et al 1978), and that of Deimos 2.0 ± 0.7 g/cm^3 (Duxbury & Veverka 1978). The difference from the mean density of Mars of 3.9 g/cm^3 lends support to the suggestion that the satellites are composed of material that did not originate in the vicinity of Mars, and albedos near 5% and reflection spectra are consistent with carbonaceous chondritic material (Pang et al 1978, Pollack et al 1978). Intact capture was considered (Pollack & Burns 1977, Mignard 1981b), but Szeto (1983) showed several seemingly insurmountable inconsistencies with the capture hypothesis independent of the impossibility of relaxing the captured satellites into the circular equatorial orbits where they are found. Despite the low density and spectral indications of carbonaceous chondrite

TABLE 1 Physical properties and orbital characteristics of the satellites.[a]

Satellite	R^b (km)	m (10^{20} kg)	ρ (g cm^{-3})	P_o (days)	P_r (days)	c (10^3 km)	e	i (deg)
Earth	6378	59742	5.515					
E1 Moon	1737.5	734.9	3.34	27.322	S	384.40	0.0549	5.15
Mars	3394	6418.5	3.933					
M1 Phobos	11.2	1.08×10^{-4}	1.90	0.3189	S	9.377	0.0151	1.082
M2 Deimos	6.3	1.80×10^{-5}	1.76	1.2624	S	23.463	0.00033	1.791
Jupiter	71492	1.8988×10^7	1.326					
JXVI Metis	20			0.2948	S	127.96	<0.004	~0
JXV Adrastea	10			0.2983	S	128.98	~0	~0
JV Almathea	90			0.4982	S	181.3	0.003	0.40
JXIV Thebe	50			0.6745		221.9	0.015	0.8
JI Io	1821	893.3	3.530	1.7691	S	421.6	0.0041	0.040
JII Europa	1565	479.7	2.99	3.5518	S	670.9	0.0101	0.470
JIII Ganymede	2634	1482	1.94	7.1546	S	1,070	0.0015	0.195
JIV Callisto	2403	1076	1.85	16.6890	S	1,883	0.007	0.281
JXIII Leda	5			238.72		11,094	0.148	27[c]
JVI Himalia	85			250.5662	0.4	11,480	0.163	28[c]
JX Lysithea	12			259.22	0.53	11,720	0.107	29[c]

JVII Elara	40				259.6528	0.207	28[c]
JXII Ananka	10				631R	0.169	147[c]
JXI Carme	15				692R	0.207	163[c]
JVIII Pasiphae	18				735R	0.378	148[c]
JIX Sinope	14				758R	0.275	153[c]
Saturn	60268	5.6850×10^6	0.687				
SXVIII Pan	10				0.5750	~0	~0
SXV Atlas	16.4				0.6019	~0	~0
SXVI Prometheus	53	0.0014	0.27		0.6130	0.0024	0.0
SX VII Pandora	43	0.0013	0.42		0.6288	0.0042	0.0
SXI Epimetheus	60	0.0055	0.63	S	0.6946	0.009	0.34
SX Janus	90	0.0198	0.65	S	0.6946	0.007	0.14
SI Mimas	198.8	0.375	1.14	S	0.9424	0.0202	1.53
SII Enceladus	249	0.73	1.12	S	1.3702	0.0042	0.02
SIII Tethys	530	6.22	1.00	S	1.8878	0.0000	1.09
SXIV Calypso(T−)	10				1.8878	~0	~0
SXIII Telesto(T+)	12				1.8878	~0	~0
SIV Dione	560	10.52	1.44	S	2.7369	0.0022	0.02
SXII Helene(T+)	16				2.7369	0.005	0.2
SV Rhea	764	23.1	1.24	S	4.5175	0.001	0.35
SVI Titan	2575	1345.5	1.881	S	15.9454	0.0292	0.33

(*continued*)

TABLE 1 (continued)

Satellite	R^b (km)	m (10^{20} kg)	ρ (g cm^{-3})	P_o (days)	P_r (days)	a (10^3 km)	e	i (deg)
SVIII Hyperion	146			21.2766	C	1481.1	0.1042	0.43
SVII Iapetus	718	15.9	1.02	79.3302	S	3651.3	0.0283	7.79[c,f]
SIX Phoebe	110			550.48R	0.4	12952	0.163	175.3[c]
Uranus	25559	8.6625×10^5	1.318					
UVI Cordelia	13			0.3350		49.752	0.000	0.1
UVII Ophelia	16			0.3764		53.764	0.010	0.1
UVIII Bianca	22			0.4346		59.165	0.001	0.2
UIX Cressida	33			0.4636		61.777	0.000	0.0
UX Desdemona	29			0.4737		62.659	0.000	0.2
UXI Juliet	42			0.4931		64.358	0.001	0.1
UXII Portia	55			0.5132		66.097	0.000	0.1
UXIII Rosalind	29			0.5585		69.927	0.000	0.3
UXIV Belinda	34			0.6235		75.255	0.000	0.0
UXV Puck	77			0.7618		86.004	0.000	0.3
UV Miranda	236	0.659	1.20	1.413	S	129.8	0.0027	4.22
UI Ariel	579	13.53	1.67	2.520	S	191.2	0.0034	0.31
UII Umbriel	585	11.72	1.49	4.144	S	266.0	0.0050	0.36
UIII Titania	789	35.27	1.71	8.706	S	435.8	0.0022	0.10

UIV Oberon	761	30.14	1.63	13.463		582.6	0.0008	0.10
UXVI Caliban	~30			580		7.17×10^5	0.0823	139.68[d]
UXVII Sycorax	~60			1290		122.1×10^5	0.509	152.67[d]
Neptune	24764	1.0278×10^6	1.683					
NIII Naiad	29			0.2944		48.227	0.000	4.74
NIV Thalassa	40			0.3115		50.075	0.000	0.21
NV Despina	74			0.3347		52.526	0.000	0.07
NVI Galatea	79			0.4287		61.953	0.000	0.05
NVII Larissa	96			0.5547		73.548	0.000	0.20
NVIII Proteus	209			1.1223		117.647	0.000	0.55
NI Triton	1353	214.7	2.054	5.8769R	S	354.76	0.000	156.834
NII Nereid	170			360.1362		5513.4	0.7512	7.23[c]
Pluto	1170	131.5	2.0					
PI Charon	586	19.0	2.24	6.3872	S	19.405	0.0076[e]	0

[a] These data are updated slightly from Yoder 1995.
[b] The satellite radii are averages if the body is very asymmetric. See Yoder (1995) for more complete descriptions of satellite shapes, for uncertainties in the determinations of all table entries, and for the primary references.
[c] Relative to the Laplacian plane.
[d] Relative to the ecliptic plane of AD 2000 (P Nicholson. Private communication).
[e] Tholen and Buie (1997).
[f] Vienne & Duriez 1995. S, synchronous; C, chaotic; R, retrograde; T, Trojan.

material, the coplanar equatorial orbits indicate that these satellites must have formed from a dissipative disk of debris orbiting the planet. If that debris were indeed carbonaceous chondritic (which is not at all certain), one possible way it could have gotten into orbit about Mars would be from the shattering of such a planetesimal that was formed in the asteroid belt region of the nebula when it collided with a denser object already in orbit about Mars. The condition here would be that the pieces would have to be sufficiently small and of sufficient number to make a dissipative disk. Samples of both Phobos and Deimos would tell us if such a contrived origin were necessary. Any scheme to capture these satellites intact and bring them into their current orbits cannot survive close inspection of the assumptions involved.

Although consensus on many details is still elusive, we have plausible origins for all of the known natural satellites. It is well known that the current distribution of satellite orbits is not the initial distribution nor are even the masses and perhaps the total number the same. The larger satellites are likely to have remained intact under the bombardment of now high-speed comets or asteroid-type planetesimals, but the smaller ones may have been shattered and possibly recollected into new bodies several times in the history of the solar system. Some of the debris from such collisions remains as rings of smaller particles. The densely cratered surfaces of many of the satellites—the Moon, in particular—are testimonies to the flux of impactors after the satellites were formed.

The changes so far described have resulted from stochastic, short-time-scale events characterized by collisions. However, the most interesting changes in the satellite systems involve the orbital modifications and changes in the satellites themselves caused by gravitational tides raised on the primaries by the satellites and raised on the satellites by the primary. Angular momentum is transferred between a spinning planet and its closer satellites, causing either a reduction in the spin rate while expanding the satellite orbits or an acceleration of the spin while the orbit contracts. The direction of the transfer depends on the relative spin and orbital angular velocities. Differential expansion of the satellite orbits leads to the establishment of the many examples of orbital resonances in which the ratio of the orbital mean angular velocities is that of two small integers. The resonances are stable against further tidal evolution up to a point. Dissipation of tidal energy within the satellites has led to striking consequences. From Table 1, we see that all of the closer satellites are rotating synchronously with their orbital motion—a natural consequence of tidal evolution. But sometimes the tidal retardation of a spin leads to chaotic tumbling when synchronous rotation is unstable, as it is for Hyperion.

Several of the satellites have surfaces much younger than their age as indicated by the paucity of impact craters, which were probably erased by indigenous processes energized by tidal dissipation. Tidal dissipation may have maintained a liquid water ocean under the ice of the Jupiter satellite, Europa. It might have softened the interiors of Ganymede to account for its partial resurfacing and perhaps an interior sufficiently hot to support the observed intrinsic magnetic field (Kivelson et al 1997). The most striking consequence of tidal dissipation in a satellite remains the remarkable volcanic activity on Io (Peale et al 1979, Morabito et al 1979, Smith et al 1979), which is the most volcanically active body in the solar system.

Whereas the dissipation in Io is ongoing, current configurations of other satellites do not imply tidal dissipation of sufficient magnitude to account for their observed resurfacing. In such cases historical configurations are inferred, in which past resonances could have forced orbital eccentricities of sufficient magnitude to cause the needed dissipation. Many of these inferred histories are uncertain; for several satellites, no history has yet been constructed that could account for current properties. Below we outline current ideas about the evolution of the satellite systems as constrained by the growing list of observational facts. There are some fascinating and unusual stories, and many of these stories are themselves evolving. We point out areas of research where plausible evolutions are still being sought to explain observed phenomenology. But first we describe briefly the most important process in this evolution—that of the tides.

3. TIDES

Every satellite or planet is distorted into a (slightly) prolate shape (i.e. exhibits a tide) by the gravitational field of another mass. The distortion is understood in terms of the gradient of the external field, where different parts of the body experience different accelerations leading to the distortion. By changing its shape, the tidally distorted body compensates the spatially varying accelerations from the external field by altering its own self-gravitational field and by internal stresses—thereby allowing the entire body to accelerate as one (See Peale 1999 for a heuristic discussion). For a spherically symmetric distribution of mass within the body acted on by the external potential, the potential due to the resulting tidal distribution of mass is proportional to the external potential evaluated at the surface. The potential decreases as $1/r^3$ outside the surface when the external potential is a second-order spherical harmonic (e.g. Love 1944). Hence,

$$V_T = -k_2 \frac{GMma_e^5}{2\mathcal{R}^5 r^5}[3(\vec{\mathcal{R}} \cdot \mathbf{r})^2 - r^2\mathcal{R}^2], \tag{3}$$

is the dominant term in the potential at $\mathbf{r}$ due to a tide raised by M located at $\vec{\mathcal{R}}$ or the potential at $\vec{\mathcal{R}}$ from a tide raised by m at $\mathbf{r}$. In Equation 3, G is the gravitational constant, a_e is the equatorial radius of the tidally distorted body, and k_2 is the potential Love number for the second-order harmonic potential. For a small satellite, k_2 is adequately approximated by its value for a homogeneous, rigid sphere (e.g. Love 1944).

$$k_2 = \frac{3/2}{1 + \frac{19\mu}{2\rho g R}} \approx \frac{3}{19} \frac{\rho g R}{\mu}, \tag{4}$$

in which μ is the rigidity, ρ is the density, g is the surface gravity acceleration, and R is the satellite radius. Because we use no other Love number, we replace the subscript 2 with a letter symbol indicating the satellite or planet to which the Love number applies. Real planets and larger satellites are not homogeneous spheres, and k_2 must be calculated numerically as part of the solution of a boundary value problem for centrally condensed and layered planets and satellites (e.g. Alterman et al 1959).

If M and m are the same mass, then

$$\vec{\mathcal{R}}(t) = \mathbf{r}(t - \Delta t) = \mathbf{r}(t) - \frac{d\mathbf{r}}{dt}\Delta t + \cdots,$$

In other words, $\vec{\mathcal{R}}$ is the position of m a short time Δt in the past relative to a coordinate system fixed in the tidally distorted body. This representation accounts for the phase lag in the response of the body relative to the tide raising potential caused by dissipation of tidal energy. In determining the forces in the equations of motion owing to the tidal potential, one does not differentiate with respect to the coordinates $\vec{\mathcal{R}} = \mathbf{r}(t - \Delta t)$. That is, one only differentiates with respect to the coordinates of m considered as the body perturbed by the potential, not as the body raising the tide (Kaula 1964).

The dissipation parameter $Q = 2\pi E_0/\Delta E$, where E_0 is the maximum energy stored in an oscillation and ΔE is the energy dissipated over a complete cycle, is related to Δt by $\omega \Delta t = 1/Q$; ω being the frequency of oscillation of a particular term in the Fourier expansion of Equation 3 that accounts for the orbital and rotational motions. Constant Δt means Q is inversely proportional to frequency, and this formulation of the tidal interaction is entirely equivalent to the expanded form used by Kaula (1964), with Q having this frequency dependence. For numerical work, this form has the advantage of being applicable to arbitrary eccentricity and inclination of the relative orbit.

The above tidal model is about the simplest conceivable, as it is based on the assumption of an equilibrium tide justified by the fact that principal frequencies of tidal oscillations are small compared with natural oscillation frequencies of typical solar system bodies. For small tidal distortions, the distortion is proportional to the distorting force, which encourages the analogy with the harmonic oscillator. For a homogeneous solid sphere, the representation of the dissipation as a simple phase lag in the response of the body to the perturbation and the assumed frequency dependence of the dissipation resulting from constant Δt are also what results in a linearly damped harmonic oscillator. For $Q \propto 1/\text{frequency}$, dissipation vanishes for low frequency and becomes very large for high frequency. Both extremes are outside the range of applicability of the model—the former because a viscouslike creep can set in for sustained strains and the latter because inertial effects must be included in a now-dynamic analysis, and the equilibrium tide will not prevail. Showman et al (1997) invoke a Maxwell model solid, which behaves elastically for high-frequency oscillations but viscously for low-frequency oscillations. The Maxwell model has the advantage that the viscosity, and hence Q, can be given a model temperature dependence, which has profound and interesting effects in the coupled orbital and thermal evolution of Ganymede to which it is applied. For limited variation in temperature, $Q \propto 1/\text{frequency}$ seems to be appropriate for liquids, but empirical studies of Q in Earth-like materials yield nearly constant values over many orders of magnitude in frequency (Knopoff 1964). Even though these tests do not approach tidal frequencies, constant Q is often assumed in calculations of time scales for tidal evolution. This assumption is probably a fair approximation if frequencies and properties of the planets and satellites involved do not change significantly over the time of evolution, or if the resulting time scales are so extreme that more elaborate models for Q would not affect the conclusions. We will adopt this practice in most of what follows, where the necessity for more elaborate models for the tides is pointed out where needed. One instance where traditional tidal models will probably have to be abandoned is in the determination of the dissipative properties of the giant gaseous planets, for it is becoming increasingly clear that the full dynamics of atmospheric response to the tide-raising potential must be considered (Ioannou & Lindzen 1993).

The simple tide is a poor representation of the tides in the oceans of Earth, where basins defined by the continents have sloshing periods that are comparable with those of the lunar and solar tides. The resultant tides are often far out of phase with the forcing potential. Despite this difficulty, the simple tide has been used for studies of the Earth-Moon history in hopes that the complicating effects of the oceans will average to something close to those of the simple tide. Formulating an evolutionary tidal model is difficult, since the effects of the ocean tides on lunar orbit evolution have changed in largely unknown ways as the continental configuration has changed. It is usually assumed that the tidal effective Q was larger in the past when the continents were joined. With the measured tidal effective Q appropriate to current expansion of the lunar orbit, the Moon would have been at the Earth surface less than 2 billion years ago. Attempts to invoke more complicated tidal models for the Earth have not produced qualitatively different plausible evolutions of the lunar orbit (Kaula 1969).

The phase lag in the response of the tidally distorted body results in misalignment of the tidal bulge with the body raising the tide and leads to the gravitational torque that transfers angular momentum between the spin of the tidally distorted body and the orbit of the body raising the tide (e.g. see Peale 1999). The angular phase lag in a periodic tidal oscillation is $1/Q$, and this lag leads to a geometric angle between an ideal tidal bulge and the direction to the tide-raising satellite of $1/2Q$ since the tide has a $180°$ geometric symmetry. The torque on a satellite in a circular equatorial orbit about a planet deduced from Equation 3 or a Fourier expansion thereof is (e.g. MacDonald 1964)

$$T = \frac{3}{2} \frac{k_2 G m_S^2 R_P^5}{a^6 Q_P}, \qquad (5)$$

where the subscripts S and P refer to satellite and planet respectively, and a is the semimajor axis of the orbit. If the satellite is rotating relative to the primary, there will be a similar expression with the indices interchanged for the torque exerted on the planet. The equal and opposite torque on the tidally distorted body results in a change in the spin angular velocity $\dot\psi$ at a rate given for a small satellite with constant Q by

$$\frac{d\dot\psi}{dt} = -\frac{45}{76}\frac{\rho n^4 R^2}{\mu Q}\,\mathrm{sign}(\dot\psi - n), \qquad (6)$$

where n is the mean orbital angular velocity (mean motion), the approximate form of Equation 4 has been used, and the moment of inertia, $2m_S R^2/5$, used is that for a homogeneous sphere.

Both the torque on the satellite by tides raised on the planet and the torque on the planet by satellite tides affect the orbital motion. However, the tidally evolved satellites will be locked in synchronous rotation, and tides raised on these satellites will not contribute to changes in the orbital angular momentum. For nearly all satellites, the spin angular momentum is such a small fraction of the orbital angular momentum that satellite effects on the orbital motion are negligible in any case. The rate of change of the orbital mean motion (mean orbital angular velocity) from the torque on an equatorial satellite in circular orbit is then (with Q_P constant) (Peale 1988)

$$\frac{dn}{dt} = -\frac{9}{2}\frac{k_P}{Q_P}\frac{m_S}{m_P}\frac{R_P^5 n^{16/3}}{[G(m_P + m_S)]^{5/3}}\,\mathrm{sign}(\dot\psi - n). \qquad (7)$$

If m_S were also spinning nonsynchronously in the same sense as its orbital motion, but with nonzero obliquity (spin axis inclined relative to the orbit normal), the tidal bulge on m_S would be carried out of the orbit plane, and there would be a component of the torque perpendicular to the spin axis. Both the spin magnitude and its direction will thus change as a result of tidal dissipation. The obliquity of the spinning satellite, if initially small, would tend to increase from the tides, if the spin is fast. This behavior can be understood if we resolve the spin angular momentum into components perpendicular and parallel to the orbit plane. The tide is decreasing the magnitude of the perpendicular component at all points in the orbit, whereas it does not reduce the parallel component when that component is pointing toward or away from M. Averaged around the orbit, the fractional change in the perpendicular component exceeds the fractional change in the parallel component, and the obliquity increases. There is an equilibrium obliquity between $0°$ and $90°$ toward which the spin axis is driven where the fractional changes in the two components are the same (Goldreich & Peale 1970). This equilibrium is close to $90°$ for fast spins, but it decreases toward $0°$ as the spin is retarded. If the spin angular velocity $\dot\psi < 2n$, the equilibrium obliquity is $0°$, so a satellite in a fixed circular orbit should approach synchronous rotation and zero obliquity simultaneously.

If the orbit is eccentric, the tidal torque averaged around the orbit vanishes at a spin angular velocity that is slightly larger than the synchronous value. From Equation 5 we see that the tidal torque on the satellite varies as m^2/r^6. One factor of m/r^3 determines the magnitude of the tide, being the difference of $1/r^2$ forces, whereas the second factor of m/r^3 comes from the differential force on the two tidal bulges. As the maximum torque occurs at the orbit periapse, the tides will try to synchronize the spin with the instantaneous value of the orbital angular velocity at this point, which is larger than n. Satellites in eccentric orbits maintain synchronous rotation in spite of a slightly faster spin favored by the tides because much larger torques on the permanent, nonaxially symmetric distribution of mass force the long axis of such a satellite to librate about the direction to the planet when the satellite is at periapse. It can only so librate while maintaining a spin that is a half-integer multiple of its orbital mean motion, where synchronous rotation is the overwhelmingly most common example. The spin of Mercury is locked at $1.5n$, but no satellite is in such a higher-order spin state. If the permanent deviation from axial symmetry is smaller than that induced by the tidal forces, the tides could win, and the endpoint of the evolution would be a spin slightly faster than synchronous. There is evidence that at least the surface ice layer on Europa may be rotating slightly faster than the synchronous rate (Geissler et al 1998, Greenberg et al 1998).

Tidal dissipation will persist in a synchronously rotating satellite in an eccentric orbit at a rate given by (Peale & Cassen 1978)

$$\frac{dE}{dt} = \frac{21}{2} \frac{k_S f}{Q_S} \frac{G m_P^2 n R_S^5 e_S^2}{a^6}, \qquad (8)$$

where the factor $f \geq 1$ has been added to account for an increase in k_S if there is a molten core (Peale et al 1979). The tide will oscillate in magnitude as the satellite-planet distance varies, and it will oscillate in a direction relative to the coordinate system fixed in the satellite because the rotation is nearly uniform, whereas the orbital motion is not. This dissipation will tend to reduce the eccentricity e, as $e \neq 0$ is the cause of the dissipation. The spin angular momentum of the satellite is conserved because of the lock into synchronous rotation. The specific orbital angular momentum $[G(m_P + m_S)a(1 - e^2)]^{1/2}$ can thus not gain angular momentum from the satellite. The orbital energy, $-G m_P m_S/2a$, must decrease if energy is dissipated in the satellite, and a must thereby decrease. But the conserved angular momentum means that e must also decrease if a decreases. At the same time, the tide raised on the planet by the satellite tends to increase the eccentricity. The greater tidal force on the satellite at the orbit periapse tends to fling the satellite to a greater apoapse distance than it would have reached without the kick–thereby increasing e. The variation in the eccentricity from the two effects is (Goldreich 1963)

$$\frac{de}{dt} = \frac{57}{8} k_P n \frac{m_S}{m_P} \frac{R_P^5}{a^5} \frac{e}{Q_P} - \frac{21}{2} k_S n \frac{m_P}{m_S} \frac{R_S^5}{a^5} \frac{e}{Q_S}. \qquad (9)$$

The dissipation in the satellite wins if $19 k_P Q_S m_S R_S \rho_S^2 / 28 k_S Q_P R_P \rho_P^2 < 1$). This situation is generally true for the giant planets, so free eccentricities of these satellites will tend to damp.

For large planets, those satellites for which the decay time constants depending on the satellite parameters (e.g. k_S, r, R, Q) are short compared with the age of the solar system, should be in nearly circular orbits and be synchronously rotating with their orbital motions with nearly zero obliquity. Table 1 shows that all the satellites that would be expected to be synchronously rotating are doing so. Some of the orbital eccentricities are forced to significant values in orbital mean motion resonances, however, and the obliquity of the Moon is 6.67°.

The tidal evolution is toward zero obliquity if one assumes that the orbit was fixed in inertial space. The orbit of the Moon is inclined by ~5° to the ecliptic and, it precesses in a retrograde sense with a period of ~18 years while maintaining a nearly constant inclination. So we find that the tides are attempting to drive the spin toward an orbit normal that is itself moving. In this case the tides bring the spin to a Cassini state (named after Cassini's laws that describe Moon rotation), in which the normal to the ecliptic, the orbit normal, and the spin vector remain coplanar as the latter two precess about the first (Colombo 1966, Peale 1969). There are either two or four possible Cassini states where the spin is fixed in the precessing frame. In the latter case, state 4 is dynamically unstable, and state 3, with the spin retrograde, is secularly unstable to tidal dissipation (Peale 1974). For the Moon, only states 2 and 3 exist, with the Moon occupying state 2, where the tides would bring it from any initial state (Peale 1974). The regular satellites of the major planets are also in precessing orbits with nearly constant inclinations to the respective planetary equators. Here inclinations are so small that the Cassini state toward which the tides drive the spins have obliquities negligibly different from 0°. Iapetus is sufficiently far from Saturn that the rate of regression of the line of nodes on the orbit plane from the solar perturbation is comparable with the rate of regression of the line of nodes on the equator plane of Saturn caused by the latter's oblateness. In this situation, the Iapetus orbit maintains nearly constant inclination to a plane called the *Laplacian plane*, which is intermediate between the equator and orbit plane. This situation leads to an interesting choice of Cassini states, that we return to below. All of the larger natural satellites have their final evolutionary spin states tabulated in the work of Peale (1977).

If the planet and satellite are of comparable size, both can approach synchronous rotation and zero obliquity as the orbit expands. Once neither satellite is rotating rapidly relative to the orbital mean motion, the tides on both bodies eventually tend to reduce the eccentricity and the obliquities of both bodies relative to their mutual orbit plane. The natural endpoint in this case is a circular orbit with both planet and satellite rotating synchronously with zero obliquity. Pluto-Charon is the only known system to have reached this final state, although there is one observation of a nonzero orbital eccentricity that has not been confirmed (Tholen & Buie 1997).

Satellites approach rotations synchronous with their orbital mean motion, and their spin vectors approach one of two Cassini states with small obliquity owing to the dissipation of tidal energy within the satellite. At the same time, tides raised on the planet expand or contract the satellite orbits with corresponding changes in the spin angular momenta of the planets. These general conclusions account for much of what we observe within the satellite systems. However, there are exceptions to these rules, and complications abound among multiple satellites in the same system. Our goal is to deduce the histories that have produced the current states

4. EARTH-MOON SYSTEM

All are familiar with the one hemisphere of the Moon that is always more or less facing toward us on Earth—the most apparent example of rotation synchronous with the mean orbital motion that results from tidal dissipation according to Equation 6. If the Moon were formed at $3.5 R_E$ (Earth radii), it would reach synchronous rotation in about 1 year from an initial period of 4 h ($\mu \sim 10^{11}$ dynes cm^{-2}, $Q_M \sim 100$). The Moon now occupies Cassini state 2, where the spin axis is fixed at constant obliquity in the frame precessing with the orbit in the plane defined by the lunar orbit normal and the normal to the ecliptic plane where it would have been brought by tidal dissipation from any initial configuration (Peale 1974). The spin axis is actually displaced from this Cassini equilibrium position by 0.26 arcseconds either because of tidal dissipation in the lunar mantle with $Q_M \approx 26$ or because of a possible liquid core-solid mantle interaction (Yoder 1981a, Dickey et al 1994). The unusually low value of the necessary Q_M compared with $Q \approx 100$ (Section 5) of Mars and the high seismic Q_M (Nakamura & Koyama 1982) make the core mantle interaction more probable (Yoder 1981a).

The origin of the Moon as the consequence of a giant impact on the Earth in the latter stages of the accretion process was discussed in Section 2 as the only probable means to arrive at the large angular momentum of the Earth-Moon system as well as the volatile and iron depletion in the Moon relative to the Earth. Still, the geochemistry of Earth's mantle is not thought to be consistent with this scenario (Drake 1986, Ringwood 1989), although a geochemistry-inspired alternative to the giant impact origin that satisfies the above constraints has not been forthcoming. In addition, there are sufficient degrees of freedom that geochemical inconsistencies can be decreased by altering details in the collision process. Despite any reservations from geochemistry, we assume that dynamical constraints and the gross chemical differences between Earth and Moon are sufficiently compelling consistencies for adoption of the impact origin as the most likely. A characteristic of the giant impact scenario is that the debris that is to accrete into the Moon will settle down into the equatorial plane for reasons discussed in Section 2. That the Moon must have formed in the equatorial plane of the Earth is the natural consequence of almost all of the numerical calculations synthesizing the impact (Benz et al 1986, Benz et al 1989, Cameron & Benz 1991, Cameron 1997, Cameron & Canup 1998a,b). Given sufficient material outside the Roche radius of $\sim 2.9 R_E$, there is no difficulty in accreting the Moon in a very short time (Stevenson 1987, Ida et al 1997), although the drastically different points of view in these last two publications—one hot and one cold—demonstrate our lack of understanding of the phenomenology between impact and final Moon formation. In the cold-accretion scenario, it appears that a lunar-sized object can accrete outside the Roche radius only for impacts by a body twice the mass of Mars with twice the current angular momentum of the current Earth-Moon system (Canup & Esposito 1996), although this minimum mass for the impactor can be relaxed if the impact occurred when Earth's formation was only ∼50 to 70% complete (Cameron & Canup 1998a,b). However, continued accretion by the Moon as the Earth completes its own accretion could lead to excessive siderophile or volatile contamination of the lunar surface layers that is not observed (Stewart & Canup, 1998). The diversity of the accretion scenarios demonstrates the inadequacy of theoretical models of the impact-generated disk.

In contrast to our ignorance of many fundamental parameters for all of the other satellites, we have detailed information on the geochemistry and ages of rocks from six sites on the Moon thanks to landings during the Apollo program. The ages of the surfaces all over the Moon can then be estimated from the calibration of the crater densities at each of the visited sites. The youngest surfaces range from 2.9 to 3.9 billion years old (Warren 1985), where repeated basaltic lava flows have covered vast areas called *maria*—usually the interiors of giant impact basins such as Mare Imbrium or Mare Serenitatis. The maria occupy only ∼17% of the lunar surface; the remainder (lunar highlands) is a thick layer (thickness is estimated from assumed isostacy of lunar highlands floating on dense mantle basalts and from the depth of sampling by large impact craters) of low density rock that is >75% plagioclase (Ca, Na-aluminosilicate). So much plagioclase on the surface is thought to result from differential crystallization in a deep (∼250-km) magma ocean with the less dense plagioclase crystals floating to the top. The minimum depth of the ocean is controversial, but it is estimated as that necessary to provide the observed plagioclase if all the available plagioclase has floated to the surface. There are additional, more subtle geochemical indications of the existence of such a magma ocean on the early Moon (Warren 1985). Very few of the returned lunar rocks are older than 4.0 billion years, where older rocks are thought to have been destroyed by a continued heavy bombardment (late heavy bombardment) that persisted until ∼500 million years after the Moon formed (Mottman 1977).

In the sections that follow, we invoke tidal dissipation (only sometimes successfully) to account for resurfacing of icy satellites with no other apparent sources of internal energy. However, one investigation of the contribution of tidal dissipation to lunar thermal history (Peale & Cassen 1978) yielded negative results. This study was motivated by the realization that the obliquity of the Moon would undergo large variations as it changed from equilibrium Cassini state 1 to a precession about and dissipative relaxation toward a still very inclined state 2 as the Moon passed through ∼$34R_E$ from the Earth during the tidally induced growth of the semimajor axis of the Moon (Ward 1975). The obliquity reached as high as 77° during this process, which led to significant tidal dissipation within the Moon. However, the growth of the semimajor axis was sufficiently rapid during this transition that the contribution to the lunar thermal budget was negligibly small (Peale & Cassen 1978). There is a possibility discussed below that tidal dissipation caused profound changes in the lunar interior during passage through an orbital resonance when the Moon was very close to the Earth.

Several authors (Darwin 1879, 1880, Gerstenkorn 1955, MacDonald 1964, Kaula 1964, Goldreich 1966) have attempted to constrain the origin of the Moon by integrating the motion approximated by Equation 7 backwards in time until the Moon was close to the Earth. The actual motion is much more complicated than implied by Equation 7. The lunar orbit has a variable inclination relative to the Earth's equator. The noncircular orbit precesses while maintaining a nearly constant inclination to the Laplacian plane, where that plane is nearly coincident with the Earth's equator plane when the Moon is close, but is now nearly coincident with the ecliptic. The Earth's spin rate is decreasing as angular momentum is transferred to the Moon and the Sun. The Earth's obliquity is changing in response to lunar and solar torques. The Earth's spin axis itself is precessing due to the torques on its oblate figure. The current rate of regression of the Moon was deduced long ago by the determination that the length of an Earth day was increasing 0.0016 s/century from the timing of solar eclipses. That rate has since been made more precise from lunar laser ranging to the Moon over the past 25 years ($da_M/dt = 3.82 \pm 0.07$ cm year^{-1}) (Dickey et al 1994).

The first thing learned from the calculations of the Earth-Moon evolution was that the Moon would have been at the surface of Earth <2 billion years ago if the tidal-effective Q_E was maintained at that value yielding the current rate of lunar orbit expansion. As the major fraction of the Earth's tidal dissipation is in the oceans and as that dissipation depends on the configuration of continental shorelines, a reasonable solution to this problem is that the changes in the continental configurations have led to a tidal effective Q today that is less than its value in the distant past. The second result was that the Moon returned to an inclination relative to the Earth's equator of ~10° instead of to the equatorial plane. As the then popular coaccretion model of lunar formation had the Moon forming in the Earth's equator plane from a debris disk, this is referred to as the "inclination problem." This result has been confirmed in a modern Hamiltonian formulation of the problem and by a symplectic integration including the complete chaotic solar system, but with the same accelerated evolution rates and no dissipation in the Moon (Touma & Wisdom 1994). The currently popular giant impact origin of the Moon also leads to formation in an equatorial disk, so the problem remains. However, it takes only another large impact on the Earth or Moon after the Moon has formed to change the orbit inclination relative to the Earth equator and thereby provide a consistent tidal evolution (e.g. Stevenson 1987). But Touma & Wisdom (1998) have another solution.

In their 1994 paper, Touma & Wisdom note that the artificially accelerated rate of tidal evolution would drag the Moon through any orbital resonances it might have encountered such that the consequences of capture in such a resonance would be missed. In addition, the evolution of the Moon is not time reversible because the Moon might be captured into an orbital resonance when approaching it from one direction, but not the other. Any viable representation of the Earth-Moon history must then involve a rate of tidal evolution reasonably close to the physical rate and must progress forward in time.

In an integration that produces the current configuration when the Moon has reached its current distance from the Earth, Touma & Wisdom (1998) start the Moon in the equatorial plane of the Earth at a separation of $3.5R_E$ with an eccentricity of 0.01, which is consistent with current expectations from the giant impact origin (e.g. Ida et al 1997). The initial obliquity of the Earth is 10°, and the initial Earth rotation period is five h. Realistic rates of tidal evolution are used in the symplectic integrations that include the entire chaotic solar system, and dissipation in the Moon is included at a variety of dissipation rates. A tidal model was used with the constant time lag as discussed in Section 3 but with the Mignard (1981a) formulation. The first strong resonance is encountered when the Moon is $\sim 4.5R_E$, where the period of the periapse motion of the lunar orbit relative to an inertial reference is ~ 1 year. This resonance is called the *evection resonance* because the same term in the disturbing function gives rise to the 1.3° amplitude, 31.8-day periodic variation in the Moon's mean longitude called the evection. Capture into the evection resonance is certain if the eccentricity is <0.07 as the resonance is approached and if the rate of tidal evolution is sufficiently slow. With the assumed parameters, capture occurs and the eccentricity grows rapidly to large values, where the maximum value reached before the system escapes the resonance is determined by the value of

$$A = \frac{k_M}{k_E} \frac{\Delta t_M}{\Delta t_E} \left(\frac{m_E}{m_M}\right)^2 \left(\frac{R_M}{R_E}\right)^3, \qquad (10)$$

where k, Δt, m, and R refer to Love number, the constant tidal time lag, mass, and radius with subscripts referring to Earth and Moon, respectively. A is a measure of the relative rates of energy dissipation in the Earth and Moon. The current value of A from the lunar laser-ranging experiment is ~ 1.1 (Dickey et al 1994). For $A = 0$ (no dissipation in the Moon), the maximum eccentricity is ~ 0.5 before escape, and the eccentricity continues to climb after escape from the resonance because of tides raised on Earth and no dissipation in the Moon (Goldreich 1963). For $A = 10$ (high dissipation in the Moon), the maximum eccentricity is only ~ 0.15. For $1 \leq A \leq 10$, the energy dissipated in the Moon in ~ 8000 years is in the range from $\sim 2 \times 10^{35}$ to 1.5×10^{36} ergs, which could lead to substantial melting (Touma & Wisdom 1998).

After escape from the evection resonance, the continued expansion of the orbit further decreases the prograde motion of the orbit periapse and twice the time derivative of the evection resonance variable plus the retrograde motion of the lunar orbit node approaches zero. The term in the Hamiltonian corresponding to this resonance has $e_M^2 i_M$ in the coefficient, but this resonance affects the inclination more than the eccentricity. Touma & Wisdom name this resonance the *eviction—*

changing the e in *evection* to i to emphasize the inclination and noting that this resonance "evicts" the Moon from an equatorial orbit. If A is not too large, the eviction resonance is approached with high eccentricity in the wrong direction for capture. Passage through the resonance leaves the eccentricity large and excites an inclination of 2–3°. If A were now to increase drastically—perhaps because the continued high eccentricity has partially melted the interior—the dissipation in the Moon becomes so high that the semimajor axis is decreased as the eccentricity is reduced. This reduction in the semimajor axis takes the system through the eviction resonance from the other direction where capture and subsequent evolution force the inclination to values between 9 and 13°. Escape from the resonance is effected because of the continued decrease in e_M, but the remnant inclination is preserved. Subsequent evolution brings the Moon to the current configuration.

This analysis of the evolution of the lunar orbit is the first in which the Moon evolves from an equatorial orbit to its current configuration from the effects of tidal dissipation alone. For this to occur, the Moon must start relatively cold so that the parameter A above is comparable with or smaller than its current value. Under this condition, the eccentricity can grow to large values in the eviction resonance. The system must pass through the eviction resonance while maintaining substantial eccentricity. The consequence of reasonably sustained high eccentricity must soften the Moon sufficiently to drastically increase the A parameter after eviction passage, where the now-enhanced dissipation within the Moon decreases the semimajor axis as well as the eccentricity. A second encounter with and capture into the eviction resonance then forces the inclination to precisely the range of values obtained by earlier workers in their time-reversal integrations of the lunar history. Although there are several special circumstances necessary for this scenario to solve the inclination problem, our ignorance of the circumstances of the initial accretion and the resulting interior properties of the Moon leaves this alternative as a possibility.

If the Moon were hot after its accretion, however, A would be large initially, and the eccentricity might never reach the necessary values for the Moon to complete its dance around the eviction. The gravitational binding energy of the Moon is $\sim 3Gm_M^2/5R_M = 1.2 \times 10^{36}$ ergs. Therefore, if the accretion is complete in <1 year, as found by Ida et al (1997) for moon-type bodies just outside the Roche radius, the accretion could conceivably produce the early magma ocean inferred by geochemical evidence (Warren 1985). The fraction of the accretional energy that is retained increases with the size of the accreting bodies if the impacts are infrequent, but most of the heat of accretion can be retained even with small particles if the accretion is sufficiently rapid. Radiation of the deposited heat is inhibited in this latter case. If the time scale for accretion is >1000 years or if accretion is inhibited by being too close to the Roche zone (Canup & Esposito 1995), the Moon might originally have been relatively cool, and the magma ocean would have formed later—perhaps through excitation of the eccentricity in the evection resonance (Touma & Wisdom 1998). The latter scheme would require a distribution of tidal heating in the Moon that would lead to the temperature exceeding the solidus first in the outer layers—a condition that might not be consistent with the distribution of tidal heating found by Peale & Cassen (1978), where the highest heating rate of a solid Moon is in the center. On the other hand, if the Moon were initially softened in the outer layers during the accretion process, while leaving the interior cool, the tidal dissipation might well be concentrated in the outer layers, and the tides could finish the job of forming the magma ocean that was begun by accretion. This latter scenario would take some improbable fine tuning, however, if the Moon is to arrive at the eviction resonance with a large eccentricity.

In his analysis of the giant impact origin of the Moon, Stevenson (1987) finds that the initial Moon must have been largely molten—not from a rapid accretion of cold solid bodies as treated by Ida et al (1997) but from the retention of the giant impact energy within the debris disk. The turbulent combination of liquid and gas in the disk of material resulting from the impact evolves so quickly that the material spreads beyond the Roche radius and forms the Moon before it has time to cool and solidify. The initial Moon would then have been at least partially molten and hot throughout. The initially small A necessary for the start of the Touma & Wisdom scenario would be precluded. On the other hand, an initially hot Moon is not consistent with the results of Solomon & Chaiken (1976), who find that the limited change in the lunar radius since the current very old surface was emplaced suggests that no more than the outer 200 km was molten and that the interior started cold. Touma & Wisdom note that such an interpretation would also cause problems with their scheme for forming the magma ocean through tidal dissipation since their Moon would also be hot throughout (if not melted in the center) after the eviction-induced heating.

From the preceding paragraphs, it is clear that our understanding of the origin and evolution of the Moon is far from complete. Any geochemical inconsistencies notwithstanding, the giant impact initial event seems to be a relatively secure explanation for the large angular momentum of the Earth-Moon system and for the Moon's lack of volatiles and iron compared to the Earth. However, details in getting from the impact to an accreted Moon outside the Roche radius are obscure. Is the Moon accreted hot or cold? The former would eliminate the Touma & Wisdom solution to the inclination problem, and it is inconsistent with the small apparent decrease in the lunar radius during the last 3.8 billion years. The latter scheme has difficulty in placing enough material outside the Roche radius without extreme values of impactor mass and angular momentum. Perhaps effort for the near future should be concentrated on understanding and possibly constraining the physics of the accretion disk. Additional processes heretofore not considered, such as spiral density waves to redistribute angular momentum more efficiently, might allow smaller impactors. As always in this research, we desire a single, self-consistent scenario that carries the Moon from its origin to the current configuration of the Earth-Moon system with the minimum number of degrees of freedom while satisfying the maximum number of observational constraints. So far we have only somewhat disjointed pieces with uncertain parameters, so that none satisfy all of the apparent constraints.

5. MARS SYSTEM

From Table 1, we see that Phobos is located well inside the corotation radius of $\sim$5.9 Martian radii (R_M), and Deimos is just outside this radius. The tides raised on Mars thus cause Phobos to be spiraling toward Mars and Deimos to be spiraling away. In fact Phobos is inside the Roche radius for a density of 1.9 g cm^{-2} and would be torn apart by tidal forces if it were a fluid. It needs only a shear strength of 10^5 dynes cm^{-2} to resist disruption (Yoder 1982)—a loose rubble pile would survive (Soter & Harris 1977, Dobrovolskis 1982).

Phobos and Deimos are synchronously rotating. Deimos would have reached this state in $<10^8$ years from an unlikely small initial spin period of 4 h, if a rigidity of 5×10^{11} dynes cm^{-2} and $Q = 100$ is assumed. Under the same assumptions, Phobos would have reached this state in $<10^5$ years at its current separation from Mars and in $<10^7$ years at its likely initial separation near the corotation radius. [See Peale (1977) for a detailed discussion of the rotation histories of all of the satellites known at that time.] We have already dismissed an intact capture origin for the satellites of Mars as untenable, given the regularity of the current orbits. A formation from accretion in a debris disk then implies that the initial orbits of Phobos and Deimos were also regular with near-zero eccentricities and inclinations to the equator plane of Mars.

The fact that Phobos and Diemos deviate considerably from spherical symmetry leads to special circumstances as they approach synchronous rotation. Both enter chaotic zones in the phase space surrounding that of stable libration about synchronous rotation where the spin axis is attitute unstable (Wisdom 1987b). This means the satellites will tumble chaotically with time scales comparable with the orbital periods until they are trapped into stable libration about the synchronous state. Tidal dissipation under these circumstances is like that for a nonsynchronous rotation, and the eccentricity will be damped at a much higher rate than it would have if the satellite were in synchronous rotation. Hence, the process of synchronizing the rotation of the satellites with the orbital motion will also effectively damp any eccentricity remaining from the accretion process. This damping precludes high eccentricities in the past and concern about collisions between Phobos and Diemos as discussed by Szeto (1983). The evolution of the system to the current configuration would then seem to require only the expansion of the orbit of Diemos from tides raised on Mars and the shrinkage of the orbit of Phobos from similar tides.

Although the effect of tides on the inclinations of the orbits to the Martian equator plane is negligible, it is of interest that the initial equatorial orbits will remain equatorial in spite of the chaotic, large amplitude variations in the obliquity of Mars (Ward 1979, Laskar & Robutel 1993, Touma & Wisdom 1993) and despite the precession of the spin axis of Mars (Goldreich 1965b). The solid angle described by the orbit normal as the satellite orbit precesses due to Mars's oblateness is an adiabatic invariant (Goldreich 1965b), as these precession rates for the Martian satellites [periods of 2.3 and 57 years for Phobos and Deimos, respectively (Peale

1977)] are fast compared with rates of change of Mars' spin axis direction relative to inertial space [timescales O (10^5 years), Touma & Wisdom 1993].

Determinations of the Q of Mars lie between about 66 and 144 (Shor 1975, Sinclair 1978, Duxbury & Callahan 1981) from observations of the secular acceleration of Phobos' orbital mean motion. If we choose a constant value of $Q_M = 100$ with $k_M = 0.14$, the orbit of Deimos could have expanded by <200 km in 4.6×10^9 years. The initial semimajor axis of Phobos would have been $\sim 5.6 R_M$ under the same assumptions. The rotation period of Mars would be essentially unaffected by the exchange of angular momentum with the satellites and would have been only ~ 10 min longer due to solar tides. Deimos has essentially its initial orbit; Phobos, having started inside the corotation radius, is consistent with the measured current value of $Q_M \approx 100$.

If we insist that both satellites started with nearly circular orbits, how then can we explain the current eccentricity of Phobos' orbit $e_P = 0.0151$? If the orbital motion is integrated backwards in time, this eccentricity grows to large values and collisions with Deimos would have been likely (Yoder 1982, Szeto 1983), even if there were no tidal dissipation in Phobos. Significant dissipation in Phobos reduces the time scale for a crossing orbit with Deimos to $<10^9$ years in the past (Yoder 1982). The current eccentricity cannot therefore be a remnant from tidal decay beginning 4.6×10^9 years ago. Yoder (1982) has identified three commensurabilities (defined when two characteristic periods in the description of the motion are in the ratio of small whole numbers) that Phobos has passed through within the past 10^9 years that provide likely gravitational excitations of the Phobos eccentricity during its inward spiral. The commensurabilities are encountered at $a = 3.8, 3.2,$ and $2.9 R_M$, where the earliest resonance was encountered only 5×10^8 years ago. The first and third are 2:1 and 3:1 commensurabilities between the rotation of Mars and the orbital mean motion, where the resonant interaction is with the axial asymmetry of Mars. At $3.2 R_M$, the 2:1 commensurability is between the apparent mean motion of the Sun and the periapse of the orbit of Phobos, where the secular motion of the latter is caused by the oblate figure of Mars. This resonance is like the evection resonance for the Moon. There is also a 3:2 spin-orbit resonance excitation of the eccentricity when $a = 4.6 R_M$, but this excitation happened so long ago that there would be no contribution to the current eccentricity. The eccentricity would have decayed after each excitation, and it plausibly arrives at the current eccentricity after the series of kicks and subsequent decays (Yoder 1982). Orbital inclination can also be excited, and, even though the resonance interaction is not as strong as it is for the eccentricities, the excited inclinations do not decay. Still, the current inclinations of the orbits are consistent with the resonance passages (Yoder 1982).

There is a condition on the dissipation in Phobos for this scenario to work. Yoder (1982) has calculated the dissipation in the satellite accounting for both the tidal dissipation caused by the eccentric orbit as discussed earlier and that caused by the forced libration of the very asymmetric satellite. This libration has an amplitude of $3.9°$ (Duxbury & Callahan 1981, Yoder 1982) and causes twice the tidal dissipation in Phobos that would occur if Phobos were nearly axially symmetric in the same eccentric orbit (Yoder 1982). Both the dissipation in Phobos and that in Mars from tides raised by Phobos damp the eccentricity. There cannot be too much damping since the series of eccentricity excitations or the current eccentricity would be less than that observed. Because the dissipation in Mars can be presumed known from the measurement of Q_M, and the magnitude of the probable excitations can be reasonably estimated from the resonance passage analysis, the current value of e_P limits the contribution by Phobos. Yoder finds that $\mu_P Q_P > 3$ to 6×10^{12} dynes cm^{-2} or, if $Q_P \sim 100$, $\mu_P \gtrsim 10^{10}$ dynes cm^{-2}, which is about that of ice. The properties of Phobos are not sufficiently well known for one to be sure that the rigidity could satisfy this constraint, but this rigidity is not unreasonable.

During the spiral of Phobos toward Mars, it is likely that it passed through the 2:1 orbital mean motion commensurability with Deimos. Such a passage would excite an eccentricity of ~ 0.002 in the orbit of Deimos if the eccentricity of Deimos were much smaller than this before resonance passage. The time of this commensurability is known if the current dissipative properties of Mars have not changed substantially since the resonance encounter. This places a lower bound on the dissipation in Deimos if the current eccentricity is the tidal remnant from an initial value of 0.002 excited by the resonance passage. Yoder (1982) finds $\mu_D Q_D (1 - \alpha_D)^2 / \alpha_D^2 \lesssim 10^{10}$ dynes cm^{-2}, where $\alpha_D = 3(B - A)/C$, with $A < B < C$ being the principal moments of inertia of Deimos. This limit may be unreasonably low, but the dissipation in Deimos may be increased if the forced libration is nearly resonant with the free libration. The enhanced amplitude of libration would lead to higher dissipation and would relax the constraint on $\mu_D Q_D$. The free libration period could be better constrained by an estimate of α_D from a more accurate determination of the shape of Deimos along with an accurate measure of its physical libration amplitude.

In any case, the Yoder hypothesis (that the satellite orbits have always been regular and current properties of the system then attributed to the effects of resonance passages by Phobos) is well supported. This hypothesis is consistent with our presumed origin from a dissipative disk of small particles. However, the necessary approximations in the developments and still uncertain dynamical properties of the satellites warrant a more thorough numerical exploration of the phase space of the system with the value of Phobos' rate of tidal orbit decay comparable with the real value.

6. JUPITER SYSTEM

The Jupiter system has four classes of satellites. The dynamical evolutions of the four small satellites closest to Jupiter, after the disk of gas and solid particles from which they formed had disappeared, are limited to the tidal retardation of their spins to synchronous rotation, unless there have been episodes of breakup by collision and reassembly as we will imply for the small satellites of Saturn. From Equation 6, the satellite Thebe at $3.1R_J$ with rocklike characteristics would slow to synchronous rotation from an initial rotation period of 4 h in less than a few thousand years. The closer small satellites are of comparable size or larger and would spin down in shorter times. All of the Galilean satellites, the next class, are observed to be synchronously rotating to within observational error, which is consistent with a time of ≤ 2.5 million years to bring all to this state. However, Io and Europa may not have the internal strength or permanent asymmetry sufficient to stabilize synchronous rotation against the averaged tidal torque (Greenberg & Weidenschilling 1984, McEwen 1986).

The two families of distant irregular satellites, four in prograde orbit with semimajor axes near $160R_J$ and four in retrograde orbits with semimajor axes near $300R_J$, are too far away even for significant evolution of the spins except for the attainment of principal axis rotation. The largest of the prograde irregular satellites, Himalia, with icelike characteristics, would require $\sim 2 \times 10^{14}$ years for significant change in an initial 4-h rotation period, and the other irregular satellites would take even longer. Thus, the only changes in the rotation states of the distant, irregular satellites are those that have been induced by occasional impacts. Light curves for several of the irregular satellites indicate rotation periods in the range of 8.3 to 13.2 h (Degewij et al 1980, Luu 1991), which fall in the middle of the range of asteroid rotation periods (e.g. Harris et al 1992).

The orbits of the irregular satellites are highly variable due to solar perturbations, but there appears to have been some orbital decay for these satellites. Two of the retrograde satellites occupy orbital resonances where their apses are locked to the apsis of Jupiter—Pasiphae (JVIII) continuously and Sinope (JIX) intermittently (Whipple & Shelus 1993, Saha & Tremaine 1993). Dissipation is required to bring the satellites to this state from an initially random orbit of the parent body. Decay of the orbits by gas drag in an extended primordial Jupiter atmosphere is not an unreasonable hypothesis, since the parent bodies for the two groups of irregular satellites may have been captured and split into the respective members of the groups from gas drag from the extended atmosphere just before the final collapse of that atmosphere (Pollack et al 1979). Capturing satellites by gas drag is tricky, however, because an atmosphere dense enough in its outskirts to capture a satellite is also dense enough to cause the satellite to spiral into the planet in a relatively short time. It appears that the time scale for atmospheric collapse would have to be less than the orbit decay time in the initial atmosphere for gas capture to work. One alternative is for capture of the parent bodies for each class of irregular satellites by three-body interactions or collisions within the sphere of influence at a time when the extended atmosphere was sufficient for some orbital decay but not so thick as to threaten the continued existence of the satellites. A second alternative would allow the parent body to decay to the resonance, which would halt the orbit decay while the atmosphere continued its collapse or was otherwise dissipated. The collision creating the family could then leave one of the resulting satellites in the resonance, a second nearly so and the other two scattered out. With the atmosphere gone, rotations imparted at the breakup would not decay by gas drag. The details of origin are likely to remain speculative, but dissipation at some stage seems necessary to leave Pasiphae and Sinope in the orbital resonance.

The Galilean satellites have much more interesting histories than do the remaining satellites that still have not been totally unraveled in a way consistent with their observed properties. The active volcanism on Io (JI) (Smith et al 1979, Morabito et al 1979) is almost certainly due to the dissipation of tidal energy sustained by the forced orbital eccentricity of Io in the Laplace orbital resonance (Peale et al 1979). But the orbital configuration resulting from a dissipative equilibrium (Yoder 1979, Yoder & Peale 1981) yields a dissipation rate that is only $\sim 1/2.5$ the measured heat flux from Io for the minimum *historical average* value of $Q_J = 6.6 \times 10^4$ (Veeder et al 1994). Observational evidence for a liquid ocean under the surface ice of Europa (JII) continues to build (Greeley et al 1998, Khurana et al 1998), but tidal dissipation—apparently the only viable source of energy to keep the ocean from freezing with the current orbital eccentricity—may not be sufficient to maintain the ocean (Cassen et al 1980a, Ross & Schubert 1987). Ganymede (JIII) is only slightly larger than Callisto (JIV), yet the former has extensive areas of grooved terrain from endogenic activity that occurred long after the ancient cratered surface of the latter was emplaced. In addition, Ganymede may be differentiated more than Callisto (Anderson et al 1996, 1997, 1998a), and it has an intrinsic magnetic field that implies a fluid conducting core (Kivelson et al 1998). One attempt to explain the dichotomy in the ages of the surfaces with the additional radioactive heat sources in the larger rocky core of Ganymede (Cassen et al 1980b) was judged unlikely adequate because of the necessary fine-tuning of unknown interior parameters (Showman et al 1997), and it almost certainly could not account for a fluid metallic core for the magnetic field generation. Tittemore (1990) considered a 3:1 Europa-Ganymede resonance where he obtained large eccentricities for both satellites ($e_E = 0.14$ and $e_G = 0.06$), which he inferred to yield sufficient heating of Ganymede to account for the resurfacing. However, Tittemore kept only the two-body Europa-Ganymede interaction, and he neglected dissipation in the satellites while obtaining the high eccentricities. His neglect of Io, presumably already locked in the 2:1 resonance with Europa, completely changes the dynamics. We shall see below that Europa's eccentricity cannot be changed arbitrarily while it is locked in a 2:1 resonance with Io, and this constraint on Europa is going to affect the Ganymede eccentricity. If satellite dissipation had been included during the evolution, the eccentricities obtained would not have been so large.

Showman et al (1997) note that significant tidal heating would have occurred in Ganymede if the system were captured into at least one Laplace-like (see below) historical orbital resonance on its way to the current configuration. Large thermal runaways generated by a rapid warming of Ganymede's mantle are the most attractive for resurfacing but are prevented by details of the model. Smaller runaways creating pockets of water or slush coupled with conduits to the surface may be capable of resurfacing. But the model dependence of these conclusions means that additional analysis is needed before a robust scenario leading to the

Ganymede-Callisto surface dichotomy can be obtained. If these implications are not sufficiently puzzling, analysis of variability of an induced magnetic field in Callisto is consistent with a conducting layer of liquid, salty water below the insulating ice crust (Khurana et al 1998). Sustaining such a layer of liquid water on Callisto over any significant time span seems totally implausible.

The apparent incomplete differentiation of Callisto is another mystery (McKinnon 1997), but it is significant that the last Callisto flyby by the Galileo spacecraft has considerably reduced the deduced $C_C/m_C R_C^2$ from ~0.4 from the two earlier encounters to 0.358 ± 0.004 (Anderson et al 1998a). McKinnon (1997) has pointed out that even a homogeneous Callisto will have $C_C/m_C R_C^2 = 0.38$ because of the compression of deeper layers and polymorphism of ice, so the latest value of $C_C/m_C R_C^2$ would indicate some differentiation. Next, the flyby data were interpreted under the assumption of hydrostatic equilibrium, and McKinnon (1997) finds that nonhydrostatic contributions to the gravitational harmonic coefficients J_2 and C_{22} could mimic an undifferentiated Callisto and by inference lead to $C_C/m_C R_C^2 = 0.358$, which still is too high. If one allows for a 3σ error as well, it seems that the conclusion that Callisto is only partially differentiated is not that secure. If there were to be no differentiation, the internal temperature and, in particular, the surface temperature could never exceed 273 K. If one assumes that all of the accretional energy must be radiated away at surface temperature T, the accretion time $t = 4\pi G \rho_C^2 R_C^3/(9\sigma T^4)$ is ~4.5×10^5 years for $T = 273$ K (DJ Stevenson, private communication), where σ is the Stefan-Boltzman constant and where only the gravitational binding energy is accounted for. This million-year accretion time scale is much longer than that obtained by McKinnon & Parmentier (1986), who find that both Ganymede and Callisto should have been substantially melted during accretion. A more detailed mapping of the Callisto gravitational field seems appropriate before theorists devote too much time to explaining incomplete differentiation.

The most striking characteristic of the Galilean satellite system is the set of orbital resonances where the orbital mean motions satisfy the relations

$$n_I - 3n_E + 2n_G = 0,$$
$$n_I - 2n_E + \dot{\varpi}_I = 0,$$
$$n_I - 2n_E + \dot{\varpi}_E = 0,$$
$$n_E - 2n_G + \dot{\varpi}_E = 0,$$
(11)

which lead to the following constraints on the mean longitudes:

$$\lambda_I - 3\lambda_E + 2\lambda_G = 180°,$$
$$\lambda_I - 2\lambda_E + \varpi_I = 0°,$$
$$\lambda_I - 2\lambda_E + \varpi_E = 180°,$$
$$\lambda_E - 2\lambda_G + \varpi_E = 0°.$$
(12)

The subscripts refer to Io, Europa, and Ganymede, respectively, and ϖ_i are the longitudes of pariapse with the dot indicating time differentiation. The Laplace relation is the first of both sets of equations. If we define $\omega_1 = n_I - 2n_E$ and $\omega_2 = n_E - 2n_G$, the Laplace resonance can be thought of as a 1:1 commensurability between ω_1 and ω_2 whose current value is 0.74° per day. The combination $\lambda_E - 2\lambda_G + \varpi_G$ is not constrained. At the conjunction of Io and Europa, Io is thus at its periapse and Europa is at its apoapse, whereas at the conjunction of Europa and Ganymede, Europa must be at its periapse and Io must be on the opposite side of Jupiter. The longitude of Ganymede's periapse is not constrained at conjunction.

Because the phase-space volume for the Laplace relation is so small compared with that available, it has been long assumed that the resonances were assembled from initially random orbits through differential tidal expansion of the orbits (T Gold, personal communication, 1962, Goldreich 1965a). Yoder (1979), with elaboration in Yoder & Peale (1981), was the first to develop a consistent analysis of tidal evolution arriving at the current configuration, where the high tidal dissipation in Io was shown to be a vital consideration in damping the amplitude of libration of the Laplace angle to the current remarkably low value of $0.066 \pm 0.013°$ (Lieske 1987). The damping of this libration to such a small value during evolution within the resonance was not possible without the high tidal dissipation in Io (Sinclair 1975).

Substitution of parameter values from Table 1 into Equation 7 shows that Io's orbit will expand more rapidly than Europa's and Europa's more rapidly than Ganymede's. Thus Yoder starts Io inside the 2:1 mean motion commensurability with Europa, so that it approaches this commensurability from a direction where capture into each of two eccentricity-type resonances corresponding to libration of the second and third angles defined by Equation 12 is certain if the respective eccentricities far from resonance are sufficiently small (e.g. Peale 1986). This condition for certain capture will surely prevail, since the secular decrease of the eccentricity e of a synchronously rotating satellite as a consequence of tidal dissipation will dominate the secular increase induced by tides raised on the planet and quickly reduce e to negligibly small values before resonance encounter (Peale et al 1980). (Time constant for decay of free eccentricity for a cold Io would be $8.2 \times 10^4 Q_I$ years with $\mu = 6.5 \times 10^{11}$ dynes cm^{-2} from Equation 9). Just after capture into the two eccentricity-type resonances at the 2:1 mean motion commensurability, $\omega_1 = n_I - 2n_E = -\dot{\varpi}_I = -\dot{\varpi}_E$ is considerably larger than the current value.

The retrograde periapse motions are a resonance effect that dominates the ordinary prograde motion from Jupiter's oblateness and from the solar perturbation. The contribution to $\dot{\varpi}$ from the resonance term is inversely proportional to the eccentricity (Peale et al 1979) such that as tides raised on Jupiter by Io continue to reduce Io's mean motion, the decrease in ω_1 forces a similar decrease in the retrograde motions of $\dot{\varpi}_I$ and $\dot{\varpi}_E$ if the resonance is to be maintained. But since this rate is inversely proportional to the eccentricites, both eccentricities are forced to higher values as tides push Io deeper into the resonance, that is, closer to exact commensurability of the mean motions. But higher eccentricities result in higher tidal dissipation in the satellites according to Equation 8, where $1 \leq f \leq 13$ in this equation for a two-layer Io model with $k_I \approx 0.035$ for a homogeneous, solid Io with rigidity between those of Earth rocks and the outer layers of the Moon (Peale et al 1979). This dissipation decreases the eccentricity at a rate given by Equation 9. Thus, tides on Jupiter are forcing Io deeper into the resonance and thereby increasing the eccentricity while tidal dissipation in Io and (much less so) in Europa from the forced eccentricities are tending to reduce the eccentricites. An equilibrium is approached where the two effects balance with e_I and e_E, ω_1, ϖ_I, and ϖ_E essentially constant as the locked pair of satellites move outward together as Io transfers angular momentum to Europa through the resonance interaction. It is easy to see that the values of the equilibrium eccentricities where the two dissipative effects balance determine the ratio $k_J Q_I / k_I f_I Q_J$ (Yoder & Peale 1981).

The locked pair of satellites continue to move away from Jupiter, leading to the eventual encounter of the 2:1 mean motion commensurability between Europa and Ganymede. Capture into the three-body Laplace resonance ($\omega_1/\omega_2 = 1$) has a probability of ~0.9 (Yoder 1979; Yoder & Peale 1981) with simultaneous capture into the eccentricity type 2:1 resonance involving Europa's periapse longitude but not that of Ganymede. Initially there are free eccentricities induced that manifest themselves as large amplitudes of libration of the angles defined in Equations 12 about their mean resonant values. Subsequent evolution of the set from continued application of the torque from the tide raised on Jupiter by Io (The torques on Europa and Ganymede from their Jupiter tides are negligible by comparison.) forces larger eccentricities for the orbits of Io and Europa as ω_1 is pushed to smaller values, while the tidal dissipation of orbital energy in Io damps the libration of the Laplace angle. The evolution is such that the amplitude of libration of the Laplace angle approaches zero as e_I and e_E approach new equilibrium values. In this state, the three satellites move out together while maintaining fixed ratios of their mean motions. Angular momentum acquired from Jupiter by Io is transferred to Europa and from Europa to Ganymede through the resonant interactions to maintain the configuration.

The miniscule amplitude of libration of the Laplace angle is consistent with the current values e_I and e_E being equilibrium values. Consequently, the balance of the dissipation effects in Io (trying to decrease e_I) to those in Jupiter (trying to increase e_I) leads to $k_I f_I / Q_I \approx 900 k_J / Q_J$ (Yoder & Peale, 1981). The 1600 Q_J year age for the Laplace resonance, deduced by Yoder & Peale if the amplitude is an evolutionary remnant, was increased to $2100 Q_J$ years in a refined analysis by Henrard (1983) and to possibly an even greater age under circumstances to be explored below (Malhotra 1991). The age could be greater than any of these constraints if the amplitude was not a remnant of the Laplace resonance evolution but had been excited by an impact on one of the satellites or if it is a libration forced by the proximity of the (2074 ± 10)-day period of free libration to the 2076-day period of a term in the solar perturbation. It may be the case at the time of this writing that the masses of the satellites have been sufficiently refined by the Galileo spacecraft encounters to determine the closeness of the two periods and thereby the expected forced libration of the Laplace angle.

The above plausible scenario for the establishment and evolution of the Laplace resonance is perhaps the simplest, but whether it actually occurred depends on the initial distribution of random orbits and on a sequence of probabilistic events. Yoder & Peale (1981) investigated passage of the system through other Laplace-like resonances where $\omega_1/\omega_2 = j/(j+1)$. In particular, the $j = 1$ resonance would most probably have been encountered, with the system temporarily captured into libration in the above simplest evolution. Yoder & Peale noted no consequence of this resonance passage except for remnant free eccentricities that would be damped by dissipation within each satellite. The current free eccentricities of the orbits of Europa and Ganymede ($e_{E_{free}} = 9 \times 10^{-5}$, $e_{G_{free}} = 0.0015$) could then be remnants of such resonance passage if relatively recent. The fact that Ganymede's free eccentricity exceeds that forced by the Laplace resonance ($e_{G_{forced}} = 0.0006$) leads to the circulation of $\lambda_E - 2\lambda_G + \varpi_G$ rather than libration.

Malhotra (1991) bypassed the analytic approximations used by Yoder & Peale by numerically integrating the system from a variety of starting points, thereby encountering several Laplace-like resonances on the way to the current 1:1 resonance. Capture into several higher-order resonances was possible, with almost certain capture into $\omega_1/\omega_2 = 1/2$. The remarkable consequence of the latter resonance is that Ganymede's eccentricity is forced to large values, where larger tidal dissipation offered the possibility of modifying Ganymede's surface after most of the cratering was complete without affecting Callisto. Showman & Malhotra (1997) extended the range of initial conditions for the integrations and found much larger excitations of Ganymede's eccentricity from passage through the $\omega_1/\omega_2 = 2/1$ and $3/2$ resonances which brightened the prospects of accounting for the dichotomy between Ganymede and Callisto. A troubling aspect of this analysis was the necessity of changing the ratio $Q_I k_J / f k_I Q_J$, either suddenly or sinusoidally in order to disrupt the $\omega_1/\omega_2 = 2/1, 3/2$ resonances. There exist arguments that are not unreasonable for varying Q_I in episodic heating and cooling of Io (Greenberg 1982, Ojakangas & Stevenson 1986) and for varying Q_J (Iaonnou & Lindzen 1993), so sudden changes in $Q_I k_J / f k_I Q_J$ should probably be abandoned. Finally, capture into the $\omega_1/\omega_2 = 2$ and $3/2$ resonances is probable only if Ganymede's eccentricity is $\lesssim 0.001$ at the time of resonance passage (Showman & Malhotra 1997). The time constant for the decay of Ganymede's free eccentricity from Equation 9 is $\sim 2.2 \times 10^{-5} \mu Q$ years, which could approach the perhaps unlikely but possible value 1.5×10^9 years if a cold Ganymede had the rigidity of rock (6.5×10^{11} dynes cm^{-2}) with $Q_G = 100$.

The $\omega_1/\omega_2 = 1/2$ resonance became unstable naturally but produced lower values of e_G. However, even with passage through the 2/1 and 3/2 Laplace resonances, Showman et al (1997) found that conditions required to achieve a thermal runaway in the Ganymede ice layer were difficult to obtain. They used a Maxwell model where Q_G had an explicit temperature dependence based on ice rheology and the orbital evolutions found by Showman & Malhotra (1997). A thermal runaway with likely extensive resurfacing required the eccentricity to reach large values in the resonance with a subsequently rapid plunge in Q_G. The rapid damping of the eccentricity with the low Q_G would deposit sufficient energy in a thermal runaway in the ice mantle to effect the resurfacing. However, radioactivity in a carbonaceous chondritic core would warm the ice too much before the resonance was approached, and the resulting higher dissipation would prevent the attainment of high eccentricity. The tidal energy would thereby be deposited more slowly over a longer period, and thermal runaway would not occur. Still, partial runaways could occur in which parts of the interior were melted or softened, and it may be possible to account for the resurfacing if conduits to the surface develop.

The authors are careful to point out the need for continued pursuit of a possible tidal origin of the Ganymede-Callisto dichotomy. First, the failure to obtain a thermal runaway in Ganymede during the resonance passage is model dependent, and motivation for modifying the assumptions may emerge in the future. Second, the tidal dissipation rates were accelerated by a factor of 1000 so the calculations could be completed in a reasonable time. Such accelerations are notorious for introducing numerical artifacts into the results if the tidal changes are not adiabatically slow. Such adiabaticity was inferred for evolution into the $\omega_1/\omega_2 = 3/2$ and 2 resonances from the ratio of the restoring torque in each resonance to the tidal torque of 10^4 and 10^2, respectively (Showman & Malhotra 1997). However, in chaotic regions of the phase space, Tittemore & Wisdom (1988) have shown that numerical artifacts develop during the integrations unless the evolution is as much as a factor of 10^3 slower than that which satisfies the ordinary adiabatic invariance criterion for representations in two degrees of freedom. If a higher dimensionality of the problem must be considered, the evolution rate may have to be that set by physical constraints to avoid the artifacts (Tittemore & Wisdom 1990). The uncertainty of whether the variations in Q_I/Q_J necessary for disrupting these resonances would actually occur and the uncertainty of the time scales if they did, the uncertain damping time of Ganymede's eccentricity, and the uninvestigated nature of many details such as a numerical verification of invariance of the results with rate of evolution mean that the continued pursuit of a means of tidally resurfacing Ganymede encouraged by Showman et al is well founded. Although it is possible that a robust tidal origin of the Ganymede-Callisto surface dichotomy can be found, it will be more difficult to maintain a molten core for Ganymede's magnetic field generation.

Critical to all of the scenarios above is sufficient torque from Jupiter to force enough orbital evolution to allow assembly of the resonances from initially random orbits. We have so far considered only a tidal torque that acts principally on Io. There is also an electromagnetic torque $N_{EMF} = \pi I R_J B_J a_I$, where $I \sim 2.8 \times 10^6 A$ (Acuna et al 1981) is the current flowing along the tube of force induced by the $\mathbf{v} \times \mathbf{B}$ electric field across Io, R_J is the Jupiter radius, and $B_J = 0.02G$ at Io's distance a_I (Goldreich & Lyndon-Bell 1969). The electromagnetic torques are not sufficient to relax the limits of $6.6 \times 10^4 \lesssim Q_J \lesssim \times 10^6$ established by the above evolutionary scheme and current configuration of the satellites (Yoder & Peale 1981). The lower bound on the *average* Q_J results from the proximity of Io to Jupiter after expanding from just outside the synchronous orbit with the Laplace resonance assumed to exist for 4.6×10^9 years. The upper bound depends on the current eccentricity of Io being nearly the equilibrium value leading to (Yoder 1979, Yoder & Peale 1981)

$$\frac{k_I}{k_J}\left(\frac{R_I}{R_J}\right)^5 \left(\frac{m_J}{M_I}\right)^2 \frac{Q_J f_I}{Q_I} = 4200, \tag{13}$$

from which, $Q_J f_I / Q_I = 1.24 \times 10^4$. From Equation 8, the tidal heating exceeds the radiogenic heating rate of Io of 6×10^{18} ergs/s if $Q_I/f_I < 370$ (Yoder & Peale 1981). This inequality is almost certainly satisfied since the Moon most probably has radiogenic heating comparable with that of Io but is largely unmelted, so the high temperature of Io needs considerable tidal heating. This bound on Q_I/f_I then imposes $Q_J < 4.6 \times 10^6$. More likely $Q_I/f_I < 100$ by comparison with other rocky bodies such as Mars and the solid Earth. This latter upper bound on Q_I/f_I leads to $Q_J < 1.2 \times 10^6$, which we nudge up to 2×10^6 to be conservative.

The upper bound on Q_J is lower than several estimates of the Q_J to be expected from turbulent viscosity in Jupiter's gaseous interior—the most extreme being $Q_J \approx 10^{13}$ (Goldreich & Nicholson 1977). A value of Q_J much larger than the above upper bound means there would be insufficient torque from Jupiter to assemble the resonances and to maintain the current hypothesized equilibrium. This insufficiency would mean that the dissipation in Io would be currently decreasing the eccentricity, increasing $\omega_{1,2}$, and dissassembling the Laplace relation and associated two-body resonances. Could the Laplace relation simply be decaying from an original state much deeper in the resonance? Libration of the Laplace angle $\sim 180°$ becomes unstable if $e_I > 0.012$, $\omega_1 < 0.14°$/day, and the time to decay from $e_I = 0.012$ to $e_I = 0.0041$ is only a few tens of millions of years with the current dissipation in Io (Yoder & Peale 1981). However, there is another stable stationary state with libration of the Laplace angle $\sim 0°$ with $\omega_1 < 0$ (Sinclair 1975). Greenberg's (1987) attempt to store the system here for subsequent decay to the current configuration requires a series of improbable events, not the least of which is the establishment of the Laplace relation at the time of satellite formation within the small amount of phase space allowed by the resonances. The unlikeliness of this scenario is in spite of possible paths between the two stationary states ($\omega_1 > 0$ and $\omega_1 < 0$, respectively), along which stable libration could apparently be maintained (Greenberg 1987).

In support of the Yoder (1979) hypothesis of an equilibrium configuration and the route thereto or modifications of that route by Malhotra (1991), two theoretical determinations of the tidal Q_J are well below the upper bound established by the observed dissipation in Io. Stevenson (1983) invokes hysteresis in the tidally induced condensation and evaporation of Helium raindrops to obtain the necessary dissipation, whereas Ioannou & Lindzen (1993) abandon the equilibrium tides used almost exclusively in the past and treat the dynamic response of Jupiter's fluid atmosphere and interior to the tidal forcing by Io to obtain Q_J as low as 10^3. This latter result depends on some parts of Jupiter being stably stratified with a Brunt-Väisälä frequency, the frequency of adiabatic oscillations of an element in the stably stratified region, the same as the tidal forcing frequency $2(\dot\psi_J - n_I)$ at some depth, where $\dot\psi_J$ is Jupiter's rotational angular velocity and n_I is Io's mean motion.

If Io's eccentricity is indeed very close to the equilibrium value, it can be used in place of the unknown interior properties of the satellite to express the energy dissipation in Io in terms of Q_J alone. If the three satellites have reached the equilibrium configuration, ω_1, ω_2, the ratio of the semimajor axes and all the forced eccentricities are nearly fixed as the system continues its expansion from the tides raised on Jupiter by Io. Conservation of angular momentum and energy requires

$$\frac{d}{dt}(L_I + L_E + L_G) = T, \qquad (14)$$

$$\frac{d}{dt}(E_I + E_E + E_G) = n_I T - H, \qquad (15)$$

where T is the torque on Io and H is the energy dissipation rate in Io. We have neglected the tidal torques on Europa and Ganymede as well as the energy dissipation therein as these are small compared with these parameters for Io. The second equation follows from the fact that dissipation in the satellites must come from the orbits because of the fixed synchronous spins. With $L = m\sqrt{Gm_J a(1-e^2)}$, $E = -Gm_J m/2a$, and $a_I^{-1} da_I/dt = a_E^{-1} da_E/dt = a_G^{-1} da_G/dt$ from the constancy of the semimajor axis ratios, we have (Lissauer et al 1984)

$$H = n_I T \left(1 - \frac{1 + \frac{m_E a_I}{m_I a_E} + \frac{m_G a_I}{m_I a_G}}{1 + \frac{m_E}{m_I}\sqrt{\frac{a_E}{a_I}} + \frac{m_G}{m_I}\sqrt{\frac{a_G}{a_I}}}\right), \qquad (16)$$

for the rate of energy dissipation in Io, where we have neglected e_i^2.

If we use the lower bound on the averaged $Q_J = 6.6 \times 10^4$ in Equation 5 for T, H in Equation 16 or $H = dE/dt$ in Equation 8 corresponds to a surface flux density of heat on Io of ~ 1000 ergs cm^{-2} sec^{-1}. The comparison of this surface flux density with a measured value of 2500 ergs cm^{-2} sec^{-1} (Veeder et al 1994) is the source of the discrepancy between the maximum dissipation rate in Jupiter and that necessary to account for the measured energy dissipation in Io. This discrepancy has led several authors to propose that the tidal heating of Io or the release of the energy from the surface may be episodic with the current values of the heat flux near a maximum of the fluctuations (e.g. Greenberg 1982, Ojakangas & Stevenson 1986). However, this maximum rate of Jupiter dissipation is based on a minimum averaged Q_J over all of history, where the minimum is derived from the proximity of Io to Jupiter. But notice that the discrepency would be removed if the current Q_J were only a factor of 2.5 smaller than the minimum averaged value, and there is no real reason for excluding a change in Q_J with time. Indeed, Ioannou & Lindzen (1993) find that the current Q_J could be even much lower than this, and their work suggests a possible mechanism for the current Q_J being much lower than it was in the ancient past. According to Ioannou & Lindzen, a low Q_J requires that some of the outer layers of Jupiter be stably stratified. It is probable that the early Jupiter was fully convective, but as it aged and cooled some layers may have become stably stratified leading to a much higher rate of dissipation of tidal energy.

In principle, Q_J can be determined by measuring the secular acceleration of Io's mean motion. There is a possibility of doing so because reasonably precise observations of eclipses of the Galilean satellites data back >300 years. Unfortunately, neither of the current determinations of dn_I/dt is consistent with the observed heating of Io in an equilibrium configuration. In this state, $(dn_I/dt)/n_I \approx -7.4 \times 10^{-11}$/year ($da_I/dt \approx 2.1$ cm/year)—if we assume $Q_J = 6.6 \times 10^4$, its minimum averaged value, and 2.5 times this value if Q_J is lowered to 2.6×10^4—to be consistent with the current measured heat flux from Io. Lieske (1987) finds $(dn_I/dt)/n_I = -0.74 \pm 0.87 \times 10^{-11}$/year—more than 1 order of magnitude too small to account for the dissipation in Io in an equilibrium configuration. Goldstein & Jacobs (1995) find $(dn_I/dt)/n_I = 4.54 \pm 0.95 \times 10^{-10}$/year, which would imply that the Laplace relation is rapidly being destroyed. In fact with zero torque from Jupiter, this rapid increase in Io's mean motion would require the surface heat flux density from Io to be ~ 6000 erg cm^{-2} s^{-1} in a steady state—2.4 times the measured value! Any torque from Jupiter would require even more dissipation in Io. It would seem that any attempt to resolve these large discrepancies of observationally estimated fractional rates of change in n_I from a value consistent with the observed dissipation in Io must await another analysis of the ancient eclipse data and the timing thereof.

Perhaps nothing in solar system science causes as much excitement today as the possibility of a current liquid ocean under the ice of Europa. The images from Voyager 1 revealed cracks and blocks of ice that were displaced and rotated, resembling the patterns on terrestrial ice flows (Smith et al 1979). Many more examples of disrupted surfaces viewed with much higher resolution by the Galileo spacecraft show lateral displacements and rotations of blocks that retain the groove patterns of the undisrupted surface, and the blocks can thereby be reassembled into original relative locations (Carr et al 1998). Gravity experiments from recent Galileo flybys of Europa imply a differentiated satellite with a low-density layer (ice and water) perhaps 150 km thick (Anderson et al 1998b), whereas the properties of the blocks discussed by Carr et al (1998) imply an ice thickness of only a few kilometers at the time of the surface disruption. Estimates of the surface age are controversial, but the disrupted surface may be no more than 10^8 years old and could be much younger—implying that the processes leading to breakup of the surface and displacement of the blocks may be ongoing (Carr et al 1998). The global patterns and superpositions of large cracks in the surface are consistent with fracture perpendicular to tidal stress fields in an ice layer that slowly shifts in longitude relative to a synchronous rotation rate, and the triple ridge pattern of these global features could result from the periodic pumping of water to the surface as the cracks open and close during the Europa eccentric orbital motion (Greenberg et al 1998). Finally, induced magnetic fields mapped by the Galileo spacecraft are consistent with a conducting, liquid ocean under the Europan ice (Khurana et al 1998).

Tidal energy deposition as a result of relatively large orbital eccentricity is perhaps the only viable energy source available to prevent such an ocean from freezing (e.g. Cassen et al 1979). However, this source may be only marginally able to maintain the ocean (Cassen et al 1980a, Ross & Schubert 1987). Substitution of Europa parameters into Equation 8 yields $dE_E/dt = 1.05 \times 10^{20} f_E/Q_E$ erg/s compared with $2.2 \times 10^{21} f_I/Q_I$ erg/s for Io. This substitution justifies its neglect in the above determination of the body independent value of the dissipation in Io as a function of Q_J. It remains to be seen whether a robust case can be made for this dissipation rate, f_E being determined by Europa's structure and rheology, and whether it can maintain the currently inferred liquid ocean beneath the ice layer.

The future of the Laplace resonance and a historical evolution of the Galilean satellite system that can account for the current heat flux from Io, the Ganymede-Callisto surface dichotomy, the Ganymede magnetic field, and the now probable liquid ocean on Europa may ultimately rest on confidence in a sufficiently low Q_J to bring it all about.

7. SATURN SYSTEM

In some respects the Saturn satellite system is the simplest in the solar system in the sense that all but 2 of its 18 satellites are regular. But it may be considered the most complicated, as those regular satellites display a rich variety of orbital resonances, satellite-ring interactions, an exotic rotation state, and, for the inner small satellites, probable repeated destruction and reassembly in a process that must be responsible for much of the debris in the rings (Smith et al 1982). Because of gravitational focusing by Saturn, Smith et al (1982) find that the cratering rate is presently twice as high on Rhea as on Iapetus and 20 times higher at Mimas. Starting with the crater density on Iapetus and estimates of cratering rates based on the flux of impacting cometary bodies, Smith et al (1982) deduce that from Dione inward all of the satellites have experienced at least one impact with enough kinetic energy to leave a crater the size of the satellite—probably disrupting the satellite. The inner satellites would have experienced such collisions many times, and the current distribution of these small satellites is most likely the result of repeated disintegrations and reassemblies. This process could account for the coorbital satellites Janus (SX) and Epimetheus (SXI). Disintegration of a parent satellite likely resulted in the F-ring along with its two shepherds, Prometheus (SXVI) and Pandora (SXVII). Telesto (SXIII) and Calypso (SXIV), the leading and trailing

"Trojans" of Tethys (SIII), are probable remnants of or reaccumulations from the last disintegration of Tethys, with a similar origin of Helene (SXII), the leading "Trojan" of Dione (SIV), from the last disintegration of Dione. Atlas (SXV) is the A-ring shepherd that could be a remnant of the A-ring parent as is Pan (SXVIII), the shepherd creating the Encke gap in the A-ring (Showalter 1991), although Atlas could equally likely have come from the F-ring parent.

From Equation 6, it is easy to show that all of the numbered satellites out to and including Iapetus should be tidally evolved to synchronous rotation. All of the large regular satellites, with the exception of Hyperion (SVIII), are observed to be synchronously rotating, and synchronous rotation for Janus and Epimetheus among the small inner satellites has been verified (Yoder 1995). Although Hyperion is tidally evolved, its very asymmetric shape and highly eccentric orbit force it to tumble chaotically on time scales comparable with the orbit period (Wisdom et al 1984). The only rotational evolution suffered by distant Phoebe, like the outer satellites of Jupiter, is an occasional change caused by impact with subsequent relaxation to principal axis rotation. Voyager observations of Phoebe indicate a rotation period of $\sim$9 h (Smith et al 1982). The Titan (SVII) synchronous rotation has been verified only recently by observation through methane windows in the atmospheric haze (Lemmon et al 1995). Although all of the synchronous satellites will also have reached a stable Cassini obliquity, the small orbital inclinations make these satellites unremarkable except for Iapetus.

At the distance of Iapetus from Saturn, the rates of its orbit precession due to the oblateness of Saturn and the solar perturbations are comparable—3.320 and 4.073 arcmin year^{-1}, respectively (Ward 1981)—so the Laplacian plane nearly bisects the angle formed by the Saturn orbit and equator planes. Ward derives the expression giving the angle of inclination of the Laplacian plane of 14.84° relative to the Saturn equator. The Iapetus orbit presses with a period of about 3200 years while maintaining a nearly constant inclination of $\sim$8° relative to the Laplacian plane (Ward 1981). In this precessing orbit, the Iapetus spin should have evolved to either Cassini state 1 or 2, where state 2 would be the end point if Iapetus is near hydrostatic equilibrium and state 1 would result from an internal strength sufficient to support moment differences comparable with those of the Moon (Peale 1977). Hydrostatic equilibrium would favor a predominantly icy satellite, whereas significant internal strength would imply a more rocklike constituency. The density of Iapetus (1.15 ± 0.08 g cm^{-3}) implies ice and perhaps hydrocarbons (Tyler et al 1982), so we might expect to find Iapetus in Cassini state 2. This prediction can be verified by the approaching long-term observations by the Cassini spacecraft since the spin axis in state 1 is near the orbit normal, whereas the obliquity will be $>8°$ in state 2 (Peale 1977).

The high inclination of the Iapetus orbit to the Laplacian plane could mean that Iapetus was a captured satellite instead of one accreted within a dissipative disk. The latter process should have produced Iapetus very close to the local Laplacian plane, and it should remain near that plane if the position of the plane changes slowly compared with the rate of precession of the line of nodes (Ward 1981).

Ward points out that the position of the Laplacian plane is influenced greatly by the mass in the disk and that a relatively rapid dissipation of the disk that violated the adiabatic condition, thereby rotating the Laplacian plane, could leave Iapetus with its present inclination while preserving its origin in a dissipative disk, as its small eccentricity seems to imply.

The leading hemisphere of Iapetus is very dark (albedo $\sim$0.04), whereas the trailing hemisphere is very bright (albedo $\sim$0.5) (Smith et al 1982). It has been suggested that this dichotomy came about because dust from Phoebe that was spiraling in due to Poynting-Robertson drag would impact only on the leading hemisphere of Iapetus because of the retrograde motion of Phoebe (Soter 1974, Hamilton 1997). Smith et al (1982) point out that the transitions between the bright and dark regions are often sharp and that craters on the bright side have dark interiors—characteristics that are not consistent with a dusting from Phoebe debris. However, Smith et al (1982) also point out difficulties with an endogenic origin of the dichotomy. Hamilton (1997) would have Iapetus being dusted on all sides during a billion-year period as it slowed to synchronous rotation. Iapetus would have then been coated with frost but with only the leading side kept dark with continued Phoebe contamination. Dark floored craters on the bright side are then the excavation down to the dark Phoebe material collected during the first phase. But it would take meticulous cratering indeed to just go down to a relatively thin layer of Phoebe dust without penetrating to the ice below. The frost layer would have to be kilometers thick as well, and the boundaries between light and dark are still too sharp for the dusting hypothesis. These caveats and the lack of a model for endogenic origin leave the origin of the albedo dichotomy on Iapetus still not understood.

Six sets of orbital mean motion resonances between satellites persist in the Saturn system, and there is a secular resonance where the line of apsides of the Rhea orbit librates about that of the Titan orbit (Greenberg 1975; Pauwels 1983). The 2:1 mean motion resonance of Mimas (SI) and Tethys is the only inclination (i-type) resonance in the solar system in which the coefficient in the term in the disturbing function controlling the resonant motion contains the product of the two orbital inclinations instead of an eccentricity (e-type), and the argument of that term contains the longitudes of the ascending nodes instead of a periapse longitude. Enceladus (SII) and Dione are in a 2:1 e-type mean motion resonance, and Titan and Hyperion are in a 4:3 e-type resonance. We have already pointed out the satellites in 1:1 coorbital resonances: Janus-Epimetheus, Telesto-Tethys-Calypso, and Helene-Dione. In the Janus-Epimetheus resonance, Epimetheus describes a horseshoe orbit in a frame rotating with the average of the two mean motions; Janus, the more massive, follows a shorter loop. Yoder et al (1983) give a clear analysis of this resonance and show how the mass of each member can be determined by the distance of closest approach during the reversal of the relative motions (see also Yoder et al 1989). Telesto and Calypso librate about the L4 and L5 Lagrange equilibrium points in the Tethys orbit in the frame rotating with the mean motion of Tethy, and Helene librates about the L4 point in the Dione orbit. There are many descriptions of such librations in tadpole-shaped trajectories about the equilibria, where the restricted three-body problem is the basis for analysis (e.g. Brown & Shook 1964). Libration in eccentricity-type resonances also has many description, but see Peale (1976) for a heuristic description of the physical mechanism of stability. Conjunctions of Enceladus and Dione always occur near the periapse of the Enceladus orbit with a corresponding forced eccentricity in the Enceladus orbit like that discussed for the Jupiter Galilean satellites. Conjunctions of Titan and Hyperion always occur near the Hyperion apoapse, and the resonance forces the Hyperion eccentricity. The mixed i-type resonance of Mimas and Tethys is considerably more complicated than the simple e-type resonances, in which both orbital inclinations are forced [but see Greenberg (1973a) for a lucid description of the physical mechanism of libration and stability]. Conjunctions of Mimas and Tethys librate about the average of the ascending node longitudes of the two orbits on the equator plane of Saturn. It is the goal of analyses of evolutionary schemes to understand how these three resonances among the somewhat larger, classical satellites came to be and at the same time to account for the exotic properties of the satellites involved. We shall see that both Mimas-Tethys and Enceladus-Dione resonances could have been assembled from initially random orbits from orbit-expanding torques, but that such an origin for the Titan-Hyperion resonance is far from robust.

We consider first the Mimas-Tethys resonance. From Equation 7, we find $dn_M/dt - 2dn_T/dt = -3.7 \times 10^{-19}/Q_S = -2.2 \times 10^{-23}$ rad sec^{-2}, where the fluid Love number for Saturn, $k_S = 0.317$ (Yoder 1995), and $Q_S = 1.7 \times 10^4$, the minimum average value that would bring Mimas from the edge of the A-ring to its present position in 4×10^9 years, are used. The negative value means the Mimas orbit is approaching that of Tethys sufficiently quickly to be captured and driven deeper into a resonance at the 2:1 commensurability—a necessary condition for a tidal origin of the resonance (e.g. Peale 1986). This conclusion assumes the same value of Q_S for both satellites. There are several slowly varying frequencies at the 2:1 commensurability of mean motions where those with the lowest order coefficients are $(2n_M - 4n_T + 2\dot{\Omega}_M)$, $(2n_M - 4n_T + \dot{\Omega}_M + \dot{\Omega}_T)$, $(2n_M - 4n_T + 2\dot{\Omega}_T)$, $(n_M - 2n_T + \dot{\varpi}_T)$, $(n_M - n_T + 2\dot{\varpi}_M)$. Mimas and Tethys are locked in the mixed i-type resonance corresponding to the second frequency. The time variations in the node and periapse positions are sufficiently rapid due to Saturn oblateness that these frequencies are actually well separated. Compare the 78.8-year libration period of the Mimas-Tethys resonance variable with the separation of the first two frequencies $\dot{\Omega}_M - \dot{\Omega}_T = -293°$ year^{-1}. This separation motivates the treatment of each resonance as isolated. But why did the system choose to occupy the second resonance?

As the Mimas-Tethys system approaches the 2:1 commensurability, $n_M - 2n_T$ is decreasing, and the resonances are encountered in the order given above if only the secular perturbations of the node and periapse are considered. (e.g. The first slowly varying frequency is the first to vanish.) However, from the Lagrange Planetary equations (e.g. Danby 1988) applied to a disturbing function that selects the resonant arguments, an approach to resonance induces the variations

$$\left(\frac{d\varpi_M}{dt}\right)_{res} = n_M \frac{m_T}{m_S}\frac{a_M}{a_T}C_1\frac{1}{e_M}; \quad \left(\frac{d\Omega_M}{dt}\right)_{res} = n_M \frac{m_T}{m_S}C_2\frac{a_M}{a_T}\frac{i_T}{i_M}, \quad (17)$$

for the simple lowest-order e and mixed i-type resonances, where C_1, C_2 are negative constants. For small e, $d\varpi/dt$ will be large and negative. The node can have a large negative motion if $i_M \ll i_T$, but the presence of i_T in the numerator means that the node motion will almost always be dominated by the nonresonant secular perturbations. It is much more likely that a small e will induce such a large retrograde motion in ϖ that a simple e-type resonance can be encountered before any of the inclination resonances and the system can automatically enter into the e-type resonance libration. However, from Equation 17 $e_M < 1.3 \times 10^{-6}$ in order for $|d\varpi_M/dt| > |d\Omega_M/dt|$. Since such a small average e_M is highly unlikely, it appears safe to assume that the Mimas-Tethys system first encountered the well separated inclination resonances in the order given (Yoder 1973; Peale 1976). We need but account for its avoidance of the first i-type resonance and capture into the second.

The current amplitude of libration of the resonance variable $2\lambda_M - 4\lambda_T + \Omega_M + \Omega_T$ corresponding to the second frequency (λs are mean longitudes) is 97°. By numerically integrating the resonance evolution backwards in time, Allen (1969) determined $i_M = 0.42°$ and $i_T = 1.05°$ at the time when the libration amplitude was 180°, that is, at the time of capture into the existing resonance about 2.2×10^8 years ago. With this value of i_M, Sinclair (1972, 1974) numerically calculated a capture probability of only 4.3% into the existing resonance, which was obtained analytically by Yoder (1973). The value of $i_M = 0.42°$ now applies, as the Mimas-Tethys system passed through the first encountered, simple i-type resonance, which yields a capture probability into this first resonance of 7.3%, numerically and analytically by the respective authors. We can thus account for Mimas-Tethys skipping the first resonance encountered and stopping in the second because the captures are probabilistic.

Although this scenario makes a nice, self-consistent story of the evolution of the Mimas-Tethys system into and within the resonance, it was developed without the benefit of modern nonlinear dynamics and high-speed numerical computations. Champenois and Vienne (1999a,b) find that secondary resonances between the libration frequency and newly discovered long-period terms in the mean longitude of Mimas introduce chaos into the system that may have been important at the time of capture. The inclination of the Mimas orbit at the time of capture may have been quite different than the above value calculated by Allen (1969). If the Tethys eccentricity before capture was much larger than it is today, the calculation of the capture probabilities is more complicated than the single, isolated resonance theory used above. Although Champenois and Vienne find that a moderate eccentricity in the Tethys orbit could increase the capture probability into the current inclination resonance, tidal damping of that eccentricity would appear to preclude much enhancement. The full richness of the Mimas-Tethys dynamical history is just beginning to be explored with modern techniques, and we might expect to find alternative routes to the current configuration and perhaps a more robust selection of the current resonance among those at the 2:1 commensurability of mean motions.

Finally, there is the puzzle of the currently large eccentricity (0.02) of Mimas. From Equation 9, the time constant for damping the Mimas eccentricity is somewhat $> 10^8$ years, where Mimas is assumed to be a homogeneous sphere of ice with rigidity $\mu = 4 \times 10^{10}$ dynes cm^{-2} and $Q_M = 100$. Mimas eccentricity would have had to have been excited only a few time constants ago for such a large remnant to have survived to the present.

Before the Voyager observations, the origin of the 2:1 Enceladus-Dione simple e-type resonance was accounted for by the same differential tidal expansion of the orbits that we used for the Mimas-Tethys resonance (e.g. Sinclair 1972, 1974; Yoder 1973; Peale 1976). Dione is $>14\times$ more massive than Enceladus, but this mass is less than the factor by which the Tethy mass exceeds that of Mimas, so the Enceladus orbit expands sufficiently fast to approach the 2:1 commensurability. The current eccentricity ($e_E = 0.0044$) is less than the maximum value that would allow certain capture ($e_E = 0.019$) (Sinclair 1972; Yoder 1973; Peale 1976), and was smaller in the past; apparently Enceladus has automatically evolved into the 2:1 libration. However, the existing resonance, in which the Enceladus eccentricity is forced, is the last resonance to be encountered in the set of resonances listed above in the Mimas-Tethys discussion. Moreover, the small inclinations, $i_E \approx i_D \approx 0.02°$, would have led to certain capture into any of the i-type resonances. The i-type resonances could have been avoided only if the inclinations were bigger in the past to make the captures probabilistic. Sinclair (1974) shows how the inclinations could be reduced from such larger values in probabilistic escapes from the inclination-type resonances to arrive at their current values prior to a probabilistic escape from the first eccentricity resonance and certain capture into the second where we find the system today. If the current eccentricity corresponding to the libration amplitude of $1.5°$ (Sinclair 1972) is carried backward to the time of capture, the average eccentricity would have been only 1.15×10^{-4} (Yoder 1973; Peale 1976). For this value of e_E, $\dot{\varpi}_{Eres} + \dot{\varpi}_{Esec} = -952°/\text{yr} \ll \dot{\Omega}_E = -152°/\text{yr}$, where Equation 17 was used. If Dione eccentricity was no smaller before capture than it is now, $\dot{\varpi}_{Dres} + \dot{\varpi}_{Dsec} = 24.7°/\text{day}$. Hence the simple e-type resonance in which we find Enceladus and Dione was librating long before any of the other resonances were encountered. This libration does not affect the evolution through the other resonances, so the conditions for avoiding capture in these resonances still apply (Sinclair 1983). Note that we would have overlapping resonances here without necessarily creating chaotic motions, since the inclinations are independent of the eccentricities as long as both are small. Recall that Io and Europa are librating simultaneously in two first-order e-type resonances at the same commensurability with no chaos. On the other hand, there may be some interesting dynamics when the system passes through the mixed e-type resonance while librating in the first-order resonance involving e_E. (Note the contrast with the situation with Io, where essentially the same eccentricity is thought to be an equilibrium value and accounts for the hot Io interior through tidal dissipation. The equilibrium eccentricity for Enceladus would be 0.03, if the satellite remained a homogeneous sphere with $\mu = 4 \times 10^{10}$ dynes cm^{-2}, $Q_E = 100$ and $Q_S = 1.7 \times 10^4$, the minimum average value.) Numerical integrations thorugh the resonances at rates sufficiently slow to avoid artifacts, with a thorough coverage of the likely available phase space, and with interpretations in terms of modern nonlinear dynamics, have not been done. In addition, the appearance of Enceladus and its surrounds strongly indicates that evolution into the simple 2:1 e-type resonance with Dione may be only part of a more exotic history experienced by this strange satellite.

Enceladus is the most geologically evolved and youthful of all of the Saturn satellites (Smith et al 1982), so its evolution must have been very special indeed. The youngest terrain is crater-free at a resolution of 4 km with an upper limit on its age of $\sim 10^9$ years (Smith et al 1982). The geometric albedo is 1—as if the surface has a fresh dusting of frost; the E-ring of Saturn, a diffuse thick ring extending from 3 to 8 Saturn radii (R_S), has its maximum surface brightness at the distance of Enceladus from Saturn. The ring particles are of the order of 1 μm in size and have relatively short lifetimes in orbit (Horanyi et al 1992). The circumstantial evidence thereby points to Enceladus as the active source of the E-ring particles to account for the concentration of particles at that distance from Saturn and to restore the losses (Pang et al 1984). Study of the dynamics of the E-ring particles—assumed to originate from the surface of Enceladus under the control of gravitational, radiation, and electromagnetic forces—leads to a distribution of particles that reproduces the observable characteristics of the ring (Horanyi et al 1992, Horanyi 1996). Such activity requires an energy source, and the only viable source proposed is again tidal dissipation within the satellite.

The apparent erasure of surface features by deposition of material from the interior implies that some part of the interior is in a molten state. According to Squyres et al (1983), the minimum heating rate required to initiate melting in Enceladus is $\sim 2 \times 10^{17}$ ergs/s, whereas current tidal heating provides only 6.2×10^{14} ergs/s, where a homogeneous body with $Q_E = 20$ and rigidity $\mu = 4 \times 10^{10}$ dynes cm^{-2} are assumed. If Q_E were a sufficiently rapidly decreasing function of tidal stress, the growth of the Enceladus eccentricity e_E to large values might trigger a melting episode (Yoder 1981b). But because the rate of energy input due to the Enceladus-Dione resonance alone is small, the time scale for eccentricity growth may be too long for this process to be important. If the eccentricity forcing were removed, a largely molten Enceladus would quickly damp the eccentricity and could freeze by conduction and radiation alone in 5×10^7 years.

Squyres et al (1983) find that the orbital eccentricity must be at least 5–7 times the current value of 0.0044 to maintain a molten interior against the cooling effects of convection, conduction, and radiation if the interior is pure-water ice, and it must be higher by a factor of 20 to initiate melting of an initially completely frozen body. The model here is of a solid-ice crust of varying thickness over a completely melted inner core. Since the equilibrium eccentricity varies as $Q_S^{-1/2}$

(Yoder & Peale 1981), the latter eccentricity of nearly 0.09 to initiate melting of a solid Enceladus could never occur unless Q_S were reduced by a factor of 10 from the minimum average used above to determine the equilibrium-forced eccentricity of 0.03. If the thermal conductivity was reduced by the inclusion of clathrate hydrates in the icy material, a significant enhancement of the present eccentricity would still be required to initiate melting, but it might be possible to maintain a molten interior and allow geologic activity with the present eccentricity. However, the conditions are too specific to be likely, and the supposition does not pass the Mimas test as pointed out by Squyres et al (1983). Mimas is comparable in size with Enceladus, it has a much larger eccentricity, and it is close to Saturn. Yet Mimas does not show any sign of tidal heating. If Enceladus contains a significant amount of ammonia along with water, the heating required is a few times less than 10^{17} ergs/s (Stevenson 1982; Squyres et al 1983). Could Enceladus have more NH_3 than Mimas, and thereby decrease, the heating requirement? Maybe, but no NH_3 has been detected on the surface.

In another attempt to account for the geologic activity, Lissauer et al (1984) noticed that Janus (SX) was just outside the 2:1 commensurability with Enceladus. Spiral density waves generated by Janus in the A-ring lead to torques on the satellite that would place it at the 2:1 commensurability only ~15 MY ago. If Janus were locked in the 2:1 eccentricity resonance that forced the Enceladus eccentricity, the relatively strong ring torques might force sufficient eccentricities on Enceladus for a melting episode. If we insert $T = 8.78 \times 10^{20}$ g cm^2 sec^2 (Lissauer & Cuzzi 1982), for an assumed ring-surface density of 50 g cm^{-2}, into Equation 16 along with the parameters for the Janus-Enceladus-Dione system assumed to be in a Laplace-type resonance, $H = 6.8 \times 10^{16}$ ergs/s for the three-body case and 4.5×10^{16} ergs/s if Dione is not involved. These values are close to those required by Squyres et al (1983) for melting Enceladus. Only Enceladus would have a significant forced eccentricity in either the two-body or the three-body cases (Lissauer et al 1984), so all the energy is deposited in Enceladus.

There are numerous problems with this scenario. First, the small mass of Janus means the 2:1 resonance is not very stable. The small width of the resonance means that fluctuations from other perturbations (e.g. from Titan) could disrupt the resonance. A Janus-Enceladus resonance is not likely to survive the encounter with Dione since the latter would induce fluctuations that were larger than the resonance width, while Enceladus was still far from the 2:1 resonance with Dione. Since the time to freeze Enceladus after the eccentricity has damped is at most $\sim 5 \times 10^7$ years (Squyres et al 1983), the resonance must have been as its peak a shorter time into the past if Enceladus is to be still active in keeping itself white and in supplying E-ring particles.

Lissauer et al (1984) point out that the ring torques are so strong that it is a puzzle that the small satellites inside Mimas have remained so close. All of these satellites would have been at the outer edge of the A ring only 50 MYA if the torques are correctly determined. Having some of the small satellites trapped in resonances with the intermediately sized satellites farther out would allow transfer of angular momentum that would permit their continued proximity to the rings, but we have already seen some of the difficulties with this scenario. Having these satellites recently formed from the debris of a catastrophic disruption of a larger body is another possibility as discussed above. But a more serious problem is that the amount of angular momentum available in the A-ring could not supply the small satellites for more than $\sim 10^8$ years (Lissauer et al 1984). This situation poses another potential problem for the Janus-Enceladus resonance in providing the energy for resurfacing, since Squyres et al (1983) argue that the resurfacing has been going on for much longer than this.

It is clear that the means by which tidal dissipation could have resurfaced Enceladus has not been secured. A careful numerical analysis of the system from the modern dynamics point of view should be undertaken with a thorough exploration of the phase space. There may be chaotic behavior and/or secondary resonances that could significantly increase the dissipation beyond that deduced by Equation 8. At the same time, there are no obvious dynamical means to account for the resurfacing of relatively small parts of Dione, Rhea, and Tethys as pointed out by Smith et al (1981, 1982).

In the Titan-Hyperion 4:3 e-type orbital resonance, conjunctions librate about the Hyperion apoapse with an amplitude of $36°$ and a period of 2 years. The distance of close approaches is thereby maximized, and Hyperion owes its continued existence to the resonance. The evolution of the Hyperion shape is easy to understand given the local dynamics. Impacts repeatedly chip away at all small satellites, but normally much of the material escaping from the surface is recollected in a relatively short time or reassembled into new satellites as must happen in the small satellite-ring region. However, anything escaping from Hyperion is likely to have sufficient initial velocity to escape the protective resonance with the giant Titan. No longer protected from close approaches to Titan, the escaped material is quickly eliminated from the region and relatively little of it reaccretes onto Hyperion (Farinella et al 1990). The remnant of this process is the flattened hamburger shape (Smith et al 1982). The large gravitational torques on the asymmetric satellite coupled with the highly eccentric orbit do not permit the normal tidal evolution to synchronous rotation. Tides slow the satellite until it enters a chaotic zone where it is condemned to tumble chaotically for its remaining lifetime in the resonance (Wisdom et al 1984).

Analytic descriptions of the Titan-Hyperion 4:3 orbital resonance are generally not adequate representations of the motion (Sinclair & Taylor 1985). The close proximity of the orbits and the high eccentricities cause very slow convergence of the series expansions. Still, it was used as the model for a simple e-type resonance, which led to an understanding of capture into such resonances as the inner orbit was expanded by tides (Greenberg 1973b). Colombo et al (1974) demonstrated capture as the Titan orbit was expanded by tides, but their conclusions are suspect because they accelerated the evolution by 10 orders of magnitude to allow the numerical computation to proceed. Such accelerations are known to introduce artifacts into the calculations (Tittemore & Wisdom 1988), although artificially high evolution rates are more likely to frustrate capture rather than to cause it. Tidal origin of the Titan-Hyperion resonance would seem to be precluded, since Q_S for Titan's tides must be >1 order of magnitude smaller than the minimum average tidal $Q_S = 1.7 \times 10^4$ established by the proximity of Mimas to Saturn (Colombo et al 1974). However, our limited knowledge of dissipative processes in gaseous planets does not prohibit Q_S having sufficient amplitude and/or frequency dependence to be low enough for Titan for significant tidal expansion of the Titan orbit while being high enough for Mimas to keep the latter close to Saturn. A careful investigation of the dynamics of the Saturn atmosphere as perturbed by the tide-raising potentials would seem appropriate (see Ioannou and Lindzen 1993).

Peale (1995), while assuming that significant tidal expansion of the Titan orbit occurred, has shown that Hyperion would suffer close approaches to Titan long before the orbits approached the 4:3 commensurability unless the satellites were trapped in a secular resonance that kept the lines of apsides nearly aligned during the tidal approach ($\varpi_H - \varpi_T$ librates about 0°). Moreover, the initial value of a_T/a_H must be $\gtrsim 0.806$ to prevent capture into other resonances before the 4:3 resonance was encountered. The current characteristics of the resonance are produced for initial conditions $(a_H, e_H, f_H, \varpi_H) = (1.171, 0.02, 36°, 0°)$ and $(a_T, e_T, f_T, \varpi_T) = (0.944, 0.025, 0°, 0°)$, where the semimajor axes are in units of the current a_T and f is the true anomaly. To move Titan from $a_T = 0.944$ to 1.0, in 4.6×10^9 years, $Q_S \approx 900$. This Q_S necessary for the 5.9% expansion in the Titan orbit is smaller than that obtained by Colombo et al (1974), who used 1.5 instead of 0.317 for the Saturn Love number. Capture into the 4:3 resonance is always certain once Hyperion is in the secular resonance with Titan. Hyperion eccentricities large enough for probabilistic capture lead to circulation of $\varpi_H - \varpi_T$ with inevitable destruction of Hyperion or escape to solar orbit.

The resonance could also be approached from initially nonresonant orbits if Titan experienced little or no tidal evolution, but Hyperion spiraled in from a larger orbit due to a nebular drag. This drag would act the same way as the solar nebular drag acts on planetesimals because of the differential orbital velocities between the gas and solid bodies. As a Saturn nebula would be rather short-lived, this drag evolution would have to be fairly rapid. But a few numerical integrations with nebular drag on Hyperion show that the libration is almost completely damped. Only if the current libration amplitude of 36° could be regenerated from the cometary collisions that chipped away at the original Hyperion could this origin of the resonance be viable. This possibility has not been investigated.

An alternative to a tidal or nebular drag origin of the 4:3 resonance would be for Hyperion to have accreted from material previously trapped within the secular and/or the 4:3 mean motion resonances. If Titan formed by runaway accretion, it would have been in existence before the smaller Hyperion could have formed and would have cleaned out all the material that was not protected by libration within the resonances. But accretion within the 4:3 resonance, in addition to having less material available, might be prevented because of the increased velocity dispersion for the nonresonant bodies induced by Titan. These bodies would crash into the resonant satellitesimals, the pieces would scatter into nonresonant orbits and eventually be eliminated from the region. Of course, other small bodies would be scattered into the resonance by the same collision process and could accrete onto the forming Hyperion. The investigation of the efficiency of accretion in the 4:3 resonance while particles perturbed by Titan into high relative velocities are bombarding the resonance zone may decide the feasibility of a primordial origin.

The tidal origin of the current 4:3 resonance suffers from the requirement that Q_S be relatively small for Titan-induced tides while requiring a high Q_S for Mimas-induced tides. The nebular drag origin suffers from the necessity to regenerate the libration amplitude that would be damped to nearly zero. The primordial origin suffers from the rapid elimination of all nearby material not in resonance and the high velocity dispersion of that material, which can lead to chipping away at the resonant material before it can accumulate into a Hyperion-sized body. It is clear that we do not yet know how to create the Titan-Hyperion orbital resonance in a robust scenario.

8. URANUS SYSTEM

The Uranian satellite system consists of 10 small satellites inside 3.4 Uranus radii (R_U) (Smith et al 1986): the classical satellites of Miranda (UV), Ariel (UI), Umbriel (UII), Titania (UIII), and Oberon (UIV), distributed from 5 to $17R_U$, as well as two recently discovered retrograde satellites (Gladman et al 1998, Nicholson et al 1998) at 278 and $477R_U$. All of the satellites are regular except the last two, although the Miranda orbital inclination of 4.22° might lead to an irregular classification. However, we shall see that this inclination is due to evolution from an initially small value within an orbital resonance (Tittemore & Wisdom 1989; Malhotra & Dermott 1990). There are no current orbital resonances among these satellites.

Like the irregular satellites of Jupiter, the evolution of the two outlying satellites, after their probable capture from close encounter or collision of heliocentric planetesimals within the Uranus sphere of influence, is limited to occasional change in orbital and rotation states from collisions with Uranus family comets with subsequent relaxation to principal axis rotation. The inner satellites are embedded in an extensive ring system, where repeated collisional disruption and reassembly have led to the current configuration of small satellites and rings (Smith et al 1986; Colwell & Esposito 1992). In particular, the shepherd satellites of the ϵ ring, Cordelia (UVI) and Ophelia (UVII), are likely products of a collisional disruption of a 100-km radius precursor satellite that simultaneously created smaller particles now confined to the ϵ ring (Colwell & Esposito 1992). Application of Equation 6 shows that the most distant inner satellite, Puck (UXV) at $3.36R_U$, slows from an initial 4-h period to synchronous rotation in $\sim 2 \times 10^4$ years, where a homogeneous sphere with rigidity $\mu = 4 \times 10^{10}$ dynes cm^{-2}, $\rho = 1$ g cm^{-3} (ice), and $Q = 100$ being assumed. The rotational decay times of the closer satellites are at most about twice this value. The most distant of the larger satellites, Oberon, would reach synchronous rotation from an initial 4-h period in $<3 \times 10^7$ years, where $\mu = 10^{11}$ dynes cm^{-2} and $Q = 100$ were assumed. The closer large satellites reach synchronous rotation in shorter times, and observations verify synchronous rotation for all five satellites (Yoder 1995).

The five larger regular satellites show a wide variety of surface characteristics. The cratering record implies that Oberon and Umbriel have very old surfaces, although the uniformly dark surface of Umbriel implies some kind of recent blanketing by fine debris whose nature and source are unknown (Smith et al 1986, Croft & Soderblom 1991). Whether Umbriel and Oberon have had extensive resurfacing is controversial (Croft & Soderblom 1991); if so, the resurfacing would have had to occur early in the accretion process to allow the preservation of the large numbers of large craters (>100 km diameter). Ariel, Miranda, and Titania show extreme to moderate resurfacing at times reasonably distant from the time of diminished bombardment. Titania has a surface age that is intermediate between that of Ariel and those of Umbriel and Oberon, and its lack of very large craters emplaced during the heavy bombardment means it has been completely resurfaced since that time (Plescia 1987a). Ariel has also been completely resurfaced since the heavy bombardment, and a second wave of resurfacing has yielded the youngest surfaces among these five satellites with the exception of Miranda (Plescia 1987a). Smith et al point out that varying crater densities on fault scarps and on smooth terrain in the bottom of grabens on Ariel implies that the resurfacing on this satellite persisted over much of the period of small crater emplacement. Ages of the various surface units on Ariel are model-dependent and very difficult to constrain, but Plescia (1987b) finds that all the new surfaces on Ariel were emplaced over a relative short period $>2.6 \times 10^9$ years ago.

Miranda has one of the most bizarre surfaces of any satellite in the solar system. Older terrain cratered densely with the smaller craters is interrupted with oval and trapezoidal patterns of concentric grooves and ridges (called *coronae*)—as if viscous material welled up from the interior at three particular spots (Pappalardo et al 1997) [but see Greenberg et al (1991) and Croft & Soderblom (1991), where the latter favor an infilling of impact-generated giant basins]. The discussion of Miranda is complicated by differing points of view on the cratering history and chronology of the various surface emplacements as pointed out in Greenberg et al (1991), but the coronae on Miranda are among the youngest surfaces in the Uranus system (Plescia, 1988). Grabens on Miranda, Ariel, and Titania imply a moderate global expansion—possibly due to internal heating after a very cold formation (Croft & Soderblom 1991). Extrapolation of the cratering flux at Oberon to Miranda implies that Miranda was probably broken up by catastrophic impact at least once in its early history (Smith et al 1986, McKinnon et al 1991, Marzari et al 1998). Perhaps such a breakup and reassembly could account for the initial resurfacing of Ariel and Titania as well.

What has caused the resurfacing of the Uranian satellites? The system is remarkable in having no orbital resonances in contrast to the satellite systems of

Saturn and Jupiter. But the resurfacing of the satellites, as well as the anomalously high 4.2° inclination of the Miranda orbit, has motivated extensive investigations of possible past orbital resonances, their tidal dissipation within, and their ultimate disruption (Peale 1988, Tittemore & Wisdom 1988, 1989, 1990, Dermott et al 1988, Malhotra & Dermott 1990, Tittemore 1990). Most have used the nominal masses for the satellites in these studies, but Peale (1988) demonstrates a variety of evolutions that depend on the satellite masses within the errors of their determinations. The 1 σ errors for the satellite masses have been reduced by about a factor of 2 (factor of 3 for Miranda) since the Peale study (Jacobson et al 1992), and the nominal masses of Miranda and Ariel would now have to be changed by $\sim 3\sigma$ in opposite directions to change the current divergence of the orbits under differential tidal expansion to convergence. Although this mass dependence of possible evolutions should be kept in mind as refined mass determinations are made, we shall assume the nominal masses adopted by the authors of tidal evolution scenarios yield the correct directions of approach to the resonances. The evolutionary scenarios have been at best only marginally or speculatively successful in obtaining sufficient tidal dissipation in the satellites to account for the resurfacing of Ariel or Miranda, but they have been so successful in accounting for the high orbital inclination of Miranda due to the 3:1 Miranda-Umbriel orbital mean motion commensurability that a reasonably robust upper bound of $Q_U/k_U \lesssim 3.9 \times 10^5$ is determined to ensure that the system passed through this resonance (Tittemore & Wisdom 1989).

Dermott (1984) remarked that since the gravitational coefficient $J_2 = 0.003343$ (Yoder 1995) for Uranus is small, the several resonances at each of the commensurabilities may not be separated sufficiently in frequency to be treated under the single resonance theory discussed by several authors (e.g. Yoder 1973; Henrard 1982; Henrard & Lemaitre 1983; Peale 1986, 1988) and used above for the Saturn and Jupiter satellite orbital resonances. The resulting possibility of widespread chaos during resonance passage meant that the predictions of the single-resonance theory discussed by these authors are not valid for the Uranian satellites. The single-resonance theory was indeed found not to apply to most of the Uranus resonances discussed, when modern numerical and analytical techniques were used (Tittemore & Wisdom 1988, 1989, 1990; Dermott et al 1988; Malhotra & Dermott 1990; Tittemore 1990). Two fundamentally new dynamical results came from the analysis of the Uranus satellite resonances: first, the rate at which a chaotic system could be carried through a resonance (i.e. the magnitude of the tidal torque) without introducing artifacts into the results of the integrations was 10 to 1000 × slower than the rate that satisfied the adiabatic invariance of the action integral in single-resonance theory (Tittemore & Wisdom 1988, 1989, 1990); second, a librating system in a stable resonance could be trapped into a secondary resonance between the libration frequency and other nearby frequencies, such as the circulation frequency of another resonance variable associated with the same mean motion commensurability, and dragged during continued tidal evolution into a chaotic zone thereby disrupting the original resonance (Tittemore & Wisdom 1989). This secondary resonance evolution can be displayed dramatically in the circular inclined or planar elliptic approximations where two degrees of freedom allow the construction of surfaces of section. But the secondary resonance capture persists in the full problem with its several degrees of freedom (Tittemore & Wisdom 1989). Malhotra & Dermott (1990) and Malhotra (1990) elaborated the theory of secondary resonances.

For quasiperiodic motion, it is clear that the results of an integration will not represent a real resonance encounter and the action of the librational motion will not be adiabatically conserved if the rate of evolution is so high that there is a significant change in the libration frequency during the time of a single libration. Hence, $\dot{\omega}\tau/\omega \ll 1$ is a necessary condition for adiabatic invariance of the action integral, and the results of the integration would normally be expected to be invariant as long as the rate of evolution was slow enough for this inequality to be satisfied. Here ω is the frequency of libration, τ is its period, and the dot indicates time differentiation. However, if the motion is chaotic from the interaction with nearby resonances, invariance of the integration results are attained for rates slower than some maximum only if a chaotic adiabatic invariant exists that is defined for a two-degree-of-freedom problem as the phase space volume enclosed by the energy surface containing the chaotic zone (Brown et al 1987). The chaotic adiabatic invariant exists if the trajectory in phase space has time to thoroughly explore the chaotic zone before there is significant change in the configuration (Tittemore & Wisdom, 1988). The rates of evolution consistent with this criterion, where results become independent of the rate of evolution, were shown to be 10 to 1000 times slower (depending on the resonance), than those allowed by the criteria for adiabatic invariance in the single resonance theory. The maximum rate of evolution in this case can be determined only by numerical experiment. In some resonances, the results of the integrations remained dependent on the rate of integration even at rates less than those allowed by the physical constraints (Tittemore & Wisdom 1990).

We now investigate the consequences of differential tidal evolution of the orbits where the Uranus Q_U is assumed the same constant for all of the satellites. Ariel's orbit expands faster than that of Miranda in spite of its greater distance from Uranus since its mass is so much larger. This expansion requires the average $Q_U/k_U \gtrsim 66{,}000$, since the two orbits would be coincident 4.6×10^9 years ago for Q_U/k_U at the lower bound. Peale (1988) has shown the important first- and second-order resonances that could have been encountered for this maximal evolution of the system. Miranda would have passed through the 4:3, 3:2, and 5:3 commensurabilities with Ariel, but close approaches between Miranda and Ariel inside the 4:3 resonance would have eliminated Miranda from the system, so the lower bound on the average Q_U/k_U would have to be increased over the above value to start the system outside the 4:3 resonance. Because the differential tidal expansions cause the orbits to diverge, all of the Miranda-Ariel resonances are approached from the wrong direction for capture (e.g. Peale 1986, Tittemore & Wisdom 1989). Still, passage through the large chaotic zone of the 5:3

commensurability leads to chaotic variations of the eccentricities and inclinations of both satellites with maxima $e_M \lesssim 0.03, e_A \lesssim 0.007, i_M \lesssim 1.5°, i_A \lesssim 0.35°$, where values of the inclinations and eccentricities before resonance encounter for both satellites were 0.005 radians and 0.005, respectively (Tittemore & Wisdom 1990). The system leaves the 5:3 resonance region with e_M and i_M about twice the initial values and e_A and i_A slightly below their initial values. There is insignificant heating of Miranda or Ariel either during or after this resonance passage as the eccentricities damp according to Equation 9 (Tittemore & Wisdom 1990). It is noteworthy that there is no chaotic adiabatic invariant for this resonance down to an evolution rate within the physical constraints, so the integrations were carried out at a rate corresponding to $Q_U/k_U = 1.1 \times 10^5$, the tentative lower bound justified below.

After leaving the 5:3 commensurability with Ariel, Miranda passes through the 3:1 commensurability with Umbriel with profoundly important consequences. For inclinations and eccentricities before resonance encounter at 0.005 radians and 0.005, respectively, for both satellites, the resonances are encountered in the order $(\lambda_M - 3\lambda_U + 2\Omega_M)$, $(\lambda_M - 3\lambda_U + \Omega_M + \Omega_U)$, $(\lambda_M - 3\lambda_U + 2\Omega_U)$, $(\lambda_M - 3\lambda_U + 2\varpi_U)$, $(\lambda_M - 3\lambda_U + \varpi_M + \varpi_U)$, $(\lambda_M - 3\lambda_U + 2\varpi_M)$. The most important event in the passage through this series of resonances at the 3:1 commensurability is capture into either the i_M^2 or the $i_M i_U$ resonance corresponding to the first two resonance variables involving the node of the Miranda orbit. In either of these resonances the Miranda inclination is driven to large values as tidal evolution of the orbits continues. As the inclination grows, the frequency of libration within the resonance increases and approaches low-order commensurabilities with the circulation frequency of the i_U^2 resonance variable third in the above list. The system can be trapped in one of these secondary resonances where subsequent evolution drags the trajectory into the chaotic zone where the system ultimately escapes the primary resonance involving the Miranda node. The value of i_M after escape from the resonance depends on the particular trajectory through the phase space, but it is always comparable with the observed large value of 4.22° (Tittemore & Wisdom 1989, 1990). There appears to be a correlation of the peak of the distribution of remnant inclinations of the Miranda orbit with the particular secondary resonance that drags the trajectory into the chaotic zone. The 2:1 secondary resonance produces inclinations that tend to be too high, the 4:1 too low, and the 3:1 close to the observed value (Malhotra & Dermott 1990). This natural explanation for the anomalously high inclination of the Miranda orbit leads one to infer that the system must have passed through this resonance. This requirement places a reasonably robust upper bound on $Q_U/k_U \lesssim 3.9 \times 10^5$ (Tittemore & Wisdom 1989).

After escape from the inclination resonances at the 3:1 commensurability of mean motions, the Miranda-Umbriel system passes into a large chaotic zone associated with the eccentricity resonances corresponding to the three resonance variables that include the longitudes of periapse. The Miranda eccentricity may reach values as large as 0.05 or 0.06 (Tittemore & Wisdom 1990) during the chaotic fluctuations, but the short time scale of these excursions leads to negligible heating of the interior. From Equation 8, the rate of energy dissipation in Miranda is $dE/dt \approx 7.4 \times 10^{18} \, e_M^2$ ergs sec^{-1}, where the rigidity of ice $\mu = 4 \times 10^{10}$ dynes cm^{-2} and $Q_M = 100$ are used. During a peak in the fluctuation of $e_M \approx 0.042$ averaged over 3 million years, a total energy deposition of $\sim 1.2 \times 10^{30}$ ergs would raise the average temperature of Miranda only ~ 1 K. If one treats a conductive body in thermal equilibrium, imposes a low thermal conductivity by assuming a methane-water clathrate outer layer (Stevenson 1982) and a low melting point by assuming that much of Miranda is made of an H_2O–NH_3 eutectic with a melting point of 175 K (Cynn et al 1988) and assumes a phase space trajectory that keeps the eccentricity high through the chaotic zone and leaves a high remnant eccentricity after escape from the resonance, and perhaps throws in some remnant heat from accretion (Squyres et al 1988), one could melt some of the interior of Miranda in the 3:1 resonance with Umbriel. But these conditions are too numerous and too special to be likely, although Croft & Soderblom (1991) argue for the H_2O–NH_3 composition.

During the excursion through the chaotic zone associated with the eccentricity resonances, Miranda inclination remains near its escape value of $\sim 4°$, albeit while undergoing small, chaotic oscillations. Most importantly, the high inclination always survives the eccentricity resonance passage to emerge as the observed remnant. Throughout the evolution through this commensurability, the eccentricity and inclination of Umbriel undergo chaotic oscillations but of small amplitude, and its remnant eccentricity and inclination are negligibly different from initial values. The remnant Miranda eccentricity could be relatively large or small, but the largest remnant eccentricity obtained by Tittemore & Wisdom (1990) for the full three-dimensional problem was only 0.02. Finally, we note in passing that the three-dimensional problem (eccentric, inclined orbits) with its interconnected chaotic zones apparently has no chaotic adiabatic invariant when Miranda inclination is large—down to evolution rates less than those imposed by physical constraints. Therefore the rate of evolution through this resonance cannot be accelerated above the physically realistic rate without introducing artifacts into the integrations of the full three-dimensional problem (Tittemore & Wisdom 1990).

The 3:1 Miranda-Umbriel resonance was the last chance to heat Miranda in a resonance, which we see can result in melting part of the interior by only a rather unlikely set of contrived circumstances. How then can one account for the youngest surface area among all of the major satellites surrounded by heavily cratered terrain? The castastrophic disruption and reassembly of Miranda has been suggested as a possible energy source for the ensuing geologic activity (Smith et al 1986, McKinnon et al 1991, Marzari et al 1998). Even the cratered terrain is younger than the old Umbriel and Oberon surfaces because large craters are missing. Calculations of both initial accretion (Squyres et al 1988) and reassembly after catastrophic disruption (Marzari et al 1998) show very short time scales for the process ($\lesssim 1000$ years). The accretional heating may create a warm, buoyant mobile zone tens of kilometers below the surface if NH_3 and CH_4 are major constituents of the ice, but little resurfacing could be initiated if pure H_2O ice dominates the interior (Squyres et al 1988). On the one hand, these conclusions are model-dependent and may be no more secure than those from the tidal heating hypothesis. On the other hand, the buoyant layer somewhat below the surface would be consistent with an upwelling of viscous material to form the coronae on the Miranda surface (Pappalardo et al 1997), but other processes are not ruled out (Greenberg et al 1991). The model dependence of schemes to resurface Miranda may mean we shall never find a secure explanation, but accretional heating during reassembly after disruption may be the most probable.

Another attempt to heat Miranda with tidal friction (Marcialis & Greenberg 1987) relies on the fact that very deformed satellites enter a zone of chaotic tumbling as they approach synchronous rotation (Wisdom et al 1984, Wisdom 1987a,b). In this scheme, a shattered early Miranda reaccretes into a sufficiently asymmetric body that it undergoes chaotic tumbling while in a reasonably high eccentricity orbit. This scheme is analogous to the chaotic tumbling of Hyperion, which, however, is still extant because its eccentricity is kept high by the 4:3 orbital resonance with Titan. Miranda would maintain chaotic tumbling only during the decay of its eccentricity and perhaps a short time after e_M was small and/or its asymmetry was relaxed before being trapped into synchronous rotation. If one assumes that any *tumbling* after the eccentricity is small is of inconsequential duration, then the total energy that can be deposited in the satellite is $\sim e_{M0}^2 E_M$, where e_{M0} is the initial value of e_M and E_M is the current orbital energy of the satellite. This result follows from the fact that the spin angular momentum of Miranda is only a few parts in 10^5 of the orbital angular momentum, so the chaotic tumbling results in only small fluctuations in the latter. The orbital angular momentum being essentially conserved on the average is $\sqrt{GM_U a_M (1 - e_M^2)}$, and we can relate de_M^2/dt to dE_M/dt through da_M/dt. If $e_{M0} = 0.1$, as assumed by Marcialis and Greenberg, the total energy available would be $\sim 0.01 E_M = 3.8 \times 10^{32}$ ergs or 5.8×10^9 ergs per gram. With a specific heat for ice of $\sim 1.3 \times 10^7$ ergs g^{-1} K^{-1}, the energy is sufficient to raise even water ice to its melting point and to melt a considerable fraction of the interior if all of the heat were contained. This energy is available if the satellite is tumbling or if it is locked in synchronous rotation during the decay of the eccentricity. The only difference between the two situations is that the eccentricity would be damped rapidly if chaotic rotation prevailed and slowly if synchronous rotation prevailed. The rapid deposition of the energy in the former case would allow insufficient time for much of the energy to escape through conduction or solid-state convection and more likely lead to internal melting.

As attractive as this scheme may seem from the energy point of view, there are several problems with it. First, it is very unlikely that Miranda could accrete gently in large pieces after being broken up by a catastrophic collision. The products of the collision would most likely suffer additional collisions and be further broken up before relaxing to a dissipative disk. From such a dissipative disk, the final eccentricity of the reaccreted satellite is likely to be small rather than 0.1 as assumed by Marcialis & Greenberg. Next, after the reaccretion there has to be sufficient time to establish the densely cratered surface on Miranda before the resurfacing took place. But the time scale for damping the eccentricity and simultaneously softening Miranda to relax to a nearly spherical shape could be as short as 6000 years (Greenberg et al 1991)—much too short for all of the geologic scenarios to have taken place. We are left without an acceptable means to account for the bizarre surface of Miranda.

If Miranda and Umbriel necessarily passed through the 3:1 mean motion commensurability, then Ariel and Umbriel passed through the 5:3, which is the most recent first- or second-order resonance to have been traversed (e.g. Peale 1988). The only treatment of passage through this resonance is the planar approximation by Tittemore & Wisdom (1988). The eccentricity resonances at the 5:3 mean motion commensurability involve a large chaotic zone separating circulation and libration of the resonance variables. The probability of not being captured in the resonance is no longer determined by a uniform distribution of random phases as in the single resonance theory (e.g. Peale 1986) because the trajectory in phase space can spend a considerable amount of time in the chaotic zone. The numerically determined probability of escape from this resonance is $\sim 30\%$ in the planar approximation (Tittemore & Wisdom 1988), where significant remnant orbital eccentricities might account for the somewhat high current eccentricities of the orbits of Umbriel and Ariel. However, at no time were eccentricities maintained in either the Ariel or the Umbriel orbit during and after the resonance passage sufficient for significant tidal heating.

Including dissipation in the satellites as well as inclination terms in the analysis could significantly change the evolutionary results of this study—the first by keeping the eccentricities at lower values than those obtained by Tittemore and Wisdom and the second by possibly forcing higher eccentricities (Tittemore & Wisdom 1988). Given the results of the planar problem, it is probably the case that a complete three-dimensional treatment of passage of Ariel and Umbriel through the 5:3 commensurability, including dissipation in the satellites, will not alter the conclusion of Tittemore and Wisdom that Ariel could not have been heated sufficiently to account for its resurfacing. But we have been surprised many times in the past, so it would be prudent to carry out the calculations to be sure.

One last attempt to resurface Ariel through tidal dissipation involves a possible 2:1 resonance between Ariel and Umbriel (Peale 1988; Tittemore & Wisdom 1990). If this resonance is to be encountered $Q_U/k_U < 1.1 \times 10^5$. If the resonance is approached with small eccentricities in both orbits, as is likely, the motion is dominated by quasiperiodic behavior, and capture into libration for both resonance variables, $((\lambda_A - 2\lambda_U + \varpi_A), (\lambda_A - 2\lambda_U + \varpi_U))$, is apparently certain (Tittemore & Wisdom 1990). No chaotic separatrices (regions in phase space separating circulation from libration) are crossed in this capture for small eccentricities. Noncapture into the resonance becomes increasingly likely for approach eccentricities >0.03 for Ariel, where the now interacting resonances at the 2:1 commensurability create substantial chaotic zones. However, approach at such a high value of eccentricity is very unlikely (Tittemore & Wisdom 1990). Upon capture into the resonance, the eccentricities grow as tides raised on Uranus force Ariel deeper into the resonance until an equilibrium eccentricity is approached that remains constant thereafter because of dissipation in the satellite. This process was discussed above for the Jupiter satellite Io. If we include the torque T_U on Umbriel in equations analogous to Equations 14 and 15 but with only two satellites, the energy dissipated in the two satellites in an equilibrium configuration is (Peale 1988)

$$H = n_A T_A \left[1 - \frac{1 + (m_U/m_A)(a_A/a_U)}{1 + (m_U/m_A)\sqrt{a_U/a_A}} \right]$$

$$+ n_U T_U \left[1 - \frac{1 + (m_A/m_U)(a_U/a_A)}{1 + (m_A/m_U)\sqrt{a_A/a_U}} \right]$$

$$= 0.249 n_A T_A, \tag{18}$$

where the symbols are analogous to those in Equation 16 and where the final form uses $T_U/T_A = (m_U/m_A)^2 (a_A/a_U)^6 = 0.0469$. Nearly all of this maximum energy dissipation is in Ariel (Peale 1988). From Equations 5, 8, and 18, the equilibrium eccentricity $e_A = 0.018$ for $Q_U/k_U = 1.1 \times 10^5$. Thus, the maximum rate of energy dissipation in Ariel at the time of the necessary disruption of the resonance is 7.7×10^{16} ergs sec^{-1} or 5.69×10^{-8} ergs g^{-1} sec^{-1} averaged over the mass of Ariel. The resonance must be disrupted when $n_A = 1.234 n_{A0}$, with n_{A0} being the current value, if the system is to reach the current configuration (Peale 1988).

If Ariel and Umbriel spent a considerable time in the resonance, then Q_U/k_U would have to be smaller to allow the resonance to begin somewhat before the necessary disruption, n_A and T_A would correspond to smaller semimajor axes, and the corresponding dissipation would have been larger. With a density of 1.67 g cm^{-3}, a water ice (1 g cm^{-3}) mantle and a rocky core (2.8 g cm^{-3}) comprising 61% of the mass would yield a radiogenic heat production rate of $\sim 0.61 \times 1.6 \times 10^{-7}$ ergs g^{-1} sec^{-1} averaged over the Ariel mass, where a time average radiogenic heating rate estimated for the Moon is used (Peale 1988). The tidal heating given above, the minimum at the end of the resonance existence, is about half of the averaged radiogenic heating rate. Although tidal heating would have been larger earlier in the resonance existence, it appears inadequate to account for any melting and resurfacing of Ariel, even if the 2:1 resonance with Umbriel had persisted.

There is a much more serious problem with the 2:1 Ariel-Umbriel resonance—there is no known way to disrupt the resonance once established (Tittemore & Wisdom 1990). If the eccentricities were allowed to increase to large values within the resonance, increasing chaos offers the possibility of escape, although capture in secondary resonances (so important in the 3:1 Miranda-Umbriel resonance) do not appear to drag the system into the chaotic zone. Still, such continued growth of the eccentricities would almost certainly result in disruption of the resonance, except dissipation in the satellites places a modest upper bound on $e_A \lesssim 0.02$. Eccentricity would increase to the equilibrium value and sit there indefinitely as the system librates for the remaining existence of the solar system. At least the secular perturbations (terms in the Hamiltonian with $\varpi_i - \varpi_j$ in the arguments) from Titania do not appear to disturb the resonance (Tittemore & Wisdom 1990). Perhaps something has been missed that will appear with an integration of the complete system in all its degrees of freedom, and that could disrupt this 2:1 resonance after it was established. For now we must assume that the system never encountered the resonance or it would still be locked within. The almost certain avoidance of this resonance and the almost certain encounter of the 3:1 commensurability between Miranda and Umbriel that so nicely accounts for the large orbital inclination of the former means $1.1 \times 10^5 < Q_U/k_U < 3.9 \times 10^5$ are apparently rigorous bounds on the dissipative properties of Uranus (Tittemore & Wisdom 1990).

Although a convincing argument for using tidal dissipation to resurface the Uranian satellites may ultimately be constructed, we have so far failed to account for any of the young surfaces on these satellites in a rigorous way. It is significant that Titania has extensive resurfacing but sits between Umbriel and Oberon whose ancient surfaces are completely undisturbed. Titania could have occupied no orbital resonances of first or second order (Peale 1988), although Tittemore (1990) looked at a possible passage of Ariel and Titania through the 4:1 mean motion commensurability. Here, the Ariel eccentricity could grow to large values that, however, could raise its temperature only ~20 K. Titania was still unaffected thermally by this resonance. Higher-order resonances are weaker because of additional factors of e or i in the coefficients of the resonance terms, and one needs to check the stability of such a resonance to perturbations by the other satellites before embracing its consequences. Still, Titania remains untouched by tidal dissipation even if it had occupied third-order resonances with Ariel (Tittemore 1990).

Titania may be telling us something about our difficulties in obtaining sufficient tidal dissipation to resurface those satellites that did occupy or pass through orbital resonances. If new surfaces on Ariel are indeed 2.6×10^9 years old (Plescia 1987b) and if internal activity can persist as long as one billion years after the initial heat pulse from disruption and reaccretion, perhaps the resurfacing is more due to the accretional heating than to tidal dissipation. Even this scheme may require an H_2O-NH_3 eutectic to lower the melting point, and it seems to fail in any case for Miranda (Squyres et al 1988). On the other hand, it is hard to believe that what we see on the surfaces of the Uranian satellites is independent of the evolution caused by the tides.

9. NEPTUNE SYSTEM

The dominant characteristic of the Neptune satellite system is the existence of the large satellite Triton (NI) in a close, circular, retrograde orbit (obliquity 156.8°). Neptune also has relatively few known satellites compared with the other major planets, and all but two of those, Triton and Neried (NII), were unknown until the Voyager spacecraft observations (Smith et al 1989). Except for Neried, there are no satellites outside the Triton orbit, and the Neried orbital eccentricity of 0.75

brings it no closer than $\sim 1.4 \times 10^6$ km from the center of Neptune because of its extremely large semimajor axis of 5.51×10^6 km ($222.6 R_N$)—well outside the Triton distance of 3.54×10^5 km ($14.3 R_N$, $R_N = 24766$ km). The newly discovered satellites, designated NIII to NVIII from the closest to the farthest, are relatively small satellites in circular, equatorial orbits with the exception of innermost Naiad (NIII), whose orbit is inclined $4.7°$ to the Neptune equator. All of the small satellites except Proteus (NVIII) are inside the radius where the orbital angular velocity matches the spin angular velocity of Neptune (corotation radius), and the inner four satellites are inside the Roche radius of $2.7 R_N$ ($\rho = 1.2$ g cm^{-3}). There are two narrow, dusty rings of material, and one wide ring among the inner satellites at $2.54 R_N$, $2.15 R_N$, and $1.69 R_N$, respectively, plus several other less well defined rings and a suspected continuous distribution of dust at very low optical depth everywhere inside $2.38 R_N$ (Smith et al 1989). The outer narrow ring contains the famous arcs, which turned out to be three distinct regions of higher optical depth in an otherwise continuous ring (Smith et al 1989).

The massive satellite Triton is blamed for most of the features of this unusual system. The retrograde orbit means that Triton was almost certainly captured intact from heliocentric orbit, most probably by colliding with a satellite already in orbit around Neptune with a mass a few percent of the Triton mass (Farinella et al 1980; McKinnon 1984; Goldreich et al 1989). Such a collision would have been sufficient to capture Triton into a very eccentric orbit extending a major fraction toward the Hill sphere boundary at $r_H \approx (m_N/3 m_\odot)^{1/3} a_N \approx 4.5 \times 10^3 R_N$ ($a_N =$ heliocentric distance) while not destroying it. For $1 - e_T \ll 1$, the most relevant evolution rate is that for a_T at fixed periapse $r_p = a_T(1 - e_T)$, $a_T^{-1} da_T/dt = -(21 n R_T^5 k_T)/(64 \mu r_P^6 Q_T)$. With plausible estimates of $k_T \approx 0.1$ and $Q_T \approx 100$, Goldreich et al (1989) find that an initial Triton orbit with semimajor axis $a_T \approx 10^3 R_N$ would damp to nearly its current circular orbit with $a_T \approx 14.3 R_N$ from tidal dissipation in Triton in ~ 4 to 5×10^8 years—comfortably less than the age of the solar system. The tremendous amount of energy dumped into Triton would melt and differentiate it completely, so Q_T was probably smaller than that assumed during most of the damping time and that time was correspondingly shorter. Consistent with this capture scenario and subsequent melting, Titan was found to have a young and active surface, although that surface activity is almost certainly solar-driven (Smith et al 1989).

While the orbit was so extended, Triton played havoc with the probably initially regular satellite system. Solar perturbations would have periodically driven the Triton periapse distance down to values as low as 5 to $8 R_N$, depending on the initial semimajor axis, before the periapse asymptotically approached the final circular radius (Goldreich et al 1989). All the satellites outside the minimum periapse distance of the elongated orbit of Triton have been either consumed by Triton, scattered into Neptune, or put onto escape trajectories—except for Neried. The latter could have escaped the fate of the others if its initial presumably circular orbit were sufficiently far from Neptune, but not without suffering repeated perturbations that could easily account for the currently very large semimajor axis, eccentricity and inclination of its irregular orbit (Goldreich et al 1989). Those satellites inside the minimum periapse distance did not survive unscathed. Current perturbations of the inner satellites by Triton are not significant, but when the Triton orbit was very eccentric an inner satellite would have suffered impulsive perturbations each time Triton passed periapse. The perturbations of a satellite with semimajor axis greater than $\sim 2R_N$ would have led to chaotic diffusion of the orbital eccentricity and inclination. Outside $\sim 3R_N$ the eccentricity would have reached values such that the satellite apoapse distance could approach the periapse distance of Triton. Thus Triton probably disposed of some fraction of the inner satellites as well, with any survivors expected to have persisting, substantial orbital inclinations (Goldreich et al 1989). The lack of such inclined orbits among the inner satellites must mean that there are no survivors among those early inner satellites.

The scenario excludes the possibility of collisions among the inner satellites (Banfield & Murray 1992). There are likely to have been several satellites inside the periapse distance of Triton by comparison with the regular systems of the other major planets. Orbit eccentricities would have been limited to maximum values ~ 0.3, since the rate of decrease of the eccentricity by tidal dissipation within the satellite exceeds Triton's ability to increase it for larger values (Banfield & Murray 1992). Two satellites with semimajor axes of $3R_N$ and $5R_N$ would have overlapping orbits that would persist in overlapping for times at least of the order of the eccentricity damping time of $\sim 10^8$ years. Recall that Triton eccentricity damps in a time scale several times this value. A collision between any of the current inner satellites of Neptune with its neighbor would lead to their mutual destruction. This follows from the relation (Stevenson et al 1986)

$$\frac{1}{2} m_i v_i^2 \sim m_s S + \frac{3}{5} \frac{G m_s^2}{\gamma R_s}, \tag{19}$$

where m_i and m_s are the masses of the impactor and satellite, respectively, v_i is the relative velocity at impact, R_s is the satellite radius, $S \sim 10^6$ erg g^{-1} is the material binding energy, and $\gamma \sim 0.1$ is a factor introduced to account for the inefficiency in converting the impact kinetic energy into kinetic energy of the fragments. The impactor kinetic energy must exceed the energy stored in material strength plus the self-gravitational energy by a sufficient amount to break up the body. From Equation 19 and eccentricities of 0.3, Naiad (NIII) could destroy all of the satellites except Proteus (NVIII), and any of the satellites NIV to NVII could destroy Proteus. This result implies that the current satellite system could not have existed prior to Triton orbit circularization (Banfield & Murray 1992). The debris from the first generation of satellites would settle into the equatorial plane in circular orbits and recollect into a second generation of inner satellites with nearly circular orbits and zero inclinations, where all memory of Triton perturbations would thereby be lost.

The current system of inner satellites is thought to be still a later generation than that first accreted after the Triton orbit circularized. Only Proteus is thought sufficiently large to have survived estimated cometary impacts (Smith et al 1989).

Banfield & Murray (1992) thus apply Equation 7 only to Proteus to estimate the lower bound on the average Neptune $Q_N = 12{,}000$ (4×10^9 years$/T_C$), where T_C is the time before the present that the Triton orbit circularized. This follows from the condition that Proteus could not have started inside the current corotation radius of $3.25 R_N$.

The evolution of the distribution of masses and orbits of the inner five satellites is completely speculative with the exception of the 4.7° inclination of the orbit of innermost satellite, Naiad. All five inner satellites are inside the corotation radius and are therefore spiraling toward Neptune at rates determined by Equation 7. Naiad is so small that its semimajor axis is decreasing from tides raised on Neptune at a rate that is slower than that of any of the other four in spite of its closer proximity to Neptune. Therefore, various mean motion resonances between Naiad and any of the other four satellites are approached from a direction (orbits approaching each other) that allows capture into and evolution within the mean motion resonances. The strongest inclination-type mean motion resonances have i_1^2, $i_1 i_2$, or i_2^2 in the coefficient of the appropriate term in the disturbing function, where i refers to the respective orbital inclinations to the Neptune equator plane for the two satellites. The importance of these resonances is that orbital inclinations are forced to grow while the system is forced deeper into the resonance by the Neptune tides, which could account for the Naiad inclination of 4.7° (Banfield & Murray 1992), as for the Uranus satellite Miranda (Section 8). Only those resonances with the Naiad inclination in the coefficient need be considered, since such a resonance increases only that inclination and not that of the other resonance member. The other satellites still have orbital inclinations near their initial very small values.

Banfield & Murray (1992) have determined capture probabilities for 35 possible inclination-type resonances between Naiad and the next four satellites. Although the individual capture probabilities are small, the probability that Naiad was captured in one of these particular resonances is ~76%. The resonances are disrupted when a secondary resonance between an adjacent primary resonance and the libration drags the system into a chaotic zone (Section 8). The inclination established within the resonance remains after the resonance is disrupted. Three resonances were found to be disrupted when the orbital inclination of Naiad was near the observed 4.7°—NIII-NV 12:10, NIII-NV 11:9, NIII-NV 10:8. The first has the greatest probability of occurrence, and it was chosen by Banfield & Murray (1992) as the best candidate for accounting for the inclination of Naiad. The total probability of occurrence of 4% includes capture into the primary resonance, capture into a 2:1 secondary resonance, and escape at the right inclination. Although this probability is not large, the fact that the probability of getting captured into 1 of 35 primary inclination resonances is 76% means that Naiad probably got caught in one, and this one is as likely as any of the others–and it matches the observations (Banfield & Murray 1992). If the NIII-NV 12:10 resonance did indeed cause the inclination in the Naiad orbit, an upper bound, $Q_N \leq 330{,}000$ (4×10^9 years$/T_C$), is established. This upper bound follows from the necessity that the NIII-NV system has passed through the 12:10 mean motion resonance. Although Thalassa (NIV) was also likely captured into inclination resonances that increase its inclination, there are numerous ways in which the likely inclination of escape was not significantly different from that observed ($\sim 0.21°$). Capture into resonances at the same commensurabilities that could affect the other satellite inclinations apparently did not occur.

Of course there is an assumption here that the distribution of masses for the inner satellites has been as it is now for most of the $\sim 4 \times 10^9$ years since the Triton orbit circularized. Any cometary fragmentation and redistribution of mass among reformed satellites after their initial formation following Triton circularization must have been confined to reasonably early times. The existence of satellites within the Roche radius is not limited to Neptune. Banfield & Murray (1992) hypothesize that they could either have formed outside the Roche radius but inside the corotation radius and be transported inwards by tidal friction or that they accreted there in spite of the opposing tidal forces through the pieces sticking together by nongravitational forces. Neither of these hypotheses has been investigated in detail to establish a self-consistent scenario. Given the uncertainty in the collisional and reformation history, any attempt to refine the history of the inner satellites must always remain nondefinitive, and limited scientific return from such an excercise is probably not worth the considerable effort involved. The conjectured possibilities for the collisional and dynamical history of the inner satellites of Neptune constructed by Banfield & Murray (1992) are representative of what might have happened.

Triton is spiraling into Neptune in its retrograde orbit. Chyba et al (1989) conjectured two possible Cassini states (Peale 1969), which fix the Triton obliquity in the frame precessing with the orbit. In Cassini state 1, Triton obliquity would be nearly zero, whereas in state 2, the obliquity would be $\sim 100°$ with vastly different rates of tidal dissipation in Triton and, hence, different rates of orbital decay for the two cases. Voyager 2 data (Smith et al 1989) revealed Triton to have a very small obliquity consistent with occupancy of state 1. In this case, Chyba et al (1989) find that Triton will reach the Roche radius in $\sim 3.6 \times 10^9$ years. If Triton is mostly solid at that time, it can continue to smaller orbital radii still intact, although it probably would not survive all the way to the surface. The breakup of Triton within the Roche radius would lead to a spectacular set of rings with initially much more total mass than those of Saturn.

10. PLUTO-CHARON SYSTEM

It was highly fortuitous that the Pluto satellite Charon was discovered sufficiently far in advance of our passing through the Pluto-Charon orbit plane that a well organized series of observations allowed a remarkably rich characterization of such a distant and otherwise obscure system. Mutual eclipses and occultations during the orbit plane passage allowed radii, masses and albedo distributions to be determined (Binzel and Hubbard 1997). Subsequent observations with the Hubble Space telescope indicate a deviation from the circular orbit expected from tidal evolution (Tholen and Buie 1997). Yet we still know relatively little about this system compared with all the others, since we have not had the benefit of close observations by spacecraft instrumentation. As a result there are perhaps fewer constraints that we must respond to in the context of the evolutionary history of the system. Still, the system displays its own unique features, and increasing knowledge of the system may provide needed constraints on the nature of the objects in the furthest reaches of the solar system.

If we assume the Pluto-Charon system is the consequence of a giant impact by a planetesimal whose mass is comparable with the initial Pluto mass (Farinella et al 1979, Dermott 1978 (unpublished); Dobrovolskis 1997), some of the debris from the impact will escape the system and some will fall back onto Pluto, but sufficient debris must end up outside the Roche radius in order to collect into the observed satellite (e.g. Cameron 1997 and references therein). The debris will settle to the equatorial plane and the orbits of individual particles will be circularized through collisional dissipation. Charon will accrete most of its mass from this disk within a few hundred years (Thompson and Stevenson 1988), and it should then end up with a nearly circular orbit with nearly zero inclination relative to the equatorial plane of Pluto. We assume, therefore, that Charon began its existence as a satellite in circular, equatorial orbit at $3R_P$ (Pluto radii). The system will be assumed to have its current angular momentum for this initial condition with the total mass and mass ratio derived from Table 1. From this initial configuration it tidally evolves to its current state of dual synchronous rotation, which has been observationally confirmed (Buie et al 1997).

From Equation 6, Charon would reach synchronous rotation at a separation of $3R_P$ from a 4-h initial period in about 25 years ($\mu = 4 \times 10^{10}$ dyne cm^{-2}, $Q_C = 100$). The actual Q should be much less than this, as Charon will have just accreted and may be partially melted. Regardless of assumptions, Charon should be locked into permanent synchronous rotation almost immediately after formation. Torques from tides raised on Charon should thus be unimportant in the subsequent orbital evolution except for helping to keep the eccentricity damped.

Integration of Equation 7 for Q_P = constant from an initial orbital period of 11.6 h at time t_i to the current period of 6.39 days at time t_f yields $t_f - t_i = 1.6 \times 10^3 Q_P/k_P \approx 1.7 \times 10^7$ years, where $\mu = 10^{11}$ and $Q_p = 100$ were assumed. The time to reach the dual synchronous rotation state is short compared with the age of the solar system. Although we expect the Pluto-Charon dual synchronous system to have a circular orbit, recent observations with the Hubble Space Telescope have indicated an orbital eccentricity, between 0.003 and 0.007 (Tholen and Buie 1997), although this determination has not been confirmed (Tholen, private communication, 1998). Two means of exciting such an eccentricity in the face of tidal damping have been proposed—direct collision of a Kuiper belt object with Charon or Pluto (Tholen and Buie 1997) and differential perturbations by passing Kuiper belt objects (Levinson and Stern 1995). The latter study finds collisional excitation very unlikely but the KBO perturbations can be sufficient, depending on the total number of Kuiper belt objects, to account for the Tholen and Buie observations. An eccentricity decay time scale e/i of about 10^7 years (Dobrovolskis 1997) was assumed. So the ultimate end state of tidal evolution for the Pluto-Charon system may be slightly prevented by continuing stochastic perturbations by passing objects. With the limited information now available for this system, there do not appear to be any major inconsistencies in or outstanding problems with our proposed origin and evolution—although we are likely to be surprised by future in situ observations by spacecraft.

11. SUMMARY

We have developed reasonably robust scenarios for the origin and evolution of all the natural satellites in the solar system. The nearly coplanar, circular orbits of the regular satellites are consistent with their origin by accretion in dissipative disks in the equatorial planes of the forming planets, the smaller ones close to their primary the result of repeated breakup and reaccretion. The distant irregular satellites as well as the retrograde Triton result naturally from capture of planetesimals from heliocentric orbit, where the necessary loss of energy effecting the capture could come from three body interactions or collisions within the planetary sphere of influence, or, less likely, from atmospheric drag. The Earth-Moon and Pluto-Charon seem to require giant impacts with the primary for their creation to account for the high specific angular momenta, and, for the Moon, the marked difference in chemical constituents from its primary. Rich interplays of the consequences of tidal evolution and past and present orbital resonances have guided our thoughts from plausible initial conditions after accretion or capture to current configurations and properties of individual satellites. Still, explanations for many observed surface characteristics, interior properties or orbital configurations remain elusive or uncertain—some simply because available techniques have not yet been applied to the problems, whereas others await identification of the proper approach.

An ingenious dynamical dance of the Moon through the evection and eviction resonances allows its evolution from an equatorial orbit to its current configuration—a solution of the inclination problem. But the uncertainties in the properties of the disk resulting from the giant impact, in the physical processes that took place, and in the accretion of the Moon within the disk, may allow initial properties of the Moon that prevent this dance. Other processes not yet investigated (with the exception of a second giant impact) and perhaps not even thought of would then be required to get the Moon from the equatorial orbit to its current orbit as tides push it away from the Earth. A reasonably robust evolution of Phobos through several orbital resonances as it spirals toward the surface of Mars leaves its remnant orbital characteristics comparable to those observed and allows a formation of both Phobos and Deimos by accretion in Mars equatorial plane. This latter accretion in a dissipative disk seems necessary to account for the nearly circular, coplanar orbits of these satellites. If the composition of the two satellites turns out to be drastically different from that of Mars, as perhaps indicated by their low densities, we should look for ways to make this material the major constituent of a dissipative disk in the equatorial plane of Mars for accretion in equatorial orbits.

Jupiter satellite Io almost certainly owes its high temperature to tidal dissipation resulting from a substantial eccentricity forced by the Laplace orbital resonance. Yet we strain to account for the Ganymede-Callisto dichotomy in surface characteristics, Ganymede's magnetic field and a sustained liquid ocean beneath the European ice with that same tidal dissipation. This frustration is in spite of clever tricks with past orbital resonances or interior properties. Understanding the dissipation of tidal energy in Jupiter should be a primary goal, as the torque from Jupiter on satellites within resonances ultimately constrains how much energy can be deposited in the satellites through their own tidal flexing. At Saturn, the existence of the Mimas-Tethys and Enceladus-Dione orbital resonances can be understood in terms of differential tidal expansion of their orbits with perhaps some fine tuning to pick the existing inclination type resonance for Mimas-Tethys among the selection at the 2/1 orbital mean motion commensurability. Such an origin for the 4/3 Titan-Hyperion resonance seems remote, but the viability of accreting Titan and Hyperion simultaneously while in the resonance has not been investigated. The real enigma is Enceladus, whose snow-white appearance and the circumstantial evidence for its being the continuing source of the E-ring particles imply a heated interior. How can such a wimpy little satellite still be warm? The forced eccentricity in the current resonance with Dione causes little tidal heating, and a past resonance with Janus is probably too weak to remain stable very long even if Enceladus were once captured. We hope for some surprises when modern nonlinear dynamics, with all degrees of freedom, is applied to the systems involving Enceladus. In contrast to these mysteries, we can be a little smug in our understanding of Hyperion's evolution to its current rotation state of chaotic tumbling.

At Uranus we are unable to account for any of the young surfaces on the satellites in a robust way from tidal heating within resonances. Especially puzzling is the moderate resurfacing of Titania, which could have participated in no orbital resonance—and it lies between Oberon and Umbriel with their apparently older surfaces. Ariel is extensively resurfaced, yet its possible orbital resonances in the past do little heating. Even a 2/1 resonance with Umbriel would probably fail to warm Ariel enough. The extreme stability of the Ariel-Umbriel 2/1 resonance means that Ariel probably started outside this resonance. We can relish the understanding of how Miranda got its high inclination within a past 3/1 resonance with Umbriel, and we can also relish the added insights into chaotic dynamics obtained during the analysis of this resonance—secondary resonances and the role of a chaotic adiabatic invariant. That Miranda must have passed through the 3/1 resonance with Umbriel to account for its orbital inclination, coupled with the exclusion of Ariel-Umbriel from the 2/1 resonance places rather tight bounds on the dissipative properties of Uranus—more so than for any other gaseous planet. The capture and subsequent decay of the orbital eccentricity of the retrograde

Neptune satellite Triton seems to yield a self-consistent history leading to the current configuration and surface properties of the satellites. The small inner satellites are second or higher generations—the products of repeated breakup and reaccretion. The high orbital inclination of Naiad in spite of its accumulation in a dissipative equatorial disk is nicely accounted for by capture into and evolution within an inclination type orbital resonance—like Miranda except here the orbits are spiraling toward the primary rather than away. Pluto-Charon have reached the endpoint of tidal evolution of dual synchronous rotation. The corresponding relaxation to circular orbit may be slightly frustrated by the perturbations of passing Kuiper belt objects.

We have observed the striking uniqueness of each satellite system within the solar system, and we have had several successes in understanding the origin of current configurations and properties of the several systems. However, we have also pointed out a significant number of remaining interesting problems that will occupy clever minds for years to come as they are resolved one by one. Greater understanding will uncover even more problems as our knowledge of the satellites is refined.

ACKNOWLEDGMENTS

It is a pleasure to thank the following colleagues for commenting on particular sections of the manuscript: R Greenberg, A Harris, H Levison, R Malhotra, N Murray, A Stern, D Stevenson, J Wisdom, and C Yoder. They detected errors and omissions and generally offered good advice. Special thanks are due MH Lee, who read the entire manuscript and offered many suggestions for improving the clarity and consistency. The author received support for this work from the Planetary Geology and Geophysics Program in NASA under grant NAG5 3646.

Visit the Annual Reviews home page at http://www.AnnualReviews.org

LITERATURE CITED

Acuna MH, Neubauer FM, Ness NF. 1981. *J. Geophys. Res.* 86:8513–21

Allen RR. 1969. *Astron. J.* 174:497–506

Alterman Z, Jarosch H, Pekeris CL. 1959. *Proc. R. Soc. London Ser. A* 252:80–95

Anderson JD, Jacobson RA, Lau EL, Sjogren WL, Schubert G, Moore WB. 1998a. *Science* 280:1573–76

Anderson JD, Lau EL, Sjogren WL, Schubert G, Moore WB. 1996. *Nature* 384:541–43

Anderson JD, Lau EL, Sjogren WL, Schubert G, Moore WB. 1997. *Nature* 387:264–66

Anderson JD, Schubert G, Jacobson RA, Lau EL, Moore WB, et al. 1998b. *Science* 281:2019–22

Banfield D, Murray N. 1992. *Icarus* 99:390–401

Beckwith SVW, Sargent AI. 1998. *Nature* 383:139–44

Benz W, Cameron AGW, Melosh HJ. 1989. *Icarus* 81:113–31

Benz W, Slattery WI, Cameron AGW. 1986. *Icarus* 66:515–35

Binzel RP, Hubbard WB. 1997. In *Pluto and Charon*, ed. SA Stern, WB Hubbard. Tucson: Univ. Arizona Press. pp. 85–102

Boss AP. 1997. *Science* 276:1836–39
Boss AP. 1998. *Astrophys. J.* 503:923–37
Boss AP, Peale SJ. 1986. In *Origin of the Moon*, ed. WK Hartmann, RJ Phillips, GJ Taylor, pp. 59–101. Houston, TX: Lunar Planet. Inst.
Brown EW, Shook CA. 1964. *Planetary Theory*, pp. 250–88. Mineola, NY: Dover
Brown R, Ott E, Grebogi C. 1987. *J. Stat. Phys.* 49:511–50
Buie MW, Tholen DJ, Wasserman LH. 1997. *Icarus* 125:233–44
Cameron AGW. 1997. *Icarus* 125:126–37
Cameron AGW, Benz W. 1991. *Icarus* 92:204–16
Cameron AGW, Canup RM. 1998a. In *Lun. Planet. Sci. Conf.* 29:1062 (Abstr.)
Cameron AGW, Canup RM. 1998b. In paper presented at Orig. Earth Moon Conf. Monterey, CA. Dec. 1–3
Cameron AGW, Ward WR. 1976. In *Proc. Lun. Planet. Sci. Conf.* 7:120–22
Canup RM, Esposito LW. 1995. *Icarus* 113:331–52
Canup RM, Esposito LW. 1996. *Icarus* 119:427–46
Carr MH, Belton MJS, Chapman CR, Davies ME, Geissler P, et al. 1998. *Nature* 391:363–65
Cassen P, Peale SJ, Reynolds RT. 1979. In *Geophys. Res. Lett.* 6:731–34
Cassen P, Peale SJ, Reynolds RT. 1980a. *Geophys. Res. Lett.* 7:987–88
Cassen P, Peale SJ, Reynolds RT. 1980b. *Icarus* 41:232–39
Champenois S, Vienne A. 1999a. *Icarus*. In press
Champenois S, Vienne A. 1999b. *Celest. Mech. Dyn. Astron.* In press
Christensen EJ, Born GH, Hildebrand CD, Williams BG. 1977. *Geophys. Res. Lett.* 4:555–57
Chyba CF, Jankowski DG, Nicholson PD. 1989. *Astron. Astrophys.* 219:L23–26
Colombo G. 1966. *Astron. J.* 71:891–96
Colombo G, Franklin FA, Shapiro II. 1974. *Astron. J.* 79:61–71
Colwell JE, Esposito LW. 1992. *J. Geophys. Res.* 97:10227–41
Croft SK, Soderblom LA. 1991. In *Uranus*, pp. 561–628. Tucson: Univ. Arizona Press
Cynn HC, Boone S, Koumvakalis A, Nicol M, Stevenson DJ. 1988. *Lun. Plan. Sci. Conf. Abs.* 19:433
Danby JMA. 1988. *Fundamentals of Celestial Mechanics*. Richmond VA: Willman-Bell, 466 pp.
Darwin GH. 1879. *Philos. Trans. R. Soc.* 170:447–530
Darwin GH. 1880. *Philos. Trans. R. Soc.* 171:713–891
Degewij J, Zellner B. Andersson LE. 1980. *Icarus* 44:520–40
Dermott SF. 1978. Pluto, Herculina, Mercury, and Venus: Their real and imaginary satellites. Unpublished
Dermott SF. 1984. In *Uranus and Neptune*, ed. J Bergstralh, pp. 377–404. NASA Conf. Publ. 2330
Dermott SF, Malhotra R, Murray CD. 1988. *Icarus* 76:295–334
Dickey JO, Bender PL, Faller JE, Newhall XX, Ricklefs RL, et al. 1994. *Science* 265:482–87
Dobrovolskis AR. 1982. *Icarus* 52:136–48
Dobrovolskis AR, Peale SJ, Harris AW. 1997. In *Pluto and Charon*, ed. AS Stern, DJ Tholen, pp. 159–90. Tucson: Univ. of Arizona Press
Dones L, Tremaine S. 1993. *Icarus* 103:67–92
Drake MJ. 1986. *J. Geophys. Res.* 92:E377–86
Duxbury TC, Callahan JD. 1981. *Lun. Planet. Sci. Conf. Abstr.* 13:191
Duxbury TC, Veverka J. 1978. *Science* 201:812–14
Farinella P, Milani A, Novili AM, Valsecchi GB. 1979. *Moon Plan.* 20:415–21
Farinella P, Milani A, Nobili AM, Valsecchi GB. 1980. *Icarus* 44:810–12
Farinella P, Paolicchi P, Strom RG, Kargel JS, Zappala V. 1990. *Icarus* 83:186–204
Geissler PE, Greenberg R, Hoppa G, Helfenstein P, McEwin A, et al. 1998. *Nature* 391:368–70
Gerstenkorn H. 1955. *Z. Astrophys.* 26:245–74

Gladman BJ, Nicholson PD, Burns JA, Kavelaars JJ, Marsden BG, et al. 1998. *Nature* 392:897–99

Goldreich P. 1963. *Mon. Not. R. Astron. Soc.* 126:257–68

Goldreich P. 1965a. *Mon. Not. R. Astron. Soc.* 130:159–81

Goldreich P. 1965b. *Astron. J.* 70:5–9

Goldreich P. 1966. *Rev. Geophys.* 5:411–39

Goldreich P, Lyndon-Bell D. 1969. *Astrophys. J.* 156:59–78

Goldreich P, Murray N, Longaretti PY, Banfield D. 1989. *Science* 245:500–4

Goldreich P, Nicholson PD. 1977. *Icarus* 30:301–4

Goldreich P, Peale SJ. 1970. *Astron. J.* 75:273–84

Goldreich P, Ward WR. 1973. *Astrophys. J.* 183:1051–61

Goldstein SJ, Jacobs KC. 1995. *Astron. J.* 110:3054–57

Greeley R, Sullivan R, Klemaszewski J, Homan K, Head JW, et al. 1998. *Icarus* 135:4–24

Greenberg R. 1973a. *Mon. Not. R. Astron. Soc.* 165:305–11

Greenberg R. 1973b. *Astron. J.* 78:338–46

Greenberg R. 1975. *Mon. Not. R. Astron. Soc.* 170:295–303

Greenberg R. 1982. In *Satellites of Jupiter*, ed. D Morrison, pp. 65–92. Tucson: Univ. of Arizona Press

Greenberg R. 1987. *Icarus* 70:334–47

Greenberg R, Croft SK, Janes DM, Kargel JS, Lebofsky LA, et al. 1991. In *Uranus*, pp. 693–735. Tucson: Univ. of Arizona Press

Greenberg R, Geissler P, Hoppa G, Tufts BR, Durda DD, et al. 1998. *Icarus* 135:64–78

Greenberg R, Weidenschilling SJ. 1984. *Icarus* 58:186–96

Greenzweig Y, Lissauer JJ. 1990. *Icarus* 87:40–77

Greenzweig Y, Lissauer JJ. 1992. *Icarus* 100:440–63

Hamilton DP. 1997. Iapetus: *Bull. Am. Astron. Soc.* 29:1010

Harris AW, Young JW, Dockweiler T, Gibson J, Poutanen M, et al. 1992. *Icarus* 95:115–47

Hartmann WK, Davis DR. 1975. *Icarus* 24:504–14

Henrard J. 1982. *Celest. Mech.* 27:3–22

Henrard J. 1983. *Icarus* 53:55–67

Henrard J, Lemaitre A. 1983. *Celest. Mech.* 30:197–218

Horanyi M. 1996. *Bull. Amer. Astron. Soc.* 28:1126

Horanyi M, Burns JA, Hamilton DP. 1992. *Icarus* 97:248–59

Ida S, Canup RM, Stewart GR. 1997. *Nature* 389:353–57

Ioannou PJ, Lindzen RS. 1993. *Astrophys. J.* 406:266–78

Jacobson RA, Campbell JK, Taylor AH, Synnott SP. 1992. *Astron. J.* 103:2068–78

Kary DM, Lissauer JJ. 1993. *Icarus* 106:288–307

Kaula WM. 1964. *Rev. Geophys.* 2:661–85

Kaula WM. 1969. *Astron. J.* 74:1108–14

Khurana KK, Kivelson MG, Stevenson DJ, Schubert G, Russel CT, et al. 1998. *Nature* 395:777–80

Kivelson MG, Khurana KK, Coroniti FV, Joy S, Russell CT, et al. 1997. *Geophys. Res. Lett.* 24:2155–61

Kivelson MG, Warnecke J, Bennett L, Joy S, Khurana KK, et al. 1998. *J. Geophys. Res.* 103:19963–72

Knopoff L. 1964. *Q. Rev. Geophys.* 2:625–60

Kokubo E, Ida S. 1998. *Icarus* 131:171–78

Laskar J, Robutel P. 1993. *Nature* 361:608–12

Lemmon MT, Karkoschka E, Tomasko M. 1995. *Icarus* 113:27–38

Levinson HF, Stern SA. 1995. *Lunar Planet. Sci.* 26:841 (Abstr.)

Lieske JH. 1987. *Astron. Astrophys.* 82:340–48

Lissauer JJ. 1987. *Icarus* 69:249–65

Lissauer JJ, Cuzzi JN. 1982. *Astron. J.* 87:1051–58

Lissauer JJ, Peale SJ, Cuzzi JN. 1984. *Icarus* 58:159–68

Love AEH. 1944. *A Treatise on the Mathematical Theory of Elasticity*. Mineola, NY: Dover

Luu J. 1991. *Astron. J.* 102:1213–25

MacDonald GJF. 1964. *Rev. Geophys.* 2:467–541

Malhotra R. 1990. *Icarus* 87:249–64
Malhotra R. 1991. *Icarus* 94:399–412
Malhotra R, Dermott RF. 1990. *Icarus* 85:444–80
Marcialis R, Greenberg R. 1987. *Nature* 328:227–29
Marzari F, Dotto E, Davis DR, Weidenschilling SJ, Vanzani V. 1998. *Astron. Astrophys.* 333:1082–91
McEwen AS. 1986. *Nature* 321:49–51
McKinnon WB. 1984. *Nature* 311:355–58
McKinnon WB. 1997. *Icarus* 130:540–43
McKinnon WB. 1989. *Astrophys. J. Lett.* 344:41–44
McKinnon WB, Chapman CR, Housen KR. 1991. In *Uranus*, pp. 629–92. Tucson: Univ. of Arizona Press
McKinnon WB, Parmentier EM. 1986. In *Satellites*, ed. JA Burns, MS Matthews. pp. 718–63. Tucson: Univ. of Arizona Press
Mignard F. 1981a. *Moon Planets* 24:189–207
Mignard F. 1981b. *Mon. Not. R. Astron. Soc.* 194:365–79
Morabito LA, Synnott SP, Kupferman PN, Collins SA. 1979. *Science* 204:972
Mottman J. 1977. *Icarus* 31:412–13
Nakamura Y, Koyama J. 1982. *J. Geophys. Res.* 87:4855–61
Nicholson PD, Burns JA, Gladman BJ, Kavelaars JJ, Marsden BG, et al. 1998. *Bull. Am. Astron. Soc.* 30:1147
Nicholson PD. 1999. In *Atrophysical disks— An BC Summer School, Astron. Soc. Pac. Conf. Ser.* Vol. 160, ed. JA Sellwood, J Goodman, p. 228–45
Ojakangas GW, Stevenson DI. 1986. *Icarus* 66:341–58
Pang KD, Pollack J, Veverka J, Lane AL, Ajello JM. 1978. *Science* 199:64–66
Pang KD, Voge CC, Rhoads JW, Ajello JM. 1984. *J. Geophys. Res.* 89:9459–70
Pappalardo RT, Reynolds SJ, Greeley R. 1997. *J. Geophys. Res.* 102:13369–79
Pauwels T. 1983. *Celest. Mech.* 30:229–47
Peale SJ. 1969. *Astron. J.* 74:483–89
Peale SJ. 1974. *Astron. J.* 79:722–44
Peale SJ. 1976. *Annu. Rev. Astron. Astrophys.* 14:215–46
Peale SJ. 1977. In *Planetary Satellites*, ed. JA Burns, pp. 87–112. Tucson: Univ. of Arizona Press
Peale SJ. 1986. In *Satellites*, ed. JA Burns, MS Matthews, pp. 159–224. Tucson: Univ. of Arizona Press
Peale SJ. 1988. *Icarus* 74:153–74
Peale SJ. 1995. *Bull. Am. Astron. Soc.* 27:1170
Peale SJ. 1999. Tides. In *Encyclopedia of Astronomy and Astrophysics*. Bristol, England: Inst. Phys. Publishing. In press
Peale SJ, Cassen P. 1978. *Icarus* 36:245–69
Peale SJ, Cassen P, Reynolds RT. 1979. *Science* 203:892–94
Peale SJ, Cassen P, Reynolds RT. 1980. *Icarus* 43:65–72
Plescia JB. 1987a. *J. Geophys. Res.* 92:14918–32
Plescia JB. 1987b. *Nature* 327:201–4
Plescia JB. 1988. *Icarus* 73:442–61
Pollack JB. 1985. In *Protostars and Planets II*, pp. 791–831. Tucson: Univ. of Arizona Press
Pollack JB, Burns JA. 1977. *Bull. Am. Astron. Soc.* 9:518–19
Pollack JB, Hubickyj O, Bodenheimer P, Lissauer JJ, Podolak M, et al. 1996. *Icarus* 124:62–85
Pollack JB, Veverka J, Pang KD, Colburn D, Lane AL, et al. 1978. *Science* 199:66–70
Pollack JB, Burns JA, Tauber ME. 1979. *Icarus* 37:587–611
Ringwood AE. 1989. *Earth Planet. Sci. Lett.* 95:208–14
Ross MN, Schubert G. 1987. *Nature* 325:133–34
Safronov VS. 1969. *Evolution of the Protoplanetary Cloud and Formation of the Earth and Planets*. Moscow: Izdatel'stvo "Nauka," English translation NASA TT F-677, 1972
Saha P, Tremaine S. 1993. *Icarus* 106:549–62
Shor VA. 1975. *Celest. Mech.* 12:61–75
Showalter MR. 1991. *Nature* 351:709–13
Showman AP, Malhotra R. 1997. *Icarus* 127:93–111

Showman AP, Stevenson DJ, Malhotra R. 1997. *Icarus* 129:367–83
Sinclair AT. 1972. *Mon. Not. R. Astron. Soc.* 160:169–87
Sinclair AT. 1974. *Mon. Not. R. Astron. Soc.* 166:165–79
Sinclair AT. 1975. *Mon. Not. R. Astron. Soc.* 171:59–72
Sinclair AT. 1978. *Vist. Astron.* 22:133–140
Sinclair AT. 1983. In *Dynamical Trapping and Evolution in the Solar System*. ed. VV Markellos, Y Kozai, D Reidel, pp. 19–25
Sinclair AT, Taylor DB. 1985. *Astron. Astrophys.* 147:241–46
Smith BA, Soderblom LA, Banfield D, Barnet C, Basilevsky AT, et al. 1989. *Science* 246:1422–49
Smith BA, Soderblom LA, Batson R, Bridges P, Inge J, et al. 1982. *Science* 215:504–37
Smith BA, Soderblom LA, Beebe R, Bliss D, Boyce JM, et al. 1986. *Science* 233:43–64
Smith BA, Soderblom LA, Beebe R, Boyce J, Briggs G, et al. 1981. *Science* 212:163–91
Smith BA, Soderblom LA, Johnson TV, Ingersoll AP, Collins SA, et al. 1979. *Science* 204:951–72
Solomon SC, Chaiken J. 1976. In *Proc. 7th Lun. Sci. Conf.* 3:3229–43
Soter, S. 1974. Presented at the IAU Planetary Satellite Conference, Cornell Univ. Ithaca, NY
Soter S, Harris AW. 1977. *Nature* 268:421–22
Squyres SW, Reynolds RT, Cassen PM, Peale SJ. 1983. *Icarus* 53:319–31
Squyres SW, Reynolds RT, Summers AL, Shung F. 1988. *J. Geophys. Res.* 93:8779–94
Stern A. 1991. *Icarus* 90:271–81
Stevenson DJ. 1982. *Nature* 298:142–44
Stevenson DJ. 1983. *J. Geophys. Res.* 88:2445–55
Stevenson DJ. 1987. *Ann. R. Earth* 15:271–315
Stevenson DJ, Harris AW, Lunine JI. 1986. In *Satellites*, ed. JA Burns, MS Matthews, pp. 39–88. Tucson: Univ. of Arizona Press
Stewart GR, Canup R. 1998. Can an early-formed Moon avoid siderophile contamination by subsequent impacts? Presented at Origin of the Earth and Moon Conference, Monterey, CA, Dec. 1–3, 1998
Stewart GR, Levison HF. 1998. *Lun. Plan. Sci. Conf. Abstr.* 29:1960
Stewart GR, Wetherill GW. 1988. *Icarus* 74:542–53
Strom SE. 1995. *Revista Mexicana de Astronomia y Astrofisica Serie de Conferencis, Vol. 1, Circumstellar Disks, Outflows and Star Formation*, Cozumel, Mexico, Nov. 28–Dec. 2, 1994, pp. 317–28
Szeto AMK. 1983. *Icarus* 55:133–68
Tancredi G, Fernández JA. 1991. *Icarus* 108: 234–42
Tholen DJ, Buie MW. 1997. *Icarus* 125:245–60
Thompson C, Stevenson DJ. 1988. *Astrophys. J.* 333:452–62
Tittemore WC. 1990. *Icarus* 87:110–39
Tittemore WC, Wisdom J. 1988. *Icarus* 74:172–230
Tittemore WC, Wisdom J. 1989. *Icarus* 78:63–89
Tittemore WC, Wisdom J. 1990. *Icarus* 85:394–443
Tolson RH, Duxbury TC, Born GH, Christensen EJ, Diehl RE, et al. 1978. *Science* 199:61–64
Touma J, Wisdom J. 1993. *Science* 259:1294–97
Touma J, Wisdom J. 1994. *Astron. J.* 108:1943–61
Touma J, Wisdom J. 1998. *Astron. J.* 115:1653–63
Toomre A. 1964. *Astrophys. J.* 139: 1217–38
Tyler GL, Eshleman VR, Anderson JD, Levy GS, Lindal GF, et al. 1982. *Science* 215:553–78
Veeder GJ, Matson DL, Johnson TV, Blaney DL, Goguen JD. 1994. *J. Geophys. Res.* 99:17095–162
Vienne A, Duriez L. 1995. *Astron. Astrophys.* 297:588–605
Ward WR. 1975. *Science* 189:377–79
Ward WR. 1976. In *Frontiers of Astrophysics*, ed. EH Avrett, pp. 1–40. Cambridge, MA: Harvard Univ. Press

Ward WR. 1979. *J. Geophys. Res.* 84:237–41
Ward WR. 1981. *Icarus* 46:97–107
Warren PH. 1985. *Annu. Rev. Earth Planet. Sci.* 13:201–40
Weidenschilling SJ. 1977. *Mon. Not. R. Astron. Soc.* 180:57–70
Weidenschilling SJ. 1995. *Icarus* 116:433–35
Wetherill GW. 1990. *Annu. Rev. Earth Planet. Sci.* 18:205–56
Wetherill GW. 1991. *Science* 253:535–38
Wetherill GW. 1992. *Icarus* 100:307–25
Wetherill GW. 1996. *Icarus* 119:219–38
Wetherill GW, Chambers JE. 1997. *Lunar Planet. Sci. Abst.* 28:1547
Whipple AL, Shelus PJ. 1993. *Icarus* 101:265–71
Whipple FL. 1972. In *From Plasma to Planet*, ed. A Elvius, pp. 211–32. New York: Wiley
Wisdom J. 1987a. *Icarus* 72:241–75
Wisdom J. 1987b. *Astron. J.* 94:1350–60
Wisdom J, Peale SJ, Mignard F. 1984. *Icarus* 58:137–52
Yoder CF. 1973. *On the Establishment and evolution of Orbit-Orbit Resonances*. PhD thesis. Univ. Calif., Santa Barbara. 303 pp.
Yoder CF. 1974. Presented at the IAU Planetary Satellite Conference, Cornell Univ. Ithaca, NY
Yoder CF. 1979. *Nature* 279:747–70
Yoder CF. 1981a. *Phil. Trans. R. Soc. Lond. A* 303:327–38
Yoder CF. 1981b. *EOS* 62:939
Yoder CF. 1982. *Icarus* 49:327–46
Yoder CF. 1984. In *Lunar Planet. Inst. Conf. Origin Moon*, pp. 6–. Houston, TX: Lunar and Planetary Institute
Yoder CF. 1995. In *Global Earth Physics: A Handbook of Physical Constants*, ed. TJ Ahrens, pp. 1–31. Washington, DC: Am. Geophys. Union
Yoder CF, Colombo G, Synnott SP, Yoder KA. 1983. *Icarus* 53:431–43
Yoder CF, Peale SJ. 1981. *Icarus* 47:1–35
Yoder CF, Synnott SP, Salo H. 1989. *Astron. J.* 98:1875–89

FAR-ULTRAVIOLET RADIATION FROM ELLIPTICAL GALAXIES

Robert W. O'Connell
Astronomy Department, University of Virginia, P.O. Box 3818, Charlottesville, Virginia 22903-0818

Key Words stellar populations, hot stars, mass loss, galaxy evolution

■ **Abstract** Far-ultraviolet radiation is a ubiquitous, if unanticipated, phenomenon in elliptical galaxies and early-type spiral bulges. It is the most variable photometric feature associated with old stellar populations. Recent observational and theoretical evidence shows that it is produced mainly by low-mass, small-envelope, helium-burning stars in extreme horizontal branch and subsequent phases of evolution. These are probably descendants of the dominant, metal rich population of the galaxies. Their lifetime UV outputs are remarkably sensitive to their physical properties and hence to the age and the helium and metal abundances of their parents. UV spectra are therefore exceptionally promising diagnostics of old stellar populations, although their calibration requires a much improved understanding of giant branch mass loss, helium enrichment, and atmospheric diffusion.

1. INTRODUCTION

Far-ultraviolet radiation was first detected from early-type galaxies by the *Orbiting Astronomical Observatory-2* in 1969. This was a major surprise because it had been expected that such old stellar populations would be entirely dark in the far-UV. To the contrary, not only did elliptical galaxies and the bulges of early-type spirals contain bright UV sources, but their energy distributions actually increased to shorter wavelengths over the range 2000 to 1200 Å, resembling the Rayleigh-Jeans tail of a hot thermal source with $T_e \gtrsim 20000$ K. The effect was therefore called the "UV-upturn," the "UV rising-branch," or, more simply, the "UVX." It was only the second new phenomenon (after X-rays from the active galaxy M87) discovered by space astronomy outside our Galaxy.

Controversy flourished over the interpretation of the UVX for the next 20 years because of the slow accumulation of high quality UV data. More recent evidence has winnowed the alternatives and strongly supports the idea that the UVX is a stellar phenomenon (as opposed to nuclear activity, for example) associated with the old, dominant, metal-rich population of early-type galaxies. It is the most variable photometric feature of old stellar populations. It appears to be produced mainly by low-mass, helium-burning stars in extreme (high temperature) horizontal branch and subsequent phases of evolution. Such objects have very thin envelopes ($M_{ENV} \lesssim 0.05\,M_\odot$) overlying their cores. On both theoretical and observational grounds, the lifetime UV outputs of these stars are exquisitely sensitive to their physical properties. They depend strongly, for instance, on helium abundance; the UV spectrum is the only observable in the integrated light of old populations with the potential to constrain their He abundances. More remarkably, changes of only a few 0.01 $M_\odot$ in the mean envelope mass of an extreme horizontal branch population can significantly affect the UV spectrum of an elliptical galaxy.

If this interpretation is correct, then far-UV observations become a uniquely delicate probe of the star formation and chemical enrichment histories of elliptical galaxies. They do, that is, once we understand the basic astrophysics of these advanced evolutionary phases and their production by their parent populations. However, this is one of the last underexplored corners of normal stellar evolution, and a complete interpretation is not yet at hand, even for nearby systems such as globular clusters where full color-magnitude diagram information is available. The key physical process involved in producing the small-envelope stars is mass loss during low-gravity phases on the red giant branch and subsequent asymptotic giant branch. Serious modeling of mass loss has only recently begun, and we so far have little intuition for the effects of population characteristics such as metal abundance. Although the interpretation of the integrated light of galaxies has heretofore relied on astrophysics established and tested in the context of local stars, it may be that the UVX problem will be the first where observations of galaxies will act as strong diagnostics of stellar evolution theory. At any rate, it is clear that to understand the controlling mechanisms of the UVX in galaxies we must conjoin integrated light observations of distant galaxies with the stellar astrophysics of globular clusters and hot field stars in our own and nearby galaxies.

There are broader ramifications of this interpretation as well. UV light acts as a tracer for stellar mass loss. As the primary source of fresh interstellar gas and dust in old populations, stellar mass loss is directly linked to a diverse set of other important phenomena, including gas recycling into young generations of stars, galactic winds, X-ray cooling flows, far-infrared interstellar emission, dust in galaxy cores, and gas-accretion fueling of nuclear black holes. The UV light also traces the production of low-mass stellar remnants. The hot UVX stars, regardless of their origin, are important distributed contributors to the interstellar ionizing radiation field of old populations. It is possible that the UVX is influenced by, and therefore reflects, galaxy dynamics. Finally, characterization of the UV light of nearby ellipticals, its separation into young or old stellar sources, and its predicted evolution is also basic to the development of realistic "K-corrections" for cosmological applications to high redshift galaxies and to interpretation of the cosmic background light.

There has been excellent progress over the last decade in understanding the UVX phenomenon, but the first question that might occur to the reader is why it took 30 years simply to identify its source. The answer lies in our historically limited capability for extragalactic UV observations, a subject we discuss in the next section. Following that, we describe the discovery of far-UV light from old populations and its basic observational characteristics, the lively debate over the leading alternative interpretations, and the confluence of theory and new observations that has led to the currently accepted interpretation. We also discuss several of the other observational opportunities presented by the generally faint UV background in galaxies. By "early-type" galaxies in this paper, we mean ellipticals, S0s, and the large bulges of spirals of types Sa and Sb, although most of the detailed analysis to date has concentrated on Es and S0s.

2. INSTRUMENTAL CONSIDERATIONS

Progress in understanding the far-ultraviolet radiation from galaxies has been more circumscribed by instrumental limitations than was the case, for instance, in extragalactic X-ray astronomy. Fewer long-lived ultraviolet facilities have been available, and most of these have not been well suited for the study of galaxies. The problems are both intrinsic and technical. Intrinsically, galaxies are faint, extended sources. For typical elliptical galaxies, incident far-UV photon rates per unit solid angle per unit wavelength are typically over 50 times smaller than in the V-band. The centers of nearby bright ellipticals produce only a few $\times 10^{-15}$ erg s^{-1} cm^{-2} Å^{-1} arcsec^{-2} at 1500 Å averaged over a 10 radius (Burstein et al 1988, Maoz et al 1996, Ohl et al 1998). The paucity of high contrast spectral features in UV hot star spectra at the spectral resolution and S/N possible for E galaxies has also hampered interpretation.

There has never been a large area UV sky survey sensitive enough to detect galaxies. The only all-sky survey yet made in the UV was by TD-1 in 1973 (Boksenberg et al 1973). This has a limit of about 9th magnitude and did not include a single galaxy or QSO. The GALEX mission (Martin et al 1997), now under development, will remedy this situation and produce a survey up to 10 magnitudes fainter. For now, however, the fact remains that the deepest survey of the UV sky is comparable to the Henry Draper catalog of stars, made around 1900. So UV astronomy, at least in this sense, is still 100 years behind optical astronomy.

The technical development of UV instrumentation has been reviewed by Boggess & Wilson (1987, spectroscopy), O'Connell (1991, imaging), Joseph (1995, detectors), and Brosch (1998, surveys). UV telescopes have been small, mostly less than 40 cm diameter. Other than the 2.4-m Hubble Space Telescope (HST), the largest UV instrument available has been the 1-m diameter *Astro* Hopkins Ultraviolet Telescope (HUT), which as a Shuttle-attached payload had an equivalent dedicated observing lifetime in 2 missions of only about 6 days (Kruk et al 1995). Observations of galaxies are difficult with the small entrance apertures available on most UV spectrometers, for example the International Ultraviolet Explorer (IUE) ($10'' \times 20''$) or the HST/Faint Object Spectrograph ($\leq 1''$), which were designed for point sources. With IUE, long exposures of typically 4–8 hours were needed to register far-UV spectra of galaxies. Newer instruments are better matched to requirements for galaxy work. HUT was the first UV spectrometer designed specifically for galaxies, with apertures as large as 19" × 197" (providing, however, only one spatial resolution element). The *Astro* Ultraviolet Imaging Telescope (UIT) experiment, designed for filter imaging in the 1230–3200 Å region, had a field of view (40') and spatial resolution (3") well matched to ground-based studies of nearby galaxies. The new Space Telescope Imaging Spectrograph (STIS) offers UV apertures up to 2" × 52", encompassing many spatial resolution elements, and can image 25" × 25" fields with UV photon-counting detectors and 0.05" resolution. The HST Advanced Camera for Surveys, scheduled for installation in 2000, has high throughput UV cameras with fields up to 30" × 30".

The quantum efficiencies of UV detectors such as cesium iodide and cesium telluride photocathodes are only modest (10–30%), and net throughputs are further compromised by the lower reflectivities and transmissions of UV optical components. The most widely used mirror coating, magnesium fluoride, has a short-wavelength cutoff near 1150 Å. To obtain response to the Lyman discontinuity at 912 Å special coatings such as silicon carbide are now available (e.g. Kruk et al 1995), though these do not achieve reflectances typical of standard coatings at longer wavelengths.

Two special requirements for far-UV observations have serious practical consequences. First is the necessity to suppress the effects of the strong geocoronal Ly-α emission line at 1216 Å. This is usually straightforward in spectrographs, but in photometers or imagers the only remedy is to use blocking filters that permit response only for $\lambda \gtrsim 1250$ Å. Second is the necessity to suppress residual filter and detector response to long-wave ($\lambda > 3000$ Å) photons. Even though this may be only a tiny fraction of peak UV response, it covers a wide wavelength range. Because cool sources, such as stars with $T_e < 7000$ K, can have optical f_λ thousands of times higher than their UV f_λ, there can be serious "red leak" contamination of UV observations. Despite considerable effort (e.g. on Wood's filters), it has not been possible to develop fully satisfactory long-wave blocking devices with good peak UV response. Therefore, red leak suppression depends on the use of "solar-blind" detectors with large photoelectron work functions, such as cesium iodide, which has very small response for $\lambda > 1800$ Å. Such detectors have been used in most UV spectrometers but were not available in the HST Wide Field Camera (WFPC2) or HST Faint Object Camera (FOC), both of which consequently required careful red leak calibrations for use shortward of 2500 Å. The effects of red leaks on HST photometry of stars and galaxies can be dramatic and have been discussed by Yi et al (1995) and Chiosi et al (1997). The requirements for simultaneous Ly-α and red leak suppression imply smaller bandwidths and lower throughputs for far-UV imaging or photometry than is typical at longer wavelengths.

Because of these technical constraints, the working "far-ultraviolet" (FUV) band covers $\sim$1250–2000 Å for imaging or photometry, extended to about 1150 Å for spectroscopy. The "mid-ultraviolet" (MUV) band covers $\sim$2000–3200 Å

(3200 Å being both the useful sensitivity limit of cesium telluride photocathodes and the short-wavelength cutoff of the Earth's atmosphere). We will call the 3200–4000 Å region accessible from the Earth's surface the "near-ultraviolet" (NUV). The 912–1150 Å region in galaxies has been explored to date only by HUT, though FUSE (launched in 1999) will also cover this range in brighter objects.

Unless noted, magnitudes quoted in this paper will be on the monochromatic system, where $m_\lambda = -2.5 \log F_\lambda - 21.1$ and F_λ is the mean incident flux in the relevant band in units of erg s^{-1} cm^{-2} Å^{-1}; the zero point is such that $m_\lambda(5500 \text{ Å}) = V$. Notation for colors will be, for instance, 1500–V $\equiv m_\lambda(1500 \text{ Å})$–V.

3. DISCOVERY AND ALTERNATIVE INTERPRETATIONS

Prior to the first UV observations, there was a widespread expectation that normal elliptical galaxies would be uninteresting in the FUV (as, later, would also be the case with the X-ray and far-infrared regions). The hottest identified stellar component of any consequence was the main sequence turnoff, with a temperature ($T_e \sim 6000$ K) too cool to produce many FUV photons. Although it was recognized that the old, metal-poor populations of globular clusters sometimes contained horizontal-branch (HB) stars with $T_e \gtrsim 10000$ K, these were thought to be absent in the clusters (e.g. 47 Tucanæ) with metal abundances nearest those of massive galaxies. The only hint of hot populations in E galaxies was the presence of [O II] emission lines, though these could plausibly be explained without stellar photoionization (Minkowski & Osterbrock 1959).

FUV radiation from galaxies was first detected by the University of Wisconsin UV photometer carried on the second *Orbiting Astronomical Observatory* (OAO-2). The experiment obtained fluxes with an entrance aperture of 10' diameter in 7 intermediate band filters extending from 4250 Å to the FUV at 1550 Å. The first announcement (Code 1969) of results for an old population was for the central bulge ($r < 900$ pc) of the Local Group Sb spiral M31. As expected, the energy distribution of M31 fell steeply between 3500 and 2500 Å but then, remarkably, began to rise again at shorter wavelengths. A more recent UV-optical spectrum of M31 is shown in Figure 1. Since the energy distributions of normal stars cooler than $T_e \sim 8500$ K (spectral type A5) decline precipitously below 1800 Å owing to absorption by metallic ionization edges (e.g. Fanelli et al 1992), the detection of any far-UV flux in galaxies implies sources with higher equivalent temperatures. After a difficult calibration process, OAO-2 photometry was ultimately published for 7 E/S0 objects and the M31 bulge (Code et al 1972, Code & Welch 1982). The OAO-2 detections of two objects were confirmed, and new detections made of another 11 E galaxies, by the *Astronomical Netherlands Satellite* (launched in 1974) using intermediate band photometry with a 2.5' × 2.5' aperture over the range 1550–3300 Å (de Boer 1982).

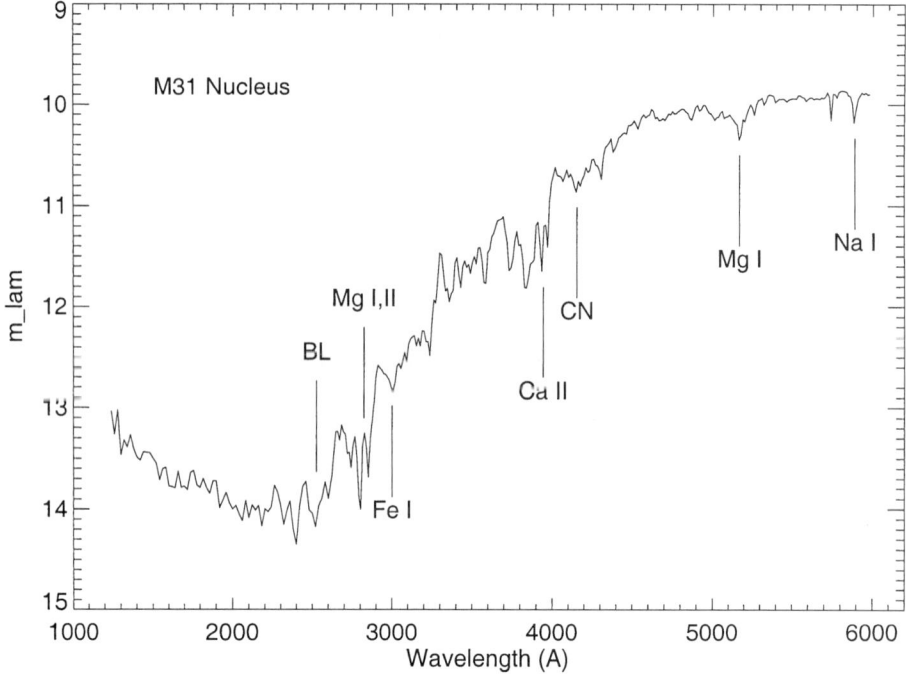

Figure 1 A composite UV-optical energy distribution for the center of the Sb galaxy M31. IUE data taken with a 10" × 20" aperture is plotted below 3200 Å, while a ground-based spectrum covering the same region is plotted above. Resolution is 20 Å below 2600 Å and 12 Å above. Irregularities in the UV spectrum below 2200 Å are mainly noise. Some of the stronger absorption line features are identified ("BL" corresponds to a strong blend of Fe and other metallic lines near 2538 Å). The "UV-upturn" is the rise in the spectrum at wavelengths shorter than 2000 Å. By simple extrapolation of the far-UV continuum slope, one finds that the upturn component contributes only about 0.3% of the V light of the galaxy. Spectrum courtesy of D Calzetti.

An immediate conclusion from the UVX observations which was emphasized by Code and his colleagues was that early-type galaxies exhibited much larger scatter in the UV than was expected from their conspicuously homogeneous behavior in the optical to near-IR (4000–20000 Å) region. The UV observations implied divergent histories at some level and were among the first indications that elliptical galaxy populations were more heterogeneous than envisioned in Baade's classic definition of Population II (Baade 1944, O'Connell 1958, O'Connell 1980, Faber et al 1995). They called into question the use of E galaxies as "standard candles" in cosmological studies. They also complicated the construction of accurate K-corrections needed to transform photometry of high redshift elliptical galaxies to standard bands in the restframe (e.g. Pence 1976, Coleman et al 1980, King & Ellis 1985, Bertola et al 1982, Kinney et al 1996).

Interpretation of the unexpected OAO-2 results was initially confused by calibration uncertainties which produced anomalously steep FUV energy distributions (in normal spirals and irregulars as well as early-type systems, see Code & Welch 1982). Code (1969) and Code et al (1972) suggested that the UVX component was nonthermal radiation from an active nucleus (AGN) or scattering of photons from massive hot stars by interstellar dust. The latter would have implied that most E/S0 galaxies contain an appreciable Population I component.

Hills (1971) pointed out that the steep rise of the M31 UV spectrum to higher photon energies was incompatible with known nonthermal sources but was closely matched by the Rayleigh-Jeans tail of a high temperature thermal source. Based on comparison with a small sample of UV-bright stars in the globular cluster M3, he proposed that the UV upturn is produced by highly evolved, hot, low-mass stars such as the central stars of planetary nebulae, now known as post asymptotic giant branch (PAGB) stars, or their hot white dwarf descendents. He did not require that these be members of a strong Population II (old, metal-poor) component but pointed out that their prominence would probably depend on metal abundance.

Tinsley (1972a) argued that the UV light arose instead from young, massive, main sequence stars and showed that a spectral synthesis model for an old galaxy with an exponentially declining star formation rate and an e-folding time of 2 Gyr could fit the OAO-2 flux for M31 observed at 1700 Å. This would imply that the UVX was related to a normal, if temporally extended, star formation process in early-type systems. Fuel for the star formation might be primordial gas consumed gradually over a Hubble time, mass loss from red giants, or material accreted from outside galaxies (Gallagher 1972, Tinsley 1972b, O'Connell 1980, Gunn et al 1981). Residual star formation histories of the type suggested by Tinsley would drastically change the predicted properties of E galaxies viewed at moderate look-back times, whereas the low-mass star interpretation would have less serious implications for spectral evolution.

It was implicit in these early studies that the hot components that dominated the far-UV light could be virtually undetectable at visible wavelengths—i.e. that the UV was providing entirely independent information about galaxies. Ignoring any contribution from the cool components to the UV light, the maximal fractional contribution of a hot component to the integrated V-band light of a galaxy will be $p_{max} \sim 10^{0.4\Delta}$, where $\Delta = (1500-V)_{hot} - (1500-V)_{obs}$. A color for a typical E galaxy is $(1500-V)_{obs} \sim +3$ while a component with an appropriate far-UV spectral slope (B0 equivalent) has $(1500-V)_{hot} \sim -4.5$, implying that $p_{max} \sim 0.001$. This is about 50 times smaller than could be directly detected in the V-band using spectral synthesis techniques.

The early workers on the UVX realized that the best tests of the alternative interpretations were (a) UV spatial structure and (b) UV spectral features observed at higher resolution. An active nucleus would be a concentrated point source, whereas a population of low-mass stars would presumably have a smooth distribution similar to that found in the optical bands for bulges and E galaxies. Young, massive-star populations would likely have a clumpy structure, similar to the OB associations found in spiral arms, and they might well be concentrated to disks. In nearer galaxies individual massive OB stars could be isolated. Spectroscopically, a UV-bright AGN would be easy to identify on the basis of broad, high-excitation emission lines. Active massive star-forming regions characteristically exhibit strong UV resonance lines of Si IV, C IV, and other species, often with P-Cygni profiles (e.g. Kinney et al 1993), whereas the spectra of hot, low-mass stars are relatively weak-lined in the 1200–2000 Å region.

Because of limited UV observing opportunities, it would not be possible to apply these tests in a definitive way until over a decade after the discovery of the UVX. Only short-duration sounding rocket or balloon experiments were available until 1978. The most productive observing facility for the study of the UVX in the period 1978–1990 was IUE (Kondo 1987). IUE's handicaps of small effective collecting area, small entrance aperture, and limited dynamic range were beautifully compensated by its record 18 year lifetime and a capability for very long integration times, and it produced an invaluable set of UVX spectra. The fact that its point spread function was smaller than its $10'' \times 20''$ entrance aperture also meant that spatial structure could be studied to a radius of $10''$. After 1990, HST and the two *Astro* missions provided new capabilities to study the UVX.

In the next two sections, we describe the basic phenomenology of far-UV sources in bright early-type galaxies, as determined by IUE and other instruments, and how this bears on the now accepted interpretation of these as low-mass stars in old stellar populations.

4. SPATIAL STRUCTURE OF THE UVX

4.1 Evidence Against Young Stars

UV imaging of early-type galaxies began in the 1970s. Early rocket and balloon experiments obtained low S/N images of the central bulge of M31 which showed that it was an extended source in the far-UV with $r \gtrsim 4'$ (Deharveng et al 1976; Carruthers et al 1978; Deharveng et al 1980). This was sufficient to exclude the AGN interpretation (in this particular case) but could not readily distinguish between the old and young star models or other types of diffuse sources. A later rocket imaging experiment by Bohlin et al (1985) provided far- and mid-UV photometry of M31 with $20''$ resolution. The UV intensity profiles of the bulge were smooth and similar to Kent's (1983) R-band profile for $r \lesssim 1.1'$. Although localized regions of massive star formation were readily detectable in the outer spiral arms, similar structures were absent in the bulge, nor could individual bright OB stars be detected there.

IUE observations of bright early-type galaxies (including M31, M32, NGC 3379, NGC 4472, M87, NGC 4552, and NGC 4649) confirmed the spatial extension of the far-UV light, even in the case of the prominent AGN of M87, and indicated that it paralleled the profile of the visible light, at least over the innermost $10''$ (Bertola et al 1980, Perola & Tarenghi 1980, Nørgaard-Nielsen & Kjærgaard 1981, Oke et al 1981, Bertola et al 1982, O'Connell et al 1986). Deharveng et al

(1982) and Welch (1982) used multiple IUE spectra to study the light distribution within the inner 15″ of the M31 bulge. They obtained a smooth profile, unlike those of star-forming regions, but found that the FUV light was slightly more concentrated to small radii within this region than MUV or B band light, producing gradients of several 0.1 mags in colors. The smooth distribution of UVX light in these cases and its similarity to the optical band profile, where old stars dominate, strongly suggested that old stars produced the UVX.

In another rocket experiment, Onaka et al (1989) and Kodaira et al (1990) obtained low-resolution, wide-field UV images of the Virgo cluster, extending the earlier photometry of Smith & Cornett (1982) at 2400 Å to 1600 Å. By comparing their total fluxes with IUE values for the nuclei, Kodaira and colleagues found evidence of large UV color gradients in five Virgo E galaxies. They attributed the blue nuclear excesses and the observed scatter in 1500–V colors to recent star formation from galactic cooling flows, though their observations were also consistent with gradients in low-mass populations.

The best available set of large area UV maps of early-type galaxies was obtained by the *Astro* UIT experiment during two Space Shuttle missions in 1990 and 1995 (Stecher et al 1997). Twenty-two ellipticals and early-type (S0-Sb) spiral bulges were imaged with good S/N at 3″ resolution, and results for 10 of these have been published (O'Connell et al 1992, Ohl et al 1998). UIT images of two Fornax cluster elliptical galaxies are shown in Figure 2. In the best cases, it was possible to obtain UV surface brightness profiles to $\mu_\lambda \sim 27$ mag arcsec^{-2}. All of

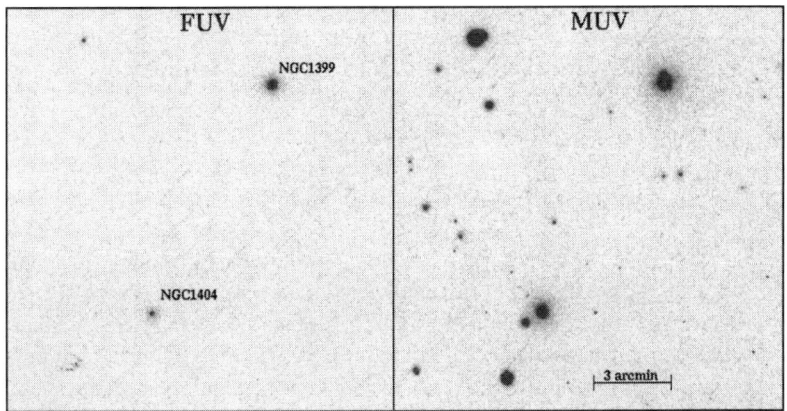

Figure 2 *Astro*/UIT images of the Fornax cluster ellipticals NGC 1399 and 1404 in broad bands in the far-UV (1500 Å) and mid-UV (2500 Å) with spatial resolution of ∼3″. The mid-UV band is dominated by the main sequence turnoff. All of the far-UV light is from the UVX component. It is smooth, without evidence for massive stars, though is more concentrated than the mid-UV light. NGC 1399 has one of the strongest UVX components yet discovered. Note that the foreground stars have mostly vanished in the FUV band; this is a pictorial representation of how unusual are the objects which make up the UVX.

the objects exhibit smooth UV profiles (except M87, in which the nonthermal jet is bright), with none of the clumpiness normally associated with recent massive-star formation, and the FUV contours are consistent in shape and orientation with optical-band isophotes. There is little evidence for dust lanes or clouds in the galaxy centers; such features should be readily detectable because of the high selective UV extinction of normal dust. In the M31 bulge, the point source detection threshold was $m_\lambda(2500 \text{ Å}) \sim 18.4$, which excluded the presence of individual main sequence stars hotter than B1 V (O'Connell et al 1992). Over 200 such objects would be expected in the central 2' of the bulge if massive stars formed with a normal initial mass function produced the FUV light. The FUV profiles of about half the sample are well fitted by de Vaucouleurs functions ($\mu \sim a + br^{0.25}$), which are characteristic of spheroids at optical wavelengths. However, the inner FUV profiles of several objects (NGC 3379, 4472, and 4649) are more consistent with an exponential function (Ohl et al 1998). Although exponentials are normally associated with disks, the FUV isophotal contours are congruent to the B-band contours, and the 3-dimensional FUV light distributions are therefore unlikely to be genuinely disklike. (Because of the large UV/optical color gradients discussed below, it is not necessarily expected that the UV profiles of objects that are true spheroids at optical wavelengths would be closely de Vaucouleurs in shape.)

High-resolution UV imaging from HST (mainly of smaller $\leq 22''$ nuclear fields with the FOC) has confirmed the absence of massive stars in the centers of M31 and M32 (King et al 1992, Bertola et al 1995, King et al 1995, Cole et al 1998, Brown et al 1998a, Lauer et al 1998) and in most UVX sources in the nuclei of 56 early-type galaxies in the 2300 Å survey of Maoz et al (1996).

The collective evidence of all these structural studies is that the far-UV light in most early-type galaxies originates in a stellar component with dynamics characteristic of the bulk of the old stellar population. Active nuclei or young massive stars are not important in most cases.

4.2 Structural Variations

However, the UV-bright population is not a simple extension of the well-studied, optically bright one. This was evident from the scatter in the ratio of UV to optical light first reported by Code et al (1972) and Code & Welch (1982), which has been amply confirmed by later observations (see Section 5.2). In addition, large internal gradients in UV/optical colors have been revealed in almost all cases studied with sufficient S/N. Five of the galaxies shown in Figure 3 from the Ohl et al (1998) UIT sample display large internal 1500–B color gradients with net changes up to ~ 1.0 mag over the region photometered. The 1500–B colors of 7 of the 8 objects become redder outward, meaning that the far-UV light is more concentrated to the galaxy centers than the optical light. Both in amplitude and sign, these changes are dramatically unlike the very mild, bluer-outward color gradients encountered in the optical and IR (e.g. Peletier et al 1990). M32 is the only object which becomes bluer in 1500–B at larger radii.

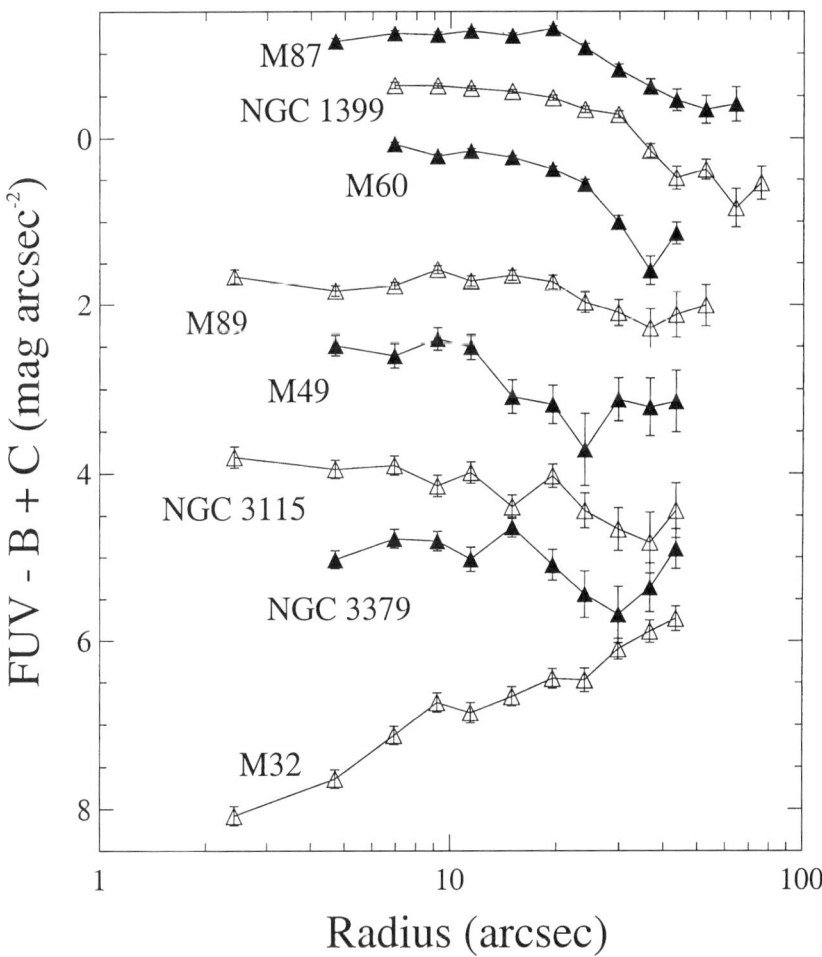

Figure 3 Radial FUV–B color profiles for 8 early-type galaxies obtained by comparing *Astro*/UIT far-UV surface photometry with B-band data from the literature. The curves have been offset for clarity and arranged in order of increasing central UVX. One sigma error bars are shown. FUV–B colors redden with increasing radius in all cases except M32, which shows a strong, reversed profile. It is the only object currently known to have this behavior. There is an interesting two-component structure in most of the profiles. Offsets in order from the top down are $C = -2.5, -2.0, -1.5, 0.0, 0.0, +1.0, +2.0,$ and $+3.5$ mag arcsec^{-2}. From Ohl et al (1998).

Internal extinction by dust cannot be responsible for these gradients. Aside from the absence of dust structures in the images and the sense of the typical gradient (implying more extinction at larger radii), the gradients are so large that significant optical-band effects would be expected, since $A(4400\text{ Å}) \sim 0.5 A(1500\text{ Å})$, where A is the total extinction in magnitudes. HUT spectroscopy also places strict limits on the amount of internal extinction (Ferguson & Davidsen 1993, Brown et al 1997). Instead, the gradients are apparently driven by a radial change in the properties of the old star population.

5. SPECTRAL AND PHOTOMETRIC CHARACTERISTICS OF THE UVX

5.1 Incidence, Spectral Shape, and Line Features

Except in cases of obscuration by a major dust lane (e.g. in edge-on S0-Sb objects), far-UV radiation has been detected in all nearby early-type systems observed with adequate S/N. As noted above (Section 3), this implies the presence of sources with $T_e > 8500$ K. Rifatto and colleagues (1995a, 1995b) have compiled all UV observations of galaxies published before 1990 and attempted to place them on a homogeneous system, which is a challenge owing to the varied types of experiments involved and the relatively low photometric precision which is typical. Their list includes 94 galaxies of type Sb or earlier with UV detections at $\lambda < 2100$ Å. The list does not include later photometry or imaging from the SCAP/FOCA balloon experiments (Milliard et al 1992, Donas et al 1995, Treyer et al 1998), *Atlas*/FAUST (Deharveng et al 1994), *Astro*/UIT (Stecher et al 1997), *Astro*/HUT (Kruk et al 1995), or the Maoz et al (1996) HST/FOC nuclear survey. Combined, these roughly double the total number of far-UV E-Sb detections, and HST is continually enlarging this sample. It is worth emphasizing that extragalactic UV observations are largely confined to relatively nearby, bright systems (except for very distant objects where the redshift brings the restframe UV into the bands accessible from the ground).

The early IUE spectra of the nuclei of bright ellipticals and spiral bulges (Johnson 1979, Bertola et al 1980, Perola & Tarenghi 1980, Nørgaard-Nielsen & Kjærgaard 1981, Oke et al 1981, Bertola et al 1982, Deharveng et al 1982, O'Connell et al 1986) showed immediately that the strong, broad emission lines characteristic of active nuclei were absent, excluding the AGN hypothesis. Signals for $\lambda \lesssim 2400$ Å were, however, very weak and subject to several kinds of detector noise (which generated some spurious claims of narrow coronal or chromospheric emission lines). Except in the brightest sources, it was necessary to average far-UV fluxes over bandwidths of ~ 50 Å.

Burstein et al (1988, hereafter BBBFL) produced the largest homogeneous set of good IUE spectra for early-type galaxies (32 objects). Two examples are shown in Figure 4. The flux rapidly declines shortward of 3300 Å. Strong absorption features from Mg I (2852 Å), Mg II (doublet at 2800 Å), Fe I (numerous lines),

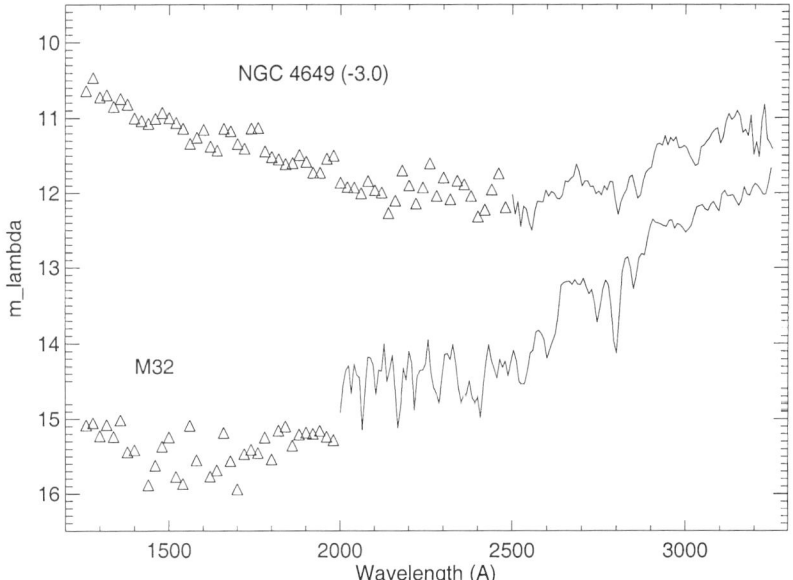

Figure 4 IUE spectra of two galaxies lying at the extremes of UVX behavior. M32 has the smallest known UV upturn, while NGC 4649 has one of the strongest. Data plotted with open triangles have 20 Å binning, while the solid line has 8 Å binning. The zero point of the NGC 4649 spectrum has been shifted −3.0 mags. Neither spectrum is corrected for redshift or foreground extinction. The FUV data for M32 are too poor to judge the slope of its UVX component. The weakness of the absorption lines near 2800 Å in NGC 4549 is caused by filling by the smooth UVX component, which contributes over 70% of the light at 2700 Å in this object. Reprocessed and recalibrated IUE spectra courtesy of RC Bohlin.

and other metallic species are easily detectable down to $\lambda \sim 2500$ Å, as are strong discontinuities caused by metallic blanketing at 2640 and 2900 Å. The mid-UV line spectrum closely resembles that of F-G dwarf stars (see Fanelli et al 1992), the spectral types expected for the main sequence turnoff in an old E-galaxy population. The flux reaches a minimum in the range 2000–2600 Å, where the S/N is almost always rather poor, then usually rises steeply again to shorter wavelengths. No maximum is detected in the rising component longward of the IUE cutoff at ∼1150 Å.

At the resolution permitted by the noise, the typical far-UV IUE spectrum is a relatively smooth continuum with an equivalent temperature $T_e \gtrsim 20000$ K. The spectral slope of the upturn is roughly constant, so that the far-UV rise begins at longer wavelengths in galaxies with brighter UVX components (e.g. NGC 4649 in Figure 4). The contribution of the UVX component to the mid-UV light can be appreciable, ranging up to 75% at 2700 Å for objects like NGC 4649 (BBBFL,

Ponder et al 1998, Dorman et al 1999), though this drops rapidly at longer wavelengths because of the steep rise in the spectrum of the cooler main sequence turnoff stars. In objects with the smallest UVX components (e.g. M32 and NGC 4382), the spectra appear to flatten below 2000 Å, rather than rise, but the S/N is too poor to estimate a temperature (BBBFL and Figure 4).

In the great majority of cases, E galaxy spectra longward of 3200 Å are quite similar to one another; the large spectral anomalies associated with the UVX are confined to the vacuum UV. This suggests that the stars responsible for the UVX are well segregated in the color-magnitude diagram from the bulk of the population. An interesting exception is M87, where anomalies are detectable up to 4000 Å (Bertola et al 1982, McNamara & O'Connell 1989); they are spatially extended and may be related to massive star formation in M87's cooling flow (see Section 8).

As originally emphasized by Tinsley (1972a), the characteristic FUV slope of UVX galaxies is consistent with star-forming models in which massive O and early B stars dominate the light. These require star formation to have occurred within the last 10–20 Myr. Tinsley did not discuss the far-UV spectral shape of her models, but later studies (e.g. Wu et al 1980, Gunn et al 1981, Nesci & Perola 1985, Rocca-Volmerange & Guiderdoni 1987, Bica & Alloin 1988, Burstein et al 1988, Ferguson et al 1991, Bruzual & Charlot 1993) would show that the young star and old, evolved star models could produce almost indistinguishable FUV energy distributions at low spectral resolution (see Figure 5). It would be necessary to consider other information, especially spectral features, to resolve the ambiguity.

Because of the limited signal-to-noise of individual spectra, IUE studies of possible far-UV absorption lines have been based on summed spectra for either the same or several objects. A serious complication is a systematic background of "fixed pattern" and camera artifact features in long-exposure IUE spectra (Crenshaw et al 1990). Welch (1982) analyzed 12 exposures of the center of M31 and detected weak absorption features at 1260 Å (Si II + S II), 1302 Å (O I + Si II + Si III), and 1335 Å (C II). The features were confirmed by BBBFL in M31, but they were weak or absent in a summed spectrum for 3 bright UVX E galaxies (BBBFL). These lines are characteristic of normal early-B stars (e.g. Fanelli et al 1992) but are considerably stronger in the stars than in any of the UVX sources. The strong Si IV (1400 Å) or C IV (1550 Å) absorption features associated with massive O stars were absent in both M31 and the summed E spectrum.

Welch (1982) and BBBFL argued that the weakness of the massive OB-star spectral features was inconsistent with recent star formation as the source of the UVX. BBBFL pointed out additional evidence in the form of continuum shapes. Although it is possible to produce a star-forming model whose spectrum matches the typical steep UVX far-UV spectral slope (see Figure 5), in fact many systems with younger populations (e.g. NGC 205 or 5102) have rather flat energy distributions in the 1200–3000 Å region. This is the signature of an aging starburst or a young starburst containing local extinction (Kinney et al 1993). Furthermore, the predicted young-star contamination of the optical band if the UVX originates in massive stars is significantly larger than limits from careful spectral synthesis

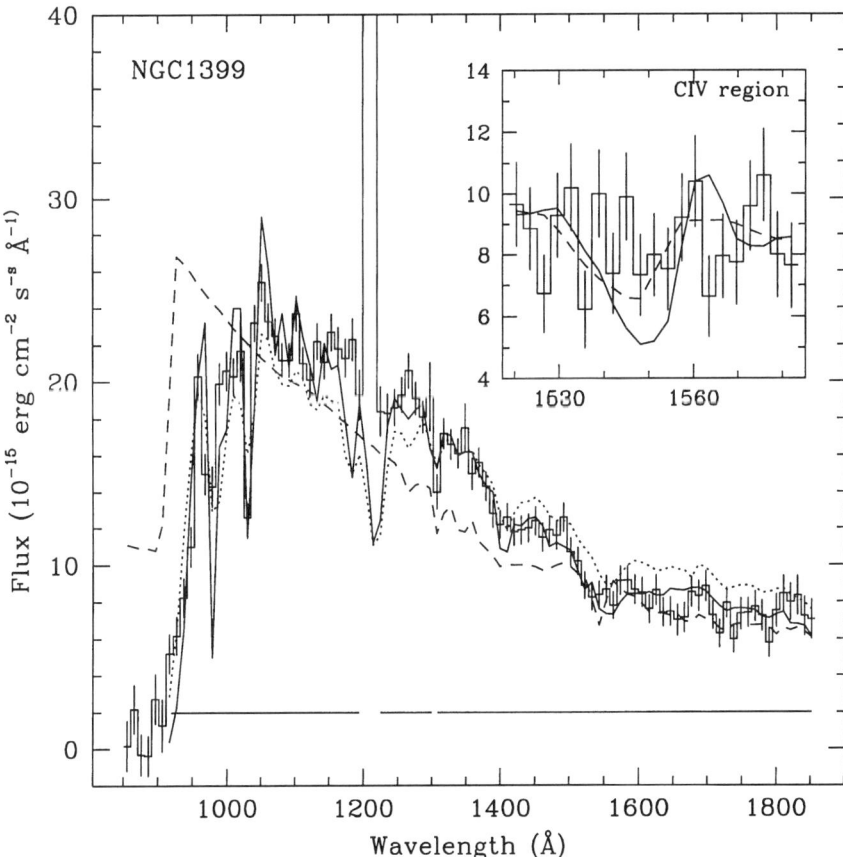

Figure 5 The *Astro*/HUT spectrum of the gE galaxy NGC 1399 in the Fornax Cluster (see Figure 2). The histogram is the observed flux in 10 Å bins. The solid line shows the best-fitting Kurucz (1991) solar abundance model atmosphere, which has $T_e = 24000$ K. The dashed line is a model from Rocca-Volmerange & Guiderdoni (1988) for an old galaxy with continuing star formation. This contains hotter starlight than is present in the galaxy. The inset shows the observed spectrum near the C IV 1550 Å doublet compared with continuous star forming models. From Ferguson et al (1991).

studies—e.g. the 2% maximum at 4000 Å set by Rose (1985) using high-resolution spectra of 12 E galaxies. The uniformity of the UV-upturn slope and the absence of warm-star effects at wavelengths longer than 2000 Å are therefore additional evidence against the involvement of massive stars in the UVX phenomenon.

The best far-UV spectra of UVX galaxies, covering the range 900–1800 Å, were obtained by *Astro*/HUT with a photon-counting detector and calibration superior to IUE's. HUT spectra of 8 early-type systems confirmed the weakness of the massive OB-star spectral features (Ferguson et al 1991, Davidsen & Ferguson

1992, Brown et al 1995, Brown et al 1997). For example, they placed an upper limit on the strength of C IV 1550 Å of $\lesssim$3 Å equivalent width in NGC 1399, which formally excluded spectral synthesis models for continuous star formation with normal metal abundances (see Figure 5). Equally important, by extending spectral coverage below 1150 Å, HUT was able to detect turnovers in the UVX spectra that place firm upper limits on their effective temperatures of $T_e \lesssim$ 25000 K (equivalent to a B0 V star). This excludes continuing star formation models with a normal initial mass function (IMF). Only contrived models (e.g. invoking a truncated IMF or an unprecedented synchronization of star formation in different galaxies) can reconcile young populations with these results. The narrow T_e range, however, also appears to require an unusual degree of "fine-tuning" in an old population interpretation.

The HUT spectra also provide excellent limits on the amount of internal interstellar extinction. Because the slope of the far-UV spectral rise is nearly at the maximum encountered among hot stars (Dean & Bruhweiler 1985, Fanelli et al 1992), there is little room for interstellar reddening. In the HUT data, there is no evidence for extinction significantly in excess of the expected Galactic foreground. Since dust is normally associated with star-forming regions, this is yet further evidence of their unimportance in UVX galaxies.

The spectroscopic evidence therefore strongly corroborates the structural evidence from the last section that massive stars are not responsible for the UV-upturn.

Recently, Bica et al (1996) have produced composite spectra for groups of E galaxies from the IUE archives. They find evidence for broad absorption features at 1400 and 1600 Å in most UVX sources, which they identify with the Ly-α satellite lines in intermediate-temperature (DA5) white dwarf stars. If true, this would be remarkable since only a very unusual population would have the requisite concentration of white dwarfs. The features are not present in the better quality HUT spectra discussed above, but Bica et al suggest they are confined only to the nuclei and have been diluted by the larger HUT entrance aperture. This can be checked with HST/STIS.

5.2 Amplitude and Correlations with Other Properties

The most striking feature of the UVX phenomenon is its large variation from object to object. As measured by the color 1500–V, the amplitude of the UVX in the centers of bright E-Sb galaxies varies from $\sim$4.5 to $\sim$2.0, which is a factor of 10 in the ratio of far-UV to visible flux. No optical-IR photometric or spectral index for normal old stellar populations exhibits a comparable range; in fact, most do not vary more than $\pm$30% (e.g. Sandage & Visvanathan 1978, Peletier et al 1990, Trager et al 1998). The UV variations are not confined to the nuclear regions observed with IUE; they are also present in all the large aperture data sets cited above (OAO, ANS, UIT, HUT, FAUST, and the various rocket/balloon experiments).

Large excursions of this kind are usually associated with an incidental component rather than with the aggregate population of a galaxy. To what extent does the UVX convey useful information about the fundamental properties of galaxies?

A key insight was provided by Faber and her colleagues (Faber 1983, BBBFL): the UVX appears to be stronger in more metal-rich galaxies. Faber (1983) found in a small sample of early-type galaxies that nuclear UV colors became bluer as the nuclear spectral line index Mg_2, which measures the Mg I + MgH absorption features near 5170 Å, increased. The Mg features are produced by the dominant old stellar population, and this correlation simultaneously links the UVX to the bulk of the galaxy while further weakening the case for massive stars (since there is no obvious reason why recent star formation would be related to metal abundance). Interestingly, the correlation is reversed in sense from the well-known dependences of (U–B) or (B–V) colors on metal abundance in old populations. Driven mainly by opacity effects in stellar envelopes and atmospheres, these become redder as abundance increases.

The correlation was confirmed by the larger sample of BBBFL. They also found significant, if weaker, correlations between 1500–V and central velocity dispersion or luminosity. A later version of the Mg_2 correlation, including data on Galactic globular clusters, is shown in Figure 6. The figure emphasizes the lack

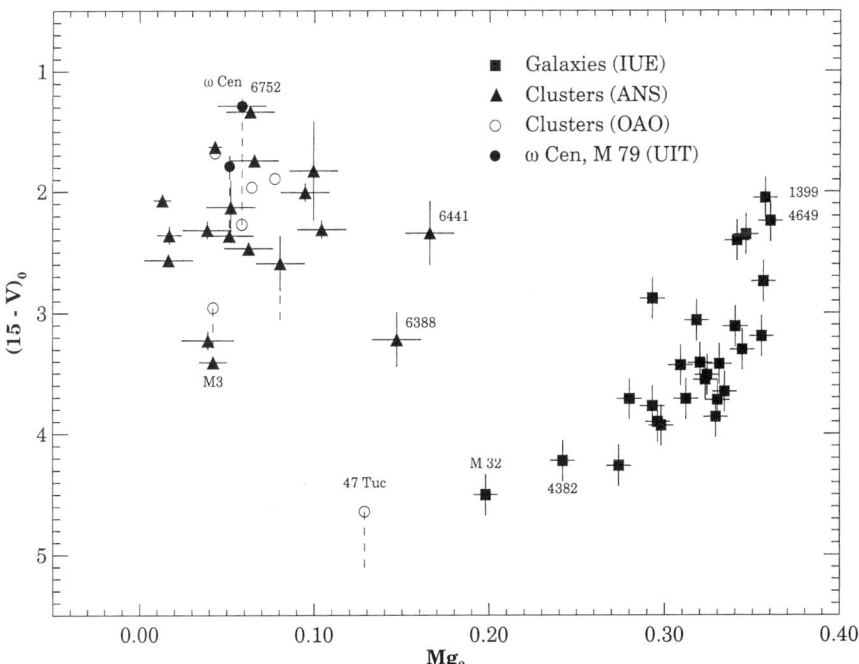

Figure 6 Amplitude of the UVX in old stellar populations, as measured by the color 1500–V, as a function of the Mg_2 line index, which measures absorption from Mg I + MgH near 5170 Å. The E galaxy data are from IUE, mostly from the study by BBBFL. The globular cluster data are from several sources, as indicated in the legend. The clusters and galaxies are clearly distinct kinds of populations. From Dorman et al (1995).

of continuity between the clusters and galaxies (see Section 6.1). Clusters with a wide range of Mg_2 index have large, if scattered, FUV fluxes. Some strong-lined clusters, e.g. 47 Tuc, are faint in the FUV while others, e.g. NGC 6388 and 6441, are bright. The galaxies with line strengths comparable to the strong-lined clusters are relatively faint in the FUV, but galaxy 1500–V colors rapidly become bluer as Mg_2 increases. The apparent correlation between 1500–V and Mg_2 is much stronger for the galaxies than the clusters. FUV behavior is only one of a number of basic spectrophotometric distinctions that show that globulars and E galaxies do not form a simple population continuum (e.g. Burstein et al 1984, Rose 1985, Ponder et al 1998).

Using the recent compilation of data for the Lick Observatory E galaxy spectral survey by Trager et al (1998), one can explore correlations between the UVX and other absorption line indices. There are good correlations, similar to that in Figure 6, between 1500–V and the Na I D lines or the CN bands at 4150 Å. But there is no correlation with a composite Fe index based on features at 5270 and 5335 Å. It is now clear that the abundances of certain light elements (N, Mg, Na) are decoupled from those of the iron peak in more luminous E galaxies (see reviews by McWilliam 1997 and Worthey 1998), although there is no clear understanding of the nucleosynthetic origin of these abundance ratio variations. The behavior of the UVX is evidently linked to that of the lighter elements.

The scatter in 1500–V at a given Mg_2 is appreciable, especially among the most metal-rich galaxies. This may indicate the influence of parameters other than metal abundance (see Section 6.3). It is also possible that the apparent correlation between UV colors and Mg_2 may not reflect smoothly varying properties but instead might arise from several discrete classes of galaxies, as discussed by DOR. There is a suggestion of grouping in Figure 6 (though this is less pronounced in the correlations with Na I and CN). Most of the galaxies have colors in the range 3.5 ± 0.5; within this group there is only a mild UV-Mg correlation. A few objects, including M32, have significantly redder colors. At the other extreme, there are four strong-lined objects with 1500–V < 3 which stand out as a distinct group.

Interestingly, Longo et al (1989) have pointed out that the strongest UV upturns occur in objects with "boxy" isophotes. Most of the systems in the middle group of Figure 6 have "disky" isophotes. The isophotal distinctions between the two groups are now known to correlate with a wider set of morphological and kinematic properties (e.g. Jaffe et al 1994, Faber et al 1997). The boxy galaxies are probably merger products (Bender 1988). It is therefore possible that the UVX is influenced by the dynamical environment of galaxies. Alternatively, all of these characteristics may be related independently to the mass of galaxies.

In low resolution photometry of 40 early-type galaxies in the Virgo cluster, Smith and Cornett (1982) detected the effects of the long-wavelength tail of the UVX component in the integrated mid-UV colors (2400–V) of E galaxies. These, however, had a significantly different color-luminosity relation than the S0 galaxies in the sample. Such potential morphological dependencies have not been carefully investigated.

The UV-Mg$_2$-Na-N correlation and the large internal UV color gradients (Section 4.2) remain the most suggestive clues linking the UVX to the global properties of galaxies. Interpretations of the UVX must accommodate such correlations, but with the caveat that we do not yet really understand the nucleosynthetic drivers of E galaxy chemistry.

6. CANDIDATE LOW MASS UVX SOURCES

On the basis of the evidence in the last two sections, old, low-mass stars are the primary sources of far-UV light in E galaxies. Interpretational effort during the last 10 years has therefore focused on the viability of various types of hot, low-mass stars and their relationship to the dominant populations of E galaxies. Since observational information on the UVX is still sparse, much of this work has been based on new generations of theoretical models for advanced stellar evolution. A fully satisfactory interpretation has not emerged, but there has been good progress in narrowing the range of possibilities. In this section and the next we review the main conclusions.

6.1 Globular Cluster-Type Populations

The presence of hot stars in old populations was, of course, not unprecedented because "blue horizontal branch" (BHB) stars had long been associated with metal-poor globular clusters (having [Fe/H] $\lesssim -1$). The most natural old-star interpretation of the UVX was therefore that it arose from the low-metallicity tail of a stellar population with a large abundance range.

Surprisingly, however, observations quickly demonstrated that the UVX could not simply be the sum of globular cluster-type populations. The first quantitative comparison between clusters and a UVX source was made using ANS data for the bulge of M31 by Wu et al (1980). In order to fit the far-UV spectrum of M31, Wu et al were forced to add an additional high temperature component to the cluster M13, which has one of the hardest UV spectra among Galactic globulars. Models by Nesci & Perola (1985) likewise showed that normal cluster BHBs could not match the galaxy IUE spectra unless a second, hotter HB component was included. Oke et al (1981), Welch (1982), and Bohlin et al (1985) all emphasized the dissimilarity between typical cluster and galaxy UV energy distributions. Compilations of ANS, IUE, HUT, and UIT data for globulars (van Albada et al 1981, Castellani & Cassatella 1987, Bica & Alloin 1988, Davidsen & Ferguson 1992, Dorman et al 1995) show that although clusters in general have total UV-bright star fractions exceeding those of galaxies (see Figure 6), galaxies can have steeper far-UV spectra than any cluster. The distinction between the two populations is quite clear in two-color diagrams such as Figure 7, where the color 1500–2500 is used to measure the slope of the UV upturn.

The temperature distribution of UV-bright sources in the clusters is evidently cooler and broader than in the galaxies. The far-UV slope in the galaxies is

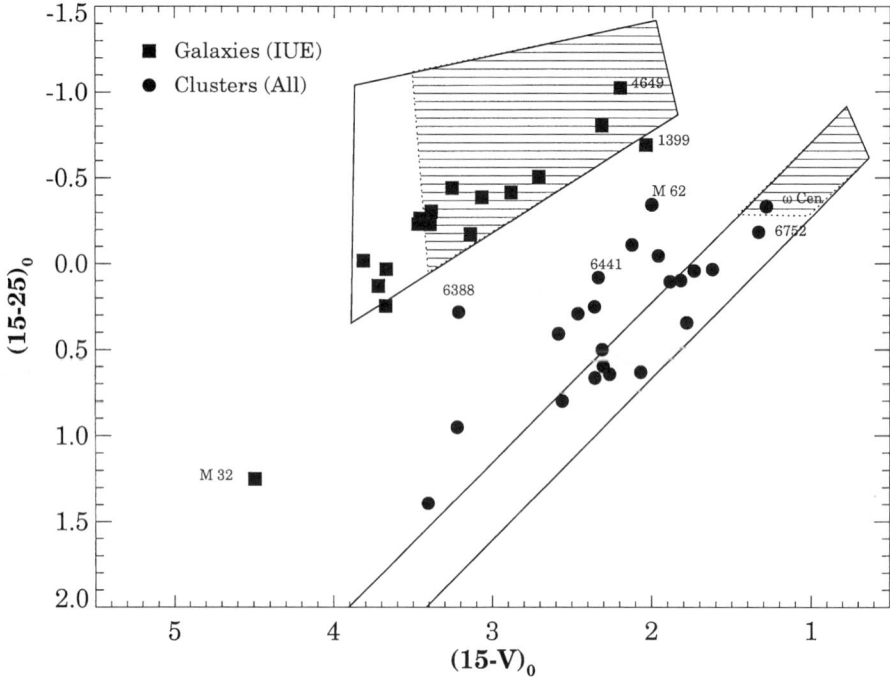

Figure 7 Broad-band UV colors for E galaxies and globular clusters compared. Data are mainly from IUE. The color 1500–2500 measures the slope of the UV-upturn component, while 1500–V measures its amplitude. Globular clusters and galaxies are clearly segregated in the diagram, with galaxies having steeper FUV spectral slopes at a given amplitude. The boxes enclose several fiducial model sets, with the upper one corresponding to $Z \gtrsim Z_\odot$ and the lower to $Z \lesssim 0.04\ Z_\odot$. The shaded regions represent models in which the hot stellar component consists mainly of EHB stars. Most of the galaxies require an EHB contribution, but most of the clusters do not. From Dorman et al (1995).

equivalent to $T_e \gtrsim 20000$ K, whereas BHBs in Galactic globulars normally do not extend beyond $T_e \sim 10000$–12000 K. Hotter stars can be found in some globulars on or above the horizontal branch, as Hills (1971) originally pointed out, but these are relatively uncommon (see de Boer 1985, 1987; and Section 6.2). Even where hotter stars are present, the mean integrated UV light of typical clusters is heavily influenced by cooler horizontal branch objects with a wide range of temperatures. Lower metallic line blanketing in the atmospheres of cluster stars on the warm HB and near the main sequence turnoff also produces larger mid- and near-UV emergent flux than for galaxies, tending to flatten the UV energy distributions.

Even if one ignores the distinctions between clusters and galaxies in UV spectral shape, the limits on metal-poor light at optical wavelengths in M31 and E galaxies are inconsistent with a large contribution from cluster-like populations in the UV (O'Connell 1976, 1980; Rose 1985; Bica & Alloin 1988; Dorman et al 1995).

The fact that the galaxy UVX does not appear to arise from a metal-poor subpopulation similar to the Galactic globular clusters does not, of course, necessarily mean that the kinds of hot stars in the galaxies differ from those in the clusters—only that the mixture of these is different.

6.2 Single Star Candidates

Since there were no ready-made local analogues for UVX populations, it was necessary to explore alternatives from a largely theoretical perspective. The seminal discussion of the various low-mass candidates for the UVX sources was presented by Greggio & Renzini (1990, hereafter GR; an updated overview is in Greggio & Renzini 1999). They discussed primarily single-star candidates, since the parameter space for binaries is much larger. Here, we also defer discussion of the possible involvement of binaries until a later section.

On the basis of the UV/optical colors, the UVX is estimated to contribute $\sim$2–3% of the bolometric luminosity in the most metal rich galaxies such as NGC 1399 and 4649 (GR). While this is not large, the challenge is to identify a mechanism for producing sufficient numbers of high-temperature stars in old populations. If the relevant evolutionary phase is short-lived compared with the lifetime of the galaxy, then the number of stars in this phase is proportional to its evolutionary lifetime. In this circumstance, the "fuel consumption theorem" shows that its contribution to the bolometric luminosity of the galaxy is proportional to the total amount of nuclear fuel consumed during the phase, which can be estimated directly from interiors models (Tinsley 1980, Renzini 1981a, Renzini & Buzzoni 1986, GR). Interesting candidates for the UVX will therefore have temperatures over $\sim$20000 K and will burn up to $\sim$0.01 $M_\odot$ of hydrogen or $\sim$0.1 $M_\odot$ of helium (GR). Dorman et al (1995, hereafter DOR) used a similar approach to estimate that the integrated monochromatic energy release at 1500 Å of suitable candidate evolutionary phases must be $E_{1500} \gtrsim 4 \times 10^{-3}$ $L_{V,\odot}$ Gyr Å^{-1}, where $L_{V,\odot} = 4.51 \times 10^{32}$ ergs s^{-1}.

The evolutionary phases of interest all occur after a low-mass star has begun moving up the red giant branch. HR diagram loci for the main types of UVX candidates that have been explored to date are illustrated in Figure 8. The general considerations are described in GR. The evolutionary trajectory of a low-mass star following He core ignition at the tip of the red giant branch is governed mainly by its envelope mass, M_{ENV}. Since the He core mass is $\sim$0.5 $M_\odot$ and is relatively insensitive to other parameters, $M_{ENV} \sim M_{TO} - \Delta M - 0.5$ $M_\odot$, where M_{TO} is the turnoff mass and ΔM is the total mass loss during the red giant phase, which, in globular cluster stars, amounts to $\sim$0.1–0.3 $M_\odot$, or 10–40% of the initial mass. The variance in mass loss leads to a scatter in M_{ENV} and hence in the initial temperature of the subsequent core He-burning stage on the "zero-age" horizontal branch (ZAHB). Envelope masses on the HB range up to $\sim$0.4 $M_\odot$. The lower

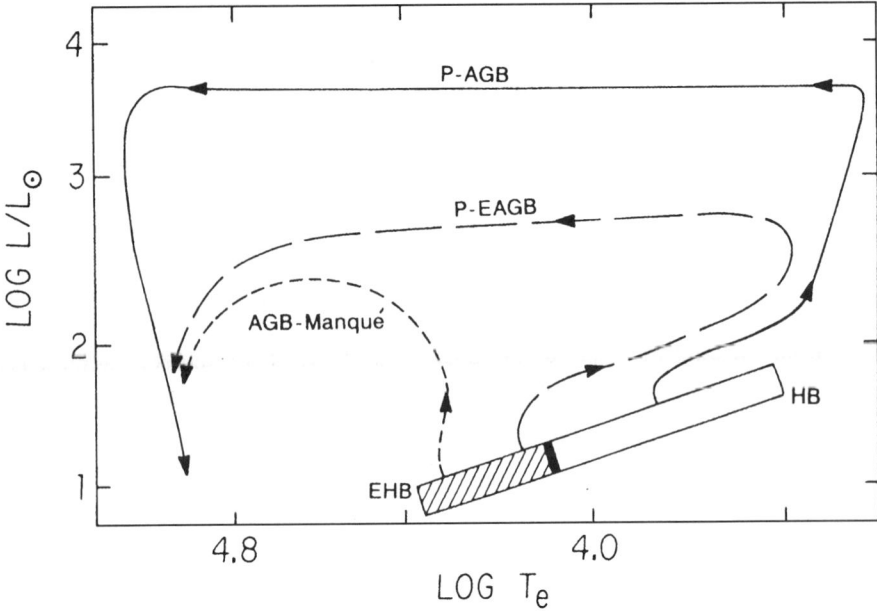

Figure 8 Schematic evolutionary tracks for the principal post-horizontal branch evolutionary phases described in Section 6. Envelope masses on the horizontal branch increase from left to right. For $Z \sim Z_\odot$, they are $M_{ENV} \sim 0.003$ $M_\odot$ at the left hand (hot) edge and ~ 0.05 $M_\odot$ at the heavy separator line, which marks the cool end of the "extreme horizontal branch" (shaded). The segment of the "P-AGB" track (solid line) which rises to the right of $\log T_e \sim 3.6$ corresponds to the AGB. Detailed evolutionary tracks for these phases are shown, for example, by Dorman et al (1993).

is M_{ENV}, the hotter is the ZAHB location (e.g. Iben & Rood 1970, Rood 1973, Sweigart 1987, Dorman 1992). Envelopes smaller than 0.05 $M_\odot$ correspond to $T_e(\text{ZAHB}) \gtrsim 14000$ K. After ~ 100 Myr, helium becomes exhausted in the center of the star, which then contains both helium-burning and hydrogen-burning shells moving outward.

If M_{ENV} is large enough, post-HB stars develop a deep convective envelope, evolve to lower temperatures, and ascend the cool asymptotic giant branch (AGB), leaving it only at high luminosity near the AGB tip, when rapid mass loss and thermal shell pulsing remove the envelope. Subsequent evolution in this case involves a rapid (10^4–10^5 yr) contraction and heating (the post-AGB or PAGB phase), in some cases with the formation of a planetary nebula, followed by cooling and fading on the white dwarf remnant sequence. Much of the pre-white dwarf time is spent at high temperatures, $T_e > 50000$ K. PAGB models for low-mass stars have been computed by Schönberner (1983), Blöcker & Schönberner (1990), and Vassiliadis & Wood (1994). The great majority of stars now near or above the main sequence turnoff in globular clusters will pass through the PAGB channel.

More exotic evolution can occur in the case of very small envelopes. For $M_{ENV} \lesssim 0.05$ M$_\odot$ the post-HB star may evolve to higher temperatures before it reaches the AGB tip or even before it approaches the cool AGB. These cases produce, respectively, post-early AGB (PEAGB) and AGB-manqué ("failed AGB") stars (see Figure 8). The first detailed models were described by Brocato et al (1990) and Caloi (1989), respectively (though similar behavior had been noted in early models by Sweigart et al 1974 and Gingold 1976). Their internal structure is similar to that of an AGB star (a double shell source) but with much thinner envelopes. They burn about the same amount of fuel as do the more familiar cool AGB stars, but they do so at much higher T_e ($\sim$25000 K). Their post-HB paths in the HR diagram can be convoluted. AGB-manqué lifetimes are $\sim 10^7$ yr, considerably longer than for the PAGB phase, after which stars evolve directly to the remnant cooling sequence. Typical E_{1500}'s for these "slow blue" post-HB phases (Horch et al 1992) are comparable to those of the hot HB phase (DOR Figure 6). Grids of such models, for a wide range of metallicities, have been computed by Castellani & Tornambè (1991), Castellani et al (1992), Horch et al (1992), Dorman et al (1993), Bertelli et al (1994), Castellani et al (1994), and Yi et al (1997a).

The least massive envelope ($\sim$0.05 M$_\odot$, if $Y \sim Y_\odot$; see Dorman et al 1993, Table 1) capable of producing a classical PAGB star yields a boundary on the ZAHB between what is now called the "extreme HB" (EHB) (to higher temperatures) and the normal HB. This is marked on Figure 8. AGB-manqué progenitors occupy the hot end of the EHB, while PEAGB progenitors occupy the end adjacent to the normal HB. Note that the normal main sequence for massive stars (not shown) crosses the HB locus at $T_e \sim 10000$ K, and that HB objects hotter than this fall below the main sequence in the classical "hot subdwarf" regime.

Another variety of hot star can be produced directly from the first ascent red giant branch if mass loss is large enough to remove the convective envelope before core He ignition. In this case, the post-RGB (PRGB) object evolves rapidly to the white-dwarf cooling sequence without passing through the HB phase. Some such objects may experience a late He-flash while on the cooling sequence as their central temperatures rise owing to gravitational core contraction. These "hot flashers" will then move to a position slightly below the EHB and follow subsequent post-EHB tracks similar to normal EHB stars. The hot flash effect was demonstrated by Castellani & Castellani (1993), and more detailed models including the secular effects of mass loss during advanced RGB evolution have been computed by D'Cruz et al (1996).

Remnants on the white dwarf cooling sequence are the inevitable descendants of all the preceding types of stars. During their early evolution, white dwarfs are still hot enough to emit UV photons, and their potential contribution to galaxy light can be estimated by integrating down the cooling curve. Magris & Bruzual (1993) and Landsman et al (1998) find that hot white dwarfs (residing on the cooling curve for $\lesssim$200 Myr) can contribute up to $\sim$10% of the far-UV light produced by their parent PAGB phases (less if their parents were EHB stars). This is too small to be of practical importance in normal circumstances. Unless one invokes an IMF truncated below $\sim$1.5 $M_\odot$, the integrated spectrum of the cooling curve also does not contain the strong Ly-α satellite features claimed by Bica et al (1996) to be present in IUE spectra of galaxies.

Finally, the much-studied "blue stragglers" (Bailyn 1995), which are warm stars lying near the main sequence but above the turnoff luminosity in star clusters, are too cool to be viable UVX candidates. They generally have temperatures below 10000 K. However, they may influence the mid-UV spectrum of old populations (Spinrad et al 1997, Landsman et al 1998).

A common characteristic of the viable UVX candidates is their extreme sensitivity to small changes in properties. Differences of only a few 0.01 $M_\odot$ in envelope mass for hot HB stars produce large changes in the type of post-HB track followed and the resulting E_{1500} (e.g. see DOR Figure 6). Likewise, models for PAGB stars show that their UV output is extraordinarily sensitive to core mass. Schönberner's (1983) 0.546 $M_\odot$ model has a lifetime 20$\times$ longer than for his 0.565 $M_\odot$ model and has E_{1500} a factor of 6.8 larger (DOR; GR Figure 3).

Likely individual examples of all these candidate types have been found in local star clusters and the Galactic field. Imaging with space telescopes has produced a fairly large sample of PAGB, EHB, AGB-manqué, and related post-HB stars in some globular clusters (e.g. in ω Cen, Whitney et al 1994, 1998; NGC 6752, Landsman et al 1996; NGC 2808, Sosin et al 1997; NGC 6338 and 6441, Rich et al 1997; M13 and M80, Ferraro et al 1998). A smaller sample of similar sources has been identified in the open clusters NGC 188 and 6791 (Liebert et al 1994, Landsman et al 1998), and Landsman et al estimate that NGC 188 and 6791 would have UV upturns in their integrated light as strong as any E galaxy. Over 1500 hot subdwarfs (sdO, sdB, and related types) are now known in the Galactic field. As first shown by Greenstein & Sargent (1974), many of these are EHB and post-EHB stars (Heber 1992, Saffer et al 1994). Kinematical studies show that some hot subdwarfs are members of the old, metal-rich disk population of the Galaxy (Thejll et al 1997). The field and open cluster examples are important cases since they demonstrate that EHB objects are not confined to low-metallicity environments. NGC 6791, in particular, has [Fe/H] $\sim$ +0.5 (Kaluzny & Rucinski 1995).

DOR summarized integrated UV outputs for the several main candidate UVX star types. PAGB tracks have $E_{1500} < 0.001$ $L_{V,\odot}$ Gyr Å^{-1}. They therefore cannot be solely responsible for the brightest UV-upturn cases, as first pointed out by GR and Castellani & Tornambè (1991). However, the EHB, PEAGB, and AGB-manqué phases burn more H + He fuel at high T_e's, by factors of $\sim$3–30, than classical PAGB stars and therefore are excellent UVX candidates.

6.3 Relationship to Global Characteristics of Parent Population

Each UVX candidate represents a potential channel to be filled by its evolving post-main sequence parent population. At least five global population parameters are known to be important in determining the occupation of the various channels.

The effects of these have been reviewed in GR and Chiosi (1996). Yi et al (1997b) nicely illustrate the effects of age, abundance, and mass-loss parameters on color-magnitude diagrams, integrated spectra, and broad-band UV colors. The single most important variable is mass loss on the giant branch, followed by helium abundance (Y).

Age As a population ages, its turnoff mass decreases, with $M_{TO} \sim 0.96\, t_{10}^{-0.2}\, M_\odot$, where t_{10} is the age in units of 10 Gyr. For a given amount of RGB mass loss, older stars will have smaller M_{ENV} and will fall at higher ZAHB temperatures. The UVX is therefore expected to increase with age, though probably in a strongly nonlinear fashion. At large enough ages, all stars evolving up the RGB will become hot EHB or PRGB objects.

Y An increase in helium abundance has important effects on post-giant branch evolution (GR, Horch et al 1992, DOR). Because of the increase in mean molecular weight, turnoff masses at a given age are smaller, which yields smaller M_{ENV} for a given amount of RGB mass loss. A higher initial helium abundance also causes stars with a given M_{ENV} to burn more of their hydrogen envelope during the core He-burning phase, producing AGB-manqué behavior for a larger range of M_{ENV}'s (Horch et al Table 1; DOR Figure 6). Increasing Y from 0.27 to 0.47 roughly quadruples the total E_{1500} for a uniform distribution of M_{ENV}'s (DOR).

Z The strong correlation between the UVX and line strengths discussed in Section 5.2 makes metal abundance effects on hot star evolution of particular interest. Based on the example of the globular clusters, one might suppose that metal abundance determines the prevalence of hot HB stars. To the contrary, theoretical models show that Z has little direct effect on either the EHB or post-EHB phases of evolution (e.g. Dorman et al 1993, DOR). Instead, these are governed mainly by M_{ENV}. Although increased metallicity does increase M_{TO} for a given age, thereby decreasing ZAHB temperatures for a given amount of RGB mass loss, hot HB stars can appear at any metal abundance as long as envelope masses are small enough. This is demonstrated in the grids of metal-rich HB models cited in Section 6.2 and was first illustrated in integrated light by Ciardullo & Demarque (1978). However, for higher metallicities, T_e for medium-envelope (0.05–0.15 $M_\odot$) stars is strongly decreased. This implies that a uniform distribution of M_{ENV} will lead to a bimodal distribution of ZAHB temperatures at higher metallicities (Dorman et al 1993, D'Cruz et al 1996).

There has been less exploration of advanced evolution with relative abundance variations among the metals. D'Cruz et al (1996) found no qualitative changes in behavior for models with [O/Fe] = +0.75. However, as discussed in Section 5.2, models incorporating variable abundance ratios among the metals would seem to be essential if the empirical line strength correlations are to be understood.

These theoretical expectations on the secondary status of metallicity effects on HB temperatures have good empirical support. The bluest UV colors in Figure 6 occur not for the most metal-poor globular clusters but for those of intermediate metallicity. Small numbers of EHB and related stars have recently been found in globular clusters with heavily populated red HBs (NGC 362, Dorman et al 1997; 47 Tucanæ, O'Connell et al 1997), and large numbers of hot HB stars are present in the relatively metal-rich clusters NGC 6388 and 6441 (with $Z \sim 0.25\, Z_\odot$, Rich et al 1997). Other clusters with EHB stars may range up to $Z \sim 3\, Z_\odot$ (NGC 6791, Liebert et al 1994).

$\Delta Y/\Delta Z$ There is good evidence from the study of emission lines in low-metallicity galaxies that helium abundance is coupled to metal abundance (e.g. Wilson & Rood 1994, Izotov & Thuan 1998). Values of $\Delta Y/\Delta Z \sim 3$–4 have been derived for low metallicity environments. If these apply to E galaxies, then the smaller effects on EHB and post-HB evolution of metallicity enhancements for $Z \gtrsim Z_\odot$ are strongly amplified by the effects of Y enhancement, as emphasized by GR. The dramatic increases in post-HB UV output found by Horch et al (1992) in metal-rich models were actually produced by the accompanying He effects (Y is increased to ~ 0.35–0.45 for $Z > 2\, Z_\odot$ in their models). Jørgensen & Thejll (1993) estimated that $\Delta Y/\Delta Z > 2.5$ is needed to produce a strong positive correlation between metal abundance and UVX above $Z_\odot$, for normal ranges of age and RGB mass loss. DOR (Section 8.3) emphasize that there is very little known about $\Delta Y/\Delta Z$ for solar abundances or above and that most available chemical evolution models suggest smaller He enhancements than for low abundances. DOR also point out that EHB stars exist in clusters and the Galactic field at moderate metallicites, and presumably moderate Ys, so that extreme Z or Y enhancements are not essential to their production.

Mass Loss Mass loss is the most important determinant of post-RGB evolution in low-mass stars but is also the most difficult to evaluate because of a paucity of both empirical evidence and theoretical exploration. RGB mass loss is usually modeled using the Reimers (1977) prescription:

$$\dot{m} = -4 \times 10^{-13}\, \eta_R\, \frac{L}{gR}\, M_\odot\, \mathrm{yr}^{-1},$$

where η_R is a mass loss efficiency parameter, L is the luminosity, g is the surface gravity, and R is the radius, with $L, g,$ and R in solar units. This formula is based on dimensional analysis rather than a well-grounded physical theory. It is consistent with the available observational data, which indicate only that mass loss increases with luminosity and decreasing surface temperature, reaching a maximum just prior to the He flash (reviewed in Dupree 1986). The Reimers prescription implicitly includes composition and age dependences through their influence on stellar structure, and hence $L, g,$ and R. Although η_R is the principal mass-loss parameter in this formulation, empirically there is always a significant spread in the effective η_R (e.g. Rood 1973), which produces a range ΔM_{ENV} on the ZAHB. There is no theory for the spread at the moment, so it appears in evolutionary synthesis models as an additional free parameter.

To produce a typical globular cluster blue horizontal branch requires $\eta_R \sim 0.2$–0.5 (e.g. Renzini 1981b, Yi et al 1997b) if cluster ages are ~ 15 Gyr. Such values are therefore regarded as "normal" and are widely used in galaxy spectral modeling. However, the globular cluster values would increase if the lower ages of ~ 12 Gyr favored by the recent Hipparchos recalibration of distance indicators are adopted (Yi et al 1999). Furthermore, there is very little evidence on whether globular cluster η_R values are appropriate in other kinds of stellar populations. It is physically plausible that η_R would increase with metal abundance, owing to grain formation, for instance. GR explored the consequences of assuming that $\eta_R \propto 1 + \frac{Z}{Z_{crit}}$, where Z_{crit} is a critical threshold. D'Cruz et al (1996) and Yi et al (1997b) considered models for a range of η_R up to 1.2, the former including self-consistently the effects of mass loss on evolution near the RGB tip and the "hot flasher" phenomenon. They find that production of EHB stars in populations with $Z \gtrsim Z_\odot$ requires $\eta_R \gtrsim 0.7$, with total mass loss of ~ 0.5 M$_\odot$ per star for ages $\gtrsim 10$ Gyr. D'Cruz et al found that a smooth distribution of η_R on the RGB leads to a strongly bimodal distribution of ZAHB temperatures if $Z \gtrsim Z_\odot$. The models imply that increases of η_R of only a factor of 2–3 over canonical globular cluster values are sufficient to produce a large population of EHB and post-EHB stars for normal ranges of age, Z, and Y.

The hot flash phenomenon is one way of routing a significant fraction of the evolving population into HB objects with $M_{ENV} < 0.05$ M$_\odot$ without the need for "fine tuning" of mass loss. D'Cruz et al (1996) showed that as long as mass loss near the RGB He flash luminosity is above a critical threshold (corresponding to $\eta_R \sim 0.7$, but not necessarily tied to the Reimers prescription), EHB stars will always be produced via the hot-flash mechanism. Neither the η_R values nor the range of values ($\Delta\eta_R$) producing EHB objects vary much with metallicity from globular cluster to supersolar values. The mechanism produces a natural concentration to a narrow range of T_e. There is also improving empirical evidence from the cluster and field samples (Section 6.2) that EHB populations do occur in nature at a wide range of metallicities. Thus, there are no obvious obstacles to the production of EHB stars through enhanced mass loss.

Although the Reimers law provides a useful schematic for exploring mass-loss effects on the UVX, a much improved physical theory is needed. Preliminary hydrodynamic models (e.g. Bowen & Willson 1991, Willson et al 1996) suggest a sudden onset of mass loss at a critical luminosity and a strong metallicity dependence, effects that may not be well modeled by a simple scaling law.

Other Parameters There are certainly other processes that can influence the production of hot stars in old populations. Sweigart (1997) showed that deep mixing in the outer envelopes of RGB stars, which results in enhanced surface He abundances, encourages the production of hot HB stars and AGB-manqué behavior. Larger mixing is presumably related to higher stellar rotation rates. The dynamical environment of galaxies could therefore influence the UVX by way of stellar spin distributions. Ferraro et al (1998) find that gaps in the hot HBs of different globular clusters occur at similar temperatures, suggesting that RGB mass loss is a multimode process. Good candidate mechanisms have not yet been identified.

6.4 Binaries and Dynamical Effects

There has been much speculation about the possible origin of hot low-mass stars through dynamical interactions, especially in binary systems through Roche-lobe mass transfer or mergers. The various recognized mechanisms have been reviewed by Bailyn (1995), while their implications for the UVX problem were most extensively discussed by GR. The mildest form of interaction occurs when a star ascending the giant branch loses part of its expanding envelope to a companion, thereby appearing with lower M_{ENV} on the ZAHB (Mengel et al 1976) but evolving normally thereafter. Based on the frequency of binaries among the sdB stars in the Galactic field (Green 1999), this process may be fairly common there. Although "fine tuning" of binary mass ratios and separations would seem to be necessary to produce small M_{ENV} without suppressing the He flash altogether, in fact the hot-flash mechanism (D'Cruz et al 1996) would mitigate this problem here as it does for normal mass loss. One reason that UV star production in binaries might depend on Z is that stellar envelopes become more inflated at higher metallicities (GR). The fact that field and cluster hot horizontal branches have gravities and luminosities consistent with single-star models suggests that more drastic interactions (e.g. mergers) are considerably more rare.

Some support for dynamical effects is provided by the observation that the extent of horizontal branch "blue tails" in Galactic globular clusters appears to correlate with cluster concentration and density (e.g. Fusi Pecci et al 1993, Buonanno et al 1997). However, other expectations for dynamical mechanisms are not met. For instance, the hot stars are not necessarily concentrated to the centers of clusters, and the system with the largest EHB/post-EHB population (ω Cen) is notably low density (e.g. Whitney et al 1994, Rich et al 1997). HST observations of 10 cluster cores (Sosin et al 1997) show that the EHB stars are not as centrally concentrated as the blue stragglers (which are almost universally agreed to be interaction products).

It is unclear how to translate the evidence for dynamical mechanisms in star clusters to the dynamical environment of galaxies, which is very different and less conducive to stellar interactions. Of course, the present field population in E galaxies may well have originated in concentrated cluster-like systems that have since disintegrated but which were responsible for establishing the binary frequency. Stellar rotation, which could depend on global dynamical characteristics of galaxies, may also influence the UVX through He mixing (Sweigart 1997). The only hint that the E galaxy UVX is related to dynamics is the connection between large UV upturns and boxy isophotes (Longo et al 1989 and Section 5.2). The core of M32 is the densest observable extragalactic system. No large radial gradients in blue light are apparent near its center at HST resolution (King et al 1992, 1995;

Bertola et al 1995; Cole et al 1998; Lauer et al 1998), though Brown et al (1998a) note that the density of resolved UV stars per unit total light increases slightly at smaller radii in both M31 and M32.

Mass transfer onto white dwarf binary companions can produce copious UV flux and has also been discussed in the context of the UVX by GR. One of the main problems is again the fine-tuning needed to ensure a sufficient but not excessive transfer. Because of the large parameter space involved, estimates are only tentative, but accreting white dwarfs seem unlikely to be major contributors to the UVX (GR). The UV spectrum of a population dominated by such objects is also expected to show fairly strong emission lines (e.g. Wu et al 1992), which are absent in E galaxy spectra.

Although dynamical interactions could certainly influence the UVX in galaxies, there is no strong evidence yet that they do so.

7. HOT LOW-MASS STARS IN ELLIPTICAL GALAXIES

7.1 Interpretation of UV Spectra and Colors

Early discussions of low-mass candidates for the UVX in the context of the observations concentrated mainly on distinguishing them from the massive star interpretation rather than from one another. Rose & Tinsley (1974) were the first to emphasize that hot PAGB stars were inevitable products of low-mass evolution and should be present in sufficient numbers to affect the integrated UV spectrum in old populations of all metallicities (assuming mass loss is not so extreme as to suppress the AGB phase altogether). O'Connell (1976) found tentative evidence for hot starlight in the 3300–4000 Å region of 3 gE galaxies, which was plausibly interpreted as from the PAGB. Following the demonstration that normal HB stars in globular cluster-like populations were not compatible with the galaxy UVX spectra (see Section 6.1), PAGB stars became the favored candidates.

Bohlin et al (1985), Renzini & Buzzoni (1986), and O'Connell et al (1986) pointed out that the extreme sensitivity of the UV output of PAGB stars to their core masses might explain the large variation in UVX strength and the Faber (1983) Mg_2 correlation. They suggested that metallicity-enhanced mass loss or age differentials drove the correlation. BBBFL, Bertelli et al (1989), and Magris & Bruzual (1993) examined the implications of PAGB models quantitatively. They found that the brightest UVX sources would require non-PAGB sources of UV light or PAGB stars with core masses smaller than those in the grid of evolutionary models by Schönberner (1983), the smallest of which was technically a PEAGB rather than a PAGB star (i.e. it did not reach the AGB tip). They also concluded that changes in Z alone could not reproduce the Faber correlation unless there were accompanying changes in t, η_R, or Y.

Shortly thereafter, two lines of evidence combined to reject PAGB stars as the principal UVX sources. First, GR and the other studies cited in Section 6.2 emphasized the shortfall in the UV output of PAGB objects compared with the strongest upturn galaxies and suggested EHB stars and their descendants as a more viable alternative. Second, HUT spectroscopy showed that the energy distributions of M31 and NGC 1399 declined shortward of 1050 Å (Ferguson et al 1991, Ferguson & Davidsen 1993). This placed an upper limit of $T_e \sim 25000$ K on the temperature of the dominant UVX stars, considerably cooler than the 50000 K characteristic of a PAGB component but entirely consistent with the EHB and post-EHB channels. Ferguson & Davidsen (1993) found significant differences between M31 and NGC 1399 which demonstrated, independent of modeling details, that the UVX is probably a composite population with the mixture of HB and other hot types varying from system to system. (Later HUT observations by Brown et al 1997 confirmed such variations in six other galaxies based on 912–1000 Å fluxes.)

EHB, PEAGB, and AGB-manqué stars have consequently become the favored candidates for the dominant UVX sources. They are energetically viable since in the brightest UVX cases only 10–20% of the evolving stars in the dominant population would need to pass through the EHB and post-EHB channels to produce the observed far-UV luminosities (GR, DOR, Brown et al 1995). DOR showed that the observed 1500–V and 2500–V colors of E galaxies were consistent with composite EHB/post-EHB and PAGB models in which the EHB channel contributes $\sim 25\%$ of the FUV light in medium-upturn systems like M31 but $\sim 75\%$ in the strongest upturns. They found that most globular clusters do not require EHB stars to explain their UV colors, consistent with independent information on color-magnitude diagrams (see Figure 7). They also found that if mass loss is left as a free parameter, 1500–V does not place useful limits on galaxy age or metal abundance, though 2500–V does. This is because the properties of the EHB/post-EHB channels are not very sensitive directly to either parameter, whereas the main sequence turnoff (which dominates for $\lambda > 2500$ Å) is. DOR found acceptable agreement with observed colors for a wide range of ages (6–20 Gyr) but only for solar or higher metallicities ($Z \sim 1-4\ Z_\odot$), again leaving mass loss unconstrained.

Brown et al (1997) analyzed HUT spectra of six E/S0 galaxies covering the 900–1800 Å region at 3 Å resolution. They were able to produce good fits to the spectra with composite EHB/post-EHB and PAGB models in which only a small fraction ($\lesssim 10\%$) of the evolving population need pass through the EHB channel (see Figure 9). Although the models formally contained very small ranges of M_{ENV} on the ZAHB (e.g. 0.021–0.046 $M_\odot$ in NGC 4649), they found that broader distributions of M_{ENV} would yield similar fits because the short wavelength FUV flux tends to be dominated by the hot AGB-manqué phases. The best fits occur for evolutionary tracks with supersolar values of Z (2–3 $Z_\odot$) and Y (0.34–0.45), which produce more flux than subsolar models below 1200Å. Interestingly, however, fits are better for atmospheres with subsolar values of Z ($\sim 0.1\ Z_\odot$). Brown and colleagues attribute this to the same processes (mainly diffusion) that create well-known abundance anomalies among Galactic hot subdwarfs (e.g. Saffer & Liebert 1995). These effects are not straightforward to analyze, but the expectation is that diffusion in high-abundance stars will tend to reduce line strengths and make

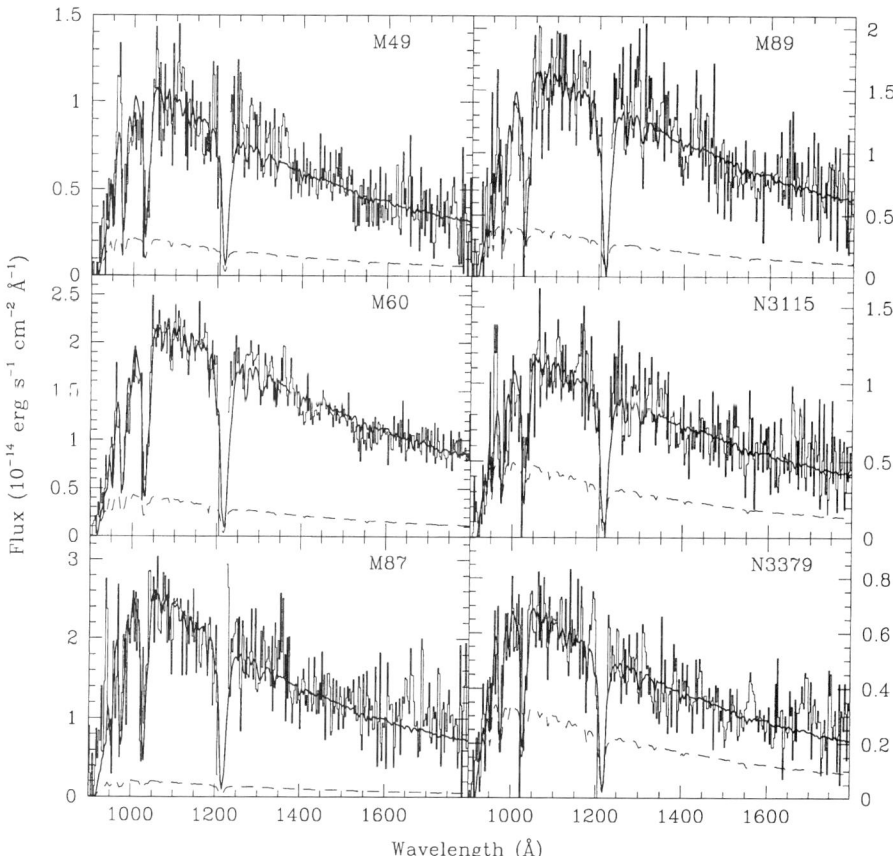

Figure 9 *Astro*/HUT far-UV spectra of 6 E/S0 galaxies compared with EHB+post-EHB+PAGB models. Fluxes are shown in 2.5 Å bins. The best-fitting composite models are shown by the solid lines. These employ evolutionary tracks with $Z \sim 2\text{--}3\ Z_\odot$ but atmospheres with $Z = 0.1\ Z_\odot$. The PAGB contribution to each model is shown by the dashed lines; this is usually considerably smaller than 50%. From Brown et al (1997).

spectra appear less metal-rich. This is a critical issue, however, since the consensus interpretation would have to be fundamentally revised if the UVX arose from a metal-poor population.

The DOR and Brown et al (1997) analyses were consistent with realistic global population models but left mass loss on the RGB as a free parameter because the physics involved are so uncertain. A large body of other evolutionary spectral synthesis studies adopt definite prescriptions for mass loss and have predicted the UV spectra of old populations with the intent of exploring how higher metal abundances might produce larger UV output (as in Figure 6). Some of these consider a fixed grid of abundance parameters, while others (e.g. the Padova group, Bressan et al 1994, 1996) include self-consistent star formation and nucleosynthetic enrichment histories to determine the abundance distribution in galaxy models of various masses. In most of these models, RGB mass loss is specified by the Reimers prescription using a fixed mean η_R value in the range $\sim$0.3–0.5, as appropriate for globular clusters (e.g. Bertelli et al 1989, Barbaro & Olivi 1989, Bruzual & Charlot 1993, Magris & Bruzual 1993, Bressan et al 1994, Lee 1994, Bertola et al 1995, Bressan et al 1996, Chiosi et al 1997, Park & Lee 1997). These model sets do not include the "hot flash" phase. It is also necessary to specify the spread of M_{ENV} on the ZAHB, ΔM_{ENV}. Since the early studies of globular cluster HBs (e.g. Rood 1973) it has been traditional to use a modified Gaussian distribution, though this is neither unique nor well justified on astrophysical grounds.

The two-color diagram in Figure 10 (from Yi et al 1997b) illustrates the general nature of the predictions from this class of models. It shows the effects of age and composition for a grid with $\eta_R = 0.7$ and $\Delta Y/\Delta Z = 3$, somewhat higher values than typically assumed in the other studies mentioned above. The color-age relation is nonmonotonic. Far-UV light reaches a minimum at about 5 Gyr, after the decay of the warm main sequence and before the EHB channel is well-filled. PAGB stars are always present but are significant in the UV light only for ages $\lesssim$5–10 Gyr, after which the EHB and post-EHB phases dominate. The EHB/post-EHB channel becomes strongly occupied only when the turnoff mass becomes small enough that the assumed RGB mass loss yields small-envelope ZAHB stars. In Figure 10 this occurs after $\sim$16 Gyr for $Z_\odot$ and $\sim$8 Gyr for 3 $Z_\odot$, yielding 1500–V $\lesssim$ 3. These threshold ages would decrease by about 5 Gyr for $\eta_R = 1$ (Yi et al 1997b, Figure 22). Although the detailed age-color relation depends on Z, the color loci for $Z \gtrsim Z_\odot$ nearly superpose, implying that the UV colors cannot be easily used to distinguish Z. Tracks for $\eta_R = 1.0$ reproduce the plotted E galaxy data better than those shown here, although these do encompass the entire color range observed. In particular, the plotted models allow significantly stronger UV-upturns (1500–V $\sim$ +0.5 and 1500–2500 $\sim$ −1) than actually observed. As indicated by the labeled points, the observed range of 1500–V (though not 1500–2500) could be produced at $t = 15$ Gyr, $\eta_R = 0.7$ by a change of Z of a factor of 2, with a concomitant change in Y. Alternatively, within the modeling uncertainty, the data could be equally well explained by changes in age at constant abundance or by a correlation between η_R and abundance.

7.2 Inferences About Chemical Abundances and Ages

The evolutionary model sets agree that the observed far-UV fluxes can be produced by old (5–15 Gyr), metal-rich ($Z \sim 1$–3 $Z_\odot$) populations given favorable, but not unreasonable, assumptions about $\Delta Y/\Delta Z$, η_R, and ΔM_{ENV}. They also agree that changes in Z alone cannot reproduce the Faber correlation, unless it is assumed that $\Delta Y/\Delta Z \gtrsim 2.5$ or that η_R increases with Z (as foreseen by GR). Models based on more conservative assumptions about mass loss require that populations with significant UVX components be older than about 10 Gyr.

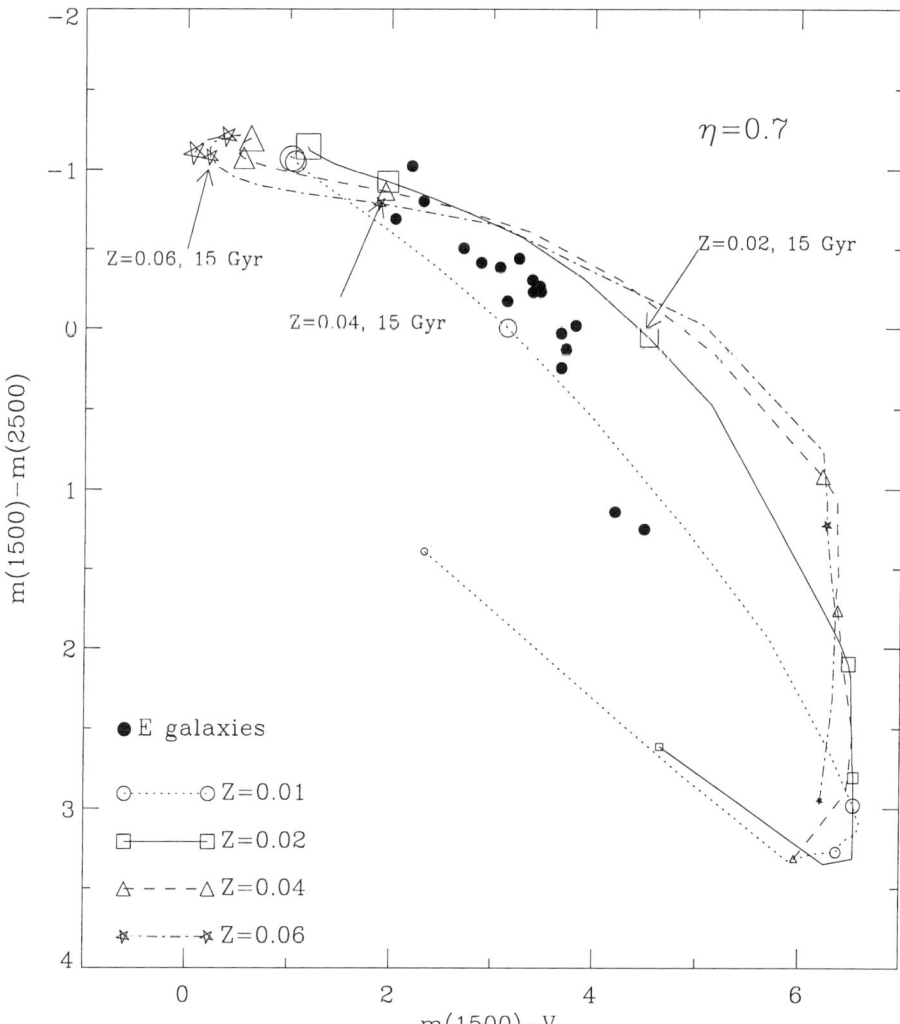

Figure 10 IUE color data (same axes as Figure 7) for E galaxies compared to theoretical models. From Yi et al (1997b). The various symbol shapes correspond to sequences for different Z, as indicated on the legend. The symbol sizes increase with age; the six models in each sequence are for 1, 5, 10, 15, 20, and 25 Gyr. All models assume a Reimers mass-loss parameter of $\eta_R = 0.7$, about twice that estimated for globular clusters. Masses on the ZAHB are assumed to be spread about the mean value (determined by η_R) in a modified Gaussian distribution with $\sigma = 0.06$ M$_\odot$. $\Delta Y / \Delta Z$ is assumed to be 3.0. See Section 7.1 for further details.

As noted in Section 6.3 and illustrated in Figure 10, the metal abundance Z, if decoupled from Y and mass loss, has only a secondary influence on the spectrum of the hot components at wavelengths above 1200 Å. There is a somewhat greater effect on the temperature distribution of light in the 900–1200 Å range (Brown et al 1997). Y can have a significant effect on UV flux production, but its influence can be masked by changes in mass loss. Furthermore, the UVX absorption line spectrum is apparently subject to atmospheric diffusion effects, as discovered by Brown et al (1997). Therefore, far-UV spectra in the 1200–2000 Å range cannot easily be used to infer abundances of the hot star populations of E galaxies.

It is also premature to try to use the UVX to age-date elliptical galaxies. The far-UV appears promising for this purpose because it is the most rapidly evolving part of a single-burst galaxy spectrum. A number of studies have noted that the "turn-on" of the UVX (which occurs at the age when M_{TO} drops to the point that the assumed RGB mass loss is able to fill the smaller envelope channels) marks an obvious spectral transition which might be used to age-date E galaxies at moderate lookback times (e.g. GR, Magris & Bruzual 1993, Chiosi 1996, Bressan et al 1996, Chiosi et al 1997, Yi et al 1999). The amplitude of the transition in UV colors is large. Unfortunately, its timing in the models is very sensitive to assumptions about mass loss and helium abundance, making results strongly model-dependent. This can be seen in the cases presented by Yi et al (1997b, 1999).

An interesting related example is the interpretation of the UVX as the product of a metal-poor subpopulation of extremely old (18–20 Gyr) stars, presumably the earliest generation to form in massive galaxies (Lee 1994, Park & Lee 1997). The models employed by Lee and Park adopt atypically small values for both total RGB mass loss (a fixed 0.22 $M_\odot$) and ΔM_{ENV} (0.02 $M_\odot$). Larger values for these parameters would significantly reduce the inferred ages. The models also do not fit the UVX spectra very well, having the flatter energy distributions for 1500–3500 Å characteristic of metal poor systems (Park & Lee 1997), even when mixed metallicities and larger effective mass loss are included (Yi et al 1999).

It is clear in general that assumptions about η_R and ΔM_{ENV} largely determine the outcome of the far-UV evolutionary synthesis models developed to date and any conclusions about t, Y, or Z that may emerge from them. This is necessarily so, given the circumstance that changes in only a few 0.01 $M_\odot$ in M_{ENV} can radically affect the UV output of stars. Age and mass loss can be traded for one another, and without a more deterministic theory of mass loss, derived ages are not reliable.

These remarks apply to the far-UV, hot-star spectra of old populations. The situation is quite different in the mid-UV region (2400–3200 Å) where the light becomes dominated by cool stars ($T_e \sim$ 5500–7500 K) near the main sequence turnoff. Turnoff light is directly sensitive to both age and abundance, whereas red giant light, which is important longward of 4500 Å, is insensitive to age. The principal obstacle to exploitation of the mid-UV as a population diagnostic is the lack of good "libraries" of mid-UV stellar energy distributions. Empirical datasets (e.g. Fanelli et al 1992) tend to be limited to solar abundance, whereas theoretical ones (e.g. Kurucz 1991) have serious shortcomings due to difficulties in treating the overwhelming UV line blanketing. HST is beginning to fill the gap, if slowly, with high quality, medium-resolution spectra (e.g. Heap et al 1998). Another technical difficulty with the mid-UV is that contamination by the long-wavelength tail of the UVX energy distribution must be removed. Although this contributes over 50% of the 2700 Å light in many cases (BBBFL, Ponder et al 1998), correction for the UVX component is straightforward because it has a smooth and well-determined shape (Dorman et al 1999).

Preliminary synthesis models of the mid-UV using theoretical stellar spectra confirm the expectation that it will be an excellent age/abundance diagnostic. DOR's experiments with fitting broad-band UV colors of E galaxies showed that 2500–V yields much more information on age and composition than does 1500–V. They estimate that $\partial(2500-V)/\partial \log Z \sim 2.7$ for old populations, a much higher sensitivity than for most optical-IR indices (e.g. Worthey 1994). The mid-UV continuum is especially useful in placing limits on the contribution of metal-poor populations to galaxy light. The available mid-UV models show, for instance, that the metal-poor fraction in E galaxies is much smaller than predicted by simple "closed box" nucleosynthetic models (e.g. Tantalo et al 1996, Worthey et al 1996). Empirically, mid-UV spectral features also strongly distinguish the populations of globular clusters and E galaxy cores from one another (e.g. Ponder et al 1998).

One of the most important applications of mid-UV stellar population analysis will be to the spectra of high redshift galaxies. Age-dating of distant objects that are passively evolving, such as LBDS 53W091, with $z = 1.55$, can constrain the earliest epoch of star formation, and hence cosmology (e.g. Spinrad et al 1997).

7.3 Resolved UV Star Populations

The HST has sufficient sensitivity to probe directly the UV star populations of nearby galaxies. Three UV imaging studies of hot stars in M31 and M32, all based on FOC observations of small fields, have appeared to date (King et al 1992, Bertola et al 1995, Brown et al 1998a). Although the two earlier studies suffered from serious calibration difficulties (see Brown et al 1998a), all three detected UV-bright stars and agree that luminous PAGB stars cannot account for more than a small fraction of the total FUV light. The photometry of Brown et al (1998a) has a detection limit of $m_\lambda(1750 Å) \sim 24.5$, which is not deep enough to reach the HB itself but does encompass PAGB, PEAGB and AGB-manqué luminosities. Brown and colleagues identify a large number of stars consistent with expectations for the descendants of EHB stars with $M_{ENV} \sim 0.002$–0.05 $M_\odot$. However, most of the UV light is produced by unresolved stars, presumably on the EHB. The integrated HUT or IUE spectra are consistent with models, normalized by the resolved samples, in which only 2% of the total population passes through the EHB channel in M31 and only 0.5% does so in M32. The lifetime of the PAGB channel, which makes up the rest of the post-HB population, is so short that few resolved objects are expected in the observed fields. Somewhat unexpectedly, the shapes of the luminosity functions for the resolved stars in M31 and M32 are similar (despite significant differences in optical absorption line spectra). M32 differs only in having fewer total stars (per unit UV surface brightness) above the detection threshold. Brown and coworkers also find that about 10% of the brighter resolved population is not explainable by existing post-HB evolutionary tracks.

Any process that reduces HB envelope masses in a significant fraction of the population can have influence extending beyond the UV region. EHB stars do not become AGB stars, and if most of the evolving population passes through the small-envelope channel, the AGB contribution to the integrated optical/IR spectrum (mainly longward of 6000 Å) of a galaxy will decrease. Changes in the AGB should also be detectable with the surface brightness fluctuation imaging method (Tonry & Schneider 1988). Ferguson & Davidsen (1993) find that the incidence of planetary nebulae, which should also decrease if the AGB population decreases, is anticorrelated with bluer 1500–V colors. This is important circumstantial evidence that small-envelope HB stars are implicated in the UVX. The correlation should be pursued with a larger sample of galaxies, and radial dependences within galaxies should be studied as well.

The existing deep imaging studies therefore provide good support for the EHB interpretation of the UVX. Imaging to the level of the EHB itself within the Local Group can probably be secured with HST/STIS and HST/ACS. Worthey (1993) has described how the surface brightness fluctuation technique can be applied to faint hot stars to extend the effective depth of such UV imaging.

7.4 Cosmic Evolution of the UVX

Since the models predict that the UVX (if dominated by the EHB or PAGB channels) should decrease as the main-sequence turnoff mass increases, there should be strong evolution of the UVX with lookback time (see Section 7.2). If E galaxies are sufficiently homogeneous, there could be a unique lookback beyond which the UVX disappears. Given the uncertainty in the models discussed above, lookback effects should probably be viewed for the present more as a valuable opportunity to refine the models than as a way to age-date the universe.

There are serious technical challenges in making restframe UV measurements at moderate redshifts. The galaxies are faint. At a redshift of 0.5 (a lookback time of 6 Gyr), the distance modulus is 43, implying that the unevolving UVX of a strong upturn source in a typical luminous elliptical would have $m_\lambda(2250$ Å$) \sim 24.5$. Simple detection of far-UV light (e.g. in broad bands) is not sufficient to distinguish a UVX component from the decaying initial burst or late star formation (see Section 5). Multiband photometry or spectroscopy is necessary. Several attempts to observe the UVX at high redshift have been made (e.g. Windhorst et al 1994), but only recently has a detection been claimed in the cluster Abell 370 ($z = 0.38$) by Brown et al (1998b). Using broad-band filters with HST/FOC, they find four cluster E galaxies to have a range of 1500–V similar to that in local galaxies. If this is UVX light, it implies a high formation redshift ($z_F > 4$) in the context of most existing models. An absence of UV evolution over the past few Gyr would be inconsistent with some classes of UVX models. It will be especially important to link changes in the UVX of distant galaxies with evolution of the initial burst at optical/IR wavelengths (now detected up to $z \sim 0.9$, e.g. Stanford et al 1998).

7.5 Summary and Key Issues

Progress on the UVX problem during the last ten years has been excellent. The theoretical, spectral, and imaging evidence has recently converged toward the view that the UVX originates from He-burning, extreme horizontal branch stars, their post-HB progeny, and post-AGB stars in the dominant, metal-rich stellar population of E galaxies. The mixture of these types apparently varies from object to object, perhaps in a systematic way with global mean metallicity or mass, but in most cases the EHB/post-EHB channels are the more important. The simplest explanation for the correlation between the UVX and optical line strengths is that the mass-loss parameter η_R increases with Z or that $\Delta Y/\Delta Z \gtrsim 2.5$.

Although evolutionary synthesis models successfully predict UV spectral properties in the ranges observed, progress in understanding the UVX, and in refining estimates of ages and abundances derived therefrom, is hampered by our lack of knowledge of two basic processes: mass loss on the giant branch and helium enrichment. Both of these are critical to the efficiency with which an old population can generate UV-bright stars. We urgently require a more complete and predictive physical theory of giant-branch mass loss. This is the highest priority for UVX theory in the near term. The question of the value of the helium enrichment parameter ($\Delta Y/\Delta Z$) near and above solar abundance also needs to be addressed. Both areas demand extensive observational programs on nearby systems as well as fundamental improvements in theoretical modeling. The same is true of diffusion in hot atmospheres, which is important to interpreting the UVX line spectrum.

These are the most serious gaps in our astrophysical understanding of the UVX, but there are other troublesome issues as well, three of which are worth mentioning:

1. The behavior of the UVX seems to be firmly linked to that of the lighter elements such as N, Mg, and Na and decoupled from the Fe-peak (Section 5.2). This adds an additional dimension to modeling space, so far unexplored, which is not at present well supported by nucleosynthetic theory (Worthey 1998).

2. The internal spatial gradients in 1500–B color discussed in Section 4.2 do not correlate with gradients in Mg_2 (Ohl et al 1998). Metallicity is evidently not the sole parameter governing the UVX. This may be related to the decoupling of the Fe-peak noted in paragraph 1 above, or it may reflect the influence of other changing parameters within galaxies, such as age or Y abundance. M32, with a large and reversed UVX gradient (see Figure 3), is an important case since there is considerable independent evidence for an intermediate age ($\lesssim 8$ Gyr) population there and possibly an age gradient in which the central regions are younger (O'Connell 1980, Freedman 1992, Rose 1994, Hardy et al 1994, Faber et al 1995, Grillmair et al 1996).

3. There has been very little work on the dependence of the UVX on galaxy morphology despite suggestions of differences between E galaxies and S0 galaxies (e.g. Smith & Cornett 1982). Bright, nearby spiral bulges could readily be studied in the UV with HST, and comparisons with E galaxies could help distinguish some of the underlying drivers of the UVX phenomenon. If, for example, bulges have a wider range of ages than E galaxies (e.g. Wyse et al 1997), then the younger ones should have smaller UV upturns than E's.

8. OTHER FAR-UV PHENOMENA IN E GALAXIES

Although the UVX produces a ubiquitous extended light background in old populations, it is at a low level and is coincident with a "dark window" in the natural sky background centered at about 2000 Å, where the sky is about 40× fainter than at any other wavelength in the optical-IR region (O'Connell 1987). The faint UV backgrounds permit isolation of other interesting phenomena that are either unique to the UV or are drowned out in the visible bands by the glare of the main sequence and giant branch stars. This includes low-luminosity active nuclei, recent star formation, blue straggler populations, gas in the 10^5–10^6 K temperature range, scattered light from dust grains, and H_2 fluorescence features near 1600 Å. As noted in Section 3, hot continuum sources that contribute as little as 0.1% of the V-band light of a galaxy can be readily detected in the UV region. In this section we briefly discuss UV observations relevant to massive star formation and nonthermal nuclei.

Recent Massive Star Formation   Identification of a minority component of massive stars in an old population depends on detection of spectral or color distortions in integrated light or on imaging of individual stars or concentrations of stars that stand out against the smooth background light. The vacuum UV is about 30–50× more sensitive to such effects than the optical/IR bands (McNamara & O'Connell 1989). Based on spectral synthesis models for constant star formation with a normal IMF (e.g. Bruzual & Charlot 1993, Cornett et al 1994), the star formation rate per unit V-band luminosity is related to far-UV color as follows:

$$\dot{S}/L_V \sim 8 \times 10^{-11}\, 10^{-0.4(1500-V)}\, M_\odot\, \text{yr}^{-1}\, L_{V,\odot}^{-1}.$$

The 1500 Å flux used to compute the color here is the part of the total far-UV flux that is attributed to young stars. The coefficient in this expression is almost independent of the period over which the star formation is assumed to have persisted, for periods over 50 Myr.

If all of the UV light in the strongest UV upturn cases (1500−V $\sim$ 2) were attributed to massive star formation, the implied normalized rate would be $\dot{S}/L_V \sim 1.3 \times 10^{-11}$, or a total rate of $\dot{S} \sim 0.25\, M_\odot\, \text{yr}^{-1}$ for a typical gE galaxy with $M_V = -21$. This is obviously a strong upper limit to massive star formation in a normal E galaxy since only a small fraction of the far-UV light can be produced by massive stars, as discussed in detail in Sections 4 and 5. If we take 20% as the upper limit on the contribution of young starlight at 1500 Å in a galaxy with a more typical UV upturn with 1500−V = 3.5, then the maximal total rate becomes $\dot{S} \lesssim 0.01$ M$_\odot$ yr^{-1}. This is a very stringent limit on the amount of continuing star formation in a typical gE galaxy.

These values can be compared with the estimated total mass loss from stars evolving up the giant branch. The "evolutionary rate" in an old population (i.e. the number of stars evolving off the main sequence per unit time) is $\sim 4 \times 10^{-11}$ yr^{-1} $L_{V,\odot}^{-1}$ (e.g. Renzini & Buzzoni 1986, DOR). If each star sheds 0.3–0.5 M$_\odot$, then the total estimated normalized mass loss rate is $\dot{m}/L_V \sim 1$–2×10^{-11} M$_\odot$ yr^{-1} $L_{V,\odot}^{-1}$, or 0.2–0.4 M$_\odot$ yr^{-1} for a galaxy with $M_V = -21$. The maximal continuing star formation rate derived from far-UV data is some 20–40× smaller. Clearly, most of the material produced by giant branch mass loss is not being recycled into new stars in normal E galaxies, at least not with a normal IMF. The UV is the key to this conclusion, since high S/N optical-band studies generally cannot exclude complete recycling (e.g. O'Connell 1980, Gunn et al 1981).

The ultimate fate of the lost red giant envelopes remains unclear. At early times the material is probably removed from galaxy interiors by high-temperature, supernovae-driven winds. In more massive galaxies, the gas forms a hot corona, which is detectable at X-ray wavelengths (e.g. Forman et al 1985). Some fraction of the corona is returned to the interior by a cooling flow (e.g. Sarazin & White 1988, David et al 1991), but the final repository of the material from the flow remains to be identified. One interesting example of young stars in a normal old population is the remarkable source P2, which is coincident with the dynamical center of M31 (King et al 1995, Lauer et al 1998). This is slightly extended and considerably bluer than the surrounding UVX population. It has the characteristics of an intermediate-age star cluster, but with $M_V = -5.7$, it can account for only a tiny fraction of recent mass loss by the bulge giants. Its massive stars may have been formed through stellar collisions. The second concentrated nuclear source in M31, denoted P1, is not at the dynamical center and is brighter at optical wavelengths. However, its UV properties are similar to those of the inner bulge. It has been suggested that this is a cannibalized galaxy nucleus in the final stages of consumption by M31. If so, it has managed to clothe itself with a UVX population similar to the bulge stars in M31.

The minority of nearby early-type galaxies that do exhibit evidence for recent star formation (including NGC 205, 5102, and 5253) have probably mostly suffered gas transfer during a recent interaction. UV observations in these cases provide a much improved picture of the massive star population and its history than do optical data (e.g. BBBFL, Wilcots et al 1990, Deharveng et al 1997, Calzetti et al 1997). By contrast with normal E galaxies, recent star formation is often found in early-type galaxies associated with massive cluster cooling flows (reviewed in Fabian 1994). Systems with UV observations include M87, Abell 2199, and NGC 1275 (Perola & Tarenghi 1980, Bertola et al 1982, Bertola et al 1986, BBBFL, McNamara & O'Connell 1989, Smith et al 1992, Dixon et al 1996). The UV is important here in placing better limits on star formation rates (always much smaller than X-ray estimates of total accretion rates) and in exploring possible anomalies in the initial mass function.

Active Nuclei   The flat energy distributions of nuclear nonthermal sources imply that the contrast of an AGN against its surroundings in an E galaxy can improve by a factor up to ∼100 in the UV compared with the optical-IR. This permits better study of known nuclei and searches for very low-luminosity activity. A number of identifications of nuclear point sources have recently been made by UV imaging either of complete samples of nearby galaxies (Maoz et al 1995, 1996) or of samples of objects with Low Ionization Nuclear Emission Region (LINER) optical spectra (Barth et al 1998). Only about 30% of the known LINERs are detected this way, and Maoz and Barth and their respective colleagues suggest that obscuration by dust reduces the visibilty of the other nuclei, at least in the disk galaxies in their samples. However, the UV brightnesses of the nonthermal nuclei support photoionization (rather than shock excitation) models for the LINER emission lines.

In the case of E galaxies with known bright nuclei (e.g. M87), the AGN contributes only a small part of the FUV light within the IUE aperture. From the UIT images of Ohl et al (1998), we find that the nucleus and jet in M87 produce only 10% of the FUV light within a radius of 10″. A similar situation applies to NGC 4278, whose nonthermal nucleus was recently detected by Moller et al (1995). These amounts are, however, sufficient to shift the active galaxies such as M87, NGC 4278, and NGC 1052 slightly in 1500–V vs Mg_2 diagrams such as Figure 6 (as first remarked by BBBFL).

The most interesting case of UV-facilitated observations of an E galaxy AGN is that of NGC 4552. This object has conspicuous radio and infrared signatures of an active nucleus and was originally observed with IUE for that reason (O'Connell et al 1986). Aside from a strong, spatially extended UV-upturn, however, there were no nuclear anomalies obvious until HST imaging was obtained by Renzini et al (1995) and Cappellari et al (1998). The HST observations show a time-variable, unresolved ($r \lesssim 0.07″$) spike of UV light which brightened by a factor of 4.5× between 1991 and 1993. Without the resolution of HST and the improved contrast offered by the UV, it would have been impossible to detect this source, which is currently the least luminous known AGN, having an Hα luminosity of only 6×10^{37} erg s^{-1}. The outburst probably corresponds to the accretion of material stripped from a single star during a close fly-by of the nuclear black hole (Cappellari et al 1998).

9. CONCLUSION

Ultraviolet observations have opened a new, and unexpectedly rich, window on old stellar populations that has revealed phenomena that are either difficult or impossible to study at longer wavelengths. The identification of the UVX component with low-mass, small-envelope stars has led to the recognition that the spectra of distant E galaxies are remarkably sensitive to what in traditional stellar population research would have been regarded as subtle astrophysical processes, including giant-branch mass loss, helium enrichment, and atmospheric diffusion. The fact that these processes are manifestly not properly understood at the moment, precluding a definitive interpretation of the UVX in terms of global population parameters, is less important than the long-term promise of UV observations as powerful probes of galaxy evolution.

ACKNOWLEDGMENTS

For comments, figures, and other help in preparing this paper, I am most grateful to Ralph Bohlin, Tom Brown, Dave Burstein, Daniela Calzetti, Jeff Cranc, Ben Dorman, Harry Ferguson, Ian Freedman, Richard de Grijs, Wayne Landsman, Ray Ohl, Alvio Renzini, Bob Rood, Ted Stecher, and Sukyoung Yi. This work has been supported in part by NASA Long Term Space Astrophysics grant NAG5-6403.

Visit the Annual Reviews home page at http://www.AnnualReviews.org

LITERATURE CITED

Baade W. 1944. *Ap. J.* 100:137–46
Bailyn CD. 1995. *Annu. Rev. Astron. Astrophys.* 33:133–62
Barbaro G, Olivi FM. 1989. *Ap. J.* 337:125–40
Barth AJ, Ho LC, Filippenko AV, Sargent WLW. 1998. *Ap. J.* 496:133–44
Bender R. 1988. *Astron. Astrophys. Lett.* 193:L7–10
Bertelli G, Bressan A, Chiosi C, Fagotto F, Nasi E. 1994. *Astron. Astrophys. Suppl.* 106:275–302
Bertelli G, Chiosi C, Bertola F. 1989. *Ap. J.* 339:889–903
Bertola F, Bressan A, Burstein D, Buson LM, Chiosi C, Di Serego Alighieri S. 1995. *Ap. J.* 438:690–94
Bertola F, Capaccioli M, Holm AV, Oke JB. 1980. *Ap. J.* 237:L65–69
Bertola F, Capaccioli M, Oke JB. 1982. *Ap. J.* 254:494–99
Bertola F, Gregg MD, Gunn JE, Oemler A. 1986. *Ap. J.* 303:624–28
Bica E, Alloin D. 1988. *Astron. Astrophys.* 192:98–106
Bica E, Bonatto C, Pastoriza MG, Alloin D. 1996. *Astron. Astrophys.* 313:405–16
Blöcker T, Schönberner D. 1990. *Astron. Astrophys. Lett.* 240:L11–14
Bohlin RC, Cornett RH, Hill JK, Hill RS, O'Connell RW, Stecher TP. 1985. *Ap. J. Lett.* 298:L37–40
Boksenberg A, Evans RG, Fowler RG, Gardner ISK, Houziaux L, et al. 1973. *MNRAS* 163:291–322
Boggess A, Wilson R. 1987. In *Exploring the Universe with IUE*, ed. Y Kondo pp. 3–20. Dordrecht: Reidel
Bowen GH, Willson LA. 1991. *Ap. J. Lett.* 375:L53–56
Bressan A, Chiosi C, Fagotto F. 1994. *Ap. J. Suppl.* 94:63–115
Bressan A, Chiosi C, Tantalo R. 1996. *Astron. Astrophys.* 311:425–45
Brocato E, Matteucci F, Mazzitelli I, Tornambé A. 1990. *Ap. J.* 349:458–70
Brosch N. 1998, *Exp. Astron.* In press
Brown TM, Ferguson HC, Davidsen AF. 1995. *Ap. J. Lett.* 454:L15–18
Brown TM, Ferguson HC, Davidsen AF, Dorman B. 1997. *Ap. J.* 482:685–707

Brown TM, Ferguson HC, Stanford SA, Deharveng JM. 1998a. *Ap. J.* 504:113–38

Brown TM, Ferguson HC, Deharveng JM, Jedrzejewski RI. 1998b. *Ap. J. Lett.* 508:L139–42

Bruzual AG, Charlot S. 1993. *Ap. J.* 405:538–53

Buonanno R, Corsi C, Bellazzini M, Ferraro FR, Fusi Pecci F. 1997. *Astron. J.* 113:706–12

Burstein D, Bertola F, Buson LM, Faber SM, Lauer TR. 1988. *Ap. J.* 328:440–62. (BBBFL)

Burstein D, Faber SM, Gaskell CM, Krumm N. 1984. *Ap. J.* 287:596–609

Caloi V. 1989. *Astron. Astrophys.* 221:27–35

Calzetti D, Meurer GR, Bohlin RC, Garnett DR, Kinney AL, et al. 1997. *Astron. J.* 114:1834–49

Cappellari M, Renzini A, Greggio L, di Serego Alighieri L, Buson M, et al. 1998. *Ap. J.* In press

Carruthers GR, Heckathorn HM, Opal CB. 1978. *Ap. J.* 225:346–56

Castellani V, Cassatella A. 1987. In *Exploring the Universe with the IUE Satellite*, ed. Y Kondo, pp. 637–54. Dordrecht: Reidel

Castellani M, Castellani V. 1993. *Ap. J.* 407:649–56

Castellani M, Castellani V, Pulone L, Tornambé A. 1994. *Astron. Astrophys.* 282:771–74

Castellani M, Limongi M, Tornambé A. 1992. *Ap. J.* 389:227–33

Castellani M, Tornambé A. 1991. *Ap. J.* 381:393–408

Chiosi C. 1996. In *From Stars to Galaxies: The Impact of Stellar Physics on Galaxy Evolution*, eds. C Leitherer, U Fritze-von Alvensleben, J Huchra, 98:181–92. San Francisco: Astron. Soc. Pacific

Chiosi C, Vallenari A, Bressan A. 1997. *Astron. Astrophys. Suppl.* 121:301–19

Ciardullo RB, Demarque P. 1978. In *The HR Diagram*, eds. AG Davis Philip, DS Hayes, (IAU Symposium No 80), pp. 345–48. Dordrecht: Reidel

Code AD. 1969. *Publ. Astron. Soc. Pac.* 81:475–87

Code AD, Welch GA. 1981. *Ap. J.* 228:95–104

Code AD, Welch GA. 1982. *Ap. J.* 256:1–12

Code AD, Welch GA, Page T. 1972. In *Scientific Results from the Orbiting Astronomical Observatory*, ed. AD Code (NASA SP-310), pp. 559–74

Cole AA, Gallagher JS, Mould JR, Clarke JT, Trauger JT, et al. 1998. *Ap. J.* 505:230–35

Coleman GD, Wu CC, Weedman DW. 1980. *Ap. J. Suppl.* 43:393–416

Cornett RH, O'Connell RW, Greason MR, Offenberg JD, Angione RJ, et al. 1994. *Ap. J.* 426:553–62

Crenshaw DM, Breugman OW, Normal DJ. 1990. *Publ. Astron. Soc. Pac.* 102:463–77

David L, Forman W, Jones C. 1991. *Ap. J.* 369:121–34

Davidsen AF, Ferguson HC. 1992. In *Physics of Nearby Galaxies: Nature or Nurture?*, eds. TX Thuan, C Balkowski, JTT Van, pp. 125–37. Paris: Editions Frontieres

D'Cruz NL, Dorman B, Rood RT, O'Connell RW. 1996. *Ap. J.* 466:359–71

Dean CA, Bruhweiler FC. 1985. *Ap. J. Suppl.* 57:133–43

de Boer KS. 1982. *Astron. Astrophys. Suppl.* 50:247–50

de Boer KS. 1985. *Astron. Astrophys.* 142:321–32

de Boer KS. 1987. In *The Second Conference on Faint Blue Stars*, eds. AGD Philip, DS Hayes, JW Liebert, (IAU Colloquium No 95), pp. 95–104. Schenectady, NY: L Davis Press

Deharveng JM, Jakobsen P, Milliard B, Laget M. 1980. *Astron. Astrophys.* 88:52–57

Deharveng JM, Jedrzejewski R, Crane P, Disney MJ, Rocca-Volmerange B. 1997. *Astron. Astrophys.* 326:528–36

Deharveng JM, Joubert M, Monnet G, Donas J. 1982. *Astron. Astrophys.* 106:16–20

Deharveng JM, Laget M, Monnet G, Vuillemin A. 1976. *Astron. Astrophys.* 50:371–75

Deharveng JM, Sasseen TP, Buat V, Bowyer S, Lampton M, Wu X. 1994. *Astron. Astrophys.* 289:715–28

Dixon WVD, Davidsen AF, Ferguson HC. 1996. *Astron. J.* 111:130–39

Donas J, Milliard B, Laget M. 1995. *Astron. Astrophys.* 303:661–72

Dorman B. 1992. *Ap. J. Suppl.* 81:221–50

Dorman B, O'Connell RW, Rood RT. 1995. *Ap. J.* 442:105–41. (DOR)

Dorman B, O'Connell RW, Rood RT. 1999. In preparation

Dorman B, Rood RT, O'Connell RW. 1993. *Ap. J.* 419:596–614

Dorman B, Shah RY, O'Connell RW, Landsman WB, Rood RT, et al. 1997. *Ap. J. Lett.* 480:L31–34

Dupree AK. 1986. *Annu. Rev. Astron. Astrophys.* 24:377–420

Faber SM. 1983. *Highlights of Astronomy* 6: 165–71

Faber SM, Trager SC, Gonzalez JJ, Worthey G. 1995. In *Stellar Populations*, ed. P van der Kruit, G Gilmore. IAU Symposium 164, pp. 249–57. Dordrecht: Kluwer

Faber SM, Tremaine S, Ajhar EA, Byun YI, Dressler A, et al. 1997. *Astron. J.* 114:1771–96

Fabian AC. 1994. *Annu. Rev. Astron. Astrophys.* 32:277–318

Fanelli MN, O'Connell RW, Burstein D, Wu CC. 1992. *Ap. J. Suppl.* 82:197–245

Ferguson HC, Davidsen AF, Kriss GA, Blair WP, Bowers CW, et al. 1991. *Ap. J.* 382:L69–73

Ferguson HC, Davidsen AF. 1993. *Ap. J.* 408: 92–107

Ferraro FR, Paltrinieri B, Fusi Pecci F, Rood RT, Dorman B. 1998. *Ap. J.* 500:311–19

Forman W, Jones C, Tucker W. 1985. *Ap. J.* 293:102–19

Freedman WL. 1992. *Astron. J.* 104:1349–59

Fusi Pecci F, Ferraro FR, Bellazzini M, Djorgovski S, Piotto G, Buonanno R. 1993. *Astron. J.* 105:1145–68

Gallagher JS. 1972. *Astron. J.* 77:568–72

Gingold RA. 1976. *Ap. J.* 204:116–30

Green EM. 1999. In *3rd Conference on Faint Blue Stars*, ed. AGD Philip. Schenectady: Davis. In press

Greenstein JL, Sargent AI. 1974. *Ap. J. Suppl.* 28:157–200

Greggio L, Renzini A. 1990. *Ap. J.* 364:35–64. (GR)

Greggio L, Renzini A. 1999. In "UV Astronomy in Italy," ed LM Buson, D DeMartino. *Mem. S. A. Italia.* In press

Grillmair CJ, Lauer TR, Worthey G, Faber SM, Freedman WL, et al. 1996. *Astron. J.* 112: 1975–87

Gunn JE, Stryker LL, Tinsley BM. 1981. *Ap. J.* 249:48–67

Hardy E, Couture J, Couture C, Joncas G. 1994. *Astron. J.* 107:195–205

Heap SR, Brown TM, Hubeny I, Landsman W, Yi S, et al. 1998. *Ap. J. Lett.* 492:L131–34

Heber U. 1992. In *Atmospheres of Early-Type Stars*, ed. U Heber, CS Jeffrey, p. 233. Berlin: Springer

Hills JG. 1971. *Astron. Astrophys.* 12:1–4

Horch E, Demarque P, Pinsonneault M. 1992. *Ap. J. Lett.* 388:L53–56

Iben I, Rood RT. 1970. *Ap. J.* 161:587–617

Izotov YI, Thuan TX. 1998. *Ap. J.* 500:188–216

Jaffe W, Ford HC, O'Connell RW, van den Bosch F, Ferrarese L. 1994. *Astron. J.* 108: 1567–78

Johnson HM. 1979. *Ap. J. Lett.* 230:L137–40

Jørgensen UG, Thejll P. 1993. *Ap. J. Lett.* 411: L67–70

Joseph CL. 1995. *Exp. Astron.* 6:97–127

Kaluzny J, Rucinski SM. 1995. *Astron. Astrophys. Suppl.* 114:1–20

Kent SM. 1983. *Ap. J.* 266:562–67

King CR, Ellis RS. 1985. *Ap. J.* 288:456–64

King IR, Deharveng JM, Albrecht R, Barbieri C, Blades JC, et al. 1992. *Ap. J. Lett.* 397: L35–38

King IR, Stanford SA, Crane P. 1995. *Astron. J.* 109:164–72

Kinney AL, Bohlin RC, Calzetti D, Panagia N, Wyse RFG. 1993. *Ap. J. Suppl.* 86:5–93

Kinney AL, Calzetti D, Bohlin RC, McQuade K, Storchi-Bergmann T, Schmitt HR. 1996. *Ap. J.* 467:38–60

Kodaira K, Watanabe T, Onaka T, Tanaka W. 1990. *Ap. J.* 363:422–34

Kondo Y, ed. 1987. *Exploring the Universe with the IUE Satellite.* Dordrecht: Reidel. 787 pp.

Kruk JW, Durrance ST, Kriss GA, Davidsen AF, Blair WP, Espey BR. 1995. *Ap. J. Lett.* 454:L1–4

Kurucz RL. 1991. In *Precision Photometry: Astrophysics of the Galaxy*, ed. AGD Philip, AR Upgren, KA Janes, pp. 27–44. Schenectady: Davis

Landsman WB, Bohlin RC, Neff SG, O'Connell RW, Roberts MS, et al. 1998. *Astron. J.* 116:789–800

Landsman WB, Sweigart AV, Bohlin RC, Neff SG, O'Connell RW, et al. 1996. *Ap. J. Lett.* 472:L93–96

Lauer TR, Faber SM, Ajhar EA, Grillmair CJ, Scowen PA. 1998. *Astron. J.* 116:2263–86

Lee YW. 1994. *Ap. J. Lett.* 430:L113–16

Liebert J, Saffer RA, Green EM. 1994. *Ap. J.* 107:1408–21

Longo G, Capaccioli M, Bender R, Busarello G. 1989. *Astron. Astrophys. Lett.* 225:L17–19

Magris G, Bruzual G. 1993. *Ap. J.* 417:102–11

Maoz D, Filippenko AV, Ho LC, Macchetto FD, Rix HW, Schneider DP. 1996. *Ap. J. Suppl.* 107:215–26

Maoz D, Filippenko AV, Ho LC, Rix HW, Bahcall JN, et al. 1995. *Ap. J.* 440:91–99

Martin C, Friedman P, Schiminovich D, Madore B, Bianchi L, et al. 1997. *Bull. Amer. Astron. Soc.* 191:#63.04

McNamara BR, O'Connell RW. 1989. *Astron. J.* 98:2018–43

McWilliam A. 1997. *Annu. Rev. Astron. Astrophys.* 35:503–56

Mengel JG, Norris J, Gross PG. 1976. *Ap. J.* 204:488–92

Milliard B, Donas J, Laget M, Armand C, Vuillemin A. 1992. *Astron. Astrophys.* 257:24–30

Minkowski R, Osterbrock D. 1959 *Ap. J.* 129:583–595

Moller P, Stiavelli M, Zeilinger WW. 1995. *MNRAS* 276:979–1002

Nesci R, Perola GC. 1985. *Astron. Astrophys.* 145:296–304

Nörgaard-Nielsen HU, Kjærgaard P. 1981. *Astron. Astrophys.* 93:290–96

O'Connell DJK, ed. 1958. *Stellar Populations*, Amsterdam: North Holland. 544 pp.

O'Connell RW. 1976. *Ap. J.* 206:370–90

O'Connell RW. 1980. *Ap. J.* 236:430–40

O'Connell RW. 1987. *Astron. J.* 94:876–82

O'Connell RW. 1991. *Adv. Space Research* 11 No. 11:71–80

O'Connell RW, Bohlin RC, Collins NR, Cornett RH, Hill JK, et al. 1992. *Ap. J. Lett.* 395:L45–L48

O'Connell RW, Dorman B, Shah RY, Rood RT, Landsman WB, et al. 1997. *Astron. J.* 114:1982–91

O'Connell RW, Thuan TX, Puschell JJ. 1986. *Ap. J. Lett.* 363:L37–L40

Ohl RG, O'Connell RW, Bohlin RC, Collins NR, Dorman B, et al. 1998. *Ap. J. Lett.* 505:L11–L14

Oke JB, Bertola F, Capccioli M. 1981. *Ap. J.* 243:453–59

Onaka T, Tanaka W, Watanabe T, Watanabe J, Yamaguchi A, et al. 1989. *Ap. J.* 342:238–49

Park JH, Lee YW. 1997. *Ap. J.* 476:28–39

Peletier RF, Davies RL, Illingworth GD, Davis LE, Cawson M. 1990. *Astron. J.* 100:1091–1142

Pence W. 1976. *Ap. J.* 203:39–51

Perola GC, Tarenghi M. 1980. *Ap. J.* 240:447–54

Ponder J, Burstein D, O'Connell RW, Rose J, Frogel JA, et al. 1998. *Astron. J.* 116:2297–314

Renzini A. 1981a. *Ann. Phys. Fr.* 6:87–102

Renzini A. 1981b. In *Effects of Mass Loss on Stellar Evolution*, eds. C Chiosi, R Stalio. pp. 319–38. Dordrecht: Reidel

Renzini A, Buzzoni A. 1986. In *Spectral Evolution of Galaxies*, eds. C Chiosi, A Renzini, pp. 195–235. Dordrecht: Reidel

Renzini A, Greggio L, di Serego Alighieri S, Cappellari M, Burstein D, Bertola F. 1995. *Nature* 378:39–41

Rich RM, Sosin C, Djorgovski SG, Piotto G,

King IR, et al. 1997. *Ap. J. Lett.* 484:L25–28
Rifatto A, Longo G, Capaccioli M. 1995a. *Astron. Astrophys. Suppl.* 109:341–45
Rifatto A, Longo G, Capaccioli M. 1995b. *Astron. Astrophys. Suppl.* 114:527–36
Rocca-Volmerange B, Guiderdoni B. 1987. *Astron. Astrophys.* 175:15–22
Rocca-Volmerange B, Guiderdoni B. 1988. *Astron. Astrophys. Suppl.* 75:93–106
Rood RT. 1973. *Ap. J.* 184:815–38
Rose JA. 1985. *Astron. J.* 90:1927–56
Rose JA. 1994. *Astron. J.* 107:206–29
Rose WK, Tinsley BM. 1974. *Ap. J.* 190:243–51
Saffer RA, Bergeron P, Koester D, Liebert J. 1994. *Ap. J.* 432:351–66
Saffer RA, Liebert J. 1995. In *Proc. of the 9th European Workshop on White Dwarfs*, ed. D Koester, K. Werner, pp. 221–32. Berlin: Springer
Sandage AR, Visvanathan N. 1978. *Ap. J.* 225:742–50
Sarazin CL, White RE. 1988. *Ap. J.* 331:102–15
Schönberner D. 1983. *Ap. J.* 272:708–14
Smith AM, Cornett RH. 1982. *Ap. J.* 261:1–11
Smith EP, O'Connell RW, Bohlin RC, Cheng KP, Cornett RH, et al. 1992. *Ap. J. Lett.* 395:L49–54
Sosin C, Dorman B, Djorgovski SG, Piotto G, Rich RM, et al. 1997. *Ap. J. Lett.* 480:L35–38
Spinrad H, Dey A, Stern D, Dunlop J, Peacock J, et al. 1997. *Ap. J.* 484:581–601
Stanford SA, Eisenhardt PR, Dickinson M. 1998. *Ap. J.* 492:461–79
Stecher TP, Cornett RH, Greason MR, Landsman WB, Hill JK, et al. 1997. *Publ. Astron. Soc. Pac.* 109:584–99
Sweigart AV. 1987. *Ap. J. Suppl.* 65:95–135
Sweigart AV. 1997. *Ap. J. Lett.* 474:L23–26
Sweigart AV, Mengel JG, Demarque P. 1974. *Astron. Astrophys.* 30:13–19
Tantalo R, Chiosi C, Bressan A, Fagotto F. 1996. *Astron. Astrophys.* 311:361–83
Thejll P, Flynn C, Williamson R, Saffer R. 1997. *Astron. Astrophys.* 317:689–93
Tinsley BM. 1972a. In *Scientific Results from the Orbiting Astronomical Observatory*, ed. AD Code, (NASA SP-310), pp. 575–81
Tinsley BM. 1972b. *Ap. J.* 178:319–36
Tinsley BM. 1980. *Fund. Cosmic Phys.* 5:287
Tonry J, Schneider DP. 1988. *Astron. J.* 96:807–15
Trager SC, Worthey G, Faber SM, Burstein D, González JJ. 1998. *Ap. J. Suppl.* 116:1–28
Treyer MA, Ellis RS, Milliard B, Donas J, Bridges TJ. 1998. *MNRAS* 300:303–14
van Albada TS, de Boer KS, Dickens RJ. 1981. *MNRAS* 195:591–606
Vassiliadis E, Wood PR. 1994. *Ap. J. Suppl.* 92:125–44
Welch GA. 1982. *Ap. J.* 259:77–88
Whitney JH, O'Connell RW, Rood RT, Dorman B, Landsman WB, et al. 1994. *Astron. J.* 108:1350–63
Whitney JH, Rood RT, O'Connell RW, D'Cruz NL, Dorman B, et al. 1998. *Ap. J.* 495:284–96
Wilcots EM, Hodge P, Eskridge PB, Bertola F, Buson L. 1990. *Ap. J.* 364:87–93
Willson LA, Bowen GH, Struck C. 1996. In *From Stars to Galaxies: The Impact of Stellar Physics on Galaxy Evolution*, eds. C Leitherer, U Fritze-von Alvensleben, J Huchra, 98:197–201. San Francisco: Astron. Soc. Pacific
Windhorst RA, Pascarelle SM, Keel WC, Bertola B, McCarthy PJ, et al. 1994. In *Frontiers of Space and Ground-Based Astronomy*, eds. EW Wamsteker, MS Longair, Y Kondo, pp. 663–67. Dordrecht: Kluwer
Wilson TL, Rood RT. 1994. *Annu. Rev. Astron. Astrophys.* 32:191–226
Worthey G. 1993. *Ap. J. Lett.* 415:L91–94
Worthey G. 1994. *Ap. J. Suppl.* 95:107–49
Worthey G. 1998. *Publ. Astron. Soc. Pac.* 110:888–99
Worthey G, Dorman B, Jones LA. 1996. *Astron. J.* 112:948–53
Wu CC, Faber SM, Gallagher JS, Peck M, Tinsley BM. 1980. *Ap. J.* 237:290–302
Wu CC, Reichert GA, Ake TB, Boggess A, Holm AV, et al. 1992. *IUE Ultraviolet Spec-*

tral Atlas of Selected Astronomical Objects. (NASA Reference Publication No. 1285).

Wyse RFG, Gilmore G, Franx M. 1997. *Annu. Rev. Astron. Astrophys.* 35:637–75

Yi S, Demarque P, Kim YC. 1997a. *Ap. J.* 482:677–84

Yi S, Demarque P, Oemler A. 1995. *Publ. Astron. Soc. Pac.* 107:273–78

Yi S, Demarque P, Oemler A. 1997b. *Ap. J.* 486:201–29

Yi S, Lee YW, Woo JH, Park JH, Demarque P, Oemler A. 1999. *Ap. J.* 513:128–41

SUBJECT INDEX

A
Absorption
  See Broad absorption lines (BALs); Broad absorption lines region (BALRs); Narrow absorption lines (NALs)
Absorption line
  diagnostics, 504–16
Abundance(s)
  abundance scale, 199–200
  analysis, 492–95, 506–12
  and calibrating stars, 206
  cluster, 204–7
  diagnostics, 499–504
  of evolved stars, 271–73
  heavy-element
    in asymptotic giant branch (AGB) stars, 252
  iron
    and cluster giants, 205
  predictions, 519–20
  in quasars, 487–524
  ratios, 494–95
  and red clump stars, 219
  and red giants, 227
  and RR Lyrae stars, 222, 228, 229
  scaled-solar, 214
  in stellar systems, 225
  and subdwarfs, 204
AC114
  dark matter in, 137, 138
  lensed images of, 133
Accretion
  and classical T Tauri (CTT) stars, 366, 380, 384, 385
  through disks, 342
  flow instabilities
    and microquasars, 438

and Ganymede and Callisto, 563
and Moon formation, 539, 551–52, 555, 556
and Phobos and Deimos, 557
and planet formation, 536–37, 538
rates
  and massive stars, 346
  and satellites of Mars, 539, 596
  and star formation, 343, 352, 353
  and stellar growth, 366–67, 368, 370
  of the terrestrial planets, 539
  and young stellar objects (YSOs), 370
Accretion disk(s), 409
  and black holes, 410, 411
  and collimated ejecta, 410
  and ejection of plasma clouds, 425
  and jet formation, 424–27, 428
  around 7 keV, 431
  and relativistic jets, 409, 438, 439
Advanced X-ray Astrophysics Facility (AXAF), 400
Age(s)
  age-color relation, 634
  globular cluster, 212–14
  and mid-UV, 637
  and UVX, 627, 634, 636–37
Air navigation
  and frequency allocation, 88

Aircraft
  interference from
    and radio astronomy, 76–77
Ambipolar diffusion
  and cores, 334
  and star formation, 326, 331, 332, 341
Andromeda nebula, 452, 455
Angular displacements
  and ejection events, 416, 417
Angular momentum
  and the Earth-Moon system, 31–33, 539, 551, 552
  and jet formation, 427
  and natural satellites, 544, 547–49, 553, 556, 564, 578
  and planet formation, 537, 538
  and star formation, 326
Arc(let)s
  mass reconstruction with, 135–38
  redshift of
    and weak lensing, 169
  spectral content of, 173–74
  spectroscopic surveys of, 166–68
  statistics of, 176–78
  and weak lensing, 129
Ariel
  satellite of Uranus, 580–84, 587–89, 596
Aspen Center for Physics, 17
Asteroids
  discovery of Icarus, 451
  Greek names for, 451n.1

649

650 SUBJECT INDEX

Astrometric measurements
  accuracy of, 97, 98
Astronomical Netherlands Satellite, 607
Astronomy
  adverse environmental impacts on, 90
  early years of, 450
  interference with, 71
  and the Mount Wilson and Palomar Observatories, 445–80
  and Project WEST FORD, 69
  radio, 99, 446, 450, 474, 476–77
    and allocation of frequencies, 65–95
  UV, 605
  X-ray, 446, 450
Astronomy Survey Committee, 18
Astrophysics
  early years of, 450
  high energy, 472, 474–80
  workshops, 17
Asymptotic giant branch (AGB) stars
  See Star(s), asymptotic giant branch (AGB)
Atacama Large Millimeter Array, 400
Atlas, 571
Atmosphere
  of Jupiter and Saturn, 537
  and planet formation, 537, 538
  and radio sources, 114
Atmospheric effects
  and weak lensing, 146
Australia
  and Motorola IRIDIUM satellites, 89–90
Autocorrelation function (ACF), 146, 148
Automated Plate Machine (APM) survey, 129

A370
  and gravitational lensing, 136
  lensed images of, 133
  lensing-cluster, 139
  radial arcs in, 137
A1942
  shear maps in, 146
A2218
  and gravitational lensing, 135
  lensing cluster, 129
  and weak lensing, 144
A2390
  lensed images of, 133
  X-ray and lensing masses in, 136

B

Baade, W, 444–80 passim
Balbus-Hawley effect, 28
Barium (Ba) isotopes
  and asymptotic giant branch (AGB) stars, 256, 257
Baum, W. A., 453–54, 456
Belfer Graduate School of Science, 12, 14–15
Betatron, 2–3
Biermann batter mechanism
  and the galactic dynamo, 39, 56–58, 61
Binaries
  and UVX, 630–31
  and UVX sources, 623
  X-ray, 409
Black hole(s), 8, 424
  accreting, 413
  and accretion disks, 410, 411
  binaries, 414, 424, 432
  and extragalactic radio sources, 109
  and jet formation, 427, 428
  mass and span, 430–31

  spin of, 409, 439
  stellar mass, 438
  and superluminal jets, 413–14
  systems
    X-ray luminosities in, 414
  X-ray binaries, 409
Born approximation
  and weak lensing, 162
Bowen, I, 446, 447–49
BP Tau
  emission line variations, 379
Broad absorption lines (BALs)
  and quasars, 489, 490, 502–4, 509, 512–14, 516–17
Broad absorption lines region (BALRs)
  and quasars, 502, 503, 509, 513, 514
Broad emission lines (BELs)
  origin of, 491–92
  and quasars, 489, 490, 494, 497, 502, 503, 516, 517, 523, 524
Broad emission lines region (BELR)
  cloud
    ionization structure for, 495, 496
  and quasars, 491, 494, 495, 497, 498, 500, 501, 503, 517
Broadcasting Satellite Service
  and radio frequency allocation, 74, 82
Business
  and the radio spectrum, 94
B335, 339–41

C

Calculating machinery
  and nuclear astrophysics, 4–5

SUBJECT INDEX   651

Calculation
  astrophysical
    and computers, 23
Callisto
  Jupiter's satellite, 561–62,
    565–66, 570, 596
Caltech, 10
Calypso, 570, 572
Camera
  HST/Advanced Camera for
    Surveys, 606
  HST Faint Object Camera
    (FOC), 606
  HST Wide Field Camera
    (WFPC2), 606
Cameron, A.G.W., 1–34
Canada-France-Hawaii
  Telescope (CFHT),
    129
Canadian Atomic Energy
  Project, 4
Canadian National Research
  Council, 12
Canadian radio astronomy
  and SSU series satellites,
    78
Carbon (C)
  burning
    and thermonuclear
      reactions, 4
  isotopes
    and stellar evolution,
      244–74 passim
    and stellar burning stages,
      6–7
Carbon monoxide
  as a tracer of molecular
    gas, 318, 319
  as a tracer of temperature,
    319–20
Carbonaceous chondrite
  See Chondrites,
    carbonaceous
Carnegie Observatories,
  445
Cassini spacecraft,
  571

Cassini state(s), 550–52
  and Iapetus, 571
Catalog(s)
  3CR radio, 477
  FK3 and FK4, 103
  Fundamental
    FK-series of, 102–3
  Hipparcos, 115–17, 119
  Humason-Mayall Redshift
    Catalog, 465–67
  Nearby Star, 200
  and the radio reference
    frame, 105
  of radio source positions,
    106
Catalogue
  Hipparcos astrometric,
    191–93, 200, 202, 209,
    216
Center for Astrophysics, 15,
  21–23
Cepheid(s), 230
  classical, 452–56
  as distance indicators, 472
  and the extragalactic
    distance scale, 191
  galactic distance scale
    from, 118
  in galaxies beyond the
    Local Group, 471
  LMC and SMC, 231
  luminosities assigned to,
    455–56
  in M31, 456
  P-L relation, 446, 452, 453,
    463
  calibration of, 454, 456,
    457, 462
  U Sgr, 462
  variables
    as distance indicators,
      216–18
Cesium (Cs)
  cesium telluride
    photocathodes, 606,
    607
  iodide, 606

isotopes
  and asymptotic giant
    branch (AGB) stars, 256,
    257
Chamaeleon I star-forming
  cloud, 396, 397
  classical T Tauri (CTT)
    stars in, 376
  ROSAT soft X-ray images
    of, 375
Chamaeleon star-forming
  clouds
  ROSAT All-Sky Survey
    around, 398
Chandra X-ray Observatory,
  400
Charon
  origin of, 538
  Pluto-Charon
    rotating synchronously,
      550
  Pluto-Charon system,
    593–95, 597
    origin of, 539
    and special circumstances,
      534
Chemical abundances
  and UVX, 634, 636–37
Chemical composition
  of stars, 452–53, 458–61
Chemical evolution
  galactic, 517–21
  of the galaxy
    and metallicity, 267–77
Chemical variations
  among the stars, 457
Chondrite(s)
  carbonaceous, 9, 393
  and the satellites of
    Mars, 539
  formation time, 279
Chondrule(s)
  formation, 392
  and remanent magnetism,
    392–93
  shocks may melt, 400
Circinus X-1, 432, 433

Civil service, 13, 14
Classical T Tauri (CTT) stars, 365, 367–69
 and accretion, 366, 380, 384, 385
 infrared designation of, 368
 intense emission from, 366
 and magnetic activity, 379–80, 384
 and outflows, 366–68, 370, 380, 389
 and X-ray emissions, 374, 384
 X-ray properties, 376
Cloudy numerical code, 490
Cloverleaf
 cluster around, 155
Clumps
 and star formation, 347–49, 352, 353
Cluster(s)
 abundances, 204–6
 distance determinations post-Hipparcos, 209, 212
 distance scale, 214
  pre-Hipparcos, 207–9
 distances, 206–12
 faintest, 469
 galaxies, 181
  as distance indicators, 472
 and galaxies, 619–23, 630
 globular
  distance determinations for stars in, 233
  distances to, 204–16
 high-redshift, 145
 lensing, 136, 145, 170, 174, 176–78
 low-mass, 182
 mass reconstruction of, 142
 and mid-UV spectral features, 637
 open, 218
  distances to, 196–98
  probing the clumpiness of, 137–38

reconstruction
 constraints from, 175–76
 and UVX, 621–23
variables
 main-sequence fitting calibration of, 226
See also Globular cluster(s)
Clustering
 source
  and weak lensing, 161
Cold Dark Matter (CDM) model, 145
Collapse
 gravitational
  and star formation, 326
 and star formation, 312–14, 332–34, 337–41, 343, 346, 350–52
Color(s)
 age-color relation, 634
 and cluster abundance, 207
 interpretation of UV spectra and, 631–34
 and luminosity, 214, 217
 and metallicity, 198, 219
 and MS fitting, 207
Columbia University, 12
Column densities, 491, 513
 and abundance analysis, 509–10
Column density
 and abundance analysis, 509–10
 gas
  and star formation, 317
  in the outer regions of dark clouds, 335
 and photoionization simulations, 495
 and star formation, 315, 316, 318, 334, 346, 347, 352
Coma
 and Hipparcos, 198
Coma Berenices
 and Hipparcos, 198

Coma cluster
 its visible and virial mass, 128
Comet Haley
 and the VEGA mission, 82, 83
Comite Consultatif International des Radiocommunications (CCIR), 67, 68, 72, 74, 79, 81, 85–88
Committee on Planetary and Lunar Exploration (COMPLEX), 18–20
Committee on Radio Astronomy Frequencies (CRAF)
 foundation of, 85–86
Committee on Radio Frequencies (CORF), 72, 80, 86
Communist Bloc
 and radio frequency allocations, 78
Computers
 and A. Cameron, 5–6
Computing technology
 evolution of, 22–25
Continuous Observation for Rotation of the Earth (CORE) program, 115
Continuum emission
 from dust, 334
Cool bottom processing, 247
Core(s)
 and ambipolar diffusion, 334
 with central objects, 335
 clumpy, 347
 and collapse, 338
 collapsing, 341
 defined, 347
 formation, 332, 333
 hot, 349, 351
 and isolated star formation, 327–31
 maps, 345

SUBJECT INDEX 653

massive, 346–48, 350, 352
massive and isolated, 348
and massive stars, 343
and mid-UV spectral
  features, 637
in the Orion clouds, 344
and star formation, 311,
  352, 353
with submillimeter
  emission, 332
in Taurus, 344
Corolis force
  and the galactic dynamo,
  60
Corona Australis cloud
  CrA IRS 7 in, 380
  IRS 5 in, 383
  See also CrA IRS5; CrA
  IRS 7
Cosmic ray(s)
  cosmic-ray heating, 328
  of gas, 320
  galactic
    ionization from, 390
  and galactic dynamos, 50,
    61
  galactic ionization, 364
  and the galactic magnetic
    field, 59
  ionization rate, 388
Cosmological Background
  Explorer (COBE) maps,
  178–79
Cosmological parameters
  and gravitational lensing,
  127, 175–78
Cosmological theory,
  463
Cosmology, 463
  and gravitational lensing,
  128
  observational, 450
  and Palomar, 451
COSPAR, 69, 78, 83
CrA IRS 5, 383, 384
CrA IRS 7, 380, 381, 383,
  399, 400

Crab Nebula, 41, 56, 57, 474
  optical polarization in, 475
Cratering
  and satellites of Uranus,
  581
  and Saturn's satellites, 570
Craters
  impact
    and natural satellites,
    544
Cyclonic motion(s)
  and galactic dynamo
  theory, 38–45, 48, 59, 60
Cygnus A, 104
Cygnus X-1, 411
  radio features, 432, 434
Cygnus X-3, 432, 433
  energy outbursts, 42
C1 Cam, 433
C10024+1654
  and weak lensing, 171
C11358+6245
  giant arc detected in, 167
C1 Cam, 433
3C 48, 476–77
3C 273
  redshift of, 477
3C 324, 155
3CR
  identifications, 478
  radio catalog, 477
  radio sources, 476

D

Dark matter
  galactic
    halos of, 163–64
  of galaxies, 162
  and halos of cluster
    galaxies, 138
  weak lensing as a probe of,
  127
  and x-ray dynamics, 136
DD Tau, 376
Deimos, 595–96
  and Mars, 557–59
  origin of, 544

Density
  and star formation, 315,
  319–22, 336, 352
Deuterium
  and radio frequency
    allocations, 81
  and solar nebula, 28
Differential Microwave
  Radiometer (DMR), 179
Differential rotation
  and galactic dynamo
  theory, 37, 39–43
Diffusion
  and dynamo theory, 37, 38,
  40, 41, 45, 50–52
Digital radio broadcasting,
  86, 87
Dione
  Saturn's satellite, 570–73,
  575–78, 596
Disk(s)
  accretion through, 342
  around black holes, 410
  circumstellar, 366
    and Class III objects,
    377
  longevity of, 365
  X-ray ionization for,
    388–90
  emission, 335
  evolution, 353
  and Herbig Be stars,
  349
  Keplerian, 335
  lifetimes, 397
  local
    HR diagram of, 200–2
  in massive regions, 346
  around massive stars, 343
  and natural satellite
    formation, 538
  and planet formation, 534,
  535
  planetesimal, 537
  protoplanetary, 364, 366
  stars, 202
  thin, 333

and young stellar object
  (YSO) magnetic activity,
  386, 387
and young stellar objects
  (YSOs), 366, 368
Distance
  to the galactic center,
    229–30
  indicators, 216–27
  to the Large Magellanic
    Cloud, 230–31
Distance scale, 463
  extragalactic, 216, 471–72
  Hubble's "remote," 471
  local, 227, 229–32
Distortion
  and gravitational lensing,
    134
DIVA, 232
DoAr 21, 376, 378, 379
Doppler boosting
  and relativistic jets, 416,
    418, 419, 423
Doppler shift(s), 322, 337
Dust
  continuum emission, 325,
    334, 351
  continuum emission maps,
    349
  continuum maps, 348
  emission, 342
  heated by photons, 320
  and star formation, 316–18
Dwarf(s), 452
  brown, 378, 397, 400
  white, 271
    binary companions, 631
    evolution, 625
    photometry for, 208
Dynamo(s)
  α-ω, 384, 385
  criticisms of the theory of,
    38, 49–55
  dynamo theory
    and galactic magnetic
      fields, 39–42
    equation, 38, 42–49

galactic, 37–62
magnetic
  and cool stars, 384
  saturation of, 60, 62

E

E-process
  environment in,
    25–26
Earth
  Earth-Moon
    and natural satellites,
      595
    and the simple tide,
      547
  Earth-Moon system,
    551–56, 595
  formation of, 31–33
  rotation
    International Earth
      Rotation Service (IERS),
      105, 114–15, 117, 121
    Monitor Earth Rotation
      and Intercompare
      Techniques (MERIT),
      105–6
Earth Exploration Service
  and World Radio
    Conference (WRC) 2000,
    94
Eastern Europe
  and Inter-Union
    Commission on the
    Allocation of Frequencies
    (IUCAF), 85
  and radio frequency
    allocations, 78
ECHO-1, 74
Einstein rings
  lensing events, 163
Elemental abundances
  and meteorites, 9–10
  and quasars, 487–524
  See also Abundance(s)
Elements
  lighter
    and UVX, 639

Ellipticity
  of galaxies
    and weak lensing,
      146–47, 150, 151, 157,
      158, 169
Emission line
  diagnosis, 490–504
  See also Broad emission
    lines (BELs); Broad
    emission lines region
    (BELR)
Enceladus
  Saturn's satellite, 572, 573,
    575–78, 596
Encke gap
  in Saturn's A-ring,
    571
Epimetheus
  Saturn's satellite,
    570–72
Equation(s)
  dynamo, 38, 42–49
  of state, 212
Eta Cha cluster, 397
Europa, 575
  Jupiter's satellite, 560, 561,
    563, 564, 568–70
  and tidal dissipation,
    544
European bloc
  and World Administrative
    Radio Conference
    (WARC) 1992, 87
European Science Foundation
  (ESF), 86
European Space Agency
  (ESA)
    Hipparcos Space
      Astrometry mission, 97,
      99, 118
Extinct radioactivities
  and solar system
    formation, 30
"Extraterrestrial Radio
  Noise," 67
1E1740.7-2942, 389, 411,
  412, 432–34

SUBJECT INDEX

F
FAME, 118, 119
Far-ultraviolet radiation
　See Ultraviolet (UV)
Field lines
　and dynamo theory,
　　45
　and galactic dynamo
　　theory, 37, 49–50, 52
Fifth Fundamental Catalog
　(FK5), 101
Filaments
　and star formation, 347,
　　349
Flares
　magnetic
　　and weak-lined T Tauri
　　(WTT) stars, 378
　radio continuum, 374
　solar, 374
　in T Tauri stars, 374–75
　X-ray
　　protostellar, 383
　　in young stellar objects
　　(YSOs), 371, 384, 387
　young stellar object (YSO),
　　400
Flaring
　and young stellar objects
　　(YSOs), 364, 365
Fornax globular clusters,
　451
Fortran, 14
Frequencies
　allocation
　　and radio astronomy,
　　65–95

G
GAIA
　See Global Astrometric
　Interferometer for
　Astrophysics (GAIA)
Galactic center
　distance to, 229–30
Galactic distance scale,
　191–233

Galaxies
　brightest in clusters, 463
　cluster
　　as distance indicators,
　　472
　and clusters, 517, 619–23,
　　630
　clusters of
　　mass distribution in,
　　135–50
　correlated polarization of,
　　162
　distant
　　weak lensing study of,
　　127
　as ellipses, 132
　elliptical
　　far-ultraviolet (FUV)
　　radiation from, 603–43
　faint
　　and magnification bias,
　　170–73
　and galaxy-galaxy lensing,
　　162–65
　and gravitational lensing,
　　132
　highly magnified
　　morphology of, 174–75
　large scale distribution of,
　　463–64
　lensed, 134
　and quasars, 489
　redshift distribution of
　　beyond B = 25, 165–73
　starburst, 351, 478–80
　and weak lensing, 126–83
Galaxy
　chemical evolution of,
　　13
　classification, 450
　clusters
　　and redshifts, 465–67,
　　469–71
　count program, 463–64
　distribution, 464
　formation, 463
　model of, 460

Fornax dwarf dE, 451
　morphology, 463
　radio waves from, 66
　relativistic jets in, 409–39
　two-point correlation
　　function, 153, 154, 156,
　　161
GALEX mission, 605
Galilean satellites, 560–63,
　569, 570, 573
Galileo spacecraft, 565,
　569
　flyby data, 563
Gamma-ray bursts
　and relativistic jets,
　　437–39
Ganymede
　Jupiter's satellite, 561–66,
　　568, 570, 596
　orbital and thermal
　　evolution of, 546
　and tidal dissipation, 544
Gas drag
　capturing satellites by, 560
　and planet formation, 537,
　　538
Gaseous atmosphere
　of Jupiter and Saturn, 537
Giant gaseous protoplanets
　(GGPPs), 29
Giants
　cluster
　　and iron abundance, 205
GL2591, 348
Global Astrometric
　Interferometer for
　Astrophysics (GAIA),
　118, 119, 232
Global Navigation Satellite
　System (GLONASS)
　and radio astronomy,
　84–85, 88–89
Global Positioning Satellites
　and radio astronomy, 84
Globular cluster(s), 453, 629
　age-dating of, 460
　"blue tails" in, 630

and discovery of the main
sequence, 456–57
and far-ultraviolet (FUV),
607
globular cluster-type
populations
and UVX, 621–23
M31, 455
main sequence, 460–61
main sequence turn-off of,
457–58
stars
and UV excess, 460
Globules
and isolated star formation,
327–31
Goddard Institute for Space
Studies, 12–14
Gordon Research
Conferences, 15–16
Gould's Belt, 398
Graduate students
at Harvard, 21–22
Gravitational lenses
probing the universe with,
126–83
GRB 990510, 437
Green Book, 80
GRO J1655-40, 430, 433,
437
accretion disk, 431
contains a Kerr black hole,
431
ejecta from, 432
and hard X-rays, 424
radio jets of, 435
superluminal motions in,
413, 414, 419, 420
GRS 1915+105, 421–23,
429–30, 433
accretion disk of, 431
ejecta from, 432
radio emission from, 434
and relativistic jets, 435,
438
superluminal motions in,
413–19

and synchrotron emission,
429–30
and X-ray flux variations,
424–26
GRS 1758-258, 411, 432,
433
and interaction of
relativistic jets with the
environment, 434

H

Hale telescope
mirror of, 446–49
See also Palomar
Halos
of cluster galaxies, 138
of galactic dark matter,
163–64
of galaxies, 162, 165
galaxy
investigation of, 181
Harvard
and A. Cameron, 21–22
Harvard College Observatory,
15, 22
Hayashi tracks
rotational evolution of stars
along, 365
HD 19445, 461
metallicity deficiencies of,
459
HD 103095, 207
HD 140283, 459, 461
HD 155555, 378
HD 283447, 376–78
HD 283472, 378
HDE 283572, 378
Helene, 571, 572
Helium (He)
abundance
and UVX, 627
burning
in stars, 243, 248
and thermonuclear
reactions, 4
enrichment, 603, 643
and UVX, 639

and giant planets, 28
helium burning stars
and far-ultraviolet
(FUV) radiation, 604
and metal abundance,
628
primordial abundance,
204
and solar evolution, 27
and star evolution,
623–26
and star formation, 11
in stars, 247, 458
and stellar evolution, 6,
245, 248–51, 258, 261,
265–67, 271, 273, 289,
293, 623–25, 626
Hendricks, D, 447, 449
Hertzsprung-Russell (HR)
diagram, 452, 455–58,
461
and classical and
weak-lined T Tauri stars,
368
defined by field stars,
198–204
and the galactic distance
scale, 191–233
of a stellar model,
242–44
and UVX, 623, 625
High-altitude platform(s)
and radio astronomy, 94
and World Radio
Conference (WRC) 2000,
93
Highly Advanced Laboratory
for Communication and
Astronomy (HALCA)
space mission, 113
Himalia
Jupiter's satellite, 560
Hipparcos astrometric
catalogue, 191–93, 200,
202, 209, 216
Hipparcos Catalog, 101,
115–17

Hipparcos satellite, 192
  and abundances, 204
  astrometric analyzes,
    216–17
  astrometry, 220
  astrometry for Galactic
    Miras, 227
  data
    and the Hyades, 197
    for Lowell
      proper-motion stars, 202
  dataset, 195
  extension of, 232
  and the
    Hertzsprung-Russell (HR)
    diagram, 191–233
  MS-fitting, 223
  optical reference frame,
    103, 108, 115–17, 119,
    121
  and the Pleiades, 197–98
  and the Solar
    Neighborhood
    population, 218
  stars, 118–19
Historian
  and scientists, 1, 2
Hopkins Ultraviolet
  Telescope (HUT)
  Astro, 605, 606, 617
  far-UV spectra, 632,
    633
  data for globulars, 621
  spectra, 637
  spectroscopy, 632
Hot-flash mechanism
  in stars, 629–30
HR diagram
  See Hertzsprung-Russell
    (HR) diagram
Hubble constant, 191, 471,
  473
Hubble diagram, 466,
  467–71, 478
Hubble, E, 445–55 passim,
  463, 464, 466, 471,
  480

Hubble Space Telescope
  (HST), 605
  data, 145
    of C10939+4713, 138
    and weak lensing, 164
  and the distance scale
    problem, 472
  and far-ultraviolet (FUV)
    detections, 614
  and gravitational distortion,
    129
  and gravitational lensing,
    135
  high spatial resolution,
    206
  HST Faint Object Camera
    (FOC), 606
  HST Wide Field Camera
    (WFPC2), 606
    observations of NGC 4552,
      642
  and the Pluto-Charon
    system, 594
Hubble Space Telescope
  (HST) images, 173, 181
  of arcs, 174
  of lensing-clusters, 137
  of quasars, 155
  and weak lensing, 169, 176
Humason, Milton, 448, 465,
  466, 471
Humason-Mayall Redshift
  Catalog, 465–67
Hyades
  distance to, 197
Hydrogen (H)
  burning
    and thermonuclear
      reactions, 4
    and frequency allocations,
      74
    and giant planets, 28
    radio astronomical
      observations of, 78
    and solar evolution, 27
    and star formation, 11
    stars burning, 457–58

and stellar evolution, 245,
  247–51, 258–60, 265–67,
  289, 293, 295, 300
and thermonuclear
  reactions, 6
and UVX, 627
Hydroxyl (OH)
  astronomical observations
    of, 87
  band(s)
    and the Geostationary
      Meteorological Satellite,
      77
    and Global Positioning
      Satellite (GPS) emissions,
      84
    in Comet Halley, 83
    and GLONASS satellites,
      88
  radio spectral lines for,
    72–74
Hyperion, 544
  chaotic tumbling of, 586
  Saturn's satellite, 571–73,
    578–80, 596
Hysteresis
  and the Oosterhoff
    dichotomy, 225–26

I

Iapetus
  and Cassini states, 550
  Saturn's satellite, 570–72
IC 348
  in Perseus, 395
IC 2602
  and Hipparcos, 198
Icarus
  discovery of, 451
Ice
  of Europa
    ocean under, 569–70
    and Iapetus, 571
Impact(s)
  giant
    and natural satellites,
      595

origin of the Moon,
538–39, 551–53
origin of the
Pluto-Charon binary
system, 539
theory of moon
formation, 32–33
Impactors
and natural satellites, 544
Inclusions
Ca-Al-rich inclusions
(CAIs), 392
formation time, 279,
284, 285
fractionated and
unknown nuclear (FUN),
29
isotopic anomalies in,
393–94
from other stellar
systems, 30
Infrared sources
classification scheme for,
331
INMARSAT
and World Radio
Conference (WRC) 2000,
93
Institutions
and rigid procedures, 13
Inter-Union Commission on
the Allocation of
Frequencies (IUCAF),
65–82 passim, 90, 92,
93
activity from 1980–1987,
85
correspondents, 78–79, 85,
86
downsizing of, 78
formation of, 69–72
and GLONASS satellites,
88, 89
preparations for World
Administrative Radio
Conference
(WARC)-1992, 86–87

and the VEGA mission in
1984, 82–83
and World Radio
Conferences (WRCs),
91
Interference
with astronomy, 71
and radio astronomy,
65–66
from satellites, 79
from satellites and aircraft
and radio astronomy,
74–78
Interferometer(s)
and reference frames,
119
space, 118
and young stellar object
(YSO) X-ray ionization,
389
Interferometric arrays
operating at millimeter
wavelengths, 324, 353
Interferometry
optical/IR, 119
radio, 104
and radio astronomy,
476
and young stellar object
(YSO) systems, 400
International Astronomical
Union (IAU), 78, 85
and the formation of the
Inter-Union Commission
on the Allocation of
Frequencies (IUCAF),
70–71
priorities for molecular
lines, 79
and radio astronomy, 68
and the radio reference
frame, 105
and reference frames, 121
and reference systems,
101
Symposium 182 *Time and
the Earth's Rotation*, 105

International Celestial
Reference Frame
(ICRF), 97, 101,
107–15, 118, 119, 121
and the dynamical
reference frame, 117
and the Hipparcos optical
reference frame, 108, 117
International Celestial
Reference System
(ICRS), 97, 99, 101,
119, 121, 193
International Civil Aviation
Organization, 77
International Earth Rotation
Service (IERS), 105,
114–15, 117, 121
International Frequency
Registration Board, 84
International Scientific Radio
Union (URSI), 65, 69,
75, 78, 82, 83, 85, 94
International
Telecommunications
Union (ITU), 65–68, 72,
74, 79, 81, 86, 90
Recommendations of,
81
World Administrative
Radio Conference
(WARC) 1992, 86–89
International Ultraviolet
Explorer (IUE), 605–6,
608, 610–11, 616, 618,
622, 642
color data for E galaxies,
635
data for globulars, 621
E galaxy data from,
619
spectra, 614–16, 637
spectra of galaxies, 626
Interstellar cloud
collapse of, 26–27
Io, 575–76
Jupiter's satellite, 560,
563–64, 567–70

SUBJECT INDEX 659

and tidal dissipation, 545,
    596
volcanism on, 561
Iodine (I)
    ^{129}I
        in meteorites, 10–11
Ionization
    ambiguities, 509
    and the broad emission line
        region (BELR), 491
    fraction
        in cores, 345–46
    ionization correction (IC)
        and abundance analysis,
            506, 508–10, 513
        and abundance ratios,
            494
    of low-mass cores, 331
    and quasars, 506, 508
    and star formation, 312,
        314, 324
    X-ray
        and young stellar objects
            (YSOs), 388–91
Iowa State College, 3
IRAM, 174
IRAS 16293, 340
IRAS sources, 435
    surveys of, 346
IRAS survey, 331
IRIDIUM satellites,
    89–90
Iron (Fe)
    abundance
        solar, 199, 205
    Fe/α
        as a clock, 520
        timescales and
            cosmology, 522, 523
    FEII/MgII
        broad emission lines
            (BEL) strengths, 516
    peaks
        and the s-process, 5, 8
Isotope(s)
    and the early solar system,
        277–99

and stellar evolution,
    239–77 passim, 289–301
^{99}Tc, 252
Isotopic anomalies
    in Ca-Al-rich inclusions,
        393–94
IUCAF
    See Inter-Union
        Commission on the
        Allocation of Frequencies
        (IUCAF)

J

Janus
    Saturn's satellite, 570–72,
        577–78
Japanese broadcasting
    satellite, 90
Japanese Meteorological
    Agency, 77
Jet(s)
    in astrophysics,
        409–10
    defined, 410
    formation, 427–28
        and accretion disk
            instabilities, 424–27
    relativistic
        in the galaxy, 409–39
        interaction with the
            environment, 434–35
Jupiter
    formation of, 536–37
    and giant planet models,
        28, 29
    satellite system of, 559–70,
        595–96
    satellites orbiting, 538

K

K-corrections
    and ultraviolet (UV) light,
        604
    and ultraviolet
        (UV)-observations, 608
Keck telescope(s), 129, 174
Kernels, 347

Kinematics
    of globules and cores,
        329
    and star formation, 322–23,
        350–51, 353
Kinetic energy
    of relativistic jets, 435
Kuiper belt objects, 594–95,
    597
7 keV, 431

L

Laplace
    Pierre-Simon, 365
Laplacian plane, 550, 553
    and Saturn's satellites,
        571–72
Large Southern Array (LSA),
    182
Lens coupling
    and weak lensing, 162
Lens parallax method,
    141
Lensing
    equations, 130–32
    and mass discrepancy,
        136–37
    probing the universe with,
        126–83
    strategies for, 157–60
Lick Observatory, 464
    E galaxy spectral survey,
        620
Lines of force
    and galactic magnetic
        fields, 41, 49, 50
Lithium (Li)
    isotopic
        and the galactic
            magnetic field, 59
    in stars, 369, 394–95,
        397–98
LkCa 15, 380
LkCa 16, 378
LkHα 92, 377
Long base-line
    interferometry, 476

Lorentz factor(s), 439
  in jets from microquasars, 438
  and relativistic jets, 423
Lorentz force(s)
  and dynamo theory, 53
  and magnetic fields, 43
Low-earth-orbit satellites
  and radio astronomy, 87
Low Ionization Nuclear Emission Region (LINER)
  optical spectra, 642
Luminosities
  for RR Lyraes and non-variable HB stars, 221
  X-ray
    and young stellar objects (YSOs), 400
Luminosity
  and abundance, 225
  bolometric
    stellar, 375
    and UVX, 623
  Cepheid period, 216–18
  and color, 214, 217
  galaxy
    as distance indicators, 472
  at the main-sequence turnoff, 209, 212
  and metallicity, 198
  and the Oosterhoff dichotomy, 225–26
  period-luminosity relation, 227, 230
  and RR Lyrae stars, 222
  of the ZAHB, 222–23
Luminous blue variables (LBVs)
  in M101, 471
Lunar formation
  theories of, 31
Lunar landings, 31

Lutz-Kelker bias
  in parallax determination, 193, 195, 197
Lutz-Kelker corrections, 217
Lyman alpha forest, 463
Lyrae
  See Star(s), Lyrae
L1495W, 395
L1527, 339
L1551 IRS5, 335, 336
L1630, 344
L1641, 344

M

Magellanic Cloud(s), 227
  Large (LMC), 462
    distance to, 230–31
  Small (SMC), 462
    distance to, 231
  stars
    Hipparcos data for, 193
Magma ocean
  on the early Moon, 552, 555, 556
Magnesium (Mg), 7
  emission lines, 503–4, 523
  FEII/MgII
    broad emission lines (BEL) strengths, 516
  isotopes
    and asymptotic giant branch (AGB) nucleosynthesis, 258, 259, 261, 262, 266, 267, 272
Magnetic activity
  and protostars, 380, 383
  as a tracer of young stellar object (YSO) populations, 394–99
  and young stellar objects (YSOs), 363–65, 367–87, 400
Magnetic field(s)
  in cores and globules, 330
  of Europa, 570

  galactic
    and dynamo theory, 37–62
  Ganymede's, 566
  geometries in young stellar objects (YSOs), 385–87
  and jet formation, 427, 429
  and radio sources, 475
  and star formation, 312, 314, 315, 323–24, 326, 333, 334, 352, 353
  strengths, 345
  and X-ray binaries, 428
  in young stellar objects (YSOs), 366–71
    tracers of, 369–74
Magnetism
  remanent
    and chondrules, 392–93
Magnification
  and gravitation lensing, 134
Magnification bias
  of bright quasar samples, 160
  and distribution of faint galaxies, 170–73
  and gravitational lensing, 141
  mass profile from, 149–50
  measuring, 155–57
  two-point correlation function, 153
  and weak lensing, 151, 161, 168, 177–78
Magnitude(s)
  and the 200-inch telescope, 452
  of the Cepheids, 456
  and redshift, 465
  scales
    and Mount Wilson and Palomar, 453–54
Main-sequence fitting, 198, 204–8, 212, 218
  analyzes 209

calibration of cluster
  variables, 226
distance determinations,
  210–11
Main-sequence location
  and abundance, 200
Main-sequence stars, 199
Main-sequence turnoff
  luminosity at, 209, 212
Maps
  CBM, 180
  projected mass density,
    158
Mars
  satellite system of, 557–59,
    595–96
  satellites of, 539, 544
Mass
  and asymptotic giant
    branch (AGB) stars,
    263–64
  density
    and gravitational
      lensing, 127
  discrepancy, 136–37
  distribution(s), 151
    cluster, 150, 176
    in clusters of galaxies,
      135–50
    in the universe,
      181
    and weak lensing,
      177
  and evolutionary phases of
    low-mass stars, 623–25
  galactic mass-metallicity
    relation, 519
  loss
    and asymptotic giant
      branch (AGB) stars, 259,
      264
    on the RGB, 633
  map, 144
  map of the universe, 129
  profile from the
    magnification bias,
    149–50

reconstruction(s), 143–46,
  149, 150, 157, 170
  with arclets, 135–38
  of clusters, 142
  from lensing inversion,
    141
  and weak lensing, 128,
    138–46, 161, 162, 165,
    181
  sheet degeneracy, 141
  transfer onto white dwarf
    binary companions, 631
  and UVX, 627
  visible and virial
    of Coma cluster, 128
Mass density
  and planet formation, 536
  projected
    and gravitational
      lensing, 139, 146, 153,
      156–58, 166, 171, 181
    maps, 158
Mass loss
  giant branch, 603, 641, 643
  red giant branch (RGB),
    628–30
  stellar, 604
  and UV flux, 636
  and UVX, 639
Medium Deep Survey
  (MDS), 164–65
Mercury, 451
  spin of, 537, 549
MERLIN, 113, 115, 117, 118,
  383, 423
  and GRS 1915+105, 418,
    422
Metal abundance(s)
  and helium, 628
  and quasars, 487–524
  and UV spectra, 633–34
  and UVX, 619, 627, 628,
    636
  See also Abundance(s)
Metallicities
  cluster, 204–6
  and evolved stars, 271–73

of stars, 200
and unevolved stars,
  273–77
Metallicity
  of calibrating stars, 206
  effects
    and the chemical
      evolution of the galaxy,
      267–77
  galactic mass-metallicity
    relation, 519
  and main-sequence fitting,
    198
  and red clump stars, 219
  and the s-process, 267–71
  scale, 204
  solar
    low-mass star of, 247–48
  of stars, 453, 457–61
  stellar, 199
  and UVX, 639
  varying
    main-sequence with, 202
Meteorite(s)
  and the early solar nebula,
    27
  and elemental abundances,
    9–10, 29–30
  exposure to energetic
    particles and shocks,
    391–94
  and extinct radioactivity,
    10
  formation time, 279, 285
  inclusion(s)
    and solar system
      environment, 29–30
  isotopes in, 400
  and isotopic ratios, 279
Meteorological satellite(s), 72
  frequency allocations in,
    74
  and radio astronomy,
    77–78
MHD equations
  and dynamo theory, 53, 54,
    56

MHD numerical simulations
of jet formation, 428
Microblazars
and relativistic jets,
437–39
Microquasars, 409–13
and new perspectives,
438–39
and quasars, 424, 438
Microwave landing system
(MLS)
and radio astronomy, 77
Mimas
satellite of Saturn, 570–80
passim, 596
Minkowski, R, 448, 449, 464,
472, 474
Miranda
satellite of Uranus, 580–89
passim, 592, 596
Miras, 230
variables, 227
Mobile satellite service(s)
frequency allocations for,
92
and World Radio
Conference (WRC) 2000,
93
Mobile services
and radio astronomy, 94
radio frequency allocations
to, 82, 84
Mobile telephone system
and radio astronomy, 87
Molecular cloud(s)
and star formation, 311–26
passim
Molecular lines
International Astronomical
Union (IAU) priorities
for, 79
Molecular spectral lines
and radio astronomy, 73
Monte Carlo calculations
and planet formation,
536
Monte Carlo technique, 325

Moon
and Cassini states, 550–52
Earth-Moon
and natural satellites,
595
and the simple tide, 547
Earth-Moon system,
551–56, 595
formation of, 31–33
origin and evolution of,
538–39
origin of, 551, 553
and radio astronomy, 74
and special circumstances,
534
Motorola
IRIDIUM satellites, 89–90
mobile telephone system
and radio astronomy, 87
Mount Wilson Catalog of
Photographic
Magnitudes in Selected
Areas, 453–54
Mount Wilson observatory,
445–80
MS0302+17
weak lensing analysis,
155
MS2137
lensed images of,
133
MS2137-23
radial arcs of, 137
Multinational corporations,
94
M3
main sequence, 456–57
and UV excess, 460
M4, 208
M5
and UV excess, 460
M13
and UV excess, 460
UV spectra, 621
M15
and UV excess, 460
M17, 350

M25, 462
M31, 449, 452, 462, 631
Cepheids in, 456
distance to, 231–32, 463,
472
energy distributions of, 632
and far-ultraviolet (FUV)
light, 611
far-ultraviolet (FUV)
spectrum of, 621
globular clusters, 455
International Ultraviolet
Explorer (IUE) spectra,
616
metal-poor light at optical
wavelengths in, 622
Sb spiral
and FUV, 607–8
second concentrated
nuclear source in, 641
and UV, 609–10
UV imaging studies of,
637–38
M32, 612, 631
core of, 630
UV imaging studies of,
637–38
UV upturn, 615
UVX gradient, 639
M81, 471
distance to, 463, 472
M82
prototype of starburst
galaxies, 478–79
M87
jet of
polarization of, 475
spectral anomalies in, 616
and UV profiles, 612
M92, 209
age of, 212, 214
distance and age of, 207
main sequence, 456–57
and UV excess, 460
M101, 471
distance to, 463, 472
stellar content of, 473

SUBJECT INDEX 663

N

Naiad
  Neptune's satellite, 590–92
Narrow absorption lines
    (NALs)
  and quasars, 489, 504–6,
    510, 512, 514–16, 523,
    524
NASA
  See National Aeronautics
    and Space Administration
    (NASA)
National Academy of
    Sciences, 18
  Committee on Radio
    Frequencies, 80
National Aeronautics and
    Space Administration
    (NASA)
  early plans of, 12
  and the National Academy
    of Sciences, 18–19
  Space Sciences Steering
    Committee, 18
  and the study of the
    origin of the solar
    system, 16
National park system model
  and the radio spectrum,
    94–95
Nearby Star catalog, 200
Nebula(e)
  distribution of, 464
  and the extragalactic
    distance scale, 191
  solar, 536–37
Nebula W50, 434–35
Neon (Ne)
  isotopes
    and asymptotic giant
      branch (AGB)
      nucleosynthesis, 258,
      259, 261, 262, 266, 267
  ^{21}Ne
    in meteoritic grains, 393
    and thermonuclear
      reactions, 7

Neptune
  debris around, 538
  formation of, 537
  satellite system of, 589–93,
    597
Neried
  Neptune's satellite, 589–90
Neutron(s)
  production, 4
  in red giant stars, 3
Neutron capture(s)
  and metallicities, 269
  reaction
    explained, 7
Neutron capture process
  See Slow-neutron-capture
    process
Neutron star(s), 8–9, 26, 414,
    424
  binaries, 411
  jet velocities, 432, 434
New Generation Space
    Telescope (NGST), 182
New York University, 12
NGC2071, 348
NGC 121, 231
NGC 188, 626
NGC 625, 478
NGC 1275
  prototype of starburst
    galaxies, 478–79
NGC 1399, 611
  Astro/Hopkins Ultraviolet
    Telescope (HUT)
    spectrum of, 617–18
  energy distributions of, 632
NGC 1404, 611
NGC 1569, 478, 480
NGC 1705, 478, 480
NGC 2244, 449
NGC 2403, 471–72
NGC 4147
  and UV excess, 460
NGC 4552
  UV-facilitated observations
    of, 642
NGC 4649, 615

NGC 6397
  distance to, 208–9
NGC 6752, 208
NGC 6791, 626
NGC 6822, 462
NGC 7790, 462
NGC 1333 star-forming
  region, 376
Nitrogen (N)
  abundances, 520–21
  isotopes
    and stellar evolution,
      244, 245, 247–49,
      258–61, 265, 267
Nova Muscae, 432
Nova Oph 93, 432
Novae, 463
  as distance indicators,
    472
Nuclear astrophysics, 4
  A. Cameron on, 25–26
Nuclei
  short-lived
    in the early solar system,
      277–99
Nucleosynthesis, 204
  in asymptotic giant branch
    (AGB) stars, 239–301
  and cluster giants, 205
Nutation
  and reference frames,
    121

O

Oberon
  satellite of Uranus, 580,
    581, 585, 589, 596
Occam's razor, 517–18
Ocean
  under the ice of Europa,
    569–70
Oosterhoff dichotomy,
    225
Ophiuchi cloud, 380, 381,
    383
  dust continuum maps of,
    348

Ophiuchus cloud complex
and star formation, 397
Orbiting Astronomical Observatory-2 (OAO-2)
and far-ultraviolet (FUV) radiation, 603, 607, 609
Orion clouds
cores in, 344
Orion giant molecular cloud, 400
Orion Nebula Cluster, 350
Outflows, 340
in the earliest stages, 332
in globules, 327
and massive stars, 343
and young stellar objects (YSOs), 366–68, 370, 380, 389
Oxygen (O)
abundance anomalies in cluster giants, 205
isotopes
and stellar evolution, 244–48, 258–61, 265, 267, 276
and thermonuclear reactions, 7

P

Palomar Observatory, 445–80
Palomar Telescope, 151
to measure cosmic shear, 159
Pan, 571
Pandora, 570
Par 1724, 378
Parallax(es)
for Cepheids, 216
determination
systematic biases in, 193, 195–96
factors
and ESA Hipparcos satellite, 192
Hipparcos, 193, 217
Hipparcos data, 193
for halo subdwarfs, 214
from Hipparcos observations
and the Pleiades, 198
measurement(s), 192, 197
temporal effects in, 196
pulsational analyzes, 217–18
statistical, 220–21
trigonometric and statistical analyzes, 226–27
trigonometric parallax measurements, 191, 219–20
Pasiphae
Jupiter's satellite, 560
Persei
and Hipparcos, 198
Perturbation theory
and large scale structures, 152
Phobos, 595–96
and Mars, 557–59
origin of, 544
Phoebe
Saturn's satellite, 571, 572
Photocathodes
cesium telluride, 606, 607
Photoionization, 498
of the broad emission line region (BELR), 491
simulation, 495–99
Photometry, 199, 204
of cluster stars, 206
Strömgren, 200
for white dwarfs, 208
Photonuclear activation curves, 3
Photonuclear reaction explained, 7
Plagioclase
on the Moon, 552
Planck-Surveyor, 179, 180, 182
Planetesimals
and planet formation, 535–37
Planets
formation of, 534–38
giant
models of the interiors of, 28–29
outer
strategy for exploration of, 19
terrestrial
accumulation of, 28
Pleiades
distance to, 197–98
Pluto
Pluto-Charon
rotating synchronously, 550
Pluto-Charon system, 593–95, 597
origin of, 539
Praesepe
and Hipparcos, 198
Precession
and reference systems, 97, 102, 103, 105, 121
Primordial field
and the galactic dynamo, 56, 58–59, 61
Prometheus, 570
Proteus
Neptune's satellite, 590–92
Protogalaxy
as a seed field, 56, 58, 61
Protostars, 367
and X rays, 380–83, 400
Pseudodisk(s), 333, 335
Pulsars, 26
Pulsation equation
for RR Lyraes, 224–25
Pycnonuclear reactions
how named, 6

Q

Quasars, 59
compact radio sources, 104
and cosmic shear, 160
discovery of, 475–78

SUBJECT INDEX

and elemental abundances, 487–524
and microquasars, 411, 413, 424, 438
and reference frames, 113–14
and shear, 155
Q2345+007, 155

R

R-process, 273, 277
  and the early solar system, 279
  and environment, 25–26
  and extinct radioactivity, 10, 11
  nuclei, 288
  and stellar evolution, 292
  and unevolved stars, 275, 276
Radar
  earth-looking, 92
  and radio astronomy, 66
  ranging
    and reference frames, 97, 104
Radio
  and hot cores, 349
Radio astronomy, 446, 450, 474, 476–77
  and allocation of frequencies, 65–95
  and interference, 65–66
  international cooperation in, 66–68
  millimeter-wave, 73
  after World War II, 66
Radio-continuum emission, 349, 384
  of protostars, 383
Radio-continuum flares, 374
  in young stellar objects (YSOs), 367
Radio emission
  of the outflows of T Tauri South, 394

and young stellar objects (YSOs), 380
Radio galaxies
  and the Hubble diagram, 467, 468
Radio reference frame, 104–7
Radio source(s)
  compact, 104, 105, 114
  extragalactic
    and reference frames, 97, 99, 101, 108, 109
    initial identification of, 472, 474–75
  shear detected around, 160
Radio spectral lines
  of greatest significance, 85
Radio spectrum
  burgeoning use of, 66
  and frequency allocation, 65, 67, 73–75
Radio transmissions
  space-based
    and radio astronomy, 83
Radio waves
  cosmic, 472
  from the galaxy, 66
Radioactive nuclei
  in the early solar system, 277–99
Radioactivities
  extinct
    and solar system formation, 10–11, 30
  original
    in the solar system, 277
Ray bundle(s)
  deformation, 131, 162
  distortion of, 127, 152
Red clump stars, 218–19
Red giant(s)
  and abundance, 227
  light
    insensitive to age, 636

Red giant branch (RGB)
  mass, 630, 634, 636
  mass loss, 628–30, 633
  and star evolution, 623–25
  and UVX, 627
Reddening
  and Cepheids, 216, 218
  cluster, 208
  and main-sequence (MS) fitting, 207
Redshift(s), 463
  distribution from lensing inversion, 168–69
  distribution of galaxies beyond $B = 25$, 165–73
  and dynamical evolution of structures, 127
  and gravitational lensing, 135, 143, 145
  high
    and quasars, 487–524
  higher
    and the Hubble diagram, 467, 469–71
  of lensed galaxies, 138
  and magnitude, 465
  papers of the Redshift Distance series, 468
  probing source
    using various lens planes, 169–70
  redshift 1
    massive clusters of galaxies at, 145
  and relativistic jets, 423
  and UVX, 638
  and weak lensing, 127, 152–58, 161, 165, 174–79, 181, 182
Reference frame(s)
  in astronomy, 97–121
  and celestial reference systems, 99–102
  dynamical, 101, 103–4, 117
  FK5, 99, 102–3

future improvements in, 118–19
Hipparcos optical, 103, 108, 115–17, 119, 121
radio, 104–7
Reference system(s)
celestial, 99–102
conventional, 100, 101
conventional celestial reference system (CCRS), 102
dynamical, 100, 103
kinematic, 99–101
of observational optical astronomy, 101–2
Relativistic ejecta
collisions with environmental material, 434
Relativity
effects
and relativistic jets, 421–23
general
possible laboratories for, 430–31
and reference frames, 100, 101, 119, 121
and relativistic jets, 411
theory
and microquasars, 439
Resonance
evection, 554–55
eviction, 554–56
Laplace, 564–65, 567, 570
and natural satellites, 560–66
and Phobos, 558
Reynold's number(s)
and galactic dynamo theory, 52, 54
Rhea, 572
Saturn's satellite, 570, 578
Roche radius, 539, 551, 556
and Neptune's satellites, 590, 593

Rocks
lunar, 552
ROSAT All-Sky Survey (RASS), 397–98
ROSAT samples
weak lensing studies of, 181
Rotation
and binarity, 340
and collapse, 338–39
in cores, 330
and star formation, 326, 333, 334, 353
RS CVn
magnetic activity, 386
Russian Space Forces
and GLONASS satellites, 88–89

S

S-process, 7
and asymptotic giant branch (AGB) stars, 254–55, 262, 277
and computers, 5
definition of, 240
elements
in stars, 252
elements produced by, 273
and metallicities, 271–73
and metallicity, 267–71
nucleosynthesis
in asymptotic giant branch (AGB) stars, 267
and unevolved stars, 275–76
Sandage period-shift effect (SPSE), 225–26
Satellite(s)
Applications Technology Satellite (ATS-F), 75
Astronomical Netherlands Satellite, 607
Belgian MLMS LEO
and radio astronomy, 91–92

broadcasting
and radio astronomy, 90
Geostationary
Meteorological Satellite
and radio astronomy, 77–78
Global Navigation Satellite System (GLONASS)
and radio astronomy, 84–85, 88–89
Global Positioning
and radio astronomy, 84
and interference, 69
interference from
and radio astronomy, 74–78
IRIDIUM, 89–90
low-earth-orbit
and radio astronomy, 87
meteorological, 72, 74
mobile satellite service
and World Radio Conference (WRC) 2000, 93
Motorola IRIDIUM, 89–90
natural
origin and evolution of, 533–97
and tides, 545–51
operators
and radio astronomy, 91
radar
and radio astronomy, 92
and radio astronomy, 74–79, 87, 94
service operators
and Recommendation 66 of World Administrative Radio Conference (WARC)-1979, 81–82
services
and radio frequency allocations to, 82, 84
survey MAP, 179

SUBJECT INDEX

Saturn
  formation of, 536–37
  satellite system of, 570–80, 596
Schmidt, B, 449
Schmidt survey telescope at Palomar, 449
Science
  and the radio spectrum, 94
Scientists
  and historians, 1, 2
Sco-Cen star-forming region, 397
Scorpius X-1, 432
  and jet velocities, 428
SCUBA, 174
Seed fields
  for the galactic dynamo, 56, 58
  and galactic dynamo theory, 40–42, 49, 55–59, 61
Serpens, 348
  cloud
    ROSAT observations of, 381
Sgr A, 73, 433, 434
Shear
  cosmic, 159, 160
    and large-scale structures, 150–62
  detected around Q2345+007, 170
  and gravitational lensing, 139, 160
  maps
    in A1942, 146
  and mass reconstruction, 145
  measuring weak, 146–49
  and quasars, 155
  reduced complex
    and gravitational lensing, 134
  and weak lensing, 150, 158, 177
Shocks
  and chondrule melting, 392–93, 400
  heat gas, 320
  and meteorites, 391
  and young stellar objects (YSOs), 394
Sinope
  Jupiter's satellite, 560
Sky survey
  Palomar Observatory-National Geographic, 464, 466
  Sloan Digital Sky Survey (SDSS), 129
  UV, 605
Slow-neutron-capture process
  and asymptotic giant branch (AGB) stars 246–257
  and ^{13}C burning, 264–67
  isotopes, 269–71
  phenomenological approach to, 254–57
Small-scale fields
  and dynamo theory, 37, 38, 52–55, 61
Smithsonian Astrophysical Observatory, 15, 21–24
Smoothed particle hydrodynamics (SPH) code
  and lunar formation, 32–33
SN 1987A, 230
Software
  and reference frame models, 114
Solar evolution, 27
Solar nebula
  processes, 26–29
Solar system
  early
    short-lived nuclei in, 277–99
  formation, 534
  and extinct radioactivities, 10–11, 30
  and nucleosynthesis in asymptotic giant branch (AGB) stars, 239–301
  origin of, 11, 26–27
  strategy for the exploration of, 20
  study of the origin of, 16
  supernova trigger for formation of, 30–31
Solar systems
  Gordon Conference on the Origins of Solar Systems, 16
Space
  emissions from
    and radio astronomy, 75
  Gordon Conference on the Chemistry and Physics of, 16
  planning for space science, 17–21
Space Interferometry Mission (SIM), 118, 119, 232–33
Space mission(s)
  and global astrometry, 118–19
  Highly Advanced Laboratory for Communication and Astronomy (HALCA), 113
Space research
  and allocation of frequencies, 65, 73, 80–81, 87
Space Science Board (SSB), 18–21
Space Sciences Summer Study of the Space Science Board, 18

Space Telecommunications
World Administrative
Radio Conference
(WARC), 73–74, 76–77,
88
Space Telescope Imaging
Spectrometer (STIS),
160
Space World Administrative
Radio Conference
(WARC)
of 1963, 68, 72–73
Spectral content
of arc(let)s, 173–74
Spectrograph
HST/Faint Object
Spectrograph, 605–6
Space Telescope Imaging
Spectrograph (STIS),
606
Spectrometers
UV, 605–6
Spectrophotometry
low-resolution,
204
Spectroscopic calibration,
199
Spectroscopic capabilities
on giant telescopes,
182
Spectroscopic surveys
of arclets, 166–68
Spectroscopy
high-resolution, 204
IR
and young stellar object
(YSO) systems, 400
of main-sequence cluster
members, 206
Spread-spectrum modulation
interference from, 83–84
and radio astronomy,
91–92
Square Kilometer Array,
400
SS 433, 409–11, 419,
432–36

Star(s)
asymptotic giant branch
(AGB), 604
contamination model,
288–89
nucleosynthesis in,
239–301
red giant, 30
sources of short-lived
nuclei, 289–99
in star evolution,
624–25
variables, 227
binary and multiple
and main-sequence (MS)
fitting, 207
blue horizontal branch
(BHB), 621, 622
as distance indicators,
219–27
"blue stragglers," 626,
630
brightest, 463
C, 251–53, 259
formation of, 262–64
calibrating
metallicity of, 206
chemical composition of,
452–53, 458–61
cluster
photometry of, 206
distance determinations
for, 233
as distance indicators,
472
EHB, 629, 630
evolved
and metallicities,
271–73
and the extragalactic
distance scale, 191
field
HR diagram defined by,
198–204
formation, 11, 173,
311–53, 394
clustered, 342–53

continuous
and Hopkins Ultraviolet
Telescope (HUT)
spectra, 618
and far-ultraviolet
(FUV), 604, 616
high-redshift
and quasars, 487, 489,
517–18, 521–22
history of, 174, 399
isolated, 326, 341–42
in M31, 610
in M87, 616
massive
and UV observations,
640–42
rates, 397
and UVX, 609, 619
via gravitational
collapse, 365–66
and the Virgo cluster,
611
fundamental, 101–3,
119
star catalogs, 101–3
Herbig Ae/Be, 342, 349,
363n.1
Hipparcos, 118–19
Hipparcos catalog of,
115–17
horizontal branch (HB),
230–31, 627–29
hot
and clusters, 630
in galaxies and clusters,
623
within the immediate Solar
Neighborhood, 200
isolated low-mass
formation of, 325–42
low-mass, 363
in elliptical galaxies,
631–40, 643
evolutionary phases of,
623–25
evolutionary trajectory
of, 623–25

and far-ultraviolet
(FUV) light, 621
and far-ultraviolet
(FUV) radiation, 603
formation and early
evolution of, 363–64
origin of, 630–31
and UVX, 609–10
Lowell proper-motion
Hipparcos data for, 202
Lyrae stars, RR, 228–31,
461
and cluster distance
determination, 208
as distance indicators,
219–27
and the extragalactic
distance scale, 191
galactic distance scale
from, 118
in M31, 231
variables, 453, 457
Wilson's absolute
magnitude of, 457
main-sequence, 199, 368
and the main sequence,
457–58
massive, 342–52
and UV-upturn, 618
UVX, 616–17, 619
metallicity of, 457–61
neutron, 8–9, 26
OB, 363n.1
spectral features, 616
observed by Hipparcos
satellite, 192–93
old
and UVX, 611, 621, 636
older, 627
post-T Tauri, 369, 394, 399
pre-main-sequence, 331,
363, 365–66, 378
red clump, 218–19, 229–31
as distance indicators,
232
red giant, 6, 244–45
of class S, 3–4

resolved UV populations,
637–38
S
and isotopes, 252
as seed fields, 56–58
single-star candidates of
UVX sources, 623–26
small-envelope, 604
subgiant
high-resolution spectra
of, 206
unevolved
and metallicities,
273 77
used to estimate
metallicities, 200–2
UV excess in, 459–60
white dwarf, 618
models, 8
X-ray discovered,
395–97
young
and UVX, 610–12
young and massive
and UVX, 609–10
See also Classical T Tauri
(CTT) stars; Clusters;
Star-forming regions;
Stellar evolution; Stellar
population(s);
Subdwarf(s); Weak-lined
T Tauri (WTT) stars
Star-forming regions, 399
X-ray images of, 375–76
young stellar objects
(YSOs) in, 365, 395–97
Statistical analysis
of parallax-selected
datasets, 195
Stellar evolution, 25, 27–28,
457–61
along the asymptotic giant
branch (AGB), 248–52
calculations, 13
and Hipparcos astrometry,
198
and Palomar, 446, 451

prior to asymptotic giant
branch (AGB) phase,
242–48
stages of, 367–68
and thermonuclear
reactions, 6
See also Star(s)
Stellar objects
See Young stellar objects
Stellar population(s)
old
and far-ultraviolet
(FUV) radiation, 603–4,
612
UV observations of,
642–43
and UVX, 619
See also Star(s)
Stellar seed fields
for the galactic dynamo,
56–58, 61
Subdwarf(s)
age of, 460
calibrating
and main-sequence
(MS)-fitting, 207
calibrators, 209
halo, 214
HR diagram for, 202–4
Hipparcos-calibrated,
212
and main-sequence
(MS)-fitting, 206, 212
metal-poor, 461
and photometry, 206
and UV excess, 460
Submillimetre Common-User
Bolometer Array
(SCUBA), 129
Sun
and X-ray emission,
364
Superbubbles, 49
Superluminal sources,
413–21
X-ray power spectrum in,
431

Supernova(e)
  as distance indicators, 472
  and extinct radioactivities, 30
  and the extragalactic distance scale, 191
  and the galactic dynamo, 58
  models for explosions, 48–49
  and seed fields, 41, 56, 57
  and star formation, 11
  trigger for formation of the solar system, 30–31
  understanding of, 25
  varied spectral features in, 10
Surveys
  weak lensing strategies for, 157–60
SVS 16, 376
Synchrotron emission, 409, 428–30, 435
  and GRS 1915+105, 426
  and relativistic jets, 411, 424, 428
Synchrotron luminosity
  and relativistic jets, 426–27
Sz 68, 379
S140, 348

T

T Tau North, 379, 380, 383
T Tau South, 383
T Tauri, 364
T Tauri South
  outflows of
    radio emission of, 394
T Tauri star(s), 370, 377
  and α-ω dynamo theory, 384
  basic picture, 365
  census of X-ray emitting, 395
  Class-II sources, 331
  and the Corona Australis cloud core, 380, 382
  in the Orion giant molecular cloud, 400
  post, 395, 399
  X-ray flares from, 367
  and X-ray luminosities, 379
  X-ray properties of, 374–76
  See also Classical T Tauri (CTT) stars; Weak-lined T Tauri (WTT) stars
Taurus
  cores in, 344
Taurus-Auriga
  classical T Tauri (CTT) stars in, 376
Taurus-Auriga cloud complex
  most active T Tauri star in, 378
Technetium
  in red giant stars of class S, 3, 4
Telescope(s)
  New Generation Space Telescope (NGST), 182
  and high-z universe, 165–75
  operating at submillimeter wavelengths, 324, 353
  Palomar Telescope, 151, 159
  Schmidt survey telescope, 449
  Ultraviolet Imaging Telescope (UIT), 606, 611–13, 621
  UV, 605
  VLT-Survey-Telescope (VST), 129
  See also Hopkins Ultraviolet Telescope (HUT); Hubble Space Telescope (HST); Hubble Space Telescope (HST) images
Telesto, 570, 572
Telluride
  cesium telluride photocathodes, 606–7
Temperature
  of a black body, 332, 334, 335
  dust
    and star formation, 316–18, 320, 325, 329
    far-ultraviolet (FUV), 607
  gas
    and star formation, 319–20, 328, 329
    and gravitational lensing, 179–80
Tethys
  Saturn's satellite, 571–75, 578, 596
Thalassa
  Neptune's satellite, 593
Thebe
  Jupiter's satellite, 560
Thermonuclear reaction(s)
  advanced stages of, 6–8
  rates, 4
  in stars, 25
Thesis advisory committees (TACs), 22
Tidal dissipation
  and Enceladus, 578
  and Ganymede and Callisto, 566
  and Io, 596
  and lunar thermal history, 552–53
  and the Moon, 555–56
  and natural satellites, 544–46, 549, 557, 561, 564
  in Phobos, 559
  and satellite of Uranus, 589
Tidal energy
  and natural satellites, 544
Tidal forces
  and the origin of the Moon, 539

SUBJECT INDEX   671

Tides
  and natural satellites,
    545–51
Titan, 586
  Saturn's satellite, 571–73,
    578–80, 596
Titania
  satellite of Uranus, 580,
    581, 589
Triton, 595
  Neptune's satellite,
    589–91, 593, 597
Troposphere
  and radio sources, 114
Tully-Fisher relation, 163–65
Turbulence
  and galactic dynamo
    theory, 37–40, 43–45, 48,
    52–54, 58
  shear-induced
    and planet formation,
      534
  and solar nebula modeling,
    28
  and star formation, 312,
    314, 322, 326, 327, 331,
    341, 349, 352, 353
TV
  high-definition, 86, 87
  and radio astronomy, 94
  high-density (HDTV)
    broadcasting satellite
      allocation, 82
TW Hya, 389, 397

U

Ultraviolet (UV)
  detectors, 606
  emission of distant
    galaxies, 173–74
  excess
    in stars, 459–60
  far-ultraviolet (FUV)
    in clusters and galaxies,
      621–22
    and early-type galaxies,
      612

and globulars and E
  galaxies, 620
hot-star spectra of old
  populations, 636
far-ultraviolet-B (FUV-B)
  for 8 early-type galaxies,
    613
far-ultraviolet (FUV) light
  and optical light in
    galaxies, 612
  spatial extension of, 610
  and white dwarfs, 625
far-ultraviolet (FUV)
  radiation
    early-type systems, 614
    from elliptical galaxies,
      603–43
    and instrumental
      limitations, 605–7
  maps of early-type
    galaxies, 611–12
mid-UV
  as a population
    diagnostic, 636–37
ultraviolet rising-branch
  (UVX)
  defined, 603
upturn(s), 603, 608, 609,
  617, 620–22, 626
  and boxy isophotes, 630
  and massive stars, 618,
    640
UVX, 611
  and accreting white
    dwarfs, 631
  and binaries, 630–31
  candidate sources,
    621–31
  cosmic evolution of,
    638–39
  and dynamical
    environment of galaxies,
      629–30, 631
  and early-type galaxies,
    608
  for 8 early-type galaxies,
    613

and far-ultraviolet
  (FUV), 615–18
and galaxies, 620–21
galaxy, 623
and globular cluster-type
  populations, 621–23
interpretation of, 603–5,
  643
large variation, 618
and lighter elements, 620
and mass loss, 629
and massive stars,
  616–17, 619
and metal abundance,
  627
and mid-UV light,
  615–16
and old stars, 611
and old stellar
  populations, 603–4
sources, 623
spacial structure of,
  610–14
spectral and photometric
  characteristics of,
    614–21
and star formation, 609
stronger in metal-rich
  galaxies, 619
study of, 610
Ultraviolet Imaging
  Telescope (UIT)
  Astro, 606, 611–13
  data for globulars, 621
Umbriel
  satellite of Uranus, 580–82,
    584, 585, 588, 589, 596
Union Radio Scientifique
  Internationale (URSI),
    67–68
Universities
  personal and academic
    freedom in, 14
University of Manitoba, 2, 3
University of Saskatchewan,
  2
Unix, 22

SUBJECT INDEX

Uranus
 debris around, 538
 formation of, 537
 satellite system of, 580–89, 596
 spin of, 537
URCA process, 7, 26

V

VEGA mission
 in 1984
  and radio astronomy, 82–83
Velocity
 and planet formation, 535–37
 velocity-distance relationship, 465–71
VENERA spacecraft
 Soviet, 82
Venus
 spin of, 537
 and the Vega mission, 82, 83
Very Large Array (VLA)
 and 1E1740.7–2942, 412
 and GRS 1915+105, 422
 and GRS J1655–40, 419
 NRAO, 376
 VLA-Faint-Images of the Radio Sky at Twenty-Centimeters (FIRST), 129
 VLA-FIRST, 182
  radio survey, 158–59
Very Long Baseline Array (VLBA)
 and GRS 1915+105, 418, 419, 422
 and GRS J1655–40, 419–20
 and reference frames, 109, 111, 113, 117, 119
 and Sgr A*, 434
 and SS433, 419

Very Long Baseline Interferometry (VLBI)
 coordinating center of the IERS, 114–15
 and GRS J1655–40, 419
 and reference frames, 97, 99, 104–9, 115, 119–21
 and reference systems, 101
 Vega, 83
Virgo cluster
 early-type galaxies in, 620
 and UV color gradients, 611
VLT-Survey-Telescope (VST), 129
Voyager
 observations of Phoebe, 571
Voyager 2
 data on Triton, 593
Voyager I
 and Europa, 569–70
V773, 378–79
V410 Tau, 378
V773 Tau, 378
V826 Tau, 378

W

Water vapor
 22 GHz transition of, 82
WD-fitting, 208
Weak-lined T Tauri (WTT)
 stars, 367–68
  in the Corona Australis cloud core, 382
  isolated, 398
  locating, 394
  and magnetic activity, 376–79, 385
  and X-ray emissions, 374
  X-ray properties, 376
WEST FORD, 69–72
Winds
 stellar
  and mass loss, 264

World Administrative Radio Conference (WARC)
 1959, 68
 1979, 79–82, 88
  Recommendation 66 of, 88, 91, 93
 1983–1988, 82
 1988 ORB 2, 84
 1992, 85–89
 and Comite Consultatif International des Radiocommunications (CCIR) Recommendations, 79
 and Inter-Union Commission on the Allocation of Frequencies (IUCAF), 79
 on Space Telecommunications, 73–74, 76–77, 88
World Radio Conferences (WRCs), 65
 1993, 1995, 1997, 90–91
 1995, 92
 1997, 92
 2000, 93
 International Telecommunications Union (ITU), 90–91

X

X-ray(s)
 astronomy
  and relativistic jets, 411
 binaries, 409, 410, 428, 429, 431, 434, 438
 catalogs
  ROSAT All-Sky Survey (RASS), 398
 and circumstellar gas and disks, 401
 effects on ambient chemistry and dust, 389, 391

emission
 and young stellar objects
  (YSOs), 384
flares
 protostellar, 383
 from T Tauri stars, 367
flux
 in GRS 1915+105,
  424–26
 and relativistic jets,
  431–32
 hard
 and GRO J1655–40, 424
images
 and mass discrepancy,
  136–37
ionization, 399
 and young stellar objects
  (YSOs), 387–90
ionization of circumstellar
 material, 364
luminosities, 400
 in black hole systems,
  414
 and young stellar objects
  (YSOs), 400
luminosity, 411
 of GRS 1915+105, 411,
  414
 and main-sequence stars,
  379
 and young stellar objects
  (YSOs), 395
and microquasars, 424
power
 of the superluminal
  sources, 430

properties
 and protostars, 381
 and protostars, 380–83
 QPOs in, 439
 solar, 364
source
 GRS 1915+105, 413–19
 and relativistic jets,
  432
 surveys of stars, 395, 397
 surveys of the sky, 365
 and T Tauri stars, 374–76
transient
 XTE J1748–288, 419,
  421
 wide-field surveys, 369
 young stellar object (YSO),
  364, 371, 375, 389, 394,
  399, 400
X-ray Multimirror Mission,
 400
XTE J1748–288, 432,
 433
 ejecta from, 432
 superluminal motions in,
  413, 419, 421

Y

Yale University, 12–13
Yeshiva University
 Belver Graduate School of
  Science at, 12, 14–15
YLW 15, 380, 381, 385
Young stellar objects (YSOs)
 census of, 365
 dispersed
  census of, 397–99

high-energy processes in,
 356–404
and magnetic activity,
 363–65, 367–84
magnetic field geometries
 in, 385–87
origin of magnetic activity
 in, 384–87, 400
outflows, 366–67
phases of evolution,
 367–69
populations
 magnetic activity
  as a tracer of,
  394–99
pre-main sequence,
 365–66
within star-forming
 regions, 395–97
and X-ray ionization,
 387–90
YY Ori stars
 accretion in, 366

Z

Zeeman effect(s), 315, 364,
 371, 380
and magnetic field
 strengths, 345
and weak-lined T Tauri
 (WTT) stars, 378
Zero-age main sequence
 (ZAMS), 364, 368, 369,
 379, 398, 399
Zeropoint
 global, 193, 196
 parallax, 193

Cumulative Indexes

CONTRIBUTING AUTHORS, VOLUMES 26–37

Alexander DR, 35:137–77
Allard F, 35:137–77
Aller LH, 33:1–17
Angel JR, 36:507–37
Antonucci R, 31:473–521
Arendt RG, 30:11–50
Arnett D, 33:115–32
Arnett WD, 27:629–700

Bachiller R, 34:111–54
Bahcall JN, 27:629–700
Bahcall NA, 26:631–86
Bai T, 27:421–67
Bailyn CD, 33:133–62
Barcons X, 30:429–56
Barnes JE, 30:705–42
Bastian TS, 36:131–88
Beck R, 34:153–204
Beckers JM, 31:13–62
Benz AO, 36:131–88
Bertelli G, 30:235–85
Bertout C, 27:351–95
Bertschinger E, 36:599–654
Bessell MS, 31:433–71
Bignami GF, 34:331–81
Binggeli B, 26:509–60
Binney J, 30:51–74
Black DC, 33:359–80
Blake GA, 36:317–68
Blandford RD, 30:311–58
Bloemen H, 27:469–516
Bodenheimer P, 26:145–97; 33:199–238
Boggess A, 27:397–420
Bolte M, 34:461–510
Bothun G, 35:267–307
Bowyer S, 29:59–88
Bradt HVD, 30:391–427

Branch D, 30:359–89; 36:17–55
Brandenburg A, 34:153–204
Bressan A, 30:235–85
Brown TM, 32:37–82
Burbidge EM, 32:1–36
Busso M, 37:239–309
Butler RP, 36:57–97

Caldeira K, 32:83–114
Cameron AGW, 26:441–72; 37:1–36
Canal R, 28:183–214
Caraveo PA, 34:331–81
Carr B, 32:531–90
Carroll SM, 30:499–542
Chanmugam G, 30:143–84
Chiosi C, 30:235–85
Colavita MM, 30:457–98
Combes F, 29:195–237
Condon JJ, 30:575–611
Conti PS, 32:227–75
Cowan JJ, 29:447–97
Cowley AP, 30:287–310

D'Antona F, 28:139–81
Davidson K, 35:1–32
Dekel A, 32:371–418
de Pater I, 28:347–99
de Vegt C, 37:97–125
de Zeeuw T, 29:239–74
Dickey JM, 28:215–61
Djorgovski S, 27:235–77
Done C, 31:716–61
Draine BT, 31:373–432
Duncan MJ, 31:265–95
Dwek E, 30:11–50

Eggen OJ, 31:1–11

Elitzur M, 30:75–112
Ellis RS, 35:389–443
Evans NJ II, 37:311–62

Fabbiano G, 27:87–138
Fabian AC, 30:429–56; 32:277–318
Feigelson ED, 37:363–408
Ferland G, 37:487–531
Ferrari A, 36:539–98
Fich M, 29:409–45
Filippenko AV, 35:309–55
Fishman GJ, 33:415–58
Fowler WA, 30:1–9
Franx M, 29:239–74; 35:637–75
Friel ED, 33:381–414
Frogel JA, 26:51–92
Fusi Pecci F, 26:199–244

Gallino R, 37:239–309
Garstang RH, 27:19–40
Gary DE, 36:131–88
Gautschy A, 33:75–113; 34:551–606
Gehrz RD, 26:377–412
Genzel R, 27:41–85
Gilliland RL, 32:37–82
Gilmore G, 27:555–627; 35:637–75
Ginzburg VL, 28:1–36
Giovanelli R, 29:499–541
Glassgold AE, 34:241–78
Goldstein ML, 33:283–325
Gosling JT, 34:35–73
Gough D, 29:627–84
Gustafsson B, 27:701–56

Haisch B, 29:275–324

675

Hamann F, 37:487–531
Harris WE, 29:543–79
Hartmann L, 34:205–39
Hartwick FDA, 28:437–89
Hauschildt PH, 35:137–77
Haxton WC, 33:459–503
Haynes MP, 29:499–541; 32:115–52
Henry RC, 29:89–127
Herbig GH, 33:19–73
Hernquist L, 30:705–42
Hickson P, 35:357–88
Higdon JC, 28:401–36
Hodge P, 27:139–59
Hollenbach DJ, 35:179–215
Holzer TE, 27:199–234
Horányi M, 34:383–418
Howard RF, 34:75–109
Hudson H, 33:239–82
Hudson HS, 26:473–507
Hummer DG, 28:303–45
Humphreys RM, 35:1–32

Impey C, 35:267–307
Isern J, 28:183–214

Johnston KJ, 37:97–125

Kahabka P, 35:69–100
Kahler SW, 30:113–41
Kennicutt RC Jr, 36:189–231
Kenyon SJ, 34:205–39
Kirshner RP, 27:629–700
Knapp GR, 36:369–433
Kondo Y, 27:397–420
Koo DC, 30:613–52
Kormendy J, 27:235–77; 33:581–624
Kovalevsky J, 36:99–129
Kron RG, 30:613–52
Kudritzki RP, 28:303–45
Kuijken K, 27:555–627
Kulkarni SR, 32:591–639
Kulsrud RM, 37:37–64
Kurtz DW, 28:607–55
Kwok S, 31:63–92

Labay J, 28:183–214
Lawrence CR, 30:653–703
Lean J, 35:33–67
Léger A, 27:161–98
Lin DNC, 33:505–40; 34:703–47
Lingenfelter RE, 28:401–36
Lissauer JJ, 31:129–74
Lockman FJ, 28:215–61
Low BC, 28:491–524
Lunine JI, 31:217–63

Macder A, 32:227–75
Majewski SR, 31:575–638
Maran SP, 27:397–420
Maraschi L, 35:445–502
Marcus PS, 31:523–73
Marcy GW, 36:57–97
Mateo M, 34:511–50
Mateo ML, 36:435–506
Mathieu RD, 32:465–530
Mathis JS, 28:37–70
Matthaeus WH, 33:283–325
Mazzitelli I, 28:139–81
McCammon D, 28:657–88
McCarthy PJ, 31:639–88
McCray R, 31:175–216
McKee CF, 31:373–432
McWilliam A, 35:503–56
Meegan CA, 33:415–58
Mellier Y, 37:127–89
Melrose DB, 29:31–57
Mendis DA, 26:11–49; 32:419–63
Meyer BS, 32:153–90
Mikkola S, 29:9–29
Mirabel IF, 34:749–92; 37:409–43
Monaghan JJ, 30:543–74
Monet DG, 26:413–40
Montmerle T, 37:363–408
Morgan WW, 26:1–9
Morris M, 34:645–701
Moss D, 34:153–204
Mushotzky RF, 31:717–61

Narayan R, 30:311–58

Narlikar JV, 29:325–62
Nordlund Å, 28:263–301

O'Connell RW, 37:603–48
Ohashi T, 30:391–427
Olszewski EW, 34:511–50
Ostriker JP, 31:689–716

Paczyński B, 34:419–59
Padmanabhan T, 29:325–62
Papaloizou JCB, 33:505–40; 34:703–47
Peale SJ, 37:533–602
Phinney ES, 32:591–639
Pinsonneault M, 35:557–605
Pounds KA, 30:391–427; 31:717–61
Press WH, 30:499–542
Puget JL, 27:161–98

Quinn T, 31:265–95

Rampino MR, 32:83–114
Rana NC, 29:129–62
Rauch M, 36:267–316
Readhead ACS, 30:653–703
Reid IN, 37:191–237
Reid MJ, 31:345–72
Renzini A, 26:199–244
Rephaeli Y, 33:541–79
Richstone D, 33:581–624
Rickett BJ, 28:561–605
Roberts DA, 33:283–325
Roberts MS, 32:115–52
Robinson B, 37:65–96
Rodonò M, 29:275–324
Rodriguez LF, 37:409–43
Rood HJ, 26:245–94
Rood RT, 32:191–226
Rosenberg M, 32:419–63
Rossi B, 29:1–8
Ryan J, 33:239–82

Saikia DJ, 26:93–144
Saio H, 33:75–113; 34:551–606

Salter CJ, 26:93–144
Sandage A, 26:509–60,
    561–630; 37:445–86
Sanders DB, 34:749–92
Sanders WT, 28:657–88
Sargent AI, 31:297–343
Savage BD, 34:279–329
Schade D, 28:437–89
Schatzman E, 34:1–34
Scott D, 32:319–70
Scoville NZ, 29:581–625
Sembach KR, 34:279–329
Serabyn E, 34:645–701
Shao M, 30:457–98
Shibazaki N, 34:607–44
Shields GA, 28:525–60
Shukurov A, 34:153–204
Silk J, 32:319–70
Sneden C, 27:279–349
Sokoloff D, 34:153–204
Spitzer L Jr, 27:1–17;
    28:71–101
Spruit HC, 28:263–301
Sramek RA, 26:295–341
Starrfield S, 35:137–77
Stern SA, 30:185–233
Stetson PB, 34:461–510
Stevenson DJ, 29:163–93
Stringfellow GS, 31:433–71

Strong KT, 29:275–324
Sturrock PA, 27:421–67
Stutzki J, 27:41–85
Suntzeff NB, 34:511–50

Tammann GA, 26:509–60;
    29:363–407; 30:359–89
Tanaka Y, 34:607–44
Teerikorpi P, 35:101–36
Telesco CM, 26:343–76
Tenorio-Tagle G, 26:145–97
Thielemann F-K,
    29:447–97
Tielens AGGM, 35:179–215
Title AM, 28:263–301
Toomre J, 29:627–84
Townes CH, 35:xiii–xliv
Tremaine S, 29:409–45
Truran JW, 29:447–97
Truran JW Jr, 27:279–349
Turner EL, 30:499–542

Ulrich M-H, 35:445–502
Urry CM, 35:445–502

Valtonen M, 29:9–29
van de Hulst HC, 36:1–16
VandenBerg DA, 34:461–510
van den Bergh S, 29:363–407

van den Heuvel EPJ,
    35:69–100
van der Klis M, 27:517–53
van Dishoeck EF, 36:317–68
van Woerden H, 35:217–66
Verbunt F, 31:93–127

Waelkens C, 36:233–66
Wagner SJ, 33:163–97
Wakker BP, 35:217–66
Wallerstein G, 36:369–433
Wasserburg GJ, 37:239–309
Waters LBFM, 36:233–66
Weidemann V, 28:103–37
Weiler KW, 26:295–341
Weissman PR, 33:327–57
Welch WJ, 31:297–343
Wheeler JC, 27:279–349
White M, 32:319–70
Wilson TL, 32:191–226
Witzel A, 33:163–97
Woolf N, 36:507–37
Woosley SE, 27:629–700
Wyse RFG, 27:555–627;
    35:637–75

Young JS, 29:581–625

Zensus JA, 35:607–36

CHAPTER TITLES, VOLUMES 26–37

Prefatory Chapters

A Morphological Life	WW Morgan	26:1–9
Dreams, Stars, and Electrons	L Spitzer Jr.	27:1–17
Notes of An Amateur Astrophysicist	VL Ginzburg	28:1–36
The Interplanetary Plasma	B Rossi	29:1–8
From Steam to Stars to the Early Universe	WA Fowler	30:1–9
Notes from a Life in the Dark	OJ Eggen	31:1–11
Watcher of the Skies	EM Burbidge	32:1–36
An Astronomical Rescue	LH Aller	33:1–17
The Desire to Understand the World	E Schatzman	34:1–34
A Physicist Courts Astronomy	CH Townes	35:xiii–xliv
Roaming Through Astrophysics	HC van de Hulst	36:1–16
Adventures in Cosmogony	AGW Cameron	37:1–36

Solar System Astrophysics

A Postencounter View of Comets	DA Mendis	26:11–49
Origin of the Solar System	AGW Cameron	26:441–72
Interaction Between the Solar Wind and the Interstellar Medium	TE Holzer	27:199–234
Radio Images of the Planets	I de Pater	28:347–99
Radioactive Dating of the Elements	JJ Cowan, F-K Thielemann, JW Truran	29:447–97
The Pluto-Charon System	SA Stern	30:185–233
Planet Formation	JJ Lissauer	31:129–74
The Atmospheres of Uranus and Neptune	JI Lunine	31:217–63
The Long-Term Dynamical Evolution of the Solar System	MJ Duncan, T Quinn	31:265–95
Jupiter's Great Red Spot and Other Vortices	PS Marcus	31:523–73
The Goldilocks Problem: Climatic Evolution and Long-Term Habitability of Terrestrial Planets	MR Rampino, K Caldeira	32:83–114
Cosmic Dusty Plasmas	DA Mendis, M Rosenberg	32:419–63
The Kuiper Belt	PR Weissman	33:327–57
Corotating and Transient Solar Wind Flows in Three Dimensions	JT Gosling	34:35–73
Charged Dust Dynamics in the Solar System	M Horányi	34:383–418
The Sun's Variable Radiation and Its Relevance for Earth	J Lean	35:33–67

Nucleosynthesis in Asymptomatic Giant Branch Stars: Relevance for Galactic Enrichment and Solar System Formation	M Busso, R Gallino, GJ Wasserburg	37:239–309
Physical Conditions in Regions of Star Formation	NJ Evans II	37:311–62
Sources of Relativistic Jets in the Galaxy	IF Mirabel, LF Rodriguez	37:409–43
The First 50 Years at Palomar: 1949-1999: The Early Years of Stellar Evolution, Cosmology, and High-Energy Astrophysics	A Sandage	37:445–86
Element Abundances in Quasistellar Objects: Star Formation and Galactic Nuclear Evolution at High Redshifts	F Hamann, G Ferland	37:487–531
Origin and Evolution of the Natural Satellites	SJ Peale	37:533–602

Solar Physics

Observed Variability of the Solar Luminosity	HS Hudson	26:473–507
Classification of Solar Flares	T Bai, PA Sturrock	27:421–67
Solar Convection	HC Spruit, Å Nordlund, AM Title	28:263–301
Equilibrium and Dynamics of Coronal Magnetic Fields	BC Low	28:491–524
Flares on the Sun and Other Stars	B Haisch, KT Strong, M Rodonó	29:275–324
Seismic Observations of the Solar Interior	D Gough, J Toomre	29:627–84
Solar Flares and Coronal Mass Ejections	SW Kahler	30:113–41
High-Energy Particles in Solar Flares	H Hudson, J Ryan	33:239–82
Magnetohydrodynamic Turbulence in the Solar Wind	ML Goldstein, DA Roberts, WH Matthaeus	33:283–325
The Solar Neutrino Problem	WC Haxton	33:459–503
Solar Active Regions as Diagnostics of Subsurface Conditions	RF Howard	34:75–109
The Sun's Variable Radiation and Its Relevance for Earth	J Lean	35:33–67
Radio Emission from Solar Flares	TS Bastian, AO Benz, DE Gary	36:131–88

Stellar Physics

Tests of Evolutionary Sequences Using Color-Magnitude Diagrams of Globular Clusters	A Renzini, F Fusi Pecci	26:199–244
Supernovae and Supernova Remnants	KW Weiler, RA Sramek	26:295–341

The Infrared Temporal Development of Classical Novae	RD Gehrz	26:377–412
Abundance Ratios as a Function of Metallicity	JC Wheeler, C Sneden, JW Truran Jr.	27:279–349
T Tauri Stars: Wild as Dust	C Bertout	27:351–95
Astrophysical Contributions of the International Ultraviolet Explorer	Y Kondo, A Boggess, SP Maran	27:397–420
Quasi-Periodic Oscillations and Noise in Low-Mass X-Ray Binaries	M van der Klis	27:517–53
Supernova 1987A	WD Arnett, JN Bahcall, RP Kirshner, SE Woosley	27:629–700
Chemical Analyses of Cool Stars	B Gustafsson	27:701–56
Masses and Evolutionary Status of White Dwarfs and Their Progenitors	V Weidemann	28:103–37
Cooling of White Dwarfs	F D'Antona, I Mazzitelli	28:139–81
The Origin of Neutron Stars in Binary Systems	R Canal, J Isern, J Labay	28:183–214
Quantitative Spectroscopy of Hot Stars	RP Kudritzki, DG Hummer	28:303–45
Rapidly Oscillating Ap Stars	DW Kurtz	28:607–55
The Search for Brown Dwarfs	DJ Stevenson	29:163–93
Flares on the Sun and Other Stars	B Haisch, KT Strong, M Rodonó	29:275–324
Galactic and Extragalactic Supernova Rates	S van den Bergh, GA Tammann	29:363–407
Radioactive Dating of the Elements	JJ Cowan, F-K Thielemann, JW Truran	29:447–97
Globular Cluster Systems in Galaxies Beyond the Local Group	WE Harris	29:543–79
Magnetic Fields of Degenerate Stars	G Chanmugam	30:143–84
New Developments in Understanding the HR Diagram	C Chiosi, G Bertelli, A Bressan	30:235–85
Evidence for Black Holes in Stellar Binary Systems	AP Cowley	30:287–310
Type Ia Supernovae as Standard Candles	D Branch, GA Tammann	30:359–89
Proto-Planetary Nebulae	S Kwok	31:63–92
Origin and Evolution of X-Ray Binaries and Binary Radio Pulsars	F Verbunt	31:93–127
Supernova 1987A Revisited	R McCray	31:175–216
The Faint End of the Stellar Luminosity Function	MS Bessell, GS Stringfellow	31:433–71
Asteroseismology	TM Brown, RL Gilliland	32:37–82

The r-, s-, and p-Processes in Nucleosynthesis	BS Meyer	32:153–90
Massive Star Populations in Nearby Galaxies	A Maeder, PS Conti	32:227–75
Pre-Main-Sequence Binary Stars	RD Mathieu	32:465–530
Binary and Millisecond Pulsars	ES Phinney, SR Kulkarni	32:591–639
Stellar Pulsations Across the HR Diagram: Part 1	A Gautschy, H Saio	33:75–113
Explosive Nucleosynthesis Revisited: Yields	D Arnett	33:115–32
Blue Stragglers and Other Stellar Anomalies: Implications for the Dynamics of Globular Clusters	CD Bailyn	33:133–62
Angular Momentum Evolution of Young Stars and Disks	P Bodenheimer	33:199–238
The Old Open Clusters of the Milky Way	ED Friel	33:381–414
The Solar Neutrino Problem	WC Haxton	33:459–503
Bipolar Molecular Outflows from Young Stars and Protostars	R Bachiller	34:111–54
The FU Orionis Phenomenon	L Hartmann, SJ Kenyon	34:205–39
Circumstellar Photochemistry	AE Glassgold	34:241–78
Geminga, Its Phenomenology, Its Fraternity, and Its Physics	GF Bignami, PA Caraveo	34:331–81
The Age of the Galactic Globular Cluster System	DA VandenBerg, M Bolte, PB Stetson	34:461–510
Old and Intermediate-Age Stellar Populations in the Magellanic Cloud	EW Olszewski, NB Suntzeff, M Mateo	34:511–50
Stellar Pulsations Across the HR Diagram: Part 2	A Gautschy, H Saio	34:551–606
X-Ray Novae	Y Tanaka, N Shibazaki	34:607–44
Eta Carina and Its Environment	K Davidson, RM Humphreys	35:1–32
Luminous Supersoft X-Ray Sources	P Kahabka, EPJ van den Heuvel	35:69–100
Model Atmospheres of Very Low Mass Stars and Brown Dwarfs	PH Hauschildt, S Starrfield, F Allard, DR Alexander	35:137–77
Optical Spectra of Supernovae	AV Filippenko	35:309–55
Abundance Ratios and Galactic Chemical Evolution	A McWilliam	35:503–56
Mixing in Stars	M Pinsonneault	35:557–605
Type Ia Supernovae and the Hubble Constant	D Branch	36:17–55

Herbig Ae/Be Stars	C Waelkens, LBFM Waters	36:233–66
Carbon Stars	G Wallerstein, GR Knapp	36:369–433
The HR Diagram and the Galactic Distance Scale After Hipparcos	IN Reid	37:191–237
High-Energy Processes in Young Stellar Objects	ED Feigelson, T Montmerle	37:363–408

Dynamical Astronomy

Recent Advances in Optical Astrometry	DG Monet	26:413–40
Quasi-Periodic Oscillations and Noise in Low-Mass X-Ray Binaries	M van der Klis	27:517–53
The Origin of Neutron Stars in Binary Systems	R Canal, J Isern, J Labay	28:183–214
The Few-Body Problem in Astrophysics	M Valtonen, S Mikkola	29:9–29
Evidence for Black Holes in Stellar Binary Systems	AP Cowley	30:287–310
Origin and Evolution of X-Ray Binaries and Binary Radio Pulsars	F Verbunt	31:93–127
Binary and Millisecond Pulsars	ES Phinney, SR Kulkarni	32:591–639
Theory of Accretion Disks I: Angular Momentum Transport Processes	JCB Papaloizou, DNC Lin	33:505–40
Geminga, Its Phenomenology, Its Fraternity, and Its Physics	GF Bignami, PA Caraveo	34:331–81
Theory of Accretion Disks II: Application to Observed Systems	DNC Lin, JCB Papaloizou	34:703–47
First Results from Hipparcos	J Kovalevsky	36:99–129
A Critical Review of Galactic Dynamos	RM Kulsrud	37:37–64

Interstellar Medium

Large-Scale Expanding Superstructures in Galaxies	G Tenorio-Tagle, P Bodenheimer	26:145–97
Supernovae and Supernova Remnants	KW Weiler, RA Sramek	26:295–341
The Orion Molecular Cloud and Star-Forming Region	R Genzel, J Stutzki	27:41–85
A New Component of the Interstellar Matter: Small Grains and Large Aromatic Molecules	JL Puget, A Léger	27:161–98
Interaction Between the Solar Wind and the Interstellar Medium	TE Holzer	27:199–234

T Tauri Stars: Wild as Dust	C Bertout	27:351–95
Diffuse Galactic Gamma-Ray Emission	H Bloemen	27:469–516
Interstellar Dust and Extinction	JS Mathis	28:37–70
Theories of the Hot Interstellar Gas	L Spitzer Jr.	28:71–101
H I in the Galaxy	JM Dickey, FJ Lockman	28:215–61
Extragalactic H II Regions	GA Shields	28:525–60
Radio Propagation Through the Turbulent Interstellar Medium	BJ Rickett	28:561–605
The Soft X-Ray Background and Its Origins	D McCammon, WT Sanders	28:657–88
Collective Plasma Radiation Processes	DB Melrose	29:31–57
Distribution of CO in the Milky Way	F Combes	29:195–237
Molecular Gas in Galaxies	JS Young, NZ Scoville	29:581–625
Dust-Gas Interactions and the Infrared Emission from Hot Astrophysical Plasmas	E Dwek, RG Arendt	30:11–50
Astronomical Masers	M Elitzur	30:75–112
Proto-Planetary Nebulae	S Kwok	31:63–92
Millimeter and Submillimeter Interferometry of Astronomical Sources	AI Sargent, WJ Welch	31:297–343
Theory of Interstellar Shocks	BT Draine, CF McKee	31:373–432
Abundances in the Interstellar Medium	TL Wilson, RT Rood	32:191–226
The Diffuse Interstellar Bands	GH Herbig	33:19–73
Angular Momentum Evolution of Young Stars and Disks	P Bodenheimer	33:199–238
Bipolar Molecular Outflows from Young Stars and Protostars	R Bachiller	34:111–54
Circumstellar Photochemistry	AE Glassgold	34:241–78
Insterstellar Abundances from Absorption-Line Observations with the Hubble Space Telescope	BD Savage, KR Sembach	34:279–329
Eta Carina and Its Environment	K Davidson, RM Humphreys	35:1–32
Dense Photodissociation Regions (PDRs)	DJ Hollenbach, AGGM Tielens	35:275–215
High-Velocity Clouds	BP Wakker, H van Woerden	35:217–66
Chemical Evolution of Star-Forming Regions	EF van Dishoeck, GA Blake	36:317–68
Reference Frames in Astronomy	KJ Johnston, C de Vegt	37:97–125

The Galaxy

The Galactic Nuclear Bulge and the Stellar Content of Spheroidal Systems	JA Frogel	26:51–92

Tests of Evolutionary Sequences Using Color-Magnitude Diagrams of Globular Clusters	A Renzini, F Fusi Pecci	26:199–244
Diffuse Galactic Gamma-Ray Emission	H Bloemen	27:469–516
Kinematics, Chemistry, and Structure of the Galaxy	G Gilmore, RFG Wyse, K Kuijken	27:555–627
H I in the Galaxy	JM Dickey, FJ Lockman	28:215–61
Chemical Evolution of the Galaxy	NC Rana	29:129–62
Distribution of CO in the Milky Way	F Combes	29:195–237
The Mass of the Galaxy	M Fich, S Tremaine	29:409–45
The Faint End of the Stellar Luminosity Function	MS Bessell, GS Stringfellow	31:433–71
Galactic Structure Surveys and the Evolution of the Milky Way	SR Majewski	31:575–638
Blue Stragglers and Other Stellar Anomalies: Implications for the Dynamics of Globular Clusters	CD Bailyn	33:133–62
The Old Open Clusters of the Milky Way	ED Friel	33:381–414
Galactic Magnetism: Recent Developments and Perspectives	R Beck, A Brandenburg, D Moss, A Shukurov, D Sokoloff	34:153–204
Gravitational Microlensing in the Local Group	B Paczyński	34:419–59
The Age of the Galactic Globular Cluster System	DA VandenBerg, M Bolte, PB Stetson	34:461–510
The Galactic Center Environment	M Morris, E Serabyn	34:645–701
High-Velocity Clouds	BP Wakker, H van Woerden	35:217–66
First Results from Hipparcos	J Kovalevsky	36:99–129

Extragalactic Astronomy and Cosmology

Polarization Properties of Extragalactic Radio Sources	DJ Saikia, CJ Salter	26:93–144
Large-Scale Expanding Superstructures in Galaxies	G Tenorio-Tagle, P Bodenheimer	26:145–97
Voids	HJ Rood	26:245–94
Supernovae and Supernova Remnants	KW Weiler, RA Sramek	26:295–341
Enhanced Star Formation and Infrared Emission in the Centers of Galaxies	CM Telesco	26:343–76
The Luminosity Function of Galaxies	B Binggeli, A Sandage, GA Tammann	26:509–60
Observational Tests of World Models	A Sandage	26:561–630

Large-Scale Structure in the Universe Indicated by Galaxy Clusters	NA Bahcall	26:631–86
X Rays From Normal Galaxies	G Fabbiano	27:87–138
Populations in Local Group Galaxies	P Hodge	27:139–59
Surface Photometry and the Structure of Elliptical Galaxies	J Kormendy, S Djorgovski	27:235–77
Astrophysical Contributions of the International Ultraviolet Explorer	Y Kondo, A Boggess, SP Maran	27:397–420
Diffuse Galactic Gamma-Ray Emission	H Bloemen	27:469–516
Supernova 1987A	WD Arnett, JN Bahcall, RP Kirshner, SE Woosley	27:629–700
Gamma-Ray Bursts	JC Higdon, RE Lingenfelter	28:401–36
The Space Distribution of Quasars	FDA Hartwick, D Schade	28:437–89
Extragalactic H II Regions	GA Shields	28:525–60
The Soft X-Ray Background and Its Origins	D McCammon, WT Sanders	28:657–88
The Cosmic Far Ultraviolet Background	S Bowyer	29:59–88
Ultraviolet Background Radiation	RC Henry	29:89–127
Structure and Dynamics of Elliptical Galaxies	T de Zeeuw, M Franx	29:239–74
Inflation for Astronomers	JV Narlikar, T Padmanabhan	29:325–62
Galactic and Extragalactic Supernova Rates	S van den Bergh, GA Tammann	29:363–407
Radioactive Dating of the Elements	JJ Cowan, F-K Thielemann, JW Truran	29:447–97
Redshift Surveys of Galaxies	R Giovanelli, MP Haynes	29:499–541
Globular Cluster Systems in Galaxies Beyond the Local Group	WE Harris	29:543–79
Molecular Gas in Galaxies	JS Young, NZ Scoville	29:581–625
Warps	J Binney	30:51–74
Astronomical Masers	M Elitzur	30:75–112
Cosmologial Applications of Gravitational Lensing	RD Blandford, R Narayan	30:311–58
Type Ia Supernovae as Standard Candles	D Branch, GA Tammann	30:359–89
The Origin of the X-Ray Background	AC Fabian, X Barcons	30:429–56
The Cosmological Constant	SM Carroll, WH Press, EL Turner	30:499–542
Smoothed Particle Hydrodynamics	JJ Monaghan	30:543–74
Radio Emission From Normal Galaxies	JJ Condon	30:575–611
Evidence for Evolution in Faint Field Galaxy Samples	DC Koo, RG Kron	30:613–52

Observations of the Isotropy of the Cosmic Microwave Background Radiation	ACS Readhead, CR Lawrence	30:653–703
Dynamics of Interacting Galaxies	JE Barnes, L Hernquist	30:705–42
Supernova 1987A Revisited	R McCray	31:175–216
Millimeter and Submillimeter Interferometry of Astronomical Sources	AI Sargent, WJ Welch	31:297–343
The Distance to the Center of the Galaxy	MJ Reid	31:345–72
The Faint End of the Stellar Luminosity Function	MS Bessell, GS Stringfellow	31:433–71
Unified Models for Active Galactic Nuclei and Quasars	R Antonucci	31:473–521
High Redshift Radio Galaxies	PJ McCarthy	31:639–88
Astronomical Tests of the Cold Dark Matter Scenario	JP Ostriker	31:689–716
X-Ray Spectra and Time Variability of Active Galactic Nuclei	RF Mushotzky, C Done, KA Pounds	31:717–61
Physical Parameters along the Hubble Sequence	MS Roberts, MP Haynes	32:115–52
Massive Star Populations in Nearby Galaxies	A Maeder, PS Conti	32:227–75
Cooling Flows in Clusters of Galaxies	AC Fabian	32:277–318
Anisotropies in the Cosmic Microwave Background	M White, D Scott, J Silk	32:319–70
Dynamics of Cosmic Flows	A Dekel	32:371–418
Baryonic Dark Matter	B Carr	32:531–90
Intraday Variablity in Quasars and BL Lac Objects	SJ Wagner, A Witzel	33:163–97
Gamma-Ray Bursts	GJ Fishman, CA Meegan	33:415–58
Comptonization of the Cosmic Microwave Background: The Sunyaev-Zeldovich Effect	Y Rephaeli	33:541–79
Inward Bound: The Search for Supermassive Black Holes in Galactic Nuclei	J Kormendy, D Richstone	33:581–624
Galactic Magnetism: Recent Developments and Perspectives	R Beck, A Brandenburg, D Moss, A Shukurov, D Sokoloff	34:153–204
Gravitational Microlensing in the Local Group	B Paczyński	34:419–59
Old and Intermediate-Age Stellar Populations in the Magellanic Cloud	EW Olszewski, NB Suntzeff, M Mateo	34:511–50
Luminous Infrared Galaxies	DB Sanders, IF Mirabel	34:749–92

Observational Selection Bias Affecting the Determination of the Extragalactic Distance Scale	P Teerikorpi	35:101–36
Low Surface Brightness Galaxies	C Impey, G Bothun	35:267–307
Optical Spectra of Supernovae	AV Filippenko	35:309–55
Compact Groups of Galaxies	P Hickson	35:357–88
Faint Blue Galaxies	RS Ellis	35:389–443
Variability of Active Galactic Nuclei	M-H Ulrich, L Maraschi, CM Urry	35:445–502
Parsec-Scale Jets in Extragalactic Radio Sources	JA Zensus	35:607–36
Galactic Bulges	RF Wyse, G Gilmore, M Franx	35:637–75
Type Ia Supernovae and the Hubble Constant	D Branch	36:17–55
Star Formation in Galaxies Along the Hubble Sequence	RC Kennicutt Jr.	36:189–231
The Lyman Alpha Forest in the Spectra of Quasistellar Objects	M Rauch	36:267–316
Dwarf Galaxies of the Local Group	ML Mateo	36:435–506
Modeling Extragalactic Jets	A Ferrari	36:539–98
Simulations of Structure Formation in the Universe	E Bertschinger	36:599–654
Probing the Universe with Weak Lensing	Y Mellier	37:127–89
Far-Ultraviolet Radiation from Elliptical Galaxies	RW O'Connell	37:603–48

High Energy Astrophysics

X Rays From Normal Galaxies	G Fabbiano	27:87–138
Gamma-Ray Bursts	JC Higdon, RE Lingenfelter	28:401–36
Collective Plasma Radiation Processes	DB Melrose	29:31–57
The Cosmic Far Ultraviolet Background	S Bowyer	29:59–88
Ultraviolet Background Radiation	RC Henry	29:89–127
The Origin of the X-Ray Background	AC Fabian, X Barcons	30:429–56
Radio Emission From Normal Galaxies	JJ Condon	30:575–611
X-Ray Spectra and Time Variability of Active Galactic Nuclei	RF Mushotzky, C Done, KA Pounds	31:717–61
Binary and Millisecond Pulsars	ES Phinney, SR Kulkarni	32:591–639
Intraday Variablity in Quasars and BL Lac Objects	SJ Wagner, A Witzel	33:163–97
High-Energy Particles in Solar Flares	H Hudson, J Ryan	33:239–82
Gamma-Ray Bursts	GJ Fishman, CA Meegan	33:415–58

Comptonization of the Cosmic Microwave Background: The Sunyaev-Zeldovich Effect	Y Rephaeli	33:541–79
X-Ray Novae	Y Tanaka, N Shibazaki	34:607–44
Luminous Supersoft X-Ray Sources	P Kahabka, EPJ van den Heuvel	35:69–100
Variability of Active Galactic Nuclei	M-H Ulrich, L Maraschi, CM Urry	35:445–502
Parsec-Scale Jets in Extragalactic Radio Sources	JA Zensus	35:607–36
Modeling Extragalactic Jets	A Ferrari	36:539–98

Instrumentation and Techniques

Recent Advances in Optical Astrometry	DG Monet	26:413–40
Astrophysical Contributions of the International Ultraviolet Explorer	Y Kondo, A Boggess, SP Maran	27:397–420
X-Ray Astronomy Missions	HVD Bradt, T Ohashi, KA Pounds	30:391–427
Long-Baseline Optical and Infrared Stellar Interferometry	M Shao, MM Colavita	30:457–98
Adaptive Optics for Astronomy: Principles, Performance, and Applications	JM Beckers	31:13–62
Frequency Allocation: The First Forty Years	B Robinson	37:65–96

New Areas of Reseach, History

The Status and Prospects for Ground-Based Observatory Sites	RH Garstang	27:19–40
X-Ray Astronomy Missions	HVD Bradt, T Ohashi, KA Pounds	30:391–427
Completing the Copernican Revolution: The Search for Other Planetary Systems	DC Black	33:359–80
Detection of Extrasolar Giant Planets	GW Marcy, RP Butler	36:57–97
Astronomical Searches for Earth-Like Planets and Signs of Life	N Woolf, JR Angel	36:507–37